Climate Change and Mango Production

Syed Atif Hasan Naqvi · Shakeel Ahmad ·
Mukhtar Ahmed
Editors

Climate Change and Mango Production

Potential Adaptation and Mitigation Options

Springer

Editors
Syed Atif Hasan Naqvi
Department of Plant Pathology
Bahauddin Zakariya University
Multan, Pakistan

Shakeel Ahmad
Institute of Agronomy
Bahauddin Zakariya University
Multan, Pakistan

Mukhtar Ahmed
Department of Agronomy
Pir Mehr Ali Shah Arid Agri University
Rawalpindi, Pakistan

ISBN 978-3-031-87034-7 ISBN 978-3-031-87035-4 (eBook)
https://doi.org/10.1007/978-3-031-87035-4

This Springer imprint is published by the registered company Springer Nature Switzerland AG
The registered company address is: Gewerbestrasse 11, 6330 Cham, Switzerland

Preface

Climate change is no longer a distant threat; it is a present reality reshaping the agricultural landscape, challenging food security, and redefining the resilience of crops worldwide. Among the myriad of crops affected, mango (*Mangifera indica* L.), often hailed as the "King of Fruits," stands at a critical juncture. Its global significance, spanning nutrition, economics, and cultural heritage, makes it a focal point for understanding the interplay between climate change and fruit crops. This book, *Climate Change and Mango Production: Potential Adaptation and Mitigation Options*, emerges as a timely and comprehensive response to the urgent need for sustainable solutions in mango cultivation amidst a rapidly changing climate.

Mango, a tropical fruit cherished for its flavor, aroma, and nutritional richness, is cultivated across diverse agroecological zones, from the arid regions of Pakistan to the humid tropics of Southeast Asia. However, its sensitivity to climatic variables, i.e., temperature fluctuations, erratic rainfall, and extreme weather events renders it highly vulnerable to climate change. Rising temperatures alter flowering patterns, while prolonged droughts and unseasonal rains exacerbate pest and disease pressures, threatening yield and quality. These challenges are compounded by the socioeconomic dependencies of millions of smallholder farmers and communities whose livelihoods are intricately tied to mango production.

This book addresses these multifaceted challenges holistically, integrating cutting-edge research, innovative technologies, and practical strategies to foster climate-resilient mango production. It aims to equip researchers, policymakers, and farmers with the knowledge and tools needed to safeguard this vital crop by bridging the gap between science and application.

The 32 chapters of this book are meticulously curated to cover every dimension of mango production in the context of climate change. The journey begins with microbiome engineering (Chap. 1), exploring how rhizospheric microbes can be harnessed to enhance stress tolerance and nutrient uptake. This sets the stage for a deeper dive into mango trees' physiological and biochemical adaptations (Chaps. 2 and 3), including the role of secondary metabolites in abiotic stress mitigation and the cytogenetic responses to environmental stressors.

The book then transitions to regional and global perspectives, evaluating the impact of climate change on mango production in diverse geographies, such as Türkiye (Chap. 4), and assessing the implications for fruit quality (Chap. 5) and nutritional composition (Chap. 6). These chapters underscore the universal relevance of climate-resilient strategies while highlighting region-specific challenges and opportunities.

Central to this book is its emphasis on practical solutions. Chapters 7–16 delve into the physiological adaptations of mango trees, grafting techniques, water management, and soil health, offering actionable insights for sustainable cultivation. The socio-economic dimensions (Chap. 11) and policy frameworks (Chap. 14) provide a macro-level perspective, emphasizing the need for institutional support and community engagement in climate adaptation efforts.

A significant portion of the book is dedicated to pest and disease management, a critical area exacerbated by climate change. Chapters 17–32 explore the dynamics of fungal diseases, insect pests, and malformation disorders, presenting advanced diagnostic tools (Chap. 29), nanotechnology-based solutions (Chap. 21), and innovative mitigation strategies (Chaps. 27 and 28). The integration of modern tools, such as forecasting models and phytosanitary measures (Chap. 26), ensures that readers are equipped with state-of-the-art knowledge to tackle emerging threats.

What sets this book apart is its seamless integration of fundamental science and applied research. Each chapter not only elucidates the underlying mechanisms of climate impacts but also translates this knowledge into practical strategies. For instance, the discussion on postharvest management (Chap. 13) and value addition highlights how climate-smart practices can reduce losses and enhance marketability, while the exploration of fruit bagging (Chap. 15) and containerized nurseries (Chap. 16) offers scalable solutions for smallholder farmers.

This book is more than a compilation of research; it is a call to action. It underscores the urgency of adopting climate-resilient practices, investing in research and innovation, and fostering collaboration among stakeholders. By presenting a roadmap for sustainable mango production, it aims to inspire a collective effort to protect this invaluable crop for future generations.

We extend our deepest gratitude to the contributors, viz., researchers, scientists, and practitioners whose expertise and dedication have made this book possible. Their insights and innovations are a testament to the power of science in addressing global challenges. We also thank Springer Nature for their unwavering support and commitment to advancing knowledge in agricultural sciences.

Whether you are a researcher, extension specialist or progressive grower seeking to deepen your understanding of mango physiology, a policymaker crafting strategies for climate adaptation, or a farmer striving to enhance productivity, this book is designed to be your guide. It is our hope that the knowledge contained within these pages will empower you to navigate the complexities of climate change and contribute to a sustainable future for mango production.

As we stand at the intersection of science and sustainability, let this book serve as a beacon of hope and a catalyst for change. Together, we can ensure that the "King of Fruits" continues to thrive in a changing world.

St. Catherines, ON, Canada — Syed Atif Hasan Naqvi
Multan, Punjab, Pakistan — Shakeel Ahmad
Rawalpindi, Punjab, Pakistan — Mukhtar Ahmed

Contents

Chapter 1
Microbiome Engineering of the Mango Rhizosphere: A Sustainable Solution for Climate-Resilient Production and Way Forward

Syed Atif Hasan Naqvi, Shakeel Ahmad, Mukhtar Ahmed, Adriene Caldeira Batista, and Ricardo Siqueira da Silva

Abstract The mango (*Mangifera indica*) rhizosphere, a dynamic interaction zone between roots, soil, and microbes, is critical in tree health and productivity. Climate change, characterized by rising temperatures, erratic rainfall, and increased pest pressures, has amplified challenges for mango cultivation, including reduced soil fertility, heightened disease susceptibility, and impaired tree physiology. Microbiome engineering of the mango rhizosphere is a promising strategy to mitigate these effects and promote sustainable, climate-resilient production systems. This chapter explores the potential of harnessing beneficial microbial communities to enhance nutrient uptake, bolster disease resistance, and improve stress tolerance in mango trees. It delves into the ecological complexity of the mango rhizosphere, highlighting the symbiotic relationships between mango roots and microbial consortia, including plant growth-promoting rhizobacteria (PGPR) and arbuscular mycorrhizal

S. A. H. Naqvi (✉)
Department of Plant Pathology, Faculty of Agricultural Sciences and Technology, Bahauddin Zakariya University, Multan, Pakistan
e-mail: atifnaqvi@bzu.edu.pk

Department of Biological Sciences, Brock University, St. Catharines, ON L2S 3A1, Canada

London Research and Development Center, Agriculture and Agri-Food Canada, Vineland Station, ON L0R 2E0, Canada

S. Ahmad
Institute of Agronomy, Faculty of Agricultural Sciences and Technology, Bahauddin Zakariya University, Multan, Pakistan

M. Ahmed
Department of Agronomy, Pir Mehr Ali Shah Arid Agriculture University, Rawalpindi 46300, Pakistan

A. C. Batista · R. S. da Silva
Department of Agronomy, Federal University of Vales of the Jequitinhonha and Mucuri, Diamantina, Minas Gerais 39100-000, Brazil

S. A. H. Naqvi et al. (eds.), *Climate Change and Mango Production*,
https://doi.org/10.1007/978-3-031-87035-4_1

fungi (AMF). Advances in metagenomics, transcriptomics, and microbial bioinformatics have enabled the identification of key functional microbes that contribute to these processes. Key areas of focus include applying bioinoculants, developing microbial consortia tailored to specific agroecological conditions, and integrating of microbiome-based strategies with sustainable orchard management practices. The chapter also addresses challenges such as scalability, compatibility with existing agricultural systems, and the need for regulatory frameworks to ensure safe implementation. By synthesizing recent advances and field applications, this chapter underscores microbiome engineering as a transformative tool to enhance the resilience and sustainability of mango production systems. The way forward involves interdisciplinary collaboration, technological innovation, and farmer engagement to operationalize microbiome engineering on a scale. As climate change continues to reshape agricultural landscapes, the engineering of rhizosphere microbiomes offers a pathway toward more sustainable and resilient mango cultivation systems.

1.1 Introduction: Climate Change Impacts on Agriculture, Focusing on Mango Production

Agriculture is extremely sensitive to climate change because of its significant sensitivity to climatic variables, resulting in substantial financial impacts (Mendelsohn 2009). Climate events such as temperature and rainfall have considerable effects on crop yields. The impact of high temperatures, precipitation changes, and fertilization from CO_2 varies depending on the crop, location, and extent of the change in parameters. Temperature increases are reported to diminish yield, but precipitation increases are likely to reduce the impact of high temperatures (Adams et al. 1998). Climate change substantially challenges agriculture, particularly for mango production in vulnerable regions (Yadav et al. 2023). Managing these difficulties requires an integrated approach which includes scientific research, technical innovation, and legislative initiatives to ensure the sustainability and resilience of mango farming systems (Prasad Pant et al. 2012; Tengsetasak et al. 2024).

The intertwined challenges of soil degradation, soil-borne diseases, and pesticide dependency represent significant barriers to sustainable mango production under climate change (Doran and Zeiss 2000). Addressing these issues requires a holistic approach integrating scientific research, sustainable farming practices, and supportive policies (Chakraborty and Newton 2011). Mango (*Mangifera indica* L.), a fruit deeply rooted in the cultural and agricultural history of South Asia, is believed to have originated in the region now known as northeastern India and Myanmar. One of the central and most widely consumed tropical fruits worldwide, mangoes represent a significant source of revenue for the economy of many nations. According to the Food and Agriculture Organization (FAO) 2023, and Khan et al. (2023), Mango (Mangifera indica L.) is a culturally and economically significant tropical fruit with global production distributed across more than 100 countries. As of 2023, India

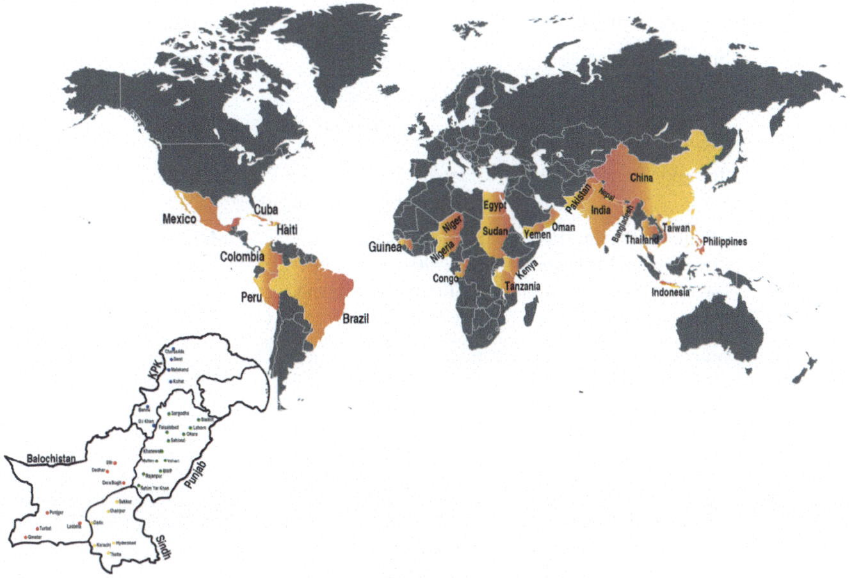

Fig. 1.1 Map of mango-producing countries in the world

leads in mango production, contributing about 44% of the global yield, followed by countries such as China, Indonesia, and Pakistan. Pakistan ranks as the fourth largest producer, accounting for 3.95% of the world's mango yield, and is the sixth largest in terms of cultivated area, highlighting its significant role in the horticultural economy. The global demand for mangoes is fueled by their unique flavor, high nutritional value, and cultural importance, particularly in South Asia, where the fruit originated (Galán Saúco 2002). However, with the rising culture of mangoes, problems are being reported with the spreading of diseases that pose serious hazards to mango production (Fig. 1.1 and Table 1.1).

1.2 Importance of Sustainable Solutions and Microbiome Engineering as a Transformative Approach

Sustainable solutions, underpinned by innovative technologies like microbiome engineering, represent the future of resilient mango production systems (Srivastava 2024). By enhancing soil health, mitigating pest and disease pressures, and improving plant stress tolerance, microbiome engineering provides a transformative pathway to address the challenges posed by climate change and resource degradation (Mendes et al. 2013). This integration of sustainability and biotechnology ensures that mango farming can thrive while preserving environmental integrity.

Table 1.1 Case studies highlighting specific challenges in mango-growing regions

Region	Specific challenges	Key impacts	References
India	- Erratic flowering due to warmer winters- Increased pest incidences (e.g., fruit flies, powdery mildew)	- Yield losses of up to 20% for varieties like *Dashehari* and *Langra*- Higher pesticide use, raising production costs	Makhmale et al. (2016)
Pakistan	- Water stress during flowering and fruiting stages- Rising soil salinity in lower Sindh	- Yield declines of 30–40% in Sindh- Stunted growth and reduced fruit size due to saline soils	Mohsin et al. (2021)
Philippines	- Cyclone damage (e.g., Typhoon Haiyan)- High humidity accelerating post-harvest losses	- Widespread tree damage and slow recovery in affected areas- Post-harvest losses estimated at 25–30%	Elauria et al. (2015)
Kenya	- Prolonged drought in arid regions (e.g., Makueni, Kitui)- Increased fruit fly infestations	- Declining yields due to limited water availability- Export losses as fruit becomes unsuitable for international markets	Griesbach, (2003)
Australia	- Heat waves causing sunburn damage- Accelerated ripening due to higher temperatures	- 20% reduction in export-grade produce in Northern Territory- Increased spoilage as harvesting schedules become misaligned	Hu et al. (2022)

1.3 Climate Change and Mango Production: Effects of Climate Change on Mango Cultivation

Climate change has significant impacts on mango cultivation, influencing growth, flowering, fruiting, and overall yield. Rising temperatures can shorten the fruit development period and enhance the synthesis of stress-related compounds, improving fruit quality in some cases. However, extreme heat, drought, and erratic rainfall can negatively affect flowering and fruit set, leading to yield reductions. High temperatures disrupt floral induction, while irregular rainfall patterns, such as pre-flowering rains, delay flowering and increase vegetative growth (Halder et al.). Additionally, water stress can influence nutrient accumulation, affecting tree productivity and fruit size. These climatic challenges underscore the need for adaptive management strategies to sustain mango production under changing environmental conditions (Legave et al. 2013).

1.3.1 Altered Temperature Regimes

Mango trees are highly sensitive to temperature changes, particularly during flowering and fruit development. Climate change has resulted in:

- **Warmer Winters**: Mango trees require specific chilling periods during winter to induce flowering. Warmer winters can disrupt this process, leading to reduced flowering and subsequent yields (Bashan et al. 2014).
- **Heat Stress**: Excessive temperatures during fruit set and maturation can impair fruit development, causing smaller fruits, sunburn damage, and decreased quality (Parmesan et al. 2022).

1.3.2 Changes in Precipitation Patterns

Erratic rainfall patterns and altered monsoon dynamics present significant challenges to mango cultivation:

- **Drought Stress**: Reduced rainfall during critical growth stages causes water scarcity, which limits nutrient uptake and stunts tree growth.
- **Flooding and Waterlogging**: Excessive rainfall can lead to root damage, soil erosion, and increased vulnerability to fungal diseases (Mahmoud Mohamed Ahmed 2012; Parmesan et al. 2022).

1.3.3 Increased Incidence of Pests and Diseases

Rising temperatures and humidity levels create favorable conditions for the proliferation of pests and diseases. Common issues include:

- **Pest Infestations**: Fruit flies and mealybugs thrive under warmer climates, increasing infestations that damage fruit and reduce marketability.
- **Fungal Diseases**: High humidity encourages the growth of fungal pathogens such as *Colletotrichum* (causing anthracnose) and *Oidium mangiferae* (powdery mildew), which can devastate flowers and fruit (Mohsin et al. 2021) (Fig. 1.2).

1.3.4 Extreme Weather Events

Extreme weather events including cyclones, hailstorms, and sudden frost events have become more frequent and intense due to climate change (Li et al. 2022). These events can cause:

- **Physical Damage**: High winds and hail can damage or destroy mango trees and fruits.

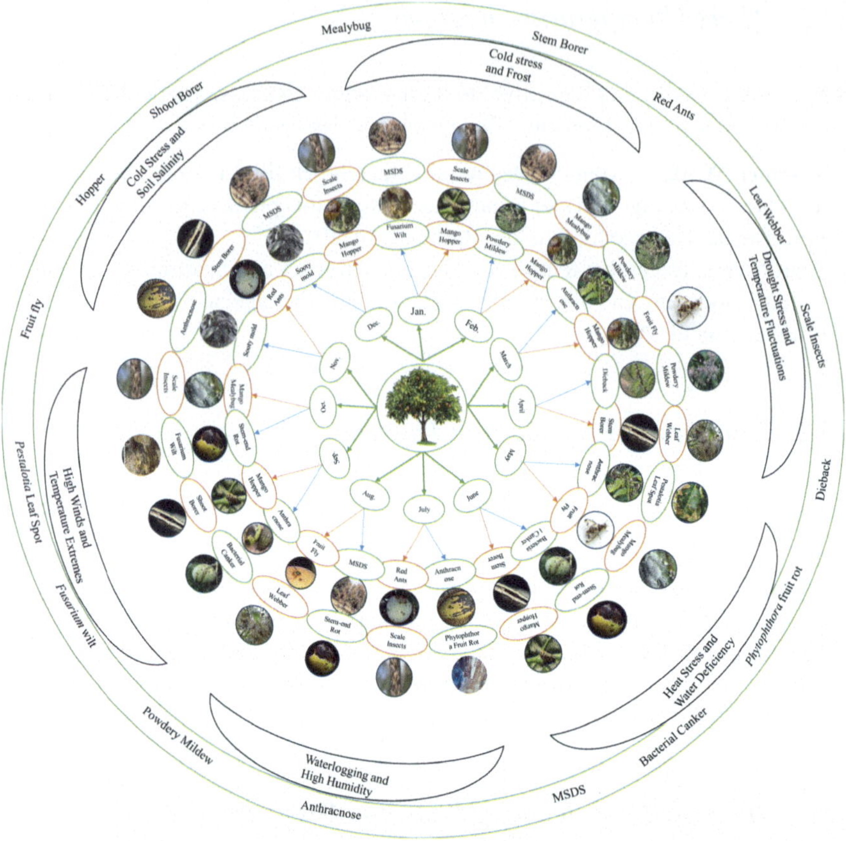

Fig. 1.2 Increased incidence of pests and diseases

- **Disrupted Harvests**: Unpredictable weather events can delay or compromise harvesting and supply chains, affecting both local and export markets (Elauria et al. 2015; Parmesan et al. 2022).

1.3.5 Decline in Fruit Quality

Climate variability impacts the quality attributes of mangoes, such as size, sweetness, and shelf life:

- Higher temperatures can accelerate ripening, reducing shelf life and increasing post-harvest losses.
- Water stress during fruit development can reduce sugar accumulation, resulting in less flavorful fruits (Chakraborty and Newton 2011; Welsh 2021).

1.4 Change in the Shifts of Mango Tree Typical Phenology to Climate-Driven Phenology

There have been reports that the mango reproductive cycle, i.e., breaking of buds' dormancy, initiation of flowers, pollination, fertilization, fruit setting, and physical and horticultural maturity has been changed regarding time e.g., previously flowering starts in the mid of February in tropical and sub-tropical countries like Pakistan and India and now due to the climate shifts this previous timings has been changes (Fig. 1.3).

1.5 The Mango Rhizosphere: An Ecological Perspective

1.5.1 Structure and Function of the Mango Tree Rhizosphere

The rhizosphere, the narrow zone of soil surrounding and influenced by plant roots, plays a critical role in the growth, health, and productivity of mango trees. This dynamic microenvironment facilitates complex interactions between the plant, soil, and microbial communities, significantly influencing nutrient acquisition, disease suppression, and stress resilience (Mohandas 2012).

1.5.1.1 Structure of the Mango Rhizosphere

1. **Root Architecture**
   - Mango trees possess a deep taproot system with extensive lateral roots, providing structural support and accessing water and nutrients from deeper soil layers.
   - Fine feeder roots in the rhizosphere are primarily responsible for nutrient uptake, secreting root exudates that shape the microbial community (Griesbach 2003; Parmesan et al. 2022).
2. **Soil Composition**
   - Roots exude organic compounds, including carbohydrates, amino acids, and secondary metabolites, enriching the soil in the rhizosphere.
   - The rhizosphere soil differs from bulk soil in physical and chemical properties and supports higher microbial diversity and activity.
3. **Microbial Communities**
   - The mango rhizosphere harbors diverse microbial populations, including bacteria, fungi, actinomycetes, and archaea. Prominent microbes include:

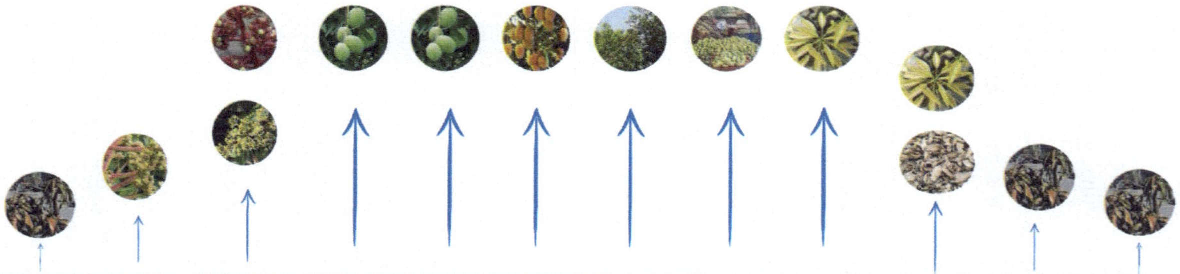

Month	January	February	March	April	May	June	July	August	September	October	November	December
Previous	D	F from mid Feb.	FF	FG	FG	FM	h	H	VG	LS and VG Oct. 15	D From 15 Nov.	D
Current	D	D	D Up to mid of march then F	FF	FG	FG	FM	h	H	Oct. 15 LS and VG	D From 15 Nov.	D

Fig. 1.3 Change in the pattern of mango phenology due to climate change

- **Plant Growth-Promoting Rhizobacteria (PGPR)**: e.g., *Pseudomonas*, *Bacillus*.
- **Mycorrhizal Fungi**: e.g., *Glomus* species, which form symbiotic associations with roots.
- **Pathogenic Microbes**: e.g., *Phytophthora* (root rot), can dominate in stressed systems (Berg et al. 2013).

1.5.1.2 Functions of the Mango Rhizosphere

1. **Nutrient Acquisition**
   - **Mineral Solubilization**: Microbial activity in the rhizosphere enhances the solubilization of nutrients like phosphorus, potassium, and micronutrients, making them available for uptake by mango roots.
   - **Nitrogen Fixation**: Certain rhizosphere bacteria (e.g., *Azotobacter* and *Rhizobium*) contribute to nitrogen availability by fixing atmospheric nitrogen (Aseri et al. 2008).
2. **Disease Suppression**
   - Beneficial microbes produce antibiotics and enzymes that inhibit the growth of soil-borne pathogens. For example, *Trichoderma* fungi suppress harmful fungi like *Fusarium* and *Phytophthora* (Asad 2022).
   - The rhizosphere's microbial diversity can outcompete pathogens for resources, reducing disease incidence (Glick 2012).
3. **Stress Mitigation**
   - Microbial communities in the rhizosphere enhance plant resilience to abiotic stresses such as drought and salinity by improving root hydraulic conductivity and producing stress-related phytohormones (e.g., abscisic acid) (Glick 2012).
4. **Carbon Cycling**
   - Mango roots secrete carbon-rich compounds that feed microbial populations. This process contributes to the soil organic matter pool, enhancing soil fertility and structure over time (Weller et al. 2002).
5. **Plant Growth Promotion**

- Rhizobacteria that support plant growth exhibit a variety of strategies that contribute to plant development. Because of their capacity to produce and release plant hormones, which include gibberellins (GAs), cytokinins, indole-3-acetic acid (IAA), and other volatiles, various PGPR can modify root structure and encourage plant development (Kejela 2024). There have been studies where the effect of PGPRs showed very promising results for breaking dormancy and inducing flowers. Santos-Villalobos (2013) reported the application of *Burkholderia caribensis* XV and *Rhizobium* sp. XXV, as plant growth-promoting rhizobacteria (PGPR) strains that were isolated from the mango rhizosphere, significantly enhanced mango growth and synchronized flowering. Improvements in morphometric, biochemical, and molecular parameters evidenced this effect. Notably, microbial inoculation upregulated the FT gene expression in treated trees compared to non-inoculated controls. Although biochemical markers such as total nitrogen and initial carbohydrate levels increased with inoculation, these parameters did not directly correlate with FT expression. Instead, they appeared to influence panicle development and flowering intensity, likely via the translocation of a putative FT protein through the phloem to receiver buds. This highlights the potential role of the FT gene as a key floral signaling molecule in mango. Interestingly, FT expression diminished sharply once flowering was established, suggesting that flower development may downregulate FT activity. Additionally, temperature played a critical role in flowering induction, as observed during a one-month period of inductive conditions (night temperatures around 10 °C and daytime temperatures near 20 °C). Bud growth during this period resulted in mixed and floral buds, with the developmental outcome influenced by exposure duration: less than two weeks for mixed buds and more than three weeks for floral buds. The study underscores the potential of inoculating mango trees with B. caribensis XV and *Rhizobium* sp. XXV as an environmentally sustainable approach to stimulate growth and flowering, offering an alternative to traditional agronomic practices. Furthermore, exploring FT and other floral regulatory genes could pave the way for developing precise, eco-friendly strategies to enhance mango production while mitigating economic and environmental challenges. Similarly, the Pre-harvest application of biocontrol formulations to manage anthracnose in mango has been reported by Ramasamy et al. (2006) where pre-harvest application of biocontrol formulations comprising *Pseudomonas fluorescens* strains FP7 and Pf1, *Bacillus subtilis* strain Bs-1, and *Saccharomyces cerevisiae* strain Sc-1, applied either biweekly or monthly with or without a chitin amendment, was evaluated for its efficacy against mango anthracnose caused by *Colletotrichum gloeosporioides*. In vitro studies demonstrated that strain FP7 significantly inhibited the growth of *C. gloeosporioides*. In field trials, the combined application of FP7 with chitin consistently reduced infection levels. Biweekly pre-harvest treatments with these biocontrol formulations also enhanced key agronomic parameters, including flower initiation, yield (average fruit number and total fruit yield), and fruit quality attributes such as total soluble solids, ascorbic acid, free acidity, and sugar content (total, reducing, and non-reducing sugars). Notably, latent symptom expression of anthracnose was delayed by 15 days under storage conditions, highlighting improved post-harvest disease management. Additionally, the highest levels of phenolic compounds, peroxidase, and polyphenol oxidase activity in mango leaves, flowers, and fruits were achieved with the FP7 + chitin treatment. These findings demonstrate the potential of integrating biocontrol formulations, particularly FP7 with chitin, into sustainable mango production systems to enhance disease resistance, yield, and fruit quality while extending post-harvest shelf life. Likewise influence of liquid plant growth promoting rhizomicrobial consortia on mango grafting was found successful by Pradeepkumar et al. (2024) wherein the application of liquid plant growth-promoting rhizomicrobial (PGPR) consortia at the nursery stage significantly enhanced mango seedling vigor and grafting success. Various liquid PGPR treatments were applied to a media mixture of soil, sand, and FYM (3:1:1 ratio), including individual and combined applications of *Azotobacter chroococcum* (T2), *Bacillus megaterium* (T3), and *Frateuria aurantia* (T4), as well as their combinations: *A. chroococcum* + *B. megaterium* (T5), *A. chroococcum* + *F. aurantia* (T6), *B. megaterium* + *F. aurantia* (T7), and *A. chroococcum* + *B. megaterium* + *F. aurantia* (T8). Among these treatments, T8 demonstrated the most pronounced effects, achieving the highest sprout success rates (82.22% and 84.44% at 30 and 60 days post-grafting, respectively), survival percentage (84.55%), sprout length (9.30 cm), sprout girth (4.76 mm), number of leaves (15.37), leaf length (14.08 cm), and leaf breadth (3.29 cm) at 120 days post-grafting. These results indicate that mango rootstocks treated with the consortium of *A. chroococcum, B. megaterium,* and *F. aurantia* exhibit superior physiological growth and grafting outcomes, underscoring the potential of PGPR consortia in sustainable mango nursery management.

1.5.2 Key Insights on Mango Rhizosphere Management

To optimize the rhizosphere's potential in mango cultivation, sustainable practices must be adopted:

- **Organic Amendments**: Adding compost or manure enriches microbial diversity, enhancing rhizosphere function.
- **Inoculation with Beneficial Microbes**: Introducing PGPR or mycorrhizal fungi can improve nutrient acquisition and disease resistance (Berendsen et al. 2012).
- **Irrigation Management**: Maintaining adequate moisture levels supports active microbial processes in the rhizosphere.
- **Reduced Chemical Use**: Minimizing synthetic fertilizers and pesticides helps preserve beneficial microbial communities (Zakaria 2021).

1.5.3 Role of Rhizospheric Microbes in Soil Health and Disease Suppression

Rhizospheric microbes are integral to maintaining soil health and suppressing diseases, especially in agricultural systems such as mango orchards. These microorganisms, which inhabit the soil surrounding plant roots, engage in complex interactions with plants and the soil environment (Kour et al. 2022; Parmesan et al. 2022). Their activities improve nutrient availability, enhance soil structure, and provide biological control against pathogens, thereby promoting sustainable crop production.

1.5.3.1 Impact of PGPR on Soil Fertility and Plant Productivity

Over the past few decades, there has been a substantial increase in interest regarding the utilization of beneficial microbes in agriculture. This heightened interest is attributed to their remarkable capacity to enhance plant resistance against both pathogens and insect pests, as well as their ability to bolster tolerance to abiotic stressors (Fernández-Lizarazo and Moreno-Fonseca 2016; Veen et al. 2019). Using environmentally friendly crop protection techniques in fruit cultivation has become key to maintaining crop yield and quality due to the increased global demand for food and the need to reduce the use of agrochemicals that harm the environment and human health. In this regard, the application of beneficial microbes emerges as a promising substitute for chemical pesticides and fungicides, promoting sustainable practices in agriculture and horticulture. Within the current climate change framework, plant-beneficial microorganisms play a pivotal role in advancing sustainability in plant productivity. As an integral component of climate-smart agricultural practices, microorganisms contribute to improved nutrient availability for plants, reciprocally obtaining nutrients through root exudates from these plants (Hamilton et al. 2016; Meena et al. 2016; Patel et al. 2015). The term "beneficial microbes" encompasses a diverse array of microorganisms associated with plants that confer positive effects on plant growth (as biofertilizers), defense (as plant protectants), and both. Noteworthy examples of successful and extensively studied biofertilizers include plant growth-promoting rhizobacteria (PGPR) (Banerjee et al. 2006), arbuscular mycorrhizal fungi (AMF) (Berruti et al. 2016), and nitrogen-fixing bacteria (NFB) (Gupta et al. 2012).

The soil serves as the natural habitat for soil microorganisms, playing a pivotal role in various soil processes that directly influence plant productivity (Meena 2015; Meena et al. 2014). Plant growth-promoting rhizobacteria (PGPRs) actively engage in numerous beneficial activities within the soil ecosystem, including the decomposition of crop residues, synthesis of soil organic matter (SOM), mineralization of SOM, immobilization of mineral nutrients, phosphate solubilization, nitrogen fixation, synthesis of phytohormones, regulation of ethylene production in roots, promotion of mycorrhizal function, reduction of heavy metal toxicity, and disease suppression. These activities collectively contribute to soil fertility improvement, ultimately enhancing overall productivity (Prasad et al. 2016; Whipps 2001). The colonization of PGPRs in the rhizosphere is contingent upon the presence of adequate soil moisture (Shrivastava et al. 2014). Rhizobacteria associated with plant roots are responsible for the synthesis of a multitude of biomolecules that become integrated into the soil, thereby contributing to improved soil health. Additionally, these bacteria act on plant residues, facilitating the decomposition and mineralization of various organic compounds (Ahmad et al. 2011, 2016; Bahadur et al. 2016; Kumar et al. 2015; Verma et al. 2015, 2014; Yadav and Sidhu 2016; Yasin et al. 2016). The resulting mineralized chemicals permeate the soil, imparting value to its composition. Beyond this process, PGPRs also synthesize growth-related chemicals that ultimately influence plant root morphology. PGPR contributes significantly to the enhancement of various plant properties during phytoremediation, encompassing improvements in biomass, contaminant uptake, plant nutrition, and health (Verma et al. 2013). These benefits extend to maintaining cell turgor, augmenting photosynthesis, reducing respiration, facilitating the transport of sugars and starches, and fortifying the plant's ability to withstand stress. The overall health and growth of plants, as indicated by grain yields, are positively influenced by the incorporation of PGPR (Verma et al. 2013). Plant Growth-Promoting Rhizobacteria (PGPR) significantly enhance the yields of various crops, including fruits and vegetables, beyond their effects on grain yield. In a study on cucumber plants, the application of PGPR strains such as *Acinetobacter radioresistens* and *Sinorhizobium meliloti* improved growth parameters like plant height, root length, and biomass, as well as fruit size, yield, and nutraceutical quality under greenhouse conditions. This demonstrates PGPR's potential to reduce reliance on chemical fertilizers and enhance sustainable production (Zapata-Sifuentes et al. 2022). Similarly, research on tomato plants revealed that PGPR strains like *Bacillus subtilis* and *Bacillus amyloliquefaciens* significantly increased fruit yield by up to 76% and improved fruit quality, including higher total soluble solids and ascorbic acid content. These results emphasize the role of PGPR in improving the productivity and quality of horticultural crops (Gashash et al. 2022). Furthermore, PGPR usage offers a substantial benefit in terms of conferring disease resistance to plants, often referred to as "biocontrol" (Lucy et al. 2004). The biocontrol efficacy of PGPRs against diseases has been previously demonstrated in tomatoes and peppers by Domenech et al. (2006). Additionally, PGPR strains prove beneficial in mitigating the adverse effects of temperature stress, combating parasitic weeds, and extending the shelf life of flowers (Ait Barka et al. 2006; Babalola et al. 2003). Within the soil environment, the growth of plants is subject to the influence of soil-borne pathogens. Plant growth-promoting rhizobacteria (PGPR) play a crucial role in safeguarding plants from pathogens and augmenting their resistance against diseases, employing various mechanisms such as antibiosis, competition, and parasitism (Beneduzi et al. 2012; Cassells and Rafferty-McArdle 2012; Khokhar 2012; Ping and Boland 2004). PGPR employs one or more of these biocontrol mechanisms to protect plants. For instance, the antibiosis mechanism involves the inhibition of pathogen growth through the production of antimicrobial compounds by bacteria (Glick et al. 1995). Similarly, PGPR diminish the availability of iron, essential for pathogen growth, thereby impeding their development. An additional significant mechanism employed by PGPR is induced systemic resistance (ISR), wherein PGPR induces changes in host-plant vulnerability, enhancing plant resistance against diseases (Khokhar 2012). Some PGPRs exhibit multiple mechanisms, enabling them to thrive not only in normal but also in stressful environments. Under stress conditions, plants respond by elevating ethylene levels, leading to increased cell and plant damage (Argueso et al. 2007). An elevated ethylene concentration can pose a threat as it triggers defoliation and various cellular processes that may adversely impact crop development (Desbrosses et al. 2009). Numerous plant growth-promoting rhizobacteria (PGPR) counteract the harmful effects of ethylene by degrading 1-aminocyclopropane-1-carboxylate (ACC), an ethylene precursor, through the production of the enzyme ACC deaminase. This enzymatic action contributes to plant growth and development by lowering ethylene levels. Furthermore, ACC deaminase-producing PGPR exhibit stress-alleviating properties, including effects on phytopathogenic bacteria, resistance to stress induced by polyaromatic hydrocarbons, salt, and drought (Glick et al. 2007). The efficacy of PGPR in promoting plant growth is contingent upon their interaction with the host plant and the soil environment, in addition to their inherent capabilities (Nadeem et al. 2014).

The pre-treatment of seeds with various plant growth-promoting rhizobacteria (PGPR) has been observed to enhance seed germination and seedling growth (Poupin et al. 2013; Rahmoune et al. 2017; Bakhshandeh et al. 2020). It is believed that, as part of this mechanism, PGPR contribute to maintaining hormonal balance, particularly the levels of auxin to cytokinin, during germination and the initial stages of plant development. This involvement plays a crucial role in shaping the genetic program governing post-embryonic roots and shoot growth (Qessaoui et al. 2019).PGPR employs diverse mechanisms to stimulate plant growth, with a notable emphasis on nutrient availability and the acquisition of mineral resources such as phosphorus, phytohormone production, the release of volatile compounds for controlling seed- and soil-borne phytopathogens, and collaboration with other plant-beneficial microorganisms to enhance resistance against various stresses (Bhattacharyya et al. 2012, 2015).

Recent findings highlight that certain rhizobacteria play a role in promoting plant growth through the release of volatile signals (Ping and Boland 2004). The identification of volatile organic compounds (VOCs) produced by rhizobacteria, including 2,3-butanediol, acetoin, terpenes, jasmonates, and others, represents a significant mechanism for stimulating plant growth. The synthesis of bioactive VOCs appears to be a strain-specific phenomenon. These VOCs, generated by plant growth-promoting rhizobacteria (PGPR), act as signaling molecules that mediate plant–microbe interactions, with concentrations sufficient to elicit plant responses when colonizing roots (Ryu et al. 2003). In maize cultivation, plant growth-promoting bacteria (PGPR) such as *Burkholderia* sp., *Azospirillum* sp., and *Bacillus* sp. enhance both total plant and grain dry weight, particularly under nitrogen-limited conditions. PGPR utilization, exemplified by P. fluorescens DR54, contributes to improved phosphorus nutrition in maize, offering potential benefits for phosphorus-deficient soils. Additionally, PGPR strains like *Bacillus amyloliquefaciens* and *Microbacterium oleovorans* exhibit biocontrol capabilities against *Fusarium verticillioides* in maize crops (Riggs et al. 2001; Shaharoona et al. 2006). Diazotrophic bacteria significantly benefit rice crops, with genera like Azospirillum and Pseudomonas contributing to increased grain production and root length. PGPR also plays a vital role in disease control in rice, particularly against *Xanthomonas* oryzae pv. oryzae, *Rhizoctonia solani*, and *Magnaporthe oryzae* (Han and Lee 2005). Wheat cultivation reaps advantages from PGPR by mitigating the necessity for chemical fertilizers. PGPR consortia, which include arbuscular mycorrhizal fungi (AMF), enhance wheat yields, encompassing an increase in grain and straw yield, by facilitating nutrient absorption. PGPR also contributes to biological control against wheat pathogens and alleviate salinity stress (Nawaz et al. 2020). In soybean crops, co-inoculation studies with rhizobia and PGPR result in increased nitrogen fixation, while biocontrol agents like Bacillus and Pseudomonas demonstrate effectiveness against soybean pathogens (Cassan et al. 2009).

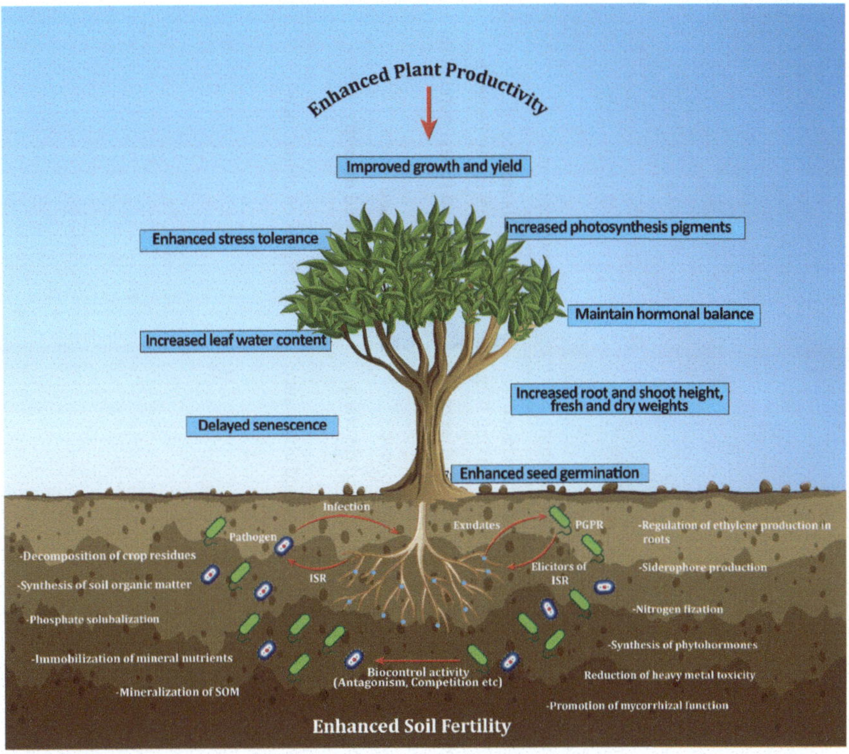

Fig. 1.4 Impact of PGPRs/ probiotics on the health of the plant

For bean cultivation, PGPR such as Pseudomonas and Bacillus enhance nodulation and nitrogen fixation (Grimes et al. 1992). Additionally, Trichoderma fungi contribute to biocontrol and iron mobilization in beans (Altomare et al. 1999). Some PGPR exhibit dual roles, simultaneously acting as biofertilizers and biocontrol agents in bean crops (Peix et al. 2001). Overall, the use of PGPR in these crops showcases their potential for sustainable agriculture by promoting plant growth, nutrient uptake, and disease resistance while reducing the reliance on chemical inputs (Fig. 1.4 and Table 1.2).

1.5.3.2 Role of Rhizospheric Microbes in Soil Health

1. **Nutrient Cycling**
    - Microorganisms in the rhizosphere are pivotal in decomposing organic matter and cycling essential nutrients like nitrogen, phosphorus, and potassium.
        - **Nitrogen Fixation**: Bacteria such as *Rhizobium* and *Azotobacter* fix atmospheric nitrogen into bioavailable forms for plants.

Table 1.2 Microbial diversity in mango orchards under varying conditions

Microbial	High organic matter	Low organic matter	High moisture	Low moisture
Bacteria	Dominance of *Bacillus*, *Pseudomonas*	Reduced *Bacillus*, increased actinobacteria	High abundance of nitrogen-fixing bacteria	Low abundance of nitrogen-fixing bacteria
Fungi	High diversity of mycorrhizal fungi	Dominance of saprophytic fungi	Increased *Aspergillus*, *Penicillium*	Reduced fungal population overall
Actinobacteria	Moderate levels	High levels	Reduced levels	Increased levels
Yeasts	High abundance of fermentative yeasts	Low abundance of fermentative yeasts	Increased diversity	Reduced diversity
Protists	High diversity, including *Amoebae*	Low diversity	Increased abundance	Reduced abundance

- **Phosphorus Solubilization**: Phosphate-solubilizing bacteria (e.g., *Pseudomonas* and *Bacillus*) convert insoluble phosphates in soil into forms that plants can absorb (Shyam et al. 2019).

2. **Improvement of Soil Structure**

   - Rhizospheric microbes secrete extracellular polysaccharides that help aggregate soil particles, enhancing soil structure, porosity, and water-holding capacity.
   - Fungi, such as mycorrhizal species, contribute to the formation of stable soil aggregates by binding particles through their hyphal networks (Bashan and De-Bashan 2010).

3. **Organic Matter Decomposition**

   - Microbes like fungi and actinomycetes break down plant residues and organic matter, enriching the soil with humus and improving fertility.

4. **Enhancing Root Growth**

   - Plant growth-promoting rhizobacteria (PGPR) produce phytohormones, such as indole-3-acetic acid (IAA) and gibberellins, which stimulate root elongation and branching.
   - Improved root architecture enhances nutrient and water uptake, indirectly contributing to soil health by maintaining a robust rhizosphere (Smith and Read 2010).

1.5.3.3 Role of Rhizospheric Microbes in Disease Suppression

Azospirillum, a prominent member of the PGPR group, has garnered significant attention for its ability to enhance crop development and yield across various environmental conditions. Its beneficial effects, such as nitrogen fixation, nitrate reduction, and phytohormone synthesis, underscore its critical role in plant–microbe interactions. Despite extensive research into these pathways, the genetic and molecular mechanisms underlying the initial interaction between *Azospirillum* and plant roots remain largely unexplored, indicating potential areas for further discovery (Smith and Read 2010). Similarly, *Pseudomonas fluorescens* strain KH-1 has demonstrated growth-promoting properties in rice plants in both controlled and natural environments, although the molecular mechanisms driving these effects are not yet fully understood. Advanced protein profiling techniques have revealed differentially expressed proteins in rice plants treated with *P. fluorescens*, offering insights into the complex molecular interactions between PGPR and their plant hosts (Ploetz 2003). Isolates such as D28 and D12, which produce indole acetic acid (IAA), ammonia, siderophores, and solubilized phosphate, have shown significant promise in promoting rice growth and inducing various plant growth-promoting traits, making them viable options for sustainable agriculture (Jain et al. 2024). The role of PGPR in enhancing crop yields is further illustrated by the ability of *Rhizobium leguminosarum trifolii* SN10 to stimulate rice growth through mechanisms such as chemotaxis, colonization, and phytohormone production, including IAA and ACC deaminase (Bhattacharjee et al.). This highlights the importance of PGPR in improving agricultural productivity, particularly in regions like the Indian subcontinent (Johnson 1999). Moreover, the antagonistic properties of rhizosphere-associated microorganisms such as Bacillus and Pseudomonas play a crucial role in promoting plant health and suppressing diseases (Gupta et al. 2022). These microorganisms enhance plant defense mechanisms by producing siderophores, solubilizing phosphate, and synthesizing IAA. Co-inoculation of these bacterial consortia has been shown to trigger defense-related enzyme activities, bolstering rice plants' resistance to pathogens and significantly improving plant health and productivity (He et al. 2024). Specific strains of *Azotobacter* spp. have also demonstrated their capacity to stimulate rice seedling growth through nitrogen fixation, siderophore production, IAA synthesis, and phosphate solubilization. These findings provide valuable information for developing effective biofertilizers for sustainable rice farming (Reddy 2015).

The molecular interactions between plants and microbes focus on the role of phenolic compounds in promoting rice plant growth and mediating defense responses. Rhizobial strains such as ANU 843 and RRE6 have been shown to induce phenolic compound synthesis in rice plants under pathogenic stress, highlighting the critical role of rhizobial-mediated defense responses in enhancing plant growth and productivity (Ploetz and Freeman 2009). In vivo, studies have demonstrated the effectiveness of co-inoculating PGPM, including *Enterobacter* sp., *Pantoea ananatis*, and *Piriformospora indica*, with fertilizers like potassium sulfate in enhancing nutrient uptake efficiency and significantly increasing rice plant tissue dry weight. These results support the use of co-inoculation techniques as a sustainable strategy for optimizing nutrient utilization and promoting plant growth in rice farming (Thind 2015). Chakraborty and Gupta (2020) isolated and identified three rice rhizosphere microorganisms, i.e., *Bacillus subtilis* DD-4, *Bacillus aryabhattai* DD-3, and *Bacillus megaterium* DD-2, from Northeast Chinese farm soil. These isolates exhibited remarkable abilities to solubilize phosphorus and potassium, produce siderophores, gibberellic acid (GA), and IAA, and enhance rice seedling growth in pot studies. Similarly Kamil et al. (2018) reported the biological control of mango die-back disease caused by *Lasiodiplodia theobromae* using *Streptomycete* and *Non-streptomycete* actinobacteria in the United Arab Emirates in a study where the authors isolated 53 actinobacterial strains from mango rhizosphere soil in the UAE, of which 66% (35 isolates) were identified as streptomycetes (SA) and 34% (18 isolates) as non-streptomycetes (NSA). Among these, 19 isolates (12 SA and 7 NSA) exhibited antagonistic activity against *L. theobromae* through the production of diffusible antifungal compounds, extracellular cell wall-degrading enzymes (CWDEs), or both. A novel mango fruit bioassay was employed to evaluate the in vivo efficacy of these isolates in reducing lesion severity caused by *L. theobromae.* Three isolates—two from the genus *Streptomyces* and one from *Micromonospora* spp.—demonstrated the strongest inhibitory effects in vitro and were further tested on mango seedlings. The antifungal activity of *Streptomyces samsunensis UAE1* was attributed to antibiosis and the production of CWDEs, such as chitinase, as well as siderophores. In contrast, *S. cavourensis UAE1* exhibited primarily antibiotic activity, while *Micromonospora tulbaghiae UAE1* relied on CWDE production. Greenhouse experiments using pre-inoculation with these promising actinobacterial isolates provided substantial disease protection in mango seedlings challenged with *L. theobromae.* The treatment with *S. samsunensis* notably reduced the number of defoliated leaves and conidia counts of the pathogen by 2- and fourfold, respectively, compared to the other two biocontrol agents. This superior performance suggests that the multifaceted antifungal mechanisms of *S. samsunensis* effectively inhibited the pathogen in planta invasion, underscoring its potential as a robust biocontrol agent for managing mango dieback disease. Endophytic fungi are known to produce bioactive secondary metabolites that enhance plant growth through various mechanisms. In a prior study, endophytic fungi were isolated from mango (*Mangifera indica* L.) trees, and the current research evaluated fifty of these isolates for their abilities to solubilize minerals, produce ammonia, and synthesize siderophores. These isolates were also found to produce phytohormones such as indole-3-acetic acid and gibberellic acid, in addition to exhibiting inhibitory effects against the plant pathogens *Colletotrichum gloeosporioides* and *Lasiodiplodia theobromae.* Among the isolates, two—*Aureobasidium pullulans* CY.OS 13 and *Aspergillus tamarii* CY.OS 144—were selected based on their promising activities and tested in dual inoculation experiments on chili plants under controlled pot-scale conditions. Seeds treated with these fungal strains, either individually or in combination, exhibited a significant ($p < 0.05$) improvement in germination rates and growth parameters, including shoot and root length, biomass, and chlorophyll content. Notably, the combination treatment led to the most pronounced enhancement in plant development. These findings suggest that the selected fungal endophytes have strong potential as biofertilizers. When used together, they could serve as an eco-friendly alternative to chemical fertilizers, benefiting human health and promoting sustainable agricultural practices (Sonsiam et al. 2024). Similarly, Talc-based bioformulations containing *Pseudomonas fluorescens, Bacillus subtilis,* and *Saccharomyces cerevisiae* were assessed for their effectiveness against mango (*Mangifera indica* L.) anthracnose caused by *Colletotrichum gloeosporioides* Penz. under endemic conditions. Preharvest aerial sprays were applied at fortnightly and monthly intervals. Among the treatments, *Pseudomonas fluorescens* (FP7) combined with chitin and applied fortnightly resulted in the highest flowering induction and improved yield attributes during the preharvest stage, along with a significant reduction in latent symptoms in the postharvest stage. The FP7 + chitin treatment also triggered substantial activity of defense-related lytic enzymes, including chitinase and β-1,3-glucanase, as confirmed by colorimetric assays and the appearance of distinct bands in native PAGE analysis. The enhanced activity of these defense enzymes likely played a key role in suppressing the anthracnose pathogen, contributing to improved yield and reduced disease severity (Vivekananthan et al. 2004) (Fig. 1.5).

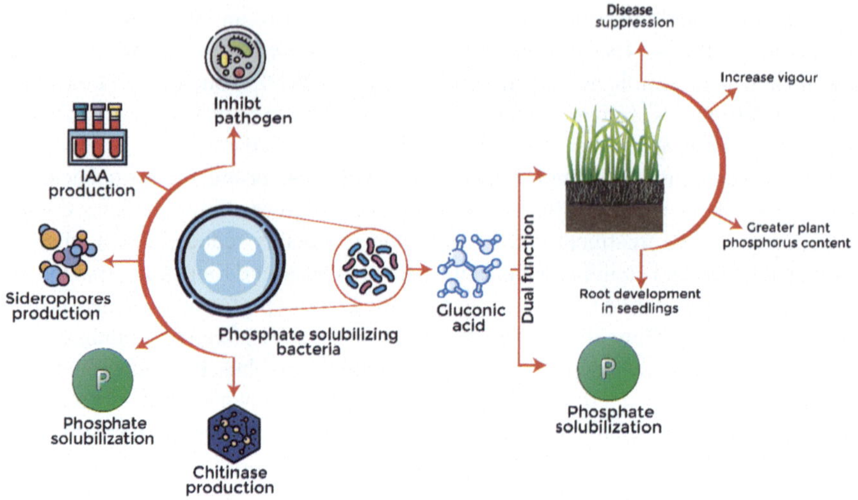

Fig. 1.5 How PGPRs play their role in the rhizosphere to help plants

1. **Antagonistic Interactions with Pathogens**
   - Beneficial microbes suppress soil-borne pathogens through direct antagonism, including:
     - **Production of Antibiotics**: Microbes like *Streptomyces* and *Pseudomonas fluorescens* produce antimicrobial compounds that inhibit fungal and bacterial pathogens.
     - **Enzymatic Degradation**: Enzymes such as chitinases and glucanases degrade the cell walls of pathogenic fungi like *Fusarium* and *Phytophthora*.
2. **Resource Competition**
   - Pathogen growth is inhibited by beneficial microorganisms competing with them for nutrients and rhizosphere space.
   - The establishment of harmful organisms in nearby areas of the root zone is prevented by this competitive exclusion.
3. **Induction of Systemic Resistance**
   - Certain microbes trigger systemic resistance mechanisms in plants, enhancing their defense against pathogens. For instance, PGPR such as *Bacillus subtilis* and mycorrhizal fungi activate signaling pathways (e.g., jasmonic acid and salicylic acid pathways) that bolster plant immunity (Mitter et al. 2019).
4. **Detoxification of Soil**
   - Microbes can degrade or immobilize toxins produced by pathogens or accumulated due to human activity, thereby restoring soil health

1.6 Microbiome Engineering: Concept and Applications

1.6.1 Definition and Scope of Microbiome Engineering

Microbiome engineering refers to the deliberate design, modification, and manipulation of microbial communities to achieve specific desired outcomes in various environments. This multidisciplinary field combines microbiology, genetics, biotechnology, and systems biology principles to enhance microbiome's functional capabilities for applications in agriculture, healthcare, industry, and environmental management (Bakker et al. 2007; Beneduzi et al. 2012).

1. **Agriculture**
   - Enhancing soil fertility and crop yield through beneficial microbial consortia.
   - Reducing the need for chemical fertilizers and pesticides by promoting natural pest resistance and nutrient cycling.
   - Developing microbial bioinoculants tailored to specific crops or environmental conditions (Bakker et al. 2007).
2. **Environmental Applications**
   - Bioremediation of polluted environments using microbes capable of degrading pollutants like plastics, oil spills, and heavy metals.
   - Engineering microbial communities to sequester carbon and mitigate climate change impacts.
   - Restoring degraded ecosystems by reintroducing or enhancing natural microbial diversity.
3. **Industrial Biotechnology**
   - Optimizing microbial fermentation processes for producing biofuels, bioplastics, and pharmaceuticals.
   - Engineering microbes for efficient waste-to-energy conversion.
4. **Synthetic Ecology**
   - Constructing synthetic microbial ecosystems for controlled and predictable outcomes in artificial or natural environments (Chaparro et al. 2012).

1.6.2 Emerging Areas

With advancements in tools such as CRISPR-Cas9, metagenomics, and machine learning, microbiome engineering is increasingly precise and targeted, offering solutions for global challenges in food security, healthcare, and environmental sustainability. This field holds transformative potential, fostering innovation across scientific and industrial domains (Agrawal et al. 2021).

1.6.3 Advances in Microbial Technologies for Rhizosphere Management

The rhizosphere, the narrow region of soil surrounding plant roots, is a hotspot of microbial activity that plays a critical role in plant health, growth, and nutrient acquisition. Recent advancements in microbial technologies have revolutionized rhizosphere management, enabling sustainable agricultural practices and improved crop productivity. Key developments include:

1.6.3.1 Bioinoculants and Probiotics

- **Beneficial Microbial Consortia**: Formulation of bioinoculants using combinations of bacteria (e.g., *Pseudomonas*, *Bacillus*, *Rhizobium*) and fungi (e.g., mycorrhizal fungi, *Trichoderma* spp.) that enhance nutrient uptake, improve soil structure, and suppress pathogens.
- **Tailored Probiotics**: Advances in microbiome profiling enable the design of crop-specific probiotics to optimize root microbiota for specific environmental conditions (Glick 2015).

1.6.3.2 Microbial Biostimulants

- **Plant Growth-Promoting Rhizobacteria (PGPR)**: Strains engineered or selected for their ability to produce growth hormones (e.g., auxins, cytokinins) or improve stress tolerance under drought, salinity, or nutrient deficiency.
- **Phosphate-Solubilizing Microorganisms (PSM)**: Use of microbes capable of mobilizing bound phosphorus, enhancing plant bioavailability.

1.6.3.3 CRISPR and Synthetic Biology

- **Engineered Microbes**: CRISPR-Cas9 and other gene-editing tools allow precise engineering of microbes to improve their functions, such as nitrogen fixation, pathogen suppression, or stress resilience (Rosenberg et al. 2013).
- **Synthetic Microbial Communities**: Construction of synthetic consortia designed to perform complementary functions in the rhizosphere, improving efficiency and stability.

1.6.4 Mechanism of Microbe Induced Immunity in Plant Against Phytopathogens

Plants exhibit varying degrees of resistance against microbial pathogens, typically triggered by chemical or biological stimuli. Induced resistance mechanisms, such as cell wall fragmentation, often fail to provide absolute control over pathogens. However, they may contribute to minimizing lesion size or quantity. In the perpetual bidirectional interaction between plants and pathogenic microbes, plants rely on their innate ability to perceive signals from potential pathogens. As sessile organisms, plants have evolved a sophisticated innate immune system encoded in their genome, comprising various traits that enable them to detect and combat pathogens. This intricate interplay between induced resistance and innate immunity underscores plant defense mechanisms' dynamic and complex nature in response to external threats (Tiwari et al. 2022). Induced Systemic Resistance (ISR) is triggered when plants interact with beneficial microorganisms, which can improve the health and growth of the associated plants. Upon activation of ISR, plants employ long-distance systemic signaling to provide protection for distal tissues, inducing rapid and strong immune responses against pathogen invasions. The transmission of ISR signaling was initially thought to be jasmonic acid- and ethylene-dependent but salicylic acid independent. However, recent research has shown that salicylic acid and jasmonic acid/ethylene signalling pathways, as well as small RNAs, play crucial roles in ISR (Yu et al. 2022). However, localized pathogen attack led to the adaptive resistance development in plants associated with priming of defense is referred to as Systemic Acquired Resistance (SAR)'. Both ISR and SAR are involved with the production of signaling compounds such as jasmonic acid, salicylic acid, ethylene, phytoalexins, some protein-related chitinases, glucanases, and so on. According to Goellner and Conrath (2008), the elicitation of plant defense resistance against pathogens can arise from either the plant's responsive reaction to the presence of beneficial microorganisms or the heightened activation of defense mechanisms induced by these associated microorganisms. Direct induction and priming are the basis for the most induced resistance phenomena (Kuć 1982).

The mechanism of microbe-induced immunity in plants against phytopathogens involves two lines of defense: PAMP-triggered immunity (PTI) and effector-triggered immunity (ETI). In the initial line of defense against pathogens, plants employ a recognition system involving their outer membrane receptors known as Pattern Recognition Receptors (PRRs). These receptors identify pathogen-derived molecules, referred to as PAMPs (Pathogen-Associated Molecular Patterns) or conserved microbe-associated molecular patterns (MAMPs), which include substances like chitin, saliva, and flagella components (Newman et al. 2013). Additionally, Boller and Felix (2009) observed that plants respond to signals originating from their own tissues due to damage caused by invading pathogens, known as Damage-Associated Molecular Patterns (DAMPs). Following the recognition of MAMPs, PRRs recruit regulatory receptor kinases to form PRR complexes, initiating a complex immune signaling cascade through receptor-like cytoplasmic kinases.

This cascade triggers a series of defensive responses, including cell wall thickening, rearrangement of the cytoskeleton, heightened reactive oxygen species (ROS) generation, and metabolic reprogramming facilitated by MAPK-mediated signaling. This orchestrated sequence of immune events, collectively known as MAMP-Triggered Immunity (MTI), serves to eliminate potential pathogenic infections (Macho et al. 2014; Hacquard et al. 2017; Nishad et al. 2020).

The plant's initial defense mechanism, involving Pattern Recognition Receptors (PRRs) and Pathogen-Associated Molecular Patterns (PAMPs) recognition, is effective in restraining the growth and development of most pathogens. However, some adaptive pathogens counteract this responsive mechanism by suppressing PAMP-Triggered Immunity (PTI), a process known as Effector-Triggered Susceptibility (ETS). In response to this evasion strategy, plants have evolved a second line of defense called Effector-Triggered Immunity (ETI), distinct from PTI in its heightened and intensified retaliatory activity. ETI culminates in hypercell death, activation of "R" genes, and callose deposition, effectively hindering the further spread of pathogens (Pastor et al. 2013). The activation of ETI is triggered by avirulence genes (avr) in pathogens encoding effector proteins. These effectors are recognized by intracellular receptors called nucleotide-binding domain leucine-rich repeats (NLR), which are encoded by resistance genes (R genes). Epigenetic modifications significantly impact plant immune responses, including DNA methylation, histone methylation, and acetylation. This epigenetic mechanism involves the expression of defense-related genes, playing a crucial role in establishing immunogenic memory. Epigenetic priming, characterized by dynamic chromatin states, exerts a transgenerational effect, making priming states heritable (Iriti and Vitalini 2021). Subsequently, plants adaptively reprogram their defense systems to counteract threats effectively. In this intricate defense scenario, the rhizosphere microbiome substantially reprograms plant defense responses (Spence et al. 2014). Whether dealing with adaptive or symbiotic rhizosphere pathogens, the mechanism involves overcoming plant immune responses and reorganization by Pattern Recognition Receptors (PRR). This underscores the multifaceted management role of microbial communities in the innate immune system, balancing the controlled accommodation of beneficial pathogens versus the elimination of harmful ones (Hacquard et al. 2017) (Fig. 1.6).

1.6.4.1 Metagenomics and Bioinformatics

- **Microbial Profiling**: High-throughput sequencing technologies enable detailed mapping of the rhizosphere microbial communities, helping identify keystone species and their functional roles.
- **Predictive Modeling**: Bioinformatics tools are used to model interactions between plants, microbes, and the soil environment, optimizing microbial applications for specific crops or conditions (Hartmann et al. 2015).

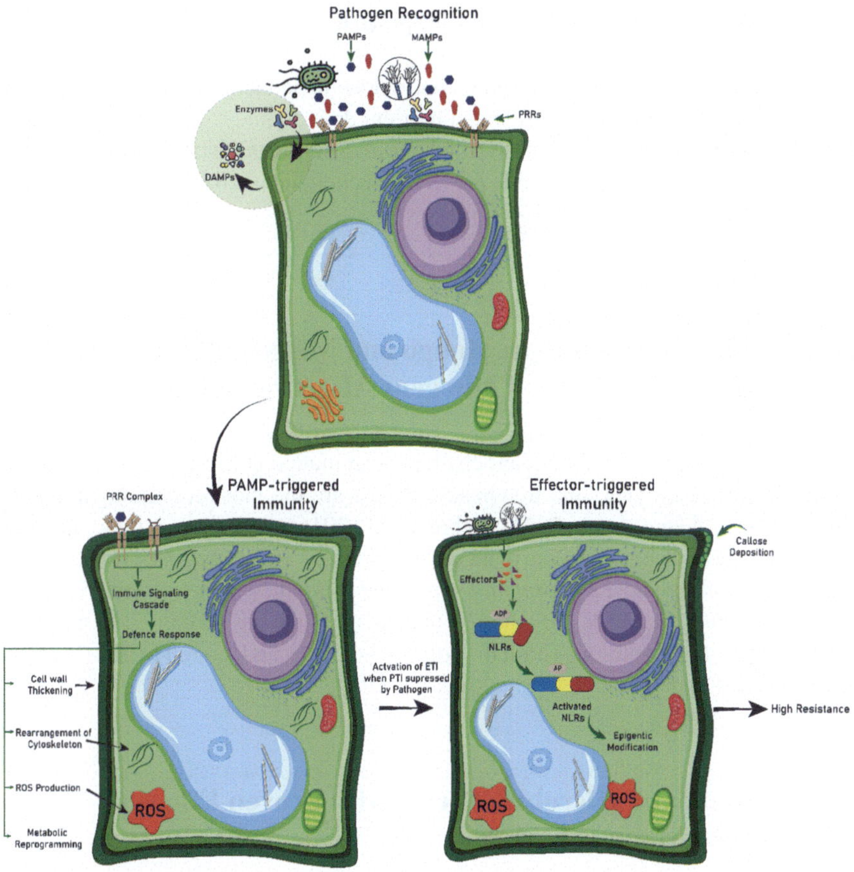

Fig. 1.6 Microbial-triggered immunity in plants to boost their response to stress

1.6.4.2 Nanotechnology and Microbial Delivery

- **Nano-carriers for Microbes**: Encapsulation of microbes in nanomaterials to protect them during application and enhance their delivery to plant roots.
- **Controlled Release Formulations**: Innovations in slow-release carriers ensure prolonged microbial activity in the rhizosphere.

1.6.4.3 Rhizosphere Manipulation for Carbon Sequestration

- Harnessing rhizosphere microbes to enhance soil organic carbon storage through optimized root exudate dynamics and microbial activity.

1.6.4.4 Microbial Biofertilizers and Biopesticides

- Development of microbial products that reduce dependence on chemical inputs by providing natural pest control and nutrient delivery.
- Biopesticides, such as those derived from *Bacillus thuringiensis* or *Metarhizium*, target soil-borne pathogens while preserving beneficial microbes (Hartmann et al. 2015; Singh et al. 2018) (Table 1.3).

1.7 Soil-Borne Diseases in Mango and Microbial Solutions: Common Soil-Borne Pathogens Affecting Mango Trees

Soil-borne pathogens are significant challenges in mango cultivation, often causing diseases that reduce yield and tree health. Below are the most common pathogens (Glick 2015; Meena et al. 2024; Nehra et al. 2016). Soil borne fungi are the most common and considered as plant destroyer e.g., *Ceratocystis fimbriata, Botryodiplodia theobromae, Lasiodiplodia theobromae* as they penetrate the vascular system from the soil and destroy the complete mango tree within days. The perpetuation of the soil-borne fungi within the soil can be better understood by the (Fig. 1.7 and Tables 1.4 and 1.5).

1.8 Examples of Microbial Agents with Proven Efficacy Against Mango Diseases

Microbial agents have shown significant efficacy in managing various mango diseases, offering sustainable alternatives to chemical treatments. For instance, *Trichoderma harzianum* and *Trichoderma viride* are effective biocontrol agents against soil-borne pathogens like *Fusarium oxysporum* and *Sclerotium rolfsii*, reducing wilt and root rot incidences. *Pseudomonas fluorescens* has been widely used for its antifungal properties, suppressing diseases like anthracnose caused by *Colletotrichum gloeosporioides*. Similarly, *Bacillus subtilis* and *Bacillus amyloliquefaciens* are known to control bacterial leaf spot and stem-end rot by producing antimicrobial compounds. *Beauveria bassiana* has shown potential in managing mango pests, indirectly reducing pathogen entry through insect wounds. These microbial agents not only control pathogens but also improve plant growth and resilience, promoting eco-friendly mango production (Glick 2015; Hartmann et al. 2015).

Table 1.3 Methods to enhance beneficial microbial populations in mango orchards

Method	Description	Benefits	Examples
Organic amendments	Application of compost, manure, or organic residues	Improves soil organic matter, promotes microbial diversity, and enhances nutrient cycling	Addition of composted farmyard manure or vermicompost
Biofertilizers	Inoculation with beneficial microorganisms such as nitrogen-fixing bacteria and phosphate solubilizers	Boosts nutrient availability to plants and reduces chemical fertilizer use	Use of *Rhizobium*, *Azotobacter*, or *Pseudomonas fluorescens*
Mulching	Use of organic or synthetic mulches around trees	Maintains soil moisture, enhances microbial habitat, and suppresses weeds	Application of straw, sugarcane bagasse, or biodegradable plastic mulch
Mycorrhizal inoculation	Introduction of arbuscular mycorrhizal (AM) fungi	Enhances nutrient uptake (especially phosphorus) and drought resistance	Use of *Glomus* spp. in mango orchards
Crop rotation and intercropping	Inclusion of legumes or other beneficial plants in cropping systems	Improves soil structure, enhances microbial interactions, and diversifies microbial populations	Intercropping mango with leguminous plants like pigeon peas or cowpeas
Reduced chemical input	Minimizing use of synthetic fertilizers and pesticides	It prevents the disruption of beneficial microbes and supports natural soil processes	Integrated pest management (IPM) and precision fertilization practices
Biopesticides	Use of microbial-based pesticides for pest control	Targets pests without harming non-target beneficial microbes	Use of *Bacillus thuringiensis* or *Beauveria bassiana*
Irrigation management	Efficient water management techniques	Prevents waterlogging or drought conditions, which can harm microbial populations	Drip irrigation systems to maintain consistent soil moisture

(continued)

Table 1.3 (continued)

Method	Description	Benefits	Examples
Cover cropping	Planting cover crops during off-seasons	Protects soil, prevents erosion, and provides a substrate for microbial growth	Use of legumes or grasses as cover crops
Microbial engineering	Application of genetically improved beneficial microbes	Enhances specific microbial functions like nitrogen fixation or pathogen resistance	CRISPR-modified *Pseudomonas* strains for enhanced pathogen control
Monitoring and soil analysis	Regular analysis of soil microbial communities	Helps track microbial diversity and guide targeted interventions	Metagenomics or soil health assays to identify microbial composition

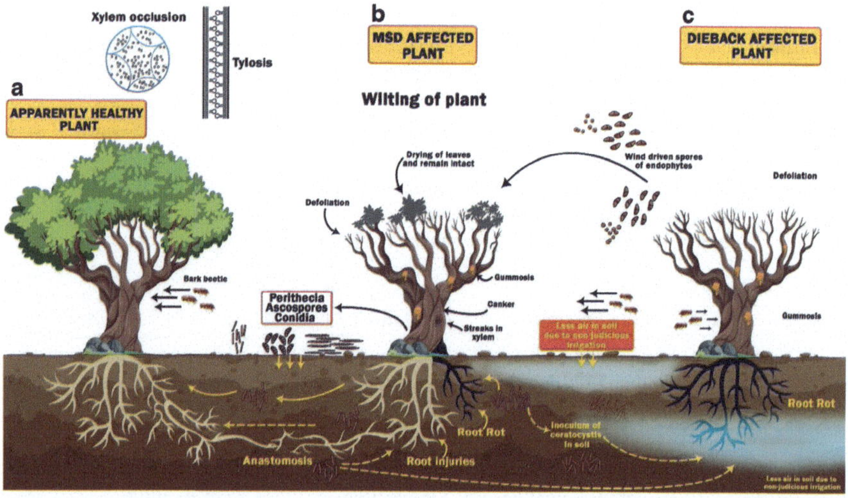

Fig. 1.7 Scheme of fungal progression in the orchards. A healthy plant in the orchard, when developing anastomosis of roots with an affected plant, gets infected with the disease, and the disease starts its progression from the roots (**a**); MSD affected plant shows complete wilting of the tree and serves as a source of inoculum in the orchards (**b**); some trees show complete wilting and while some show partial wilting of the trees in the orchards may be of one limb or two limbs at a time are severely pruned to the disease and needs the attention of the growers (**c**)

Table 1.4 Soil-borne pathogens, their causal agents, and symptoms

Pathogen	Disease caused	Symptoms
Fusarium oxysporum	Fusarium wilt	Wilting, yellowing of leaves, vascular discoloration
Phytophthora palmivora	Phytophthora root rot and collar rot	Water-soaked lesions on roots stem cankers, and eventual plant collapse
Ceratocystis fimbriata	Mango wilt	Sudden wilting, gummosis, vascular browning, and death of branches
Macrophomina phaseolina	Charcoal rot	Root necrosis, leaf drop, and black fungal sclerotia in tissues
Lasiodiplodia theobromae	Stem-end rot and root rot	Dieback, black discoloration of vascular tissues, and stem cankers
Pythium spp.	Pythium root rot	Root decay, stunted growth, and water-soaked lesions on root systems
Sclerotium rolfsii	Southern blight	Collar rot, white mycelial growth, and sclerotia formation on the soil
Rhizoctonia solani	Root and stem rot	Stem cankers, damping-off in seedlings, and brown root lesions
Verticillium dahliae	Verticillium wilt	Yellowing of leaves, unilateral leaf scorch, and vascular discoloration
Armillaria mellea	Armillaria root rot	Mushrooms near roots, mycelial fans under bark, and tree decline

1.9 Integrating Cultural Practices for Enhanced Effectiveness

1.9.1 Importance of Revised Cultural Practices in Climate Change Scenarios

Climate change presents one of the most significant challenges of the twenty-first century, impacting ecosystems, economies, and societies globally. Addressing this crisis requires not only technological and policy solutions but also a profound shift in cultural practices. This essay explores the importance of revising cultural practices to mitigate the effects of climate change and adapt to its consequences, drawing on a range of scholarly sources.

Table 1.5 Traditional versus microbiome-based management strategies

Aspect	Traditional management strategies	Microbiome-based management strategies
Definition	Conventional practices for disease and nutrient management often rely on chemicals or mechanical methods	Utilizes microbial communities and their interactions to enhance plant health and productivity
Approach	Reactive: focuses on treating symptoms of diseases or deficiencies	Proactive: enhances beneficial microbial populations to prevent issues
Nutrient management	Use of chemical fertilizers	Application of microbial biofertilizers (e.g., nitrogen-fixing or phosphate-solubilizing bacteria)
Pest and disease control	Chemical pesticides and fungicides	Biocontrol agents like *Trichoderma*, *Bacillus*, or *Pseudomonas*
Soil health	Degradation over time due to chemical buildup	Improves soil structure and fertility through microbial activity and organic matter cycling
Sustainability	Unsustainable due to overuse of inputs and potential environmental harm	Sustainable; promotes natural soil and ecosystem balance
Stress tolerance	Limited ability to enhance plant stress resilience	Microbes improve plant tolerance to drought, salinity, and other stresses
Cost	High due to repeated applications of chemicals and labor	Lower long-term cost due to reduced dependency on external inputs
Environmental impact	Negative impact, including pollution and harm to non-target organisms	Eco-friendly; reduces pollution and supports biodiversity
Ease of adoption	Widespread familiarity and established markets for traditional products	Requires awareness, training, and initial investment in microbiome technologies
Target specificity	Broad-spectrum and may harm beneficial organisms	Specific to pathogens or plants, preserving beneficial microbes
Technology use	Basic tools and chemical application	Advanced tools like CRISPR, metagenomics, and microbial formulations
Adaptability	Limited adaptability to varying environmental conditions	Microbes can be tailored to specific crops and environmental conditions

(continued)

Table 1.5 (continued)

Aspect	Traditional management strategies	Microbiome-based management strategies
Carbon footprint	High due to chemical synthesis and energy use	Low; promotes carbon sequestration and uses natural processes
Consumer preference	Increasing concerns over chemical residues in food	Aligns with consumer demand for organic and residue-free produce

1.9.2 Shifting Consumption Patterns

a. **Sustainable Diets:** Cultural practices around food consumption significantly impact the environment. Adopting sustainable diets, which emphasize plant-based foods over animal products, can reduce greenhouse gas emissions, land use, and water consumption (Smith and Read 2010). This shift necessitates changes in cultural norms and values regarding food.

b. **Reduced Waste:** Cultural practices of excessive consumption and waste contribute to environmental degradation. Embracing a culture of sustainability, where reducing, reusing, and recycling are prioritized can mitigate the environmental impact. Policies and educational campaigns aimed at changing consumer behavior are essential in fostering such cultural shifts (Ploetz 2003).

1.9.3 Redefining Progress and Success

a. **Beyond GDP:** Traditional measures of success, such as Gross Domestic Product (GDP), do not account for environmental degradation. Promoting cultural values that prioritize ecological health and well-being over mere economic growth is crucial. Alternative metrics like the Genuine Progress Indicator (GPI) or Gross National Happiness (GNH) incorporate environmental and social factors, encouraging more sustainable development models (Diedhiou et al. 2014; Johnson 1999; Reddy 2015; Vikas Gupta et al. 2013****).

b. **Community and Well-being:** Cultures that value community well-being and environmental stewardship over individual wealth can foster more resilient societies. Encouraging cultural narratives that highlight the interconnectedness of human and environmental health can drive collective action against climate change (Ploetz 2003; Reddy 2015).

1.9.4 Adapting Traditional Knowledge

a. **Indigenous Practices:** Indigenous communities often possess traditional knowledge systems that are inherently sustainable and resilient to climate change. Integrating this knowledge with modern practices can enhance adaptive capacities and environmental management. Respecting and revitalizing Indigenous cultural practices are vital for biodiversity conservation and climate resilience (Chakraborty and Gupta 2020; Mahmoud Mohamed Ahmed 2012; Thind 2015).

   b. **Agricultural Practices:** Revising agricultural practices incorporating traditional methods such as crop rotation, agroforestry, and permaculture can improve soil health and reduce emissions. These practices, rooted in cultural heritage, offer sustainable alternatives to industrial agriculture (Naqvi 2004).

1.9.5 Promoting Environmental Education

a. **Education for Sustainability:** Integrating environmental education into curricula at all levels can cultivate a culture of sustainability. Teaching young people about climate change, its impacts, and mitigation strategies fosters a generation that values and acts to protect the environment (Glick 2015; Gurr et al. 2003; Meena et al. 2014).

   b. **Public Awareness Campaigns:** Cultural change requires widespread public engagement. Campaigns that raise awareness about climate change and promote sustainable practices can shift public attitudes and behaviors. Media, art, and community initiatives are crucial in disseminating information and inspiring action (Expósito et al.; Glick 2015; Pimentel 2005).

1.9.6 Encouraging Civic Participation

a. **Grassroots Movements:** Civic participation in environmental movements can drive cultural change. Grassroots organizations and community-led initiatives often lead the way in advocating for sustainable policies and practices. Supporting these movements can amplify their impact and foster a culture of environmental activism (Chaparro et al. 2012).

   b. **Policy Engagement:** Encouraging individuals to engage in policy processes and advocate for climate action is essential. Cultural practices that promote civic responsibility and participation can influence governmental policies and corporate practices toward more sustainable outcomes (Smith and Read 2010).

1.9.7 *Environmental and Economic Impacts*

1.9.7.1 Reducing Reliance on Chemical Pesticides and Fungicides

The widespread use of chemical pesticides and fungicides in agriculture has raised significant environmental and health concerns. These chemicals can lead to soil degradation, water contamination, and harm to non-target organisms, including beneficial insects and humans (Singh and Trivedi 2017). Therefore, reducing reliance on chemical pesticides and fungicides is essential for sustainable agriculture. This essay discusses various strategies to achieve this goal, supported by references from credible sources (Sylvia et al. 2005).

1.10 Integrated Pest Management (IPM)

a. **Principles of IPM:** Integrated Pest Management (IPM) is a holistic approach that combines biological, cultural, physical, and chemical methods to control pests and diseases (Rushchitskaya et al. 2024). The emphasis is on using environmentally friendly practices and reducing chemical inputs. IPM includes monitoring pest populations, using pest-resistant crop varieties, and employing biological control agents (Naik et al. 2008).

    b. **Benefits of IPM:** Studies have shown that IPM can effectively reduce pest populations while minimizing the use of chemical pesticides. For instance, implementing IPM strategies in rice cultivation in Asia has led to significant reductions in pesticide use without compromising yield (Jaiswal et al. 2022).

1.10.1 *Biological Control*

a. **Natural Predators:** Utilizing natural predators, such as ladybugs, spiders, and predatory mites, can help control pest populations. These biological control agents prey on harmful insects, reducing the need for chemical pesticides (Tilman et al. 2002).

    b. **Microbial Biocontrol Agents:** Bacteria, fungi, and viruses can act as biocontrol agents against specific pests and diseases. For example, *Bacillus thuringiensis* (Bt) is a bacterium that produces toxins lethal to certain insects but is safe for humans and other non-target organisms (Klein et al. 2018).

1.10.2 Cultural Practices

a. **Crop Rotation:** Rotating crops disrupts the life cycles of pests and reduces their populations. Different crops can break the habitat continuity required for pest proliferation, thus diminishing the need for chemical interventions (Lugtenberg and Kamilova 2009).

   b. **Intercropping and Companion Planting:** Planting different crops together can reduce pest incidence. Certain plants can repel pests or attract beneficial insects that prey on pests. For example, marigolds can deter nematodes, and flowering plants can attract pollinators and predators of pests (Altieri 2000). Controlled irrigation can also help the spread of the soil borne pathogen which exists in the soil (Fig. 1.8).

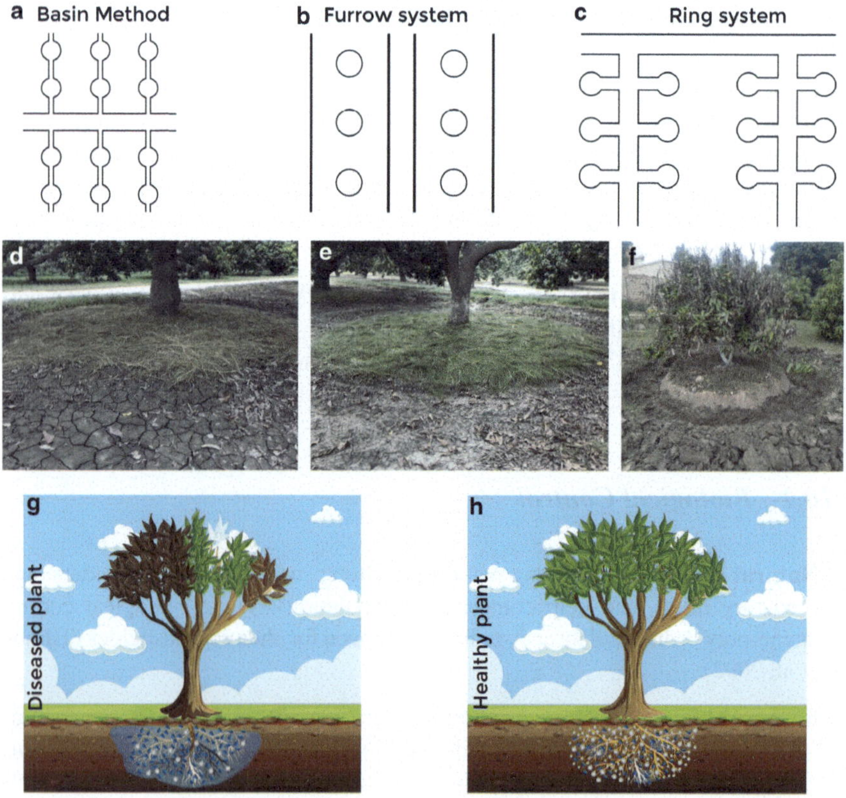

Fig. 1.8 Different methods adopted to mitigate the spread and development of disease in mango orchards

1.10.3 Mechanical and Physical Control

a. **Physical Barriers:** Physical barriers such as row covers, nets, and traps can protect crops from pest attacks. These methods are particularly effective in small-scale farming and organic agriculture (Sustainable Agriculture Research and Education (Berkes 2017).

b. **Soil Solarization:** Soil solarization involves covering the soil with clear plastic to trap solar energy and heating the soil to temperatures that kill soil-borne pests and pathogens. This method is effective in warm climates and can reduce the need for chemical soil treatments (Brown and Kasser 2005).

1.10.4 Organic Farming Practices

a. **Organic Amendments:** Organic farming relies on natural amendments such as compost, manure, and green manure to enhance soil fertility and health. Healthy soil supports robust plants that are more resistant to pests and diseases (Buchmann-Duck and Beazley 2020).

b. **Botanical Pesticides:** Organic farming uses botanical pesticides derived from plants with natural pesticidal properties, such as neem oil and pyrethrin. These substances are less harmful to non-target organisms and degrade more quickly than synthetic chemicals (Costanza et al. 2014).

1.10.5 Genetic Approaches

a. **Breeding for Resistance:** Developing and planting pest-resistant crop varieties can reduce the need for chemical treatments. Modern breeding techniques and genetic modification can enhance crop resistance to specific pests and diseases (Moser and Dilling 2011).

b. **RNA Interference (RNAi):** RNAi technology can be used to target and silence specific genes in pests, leading to their death or reduced fitness. This method offers a highly targeted approach to pest control with minimal impact on non-target species (Nisbet and Markowitz 2016; Patwardhan et al. 2012; Springmann et al. 2018).

1.11 Challenges and Future Directions

The mango (*Mangifera indica* L.) is a tropical fruit of significant economic and nutritional value. However, climate change poses serious threats to mango production, including increased pests and diseases, altered rainfall patterns, and extreme temperature fluctuations (Wals 2015; Zeevi et al. 2015). Microbiome engineering of the mango rhizosphere holds great promise for improving climate resilience. Research has demonstrated that manipulating the microbial community in the root zone can enhance drought tolerance and overall plant health under climate stress. For example, studies have shown that inoculating mango trees with specific plant growth-promoting rhizobacteria (PGPR), such as *Bacillus* and *Pseudomonas* species, can improve water use efficiency and promote better recovery from drought . This approach could significantly bolster mango production in regions affected by climate variability (de Vries et al. 2020).

a. **Diversity and Dynamics:** The mango rhizosphere hosts a diverse and dynamic microbial community, including bacteria, fungi, archaea, and viruses. These microorganisms interact in complex networks, making it challenging to manipulate specific microbial populations without unintended consequences (Ehler 2006).

    b. **Microbial Interactions:** The interactions between different microbial species and between microbes and the mango plant can be synergistic, antagonistic, or neutral. Understanding and predicting these interactions is difficult but essential for effective microbiome engineering (Isman 2006).

1.11.1 Technical and Methodological Barriers

a. **Culturing Difficulties:** Many rhizosphere microorganisms are difficult to culture in the laboratory, which limits the ability to study and manipulate them directly. Advances in culturing techniques and high-throughput sequencing are needed to overcome this barrier (Katan 1981).

    b. **Delivery Methods:** Effective methods for delivering engineered microbes to the mango rhizosphere are lacking. Ensuring that introduced microbes establish and persist in the rhizosphere environment is a significant technical challenge (Liebman 2000; Pimentel and Burgess 2014; Wang et al. 2014).

1.11.2 Environmental and Ecological Concerns

a. **Ecosystem Impact:** Introducing engineered or non-native microbes into the mango rhizosphere may disrupt existing microbial communities and ecological balance. The potential ecological consequences need a thorough assessment (Pierre et al. 2023).

   b. **Horizontal Gene Transfer:** The possibility of horizontal gene transfer between engineered microbes and native soil microorganisms raises concerns about the spread of engineered traits and their ecological impacts (Pretty and Pervez Bharucha 2015).

1.11.3 Regulatory and Social Barriers

a. **Regulatory Frameworks:** The regulatory landscape for microbiome engineering is still evolving. Clear guidelines and standards are needed to ensure the safety and efficacy of microbiome-based interventions (Altieri et al. 2005).

   b. **Public Perception:** Public acceptance of microbiome engineering technologies is crucial. Effective communication and engagement strategies are necessary to address concerns and build trust among stakeholders (Van Lenteren 2012).

1.11.4 Future Directions

1.11.4.1 Advancing Scientific Understanding

a. **Functional Characterization:** Comprehensive functional characterization of the mango rhizosphere microbiome is needed to identify key microbial players and their roles in promoting plant health and resilience. Metagenomics, meta transcriptomics, and metabolomics can provide valuable insights (Bender et al. 2016).
b. **Host-Microbe Interactions:** Investigating the molecular mechanisms underlying host-microbe interactions can inform the design of targeted microbiome interventions. Studies on root exudates and their influence on microbial communities are particularly important (Cappellano and Kurowska-Pysz 2020).

1.11.4.2 Developing Effective Interventions

a. **Biocontrol Agents:** Identifying and deploying beneficial microbes as biocontrol agents can help manage pests and diseases. These agents can be naturally occurring or engineered to enhance their efficacy and resilience (Mammola et al. 2020).

b. **Biofertilizers:** Microbes that promote nutrient uptake and soil health can be used as biofertilizers. Formulating effective microbial consortia that work synergistically with mango plants is a promising approach (Nosheen et al. 2021).

1.11.4.3 Improving Delivery and Establishment

a. **Formulation Technologies:** Developing advanced formulation technologies, such as encapsulation and slow-release carriers, can enhance the delivery and establishment of beneficial microbes in the rhizosphere (Knight et al. 2018).

b. **Field Trials:** Conducting large-scale field trials under diverse environmental conditions is essential to evaluate the performance and stability of microbiome interventions. These trials can provide critical data for optimizing practices and formulations (Miranda and Zolas 2017).

1.11.4.4 Addressing Regulatory and Ethical Issues

a. **Policy Development:** Collaborative efforts between scientists, regulators, and policymakers are needed to develop clear and science-based regulatory frameworks. These policies should balance innovation with safety and environmental protection (Pérez-Jaramillo et al. 2016).

b. **Stakeholder Engagement:** Engaging with farmers, consumers, and other stakeholders through transparent communication and participatory approaches can build public trust and support for microbiome engineering initiatives (Sahoo et al. 2017).

1.11.5 Role of Government Policies and Incentives in Promoting Microbiome-Based Approaches

The microbiome has emerged as a critical health, agriculture, and environmental sustainability factor. As research continues to uncover the potential benefits of microbiome-based approaches, the role of government policies and incentives becomes increasingly important in promoting these innovative solutions (Schlaeppi and Bulgarelli 2015). This essay examines how government interventions can facilitate the adoption and development of microbiome-based approaches, supported by references from credible sources (Stewart 2012).

1.11.5.1 Research and Development Funding

a. **Grant Programs:** Governments can promote microbiome research by establishing grant programs that provide funding for basic and applied research. Such programs can support universities, research institutions, and private companies in exploring microbiome applications. For example, the National Institutes of Health (NIH) in the United States has invested heavily in microbiome research through initiatives like the Human Microbiome Project (Vandermeer 2009).

b. **Public–Private Partnerships:** Encouraging collaboration between public institutions and private enterprises can accelerate innovation. Public–private partnerships (PPPs) can leverage the strengths of both sectors, combining public funding with private sector expertise and resources (Bais et al. 2006).

1.11.5.2 Regulatory Frameworks

a. **Clear Guidelines and Standards:** Developing clear regulatory guidelines and standards for microbiome-based products and therapies is crucial. These regulations ensure safety and efficacy while allowing new products to enter the market. For example, the European Medicines Agency (EMA) has begun to establish frameworks for evaluating microbiome therapeutics (Bashan et al. 2014).
b. **Fast-Track Approval Processes:** Implementing fast-track approval processes for microbiome-based innovations can reduce the time and cost of bringing new products to market. This can be particularly important for addressing urgent public health needs (Berendsen et al. 2012).

1.11.5.3 Educational and Outreach Programs

a. **Public Awareness Campaigns:** Raising the general understanding of the potential benefits of microbiome-based treatments through educational initiatives may raise acceptability and demand. Informing the public on the potential health, agricultural, and environmental benefits helps generate a favorable atmosphere for innovation (Glick 2012).

b. **Training and Workforce Development:** Investing in education and training programs to develop a skilled workforce is essential for advancing microbiome science. Governments can support academic programs, professional development courses, and vocational training that focus on microbiome research and its applications (De Lorenzo et al. 2018).

1.11.5.4 Support for Sustainable Practices

a. **Agricultural Policies:** Governments can promote microbiome-based approaches in agriculture by supporting sustainable farming practices that enhance soil health and crop resilience. Policies that encourage the use of biofertilizers, biopesticides, and crop rotations can improve agricultural sustainability (Jefferson et al. 2014).

b. **Environmental Regulations:** Implementing environmental regulations that reduce harmful chemicals and promote biodiversity can create a favorable environment for microbiome-based solutions. Such regulations can drive the adoption of environmentally friendly practices that benefit the microbiome (Committee et al. 2020).

1.12 Conclusion

Microbiome engineering of the mango rhizosphere presents a transformative approach to achieving sustainable and climate-resilient agricultural practices. By leveraging the intricate interactions between plants and their associated microbiota, this method offers the potential to enhance mango productivity under the increasingly unpredictable and adverse environmental conditions associated with climate change. Integrating beneficial microbial consortia into the rhizosphere has been shown to promote nutrient acquisition, improve soil health, and bolster plant defenses against abiotic and biotic stresses. This biotechnological intervention not only reduces dependence on chemical fertilizers and pesticides but also aligns with global efforts to reduce agriculture's environmental footprint.. Additionally, microbiome engineering fosters a circular, regenerative approach to resource management, improving the ecological balance of mango orchards while maintaining their economic viability. Further research and field validation are necessary to optimize the formulation and application of microbial inoculants tailored to specific soil and climatic conditions. Collaborations between scientists, growers, and policymakers will be critical to scaling this solution, ensuring its accessibility and long-term impact. In conclusion, microbiome engineering represents a promising avenue for creating a resilient, sustainable mango production system. It embodies a paradigm shift toward harnessing ecological principles to meet the dual challenges of food security and environmental sustainability, ultimately paving the way for a more resilient agricultural future.

References

Adams RM, Hurd BH, Lenhart S, Leary N (1998) Effects of global climate change on agriculture: an interpretative review. Clim Res 11(1):19–30

Agrawal N, Kumar V, Shahi SK (2021) Biodegradation and detoxification of phenanthrene in vitro and in vivo conditions by a newly isolated ligninolytic fungus Coriolopsis byrsina strain APC5 and characterization of their metabolites for environmental safety. Environ Sci Pollut Res 1–16

Ahmad I, Khan MSA, Aqil F, Singh M (2011) Microbial applications in agriculture and the environment: a broad perspective. In: Microbes and microbial technology: agricultural and environmental applications, pp 1–27

Ahmad M, Nadeem SM, Naveed M, Zahir ZA (2016) Potassium-solubilizing bacteria and their application in agriculture. In: Potassium solubilizing microorganisms for sustainable agriculture, pp 293–313

Ait Barka E, Nowak J, Clément C (2006) Enhancement of chilling resistance of inoculated grapevine plantlets with a plant growth-promoting rhizobacterium, *Burkholderia phytofirmans* strain PsJN. Appl Environ Microbiol 72(11):7246–7252

Altomare C, Norvell WA, Björkman TH, Harman G (1999) Solubilization of phosphates and micronutrients by the plant-growth-promoting and biocontrol fungus *Trichoderma harzianum* Rifai 1295-22. Appl Environ Microbiol 65(7):2926–2933

Mahmoud Mohamed Ahmed Y (2012) Studies on mango soilborne diseases with special reference to Phytophtora root rot

Aktar W, Sengupta D, Chowdhury A (2009) Impact of pesticides use in agriculture: their benefits and hazards. Interdiscip Toxicol 2(1):1–12

Altieri MA, Nicholls CI, Fritz MA (2005) Manage insects on your farm: a guide to ecological strategies. Sustainable Agriculture Research and Education (SARE)

Altieri MA (2000) Agroecology: principles and strategies for designing sustainable farming systems. In: Agroecology in action

Asad SA (2022) Mechanisms of action and biocontrol potential of Trichoderma against fungal plant diseases—a review. Ecol Compl 49:100978

Aseri G, Jain N, Panwar J, Rao A, Meghwal P (2008) Biofertilizers improve plant growth, fruit yield, nutrition, metabolism and rhizosphere enzyme activities of pomegranate (*Punica granatum* L.) in Indian Thar Desert. Sci Horticul 117(2):130–135

Babalola OO, Osir EO, Sanni AI, Odhiambo GD, Bulimo WD (2003) Amplification of 1-amino-cyclopropane-1-carboxylic (ACC) deaminase from plant growth promoting rhizobacteria in Striga-infested soil. Afr J Biotechnol 2(6):157–160

Bahadur I, Maurya BR, Kumar A, Meena VS, Raghuwanshi R (2016) Towards the soil sustainability and potassium-solubilizing microorganisms. In: Potassium solubilizing microorganisms for sustainable agriculture, pp 255–266

Bais HP, Weir TL, Perry LG, Gilroy S, Vivanco JM (2006) The role of root exudates in rhizosphere interactions with plants and other organisms. Annu Rev Plant Biol 57(1):233–266

Bakhshandeh E, Pirdashti H, Shahsavarpour Lendeh K, Gilani Z, Yaghoubi Khanghahi M, Crecchio C (2020) Effects of plant growth promoting microorganisms inoculums on mineral nutrition, growth and productivity of rice (*Oryza sativa* L.). J Plant Nutr 43(11):1643–1660

Bakker PA, Pieterse CM, Van Loon L (2007) Induced systemic resistance by fluorescent *Pseudomonas* spp. Phytopathology 97(2):239–243

Banerjee MR, Yesmin L, Vessey JK, Rai M (2006) Plant-growth-promoting rhizobacteria as biofertilizers and biopesticides. Handbook of microbial biofertilizers. Food Products Press, New York, pp 137–181

Bashan Y, de-Bashan LE, Prabhu S, Hernandez J-P (2014) Advances in plant growth-promoting bacterial inoculant technology: formulations and practical perspectives (1998–2013). Plant Soil 378:1–33

Bashan Y, De-Bashan LE (2010) How the plant growth-promoting bacterium *Azospirillum* promotes plant growth—a critical assessment. Adv Agron 108:77–136

Baum JA, Bogaert T, Clinton W, Heck GR, Feldmann P, Ilagan O, Johnson S, Plaetinck G, Munyikwa T, Pleau M (2007) Control of coleopteran insect pests through RNA interference. Nat Biotechnol 25(11):1322–1326

Bender SF, Wagg C, van der Heijden MG (2016) An underground revolution: biodiversity and soil ecological engineering for agricultural sustainability. Trends Ecol Evol 31(6):440–452

Beneduzi A, Ambrosini A, Passaglia LM (2012) Plant growth-promoting rhizobacteria (PGPR): their potential as antagonists and biocontrol agents. Genet Mol Biol 35:1044–1051

Berendsen RL, Pieterse CM, Bakker PA (2012) The rhizosphere microbiome and plant health. Trends Plant Sci 17(8):478–486
Berg G, Alavi M, Schmid M, Hartmann A (2013) The rhizosphere as a reservoir for opportunistic human pathogenic bacteria. Mol Microbial Ecol Rhizosphere 1:1209–1216
Berkes F (2017) Sacred ecology. Routledge
Berruti A, Lumini E, Balestrini R, Bianciotto V (2016) Arbuscular mycorrhizal fungi as natural biofertilizers: let's benefit from past successes. Front Microbiol 6:1559
Bhattacharyya PN, Jha DK (2012) Plant growth-promoting rhizobacteria (PGPR): emergence in agriculture. World J Microbiol Biotechnol 28:1327–1350
Bhattacharyya R, Das TK, Sudhishri S, Dudwal B, Sharma AR, Bhatia A, Singh G (2015) Conservation agriculture effects on soil organic carbon accumulation and crop productivity under a rice–wheat cropping system in the western Indo-Gangetic Plains. Eur J Agron 70:11–21
Bhattacharjee RB, Jourand P, Singh A, Mukhopadhyay SN Indole acetic acid and ACC deaminase-producing Rhizobium leguminosarum bv. trifoUi SNIO promote rice growth, and in the process undergo colonization and chemotaxis. Habilitation à Diriger des Recherches 62
Bicas JL, Molina G, Dionísio AP, Barros FFC, Wagner R, Maróstica MR Jr, Pastore GM (2011) Volatile constituents of exotic fruits from Brazil. Food Res Int 44(7):1843–1855
Boller T, Felix G (2009) A renaissance of elicitors: perception of microbe-associated molecular patterns and danger signals by pattern-recognition receptors. Annu Rev Plant Biol 60(1):379–406
Bravo A, Likitvivatanavong S, Gill SS, Soberón M (2011) Bacillus thuringiensis: a story of a successful bioinsecticide. Insect Biochem Mol Biol 41(7):423–431
Brown KW, Kasser T (2005) Are psychological and ecological well-being compatible? The role of values, mindfulness, and lifestyle. Soc Indic Res 74:349–368
Buchmann-Duck J, Beazley KF (2020) An urgent call for circular economy advocates to acknowledge its limitations in conserving biodiversity. Sci Total Environ 727:138602
Cappellano F, Kurowska-Pysz J (2020) The mission-oriented approach for (cross-border) regional development. Sustainability 12(12):5181
Cassan F, Maiale S, Masciarelli O, Vidal A, Luna V, Ruiz O (2009) Cadaverine production by *Azospirillum brasilense* and its possible role in plant growth promotion and osmotic stress mitigation. Eur J Soil Biol 45(1):12–19
Cassells AC, Rafferty-McArdle SM (2012) Priming of plant defences by PGPR against fungal and bacterial plant foliar pathogens. Bact Agrobiol: Stress Manag 1–26
Chakraborty N, Gupta NS (2020) Microbial bio-pesticides: the major component for sustainable agriculture. TTPP 483
Chakraborty S, Newton AC (2011) Climate change, plant diseases and food security: an overview. Plant Pathol 60(1):2–14
Chaparro JM, Sheflin AM, Manter DK, Vivanco JM (2012) Manipulating the soil microbiome to increase soil health and plant fertility. Biol Fertil Soils 48:489–499
Committee ES, More S, Bampidis V, Benford D, Bragard C, Halldorsson T, Hernández-Jerez A, Susanne HB, Koutsoumanis K, Machera K (2020) Evaluation of existing guidelines for their adequacy for the microbial characterisation and environmental risk assessment of microorganisms obtained through synthetic biology. EFSA J 18(10):e06263
Costanza R, Kubiszewski I, Giovannini E, Lovins H, McGlade J, Pickett KE, Ragnarsdóttir KV, Roberts D, De Vogli R, Wilkinson R (2014) Development: time to leave GDP behind. Nature 505(7483):283–285
De Lorenzo V, Prather KL, Chen GQ, O'Day E, von Kameke C, Oyarzún DA, Hosta-Rigau L, Alsafar H, Cao C, Ji W (2018) The power of synthetic biology for bioproduction, remediation and pollution control: the UN's sustainable development Goals will inevitably require the application of molecular biology and biotechnology on a global scale. EMBO Rep 19(4):e45658

de los Santos-Villalobos S, de Folter S, Délano-Frier JP, Gómez-Lim MA, Guzmán-Ortiz DA, Pena-Cabriales JJ (2013) Growth promotion and flowering induction in mango (*Mangifera indica* L. cv "Ataulfo") trees by Burkholderia and Rhizobium inoculation: morphometric, biochemical, and molecular events. J Plant Growth Regul 32:615–627

Desbrosses G, Contesto C, Varoquaux F, Galland M, Touraine B (2009) PGPR-arabidopsis interactions is a useful system to study signaling pathways involved in plant developmental control. Plant Signal Behav 4(4):319–321

de Vries FT, Griffiths RI, Knight CG, Nicolitch O, Williams A (2020) Harnessing rhizosphere microbiomes for drought-resilient crop production. Science 368(6488):270–274

Diedhiou P, Mbaye N, Faye R, Samb P (2014) Field tests of fungicides against post-harvest rot of mangoes in Senegal. Int J Sci Environ Technol 3(2):597–606

Domenech J, Reddy MS, Kloepper JW, Ramos B, Gutierrez-Mañero J (2006) Combined application of the biological product LS213 with *Bacillus*, *Pseudomonas* or *Chryseobacterium* for growth promotion and biological control of soil-borne diseases in pepper and tomato. BioControl 51:245–258

Doran JW, Zeiss MR (2000) Soil health and sustainability: managing the biotic component of soil quality. Appl Soil Ecol 15(1):3–11

Ehler LE (2006) Integrated pest management (IPM): definition, historical development and implementation, and the other IPM. Pest Manag Sci 62(9):787–789

Elauria MM, Manilay AA, Abrigo GNA (2015) Adoption and socio-economic impacts of the pre-harvest disease management technology for carabao mango in the Philippines. J Econ Manage Agric Dev 1(1):1–14

Expósito RG, de Bruijn I, Jack A, Postma J, Raaijmakers JM Successional changes in rhizosphere microbiome composition and functions during induction of the soil immune response to a fungal root pathogen. In: Microbiome dynamics of disease suppressive soils, p 89

Fernández-Lizarazo JC, Moreno-Fonseca LP (2016) Mechanisms for tolerance to water-deficit stress in plants inoculated with arbuscular mycorrhizal fungi. A review. Agronomía Colombiana 34(2):179–189

Galán Saúco V (2002) Mango production and world market: current situation and future prospects. VII International Mango Symposium, p 645

Gashash EA, Osman NA, Alsahli AA, Hewait HM, Ashmawi AE, Alshallash KS, El-Taher AM, Azab ES, Abd El-Raouf HS, Ibrahim MF (2022) Effects of plant-growth-promoting rhizobacteria (PGPR) and cyanobacteria on botanical characteristics of tomato (*Solanum lycopersicon* L.) plants. Plants 11(20):2732

Glick BR (2012) Plant growth-promoting bacteria: mechanisms and applications. Scientifica 2012(1):963401

Glick BR (2015) Beneficial plant-bacterial interactions, vol 243. Springer

Glick BR, Karaturovíc DM, Newell PC (1995) A novel procedure for rapid isolation of plant growth promoting pseudomonads. Can J Microbiol 41(6):533–536

Glick BR, Todorovic B, Czarny J, Cheng Z, Duan J, McConkey B (2007) Promotion of plant growth by bacterial ACC deaminase. Crit Rev Plant Sci 26(5–6):227–242

Goellner K, Conrath U (2008) Priming: it's all the world to induced disease resistance. In: Sustainable disease management in a European context, pp 233–242

Griesbach J (2003) Mango growing in Kenya. World Agroforestry Centre

Grimes DW, Wiley PL, Sheesley WR (1992) Alfalfa yield and plant water relations with variable irrigation. Crop Science 32(6):1381–1387

Gurr GM, Wratten SD, Luna, JM (2003) Multi-function agricultural biodiversity: pest management and other benefits. Basic Appl Ecol 4(2):107–116

Gupta G, Panwar J, Akhtar MS, Jha PN (2012) Endophytic nitrogen-fixing bacteria as biofertilizer. Sustain Agric Rev 11:183–221

Gupta A, Singh UB, Sahu PK, Paul S, Kumar A, Malviya D, Singh S, Kuppusamy P, Singh P, Paul D (2022) Linking soil microbial diversity to modern agriculture practices: a review. Int J Environ Res Public Health 19(5):3141

Hacquard S, Spaepen S, Garrido-Oter R, Schulze-Lefert P (2017) Interplay between innate immunity and the plant microbiota. Annu Rev Phytopathol 55(1):565–589

Halder R, Varma S, Singh M, Dahiya A From orchard to table: understanding climate change impacts on mango production in India: a review

Hamilton CE, Bever JD, Labbé J, Yang X, Yin H (2016) Mitigating climate change through managing constructed-microbial communities in agriculture. Agric Ecosyst Environ 216:304–308

Han HS, Lee KD (2005) Physiological responses of soybean-inoculation of *Bradyrhizobium japonicum* with PGPR in saline soil conditions. Res J Agric Biol Sci 1(3):216–221

Hartmann M, Frey B, Mayer J, Mäder P, Widmer F (2015) Distinct soil microbial diversity under long-term organic and conventional farming. ISME J 9(5):1177–1194

He S, Li L, Lv M, Wang R, Wang L, Yu S, Gao Z, Li X (2024) PGPR: key to enhancing crop productivity and achieving sustainable agriculture. Curr Microbiol 81(11):1–17

Hu K, Awange J, Kuhn M, Zerihun A (2022) Irrigated agriculture potential of Australia's northern territory inferred from spatial assessment of groundwater availability and crop evapotranspiration. Agric Water Manag 264:107466

Iriti M, Vitalini S (2021) Plant immunity and crop yield: a sustainable approach in agri-food systems. Vaccines 9(2):121

Isman MB (2006) Botanical insecticides, deterrents, and repellents in modern agriculture and an increasingly regulated world. Annu Rev Entomol 51(1):45–66

Jain R, Mathur A, Singh D, Rohillad A (2024) Fostering sustainable plant growth with Rhizozpheric allies: a review. Asian Biotechnol Dev Rev 26(1)

Jaiswal DK, Krishna R, Chouhan GK, de Araujo Pereira AP, Ade AB, Prakash S, Verma SK, Prasad R, Yadav J, Verma JP (2022) Bio-fortification of minerals in crops: current scenario and future prospects for sustainable agriculture and human health. Plant Growth Regul 98(1):5–22

Jefferson C, Lentzos F, Marris C (2014) Synthetic biology and biosecurity: challenging the "myths." Front Public Health 2:115

Johnson GI (1999) Postharvest Technology in Asia: ACIAR's framework for collaborative research and development. ACIAR PROCEEDINGS

Kamil FH, Saeed EE, El-Tarabily KA, AbuQamar SF (2018) Biological control of mango dieback disease caused by *Lasiodiplodia theobromae* using streptomycete and non-streptomycete actinobacteria in the United Arab Emirates. Front Microbiol 9:829

Katan J (1981) Solar heating (solarization) of soil for control of soilborne pests

Kejela T (2024) Phytohormone-producing rhizobacteria and their role in plant growth

Khan MW, Hussain Z, Ghafoor I (2023) Mango fruit orchards infestation: a case study on mango fruits in Pakistan

Khokhar SN (2012) Crop inoculants-an underexplored potential in agriculture

Klein EY, Van Boeckel TP, Martinez EM, Pant S, Gandra S, Levin SA, Goossens H, Laxminarayan R (2018) Global increase and geographic convergence in antibiotic consumption between 2000 and 2015. Proc Natl Acad Sci 115(15):E3463–E3470

Knight R, Vrbanac A, Taylor BC, Aksenov A, Callewaert C, Debelius J, Gonzalez A, Kosciolek T, McCall L-I, McDonald D (2018) Best practices for analysing microbiomes. Nat Rev Microbiol 16(7):410–422

Kour D, Khan SS, Kaur T, Kour H, Singh G, Yadav A, Yadav AN (2022) Drought adaptive microbes as bioinoculants for the horticultural crops. Heliyon 8(5)

Kuć J (1982) Induced immunity to plant disease. Bioscience 32(11):854–860

Kumar A, Bahadur I, Maurya B, Raghuwanshi R, Meena V, Singh D, Dixit J (2015) Does a plant growth-promoting rhizobacteria enhance agricultural sustainability. J Pure Appl Microbiol 9(1):715–724

Legave J-M, Normand F, Lauri P-E (2013) Climate change and its probable effects on mango production and cultivation. In: X international mango symposium, p 1075

Li AK, Mao Y, Manjanna S, Liu S, Dhanoa J, Mehta B, Edwards VM, Cladera F, Hsieh MA, Le Men M (2022) Towards understanding underwater weather events in rivers using autonomous surface vehicles. OCEANS 2022, Hampton Roads

Liebman (2000) Integration of soil, crop and weed management in low-external-input farming systems. Weed Res 40(1):27–47

Lucy M, Reed E, Glick BR (2004) Applications of free living plant growth-promoting rhizobacteria. Antonie Van Leeuwenhoek 86:1–25

Lugtenberg B, Kamilova F (2009) Plant-growth-promoting rhizobacteria. Annu Rev Microbiol 63(1):541–556

Macho AP, Schwessinger B, Ntoukakis V, Brutus A, Segonzac C, Roy S, Kadota Y, Oh MH, Sklenar J, Derbyshire P, Lozano-Durán R (2014) A bacterial tyrosine phosphatase inhibits plant pattern recognition receptor activation. Science 343(6178):1509–1512

Makhmale S, Bhutada P, Yadav L, Yadav B (2016) Impact of climate change on phenology of mango-the case study. Ecol Environ Conserv 22(9):S127–S132

Mammola S, Riccardi N, Prié V, Correia R, Cardoso P, Lopes-Lima M, Sousa R (2020) Towards a taxonomically unbiased European Union biodiversity strategy for 2030. Proc R Soc B 287(1940):20202166

Meena RS (2015) The needs of healthy soils for a healthy world. J Clean Prod 102:560–561

Meena VS, Maurya B, Verma JP (2014) Does a rhizospheric microorganism enhance K+ availability in agricultural soils? Microbiol Res 169(5–6):337–347

Meena VS, Maurya B, Meena SK, Meena RK, Kumar A, Verma J, Singh N (2016) Can *Bacillus* species enhance nutrient availability in agricultural soils? Bacilli Agrobiotechnol 367–395

Meena R, Kumari T, Solanki V, Partel V, Singh S, Sinha R (2024) Soil, water, and biodiversity conservation through agroforestry for crop production. In: Agroforestry to combat global challenges: current prospects and future challenges. Springer, pp 345–366

Mendelsohn R (2009) The impact of climate change on agriculture in developing countries. J Nat Resour Policy Res 1(1):5–19

Mendes R, Garbeva P, Raaijmakers JM (2013) The rhizosphere microbiome: significance of plant beneficial, plant pathogenic, and human pathogenic microorganisms. FEMS Microbiol Rev 37(5):634–663

Miranda J, Zolas N (2017) Measuring the impact of household innovation using administrative data. In Measuring and accounting for innovation in the 21st century. University of Chicago Press.

Mitter B, Brader G, Pfaffenbichler N, Sessitsch A (2019) Next generation microbiome applications for crop production—limitations and the need of knowledge-based solutions. Curr Opin Microbiol 49:59–65

Mohandas S (2012) Arbuscular mycorrhizal fungi benefit mango (*Mangifera indica* L.) plant growth in the field. Sci Hortic 143:43–48

Mohsin M, Mehmood S, Jamal F, Anwer MM (2021) Modeling and forecasting of declining area and production of mango orchards in district Bahawalpur, Pakistan: a case of tehsil ahmedpur east. J Agric Res 59(2):197–212

Moser SC, Dilling L (2011) Communicating change science: closing action climate. In: The Oxford handbook of climate change and society, p 161

Nadeem SM, Ahmad M, Zahir ZA, Javaid A, Ashraf M (2014) The role of mycorrhizae and plant growth promoting rhizobacteria (PGPR) in improving crop productivity under stressful environments. Biotechnol Adv 32(2):429–448

Naik PR, Sahoo N, Goswami D, Ayyadurai N, Sakthivel N (2008) Genetic and functional diversity among fluorescent pseudomonads isolated from the rhizosphere of banana. Microb Ecol 56:492–504

Naqvi S (2004) Diseases of fruits and vegetables. Springer

Nawaz A, Shahbaz M, Asadullah IA, Marghoob MU, Imtiaz M, Mubeen F (2020) Potential of salt tolerant PGPR in growth and yield augmentation of wheat (*Triticum aestivum* L.) under saline conditions. Front Microbiol 11:2019

Nehra V, Saharan BS, Choudhary M (2016) Evaluation of Brevibacillus brevis as a potential plant growth promoting rhizobacteria for cotton (*Gossypium hirsutum*) crop. Springerplus 5:1–10

Newman MA, Sundelin T, Nielsen JT, Erbs G (2013) MAMP (microbe-associated molecular pattern) triggered immunity in plants. Front Plant Sci 4:139

Nisbet MC, Markowitz E (2016) Strategic science communication on environmental issues. Commissioned White Paper in Support of the Alan Leshner Leadership Institute. American Association for the Advancement of Science. https://www.aaas.org/sites/default/files/content_files/NisbetMarkowitz_StrategicSciCommOnEnvironmentalIssues_WhitePaper.pdf

Nishad R, Ahmed T, Rahman VJ, Kareem A (2020) Modulation of plant defense system in response to microbial interactions. Front Microbiol 11:1298

Nosheen S, Ajmal I, Song Y (2021) Microbes as biofertilizers, a potential approach for sustainable crop production. Sustainability 13(4):1868

Pastor V, Luna E, Mauch-Mani B, Ton J, Flors V (2013) Primed plants do not forget. Environ Exp Bot 94:46–56

Paquet J-C, Claus SP, Cordaillat-Simmons M, Mazier W, Rawadi G, Rinaldi L, Elustondo F, Rouanet A (2021) Entering first-in-human clinical study with a single-strain live biotherapeutic product: input and feedback gained from the EMA and the FDA. Front Med 8:716266

Parmesan C, Morecroft MD, Trisurat Y (2022) Climate change 2022: impacts, adaptation and vulnerability GIEC

Patel JS, Singh A, Singh HB, Sarma BK (2015) Plant genotype, microbial recruitment and nutritional security. Front Plant Sci 6:608

Patwardhan A, Hallegatte S, Yanda PZ, O'Brien K, Wilbanks T, Maskrey A, Pelling M, Oswald-Spring U, Oki T (2012) Toward a sustainable and resilient future. Cambridge University Press

Peix A, Rivas-Boyero AA, Mateos PF, Rodríguez-Barrueco C, Martınez-Molina E, Velazquez E (2001) Growth promotion of chickpea and barley by a phosphate solubilizing strain of *Mesorhizobium mediterraneum* under growth chamber conditions. Soil Biol Biochem 33(1):103–110

Pérez-Jaramillo JE, Mendes R, Raaijmakers JM (2016) Impact of plant domestication on rhizosphere microbiome assembly and functions. Plant Mol Biol 90:635–644

Pierre JF, Jacobsen KL, Latournerie-Moreno L, Torres-Cab WJ, Chan-Canché R, Ruiz-Sánchez E (2023) A review of the impact of maize-legume intercrops on the diversity and abundance of entomophagous and phytophagous insects. PeerJ 11:e15640

Ping L, Boland W (2004) Signals from the underground: bacterial volatiles promote growth in Arabidopsis. Trends Plant Sci 9(6):263–266

Pimentel D (2005) Environmental and economic costs of the application of pesticides primarily in the United States. Environ Dev Sustain 7:229–252

Pimentel D, Burgess M (2014) An environmental, energetic and economic comparison of organic and conventional farming systems. Integr Pest Manage Pesticide Probl 3:141–166

Ploetz R, Freeman S (2009) Foliar, floral and soilborne diseases. In: The mango: botany, production and uses. CABI Wallingford, UK, pp 231–302

Ploetz R (2003) Diseases of mango. In: Diseases of tropical fruit crops, pp 327–363

Poupin MJ, Timmermann T, Vega A, Zuñiga A, González B (2013) Effects of the plant growth-promoting bacterium *Burkholderia phytofirmans* PsJN throughout the life cycle of *Arabidopsis thaliana*. PLoS One 8(7):e69435

Pradeepkumar D, Sridhar R, Shivakumar BS, Nandish MS, Ganapathi M (2024) Influence of liquid plant growth promoting rhizomicrobial consortia on graft success in mango (*Mangifera indica* L.). Environ Ecol 42(3A):1113–1117

Prasad R, Dhyani S, Newaj R, Kumar S, Tripathi V (2016) Contribution of advanced agroforestry research in sustaining soil quality for increased food production and food security. J Soil Water Conserv 15(1):31–39

Prasad Pant L, Hambly-Odame H, Hall A, Sulaiman VR (2012) Beyond the supply chains of technology and commodity: challenges to strengthening mango innovation systems in Andhra Pradesh, India. World J Sci Technol Sustain Dev 9(3):175–193

Pretty J, Pervez Bharucha Z (2015) Integrated pest management for sustainable intensification of agriculture in Asia and Africa. Insects 6(1):152–182

Qessaoui R, Bouharroud R, Furze JN, El Aalaoui M, Akroud H, Amarraque A, Vaerenbergh JV, Tahzima R, Mayad EH, Chebli B (2019) Applications of new rhizobacteria *Pseudomonas isolates* in agroecology via fundamental processes complementing plant growth. Sci Rep 9(1):12832

Rahmoune B, Morsli A, Khelifi-Slaoui M, Khelifi L, Strueh A, Erban A, Kopka J, Prell J, van Dongen JT (2017) Isolation and characterization of three new PGPR and their effects on the growth of *Arabidopsis* and *Datura* plants. J Plant Interact 12(1):1–6

Ramasamy Vivekananthan RV, Muthusamy Ravi MR, Ayyasamy Ramanathan AR, Neelagandapillai Kumar NK, Ramasamy Samiyappan RS (2006) Pre-harvest application of a new biological control formulation induces resistance to postharvest anthracnose and enhances fruit yield in mango, pp 126–138

Reddy PP (2015) Plant protection in tropical root and tuber crops

Riggs PJ, Chelius MK, Iniguez AL, Kaeppler SM, Triplett EW (2001) Enhanced maize productivity by inoculation with diazotrophic bacteria. Funct Plant Biol 28(9):829–36

Rosenberg E, DeLong EF, Lory S, Stackebrandt E, Thompson F (2013) The prokaryotes: prokaryotic communities and ecophysiology. Springer, Berlin Heidelberg

Rushchitskaya O, Kulikova E, Kot E, Kruzhkova T (2024) Sustainable practices and technological innovations transforming agribusiness dynamics. In: E3S web of conferences

Ryu CM, Farag MA, Hu CH, Reddy MS, Wei HX, Paré PW, Kloepper JW (2003) Bacterial volatiles promote growth in *Arabidopsis*. Proc Natl Acad Sci 100(8):4927–4932

Sahoo S, Sarangi S, Kerry RG (2017) Bioprospecting of endophytes for agricultural and environmental sustainability. In: Microbial biotechnology: vol 1. Applications in Agriculture and Environment, pp 429–458

Schlaeppi K, Bulgarelli D (2015) The plant microbiome at work. Mol Plant Microbe Interact 28(3):212–217

Shaharoona B, Arshad M, Zahir ZA (2006) Effect of plant growth promoting rhizobacteria containing ACC-deaminase on maize (*Zea mays* L.) growth under axenic conditions and on nodulation in mung bean (*Vigna radiata* L.). Lett Appl Microbiol 42(2):155–159

Shrivastava S, Prasad R, Varma A (2014) Anatomy of root from eyes of a microbiologist. In: Root engineering: basic and applied concepts, pp 3–22

Shyam R, Prakash S, Prasad J (2019) Conservation practice, irrigation and fertilizer dose influence yield and quality of mango cv. Amrapali under high density orchard in Kosi zone of Bihar. Curr J Appl Sci Technol 37(6):1–6

Singh A, Sharma D, Kumar R, Kumar A, Yadav R, Gupta S (2018) Soil salinity management in fruit crops: a review of options and challenges. In: Engineering practices for management of soil salinity: agricultural, physiological, and adaptive approaches, pp 39–85

Singh BK, Trivedi P (2017) Microbiome and the future for food and nutrient security. Microb Biotechnol 10(1):50

Smith SE, Read DJ (2010) Mycorrhizal symbiosis. Academic

Sonsiam K, Sonesouphap O, Sangkaew A, Chulaka P, Sihanonth P, Yompakdee C (2024) Roles of endophytic fungi isolated from *Mangifera indica* L. in promoting plant growth. J Microbiol Biotechnol 34(9):1857

Spence C, Alff E, Johnson C, Ramos C, Donofrio N, Sundaresan V, Bais H (2014) Natural rice rhizospheric microbes suppress rice blast infections. BMC plant biology 14:1–7

Springmann M, Clark M, Mason-D'Croz D, Wiebe K, Bodirsky BL, Lassaletta L, De Vries W, Vermeulen SJ, Herrero M, Carlson KM (2018) Options for keeping the food system within environmental limits. Nature 562(7728):519–525

Srivastava AK (2024) Challenges of plant nutrition and climate change: focus on fruit crops. In: Silicon advances for sustainable agriculture and human health: increased nutrition and disease prevention. Springer, pp 1–40

Stewart EJ (2012) Growing unculturable bacteria. J Bacteriol 194(16):4151–4160

Sylvia DM, Fuhrmann JJ, Hartel PG, Zuberer DA (2005) Principles and applications of soil microbiology. Pearson

Tengsetasak P, Tongkoom K, Yomkerd J, Susawaengsup C, Khongdee N, Chatsungnoen T, Dangtungee R, Bhuyar P (2024) Sustainable strategies for fresh Mangosteen: adapting to climate challenges. Earth Syst Environ 1–19

Thind TS (2015) Perspectives on crop protection in India. Outlooks Pest Manage 26(3):121–127

Tilman D, Cassman KG, Matson PA, Naylor R, Polasky S (2002) Agricultural sustainability and intensive production practices. Nature 418(6898):671–677

Tiwari JK, Bairwa A, Bhatia N, Zinta R, Kaushal N, Kumar V, Sharma AK, Sharma S, Choudhary B, Luthra SK, Buckseth T (2022) Resistance evaluation for native potato accessions against late blight disease and potato cyst nematodes by molecular markers and phenotypic screening in India. Life 13(1):33

Van Lenteren JC (2012) The state of commercial augmentative biological control: plenty of natural enemies, but a frustrating lack of uptake. Biocontrol 57(1):1–20

Vandermeer JH (2009) The ecology of agroecosystems. Jones & Bartlett Publishers

Veen G, Wubs EJ, Bardgett RD, Barrios E, Bradford MA, Carvalho S, De Deyn GB, de Vries FT, Giller KE, Kleijn D (2019) Applying the aboveground-belowground interaction concept in agriculture: spatio-temporal scales matter. Front Ecol Evol 7:300

Verma JP, Yadav J, Tiwari KN, Kumar A (2013) Effect of indigenous Mesorhizobium spp. and plant growth promoting rhizobacteria on yields and nutrients uptake of chickpea (*Cicer arietinum* L.) under sustainable agriculture. Ecol Eng 51:282–286

Verma R, Maurya B, Meena VS (2014) Integrated effect of bio-organics with chemical fertilizer on growth, yield and quality of cabbage (*Brassica oleracea* var *capitata*). Indian J Agric Sci 84(8):914–919

Verma JP, Jaiswal DK, Meena VS, Kumar A, Meena R (2015) Issues and challenges about sustainable agriculture production for management of natural resources to sustain soil fertility and health

Vikas Gupta VG, Razdan V, Kumar D, Vishal Gupta VG (2013) Status of soil borne diseases of brinjal and the frequency of associated pathogens

Vivekananthan R, Ravi M, Ramanathan A, Samiyappan R (2004) Lytic enzymes induced by Pseudomonas fluorescens and other biocontrol organisms mediate defence against the anthracnose pathogen in mango. World J Microbiol Biotechnol 20:235–244

Vivekananthan R, Ravi M, Ramanathan A, Kumar N, Samiyappan R (2006) Pre-harvest application of a new biocontrol formulation induces resistance to post-harvest anthracnose and enhances fruit yield in mango. Phytopathol Mediterr 45(2):126–138

Wals AE (2015) Beyond unreasonable doubt. Education and learning for socio-ecological sustainability in the Anthropocene. Wageningen University

Wang Y, Zhang L, Li Y, Liu Y, Han L, Zhu Z, Wang F, Peng Y (2014) Expression of Cry1Ab protein in a marker-free transgenic Bt rice line and its efficacy in controlling a target pest, Chilo suppressalis (Lepidoptera: Crambidae). Environ Entomol 43(2):528–536

Weller DM, Raaijmakers JM, Gardener BBM, Thomashow LS (2002) Microbial populations responsible for specific soil suppressiveness to plant pathogens. Annu Rev Phytopathol 40(1):309–348

Welsh C (2021) The effects of climate change in Africa. Center for Strategic and International Studies (CSIS)

Whipps JM (2001) Microbial interactions and biocontrol in the rhizosphere. J Exp Bot 52(suppl_1):487–511

Yadav S, Korat J, Yadav S, Mondal K, Kumar A, Kumar S (2023) Impacts of climate change on fruit crops: a comprehensive review of physiological, phenological, and pest-related responses. Int J Environ Clim Change 13(11):363–371

Yadav BK, Sidhu AS (2016) Dynamics of potassium and their bioavailability for plant nutrition. In: Potassium solubilizing microorganisms for sustainable agriculture, pp 187–201

Yasin M, Munir I, Faisal M (2016) Can *Bacillus* spp. enhance K+ uptake in crop species. In: Potassium solubilizing microorganisms for sustainable agriculture, pp 163–170

Yu Y, Gui Y, Li Z, Jiang C, Guo J, Niu D (2022) Induced systemic resistance for improving plant immunity by beneficial microbes. Plants 11(3):386

Zakaria L (2021) Diversity of Colletotrichum species associated with anthracnose disease in tropical fruit crops—a review. Agriculture 11(4):297

Zapata-Sifuentes G, Hernandez-Montiel LG, Saenz-Mata J, Fortis-Hernandez M, Blanco-Contreras E, Chiquito-Contreras RG, Preciado-Rangel P (2022) Plant growth-promoting rhizobacteria improve growth and fruit quality of cucumber under greenhouse conditions. Plants 11(12):1612

Zeevi D, Korem T, Zmora N, Israeli D, Rothschild D, Weinberger A, Ben-Yacov O, Lador D, Avnit-Sagi T, Lotan-Pompan M (2015) Personalized nutrition by prediction of glycemic responses. Cell 163(5):1079–1094

Chapter 2
Impact of Climate Change on Secondary Metabolic Profiles of Plants: Abiotic Stress Adaptation by Modulating Secondary Metabolites

Muhammad Junaid Rao, Duan Mingzheng, Syed Atif Hasan Naqvi, and Bingsong Zheng

Abstract Climate change, driven by human activities, is causing global impacts like temperature increases, greenhouse gases, air pollutants, altered solar radiation, UV-B, and changes in precipitation patterns. These changes lead to more frequent droughts, uneven temperature, salinity, and heavy metal, significantly affecting agriculture by reducing crop yields and quality, which has major socio-economic implications. Such abiotic stresses adversely affect plant growth, development, and metabolic activities, consequently reducing crop yields and quality. With the future climate changes being unpredictable, it is crucial to modify existing agricultural practices to cope with these changes. Abiotic stresses can modify plant metabolism, either inhibiting or enhancing the production of secondary metabolites. While the essential roles of these compounds in plant adaptation and defense are recognized, their reaction to climate change remains largely unexplored. As climate change impacts intensify, understanding the regulatory effects on plant secondary metabolism is becoming more critical. Research continues on how individual climate change factors like high temperatures, increased salinity levels, drought, and elevated ultraviolet-B radiation, along with their combined effects,

M. J. Rao (✉) · B. Zheng (✉)
National Key Laboratory for Development and Utilization of Forest Food Resources, Zhejiang A & F University, Hangzhou 311300, Zhejiang, China
e-mail: mjr@zafu.edu.cn

B. Zheng
e-mail: bszheng@zafu.edu.cn

D. Mingzheng
College of Agronomy and Life Sciences, Zhaotong University, Zhaotong, China

S. A. H. Naqvi
Department of Plant Pathology, Faculty of Agricultural Sciences and Technology, Bahauddin Zakariya University, Multan, Pakistan

S. A. H. Naqvi et al. (eds.), *Climate Change and Mango Production*,
https://doi.org/10.1007/978-3-031-87035-4_2

influence secondary metabolites such as phenolics, flavonoids, terpenes, alkaloids, etc. of mango plants. It is vital to grasp how these secondary metabolites contribute to the resilience of certain plants under future conditions. The book chapter aims to compile recent research on how climate change impacts the secondary metabolism of plants, particularly mangoes. It discusses the variations in compound concentrations caused by climate factors, providing essential insights into plant responses and performance under changing environmental conditions.

Keywords Climate change · Secondary metabolites · Mango · Environmental stressors · Flavonoids · Drought · Temperature · Light · Heavy metals

2.1 Introduction

Atmospheric levels of carbon dioxide (CO_2) have soared to unprecedented levels since the industrial era began, recording an average concentration of 416.45 μmol mol−[1] in 2021, with projections suggesting a rise to 700 μmol mol^{-1} by the year 2100. This escalation in CO_2, alongside other greenhouse gases, has contributed to a temperature rise of 1.59 °C on land and 0.88 °C over oceans, with potential increases reaching up to 5.7 °C by century's end under scenarios of high greenhouse gas emissions (Change 2013). Additionally, pollutants like chlorofluorocarbons and nitrous oxide (N_2O) have diminished stratospheric ozone levels, enhancing the penetration of high-energy ultraviolet-B (UVB) radiation to the Earth's surface, which also alters other light properties such as intensity and quality (Ravishankara et al. 2009). These environmental stressors are anticipated to significantly impact summer rainfall patterns. Collectively, these changes are detrimental to plant health, potentially causing significant reductions in both vegetative growth and crop yields, thereby decreasing agricultural productivity. This book chapter discusses the impacts of various environmental stressors such as increased temperatures, drought conditions, and UVB exposure on the secondary metabolites in plants (Saibo et al. 2009; Qaderi et al. 2023). It is noted that rising CO_2 levels and other greenhouse gases have led to higher global temperatures. The interconnection between drought and elevated temperatures is evident as higher temperatures increase water demand, leading to drought conditions which in turn escalate temperatures further (Mukherjee et al. 2018). Research indicates that these climate change components are interconnected and their effects could intensify throughout the century. With the global population expected to reach 10.9 billion by 2100, enhancing agricultural output through innovative mitigation strategies is crucial. Understanding plant responses to these climate change factors is essential for developing effective agricultural interventions (Jones and Warner 2016).

Mango, scientifically known as *Mangifera indica* L., is often referred to as the "King of fruits" due to its delightful flavor (Tharanathan et al. 2006). It is a member of the Anacardiaceae family within the Sapindales order. Extensive research has been conducted on mangoes to explore the structure and function of mango leaves, the pathways of mango metabolism, the transcriptome of the epidermal peel, and the progression of mango fruit from the flowering stage to ripeness. Plant developmental stages such as growth, flowering, fruiting, and the quality of the fruit are significantly affected by aforementioned climate variations (Tharanathan et al. 2006). Key climate factors impacting plant life include erratic rainfall and unpredictable spikes in temperature, which are likely to decrease agricultural output (Francini and Sebastiani 2019). During the flowering phase, moisture stress and elevated temperatures can severely affect the quality of pollen and ovules, impacting fruit production and overall yield. In papaya, high temperatures lead to the sterility of stigma and stamens, resulting in flower drop and sex alteration in both female and hermaphrodite flowers (Era et al. 2019). Low temperatures during the flowering period commonly cause flower drop in mango, guava, litchi, and other fruits. Research indicates a significant decline in the population and pollination activities of key pollinators such as bees, moths, and butterflies due to climate change (Abrol and Abrol 2012).

Increased temperatures during the fruit development phase can lead to various physiological issues like spongy tissue formation and black tip in mangoes, fruit cracking, and pulp hardening in citrus fruits (Arbona et al. 2015; Liu et al. 2023). Such conditions also favor the proliferation of beetles, bugs, and other pests, as well as diseases like mildew and blight. Therefore, breeding new plant varieties that can withstand multiple stress factors such as drought, flooding, and salinity are crucial for maintaining agricultural productivity.

Plants synthesize a variety of secondary metabolites in response to abiotic stressors including drought, salinity, UV radiation, and temperature fluctuations, which help them adapt to environmental changes (Rao et al. 2018, 2021; Khanum et al. 2020; Qaderi et al. 2023). Changes in factors like ozone, UV radiation, and temperature influence the production of these internal plant metabolites for better adaptation. Additionally, plants have evolved to produce numerous metabolites that are unique to specific species and influenced by environmental conditions (Hussain et al. 2019). These metabolites, such as polyamines, flavonoids, jasmonic acid, methyl-jasmonate, and glycine betaine, play a crucial role in protecting plants during abiotic stress. To manage these stresses, plants employ various strategies such as neutralizing reactive oxygen species, generating antioxidants, maintaining membrane integrity, and regulating or adapting their solute concentrations (Rao et al. 2020). This book chapter highlights the importance of mango plant metabolites in adaptation, particularly focusing on secondary compounds in relation to key global environmental changes like increased light, drought, flooding, UV radiation, temperature, and cold stress.

2.2 Climate Change Influence on Mangoes and Plant Secondary Metabolites

Climate change is increasingly acknowledged worldwide as a significant issue due to its adverse effects on mango production (Khanum et al. 2020). Water scarcity and food security are becoming prominent global issues. Specifically, the cultivation and production of mangoes are being severely affected by climate change, presenting numerous challenges in terms of growth, yield, and quality due to abiotic stresses such as extreme heat, cold, UV, and drought conditions (Rao et al. 2019; Khanum et al. 2020; Liu et al. 2023). With rising temperatures, more unpredictable weather patterns, and more frequent extreme weather events, mango farmers are encountering numerous obstacles at every stage of mango cultivation (Abrol and Abrol 2012; Jones and Warner 2016; Francini and Sebastiani 2019). During exposure to cold stress, a comprehensive activation of various genetic pathways was observed in mango leaves, particularly those associated with photosynthesis and the biosynthesis of secondary metabolites (Abdel-Aziz et al. 2023). Specifically, genes encoding for antenna proteins, which play a crucial role in light absorption and energy transfer during photosynthesis, exhibited significant upregulation (Wang et al. 2021). Concurrently, there was a marked enhancement in the biosynthetic pathways of several classes of phytochemicals including flavonoids, terpenoids (comprising both di- and sesquiterpenoids), and alkaloids.

Plant secondary metabolites (SMs) are crucial for plant survival and fostering ecological relationships with other species. Besides yielding numerous valuable natural products, these metabolites shield plants from pathogens and environmental stressors (Qaderi et al. 2023). Due to their immobile nature, plants rely on accumulating these bioactive compounds for protection. Secondary metabolites serve as deterrents to herbivores, barriers to pathogen entry, and agents reducing oxidative stress (Croteau et al. 2000). The production of SMs is highly sensitive to environmental conditions such as light, temperature, soil moisture, soil fertility, drought, and salinity (Al Hassan et al. 2017; Rao et al. 2020). Often, altering a single environmental factor can change the levels of secondary metabolites, even if other conditions remain stable.

This genetic activation led to a pronounced accumulation of these secondary metabolites, suggesting a protective or adaptive response of mango leaves to low temperature conditions (Abdel-Aziz et al. 2023). Flavonoids, known for their antioxidant properties, likely mitigate oxidative stress induced by cold temperatures, whereas terpenoids and alkaloids may function in stabilizing cellular structures and interacting with signaling molecules that mediate stress responses (Shen et al. 2022). Additionally, the increased levels of lignans and coumarins, which are also known for their antioxidative and photoprotective effects, further imply a multifaceted biochemical strategy employed by mango leaves to combat cold stress. The observed biochemical responses not only highlight the resilience of mango leaves but also underscore the complex molecular mechanisms plants deploy to endure abiotic stresses. These findings provide valuable insights into the adaptive strategies of tropical and subtropical flora facing sudden temperature fluctuations, potentially guiding future agricultural practices and genetic engineering efforts aimed at enhancing cold stress tolerance in crop plants.

In the subsequent section, we provide a detailed exploration of the adaptive mechanisms of plant secondary metabolites when subjected to various environmental stressors. This analysis encompasses a range of abiotic stresses including extreme temperature variations (both high and low), prolonged periods of drought, elevated salinity levels, exposure to ultraviolet radiation, air pollutants, and the presence of heavy metals in the environment. Each of these stressors can significantly impact plant physiology and biochemistry, prompting distinct responses in the synthesis and accumulation of secondary metabolites (Compean and Ynalvez 2014; Anwar et al. 2021). These compounds play crucial roles in plant defense and adaptation, enabling survival and resilience in adverse conditions (Bennett and Wallsgrove 1994). The synthesis pathways of these metabolites are intricately linked to the plant's immediate environmental conditions and stress exposure, illustrating a complex interaction between biochemical networks and ecological pressures.

2.3 Secondary Metabolic Response of Mangoes and Other Plants to Different Environmental Stresses

Accumulation of secondary metabolites in mangoes and other plants is often influenced by their growth conditions and associated metabolic pathways, typically triggered by various stress factors (Kabera et al. 2014; Khanum et al. 2020). These stressors include environmental challenges like extreme temperatures, drought, high salinity, alkalinity, UV exposure, and air pollutants, all of which can significantly harm mangoes health by increasing the reactive oxygen species which causes oxidative damage; in response plant activates it secondary metabolic pathways to accumulation fortified antioxidant metabolites to alleviate the oxidative stress (Fig. 2.1). When mangoes face these abiotic stresses, they undergo common physiological responses that may cause cells to lose water, leading to increased osmotic pressure and the movement of water from the cytoplasm into vacuoles. Such external stresses can disrupt certain growth and developmental processes in mangoes, affecting their capacity to produce SMs (Khanum et al. 2020; Liu et al. 2023; Muthuramalingam et al. 2023; Hou et al. 2024). This, in turn, results in variations in their phytochemical compositions, crucial for generating bioactive compounds. Essentially, the production of SMs is a natural response of mangoes to environmental stress, playing a critical role in their adaptation and survival through modifying their interactions with the environment and other organisms (Francini and Sebastiani 2019; Muthuramalingam et al. 2023). In the following discussion, we will explore how SM synthesis is linked with major abiotic stresses.

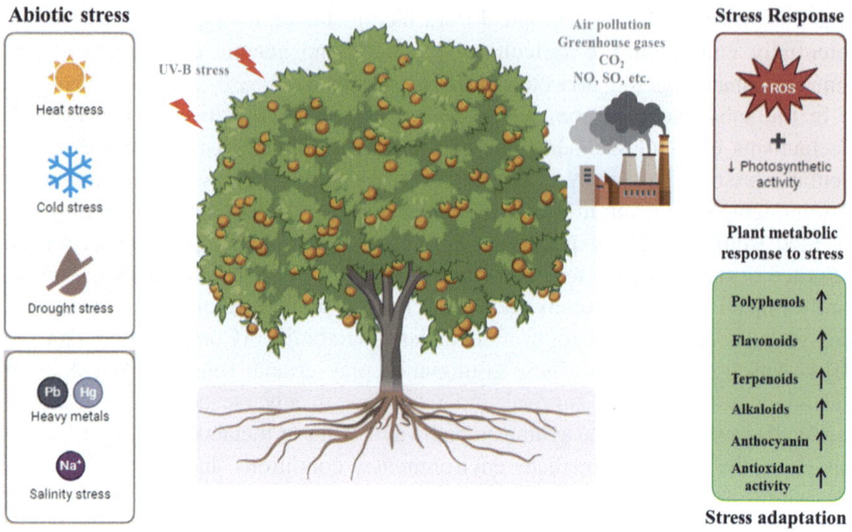

Fig. 2.1 Abiotic stress affects mango trees and plants secondary metabolic response to acclimatize to environmental stress

2.3.1 Effects of Temperature on the Secondary Metabolites of Mangoes and Other Plants

Mangoes exhibit optimal growth and development within a temperature range of 24–26.7 °C. However, when ambient temperatures fall below 10–12 °C, these tropical plants begin to suffer from cold or chilling stress, which significantly impacts their physiological and biochemical processes (Vega-Alvarez et al. 2020). Particularly susceptible to colder climates ranging between 0–15 °C, mango trees display a variety of stress symptoms including pitting, rotting, darkening of lenticels, uneven coloration, and a reduction in fragrance. These manifestations are generally associated with the degradation of the cell wall membrane, a decline in energy production, and an increase in the levels of reactive oxygen species (Li et al. 2014). Further exposure of mango leaves to temperatures within the 5–10 °C range results in a substantial decrease in relative water content (RWC), membrane stability index (MSI), and total free amino acids (TFAA), coupled with an increase in electrolyte leakage percentage (EC%) (Cantre et al. 2017). These changes underscore the critical need for a comprehensive understanding of the mango's physiological responses to low temperature stress (Cantre et al. 2017; Liu et al. 2023). Such knowledge is essential for the development of mango varieties that are more resilient to cold environments and for the formulation of innovative protective strategies.

Research in this area has revealed that mature mango fruits, especially those that are yellow or transitioning to yellow, exhibit a lower susceptibility to chilling injuries compared to their green counterparts (Cantre et al. 2017; Liu et al.

2023). This resilience is likely attributable to enhanced antioxidant activities in the ripening fruit. In-depth studies conducted by Sivankalyani and colleagues have shown that the peels of red mangoes, which accumulate higher levels of anthocyanins and flavonoids during the ripening process, provide better protection against cold stress (Sudheeran et al. 2018). In contrast, green mangoes are more prone to severe chilling injuries, characterized by black spots and pitting.

Transcriptomic analyses of mangoes subjected to cold stress at 5 °C have highlighted significant upregulation in the α-linolenic acid metabolic pathway (Sivankalyani et al. 2016). This pathway plays a pivotal role in the biosynthesis of methyl jasmonate (MeJA) and oxylipin volatile compounds, which are crucial for the plant's response to stress (Zhao et al. 2009; Sudheeran et al. 2018). Additionally, the application of quarantine hot water treatment (HWT) to 'Keitt' mangoes has been explored for its potential to enhance cold stress tolerance (Vega-Alvarez et al. 2020). This treatment appears to activate several protective mechanisms, including the induction of heat shock proteins (HSPs), enzymes involved in energy metabolism, and the synthesis of secondary metabolites such as phenylpropanoids and carotenoids (Sivankalyani et al. 2016; Sudheeran et al. 2018). Furthermore, it stimulates the activity of antioxidant enzymes, hormone metabolism pathways, and pathogenesis-related (PR) proteins, as well as enzymes crucial for cell wall and chloroplast metabolism (Sivankalyani et al. 2016). These findings collectively contribute to a deeper understanding of the molecular and physiological mechanisms underlying the response of mangoes to cold stress. They also provide valuable insights into potential agricultural practices and biotechnological interventions aimed at mitigating the adverse effects of cold stress on mango cultivation, thereby ensuring the sustainability and productivity of this important fruit crop in varying climatic conditions.

The average global temperature rose by approximately 0.47 degrees Celsius during the last century. It is predicted that global warming will significantly impact the production of mango secondary metabolites (Liu et al. 2023; Hou et al. 2024). Both high and low temperatures serve as abiotic stresses that can threaten mango survival, prompting them to modify their metabolic processes to boost essential metabolites and cope with these adverse conditions (Liu et al. 2023). Elevated temperatures are known to increase the production of alkaloids, whereas the synthesis of certain alkaloids like morphinane, phthalisoquinoline, and benzylisoquinoline is reduced under cooler conditions in species like *Papaver somniferum*. For instance, the alkaloid hydroxycamptothecin, which aids in heat-shock tolerance, was observed to multiply six times in *Camptotheca acuminata* seedlings after two hours at 40 °C (Zu et al. 2003). Similarly, *Lupinus angustifolius* varieties significantly increase alkaloid production when subjected to high temperatures (Jansen et al. 2009). On the other hand, *Catharanthus roseus* exhibits a decrease in vindoline and catharanthine levels under cold conditions (Dutta et al. 2007). Temperature also affects the production of terpenoids in trees; for example, the emission of isoprenes from *Quercus rubra* and *Quercus alba* doubles under high temperatures compared to lower ones. High temperatures lead to an accumulation of terpenes in *Daucus carota*, except for α-terpinolene, which decreases as temperatures rise (Zhao et al. 2016). Zea mays seedlings show increased anthocyanin levels under prolonged and severe cold. *Rhodiola rosea* also adjusts its SM synthesis and accumulates more SMs when exposed to cold temperatures combined with heavy metal stress (Chan et al. 2010). Phenylamides, which scavenge reactive oxygen species during stress, are produced in tobacco and beans under conditions of heat shock and water stress. Cold temperatures boost the production of phenolic compounds, which integrate into the plant cell wall as lignin or suberin (Li et al. 2019). Plants adapted to cold climates tend to produce higher levels of chlorogenic acid. Additionally, temperature variations affect the production of ginsenosides in the root hairs of Panax ginseng, while *Melastoma malabathricum* produces elevated levels of anthocyanin in cooler temperatures.

Temperature increases have been observed and are projected to continue escalating throughout this century. Previous research indicates that high temperatures (HT) negatively impact on mangoes yields by reducing photosynthesis and increasing both transpiration and stomatal conductance (Bhardwaj et al. 2015; Zhao et al. 2016; Liu et al. 2023). HT can also cause superficial damage to mangoes, such as scorching of leaves and stems, as well as leaf drop and aging (Sandip et al. 2015; Wang et al. 2021). Temperature plays a crucial role in critical life processes like seed germination and flowering; extreme temperature events have historically led to significant socioeconomic impacts and continue to threaten global food security (Rajan 2012; Sandip et al. 2015). It is estimated that HT could result in a significant yield reduction in mango, the king of fruits, and produces low quality and unmarketable fruits (Rajan 2012; Liu et al. 2021, 2023; Hou et al. 2024). Additionally, brief periods of heat stress can disrupt the biochemistry of anther development in crops like barley and Arabidopsis, causing irreversible male sterility (Yang et al. 2007).

All cellular components in mangoes simultaneously detect temperature changes through thermosensors, which trigger subsequent biochemical reactions (Khanum et al. 2020). Although the precise mechanisms remain somewhat elusive, it is widely recognized that the heat shock transcription factors, specifically HsfA1s, play a pivotal role in the pathway that manages the mangoes response to heat stress (Muthuramalingam et al. 2023). Conversely, mangoes are also vulnerable to cold stress, which involves temperatures below 15 °C, including sub-zero conditions. Such extreme temperatures can adversely impact plant growth and spread, leading to significant mango losses. Changes in temperature prompt cellular reactions that cause various alterations in mango plant metabolism (Wang et al. 2021; Liu et al. 2023). Temperature stress may lead to the denaturation of proteins, liquefaction of lipids, and damage to cellular membranes. There is a reduction in energy and protein metabolism to allow for the buildup of compounds that protect the plant, suggesting a possible dependence of these protective mechanisms on temperature variations (Muthuramalingam et al. 2023).

Thermal reactions happen swiftly, with the majority of changes in metabolites related to temperature occurring within the initial 30 min of exposure. The shift of carbon allocation from growth to defense mechanisms suggests potential detrimental effects; although the plant may continue to live, it would be smaller and under stress, leading to reduced output. Additionally, there is growing evidence that extreme temperatures can cause damage to cellular membranes, leading to increased production of highly toxic reactive oxygen species (Rajan 2012; Hou et al. 2024). This often results in lipid peroxidation, which is commonly used to measure stress-induced oxidative harm. Thermo-tolerance is achieved through the combined actions of various compatible solutes, including hormones and different plant secondary metabolites. While thermal signaling can involve both cold and heat stress, the pathways for these responses might differ. In studies with Arabidopsis, out of the analyzed metabolites, 31% showed no reaction to temperature extremes, 4% only to heat, 19% to both, and 38% only to cold. It's crucial to recognize that an increase in a particular secondary metabolite in response to high temperatures does not necessarily imply a decrease under cold stress conditions, and these extremes should be considered separately (Szymańska et al. 2017; Jan et al. 2021).

2.3.2 Impact of Water Alterations on Mango and Other Plants Metabolic Mechanisms

Drought stress arises from reduced water availability due to factors like low rainfall or high rates of evaporation (Naser and Mohammad 2016). This condition often diminishes plant growth and crop yields (Hou et al. 2024). Globally, agriculture, which consume about 70% of the planet's accessible water and relies heavily on irrigation for 40% of its production, faces significant economic threats from varying water supplies (Jan et al. 2021; Hou et al. 2024). While drought stress can be detrimental, plants may slowly acclimate over weeks, adjusting their metabolism and experiencing stunted growth. Each species reacts uniquely to water deficit, with over 30 known metabolites reacting to water shortages (Khanum et al. 2020; Jan et al. 2021; Hou et al. 2024). Critical for survival, these responses are vital, particularly in crops that generally lack drought resistance. Initially, drought stress disrupts the osmotic balance within plants including mangoes, though many can adjust by maintaining cellular turgor pressure (Liu et al. 2021). Another key response involves solute accumulation, including protective secondary metabolites, triggered by reduced CO_2 intake which affects the Calvin cycle. These substances, such as proline and polyols, help protect vital proteins from degradation due to osmotic stress. Research shows that drought conditions often lead to an increase in flavonoids and other phenolic compounds, particularly in drought-resistant mango and citrus varieties (Hussain et al. 2018; Khanum et al. 2020; Rao et al. 2020; Hou et al. 2024). For instance, under drought mangoes showed an increase in total phenolic content, unlike its sensitive counterparts (Khanum et al. 2020; Hamdy et al. 2022). Similarly, increases were noted in other resilient species like mangoes, cumin, peanut, and milfacadas—a native plant of southern Brazil adapted to arid conditions. In mango varieties, drought-stressed plants recorded a 10% rise in metabolic pathways especially in phenolic content per plant compared to controls (Khanum et al. 2020; Hamdy et al. 2022; Muthuramalingam et al. 2023). Conversely, in drought-sensitive varieties like cotton, tea, lemon, and developing oat seeds, phenolic levels declined, indicating their vulnerability to drought conditions (Hussain et al. 2018; Francini and Sebastiani 2019).

Drought, a widespread and complex form of abiotic stress, affects mangoes and other plants in dry and semi-dry regions globally (Jan et al. 2021; Muthuramalingam et al. 2023; Qaderi et al. 2023). This environmental stressor triggers a range of changes in mango—morphological, physiological, biochemical, ecological, and genetic—that can negatively impact both the quality and the yield of plant biomass (Muthuramalingam et al. 2023). Over evolutionary periods, plants have developed mechanisms to cope with drought by accumulating secondary metabolites like terpenes, alkaloids, and phenolic complexes through mechanisms triggered by ionic or osmotic stress. However, this increase in SMs often results in decreased plant biomass. For example, exposure to drought stress leads to higher levels of phenolic compounds in *Hypericum brasilience*, *Mangifera indica* L. and *Pisum sativum* (Tharanathan et al. 2006; Francini and Sebastiani 2019; Hou et al. 2024). These phenolic compounds generally build up under drought conditions due to alterations in the phenylpropanoid pathway, where the majority of crucial pathway genes are activated by the stress (Cin et al. 2011; Hou et al. 2024). Drought stress, for instance, triggers the activation of the *PAL* gene in lettuce and boosts the expression of genes involved in flavonoid biosynthesis in citrus (Rao et al. 2023; Hou et al. 2024). Likewise, in Salvia officinalis, terpene levels rise under drought conditions, though overall biomass diminishes (Jan et al. 2021). The oxidative stress induced by drought also stimulates the synthesis of flavonoids, which serve as antioxidants and help shield plants from the detrimental impacts of water shortages (Francini and Sebastiani 2019; Hussain et al. 2019; Khanum et al. 2020; Liu et al. 2021).

For instance, under conditions of drought stress, the levels of kaempferol and quercetin in tomato, citrus and mango plants are observed to rise; these substances are capable of neutralizing the hydrogen peroxide molecules formed during such stress (Sánchez-Rodríguez et al. 2010; Khanum et al. 2020; Rao et al. 2023; Hou et al. 2024). However, some secondary metabolites decrease in presence of water shortage, such as saponins in Chenopodium quinoa. In laboratory settings, researchers can simulate and examine drought stress to study its correlation with the accumulation of SMs; by adjusting the culture medium—which includes nutrients, carbon sources, and osmotic stabilizers—it's possible to mimic drought conditions and study their impact on the metabolic pathways that lead to the buildup of SMs (Khanum et al. 2020; Jan et al. 2021; Hou et al. 2024). For instance, altering the culture medium can notably influence the production of camptothecin in *Nothapodytes nimmoniana* and *Ophiorrhiza mungos*, phenolic compounds in *Bellis perennis*, and tropane alkaloids in *Brugmansia candida*. Studies have also explored the impact of drought on SMs in conifer plants; notably, total terpenes in the seedlings of *Pinus sylvestris* and *Pinus abies* were reported to increase by approximately 32–39% and 35–45% respectively in drought-stressed plants compared to their unstressed counterparts (Kumar et al. 2019; Jan et al. 2021).

Terpenoids display significant variability in their response to drought conditions, terpenes metabolism activates following drought stress in plants including mangoes (Jan et al. 2021; Datir and Regan 2022). In thyme, the rise in terpene levels is temporary, peaking about two weeks after drought stress begins, before stabilizing and potentially diminishing over time (Jan et al. 2021). Variations in growth phases and testing intervals might explain some of the differences observed across studies. Mango fruit trees, activates several secondary metabolic pathways and accumulates significant amount of secondary metabolites such as flavonoids, phenolic, anthocyanins terpenoids to acclimatize the drought conditions (Khanum et al. 2020). As antioxidant defenses become compromised under water deficit, plants may boost xanthophyll pigments to dissipate excess energy thermally. This adaptation helps protect thylakoid membranes by preventing reactive oxygen species production (Liu et al. 2021; Datir and Regan 2022; Hamdy et al. 2022). However, in severe drought conditions, the increase in zeaxanthin may not be enough to maintain the photochemistry of PSII, leading to enzymatic dysfunction and photoinhibition (Wang et al. 2021).

2.3.3 Light Stress (UV-B) Effect on Mango Metabolism

Light is a fundamental component for plant photosynthesis, serving as the primary energy source that drives the complex biochemical processes within plants (Rao et al. 2019). Photosynthetic systems, particularly sensitive to environmental stressors, are frequently studied to understand the effects of UV-B radiation (wavelength 280–320 nm) (Dias et al. 2020). This type of radiation is known to adversely affect the photosynthetic machinery, leading to the degradation of thylakoid membranes, damage to photosystem II (PSII), reduced CO_2 assimilation, and diminished oxygen production, among other detrimental impacts (Blankenship 2008). Historically, UV-B radiation has been acknowledged as a significant environmental stressor that impairs plant growth through mechanisms such as stunted growth, reduced biomass, lowered photosynthetic efficiency, and disruptions in DNA integrity and repair processes.

In a detailed experiment, a group of adult 'Tainong No. 1' mango trees (*Mangifera indica*), numbering nine, were exposed to simulated UV-B radiation at varying intensities of 24 and 96 kJ/(m^2 d) within their natural environment (Wang et al. 2021). The study meticulously recorded changes in photochemical reactions, the functionality of key enzymes involved in carbon assimilation, and patterns of gene expression. The results from the study revealed that exposure to 96 kJ/(m^2 d) of UV-B radiation resulted in a notable decline in fruit yield, reduction in soluble sugars, decrease in sugar-acid ratio, and a lower vitamin C content when compared to the control group. In contrast, the 24 kJ/(m^2 d) treatment did not exhibit significant alterations in these parameters (Wang et al. 2021).

Further analysis showed that after 20 or 40 days of exposure, critical parameters such as the net photosynthetic rate (Pn), stomatal conductance (Sc), transpiration rate (Tr), internal CO2 concentration (Ci), and the chlorophyll a/b ratio were significantly reduced under the 96 kJ/(m2·d) treatment compared to the control conditions (Wang et al. 2021). On the other hand, the levels of chlorophyll a, chlorophyll b, carotenoids, Hill reaction activity, and the photochemical quenching coefficient (qP) along with Rubisco activities showed a significant increase. Remarkably, under the 24 kJ/(m^2 d) radiation treatment, both Hill reaction and Rubisco activities not only remained stable but increased by 350% and 30.8%, respectively, suggesting an adaptive response or resilience at lower radiation intensities. However, the photosynthetic pigment levels, along with Pn, Sc, Tr, Ci, remained comparable to those observed in the control group (Wang et al. 2021). Moreover, the study highlighted that the gene encoding the Rubisco large subunit (rbcL) experienced a suppression in its expression under the higher radiation regime of 96 kJ/(m^2 d) (Wang et al. 2021). The findings underscore the nuanced responses of mango trees to different intensities of UV-B radiation, illustrating a complex interplay between radiation-induced stress and the physiological adaptations of the plants.

Recently, the effects of UV-B radiation at an intensity of 96 kJ·m^{-2} d$-^{1}$ were examined on the 'Tainong 1' mango tree. Observations revealed that the flavonoid concentrations in the treated group were substantially elevated when compared to the untreated control group (Shui et al. 2024). The period leading up to 60 days after flowering (DAF) was marked by a notable increase in enzyme activity, which was facilitated through the upregulation of specific genes including CHS, R02446, and CYP98A. This genetic activation led to the synthesis of a range of flavonoids, namely gallochol, leucoside, kaempferoside, quercetin, isoquercetin, and luteolin. These compounds played a pivotal role in the scavenging of reactive oxygen species (ROS), thereby mitigating the oxidative stress induced by the enhanced UV-B radiation (Shui et al. 2024). The study highlights the critical antioxidant role of flavonoids in countering the adverse effects of increased UV-B exposure. Specifically, before reaching 60 DAF, the mango pulp effectively neutralized ROS by augmenting the synthesis of flavonoids. This biochemical adjustment not only preserved the normal physiological functions of the mango fruit but also improved the nutritional quality of the pulp during this crucial developmental phase (Shui et al. 2024). However, beyond 60 DAF, the continued exposure to intense UV-B radiation resulted in an excessive production of ROS within the fruit pulp. This surge overwhelmed the flavonoid-mediated defensive mechanisms, leading to significant oxidative damage within the cellular structures of the pulp. Table 2.1 provides a comprehensive overview of how climate change factors influence the secondary metabolic profile of mango plants, highlighting the adaptive responses through secondary metabolites.

Flavonoids, beyond their role as antioxidants, also function as signaling molecules, UV filters, and agents for ROS neutralization. These flavonoids are integral in forming the primary defense against UV-B radiation (Xi et al. 2014; Shen et al. 2022). Furthermore, Dias et al. (2020) noted that in olives, alterations in flavonoid

Table 2.1 Climate factors influence on secondary metabolites of plants

Climate change factor	Abiotic stress	Impact on mango plants	Modulation of secondary metabolites	References
Rising temperatures	Heat stress	• Increased respiration rates • Reduced photosynthetic efficiency • Altered fruit development	• Enhanced synthesis of heat shock proteins (HSPs) • Increased production of phenolic compounds and flavonoids for stress protection	Khanum et al. (2020)
Drought conditions	Drought stress	• Reduced leaf water potential • Stomatal closure • Leaf wilting and drop	• Accumulation of osmoprotectants (e.g., proline) • Increased synthesis of flavonoids and anthocyanins to enhance drought tolerance	Rao et al. (2024)
Increased CO_2 levels	Elevated CO_2	• Enhanced growth and yield under optimal conditions • Potential nutrient dilution	• Altered secondary metabolite profiles, including increased phenolic content • Changes in flavonoid composition affecting fruit quality	Zhao et al. (2019)
Salinity	Salinity stress	• Ion toxicity (Na^+ and Cl– accumulation) • Reduced nutrient uptake	• Increased synthesis of compatible solutes (e.g., glycine betaine) • Enhanced production of secondary metabolites like flavonoids and tannins for stress mitigation	Perveen et al. (2024)

(continued)

Table 2.1 (continued)

Climate change factor	Abiotic stress	Impact on mango plants	Modulation of secondary metabolites	References
Extreme weather events	Flooding	• Root oxygen deficiency • Increased susceptibility to diseases	• Enhanced synthesis of phenolic compounds for root protection • Increased antioxidant enzyme activity to combat oxidative stress	Deena Dayalan et al. (2024)
UV radiation	UV stress	• Leaf damage and reduced photosynthetic efficiency	• Increased synthesis of UV-absorbing compounds (e.g., flavonoids, phenolics) • Enhanced antioxidant activity to mitigate oxidative stress	Zhu et al. (2023), Fatima et al. (2024)

metabolism constitute a fundamental response to UV-B stress, underscoring the universal importance of these compounds in plant defense mechanisms (Dias et al. 2020). Additionally, recent studies have further confirmed that flavonoids are indispensable for the effective removal of ROS, highlighting their essential role in maintaining cellular homeostasis and protecting against environmental stressors (Rao et al. 2022). This investigation thus provides substantial evidence that flavonoids are crucial in enhancing plant resilience against UV-B radiation, suggesting potential strategies for agricultural practices to enhance crop protection and yield through biofortification and genetic engineering.

2.3.4 Effects of Salinity on the Secondary Metabolites

Soil salinization is increasingly recognized as a significant global challenge that adversely affects the productivity of horticultural crops. Salinity significantly impacts plants by inducing cellular dehydration, which leads to either an increase or decrease in the production of specific secondary metabolites (Laxmi Kamal et al. 2021). This issue is being intensified by factors such as climate change and rapid urbanization. This salt stress serves as a trigger, enhancing the accumulation of SMs that help protect plant cells from the oxidative harm caused by ionic build-up at both cellular and subcellular levels. Consequently, these SMs mitigate the adverse effects of salinity (Jan et al. 2021; Laxmi Kamal et al. 2021; Muthuramalingam et al. 2023). In response to this growing concern, the development of salt-tolerant rootstock varieties emerges as a pivotal strategy for mitigating stress in mango and other perennial fruit crops (Laxmi Kamal et al. 2021).

Salinity significantly impacts plants by causing cellular dehydration, a major type of abiotic stress (Laxmi Kamal et al. 2021). The salinity results from ionic or osmotic disturbances, leading to either an increase or decrease in the accumulation of specific secondary metabolites in plants (Jan et al. 2021). These SMs serve as protective agents against the oxidative damage brought on by ion accumulation at both cellular and subcellular levels, mitigating the adverse effects of salinity. The flexibility of plant genotypes in various saline environments allows for the synthesis of crucial SMs, essential for plant survival under such conditions. Salinity influences plants physiologically, biochemically, morphologically, and through the biosynthesis of SM profiles by triggering stress and defense response pathways that generate reactive oxygen species, which play a key role in SM modulation.

Salinity is a significant abiotic stressor for crop plants, primarily due to the increased uptake of Na^+ and Cl− ions, which leads to ionic toxicity and osmotic stress. This imbalance causes a deficiency in K^+ and Ca^{2+}, leading to nutritional issues. Mango trees, known for their sensitivity to salt, often exhibit symptoms like leaf curling and tip burn, and in extreme cases, can experience stunted growth, leaf drop, and even death. Research has shown that salinity restricts the growth of seedlings, reduces chlorophyll content, decreases CO_2 absorption, and hampers nutrient uptake (Laxmi Kamal et al. 2021). Kishor and colleagues have suggested that treatments with paclobutrazol (PBZ) can enhance salinity tolerance in mangoes by increasing photosynthetic pigment levels and K^+ absorption, improving water potential, and reducing membrane damage and the uptake of harmful ions (Kishor et al. 2009). The response to salt stress varies among different mango rootstocks; for example, "Olour" and "Turpentine" rootstocks have shown tolerance due to their ability to limit Cl− and Na^+ ion uptake, likely through increased proline levels (Pandey et al. 2014). Another study indicated that the 'Gomera-3' genotype is more sensitive to salinity compared to the '13/1' genotype, which displays enhanced physiological mechanisms such as increased foliar K^+ and proline levels and delayed leaf shedding (Mahouachi 2018). Recent research has explored the use of sodium silicate pentahydrate and glycine betaine (GB) to mitigate the adverse effects of salinity on the 'El-Gahrawey' mango cultivar (Shalan 2020). The foliar application of Si and GB significantly enhanced growth parameters, leaf mineral content, and total phenolic content, improving the cultivar's resilience to salinity stress (Shalan 2020). These findings underscore the importance of understanding the molecular and physiological responses of mango trees to effectively manage and improve their resistance to abiotic stresses like salinity.

In an experiment, seven mango seedlings subjected to saline conditions (200 mM NaCl, Electrical Conductivity: >4.0 dSm^{-1}, pH 8.5) has demonstrated that elevated levels of proline (19.07 μg $g-^{1}$ Fresh Weight in Bappakai), glycine betaine (55.11 μg g^{-1} Fresh Weight in 13–1), and total sugars (17.33 mg $g-^{1}$

Fresh Weight in Kurukkan) play crucial roles as adaptive mechanisms to manage osmotic stress under suboptimal conditions (Laxmi Kamal et al. 2021). Furthermore, significant increases in non-enzymatic antioxidants such as tannins (17.18 mg g^{-1}), phenols (18.68 mg g−[1]), and anthocyanins (1.59 mg g^{-1}) were observed in the seedlings of "13–1", a variety known for its salt resistance originating from Israel (Laxmi Kamal et al. 2021). These biochemical compounds are essential for the scavenging of reactive oxygen species, thereby protecting cellular structures from oxidative damage. This study also highlighted the enhanced activity of antioxidant enzymes like superoxide dismutase and catalase, particularly in Indian polyembryonic cultivars (Kurukkan and Nekkare), underscoring their potential utility as rootstocks capable of countering oxidative stress (Laxmi Kamal et al. 2021). These findings not only emphasize the importance of osmotic stress relief and redox balance but also align the salt tolerance capabilities of the Indian cultivars Nekkare and Kurukkan with other globally recognized salt-tolerant rootstocks. This research contributes to the broader understanding of physiological and biochemical strategies that enhance plant resilience to salinity, offering promising avenues for the cultivation of fruit crops in salt-affected regions.

2.3.5 Impact of Heavy Metals on Secondary Metabolites in Plants

Heavy metals are increasingly acknowledged as significant abiotic stressors affecting plant physiology due to their widespread use in agriculture and industry, coupled with their high potential for bioaccumulation and environmental persistence (Jan et al. 2021). Throughout the progression of the study into late April, a detailed analysis indicated that the concentrations of these metals varied significantly, with manganese (Mn) levels ranging from 0.30 to 1.30 mg/L, iron (Fe) from 0.80 to 1.60 mg/L, copper (Cu) from 0.09 to 0.20 mg/L, and lead (Pb) from 0.33 to 0.50 mg/L. During this investigative phase, the highest recorded concentrations were observed in Binta Sugar for Mn, Julie mango for Fe, Binta Sugar for Cu, and both Sherry and Binta Sugar varieties for Pb (Dangana Mohammed Chata et al. 2018).

Extensive research has been conducted on how heavy metals influence plant growth and overall physiological responses; however, there is a relatively lesser focus on their effects on the biosynthesis of SMs (Khanum et al. 2020; Muthuramalingam et al. 2023). The alteration in plant metabolic processes due to heavy metal stress can significantly affect the production of essential biochemicals such as sugars, photosynthetic pigments, and proteins (Dangana Mohammed Chata et al. 2018; Datir and Regan 2022; Abdel-Aziz et al. 2023). This is primarily through the inhibition of key enzymes involved in these biosynthetic pathways. Furthermore, heavy metals are known to modify the synthesis of bioactive compounds by targeting specific pathways within secondary metabolism. These findings were meticulously compared to the established safety standards set by the World Health Organization (WHO) and the Food and Agriculture Organization (FAO) in 2011 (Dangana Mohammed Chata et al. 2018). The comparison revealed that while the levels of Mn, Fe, and Cu remained within permissible safety limits, the concentrations of Pb significantly exceeded these thresholds, posing potential risks.

In specialized studies, such as those involving hairy root cultures of Brugmansia candida, it has been demonstrated that exposure to cadmium chloride ($CdCl_2$) or silver nitrate ($AgNO_3$) can lead to an excessive production of tropane alkaloids like hyoscyamine and scopolamine (Lajayer et al. 2017). Similarly, in cell cultures of Taxus species, the synthesis of taxol, an important anticancer compound, is enhanced when treated with lanthanum, a rare earth metal. Additionally, the presence of copper ions (Cu^{2+}) has been shown to significantly enhance the production of betalains in Beta vulgaris (Singh and Sinha 2005). This enhancement is also observed with cobalt (Co^{2+}), which stimulates the synthesis of secondary metabolites in this species. Moreover, zinc treatment at a concentration of 900 μM has been found to increase the production of lepidine in Lepidium sativum cultures, although copper has proven to be more effective in enhancing the yield of these secondary metabolites (Saba et al. 2000).

The impact of heavy metals on the biosynthesis of SMs in plants has been increasingly recognized as a critical area of study within the field of plant physiology and biochemistry (Singh and Sinha 2005; Lajayer et al. 2017; Dangana Mohammed Chata et al. 2018). A recent research focuses on the response of secondary metabolites to varying concentrations of heavy metals, specifically manganese (Mn), iron (Fe), copper (Cu), and lead (Pb). The analysis revealed that the concentrations of these metals in the plant tissues ranged significantly, with Mn concentrations varying from 0.10 to 1.30 mg/L, Fe from 0.59 to 1.84 mg/L, Cu from 0.03 to 0.16 mg/L, and Pb from 0.50 to 0.80 mg/L (Dangana Mohammed Chata et al. 2018). Notably, the highest average concentrations of Mn and Pb were detected in the Binta Sugar and Kerosene mango varieties, suggesting a varietal dependency in metal accumulation. Moreover, the Kerosene variety also demonstrated elevated levels of Fe and Cu, particularly during the late February sampling period, indicating possible seasonal influences on metal uptake and accumulation (Dangana Mohammed Chata et al. 2018). This observation is crucial as it suggests that certain plant varieties might possess intrinsic mechanisms that facilitate higher uptake and accumulation of specific metals under certain environmental conditions.

It is well-documented that metals such as Fe, silver (Ag), nickel (Ni), and cobalt (Co) can induce the production of secondary metabolites in various plant species (Muthuramalingam et al. 2023). These secondary metabolites play essential roles in plant defense mechanisms against biotic and abiotic stresses and can influence plant growth and productivity. For instance, a detailed examination of *Brassica juncea* demonstrated that the accumulation of metals like Fe, chromium (Cr), zinc (Zn), and Mn could enhance the plant's oil content by up to 35% (Singh and Sinha 2005). This finding is significant as it highlights the potential of using metal stress to enhance the yield of valuable plant-derived products. Additionally, studies focusing on cadmium (Cd^{2+}) and copper (Cu^{2+}) have reported peak accumulations of specific secondary metabolites such as shikonin and digitalin (Tatsumi et al. 2016). These compounds are of considerable interest due to their pharmacological properties and commercial value (Muthuramalingam et al. 2023). The ability of plants to increase the production of such metabolites under metal stress could be harnessed to develop innovative agricultural practices aimed at producing high-value biochemicals in a sustainable manner.

In conclusion, this research underscores the complex interactions between heavy metal exposure and secondary metabolite production in plants. It highlights the potential for selective breeding and genetic engineering to enhance the ability of plants to tolerate metal stress and simultaneously increase the biosynthesis of economically important secondary metabolites. Further studies are required to elucidate the underlying molecular mechanisms and to optimize conditions for maximizing these beneficial effects in different plant species. There is imperative need to monitor and regulate the presence of heavy metals in food products, particularly in regions where agricultural practices might lead to elevated contamination risks. Ensuring the adherence to safety standards for heavy metals is crucial not only for safeguarding public health but also for maintaining the nutritional integrity of food supplies. As such, continuous monitoring and rigorous assessment of these metals are essential to prevent potential health hazards and to enhance food safety protocols, thereby ensuring the well-being and health of the population.

2.3.6 Air Pollution Effects on Plant Secondary Metabolites

Air pollution is increasingly recognized as a significant global problem. Among human-made sources, vehicles are the primary contributors, accounting for 70% of total air pollution, others are industrial activity, and factory emissions. They release substantial amounts of pollutants into the environment, adversely affecting air quality (Mukherjee et al. 2019). To mitigate these effects, environmentalists and policymakers have highlighted the importance of using trees to monitor air pollution. Prominent mango varieties in Malda district are Fazli, Aswina, Laxmanbhog, Langra, and Himsagar, grown across national highway, which have been chosen for this bio-monitoring study. The leave samples showed that the highest monthly average nitrogen oxides (NO) concentration was 22.04 $\mu g/m^3$ in December, with the lowest at 19.04 $\mu g/m^3$ in June 2016. Sulfur oxides (SO) concentrations peaked at 4.38 $\mu g/m^3$ in October and dipped to 2.00 $\mu g/m^3$ in June and August. The highest monthly average particulate matter (PM10) concentration was recorded in January at 124.43 $\mu g/m^3$, exceeding the National Ambient Air Quality Standards (NAAQS) limit of 100 $\mu g/m^3$ (Das and Sarkar 2022). Among the varieties, Fazli exhibited the highest average Ascorbic Acid (ASC), total Chlorophyll (tchl) content, and relative water content (RWC). The pH of leaf extracts varied, with Langra at 4.93 ± 0.70 and Himsagar at 6.46 ± 0.33. Among the mango varieties studied, Fazli stood out with the highest average levels of ASC, Tchl content, and RWC, indicating a robust physiological adaptation to environmental stressors. Notably, Fazli, with the highest air pollution tolerance index (APTI) value of 22.12 ± 0.26, was identified as a highly tolerant species, capable of withstanding higher levels of air pollutants (Das and Sarkar 2022). In contrast, varieties such as Aswina, Laxmanbhog, Langra, and Himsagar were categorized as sensitive, demonstrating lower tolerance to the air pollutants measured in this study (Das and Sarkar 2022). This differential response underscores the importance of selecting appropriate plant species for bio-monitoring and environmental management practices aimed at mitigating air pollution.

In another study, the biochemical alterations were measured within the leaves of the mango tree (*Mangifera indica*), a species prevalent in West Bengal, India (Mukherjee et al. 2019). The samples were collected from four urban locations in Kolkata, noted for their elevated pollution levels, alongside a comparison sample from a rural site known for its relatively lower pollution levels. The analysis of these samples disclosed significant biochemical changes in the foliage from areas with higher pollution. Notably, there was a significant upregulation in the levels of phenol, proline, malondialdehyde, and cellulose, coupled with a decrease in chlorophyll content in mango leaves (Mukherjee et al. 2019). These findings suggest that air pollution precipitates a series of biochemical adaptations in mango plants, serving as a potential defense mechanism against environmental stress (Zhao et al. 2016; Mukherjee et al. 2019). The alterations in these biochemical markers, therefore, could potentially serve as reliable indicators of air quality, providing insights into the extent of pollution and its impact on plant biology.

The environmental conditions play a pivotal role in influencing the accumulation of polyphenolic compounds in various parts of the mango plant, including the fruit pulp, peel, seed kernel, leaves, and stem bark (Masibo and He 2008). It is particularly interesting to note that the leaves accumulate a range of substances such as galloyl derivatives, hydroxy benzoyl esters, and epicatechin, with a significant concentration of mangiferin—a robust polyphenol—predominantly found in the bark and leaves (Masibo and He 2008; Mukherjee et al. 2019). The outcomes of this study indicate that elevated pollution levels catalyze the synthesis of these phenolic compounds, likely as a biological response to environmental stressors. These phenolic compounds are crucial as they function as antioxidants, protecting essential cellular components such as lipids, proteins, DNA, and RNA from oxidative damage (Rao and Zheng 2025; Rao et al. 2022). This protective mechanism enhances the plant's resilience and ability to withstand adverse environmental conditions.

Moreover, mango plants grown under air pollution environmental stress leads to an increase in cellulose accumulation in the leaves of the mango plants (Mukherjee et al. 2019). This enhanced deposition of cellulose within the cell walls results in the formation of thicker secondary walls, a critical adaptation that fortifies the plant's structural integrity and resistance to various forms of stress (Mukherjee et al. 2019; Khanum et al. 2020; Muthuramalingam et al. 2023). This adaptation not only plays a significant role in the mechanical support of the plant but also in its overall ability to survive and thrive under challenging environmental conditions (Das and Sarkar 2022). Through these observations, it becomes evident that the biochemical and structural adaptations of plants are intricately linked to their environmental exposures, highlighting the dynamic interplay between biological processes and ecological factors.

2.4 Future Perspectives and Concluding Remarks

Climate change influences metabolic transformations in mango plants, affecting both their primary and secondary metabolic processes. Research on how multiple stress factors simultaneously impact plants is limited, despite the known responses of plants to these factors in their natural environments and their distinct impacts compared to individual stressors. For example, elevated CO_2 levels might alleviate drought stress, but when combined with high temperatures, the effects are more harmful than either stressor alone. Studies have demonstrated that mango plants can become more resilient by producing antioxidant metabolites such as phenolic, flavonoids, etc., when exposed to single stress factors, but their resistance is compromised under multiple environmental stresses.

Plant stress responses are complex and result from evolutionary adaptations over millennia. The resilience of mango plants to abiotic stress involves multiple factors. This chapter indicates a general rise in plant secondary metabolites under various stress conditions, although elevated CO_2 alone has shown to reduce these compounds by decreasing protein and nutritional levels in crops, potentially hindering key biosynthetic pathways due to limited availability of precursor metabolites and reduced activity of crucial enzymes like PAL.

Interestingly, elevated CO_2 has also been found to increase toxic substances in some plants, which could potentially worsen their impact on human health. However, it's unclear if CO_2 would have the same effect when present with other abiotic stresses. The increase in SMs is crucial for mango plant survival as these compounds play defensive roles against a wide array of environmental stresses. Depending on agricultural goals, the increase in SMs can be seen as either beneficial or detrimental. Certain SMs, such as phenolic compounds and flavonoids, serve as antioxidants and are vital for human health. They are also utilized in treating diseases like ovarian and breast cancers, heart disease, arthritis, and other inflammatory conditions and even in medical applications. It is therefore essential to explore how these metabolites react under multiple stress conditions to better understand the potential impacts of climate change on mango plant metabolism. This knowledge is crucial for addressing one of the major global challenges: enhancing the nutritional value and yield of mango crops while reducing the presence of harmful ROS. Future research should also focus on increasing the bio-fortification of antioxidant metabolites such as phenolic and flavonoids in mango plants, under stress conditions and employing advanced techniques to assess the impact of external application of SMs on plant tolerance to various stresses.

References

Abdel-Aziz HF, Hamdy AE, Sharaf A, Abd El-Wahed AEN, Elnaggar IA, Seleiman MF, Omar M, Al-Saif AM, Shahid MA, Sharaf M (2023) Effects of fogging system and nitric oxide on growth and yield of 'Naomi' mango trees exposed to frost stress. Life 13

Abrol DP (2012) Climate change and pollinators. Pollination biology: biodiversity conservation and agricultural production. Springer, Dordrecht, pp 479–508

Anwar M, Chen L, Xiao Y, Wu J, Zeng L, Li H, Wu Q, Hu Z (2021) Recent advanced metabolic and genetic engineering of phenylpropanoid biosynthetic pathways. Int J Mol Sci 22:9544

Arbona V, De Ollas CJ, Argamasilla R, López-Climent MF, Gómez-Cadenas A (2015) Hormone and metabolite traits related to abiotic stress tolerance in citrus. Acta Horticulturae 1275–1281

Bennett RN, Wallsgrove RM (1994) Secondary metabolites in plant defence mechanisms. New Phytol 127:617–633

Bhardwaj AR, Joshi G, Kukreja B, Malik V, Arora P, Pandey R, Shukla RN, Bankar KG, Katiyar-agarwal S, Goel S et al (2015) Global insights into high temperature and drought stress regulated genes by RNA-Seq in economically important oilseed crop *Brassica juncea*. BMC Plant Biol 15:1–15

Blankenship RE (2008) Photosynthetic pigments: structure and spectroscopy. Mol Mech Photosynth 2:41–58

Cantre D, Herremans E, Verboven P, Ampofo-Asiama J, Hertog ML, Nicolai BM (2017) Tissue breakdown of mango (*Mangifera indica* L. cv. Carabao) due to chilling injury. Postharvest Biol Technol 125:99–111

Chan LK, Koay SS, Boey PL, Bhatt A (2010) Effects of abiotic stress on biomass and anthocyanin production in cell cultures of *Melastoma malabathricum*. Biol Res 43:127–135

Change IC (2013) The physical science basis. In: Contribution of working Group I to the fourth assessment report of the intergovernmental panel on climate change (IPCC).

Chata DM, Iliya AY, Chidawa MS, Emmanuel MB, Juliana OO, Chibuzor EK (2018) Determination of heavy metals in four mango fruit varieties sold in Minna modern market, Niger state, Nigeria. Int J Biol Environ Eng 1:24–29

Cin VD, Tieman DM, Tohge T, McQuinn R, de Vos RCH, Osorio S, Schmelz EA, Taylor MG, Smits-Kroon MT, Schuurink RC et al (2011) Identification of genes in the phenylalanine metabolic pathway by ectopic expression of a MYB transcription factor in tomato fruit. Plant Cell 23:2738–2753

Compean KL, Ynalvez RA (2014) Antimicrobial activity of plant secondary metabolites: a review. Res J Med Plants 8:204–213

Croteau R, Kutchan TM, Lewis NG (2000) Natural products (secondary metabolites). Biochem Mol Biol Plants 24:1250–1319

Das S, Sarkar A (2022) Utilization of mango plants (*Mangifera indica* L.) as in situ bio-monitoring tool against vehicular air pollution along National and State Highways: a case study from Malda district, West Bengal. India. J Biol Today's World 11:1–4

Datir S, Regan S (2022) Advances in physiological, transcriptomic, proteomic, metabolomic, and molecular genetic approaches for enhancing mango fruit quality. J Agric Food Chem 71:20–34

Deena Dayalan M, Ramesh Kumar A, Senthilkumar S, Srivignesh S, Manivannan S (2024) Impacts of flooding stress in fruit crops and its adaptation strategies. Interdiscip Approaches Agric For 39

Dias MC, Pinto DCGA, Freitas H, Santos C, Silva AMS (2020) The antioxidant system in *Olea europaea* to enhanced UV-B radiation also depends on flavonoids and secoiridoids. Phytochemistry 170:112199

Dutta A, Sen J, Deswal R (2007) Downregulation of terpenoid indole alkaloid biosynthetic pathway by low temperature and cloning of an AP2 type C-repeat binding factor (CBF) from *Catharanthus roseus* (L.). G. Don. Plant Cell Rep 26:1869–1878

Era FM, Raihan MS, Islam AKMA (2019) Sex expression in papaya: morphological marker, molecular genetics and environments. Fund Appl Agric 4:978–994

Fatima F, Basit A, Osaidullah, Mohamed HI (2024) Role of UV radiation management strategies: towards mitigating postharvest losses, quality, phenolic and antioxidant activity, and ripening rate of mango (*Mangifera indica* L.) cultivars. Appl Fruit Sci 1–9

Francini A, Sebastiani L (2019) Abiotic stress effects on performance of horticultural crops. Horticulturae 5:67

Hamdy AE, Abdel-Aziz HF, El-Khamissi H, AlJwaizea NI, El-Yazied AA, Selim S, Tawfik MM, AlHarbi K, Ali MSM, Elkelish A (2022) Kaolin improves photosynthetic pigments, antioxidant content, and decreases sunburn of mangoes: field study. Agronomy 12:1535

Hassan MA, Chaura J, Donat-Torres MP, Boscaiu M, Vicente O (2017) Antioxidant responses under salinity and drought in three closely related wild monocots with different ecological optima. AoB Plants 9: plx009

Hou X, Kong Y, Teng Z, Yang C, Li Y, Zhu Z (2024) Integrating genes and metabolites: unraveling mango's drought resilience mechanisms. BMC Plant Biol 1–15

Hussain S, Khalid MF, Saqib M, Ahmad S, Zafar W, Rao MJ, Morillon R, Anjum MA (2018) Drought tolerance in citrus rootstocks is associated with better antioxidant defense mechanism. Acta Physiol Plant 40:135

Hussain S, Rao MJ, Anjum MA, Ejaz S, Zakir I, Ali MA, Ahmad N, Ahmad S (2019) Oxidative stress and antioxidant defense in plants under drought conditions. In: Hasanuzzaman M, Hakeem KR, Nahar K, Alharby HF (eds) Plant abiotic stress tolerance: agronomic, molecular and biotechnological approaches. Springer, Switzerland, pp 207–219

Jan R, Asaf S, Numan M, Kim K (2021) Plant secondary metabolite biosynthesis and transcriptional regulation in response to biotic and abiotic stress conditions. Agronomy 11:968

Jansen G, Jürgens H, Ordon F (2009) Effects of temperature on the alkaloid content of seeds of *Lupinus angustifolius* cultivars. J Agron Crop Sci 195:172–177

Jones GA, Warner KJ (2016) The 21st century population-energy-climate nexus. Energy Policy 93:206–212

Kabera JN, Semana E, Mussa AR, He X (2014) Plant secondary metabolites: biosynthesis, classification, function, and pharmacological properties. J Pharm Pharmacol 2:377–392

Khanum Z, Tiznado-Hernández ME, Ali A, Musharraf SG, Shakeel M, Khan IA (2020) Adaptation mechanism of mango fruit (*Mangifera indica* L. cv. Chaunsa White) to heat suggests modulation in several metabolic pathways. RSC Adv 10:35531–35544

Kishor A, Srivastav M, Dubey AK, Singh AK, Sairam RK, Pandey RN, Dahuja A, Sharma RR (2009) Paclobutrazol minimises the effects of salt stress in mango (*Mangifera indica* L.). J Hortic Sci Biotechnol 84:459–465

Kumar R, Berwal MK, Saroj PL (2019) Morphological, physiological, biochemical and molecular facet of drought stress in horticultural crops. J Bio-Res Stress Manag 10:545–560

Lajayer BA, Ghorbanpour M, Nikabadi S (2017) Heavy metals in contaminated environment: destiny of secondary metabolite biosynthesis, oxidative status and phytoextraction in medicinal plants. Ecotoxicol Environ Saf 145:377–390

Laxmi Kamal A, Kumar V, Muthukumar M, Bajpai A (2021) Morphological indicators of salinity stress and their relation with osmolyte associated redox regulation in mango cultivars. J Plant Biochem Biotechnol 30:918–929

Li P, Zheng X, Liu Y, Zhu Y (2014) Pre-storage application of oxalic acid alleviates chilling injury in mango fruit by modulating proline metabolism and energy status under chilling stress. Food Chem 142:72–78

Li X, Li M, Zhou B, Yang Y, Wei Q, Zhang J (2019) Transcriptome analysis provides insights into the stress response crosstalk in apple (Malus × domestica) subjected to drought, cold and high salinity. Sci Rep 1–10

Liu X, Peng Y, Yang Q, Wang X, Cui N (2021) Determining optimal deficit irrigation and fertilization to increase mango yield, quality, and WUE in a dry hot environment based on TOPSIS. Agric Water Manag 245:106650

Liu X, Xiao Y, Zi J, Yan J, Li C, Du C, Wan J, Wu H, Zheng B, Wang S, et al (2023) Differential effects of low and high temperature stress on pollen germination and tube length of mango (*Mangifera indica* L.) genotypes. Sci Rep 13:611

Mahouachi J (2018) Long-term salt stress influence on vegetative growth and foliar nutrient changes in mango (*Mangifera indica* L.) seedlings. Sci Hortic 234:95–100

Masibo M, He Q (2008) Major mango polyphenols and their potential significance to human health. Compr Rev Food Sci Food Saf 7:309–319

Mukherjee S, Mishra A, Trenberth KE (2018) Climate change and drought: a perspective on drought indices. Curr Clim Change Rep 4:145–163

Mukherjee S, Chakraborty A, Mondal S, Saha S, Haque A, Paul S (2019) Assessment of common plant parameters as biomarkers of air pollution. Environ Monit Assess 191:1–8

Muthuramalingam P, Muthamil S, Shilpha J, Venkatramanan V, Priya A, Kim J, Shin Y, Chen JT, Baskar V, Park K, et al (2023) Molecular insights into abiotic stresses in mango. Plants 12

Naser TZ, Mohammad B (2016) Physiological and morphological response to drought stress in seedlings of ten citrus species. Trees

Pandey P, Singh AK, Dubey AK, Dahuja A (2014) Biochemical and salt ion uptake responses of seven mango (*Mangifera indica* L.) rootstocks to NaCl stress. J Hortic Sci Biotechnol 89:367–372

Perveen N, Dinesh MR, Sankaran M, Lakkireddy V, Shivashankara KS, Venugopal R, Mir H (2024) Phenolics signatures in response to salinity stress provide novel insights into physiological basis of salt tolerance in mango (*Mangifera indica* L.). J Plant Growth Regul 1–20

Qaderi MM, Martel AB, Strugnell CA (2023) Environmental factors regulate plant secondary metabolites. Plants 12:447

Rajan S (2012) Phenological responses to temperature and rainfall: a case study of mango. Trop Fruit Tree Species Clim Change 71

Rao MJ, Ding F, Wang N, Deng X, Xu Q (2018) Metabolic mechanisms of host species against citrus Huanglongbing (Greening Disease). Crit Rev Plant Sci 37:496–511

Rao MJ, Xu Y, Huang Y, Tang X, Deng X, Xu Q (2019) Ectopic expression of citrus UDP-GLUCOSYL TRANSFERASE gene enhances anthocyanin and proanthocyanidins contents and confers high light tolerance in *Arabidopsis*. BMC Plant Biol 19:603

Rao MJ, Xu Y, Tang X, Huang Y, Liu J, Deng X (2020) *CsCYT75B1*, a citrus CYTOCHROME P450 gene, is involved in accumulation of antioxidant flavonoids and induces drought tolerance in transgenic *Arabidopsis*. Antioxidants 9:161

Rao MJ, Wu S, Duan M, Wang L (2021) Antioxidant metabolites in primitive, wild, and cultivated citrus and their role in stress tolerance. Molecules 26:5801

Rao MJ, Duan M, Yang M, Li M, Wang L (2022) Sugarcane rind secondary metabolites and their antioxidant activities in eleven cultivated sugarcane varieties. Sugar Tech 24:1570–1582

Rao MJ, Duan M, Eman M, Yuan H, Sharma A, Zheng B (2024) Comparative analysis of citrus species' flavonoid metabolism, gene expression profiling, and their antioxidant capacity under drought stress. Antioxidants 13:1149

Rao MJ, Feng B, Ahmad MH, Tahir M, Aslam MZ, Khalid MF, Hussain S, Zhong R, Ali Q, Xu Q, et al (2023) LC-MS/MS-based metabolomics approach identified novel antioxidant flavonoids associated with drought tolerance in citrus species. Front Plant Sci 14

Ravishankara AR, Daniel JS, Portmann RW (2009) Nitrous oxide (N_2O): the dominant ozone-depleting substance emitted in the 21st century. Science 326:123–125

Saba PD, Iqbal M, Srivastava PS (2000) Effect of $ZnSO_4$ and $CuSO_4$ on regeneration and lepidine content in *Lepidium sativum* L. Biol Plant 43:253–256

Saibo NJM, Lourenço T, Oliveira MM (2009) Transcription factors and regulation of photosynthetic and related metabolism under environmental stresses. Ann Bot 103:609–623

Sánchez-Rodríguez E, Rubio-Wilhelmi M, Cervilla LM, Blasco B, Rios JJ, Rosales MA, Romero L, Ruiz JM (2010) Genotypic differences in some physiological parameters symptomatic for oxidative stress under moderate drought in tomato plants. Plant Sci 178:30–40

Sandip M, Makwana AN, Barad AV, Nawade BD (2015) Physiology of flowering—the case of mango. Int J Appl Res 1:1008–1012

Shalan AM (2020) Inducing salinity tolerance in mango (*Mangifera indica* L.) cv. “El-Gahrawey” by sodium silicate pentahydrate and glycine betaine. تعزيز تحمل شتلات المانجو صنف "الجحراوى" للملوحة بواسطة سليكات الصوديوم والجليسين بيتين. J Plant Prod 11:541–549

Shen N, Wang T, Gan Q, Liu S, Wang L, Jin B (2022) Plant flavonoids: classification, distribution, biosynthesis, and antioxidant activity. Food Chem 383:132531

Shui X, Chen T, Qian M, Peng J, Du J, Zhou K, Liu F (2024) The antioxidant response mechanism of flavonoids in ‘Tainong 1’ mango pulp under enhanced UV-B radiation. Cogent Food Agric 10:2301273

Singh S, Sinha S (2005) Accumulation of metals and its effects in *Brassica juncea* (L.) Czern. (cv. Rohini) grown on various amendments of tannery waste. Ecotoxicol Environ Saf 62:118–127

Sivankalyani V, Feygenberg O, Diskin S, Wright B, Alkan N (2016) Increased anthocyanin and flavonoids in mango fruit peel are associated with cold and pathogen resistance. Postharvest Biol Technol 111:132–139

Sudheeran PK, Feygenberg O, Maurer D, Alkan N (2018) Improved cold tolerance of mango fruit with enhanced anthocyanin and flavonoid contents. Molecules 23:1832

Szymańska R, Ślesak I, Orzechowska A, Kruk J (2017) Physiological and biochemical responses to high light and temperature stress in plants. Environ Exp Bot 139:165–177

Tatsumi K, Yano M, Kaminade K, Sugiyama A, Sato M, Toyooka K, Aoyama T, Sato F, Yazaki K (2016) Characterization of shikonin derivative secretion in *Lithospermum erythrorhizon* hairy roots as a model of lipid-soluble metabolite secretion from plants. Front Plant Sci 7:1066

Tharanathan RN, Yashoda HM, Prabha TN (2006) Mango (*Mangifera indica* L.), “The king of fruits”—an overview. Food Rev Int 22:95–123

Vega-Alvarez M, Salazar-Salas NY, López-Angulo G, Pineda-Hidalgo KV, López-López ME, Vega-García MO, Delgado-Vargas F, López-Valenzuela JA (2020) Metabolomic changes in mango fruit peel associated with chilling injury tolerance induced by quarantine hot water treatment. Postharvest Biol Technol 169:111299

Wang H, Guo Y, Zhu J, Yue K, Zhou K (2021) Characteristics of mango leaf photosynthetic inhibition by enhanced UV-B radiation. Horticulturae 7:557

Yang C, Xu Z, Song J, Conner K, Vizcay Barrena G, Wilson ZA (2007) Arabidopsis MYB26/MALE STERILE35 regulates secondary thickening in the endothecium and is essential for anther dehiscence. Plant Cell 19:534–548

Rao MJ, Zheng B (2025) The Role of Polyphenols in Abiotic Stress Tolerance and Their Antioxidant Properties to Scavenge Reactive Oxygen Species and Free Radicals. Antioxidants 14:74.

Zhao Z, Cao J, Jiang W, Gu Y, Zhao Y (2009) Maturity-related chilling tolerance in mango fruit and the antioxidant capacity involved. J Sci Food Agric 89:304–309

Zhao YH, Jia X, Wang WK, Liu T, Huang SP, Yang MY (2016) Growth under elevated air temperature alters secondary metabolites in *Robinia pseudoacacia* L. seedlings in Cd-and Pb-contaminated soils. Sci Total Environ 565:586–594

Zhao X, Zhou N, Lai S, Frei M, Wang Y, Yang L (2019) Elevated CO_2 improves lodging resistance of rice by changing physicochemical properties of the basal internodes. Sci Total Environ 647:223–231

Zhu W, Wu H, Yang C, Shi B, Zheng B, Ma X, Zhou K, Qian M (2023) Postharvest light-induced flavonoids accumulation in mango (*Mangifera indica* L.) peel is associated with the up-regulation of flavonoids-related and light signal pathway genes. Front Plant Sci 14:1136281

Zu YG, Tang ZH, Yu JH, Liu SG, Wang W, Guo XR (2003) Different responses of camptothecin and 10-hydroxycamptothecin to heat shock in *Camptotheca acuminata* seedlings. Acta Bot Sin 45:809–814

Chapter 3
Effect of Climate Change on Cytogenetics and Some Biological Activity in *Mangifera* Sp.

Hülya Doğan, Halil Erhan Eroğlu, and Hatice Baş

Abstract The sector most affected by climate change is agriculture. Greenhouse gases are the most fundamental factors in climate change. Reducing the emissions of these gases also affects agricultural production. The temperature increases due to global warming rise every year. The change in temperature affects the photosynthesis level of plants, which in turn affects the synthesis of antioxidant compounds and the quality of the product after harvest. Increasing ozone levels cause changes in important quality parameters. Different types of visual damage and physiological disorders are seen in plants. Temperatures above optimum temperatures cause irreversible damage to plant growth and development. High temperatures cause plants to develop faster, thus shortening the vegetation period, resulting in lower yields. Temperature affects physiological and biochemical processes in plants. In many plants, growth is slow at temperatures above 40 °C and below 10 °C due to reduced or inhibited photosynthesis. The physiological processes most affected by temperatures above or below optimum growth temperatures for plants are photosynthesis, growth and respiration. Similarly, biochemical changes such as changes in the viscosity, permeability and fluidity of the cell membrane also occur. In addition, there are changes in the synthesis of antioxidant enzymes and activities. In this review study, it was aimed to evaluate the general cytogenetics and phytochemicals of the plant in mango cultivation along with global climate change.

Keywords Chromosome · Antioxidant activity · Essential oil · Flavonoid compounds

H. Doğan (✉)
Department of Agriculture and Food, Hemp Research Institute, Yozgat Bozok University, Yozgat, Türkiye
e-mail: hulya.dogan@bozok.edu.tr

H. E. Eroğlu · H. Baş
Department of Biology, Faculty of Arts and Sciences, Yozgat Bozok University, Yozgat, Türkiye

S. A. H. Naqvi et al. (eds.), *Climate Change and Mango Production*,
https://doi.org/10.1007/978-3-031-87035-4_3

Abbreviations

DPPH	1,1- Diphenyl-2-picryhydrazyl radical
TPC	Total phenolic content
FRAP	Ferric reducing antioxidant potential assay
ROS	Reactive oxygen species

3.1 Introduction

Climate change is the most important result of global warming. Global warming is the greenhouse effect and the increase in temperature on earth due to the heat-absorbing gases released on earth acting permeable to sunlight and less permeable to the radiation emitted back (Zandalinas et al. 2021; Fernihough and O'Rourke, 2021). The increase in temperature on earth is changing the climate. In many regions, the four-season transitions are evolving into two seasons, and sudden natural events that can be described as disasters are experienced. Especially natural events such as untimely heavy rains, floods, and droughts affect ecosystems, destroying the tradition of human activity or creating great pressure on it. The most important human activities affected by climate change are food production. Since plant production is a production method directly related to climate, sudden natural events, and droughts necessitate radical changes in production. Changes in seasonal norms; cause the decrease in agricultural areas, the replacement of traditionally produced species with different species, the excessive use of protective chemicals, and the application of different technological methods to make products more durable and sustainable (Rojas-Downing et al., 2017; Downing et al., 2017; Rezazadeh et al., 2011; Sauco, 2009; Cdjoe et al., 2020; Resida et al., 2019; Ronchi et al., 1992). We know that this change has profound effects on ecosystems, species and biodiversity. The effects of global climate change include habitat changes and the risk of extinction of some plant species over time. There are 69 species of *Mangifera. Mangifera indica* L. is a medicinal plant used by indigenous people in tropical regions. They are members of the Anacardiaceae family and belong to the Mangifera genus. Mango is a fruit native to South and Southeast Asia and is now grown all over the world (Akhter et al. 2017). Mangoes have a superior nutritional profile (vitamins, fibers, and minerals) as well as a high concentration of bioactive compounds. There is a great diversity among fruits in fruit weight, shape, color, composition and bioactive components in plants. They are influenced by the district where they are grown, soil, climate and cultural practices. During its development, it undergoes various amendments for instance ripening, cell wall transformation and tissue detente, impairment of chlorophylls, improvement of pigments, transformation of starch to sugars, pile up of aromas and flavours (Maldonado-Celis et al. 2019). The mango fruit is important, but other parts of the plant, such as the seed, bark, flowers, and leaves are significant sources of bioactive components with probable health promoting activity. Mangos antioxidant effects, like vitamin C, are related to its capacity to ROS, protecting the human body against diseases associated with oxidative stress. In addition, carotenoids affect gene expression regulation, which explains part of the association between high carotenoid intake and lower risk of certain diseases. Several phenolic components have been identified in mango fruit peel, flesh, and seeds. Flavonols quercetin and kaempferol have biological activities including antibacterial, antithrombotic, vasodilatory, anti-inflammatory, and anti-carcinogenic (Mieszczakowska-Frąc et al. 2016). Mango fruits, which can now be found in almost all chain markets, have a high antioxidant capacity due to their ascorbic acid, carotenoid, and phenolic substance content (Ribeiro et al. 2008). In this review study, it was aimed to evaluate the general cytogenetics and phytochemicals of the plant in mango cultivation along with global climate change.

3.2 Cytogenetics

The chromosome number databases show that the genus *Mangifera* have the stable chromosome number ($2n = 40$) (Table 3.1). Mukherjee (1950a, b) suggested that there were the partial allopolyploidy in mango. However, according to some authors showing genetic markers were inherited disomically, mango had a diploid structure (Duval et al. 2005; Schnell et al. 2006).

3.2.1 Mangifera indica

The chromosome number is $2n = 40$. The chromosomes are small and the chromosome lengths vary between 0.4 and 2.0 μm. The chromosomes are divided into three groups, namely Group-I consisting of long chromosomes in 1.5–2.0 μm, Group-II consisting of medium chromosomes in 1.0–1.5 μm, and Group-III consisting of

Table 3.1 The chromosome numbers of genus *Mangifera*

Species	2*n*	References
Mangifera indica	40 24*	Mukherjee, 1950a Singhal and Gill, 1990 Pierozzi and Rossetto, 2011 Sarkar, 1992
Mangifera caesia	40	Mukherjee, 1957
Mangifera sylvatica	40	Mukherjee, 1950a
Mangifera caloneura	40	Mukherjee, 1950a
Mangifera foetida	40	Mukherjee, 1957
Mangifera khasiana	40	Kumar and Subramaniam, 1987

[*] Probably wrong

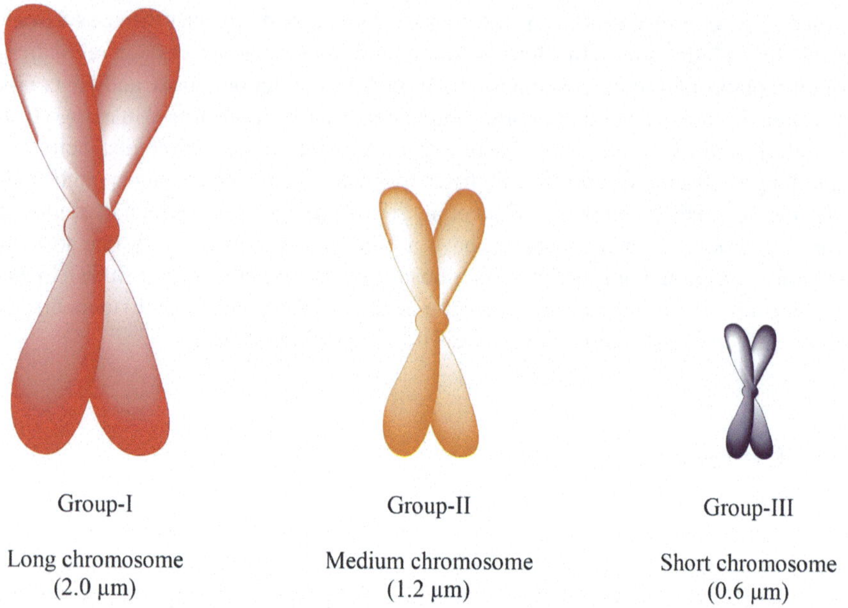

Fig. 3.1 The chromosomal groups in *Mangifera indica*. Group-I consists of long chromosomes (1.5–2.0 µm). Group-II consists of medium chromosomes (1.0–1.5 µm). Group-III consists of short chromosomes (0.6–1.0 µm)

short chromosomes in 0.6–1.0 µm. The chromosome numbers of the groups differ according to the varieties. There is satellite or secondary constriction in some chromosomes (Mukherjee 1950a). In Fig. 3.1, an ideogram is drawn showing one chromosome from each group. In addition, the chromosome lengths vary between 0.87 and 1.83 µm. The karyotype formulae are 8 m + 10sm + 2sm in Stahl variety and 7 m + 11sm + 2sm in IAC-140 variety. The total haploid chromosome length and mean chromosome length are 28.08 µm and 1.40 µm in Stahl variety and 24.37 µm and 1.22 µm in IAC-140 variety. There is satellite or secondary constriction in some chromosomes (Pierozzi and Rossetto 2011).

3.2.2 *Mangifera caesia*

The chromosome number is $2n = 40$. The chromosomes are small and the chromosome lengths vary between 0.6 and 2.0 µm. The chromosomes are divided into three groups, namely Group-I consisting of 4 long chromosomes in 1.5–2.0 µm, Group-II consisting of 10 medium chromosomes in 1.0–1.5 µm, and Group-III consisting of 26 short chromosomes in 0.6–1.0 µm. The total haploid chromosome length and mean chromosome length are 19.5 µm and 0.98 µm, respectively. There is secondary constriction in some chromosomes (Mukherjee 1957).

3.2.3 *Mangifera sylvatica*

The chromosome number is $2n = 40$. The chromosomes are small and the chromosome lengths vary between 0.4 and 1.5 μm. The chromosomes are divided into two groups, namely Group-I consisting of 10 medium chromosomes in 1.0–1.5 μm and Group-II consisting of 30 short chromosomes in 0.4–1.0 μm. The total haploid chromosome length and mean chromosome length are 17.5 μm and 0.88 μm, respectively. There is satellite or secondary constriction in some chromosomes (Mukherjee 1950a).

3.2.4 *Mangifera caloneura*

The chromosome number is $2n = 40$ (Mukherjee 1950a).

3.2.5 *Mangifera foetida*

The chromosome number is $2n = 40$. The chromosomes are small and the chromosome lengths vary between 0.6 and 1.6 μm (Mukherjee 1957).

3.2.6 *Mangifera khasiana*

The chromosome number is $2n = 40$ (Kumar and Subramaniam 1987).

3.3 Polyploidy as an Adaptation to Climate Change

Polyploidy, which occurs as a result of genome duplications and is a very common mechanism in the plant kingdom, is one of the main sources of speciation (Saúco et al., 2001). Although most of the natural and cultivated forms of mango were diploid ($2n = 2x = 40$), some polyploids were also recorded (García-García et al., 2020). The spontaneous autopolyploidy were reported for the first time in Manga blanca or Gomera-1 (G-1 mango) (Saúco et al. 2001). In addition, new natural and artificial polyploids were later demonstrated (Grajal Martín et al. 2007; Sauco 2009). For example, Rezazadeh et al. (2011) showed that there were tetraploidy ($2n = 4x = 80$) in somatic hybrid embryo lines. In summary, we can say that the high polyploidy rate in the plant kingdom is also common in genus *Mangifera*. In genus, the polyploidization is likely to be related to global climate change. Because the number of polyploid species increased during major climate change periods such as the end of the Cretaceous or the transition from Paleocene to Eocene, which caused mass extinctions. In addition, the duplicated genome provided a great deal of survival mechanisms for polyploid species to adapt to climate change (Eroğlu 2022).

The polyploid plants show higher adaptability to adverse conditions, and especially tetraploid plants are more resistant to direct climate change stresses such as light and temperature (Oustric et al. 2017). In this case, polyploid mango species will have more biomass with larger cells and tissues, increased crop size and yield, and resistance to biotic and abiotic stresses, including those associated with climate change. For example, in high water content environments, tetraploid mangoes with larger stomata cells, mesophyll cells, and chloroplasts showed higher leaf elasticity and lower water loss than diploid mangoes (Saúco et al. 2001; Barceló-Anguiano et al. 2021). Tetraploid mangoes, which develop from embryos resulting from spontaneous somatic duplication in the nucella, had larger fruit size and more advanced photoprotection (García-García et al. 2020). In polyploid Gomera-1, water loss from leaves was less compared to diploids, and resistance to drought caused by climate change can be achieved by polyploids with physiological potential that save water (Perera-Castro et al. 2023).

As a result, the above-mentioned polyploidy advantages can be increased. In addition, some polyploid mango species are likely to be more resistant to future climate change-induced crop production reductions.

3.4 Impact of Climate Change on Phytochemicals in Mango Parts

Various epidemiological studies have shown that phytochemicals are linked to biological activities that provide benefits to the human body. According to the literature, it supports the biological activity of natural plant forms of phytochemicals. These have been used in traditional medicine for a long time (Prakash et al. 2007). *Mangifera indica* L. has been recognized economically important tropical fruit crops. Mango is quite rich in phytochemicals and nutrients. Mango is an evergreen tree. Besides being a famous fruit, it also has many traditional medicinal resources. Mango fruit contains antioxidant pigments such as carotenoids, polyphenol compounds and polyunsaturated fatty acids, as well as antioxidant substances such as carotenoids, provitamin A, beta-carotene, lutein, quercetin, kaempferol, gallic acid, caffeic acid and tannin in abundance in the peel of the fruit (Chauhan and Patil, 2013). These compounds found in mango have a high free radical scavenging effect and therefore are preferred in the treatment of many diseases. The economically important part is the fruits, but during pruning, product residues such as bark, flowers, leaves, and stems are formed, and farmers experience disposal difficulties. Mango leaves are potential mineral sources. The main macromolecule detected in mango leaves is protein (Table 3.2). Mango leaves could be used as another resource of animal feed to alleviate food shortages for

Table 3.2 Bioactive components of some mango species

Species	Variety	Bioactive compounds	References
Mangifera indica L	Mango leaves	Apigenin; ferulic acid; quercetin; manindicins B; mangiferin; norathyriol gallic acid; caffeic acid; oleic acid; acetaldehyde; 2-hydroxyacetophenone; 2-furanmethanol; furfural; phenol; acarbose; manindicin A; o-catechol; pyrogallol; hydroquinone	Gu et al. 2019; Martínez-Bernett et al. 2016; Ronchi et al. 2015
	Mango pulp	Cyanidin; Catechin; Kaempferol; Myricetin; Proanthocyanidin dimers; Proanthocyanidin trimers (mg); Proanthocyanidin; Gallic acid	Lebaka et al. 2021
	Mango seed kernel	Ellagic acid; gallic acid; caffeic acid; cinnamic acid; ferulic acid; mangiferin; tannins; coumarin; vanillin	Lebaka et al. 2021
	Mango peel	Mangiferin gallate; quercetin;isomangiferin; isomangiferin gallate; rhamnetin 3–0 galactoside/ glucoside;mangiferin;	Lebaka et al. 2021
Mangifera foetida L	Macang leaves	Cis-anethole; scoparone; chromen; xanthen; quercitrin dan morin (flavon); daidzein dan sophoricoside (isoflavon); naringenin (flavanon); sophoraflavone a; diosmetin; cymene; 2-phenoxyaniline; ethynyl imidazolyl porphyrin; acetic acid; phenylhexanoic acid; piperidinepropionic acid; nootkatone; hydroquinone	Resida and Roza, 2019
Mangifera sylvatica L	Wild mango pulp	Mangiferin; quercetin; kaempferol; p-coumaricacid; gallic acid; ellagic acid	Akhter et al. 2021

livestock in developing countries. Mango leaves are used in folk medicine (Lebaka et al. 2021).

3.5 Impact of Climate Change on Antioxidant Activities and Chemical Composition of the Mango Parts

3.5.1 Antioxidant Activities

Free radicals are atoms or molecules that have one or more unpaired electrons in their structure. These radical species are unstable, short-lived, and have a low molecular weight. Free radicals damage the membranes, proteins, fats, and DNA of body cells involved in metabolism. They also cause diabetes, cancer, liver cirrhosis, coronary artery damage, and many other diseases (Tunca et al., 2018). Plants, which act as primary producers in the food chain, are the main source of many natural compounds on earth. In addition, these plants show qualitative and quantitative changes due to their infinite genetic differences. They therefore are widely used in various fields such as food, medicine, pharmacology, biology, cosmetics, and the paint industry (Schraml and Grillari 2012). Different studies have reported that mango leaves have antioxidant capacities due to the flavonoids and phenolics they contain (Kumar et al. 2020). The extract from mango leaves (main compounds) identified kaempferol-3-O-rutinoside, mangiferin, neomangiferin, quercetin and isoquercitrin. These components directly contributed to the antioxidant activity of mango leaves (Wu et al. 2020).

The highest antioxidant activity in the study conducted on wild mangoes (DPPH) were obtained from the leave, 0.88 ppm and the peel, 33.24 ppm (Fitmawati et al. 2020).

If the IC_{50} value is less than 50, 50–100 ppm gives a very strong antioxidant value; however, the IC_{50} value is weak in the range of 100–250 ppm and the $IC_{50} < 250$ ppm is ineffective (Islam et al. 2010). The antioxidants found in mangoes are essential for strengthening the immune system in the body. The study reported that leaves extracts (wild mango) are 69.67% phagocytic activity. Ascorbic acid found in the leaves and peel of mangoes, can heal chronic wounds and inflammations (Fitmawati et al. 2018). Mangoes are a rich source of polyphenols, a group of organic micronutrients found in plants that provide certain health benefits (Okwu and Ezenagu, 2008). In the research conducted, the highest phenolic compounds of mangoes were found in the peel (100.65 mg/g) and in the leaves (92.48 mg/g) (Fitmawati et al. 2020). Nowadays, herbal tea is made from mango leaves. Research has shown that mango leaves are the source of mangiferin and other phenolic contents (Ali et al. 2012; Cornelia and Sutisna 2019).

According to the study, the highest total phenolic contents were obtained from ripe green mango (142.57 μg GAE/g sample), respectively. Again, the highest DPPH activity from ripe green extracts was determined as 408.21 μg TE/g sample and metal chelating activity as 93.68%. As a result, the combined or enhancing effects of gallic acid and mangiferin contents may be responsible for the activities. The highest mangiferin content from ripe mango fruits (31.72 μg/g sample) may be the main contribution of the highest FRAP activity (868.29 μg TE/g sample) (Sulaiman and Ooi 2012). In recent years, global agricultural activities have been negatively affected by the increase in temperature due to global climate change, changes in precipitation amounts and regimes, and the increasing frequency and severity of extreme weather events (Algataa, 2020). With the change in climate, the weather conditions in the study areas have changed. Climate change can cause changes in temperature, precipitation, and season length. It has changed and shortened the growing season of many crops. Studies show that climate change has a significant effect on the yield of *Magnifera indica* (Kumar et al. 2011).

3.5.2 Essential Oil and Fatty Acid

Mango is a plant species with a pleasant aroma that varies depending on where it is grown, the variety, and the climatic conditions. Distillation method is used to obtain essential oil from mango leaves. Mango essential oil was found to contain 46.98% monoterpenes, 38.17% sesquiterpenes, and 10.67% other components and 4.19% oxygenated hydrocarbons. According to the study, the presence of seven chemical compounds namely α-pinene, camphene, α-copaene, pinene, beta -elemen, α-gurjunene and α-humulene were found as essential oil components in mango leaves (Wetungu et al. 2015). In another study, it was determined that it was 70.3% sesquiterpene and the most important compounds were α-gurjunene, beta –caryophyllene, δ-3 carene, and beta -selinene (D[Zbreve]Ami´c et al. 2010). Another study, the essential oil obtained from the leaf extract showed selinene, caryophyllene oxide and humulene epoxide II as the main components (Fontenelle et al. 2017). Lipids are known for their high energy value and beneficial properties.

The oil content in mango seeds is between 8.15–13.16%. It is nutritious and beneficial for health (Solís-Fuentes and del Carmen Durán-de-Bazúa 2011). Its fixed oil is comparable to that of vegetable butter due to the presence of stearic and oleic acids (Muchiri et al. 2012). In some mango varieties, unsaturated fatty acids (linoleic acid) are twice as high as saturated fatty acids (Arogba 1999).

High content of stearic and oleic acids in total fatty acid form of mango seed. It has higher stability than oils containing polyunsaturated fatty acids. Mango seed oil is suitable for use as cooking oil. (Solís-Fuentes and del Carmen Durán-de-Bazúa 2011). It has been determined that the total phenolic substance content and induction time of the oil of this plant are higher than many vegetable oils (Nadeem et al. 2016).

3.5.3 Antibacterial Activities

Antibacterial activity tests are used to determine the effect of a substance under investigation on bacterial growth (Zorofchian Moghadamtousi et al. 2014). Bacterial infections have often caused dangerous infections in humans. According to many sources, research is rapidly continuing to develop antibacterial agents. Medicinal and aromatic plants contain many phytochemicals that help fight bacterial infections. The disk diffusion method is the most common method used to measure antibacterial activity. Antibacterial activity is determined by measuring the zones formed around the discs (Al Akeel et al. 2018). Methanol extracts (mango leaves) showed antibacterial activity against Gram positive bacteria. According to the results obtained from current research, it has been determined that mango fruits and leaves can be used in the treatment of skin diseases usually caused by *S. aureus* infection (Maharaj et al. 2022). Nowadays, as awareness of the harms of synthetic substances increases, people are turning to nature and synthetic antioxidants are becoming less preferred. Besides, the number of alternatives that can be used against bacteria that have become resistant to existing antibiotics is decreasing day by day. This creates a significant problem in the antibiotic treatment of microbial diseases. Antibacterial activity in mango leaves may indicate the presence of tannins, phenols and flavonoids. These compounds are good antibacterial agents which may indicate the activity in mango leaves (Seleshe and Kang 2019; Cdjoe et al. 2020).

3.6 Conclusion

Low temperatures cause poor growth and development in plants, especially in tropical and subtropical regions around the world. Flowering phenology in plants is a complex trait governed by various environmental factors, shoot maturity and tree age, and the genetic makeup of cultivars.

Projected changes in climate change have the potential to alter tree phenology through significant disruptions in the timing of fruit, seed and flower availability. Important environmental cues that influence the phenology of tropical and subtropical plants include humidity, temperature, evaporation, and photoperiod. Mango trees develop more flowers and fruits when there is growth regulation caused by dry, cool weather conditions. Bioclimatic variables are being analyzed with the research methods used today. In addition, genetic and cytogenetic studies suggest that some polyploidy mango species are likely to be more resistant to future climate change-induced crop production disruptions. Although there are many studies in this area, more analysis is needed to determine the sensitivity of the species by taking into account more environmental and climatic variables. More studies should be done, especially to determine the changes in the active substances in plants due to global climate change. Studies have shown that the high antioxidant capacity value of the mango plant leaves also has a good value from a medical perspective. It has been determined that the volatile oil and fixed oil values are at least as good as cooking oil. In this respect, these changes in these tropical plants exposed to climate change need to be investigated further.

References

Akhter S, McDonald MA, van Breugel P, Sohel S, Kjaer ED, Mariott R (2017) Habitat distribution modelling to identify areas of high conservation value under climate change for *Mangifera sylvatica* Roxb. of Bangladesh. Land Use Policy 60:223–232

Akhter S, McDonald M, Marriott R, Jashimuddin M, Bashirul-Al-Mamun M (2021) Morphological, nutritional and medicinal traits of wild mango (*Mangifera sylvatica* Roxb.): Implications for increased use and options for cultivar development. Trees For People 6:100126

Al Akeel R, Mateen A, Alharbi KK, Alyousef AA, Al-Mandeel HM, Syed R (2018) Purification and MIC analysis of antimicrobial proteins from *Cucumis sativus* L. seeds. BMC Complement Altern Med 18:1–6

Algataa B (2020) Analysis of the impact of climate change and storage methods on the quality of olive oil in Libya. Int J Sci Res Publ 10(5):968–972

Ali MR, Yong MJ, Gyawali R et al (2012) Mango (*Mangifera indica* L.) peel extracts inhibit proliferation of HeLa human Cervical Carcinoma Cell via induction of Apoptosis. J Korean Soc Appl Biol Chem 55:397–405

Arogba SS (1999) The performance of processed mango (*Mangifera indica*) kernel flour in a model food system. Bioresour Technol 70:277–281

Barceló-Anguiano M, Holbrook NM, Hormaza JI, Losada JM (2021) Changes in ploidy affect vascular allometry and hydraulic function in *Mangifera indica* trees. Plant J 108:541–554

Chauhan A, Patil V (2013) Effect of packaging material on storage ability of mango milk powder and the quality of reconstituted mango milk drink. Powder Technol 239:86–93

Cornelia M, Sutisna JA (2019) Pemanfaatan daun mangga arum manis (*Mangifera indica* L.) sebagai minuman teh celup. FaST—Jurnal Sains Dan Teknologi 3 (1):71–81

Cudjoe E, Donu D, Okonu RE, Amponsah JA (2020) Amoah LE (2020) The in vitro antiplasmodial activities of aqueous extracts of selected Ghanaian herbal plants. J Parasitol Res 1:5041919

D[Zbreve]Ami´c AM, Marin PD, Gbolade AA, Risti´c MS (2010) Chemical composition of *Mangifera indica* essential oil from Nigeria. J Essent Oil Res 22:123–125

Duval MF, Bunel J, Sitbon C, Risterucci AM (2005) Development of microsatellite markers for mango (*Mangifera indica* L.). Mol Ecol Notes 5:824–826

Eroğlu HE (2022) The polyploidy mechanism as response to global climate change. In: Doğan H, Fidan H (eds) Global climate change: agriculture and food science perspective. İksad Publishing House, Ankara, pp 135–147

Fernihough A, O'rourke KH (2021) Coal and the European industrial revolution. Econ J 131(635):1135–1149

Fitmawati F, Juliantari E, Saputra A, et al (2018) The potential of wild mango leaves from Sumatera as the Immunostimulant agents. Biosaintifika: J Biol Biol Educ 10(3):671–677

Fitmawati F, Resida E, Kholifah SN, Roza RM, Almurdani M, Emrizal E (2020) Phytochemical screening and antioxidant profiling of Sumatran wild mangoes (*Mangifera* spp.): a potential source for medicine anti degenerative effects. F1000Research, 9

Fontenelle RODS, Sobrinho CAN, Soares BV, De Aguiar FLL, De Brito EHS, Cavalcante CSDP, Rocha MFG, De Morais SM (2017) Effect of essential oils from *Mangifera indica* L. cultivars on the antifungal susceptibility of *Candida* spp. strains isolated from dogs. Rev. Bras Saúde Produção Anim 18:337–346

Ronchi SN, Brasil GA, do Nascimento AM, de Lima EM, Scherer R, Costa HB, de Andrade TU (2015) Phytochemical and in vitro and in vivo biological investigation on the antihypertensive activity of mango leaves (*Mangifera indica* L.). Ther Adv Cardiovasc Dis 9(5):244–256

García-García AL, Grajal-Martín MJ, González-Rodríguez ÁM (2020) Polyploidization enhances photoprotection in the first stages of *Mangifera indica*. Sci Hortic 264:109198

Grajal Martín MJ, Fernández Galván D, Saúco VG, Juárez J, Navarro L (2007) Filipino un nuevo tetraploide de mango. Congreso Nacional de Ciencias Hortícolas, Albacete, Spain, p 51

Gu C, Yang M, Zhou Z, Khan A, Cao J, Cheng G (2019) Purification and characterization of four benzophenone derivatives from Mangifera indica L. leaves and their antioxidant, immunosuppressive and -glucosidase inhibitory activities. J. Funct. Foods. 52:709–714

Islam MR, Mannan MA, Kabir MHB, Islam A, Olival KJ (2010) Analgesic, anti-inflammatory and antimicrobial effects of ethanol extracts of mango leaves. J Bangladesh Agric Univ 8(2)

Kumar V, Subramaniam B (1987) Chromosome atlas of flowering plants of the Indian subcontinent, vol 1. Dicotyledons. Botanical Survey of India, Calcutta, p 464

Kumar V, Sharma Y, Chauhan S (2011) Impact of climate change on the growth and production of *Saccharum offcinarum* and *Magnifera indica*. Int J Sci Tech Manag 2(1):42–47

Kumar Y, Kumar V, Sangeeta, (2020) Comparative antioxidant capacity of plant leaves and herbs with their antioxidative potential in meat system under accelerated oxidation conditions. J Food Meas Charact 14:3250–3262

Lebaka VR, Wee YJ, Ye W, Korivi M (2021) Nutritional composition and bioactive compounds in three different parts of mango fruit. Int J Environ Res Public Health 18(2):741

Maldonado-Celis ME, Yahia EM, Bedoya R, Landázuri P, Loango N, Aguillón J, Guerrero Ospina JC (2019) Chemical composition of mango (*Mangifera indica* L.) fruit: Nutritional and phytochemical compounds. Front Plant Sci 10:1073

Martínez-Bernett D, Silva-Granados A, Correa-Torres SN, Herrera A (2016) Chromatographic analysis of phytochemicals components present in *Mangifera indica* leaves for the synthesis of silver nanoparticles by $AgNO_3$ reduction. J Phys Conf Ser 687:12033

Maharaj A, Naidoo Y, Dewir YH, Rihan H (2022) Phytochemical screening, and antibacterial and antioxidant activities of *Mangifera indica* L. leaves. Horticulturae 8(10):909

Mieszczakowska-Frąc M, Dyki B, Konopacka D (2016) Effects of ultrasound on polyphenol retention in apples after the application of predrying treatments in liquid medium. Food Bioprocess Technol 9:543–552

Muchiri DR, Mahungu SM, Gituanja SN (2012) Studies on mango (*Mangifera indica*, L.) kernel fat of some Kenyan varieties in Meru. J Am Oil Chem Soc 89:1567–1575

Mukherjee SK (1950a) Cytological investigation on the mango (*Mangifera indica* L.) and the allied Indian species. Proc Nat Inst Sci: India 16:287–303

Mukherjee SK (1950b) Mango: Its allopolyploid nature. Nature 4213:196–197

Mukherjee SK (1957) Cytology of some Malayan species of *Mangifera* L. Cytologia 22:239–241

Nadeem M, Imran M, Khalique A (2016) Promising features of mango (*Mangifera indica* L.) kernel oil: a review. J Food Sci Technol 53:2185–2195

Okwu D (2008) Ezenagu V (2008) Evaluation of the phytochemical composition of mango (*Mangifera indica* Linn) stem bark and leaves. Int J Chem Sci 6(2):705–716

Oustric J, Morillon R, Luro F, Herbette S, Lourkisti R, Giannettini J, Berti L, Santini J (2017) Tetraploid *Carrizo citrange* rootstock (*Citrus sinensis* Osb. x *Poncirus trifoliata* L. Raf.) enhances natural chilling stress tolerance of common clementine (*Citrus clementina* Hort. ex Tan). J Plant Physiol 214:108–115

Perera-Castro AV, Hernández B, Grajal-Martín MJ, González-Rodríguez ÁM (2023) Assessment of drought stress tolerance of *Mangifera indica* L. autotetraploids. Agronomy 13(1):277

Pierozzi NI, Rossetto CJ (2011) Chromosome characterization of two varieties of *Mangifera indica* L. Rev Bras Frutic 33:546–551

Prakash D, Suri S, Upadhyay G, Singh B (2007) Total phenol, antioxidant and free radical scavenging activities of some medicinal plants. Int J Food Sci Nutr 58:18–28

Resida E, Roza RM (2019) The secondary metabolite diversity analysis of three *Mangifera foetida* L. varieties based on liquid chromatography-mass spectrometry (LC-MS). J Phys: Conf Ser 1351(1):012027

Rezazadeh R, Harrison DK, Williams RR (2011) Intraspecific somatic hybridization of mango (*Mangifera indica* L.) through protoplast fusion. J Appl Hortic Lucknow 13:101–107

Ribeiro SMR, Barbosa LCA, Queiroz JH, Knödler M Schieber A (2008) Phenolic compounds and antioxidant capacity of Brazilian Mango (*Mangifera indica* L.) varieties. Food Chem 110(3):620–626

Rojas-Downing MM, Nejadhashemi AP, Harrigan T, Woznicki S A (2017) Climate change and livestock: impacts, adaptation, and mitigation. Clim Risk Manage 16:145–163

Ronchi SN, Brasil GA, Nascimento AMD, De Lima EM, Scherer R, Costa HB, Romão W, Boëchat GAP, Lenz D, Sarkar AK (1992) An interesting observation on the karyotype studies of the parasite, *Macrosolen cochinchinensis* (Lour.) van Tiegh., growing on different hosts. Proceedings of indian science congress association 79(3: VIII):124–125

Saúco VG (2009) El Cultivo del Mango; Ediciones Mundi-Prensa, Madrid, España

Saúco VG, Martín MJG, Galván DF, Torres AC, Juárez J, Navarro L (2001) Occurrence of spontaneous tetraploid nucellar mango plants. HortScience 36(4):755–757

Schnell RJ, Brown JS, Olano CT, Meerow AW, Campbell RJ, Kuhn DN (2006) Mango genetic diversity analysis and pedigree inferences for Florida cultivars using microsatellite markers. J Am Soc Hortic Sci 131:214–224

Schraml E, Grillari J (2012) From cellular senescence to age-associated diseases: the miRNA connection. Longev Health 1:10

Seleshe S, Kang SN (2019) In vitro antimicrobial activity of different solvent extracts from *Moringa stenopetala* leaves. Prev Nutr Food Sci 24(1):70

Singhal VK, Gill BS (1990) Chromosomal studies in some members of Anacardiaceae. J Cytol Genet 25:36–42

Solís-Fuentes JA, del Carmen Durán-de-Bazúa M (2011) Mango (*Mangifera indica* L.) seed and its fats. In Nuts and Seeds in Health and Disease Prevention; Elsevier: Amsterdam, The Netherlands.741–748

Sulaiman SF, Ooi KL (2012) Polyphenolic and vitamin C contents and antioxidant activities of aqueous extracts from mature-green and ripe fruit fleshes of *Mangifera* sp. J Agric Food Chem 60(47):11832–11838

Tunca ZŞ, Topal A, Karaduman Y (2018) Buğdaya Çavdar Kromozom Parçacıklarının Aktarılması (Translokasyonları). Bahri Dağdaş Bitkisel Araştırma Dergisi 7(2):55–65

Wetungu M, Tarus P, Segor F, Cheseto X, Omolo M (2015) Essential oil chemistry of some Mangifera indica varieties from Kenya. Am J Essent Oils Nat Prod 3:1–6

Wu L, Wu W, Cai Y, Li C, Wang L (2020) HPLC fingerprinting-based multivariate analysis of phenolic compounds in mango leaves varieties: correlation to their antioxidant activity and in silico–glucoidase inhibitory ability. J Pharm Biomed Anal 191:113616

Zandalinas SI, Fritschi FB, Mittler R (2021) Global warming, climate change, and environmental pollution: recipe for a multifactorial stress combination disaster. Trends Plant Sci 26(6):588–599

Zorofchian Moghadamtousi S, Abdul Kadir H, Hassandarvish P, Tajik H, Abubakar S, Zandi K (2014) A review on antibacterial, antiviral, and antifungal activity of curcumin. BioMed Res Int 1:186864

Chapter 4
Evaluation of Mango Production Under Climate Change in Türkiye

Mustafa Kan, Kadir Akan, Melike Bahçeci, and Ayşe Feyza Yaman

Abstract Turkiye, traditionally known for its diverse agricultural landscape, has seen an emerging interest in mango cultivation due to favorable climatic conditions in its southern regions. However, climate change poses significant challenges and opportunities for mango production in the country. Shifts in temperature, altered precipitation patterns, and the increasing frequency of extreme weather events are influencing mango phenology, flowering, and fruiting dynamics. Rising temperatures, particularly during critical growth stages, may lead to premature flowering or fruit drop, impacting yield and quality. Meanwhile, erratic rainfall and drought conditions exacerbate water stress and susceptibility to pests and diseases. Despite these challenges, the adaptation of climate-smart agricultural practices such as optimized irrigation, the use of heat- and drought-tolerant cultivars, and the adoption of protective measures like fruit bagging are paving the way for sustainable mango production. This evaluation highlights the necessity for targeted research on varietal adaptation, soil management, and microclimatic interventions in Turkiye. Furthermore, climate-resilient strategies combined with modern cultivation techniques could enhance mango productivity and market potential in the region, aligning with the global demand for this tropical fruit.

Keywords Mango production · Climate change · Türkiye · Sustainable agriculture

M. Kan (✉) · A. F. Yaman
Agriculture Faculty, Department of Agricultural Economics, Kırşehir Ahi Evran University, Kırşehir, Türkiye
e-mail: mustafa.kan@ahievran.edu.tr

A. F. Yaman
e-mail: aysseeyamann@gmail.com

K. Akan
Agriculture Faculty, Department of Plant Protection, Kırşehir Ahi Evran University, Kırşehir, Türkiye
e-mail: kadir.akan@ahievran.edu.tr

M. Bahçeci
Ministry of Agriculture and Forestry, International Agricultural Research and Training Center, Menemen, Türkiye
e-mail: melikebahceci0721@gmail.com

S. A. H. Naqvi et al. (eds.), *Climate Change and Mango Production*,
https://doi.org/10.1007/978-3-031-87035-4_4

4.1 Introduction

In order to limit the negative impacts of climate change on agricultural production and ensure sustainable agricultural production and adaptation to climate change, the 13th Sustainable Development Goal of the United Nations (UN) has been identified as "climate action". The main objectives of this action plan are to strengthen resilience and adaptive capacity to climate-related hazards and natural disasters in all countries and to integrate climate change-related measures into national policies, strategies, and plans to improve education, awareness-raising, and all-encompassing focus and institutional capacity on mitigating climate change impacts, adaptation to climate change, mitigation, forecasting, and early warning. In the agriculture sector, these goals are; ending hunger, ensuring food security, and promoting sustainable agriculture; resilience of plant patterns to stress factors; and adaptation of production systems to climate risks and foreseeable natural disasters; adaptation of agricultural systems to climate change; ensuring sustainable food production and food security; increasing production with increasing population; and protecting natural ecosystems (UN 2015).

It has been stated that one of the most important impacts of climate change will be on the "water cycle" (Douville et al. 2021; Wang and Liu 2023). By 2025, 1.8 billion people will face what is called "absolute water scarcity" and two-thirds of the global population will suffer from water stress (UNEP 2024). Turkiye is predicted to experience the negative impact of climate change, especially on water resources. Approximately one third of Turkiye's 78 million hectares of surface area, 24 million hectares, is arable agricultural land, and the economically irrigable area of this area is reported to be 8.5 million hectares. Currently, about 81.9% of the 8.5 million hectares of economically irrigable agricultural land in Turkiye can be irrigated (DSİ 2024). In Turkiye, which is experiencing water stress, significant climatic changes are expected, such as a decrease in rain and snowfall relative to long-term averages, increasing air temperatures increasing evaporation, frequent recurrence, and prolonged duration of heat waves and drought events (TOB 2021). The Mediterranean Region, including Turkiye, is among the regions most affected by global warming and climate change and is expected to be increasingly affected (IPCC 2013). It is estimated that the warming levels of the Balkans, Iberian Peninsula, North Africa, and Turkiye will reach approximately 50% higher values than annual global warming rates (IPCC 2023). This situation is considered an important danger indicator for Turkiye, which is located in the Mediterranean climate zone, especially in the Mediterranean Region.

Within the framework of the EU Program for Enhancing Dialog between Civil Society Organizations-II, TEMA Turkiye is supporting the Turkish Foundation for Combating Erosion. In the focus group meetings conducted in the Mediterranean Region within the scope of the project "Enhancing Civil Society Participation in Climate Change Policy Making Processes in Turkiye" jointly implemented by the Foundation for Reforestation and Protection of Natural Habitats (TEMA) and the

World Wildlife Fund (WWF-Turkiye), the participants reported that the most important problem in the Mediterranean Region is the "above-seasonal temperatures and the consequent decrease in water resources and drought" (TEMA-WWF 2015).

Turkiye has different advantages for the cultivation of many temperate and subtropical fruit species due to its favorable ecological conditions (Dal and Gübbük 2023). It is known that many tropical and subtropical species such as banana, avocado, pineapple, mango, guava, kiwi, and pitaya have habitat zones in some local production areas, especially in the Mediterranean Region (Köseoğlu et al. 1988; Bayram et al. 2006; Bayram and Arslan 2007; Şahin and Kendirli 2012; Bayazit et al. 2021; Gübbük 2018; Dal and Gübbük 2023; Gözalan 2023; Tunç and Yılmaz 2023). Many of these fruits are still grown under cover in the Mediterranean region. One of these products is "mango (*Mangifera indica* L.)". The first varieties of mango, which were brought to Turkiye by different means, were brought from the USA in 2012. After 2012, seedlings were brought from different sources and began to be grown in open and greenhouses in the Alanya, Anamur, and Gazipaşa districts. After 2018, greenhouse cultivation was emphasized, and it was started to be planted between banana gardens in the open (Güven 2022). Examples of greenhouse mango production are available from Spain, Japan, Korea, and Malaysia (Akinaga and Hasbullah 2002; Medina et al. 2009; Lim et al. 2016; Shiroma et al. 2019; Zul Helmey et al. 2022). For this reason, greenhouse mango production activities in the Mediterranean region continue rapidly in Turkiye.

In Turkiye, the production pattern has changed in the cultivation of different tropical plants, especially in the production areas of the Mediterranean region, as citrus and vegetable production areas have shrunk due to low production costs and high economic returns, and tropical and subtropical crops are cultivated in these areas (Gözalan 2023). Although the change in crop production patterns suggests that producers' profitability has increased, one of the important issues to be considered in the production of these newly introduced species is the water consumption of plants. Considering the changing climatic conditions, it is extremely important to determine the water consumption values of tropical and subtropical fruits that have started to occur in the agricultural product pattern of Turkiye, which has limited water resources. It is necessary to prefer new plant species that have high economic returns and that do not harm the country's natural resources (such as fresh water reserves).

In recent years, environmental awareness has been observed to be increasing daily to create a sustainable production system to ensure food safety by calculating carbon and water footprints in the production of many products at the global level. This new situation affects fresh fruit producers, retailers and consumers at different levels. The debate on carbon and water footprinting has gained more visibility with the publication of ISO 14067 (ISO 2013) and 14046 (ISO 2014) standards, which promote certification activities in the food production chain (NMB 2010). Both standards are based on life cycle assessment (LCA), as described in ISO 14040 (not listed in ISO 2006) and ISO 14044 (not listed in ISO 2006b). ISO 14067 relates to climate change for carbon footprint studies, and ISO 14046 relates to water-related categories such as water scarcity, eutrophication, and toxicity for water footprint assessments. When studies on mango water requirements are examined; Dyvik (2024) reported that 1,800 L of water are needed for 1 kg of mango production. A comparison for a better understanding of the amount of consumption shows that this value for 1 kg production is 214 L for tomatoes, 1222 L for corn, 4325 L for chicken meat, and 15,400 L for beef. There is no detailed research on water use in mango production in Turkiye. When studies on the annual water consumption of mango trees are examined, it is found that 979 mm/m^2 in Saudi Arabia (Mohammad et al. 2015) 1198 mm/m^2 in Israel (Levin et al. 2018) 1197 mm/m^2 in South Africa (Mostert and Hoffman, 1997), 545 mm/m^2 in Spain (Durán Zuazo et al. 2011), 416.6–532.6 mm/m^2 in terraced gardens in Spain (Durán Zuazo et al. 2018), 403–420 mm in Brazil (De Souza et al. 2016), and 551.6 mm is in another study in Brazil (De Azevedo et al. 2003).

Dal and Gübbük (2023) evaluated the adaptation results of passiflora, pitaya, guava, guava, longan, litchi, and mango in their study of different tropical and subtropical fruit species adapted to Turkiye. As a result of the study, among the species tested in Antalya province, mango was determined to be the most sensitive species to cold and sunburn from the first years of planting. In addition, mango was found to have the lowest adaptation ability among the species tested in this study. This situation shows that mango production in tropical fruit cultivation in the region takes a long time to adapt to the region. Gözalan et al. (2024) in their study on mango cultivation (plantations) in Antalya province reported that there is no potential for efficient mango cultivation in open areas in Antalya province when the temperature demand of the mango plant and the daily minimum temperature frequencies and extreme minimum temperatures measured at meteorological stations in Antalya province are examined. In this case, the farmer reported that the only viable alternative for mango cultivation was greenhouse production.

In this study, the production of mango, a tropical fruit, which has started in recent years in the Mediterranean Region of Turkiye, especially in Antalya and Mersin provinces, was evaluated in view of the climate change in Turkiye. For this purpose, surveys were conducted on the perspectives of producers in Antalya and Mersin provinces on mango production, reasons for their production, and expectations for the future. Within the framework of the findings, recommendations for the future of mango production in the Mediterranean Region of Turkiye were developed.

4.2 Material and Methods

When the data from the Farmer Registration System of the Ministry of Agriculture and Forestry and literature studies on the subject in Turkiye were analyzed, it was determined that the provinces with the highest mango cultivation in Turkiye are Antalya and Mersin provinces. For this reason, the main content of the study consists of survey results obtained from mango producers in these provinces (Fig. 4.1). In addition, reports, articles, and statistics published as a result of relevant studies conducted on mango in the region constitute secondary sources of the study.

Fig. 4.1 Map of Antalya and Mersin Provinces (Turkiye)

The Antalya and Mersin provinces are located in the Mediterranean Region of Turkiye, and they have Mediterranean climate characteristics. The Mediterranean climate is a subtropical climate with hot, dry, and sunny summers and mild and moderately rainy winters (MGM 2024). Because the climate type of region creates the most suitable areas for tropical fruit cultivation in Turkiye, mango production areas have expanded to these provinces. According to the Farmer Registration System (FRS) data of the Antalya and Mersin Provincial Directorates of Agriculture and Forestry for the year 2024, 99 producers in Antalya produce mango on 59.29 ha and 19 producers in Mersin produce mango on 8.39 ha. These producers constitute the main population of the study (Table 4.1).

In order to reach a sample size that can represent the main population of mango producers in the research area, it was aimed to reach at least 30% of the total 118 mango producers in both provinces. As a result of the study, a face-to-face survey was conducted with a total of 40 mango producers in Antalya and Mersin provinces. This rate constitutes 33.90% of the total population, and the data obtained can represent the total population. The study was carried out on a total of 67.68 ha of mango production area in Antalya and Mersin provinces, 59.29 and 8.39 ha, respectively. The survey questionnaire consisted of structured and semi-structured questions and aimed to determine the socio-demographic and economic status of producers and their tendencies regarding mango production. The necessary permissions were obtained from Kırşehir Ahi Evran University Science and Engineering Sciences Research and Publication Ethics Committee for the conduct of the survey (permission number 2024/01/01 dated 13.08.2024).

Tables were created for the data obtained in the research, and frequencies and percentages of the data were determined. Chi-Square statistical analysis was used to analyze the discrete data with the help of the created cross-tables, and it was determined whether the relationships (dependency) between the variables were statistically significant or not. In continuous data, t-tests for two variables from parametric tests were analyzed (Düzgüneş et al. 1983). For statistical analyses, producers were divided into 2 groups. The producers in the 1st group are the producers with mango

Table 4.1 Number of mango producers and production areas (Ha) in the research area (2024)

Province/Districts names	Number of farmers	Production area (Ha)
Antalya	*99*	*59.29*
Aksu	2	10.00
Alanya	47	21.41
Finike	2	0.42
Gazipaşa	19	4.95
Kumluca	1	0.30
Manavgat	20	17.61
Muratpaşa	1	0.01
Serik	7	4.41
Mersin	*19*	*8.39*
Anamur	6	1.25
Bozyazı	5	0.80
Çamlıyayla	1	0.01
Erdemli	5	5,58
Silifke	2	0.75
TOPLAM	*118*	*67.68*

production area below 0,35 ha and the producers in the 2nd group are the producers with mango production area of 0,35 ha and above. Statistical analyses were made according to this grouping.

4.3 Results and Discussion

In the Mediterranean region of Turkiye, especially in the production areas of different tropical fruits, the production of these fruits is becoming increasingly widespread, which has created an increasing interest of producers in these new species. The fact that the Research Institutes of the Ministry of Agriculture and Forestry and Universities are working on the adaptation of tiki species in the region is another factor that has increased this interest. Among the new species that can be cultivated in tropical and subtropical regions, mango is one of the new species that can be grown in these regions, and the investments and studies carried out by both the public and private sectors in this new field have an important effect on the increase in the production area every day. In this research conducted in Antalya and Mersin provinces, the findings related to the socio-demographic and economic characteristics of mango producers are presented in Table 4.2. When Table 4.2 is examined, it is determined that the producers in the area where the survey was conducted are generally male, and their average age is close to 50 years. When the education level of the producers is analyzed, it is determined that they have a higher level of education than producers engaged in plant production in Turkiye. The fact that 45% of the interviewed producers graduated from vocational high school or undergraduate level was evaluated as the fact that producers of this product constitute an educated segment with a higher education level. This situation also reveals producers' income levels. It was determined that 97.50% of the interviewed producers had an income above the middle level, and even 40.00% of them had a good income. Again, when the attitudes of the interviewed producers toward risk-taking were analyzed, 80.00% stated that they like to take risks.

When the socio-demographic characteristics of the producers were analyzed according to the mango land size groupings, no statistically significant difference was determined except for the income levels of the producers. Income levels of Group 2 producers with large mango areas were higher than income levels of Group

Table 4.2 Socio-demographic and economic status of mango producers by groups

		1. Grup		2. Grup		Total		T-test/chi square test
		%	Mean	%	Mean	%	Mean	
Sex	Male	94.74		100.00		97.50		1.13
	Female	5.26		0.00		2.50		
Age			46.84		50.71		48.88	− 1.20
Education level	Illiterate	0.00		0.00		0.00		5.25
	Literate	0.00		0.00		0.00		
	Primary school	21.05		19.05		20.00		
	Secondary school	15.79		0.00		7.50		
	High school	31.58		23.81		27.50		
	Vacational School	5.26		4.76		5.00		
	Bachelor's degree	26.32		52.38		40.00		
Social security	No social security	0.00		0.00		0.00		5.73
	Retirement fund	21.05		42.86		32.50		
	SSK (social security institution)	36.84		19.05		27.50		
	BAGKUR (Worker)	36.84		28.57		32.50		
	BAGKUR (Farmer)	5.26		0.00		2.50		
	Other	0.00		9.52		5.00		
Income level	Poor	0.00		0.00		0.00		7.05**
	Mid-poor	0.00		4.76		2.50		
	Mid-good	78.95		38.10		57.50		
	Good	21.05		57.14		40.00		
Risk status	Risk lover	78.95		80.95		80.00		0.56
	Risk notr	10.53		4.76		7.50		
	Risk averse	10.53		14.29		12.50		

Statistically significant at * 90%, ** 95%, and *** 99% levels

1 producers, and the difference was statistically significant at the 95% confidence level (Table 4.2).

The interviewed mango producers had a high level of education, were open to innovations, were in search of alternative crops, and were risk-loving. This shows that the characteristics of mango producers significantly overlap with those of the group called "early adopters" who adopt innovations in crop production. Akçaöz et al. (2006) conducted a study on risk management and agricultural insurance with producers in Antalya province and reported that 39.9% of the producers liked risk. In a study conducted by Özer and İkikat Tümer (2020) with lemon producers in Mersin province, 30.7% of the producers reported liking the risk. When the results of the studies carried out on similar subjects are compared with the results of this study, the fact that the rate of risk-loving producers is very high is accepted as another indicator that these producers are among the early adopters of innovations. However, the most important issue to be considered here is related to the occupations of the interviewed producers other than crop production. When we look at the main occupational groups of producers, we find that the number of those engaged in crop production as their main occupation is very small. A significant number of the producers surveyed are businessmen, architects, engineers, and physicians. This group is generally the upper-income group in Turkiye, and it is considered that this group has an important share in the increase in mango investments in the region.

When the land assets and production patterns of the producers interviewed in this research were analyzed, it was determined that the producers had an average production area of 3 ha per agricultural holding. Most of these production areas are owned by farmers, and the area cultivated by renting and sharecropping is very small (Table 4.3). When the production pattern is analyzed, it is determined that there is a tropical fruit-dominated production pattern, especially banana and avocado. Mango production ranks second, followed by banana production (Table 4.3).

Mango production has recently started in Turkiye. In particular, after 2012, seedlings were brought from different plant production sources and began to be grown in open and greenhouses in the Alanya, Anamur, and Gazipaşa districts. After 2018, greenhouse cultivation has been emphasized in particular, and it has started to be planted between banana gardens in open areas. There are 2 mango varieties that have been completely adapted in Turkiye. Among these, the Kent variety was registered at the Western Mediterranean Agricultural Research Institute (BATEM), and adaptation studies of the Irwin variety have been completed. Apart from these, the existence of Keitt, Nam doi mai, Osteen, Honey Gold, Tommy Atkins, Gleen, Palmer, Valencia pride, and R2E2 varieties, which were brought to Turkiye by different means and adaptation studies have been carried out in different production regions of the world, has also been reported (Güven 2021). As a result of a survey conducted in the research area, the mango varieties grown by producers and the systems applied in the garden establishment are given in Table 4.4. Because mango, which is sensitive to climate, i.e. cold, is suitable for greenhouse cultivation, most of the production is cultivated under cover, except in some microclimate areas. It was determined that mango production in the research area is mostly carried out under cover. According to the survey results, 92.44% of the current production area consists of greenhouses.

Table 4.3 Land asset and land use status of mango producers in the research area

	Average area (Ha) per farm	Total area (Ha)
Own area	2.80	111.99
Irrigated	2.10	83.99
Rainfed	0.70	28.00
Rented area	0.65	26.00
Irrigated	0.65	26.00
Rainfed	0.00	0.00
Sharecropping area	0.13	5.30
Irrigated	0.13	5.30
Rainfed	0.00	0.00
Mango-greenhouse	0.82	32.65
Mango-open field	0.06	2.49
Banana-greenhouse	0.67	26.80
Banana-open field	0.13	5.10
Dragon fruit	0.05	2.00
Vegetable-green house	0.02	0.90
Avacado	0.82	32.80
Other crops	1.01	40.55

The average age of trees is generally below 3 years and studies on this subject indicate that mango starts to be productive after 3 years (Güven 2021). Therefore, the number of productive trees is limited. When Tables 4.4 and 4.5 are analyzed together, it is determined that only 24.24% of the total number of trees is productive. Although there were 1022 trees per hectare in the greenhouse, there were 324 trees in the open field. Although the yield per tree varied according to the age of the tree and the variety, when Table 4.5 was analyzed, the average is 27–28 trees and this amount is 34–35 trees in the open field. In Turkiye, mango sales by producers are not at the grain level but in kilograms. On the other hand, grains are used for sales to consumers. In the research conducted, it was determined that the average selling price of mango was $6.80/kg (Table 4.5). In India, one of the highest producing countries in the world, the retail price range of mango per kilogram in 2024 is US$7.84/kg to $14.69/kg per kilogram, and the wholesale price is US$5.49/kg to US$10.29/kg. The prices were reported to be $2.59–4.31/kg and $1.81–3.02/kg in China, $0.59–1.47/kg and $0.41–1.03/kg in Thailand, respectively. In the same study, pricing for Turkiye was reported to be 3.42–6.34$/kg and 2.39–4.44$/kg (Selina Wamucii 2024).

Table 4.6 provides information on the reasons why mango producers in the research area are interested in mango production. When the table is examined, it is determined that the most important reason for producers to choose this product is the high interest of producers in trying new, untested alternative plant products. The second reason is that the production of this product has become popular in the

Table 4.4 Mango production statistics by mango varieties and production systems

Production system/ varieties		Inter-row (m)	Intra-row (m)	Production area (m^2)		Tree age (Year)	Number of trees	Number of trees per m^2
		Mean	Mean	Sum	Mean	Mean	Mean	Mean
Greenhouse	Keitt	3.05	3.20	134,500.00	6725.00	2.90	636.75	0.114
	Kent	3.41	3.38	74,200.00	4637.50	2.78	535.94	0.085
	Naomi	3.11	3.28	55,750.00	6194.44	2.33	558.22	0.109
	Irwin	2.83	3.25	23,300.00	3883.33	2.25	380.00	0.083
	Osteen	3.61	3.83	13,500.00	1500.00	2.78	142.11	0.092
	Tommy Atkins	2.00	2.00	7000.00	7000.00	2.00	1600.00	0.230
	Maya	3.00	3.00	4000.00	4000.00	1.00	500.00	0.130
	Haden	3.00	3.00	3000.00	3000.00	2.00	300.00	0.100
	Alphonso	3.00	3.00	1000.00	1000.00	3.00	45.00	0.050
	Palmer	3.00	3.00	250.00	250.00	3.00	10.00	0.040
Open field	Irwin	4.25	4.25	11,900.00	2975.00	3.00	83.75	0.028
	Keitt	4.33	4.50	11,000.00	1833.33	3.33	50.83	0.037
	Osteen	4.00	4.50	2500.00	1250.00	2.00	65.00	0.055
	Palmer	3.00	3.00	500.00	500.00	5.00	20.00	0.040

Table 4.5 Mango yield and selling price statistics of mango varieties and production systems

		Production area (m^2)		Tree age (Year)	Number of trees	Yield (number /tree)	Selling price (₺/Kg)	Selling price ($[a]/Kg)
		Sum	Mean	Mean	Mean	Mean	Mean	Mean
Greenhouse	Keitt	64,500.00	9214.29	4.43	510.00	31.29	232.86	6.88
	Kent	6700.00	2233.33	6.33	243.33	39.33	210.00	6.21
	Naomi	2000.00	2000.00	4.00	300.00	20.00	220.00	6.50
	Osteen	1500.00	750.00	5.00	30.00	10.00	245.00	7.24
	Alphonso	1000.00	1000.00	3.00	45.00	20.00	240.00	7.09
	Irwin	800.00	800.00	6.00	30.00	15.00	230.00	6.80
Open field	Tommy Atkins	4000.00	4000.00	9.00	140.00	80.00	300.00	8.87
	Keitt	1500.00	1500.00	6.00	100.00	15.00	200.00	5.91
	Irwin	1000.00	1000.00	5.00	10.00	8.00	200.00	5.91

[a] Average of August 2024 and September 2024 US Dollar selling rate of the Central Bank of the Republic of Turkiye (1$ = 33.83₺)

region and attracts producers. As is known from the socio-demographic structure of producers, the fact that producers are risk-takers (Table 4.1) is the most important driving force that pushes producers to produce or try new plant products. Therefore, it is important for producers to be pioneers in the region in selecting mango production and to be the first to seize the opportunities brought by new crops. In addition to producers' characteristics, factors that influence their choice of such crops are important. In recent years, the serious efforts of both the public and private sector in mango production have been an important factor in increasing production. In particular, the ease of access to information through the internet, the practices of pioneer farmers and attention-grabbing advertisements by the private sector have been effective in increasing mango production in Turkiye.

Although the high level of curiosity among producers who started mango production is an important factor, having knowledge is one of the most important factors in agricultural production. Although the producers started mango production, they were also tried to determine their level of knowledge about mango production. The results obtained are given in Fig. 4.2. Figure 4.2 shows that more than half of the producers does not have enough information about mango production. This indicates the existence of an important problem, especially for producers. In addition to the high dependence of producers on external sources of information produced by other countries, this increases the likelihood of errors in production, which increases the likelihood of significant losses both individually and nationally. Although unconscious agricultural production incurs significant economic losses individually, it may cause different problems, especially the unconscious use of natural resources such as water and soil. It has been reported that widespread greenhouse gas production may also put significant investments at risk. In the study conducted by Oğuz and Oğuz

Table 4.6 Reasons why producers prefer mango production

	First choice		Second choice		Third choice	
	Count	%	Count	%	Count	%
Because he/she likes to try new things	20	50.00	7	24.14	3	21.43
More profitable	10	25.00	5	17.24	3	21.43
Required less water	5	12.50	1	3.45	2	14.29
Less labor demand	2	5.00	4	13.79	1	7.14
Because it is popular	1	2.50	8	27.59	1	7.14
Lower production costs	0	0.00	1	3.45	1	7.14
Lowering production risk	0	0.00	0	0.00	1	7.14
Lower marketing risk	0	0.00	2	6.90	1	7.14
Fewer disease pests	0	0.00	0	0.00	0	0.00
Other	2	5.00	1	3.45	1	7.14

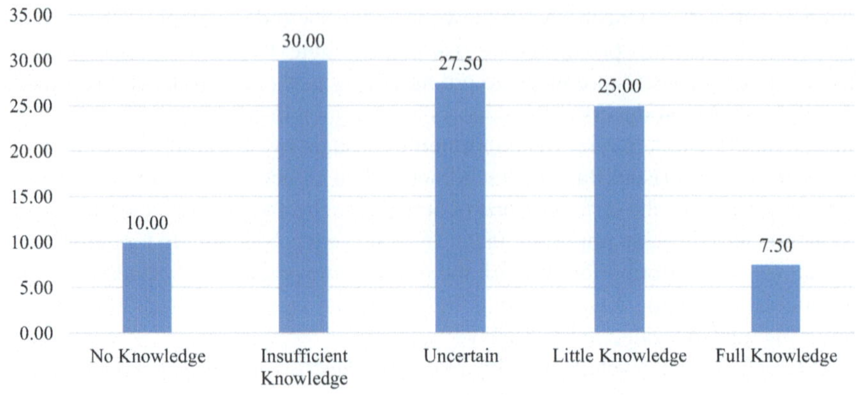

Fig. 4.2 Knowledge level of the producers on mango production (%)

(2021) on fruit cultivation under greenhouse conditions, greenhouse fruit cultivation is an expensive investment. Therefore, the marketing and sale of such products should be performed correctly. In addition, in this study, it was emphasized that it is important for producers to have sufficient knowledge and equipment to operate such production systems.

The products that producers gave up for mango production are presented in Fig. 4.3. When Fig. 4.3 is examined, it is determined that mango production has replaced banana and avocado production in the region. In the Mediterranean region of Turkiye, banana cultivation, especially in terms of greenhouse fruit growing, has been carried out for many years. In recent years, banana gardens have been replaced by avocado gardens, and it has been determined that mango is an alternative to banana and avocado. Due to the decrease in the price per kg of banana and the fact that banana consumes more water than mango, interest in mango has increased.

The reasons for abandoning products that were abandoned for mango production are presented in Fig. 4.4. When Fig. 4.4 is examined, it is reported that the first

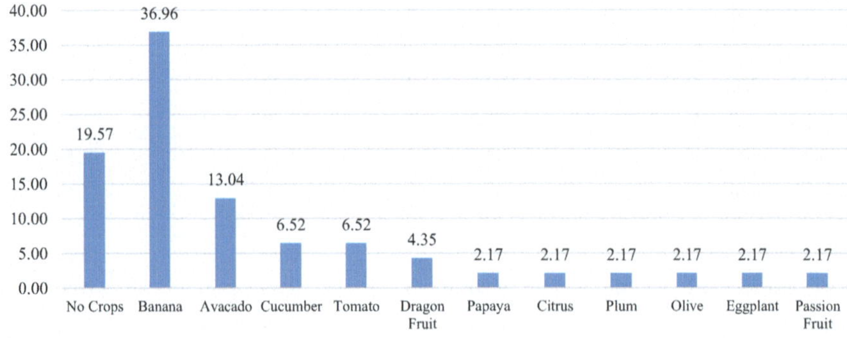

Fig. 4.3 Abandoned crops for mango production (%)

reason for abandoning the production of these products is that these products are less profitable compared to mango and that these products require more water than mango. Since there is no comparative study on the profitability and cost analysis of mango in Turkiye, when the literature conducted on the subject is examined, it is evaluated that it is parallel to the answers given by the producers on this subject. In a study conducted by Al-Hebshi (n.d.) in Yemen, they conducted an economic analysis of banana and mango production and compared them. As a result, they reported the rates of mango and banana production as 71% and banana as 51% in terms of internal profitability. The Benefit/Cost (B/C) ratio was calculated to be 10,690 for mango and 2159 for banana. The water usage rate per hectare was calculated as 34,000 m^3 for banana production and 24,000 m^3 for mango production. It has been reported that mango production saves 10,000 m^3 of water per hectare compared to banana production. In their value chain analysis of mango and banana production in India, the B/C ratio in production was determined to be 1.6 for Jalgaon banana producers, while this ratio was calculated to be 2.38 for mango producers in Karnataka and 2.25 for Maharashtra mango producers. In the same study, it was reported that water is used a lot in banana production, and water needs to be given 70–75 times a year, which is approximately 1800–2000 mm. However, it is also stated that 45% water saving is achieved, especially with the drip irrigation system. It has been reported that mango production is adapted to tropical and subtropical climates and requires sufficient rain during the growth period (June-October) and a dry period during the flowering period (November). In mango production, it is among the other issues stated that it can be grown in areas where there is a rainfall of 750–2500 mm between June and September and then a dry season lasting eight months, and that orchards should be irrigated in production areas that receive less than 750 mm of rainfall per year. It is also stated that water availability and sustainability are important aspects of this production (Wardhan et al. 2022). When the studies conducted are evaluated together, it can be concluded that since studies on water consumption are conducted in open areas and tropical regions, rain-dependent production systems are more common. In the Mediterranean Region of Türkiye, the preference for mango instead of banana in greenhouses suggests that it saves water. However, the significant amount of water required for mango should not be ignored.

The problems experienced by producers in terms of mango production are presented in Fig. 4.5. When Fig. 4.5 is examined, the primary problem in mango production is the supply of seedlings. In Turkiye, the high prices of seedlings are at the top of the complaint list. The main reasons for this situation can be explained by the fact that mango production and seedling production are still not widespread; therefore, there are not enough investments in seedlings in the country. Among the most important production problems, access to technical information and/or consultancy services is limited. The lack of sufficient information about mango production is still not widespread in the country is one of the factors that makes investing in this production area risky. Therefore, the problems experienced by producers in this regard are clearly depicted in Fig. 4.5. Another striking problem is the presence of diseases and pests in mango production. Remarkably, the issue of water consumption

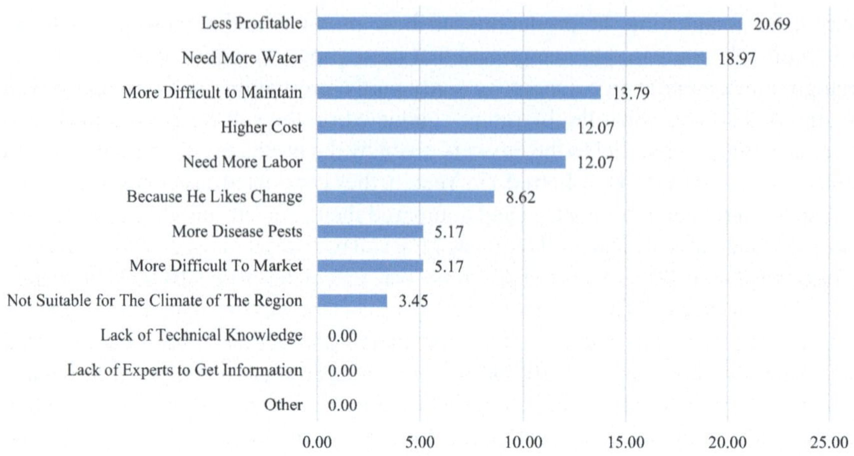

Fig. 4.4 Reasons of the producers not to product these crops (%)

is shown as a problem in mango production by producers in the last ranks and is not considered a significant problem (Fig. 4.5).

In the research area, mango producers were asked to evaluate mango production according to different criteria. When Table 4.7 is examined, the first evaluation of producers is whether mango is a suitable product to be produced in Antalya and Mersin provinces. It was determined that 85% of producers agreed that mango is a suitable product for the region. Another issue is the water demand of mangoes. The water demand of the product was asked to be evaluated in mango production,

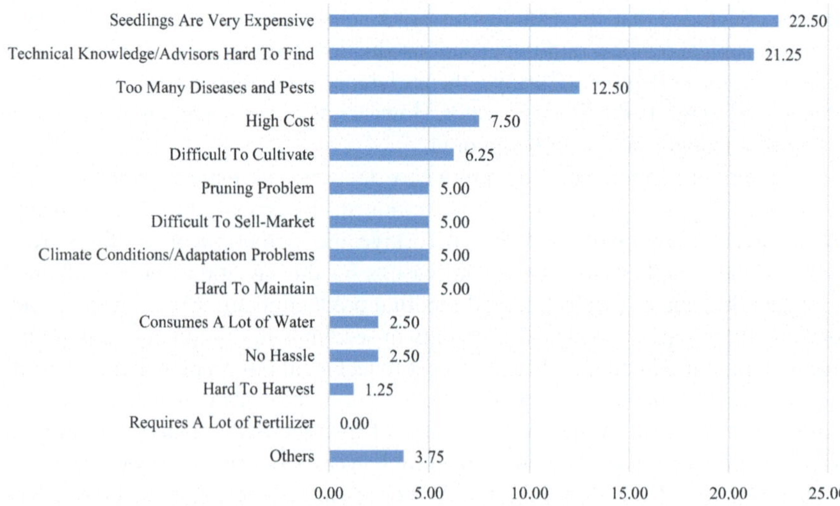

Fig. 4.5 The most challenging aspects of mango production (%)

and more than 60% of the producers declared that mango water consumption is low. It was determined that the opinions of producers on both the adaptation of mango to the region and the desires of mango to be grown did not have a statistically significant difference among the different groups. However, when asked whether producers could carry out mango production in accordance with the technique, it was determined that almost 50% of producers could not carry out the production in accordance with the technique. This situation was more common in Group 1, which included small producers, and it was determined that this situation was statistically significant at the 90% confidence level. There is a tendency to think that mango producers with larger production areas have better practices in this regard than other producers.

One of the most debated issues in Turkiye is the cultivation of tropical fruits, especially in greenhouses in the Mediterranean Region. There is a great public debate on this issue, and it is particularly true that tropical fruit cultivation "creates a significant negative impact" on limited water resources. When producers were asked

Table 4.7 Producers' thoughts on different aspects of mango production

		1. Group		2. Group		Chi-square/ likelihood ratio
		Count	%	Count	%	
Is mango a suitable product for the region?	Definitely not suitable	2	10.53	0	0.00	3.46
	Not appropriate	0	0.00	1	4.76	
	Uncertain	1	5.26	2	9.52	
	Suitable plant	8	42.11	8	38.10	
	Very suitable plants	8	42.11	10	47.62	
Water consumption of mangoes	Does not want water	3	15.79	3	14.29	1.08
	Reduced water demand	8	42.11	11	52.38	
	Not sufficient information	5	26.32	3	14.29	
	Water consuming	2	10.53	3	14.29	
	Consuming more water	1	5.26	1	4.76	
Do you think you are doing your current mango production properly?	Absolutely no	4	21.05	4	19.05	7.84*
	No	7	36.84	3	14.29	
	Uncertain	6	31.58	4	19.05	
	Yes	1	5.26	4	19.05	
	Absolutely yes	1	5.26	6	28.57	

Statistically significant at * 90%, ** 95%, *** 99% levels

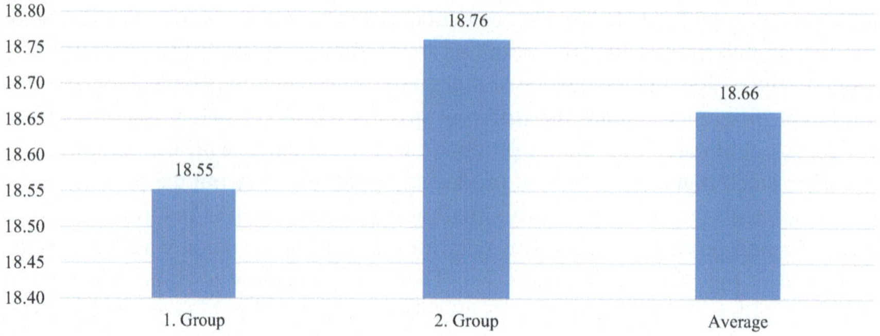

Fig. 4.6 Irrigation cost in total cost of mango production by groups

about their thoughts on water consumption in mango production, they stated that "on average, irrigation costs constitute approximately 18.66% of the total production cost." According to the grouping, the difference in irrigation costs was not statistically significant (t-test:-0.043, p:0.97) (Fig. 4.6).

The issue of "water consumption" is one of the topics discussed in the Mediterranean Region of Turkiye in terms of greenhouse fruit production. In their study on the status and problems of banana cultivation in Turkiye and their solution suggestions, Pınar et al. (2007) stated that the problems of water used in particular were not suitable for plant production (inappropriate salt, pH level), application without fully calculating plant water consumption, and excessive water application while balancing the humidity and temperature of the growing environment in many greenhouses. No detailed research has been conducted on water use in mango production in Turkiye. On the other hand, when studies on the annual water consumption of mango trees are examined, it is stated that the water needed for mango production is 979 mm/m^2 in Saudi Arabia (Mohammad et al. 2015), 1198 mm/m^2 in Israel (Levin et al. 2018), 1197 mm/m^2 in South Africa (Mostert and Hoffman 1997), 545 mm/m^2 in Spain (Durán Zuazo et al. 2011), 416.6–532.6 mm/m^2 in terraced gardens again in Spain (Durán Zuazo et al. 2018), 403–420 mm in a study conducted in Brazil (De Souza et al. 2016) and 551.6 mm in another study conducted in Brazil (De Azevedo et al. 2003).

4.4 Conclusion

The Mediterranean Basin is among the regions expected to be most affected by climate change. Therefore, there is a significant segment who sees any attempt to increase water use as dangerous for the future of the region. It is important for the Research Institutes affiliated with the Ministry of Agriculture and Forestry, in particular, to conduct studies on the adaptation of tropical and subtropical fruits, which are not currently produced or are produced in limited quantities in Turkiye, to the region to apply the appropriate varieties and production techniques and to guide producers correctly. However, it is striking that there are still no sufficient studies on water consumption, appropriate irrigation systems, and economic analyses of such fruits in greenhouse environments, and it is more of a necessity than a need to conduct research in this area as soon as possible. In particular, in all media environments, there are significant concerns that the production of tropical and subtropical products may have negative effects on water resources in the region, as well as the existence of striking warnings, which raise significant concerns about whether it is appropriate to produce such fruits.

Supporting plant production in Turkiye is an important issue, and there is a need for planning and sensitivity to supporting species that are not produced in Turkiye or are produced in limited quantities. Among the study topics, there is a need for priority and national level studies on plant water consumption, identification of problems that may arise from diseases and pests specific to new species, determination of varieties with appropriate yield and quality, and adaptation studies, risks that may arise from the production and sale of seedlings that are not controlled in terms of plant health in the market, delivery of sufficient information and techniques to producers, creation of resources from which producers can receive consultancy, and market and economic analyses of production.

It is known that today, instead of "just production", the concepts of "sustainable production" and even "responsible consumption" have come to the fore. Tools such as carbon footprint and water footprint that support these concepts are becoming more prominent in the production process. For this reason, it is important to actively use such tools not only for tropical fruits but also in every system in which intensive agriculture is carried out and to make these calculations from production to consumption. More studies on carbon neutral production systems with low water footprints and even supply chains. Mango production in Turkiye is a tropical fruit species that is being tried to be grown in the Mediterranean Region within the framework of these uncertainties. In our world, where food crises are experienced more frequently, food inflation is a significant problem, and concepts such as hunger and food security are still important problems for the whole world. The issue of responsible production and consumption must be a priority in the studies carried out to meet the needs within the framework of the United Nations Sustainable Development Goals. Especially when the production planning on which the Ministry of Agriculture and Forestry has been conducting its master studies in recent years and the support policies planned to be implemented in accordance with this goal are taken into consideration, it is obvious that such activities should be added to the agenda for discussion.

References

Akçaöz H, Özkan B, Kızılay H (2006) Antalya ilinde tarımsal üretimde risk yönetimi ve tarım sigortası uygulamaları. Tekirdağ Ziraat Fakültesi Dergisi 3:93–103

Akinaga T, Hasbullah R (2002) Mango production using plastic greenhouse in Okinawa. Acta Hortic 575:745–749

Al-Hebshi MA (n.d.) The economics of banana and mango production in Wadi Surdud, Yemen: a comparative study. https://www.academia.edu/14434531/The_Economics_of_Banana_and_Mango_Production_in_Wadi_Surdud_Yemen_A_Comparative_Study

Bayazit S, Çalışkan O, Kılıç D (2021) Akdeniz Bölgesi'nde örtüaltı meyve yetiştiriciliği. Bahçe 50:59–70

Bayram S, Arslan M (2007) Düşük ve yüksek sıcaklıkların avokado yetiştiriciliği üzerine etkisi. Derim 24:9–19

Bayram S, Arslan MA, Turgutoğlu E (2006) Türkiye'de avokado yetiştiriciliğinin gelişimi, önemi ve önerilen bazı çeşitler. Derim 23:1–13

Dal B, Gübbük H (2023) Türkiye'ye adaptasyon sağlamış tropik meyve türleriyle ilgili çalışmalardan ilk izlenimler. Meyve Bilimi 10:146–151

De Azevedo PV, Da Silva BB, Da Silva BPR (2003) Water requirements of irrigated mango orchards in northeast Brazil. Agric Water Manag 58:241–254

De Souza PPJO, Rodrigues CJ, De Sousa LAM, Lima TR, Rocha PEJ, Ortega FS (2016) Water requirements estimate for the reproductive period of mango orchards in the northeast of the state of Pará, Brazil. Rev Bras Frutic 38

Devlet Su İşleri (DSİ) (2024) T.C. Tarım ve Orman Bakanlığı, Devlet Su İşleri Genel Müdürlüğü. https://www.dsi.gov.tr/Sayfa/Detay/720

Douville H, Raghavan K, Renwick J, Allan RP, Arias PA, Barlow M, Cerezo-Mota R, Cherchi A, Gan TY, Gergis J, Jiang D, Khan A, Pokam Mba W, Rosenfeld D, Tierney J, Zolina O (2021) Water cycle changes. In: Masson-Delmotte V, Zhai P, Pirani A, Connors SL, Péan C, Berger S, Caud N, Chen Y, Goldfarb L, Gomis MI, Huang M, Leitzell K, Lonnoy E, Matthews JBR, Maycock TK, Waterfield T, Yelekçi O, Yu R, Zhou B (eds), Climate change 2021: the physical science basis. Contribution of working Group I to the sixth assessment report of the intergovernmental panel on climate change. Cambridge University Press, Cambridge, pp 1055–1210

Durán Zuazo VH, Rodríguez Pleguezuelo CR, Franco TD (2011) Impact of sustained deficit irrigation on tree growth, mineral nutrition, fruit yield and quality of mango in Spain. Fruits 66:257–268

Durán Zuazo VH, Rodríguez Pleguezuelo CR, Gálvez Ruiz B, Gutiérrez Gordillo S, García-Tejero IF (2018) Water use and fruit yield of mango (*Mangifera indica* L.) grown in a subtropical Mediterranean climate. Int J Fruit Sci 19:136–150

Düzgüneş O, Kesici T, Gürbüz F (1983) İstatistik Metodları I. Ankara Üniversitesi Ziraat Fakültesi Yayın No: 861, Ankara

Dyvik EH (2024) Water footprint of food and consumer goods 2024, by product. https://www.statista.com/statistics/264516/water-footprint-of-food-and-consumer-goods/

Gözalan S, Coşkun S, Öztekinci M (2024) Tropikal meyve ağaçlarından mangonun açık alanda yetişme potansiyeli: Antalya ili örneği. J Hist School 71:2142–2171. https://doi.org/10.29228/Joh.76865

Gözalan S (2023) Akdeniz Bölgesi'nde avokado ağacının yetişebileceği uygun alanların belirlenmesi (Yayımlanmamış doktora tezi). Karabük Üniversitesi Lisansüstü Eğitim Enstitüsü, Karabük

Gübbük H, Altınkaya L, Balkıç R (2018) Banana: a very profitable tropical crop for Turkey. Chron Hortic 57:20–25

Güven D (2022) Dünya'da ve Türkiye'de Mango Yetiştiriciliği, pp 108–175, in: Tropik Meyve Ders Notları. T.C. Tarım ve Orman Bakanlığı, Antalya İl Tarım ve Orman Müdürlüğü. https://antalya.tarimorman.gov.tr/Belgeler/Tropik%20Meyve%20Yeti%C5%9Ftiricili%C4%9Fi%20Ders%20Notlar%C4%B1.pdf

IPCC (2013) Climate change 2013: the physical science basis. In: Stocker TF, Qin D, Plattner GK, Tignor M, Allen SK, Boschung J, Nauels A, Xia Y, Bex V, Midgley PM (eds) Contribution of working Group I to the fifth assessment report of the intergovernmental panel on climate change. Cambridge University Press, Cambridge, United Kingdom and New York, NY, USA, p 1535. https://doi.org/10.1017/CBO9781107415324

IPCC (2023) Summary for policymakers. In: Lee H, Romero J (eds) Climate change 2023: synthesis report. Contribution of working Groups I, II and III to the sixth assessment report of the intergovernmental panel on climate change [Core Writing Team]. IPCC, Geneva, Switzerland, pp 1–34. https://doi.org/10.59327/IPCC/AR6-9789291691647.001

Köseoğlu AT, Onur C, Uludağ N, Arpacıoğlu A (1988) Akdeniz Bölgesinde muz yetiştirilen alanlarda toprak-bitki ilişkilerinin belirlenmesi. Akdeniz Üniversitesi Ziraat Fakültesi Dergisi 1:53–56

Levin AG, Peres M, Noy M, Love C, Gal Y, Naor A (2018) The response of field-grown mango (cv. Keitt) trees to regulated deficit irrigation at three phenological stages. Irrig Sci 36:25–35. https://doi.org/10.1007/s00271-017-0557-5

Lim CK, Moon DG, Seong KC, Kim CH, Jung YB, Park KS, Song SY (2016) Selection of mango (*Mangifera indica* L.) cultivars suitable for fruit production from greenhouse conditions in Jeju Island. Korea. Korean J Breed Sci 48:126–132

Medina D, Gómeza R, Guirado E (2009) Preliminary observations about Tommy Atkins production outdoors and under plastic greenhouse in southern mainland Spain. Acta Hortic 820:443–445

MGM (2024) Meteoroloji Sözlüğü. T.C. Çevre Şehircilik ve İklim Değişikliği Bakanlığı, Meteoroloji Genel Müdürlüğü. https://www.mgm.gov.tr/genel/meteorolojisozlugu.aspx

Mohammad FS, Alamoud AI, Mahmoud SH (2015) Water requirements and water use of mango orchards in Jazan Region, Saudi Arabia. J Anim Plant Sci 25:1008–1015

Mostert PG, Hoffman JE (1997) Water requirements and irrigation of mature mango trees. Acta Hortic 455:331–338. https://doi.org/10.17660/ActaHortic.1997.455.43

Oğuz Hİ, Oğuz İ (2021) Greenhouse cultivation fruits in the Mediterranean Region of Turkey. In: Pakyürek M (ed) Recent headways in pomology, Chap. 4. Iksad Publishing House, Ankara, pp 93–127

Özer D, İkikat Tümer E (2020) Limon üreticilerinin risk karşısındaki tutum ve davranışlarını etkileyen faktörlerin analizi: Mersin ili Erdemli ilçesi örneği. Türk Tarım Ve Doğa Bilimleri Dergisi 7:988–996. https://doi.org/10.30910/turkjans.735929

Pınar H, Türkay C, Canan İ (2007) Türkiye'de muz yetiştiriciliği, sorunları ve çözüm önerileri. Alatarım 6:15–20

Şahin G, Kendirli B (2012) Türkiye'de örtüaltı meyve yetiştiriciliği. Akdeniz Üniversitesi Ziraat Fakültesi Dergisi 25:9–15

Shiroma Y, Afuso H, Suwa R, Atsushi K, Yuki T, Takeshi K, Itaru N, Shiro T, Gauri M (2019) Development of higher yield and high-quality mango production system based on internet of things. Electron Commun Jpn 102:33–41

TEMA and WWF (2015) İklim Değişikliğinin Yerel Etkileri Raporu. https://cdn-tema.mncdn.com/Uploads/Cms/iklim-degisikliginin-yerel-etkiler-raporu.pdf

TOB (2021) İklim Değişikliği ve Tarım Değerlendirme Raporu. T.C. Tarım ve Orman Bakanlığı Tarım Reformu Genel Müdürlüğü. https://www.tarimorman.gov.tr/TRGM/Belgeler/IKLIM%20DEGISIKLIGI%20VE%20TARIM%20DEGERLENDIRME%20RAPORU.pdf

Tunç Y, Yılmaz KU (2023) Türkiye'de yetiştiriciliği yapılan bazı subtropik iklim meyvelerinin üretim projeksiyonu. Erciyes Tarım ve Hayvan Bilimleri Dergisi 6:17–22. https://doi.org/10.55257/ethabd.1216677

UNEP (2024) As shortages mount, countries hunt for novel sources of water. United Nations Environment Programme (UNEP). https://www.unep.org/news-and-stories/story/shortages-mount-countries-hunt-novel-sources-water

United Nations (UN) (2015) The 17 goals. Department of Economic and Social Affairs, Sustainable Development. https://sdgs.un.org/goals

Selina Wamucii (2024) Mango prices in China, India, Thailand and Turkiye. https://www.selinawamucii.com/

Wang X, Liu L (2023) The impacts of climate change on the hydrological cycle and water resource management. Water 15:2342. https://doi.org/10.3390/w15132342

Wardhan H, Das S, Gulati A (2022) Banana and mango value chains. In: Gulati A, Ganguly K, Wardhan H (eds) Agricultural value chains in India. India Studies in Business and Economics. Springer, Singapore, pp 1–34. https://doi.org/10.1007/978-981-33-4268-2_4

WSSD (2002) World Summit on sustainable development, implementation report, Johannesburg, 26 Ağustos–3 Eylül 2002. Güney Afrika, www.johannesburgsummit.org

Zul Helmey MS, Muhamad Hafiz MH, Wan Mahfuzah WI, Shaidatul Azdawiyah AT, Mohd Aziz R, Mohd Farid AF (2022) Evaluation of growth, yield and fruit quality of Harumanis mango under greenhouse and open field conditions. Trans Malaysian Soc Plant Physiol 30:220–225

Chapter 5
Mango Quality Under Changing Climate Scenario

Muhammad Tauseef Sultan, Aliza Nayyar, Muhammad Maaz, and Ahmad Mujtaba Noman

Abstract Climate change presents critical challenges to mango (*Mangifera indica*) production, directly impacting fruit quality through rising temperatures, erratic rainfall, and increased pest and disease pressure. These environmental changes contribute to physiological disorders such as fruit cracking, sunburn, and uneven ripening, reducing the market value and consumer appeal of mangoes. Adaptive strategies are essential to sustain mango quality under these changing conditions. Pre-harvest practices like fruit bagging provide a protective barrier against environmental stressors, pests, and diseases while enhancing fruit appearance and reducing chemical pesticide use. Developing biodegradable bagging materials offers a sustainable alternative to conventional options. Additionally, breeding programs focused on climate-resilient mango cultivars with heat and drought tolerance and improved disease resistance are vital for long-term sustainability. Post-harvest interventions, such as advanced storage techniques and modified atmospheric packaging, play a critical role in preserving mango quality during distribution and extending shelf life. Agroforestry systems, optimized irrigation schedules, and predictive climate models further aid in adapting mango cultivation to climate variability.

This chapter highlights the need for integrated strategies combining traditional knowledge and scientific innovations to address the multifaceted impacts of climate change on mango quality. Scalable solutions, farmer training, and economic feasibility are pivotal for ensuring sustainable mango production for local and global markets.

Keywords Mango quality · Climate resilience · Sustainable agriculture · Post-harvest management

---

M. T. Sultan (✉) · A. Nayyar · M. Maaz · A. M. Noman
Faculty of Food Science and Nutrition, Bahauddin Zakariya University, Multan, Pakistan
e-mail: tauseefsultan@bzu.edu.pk

S. A. H. Naqvi et al. (eds.), *Climate Change and Mango Production*,
https://doi.org/10.1007/978-3-031-87035-4_5

5.1 Introduction

The main cause of the 50-year global warming trend has been human-caused greenhouse gas emissions, mostly from utilising fossil fuels like natural gas, coal, and oil. Additionally, by releasing greenhouse gases and other carbon dioxide into the atmosphere, operations like deforestation and agriculture greatly contribute to these effluents. These gases are the cause of many natural instabilities, most notably climate change. Long-term patterns in precipitation, temperature, and other environmental variables including humidity and pressure in the air are indicative of climate change (US Global Change Research Program 2009). A multifaceted worldwide issue, the changing climate affects many facets of socio-political, socio-economic systems, ecological and the environment (Abbass et al. 2022). Given that climate change is now one of the biggest environmental issues the world is facing, an evaluation of its consequences on agriculture is required (Kalra et al. 2007).

Agriculture is the largest global sector which required freshwater resources, making it particularly susceptible to climate change. In many developing countries, agriculture is the primary source of livelihood and employment for majority population and significantly contributes to national GDP. Therefore, any reduction in agricultural output due to future climate change could severely impact food security and worsen the living conditions of the rural and urban communities. As an agro-based economy, Pakistan faces significant intimidations from global climate changes. The various agro-ecological zones of the nation will experience different consequences from climate change in terms of agriculture and natural resources. Rising temperatures have the potential to hasten deglaciation in the arid mountain regions of the West, thereby impacting water resources that are essential for energy and agricultural production. Numerous both human and natural activities have already placed these areas under extreme stress, which has resulted in continued environmental degradation (Ullah et al. 2017; Calzadilla et al. 2013).

5.2 Mango Quality

Mango is a popular fruit in various continents, including Asia, Africa and Central and South America. Mango (Mangifera indica Linn) is widely recognized for its nutritional aspects, being abundant in minerals, fiber, vitamins, and provitamins. This fruit is referred as one of the primary exotic fruits, often called the "peach of the tropics," The sweetness, refreshing quality, and aromatic nature of mango contribute to its global commercial popularity. Mango is also celebrated as the national fruit of India, Pakistan, and Philippines, and recognized as national tree of Bangladesh. Globally, Asian countries are the leading producers of mango; moreover, numerous mango varieties are cultivated in various countries around the world. Among the thousands of mango varieties produced globally, the most commercially dominant cultivars include Alphonso, Kent, Tommy Atkins, Ataulfo, Haden and Keitt. India is the top producer, accounting for over 40% of the world's mango production. India boasts over 1000 varieties of mangoes, but only about 3% of these dominate the trade and export market. Amongst, some notable varieties are Chausa, Banganpalli, Dasheri, Kesar, Alphonso, Langra, Raspuri and Totapuri (Thakor 2019; Mitra 2014; Tharanathan et al. 2006).

Mango cultivation in Pakistan dates back over two thousand years, making the country one of the top producers globally, ranking sixth after India, China, Mexico, Indonesia, and Thailand. The climate in Sindh region warms up about a month earlier than Punjab, allowing cultivating early mango varieties (Memon 2014). Mangoes are rich in antioxidants and nutrients because they contain a range of bioactive substances, including as polyphenols, β-carotene, and vitamin C. Mango pulp's chemical makeup is influenced by a number of variables, including maturity stage, variety, and growing site. Starch concentration rises from 1 to 14% throughout fruit growth; nevertheless, by the end of the maturity process, both reduced and non-reducing sugars increased. A number of psychological, biological, and sensory changes occur during ripening, producing soft, tasty fruit with appealing characteristics (Tharanathan et al. 2006).

Given that the ripening stage of a mango might affect its quality, the fruit quality characteristics of six Thai mango cultivars Nuan Namdokmai Si-Thong Chan, Namdokmai No. 4, Okrong, Thongdum, and Yaigrum—were investigated. Based on the weight and size of each individual fruit, the Namdokmai Si-Thong and Namdokmai No. 4 types yielded massive fruits with an average weight that was 401 g and 414 g, respectively. Thongdum and Nuan Chan produced medium-size fruits, with average weights of 311 g and 287 g, respectively, whereas varieties Okrong and Yaigrum produced little fruits, with mean weights of 152 g and 174 g, respectively. Mango fruit contains sucrose as its main sugar, followed by fructose. The acidity level of the small-sized cultivars is rather low. The identified primary organic acids found in mango fruit are citric, malic and ascorbic acids. In addition to this, the Thai cultivars had greater and lower total soluble solids and titratable levels, respectively. The cultivars of Thongdum and Yaigrum had the greatest TSS, whilst Namdokmai No. 4 had the lowest. Thongdum contained greater amounts of carotenoid (15 mg/100 g dry weight; DW) than other cultivars. The total flavonoid concentration was greater in the cultivars Okrong and Namdokmai No. 4. The range of the total phenolic compounds was 7.9–21.8 mg/100 g DW (Table 5.1) (Rumainum et al. 2018).

5.3 Impact of Climate Change on Mango Quality

Recent years have seen a significant variation in global climate change and climatic unpredictability due to massive carbon emissions into the atmosphere and the phenomenon's more noticeable effects on fruit crops due to their perennial nature. According to climate projection models (CPMs), the depletion of water resources with drastic climate variability, increased frequency of drought events, and a five-fold increase in water demand deficits during the twenty-first century are expected,

Table 5.1 Fruit quality of six different mango cultivars

Mango cultivars	Fresh weight (g)	Fructose (g/ 100 g DW)	Glucose (g/ 100 g DW)	Sucrose (g/ 100 g DW)	TSS (°Brix)	TA (%)
Namdokmai No.4	401.9	13.00	5.81	19.04	18.9	0.4
Yaigrum	174.3	13.08	5.35	28.13	23.1	0.1
Nuan Chan	286.5	9.80	1.72	48.95	21.2	0.1
Okrong	152.9	17.33	4.18	31.58	20.3	0.2
Namdokmai Si-Thong	414.3	19.47	6.28	31.24	20.1	0.3
Thongdum	311.2	13.06	5.39	55.40	23.5	0.3

TSS = total soluble solids, TA = titratable acids
Source (Rumainum et al., 2018)

imposing significant impact andserious threat on agricultural productivity and global food security. The biochemistry, physiology, biotic stressors and floral biology of crops is impacted by altered climate conditions, ultimately led to decline in fruit crop yield and quality. Thus, it is a significant challenge to global scientists and they have been tried to cope up such issues through various approaches (Rajatiya et al. 2018; Neupane et al. 2022).

Mangos are one of the most popular, extensively consumed, and historically farmed fruits, and they are grown in over 100 countries. This fruit, which is now the second- most produced tropical fruit worldwide, has a high cropping capacity under climate change scenarios. Additionally, it ranks as the sixth most significant fruit crop globally. Mango fruit seems to be one of the most essential food sources for the world's growing population. However, mango orchards suffer with a variety of problems that eventually reduce yield and exports (Ahmad et al. 2018).

Mangos are cultivated from Australia to Spain since they are well suited to both tropical and subtropical climates. The crop has evolved to a variety of growing environments with varying rainfall patterns, temperatures, and soil types. Nonetheless, regions with a cool or dry spell before flowering, lots of soil moisture, and fairly high temperatures (30–33 °C) during fruit development are the ideal for commercial production (Laxman et al. 2016).

Variations in climatic factors cause stress that leads to morphological, biochemical, anatomical, and physiological changes in mango plants, impacting the yield, growth, and quality. For example, drastic temperature changes combined with unseasonal rain can reduce both the quality and yield of mango fruits. The alterations in the plant's physical characteristics (such as canopy and vigor), reproductive traits (fruiting potential and fruit size), and quality traits (shelf life, susceptibility to pests and diseases, juice content) contribute to these reductions. Additionally, exposure to high temperatures and low humidity decreases the efficiency of photosynthesis, transpiration and water potential in mango leaves (Antwi-Boasiako et al. 2024).

One of the many negative consequences of climate change is the alteration of the natural environment and vegetation. One of the well-known consequences of global warming is the "change in phenotype," or altered periods for plant development. Climate change causes changes in the duration of the vegetative and reproductive phases. Because the vegetative stage was shortened by the high temperatures, most fruit crops need fewer days to blossom. Additionally, since cell wall enzymes are less active at higher temperatures, ripeness is postponed throughout growth and development. It also has an impact on the chemical makeup of fruits due to fluctuating precipitation and trends in moisture stress. A temperature rise of 0.7–1.0 °C may alter the region that is now best suited for producing high-quality Dashehari and Alphonso mango cultivars. Variations in blooming times provide ecological challenges in temperate regions. Compared to species located at higher latitudes, warming is more detrimental to tropical insects. Despite rising atmospheric CO_2, the future supply of food is uncertain because of global warming and altered precipitation. The problem will worsen with decreased plant variety and region appropriateness. Adaptation and mitigation strategies should be included in a strategy based on a strategic scientific evaluation of the significance of such risks to the world's fruit production (Haokip et al. 2020).

The climate change has an impact on soil composition, thus affecting the sixth-most significant fruit crop in the world. The mango (*Mangifera indica* L.), is probably vulnerable to increased soil natural matter mineralization and limited plant nutrient supplies, particularly boron (B) and zinc (Zn), in the result of climate change. The study of Ahmad et al. (2018) find out the best way to formulate and apply nutrients to potentially address nutritional deficiencies in mango trees growing in Multan, Pakistan, of the planet's hottest and driest locations. Twenty-year-old mango trees were evaluated for their yield and physical (quality) responses to seven distinct foliar and soil Zn and B treatment regimens.

The Nam Dok Mai mango is one of Thailand's most significant economic crops, is also impacted by climate change. The relationship between Nam Dok Mai mango yield and climate was investigated. The association between the mango quality, mango yields, diseases, and climate in both the on-season and off-season crops was determined by using the correlation coefficient. The mango yields and climate data were measured and gathered from the chosen mango orchards in three climatically distinct regions, i.e., northern, eastern, and northeastern. The findings showed a stronger correlation between temperature and relative humidity with mango yields and quality during the off-season than on-season. Mango starch is significantly impacted by high temperature and relative humidity during on and off growing season (Trongratsameethong et al. 2021).

5.3.1 Impact of Temperature on Mango Quality

Global warming and climate change have become critical issues due to rise in human activities because the planet is witnessing an extraordinary increase in surface temperatures. Over the past two decades, the concern regarding climate change has increased due to the major contribution of human activities. The global average surface temperature increased by roughly 0.74 °C in the course of the twentieth century, while predictions for the twenty-first century indicate a rise of 1.1 to 6.4 °C (Talib et al. 2020).The temperature significantly affecting the mango's growth cycle, flowering timing and frequency, fruit development, flavor, and appearance in nearly all production regions (Makhmale et al. 2016).Generally, most mango varieties can tolerate temperatures between 1–2 °C if they are not actively growing and the cold should not last for more than a few hours. The emergence of inflorescence begins immediately after the coldest part of winter; however, the mango growth requires relatively higher temperature conditions. Mango trees are capable of enduring a broad temperature range from 0 °C to 48 °C without significant harm. However, exposure to low temperatures for more than six hours can be fatal to the plant. On the other hand, young trees that are in vigorous growth can be severely damaged at 0 °C (Rajan 2012).

China is a major mango grower, and the nation's economy and agricultural sector both greatly benefit from the country's expanding mango business. Mango agriculture in China is confronted with obstacles due to global climate change, which includes the occurrence of severe weather conditions such as late-spring freezing temperatures and dry winds in the spring. Research on the effects of extreme temperatures on various mango pollen characteristics is crucial. In light of this, studies were done on four popularly produced mango types in different South Chinese provinces: Dashehari, Renong No. 1, Apple mango and Jinhuang. Pollen tube length (PTL), pollen rate of germination (PGR), and their fundamental temperature under various temperature stress conditions were the main subjects of the study. The pollen growing rate and pollen tube distance end to end were highest when grown at 24 °C–30 °C for three hours, according to preliminary experimental data on pollen germination and development of pollen tubes across different mango types at varied temperatures. Furthermore, below 20 °C, pollen germination proved difficult. Most mango types showed a rise in germination rate with temperature, peaking at 28 °C with an average germinating rate of 28.3%.The germination rate slower by 10% at temperatures higher than 32 °C after this point (Table 5.2). Corresponding to this pattern, pollen tube growth also decreased with temperature, from an average size of 50.1 μm at 28 °C to 34.7 μm at 34 °C (Table 5.3) (Liu et al. 2023).

Along with this, other researchers have proved the impact of different climate conditions on pollen features. Pollen may flourish in a warm climate, while low temperature may interfere with the pollen tube's capacity to reach the ovule and stop pollen formation. Despite the ability of mangos to self- and cross-pollinate, the effectiveness of cross-pollination techniques have been reported with varied results globally. Pollen grains lose their viability and develop other defects, when exposed to temperature below 17 °C. The meiotic recalculated phase, known as initial stage of

Table 5.2 Maximum pollen germination % and cardinal temperatures for pollen germination of four mangoes genotypes in response to different temperature stress

Mango varieties	Maximum pollen germination (%)	Cardinal temperature		
		T_{min}	T_{opt}	T_{max}
Deshehari	32.1	21.9	28.3	32.6
Renong No. 1	35.5	22.1	30.6	33.7
Jinhuang	34.3	22.8	30.4	34.3
Apple mango	29.1	20.3	26.7	30.4

Table 5.3 Maximum pollen tube length and cardinal temperatures for pollen tube length of four mangoes genotypes in response under different temperature stress

Mango varieties	Maximum pollen tube length (μm)	Cardinal temperature		
		T_{min}	T_{opt}	T_{max}
Deshehari	51.2	20.7	28.3	30.3
Renong No. 1	54.8	20.5	30.4	33.8
Jinhuang	56.6	21.2	32.1	34.4
Apple mango	52.0	20.3	27.9	30.2

micro sporogenesis, is greatly impacted by temperatures below 10 °C. Furthermore, low temperature adversely affects the germination and growth of pollen tubes and growth completely ceases at temperatures below 15 °C (Halder et al. 2024).

5.3.2 *Impact of Rainfall and Humidity on Mango Quality*

Rainfall and humidity are other important factors substantially affecting growth and quality of mango. Unpredictable rainfall throughout the flowering and pre-flowering stages might result in limited pollinators' activity and poor fruit set. In changing climates, storms during later fruit development can significantly impact the harvest. The flavor and look of mature mangoes are adversely affected by altered rainfall patterns, and unseasonal rains can encourage pests and lower yields. Mangoes prefer warm to hot weather during fruiting and do best in areas with little to no rainfall and humid during flowering, and fruit setting, and harvesting (Makhmale et al. 2016).

Mango cultivation is favored in regions with seasonal rainfall of 100 mm or more, as vegetative growth slows during dry periods. Both the amount and distribution of rainfall are crucial for mango production. Mangoes can grow naturally in regions with low to large rainfall (250–3000 mm), but sufficient rainfall is also necessary for fruit growth and maturity. Mangoes can be cultivated in regions where the average annual rainfall is more than 250 mm with little to no irrigation. In regions with high rainfall, mature fruits are susceptible to diseases such as anthracnose, causing the peel to blackish, and affecting its appearance (Rajan 2012).

Mango yields decrease when pre-flowering and flowering periods coincide with rains, as this impedes fruit set and pollination activities. Additionally, storms can cause significant harvest loss during the later stages of fruit development. Soil water content affects fruit size, yield, and quality parameters such as soluble solids, sugar, starch, and vitamin C levels. Various studies have confirmed the essential role of irrigation in rain fed conditions to achieve higher yields and better-quality mango fruits in a rapidly changing climate. Climate variability also influences fruit quality and the appearance of mature mangoes, partially due to higher incidences of pests and diseases like anthracnose under conditions of high humidity and rainfall. During such periods, mangoes may lose their aesthetic value through peel blackening. Contrastingly, low humidity and rainfall are beneficial for fruit setting and harvesting and help reduce biotic stresses. The mango genetics is highly diverse, which is beneficial for breeding and developing resistant climate change varieties. Varietal differences among mango genotypes in terms of productivity under wet and dry conditions can be exploited to mitigate the effects of climate variability on mango yields and quality. The mango yields and quality can be enhanced in emerging production regions, having mango cultivars compatibility with particular climate, thus proves efficient in mitigation of climate change effects (Antwi-Boasiako et al. 2024).

The shifting climate patterns can lead to storms during the final stages of fruit maturation, which might destroy a significant portion of the crop. The look and flavour of mature mango fruits can be significantly impacted by variations in rainfall patterns. Unseasonal rainfall can also draw pests, which lowers fruit output (Halder et al. 2024).

5.3.3 Impact of Flooding on Mango Quality

Flooding has been observed to decrease the photosynthetic rate in the leaves of many tree species. Flooding can significantly affect the quality of mangoes by disrupting the physiological processes of the trees. Climate change caused by global warming can increase the chance of flooding, which might potentially harm many mango crops located in tropical areas (Hara et al. 2014). The fruit trees are perennial crops that need several years to bear fruit and after flooding fruit production fields rarely returned to their normal position. It seems that regions that experience frequent flooding pose difficulties for the sustainable cultivation of fruit, and even in regions that do not experience flooding, the potential for future flooding contributes to a decrease in the planting of fruit trees (Schaffer et al. 2006).

The study looked at how seedlings and grafted mango trees differed in their ability to withstand flooding. Three scenarios—"30-flood," where trees were plunged up to 30 cm below the soil's surface, and "10-flood," where trees were submerged up to 10 cm below the soil's surface—were assessed. Each of the three situations was an untreated control. The grafted trees were flooded up to 5 cm above the graft union after being treated for 30 floods. Following the treatment, the amount of starch inside of seedlings exposed to both the 10- and 30-floods, as well as grafting trees subjected to the 10-flood, rapidly recovered. Even three months after treatment, grafted trees exposed to the 30-flood did not exhibit any signs of recovery. The grafting plants in the 30-flood conditions continued to have a significantly smaller amount of sugar than with all other treatments, but there were no discernible differences in the amount of sugar of the seedling across the treatments. Flood-tolerant tree species have high concentrations of root starch, and one reason for this is probably the continuous flow of carbohydrates produced by photosynthetic processes. Therefore, it was believed that the lower starch content in grafted plants that were exposed to the 10- or 30-flooding treatments at the end of their treatment compared to the control was caused by a reduction in photosynthates given to the roots (Saeki and Iwasaki 2020).

Flooding has a negative impact on mango development and gas exchange, but in order for the trees to withstand extended flooding stress, they generate hypertrophic lenticels on stems just above the waterline. Because of the harmful ions that create nutrient imbalances and lower development, fruit bearing, and yield, saline circumstances are not ideal for the successful growing of mangos. On the other hand, different rootstock cultivars have shown varying degrees of tolerance to saline stress (Laxman et al. 2016).

5.4 Conclusion

Climate change poses significant challenges to mango production, threatening fruit quality, marketability, and the livelihoods of growers. This chapter emphasizes the multifaceted impacts of rising temperatures, erratic rainfall patterns, and heightened pest and disease pressure on mango quality, leading to physiological disorders such as fruit cracking, sunburn, and uneven ripening. These issues demand urgent and adaptive solutions to sustain mango production under shifting climatic conditions. Pre-harvest interventions like fruit bagging, especially with sustainable biodegradable materials, offer immediate and effective protection against environmental stressors while reducing chemical pesticide reliance. Simultaneously, breeding programs aimed at developing climate-resilient mango cultivars with enhanced heat and drought tolerance and improved disease resistance are critical for ensuring long-term sustainability. Post-harvest technologies, including advanced storage systems and modified atmospheric packaging, are equally essential for maintaining fruit quality and extending shelf life in increasingly unpredictable climates. Integrated strategies that combine traditional knowledge with scientific innovations, such as agroforestry, optimized irrigation, and predictive climate models, provide a comprehensive approach to addressing climate variability. To ensure these solutions are scalable and accessible, farmer training programs, robust extension services, and economic feasibility must remain at the forefront of adaptation efforts. By adopting a holistic and inclusive approach, the mango industry can mitigate the adverse impacts of climate change, securing sustainable production systems that meet the demands of both local and global markets while preserving the livelihoods of farming communities.

References

Abbass K, Qasim MZ, Song H, Murshed M, Mahmood H, Younis I (2022) A review of the global climate change impacts, adaptation, and sustainable mitigation measures. Environ Sci Pollut Res 29:42539–42559

Ahmad I, Bibi F, Ullah H, Munir TM (2018) Mango fruit yield and critical quality parameters respond to foliar and soil applications of zinc and boron. Plants 7:97

Antwi-Boasiako A, Amponsah P, Opoku JA, Coulibaly D, Mintah P (2024) Increasing mango production efficiency under the fast-changing climate

Calzadilla A, Rehdanz K, Betts R, Falloon P, Wiltshire A, Tol RS (2013) Climate change impacts on global agriculture. Clim Change 120:357–374

Halder R, Varma S, Singh M, Dahiya A (2024) From orchard to table: understanding climate change impacts on mango production in India: a review. Int J Adv Biochem Res 8(4):472–478

Haokip SW, Shankar K, Lalrinngheta J (2020) Climate change and its impact on fruit crops. J Pharmacogn Phytochem 9:435–438

Hara Y, Yukimatsu H, Thaitakoo D, Tsuchiya K (2014) Agricultural landscape changes and its resilience in response to the 2011 serious floods in the urban fringe of Bangkok. J Rural Plan 33:191–196

Kalra N, Chander S, Pathak H, Aggarwal PK, Gupta NC, Sehgal M, Chakraborty D (2007) Impacts of climate change on agriculture. Outlook Agric 36:109–118

Laxman RH, Annapoornamma CJ, Biradar G (2016) Mango. Abiotic stress physiology of horticultural crops, pp 169–181

Liu X, Xiao Y, Zi J, Yan J, Li C, Du C, Liang Q (2023) Differential effects of low and high temperature stress on pollen germination and tube length of mango (*Mangifera indica* L.) genotypes. Sci Rep 13:611

Makhmale S, Bhutada P, Yadav L, Yadav BK (2016) Impact of climate change on phenology of mango–The case study. Ecol Environ Conserv 22–S132

Mitra SK (2014) Mango production in the world–present situation and future prospect. In: XXIX International Horticultural Congress on Horticulture: Sustaining Lives, Livelihoods and Landscapes (IHC2014): IV, vol 1111, pp 287–296

Neupane D, Adhikari P, Bhattarai D, Rana B, Ahmed Z, Sharma U, Adhikari D (2022) Does climate change affect the yield of the top three cereals and food security in the world? Earth 3:45–71

Rajan S (2012) Phenological responses to temperature and rainfall: a case study of mango. Trop Fruit Tree Species Clim Change. 71

Rajatiya J, Varu DK, Gohil P, Solanki M, Halepotara F, Gohil M, Solanki R (2018) Climate change: impact, mitigation and adaptation in fruit crops. Int J Pure Appl Biosci 6:1161–1169

Rumainum IM, Worarad K, Srilaong V, Yamane K (2018) Fruit quality and antioxidant capacity of six Thai mango cultivars. Agric Nat Resour 52:208–214

Saeki A, Iwasaki N (2020) The submergence of the graft union causes the death of grafted mango trees (*Mangifera indica* L.) under flooding. Agronomy 10:1121

Schaffer B, Davies FS, Crane JH (2006) Responses of subtropical and tropical fruit trees to flooding in calcareous soil. HortScience 41:549–555

Talib SAA, Hassan MHM, Rashid MA, Sabdin ZHM, Rashid MZA, Ibrahim WMW, Mustaffa MAO (2020) Effects of environmental temperature and precipitation pattern on growth stages of *Mangifera indica* cv. Harumanis Mango. J Agric Sci 12:26

Thakor NJ (2019) Indian mango–production and export scenario. Peach 18:0–12

Tharanathan RN, Yashoda HM, Prabha TN (2006) Mango (*Mangifera indica* L.), "The King of Fruits"—an overview. Food Rev Int 22:95–123

Trongratsameethong A, Kumpoun W, Chantrasri P, Chuttong B, Tiyayon C (2021) Climate change and Nam Dok Mai mango yield: a process management. RICE J Creative Entrep Manage 2:39–53
Ullah S (2017) Climate change impact on agriculture of Pakistan—a leading agent to food security. Int J Environ Sci Nat Resour 6:76–79
US Global Change Research Program (2009) Global climate change impacts in the United States. Cambridge University Press

Chapter 6
Chemical Composition Different Varieties of Mangoes: Nutritional and Phytochemical Perspectives

Hassan Raza and Muhammad Tauseef Sultan

Abstract Mangoes (*Mangifera indica*) are esteemed for their succulent taste and rich nutritional profile, offering an array of health benefits that make them a popular choice among fruit enthusiasts. From their juicy flesh to their vibrant colors, mangoes not only tantalize the taste buds but also provide a treasure trove of essential nutrients, vitamins, and minerals. It is considered the king of fruits due to its chemical makeup, which places it fifth in terms of total production and the second most marketed tropical fruit worldwide (FAOSTAT (2023) Crops and Livestock Products). In 2022, there were over 59 million mangoes produced worldwide. India was the top mango grower, producing about 23 million tonnes year, followed by Indonesia, China, Pakistan, Mexico, and Brazil. With an annual percentage of over 38% of the world's mango exports, Mexico is the leading exporter of mangoes (FAOSTAT (2023) Crops and Livestock Products). The mango production in Pakistan has been fluctuating for the past few years but in the year 2022–23, mango production was around 2.70 million tons. In this rate, Punjab contributed around 78% of the produced fruit while the area percentage of Punjab was around 62% (PBS 2023) (Gallup Pakistan Analysis of MNFS&R and PBS data).

Keywords *Mangifera indica* · Nutritional profile · Phytochemeistry

Mangoes have been cherished for their versatility and delicious flavor in the culinary applications. Other than their delightful taste, the fruit is renowned for its nutritional richness, which offers a harmonious blend of minerals, vitamins, and antioxidants. Mango if enjoyed fresh, incorporated in the savory dishes or blended into smoothies, it is a nutritious and delicious addition to any diet (Kirker and Newman 2024).

H. Raza · M. T. Sultan (✉)
Faculty of Food Science and Nutrition, Bahauddin Zakariya University, Multan, Pakistan
e-mail: tauseefsultan@bzu.edu.pk

H. Raza
e-mail: hassanraza@student.bzu.edu.pk

S. A. H. Naqvi et al. (eds.), *Climate Change and Mango Production*,
https://doi.org/10.1007/978-3-031-87035-4_6

Mangoes boast well-composed nutritional composition, which consists of carbohydrates, minerals, vitamins, and phytochemicals. They are rich dietary fiber source, which promotes satiety and helps in digestion. Moreover, mangoes give essential vitamins like vitamin B6, vitamin K, vitamin A, and vitamin C, with minerals like folate, potassium, and magnesium. Mango revered for its aroma, taste, and nutrition, fruit includes pulp, kernel, and peel. The pulp, which is abundant in sugars, pectin, amino acids, anthocyanins, vitamins, and polyphenols, is most consumable part. Kernel and peel portion despite being discarded, are significant nutritionally. The fruit peel contains compounds such as mangiferin, β-carotene, and protocatechuic acids which offers the treatment option as anti-diabetic, antimicrobial, anti-carcinogenic and anti-inflammatory agents. The kernel on the other hand, has higher content of polyphenols and antioxidants, and it is also used for extraction of oil (Lebaka et al. 2021).

Texture also plays integral role in the enjoyment of mangoes, from stringy fibrous varieties to creamy ones. Many varieties like Irwin, Keitt, and Mallika are developed particularly to possess minimal fibres. On the other hand for traditionalists, the fibrous part is a vital portion of visceral mango-eating experience (Kirker and Newma 2024).

6.1 Comprehensive Overview of Mango Fruit Composition

Mango fruit is well known for having a high phytochemical content and a complex nutritional profile, which provide a host of health advantages. This publication offers a thorough update on the nutritional components, phytochemical substances, and dynamic changes that occur during growth and postharvest stages of mango fruit. Mango fruit experiences physiological, structural changes and biochemical that affect the nutritional and phytochemical makeup of the fruit as it develops and reaches maturity. Softening and alterations in flavour, fragrance, and antioxidant capability result from these changes. Postharvest management procedures further influence the total concentration of vitamin C, carotenoids, antioxidant activity, phenolic compounds and organoleptic qualities (Abbasi et al. 2015).

Essential macronutrients including proteins, carbs, lipids, amino acids, other organic acids, and fatty acids are found in mango fruit methionine, arginine, phenylalanine, Lysine, valine, cysteine, and leucine are among the noteworthy amino acids that are present. When a fruit ripens, its lipid composition—especially that of omega-3 and omega-6 fatty acids—increases dramatically. Pectins and cellulose are examples of structural carbohydrates that add to the fruit's texture and nutritional content (Maldonado et al. 2019).

Mangoes are rich in micronutrients, including vitamins and minerals, crucial for overall health and well-being. Vitamin C, and A, along with various minerals play pivotal roles in immune function, vision, and metabolism regulation. The phytochemical compounds in mango fruit, like phenolics, pigments, polyphenols, and volatile constituents, contribute significantly to its antioxidant and therapeutic properties. Notable pigments include chlorophylls a &b and carotenoids. However, the distinctive fruity acidity is imparted by organic acids like as citric and malic acids. The

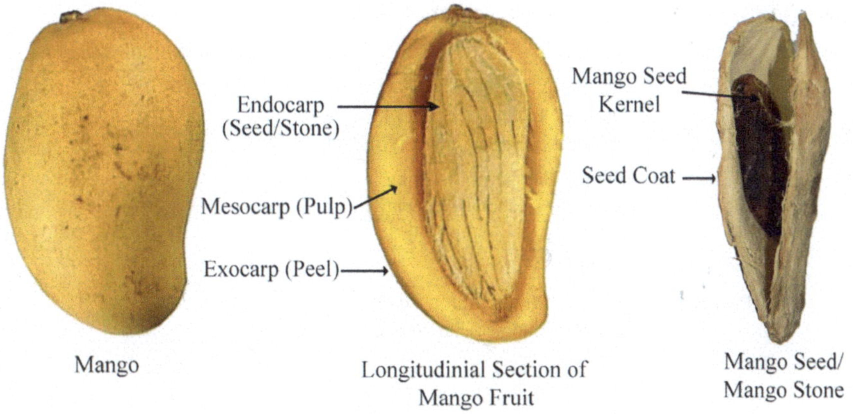

Fig. 6.1 Fruit and seed kernel of mango

fruit's flavour and aromatic profile are enhanced by volatile components (Hussain et al. 2021).

The mango fruit is characterised by three unique tissue layers: the hard wooden endocarp, which contains the seed kernel, (2) the mushy mesocarp, also known as the pulp, and (3) the pericarp, also known as peel (Fig. 6.1). Mango fruit seeds can be either monoembryonic, meaning they only yield one seedling, or polyembryonic, meaning they yield several plants. The majority of mango fruit varieties in India, the country that produces the majority of the fruit, are monoembryonic, but there are plenty of polyembryonic varieties in Myanmar, the Philippines, Thailand, and Indonesia. According to the fruit variety, the section of the seed makes up between 12 and 25% of the total weight of the fruit, and the kernel makes up between 45 and 85% of the seed portion, or around 20% of the entire fruit (Solís and Durán 2011) Tables 6.1, 6.2, 6.3 and 6.4.

6.2 Phyto-Chemistry of Mangoes

6.2.1 Phytochemical Compounds and Functional Properties of Mangoes

Mango, family Anacardiaceae, is rich in multiple bioactive phytochemical. These compounds offer multiple health benefits and functional properties.

Mangoes' bioactive substances have additive and synergistic effects, which indicate that the fruit has significant potential to enhance health and lower the risk of

Table 6.1 Top mango producers globally (2023)

Rank No.	Country	Variety	Production (Million Tons/Year)
1	India	Alphonso, Chaunsa, Langra, Dasheri, Jauhari, Lucknowi, Kesar, Totapuri, Sindhuri, Kalmi, Badami	25
2	China	Zillate, Bambaroo, Mallika, Spooner, Deshehari, Edward, Saigon	3.8
3	Indonesia	Gedong, Arumanis	3.6
4	Pakistan	Dasehri, Anwar Ratol, Sindhri, Langra, Chaunsa, Samar Bahist, Fajri	2.7
5	Mexico	Ataulfo, Keitt, Tommy Atkins, Haden, Kent	2.4
6	Brazil	Espada, Uba, Rosa, Bourbon,	2.1
7	Malawi	Dodo, Keitt, Alphonso, Tommy Aktins	1.7
8	Thailand	Brahm Kai Meu	1.6
9	Bangladesh	Gopalbhog, Amropali Khirsapat, Langra, Fasli, Himsagar,	1.5
10	Vietnam	Sherri, Kerosene	1.4

Source FAOSTAT (2023)

chronic diseases (Kabir et al. 2017). Polyphenols present in mango, act as antioxidants, gave protection to human cells from oxidative stress-triggered damage, by mitigating the DNA damage, lipid peroxidationalong with degenerative diseases. Mango pulp polyphenolic compounds include gallic acids, mangiferin, quercetin, gallotannins, isoquercetin, β-glucogallin and ellagic acid (Yadav et al. 2022). Previous studies identified catechin, leucocyanidin, epicatechin, quercitrin, and chlorogenic acid as main mango phenolics. As primary flavonoids mango usually have catechin and quercetin. In mango fruit pigments accumulation varies by the difference in cultivar, but generally mango has a high carotenoid content in the mesocarp tissue, contributing to its intense yellow color. Mango is also abundant in carotenoids, which contribute to its vibrant yellow color and also providing additional health benefits (Ediriweera et al. 2017).

Mango leaves are also rich in various phytochemicals compounds like phenolic acids, mangiferin, flavonoids, tocopherols, benzophenones, carotenoids and ascorbic acid, and. Mango leaves extracts have indicated multiple biological activities, like antioxidant, anti-cancer, antimicrobial, anti-diarrheal, anti-diabetic, lipid-lowering, anti-obesity, and hepatoprotective activity (Kumar et al. 2021).

Different studies have revealed the presence of up-to 270 volatile compounds present in mango fruit varieties. These chemical compounds also include esters which give the fruit its fruity flavour. Some varieties of the mango are name for their scent like variety from Puerto Rican 'Turpentine'. Different terpene category chemicals are

Table 6.2 Mango pulp proximate composition of different varieties and origin

Variety (country origin)	Moisture content (%)	Crude protein (g/100 g)	Ash (g/100 g)	Crude fiber (%)	Carbohydrates (%)	References
Apple, Ngowe, Kent, (Kenya)	79.96–85.84	0.06–0.12	0.06–0.12	0.18–0.41	13.51–16.74	Okoth et al. (2013)
Pakistan (Dasehri, Anwar Ratul, Langra, Sindhri)	81.45–84.32	0.72–0.84	0.23–0.45	1.60–1.82	14.12–16.1	Lebaka et al. (2021)
Hindi mango, Julie mango, Local mango (Nigeria)	77.84–82.34	1.3–1.65	0.32–0.66	0.84–1.12	16.23–19.84	Ubwa et al. (2014)
Chaunsa, Lnagra, Alphonso (India)	78.8–82.9	0.36–0.41	0.36–0.54	0.86–1.08	16.23–17.20	Rajasekaran and Soundarapandian (2023)

Table 6.3 Mango pulp chemical composition with respect to raw and ripe

Parameter	Raw	Ripe
Protein %	1.45–1.75	1.75–2.09
Carbohydrate %	21.50–26.54	20.90–28.23
Moisture %	66–70.2	72.5–75.32
Fat %	2.17–2.50	2.23–2.66
Crude fiber %	3.27–3.80	5.81–7.41
Vitamin C (μg/g)	190–313	350–393
Carotenoids (μg/g)	365–495	1400–3950

Source Ajila et al. (2007)

found in the fruit thus giving it particular turpentine aroma. The pungent *Mangifera odorata* 'funky' smell is considered by few as attractive attribute of this variety. Nam Dok Mai, a variety cultivated in Thailand is fragrant (Kirker and Newman 2024).

Table 6.4 Mango seed kernel proximate composition of different varieties and origin

Variety (country origin)	Crude protein (%)	Fat/oil (%)	Ash (%)	Crude fiber (%)	Carbohydrates (%)	References
Apple, Ngowe, Kent, and Sabine (Kenya)	6.72–9.2	9.85–13.09	2.65–3.72	1.79–2.84	72.84–75.93	Mutua et al. (2017)
Nigeria (Kibangou, Julie mango)	6.39–10.45	13.09–18.05	1.19–3.25	2.03–2.45	63.25–65.59	Nzikou et al. (2010), Ifesan (2017)
Pakistan (Dasehri, Anwar Ratul, Langra, Sindhri)	6.93–10.06	6.33–9.88	2.62–3.32	2.40–2.70	70.12–74.14	Zahoor et al. (2023), Idris (2012)
India (Chaunsa, Lnagra, Alphonso)	6.5	10.3	2.65	2.01	77.84	Kaur and Brar (2017)
Egypt (Awees, Alfons, Hendi)	7.54	11.2	2.72	2.49	68.23	Abd-Elaziz (2018)

6.2.2 *Mango Peel, Nutritional Profile and Industrial Utilization*

Mangoes are the highly cultivated fruits across the world that comprise of diverse mango varieties. Sindhari, Anwar Retual, Dusehri, Langra, Chaunsa, Kala Chaunsa, Sufaid Chaunsa, and Fajri are the different varieties of mangoes, exhibiting unique effect on the health of consumers. Pakistan is among the top five high mango producing countries that accounts for approximately 2782 in 1000 metric tons production in the annual year 2022, holding 12% of the GDP of Pakistan (Statista 2022). Mangoes, the king of fruits, are produced in the different regions of Pakistan including Multan, Hyderabad, Mirpur, Sukkur Khas, Karachi, Muzaffargarh, Nawab Shah, and Faisalabad. Multan is famous for its mango production and is ranked six among all the cities of Pakistan (Hussain et al. 2023).

The unique phytochemistry, nutritional profile, and sensory attributes (sweet and soothing taste, flavor, aroma, and color) of mangoes make it effective and acceptable for the consumers, maintaining the nutritional status as well as providing other health-promoting effects (Choudhary et al. 2023). Numerous bioactive components including carotenoids, flavonoids, polyphenols, and antioxidants (vitamin C, vitamin E, vitamin A, zinc, and selenium) are identified that demonstrate therapeutic properties for different disorders by scavenging free radicals, involved in the pathogenesis of metabolic and cardiovascular disorders (Jamil et al. 2015).

Mango peels are the recycled products that are constantly wasted due to the poor awareness about the phytochemical and nutritional composition. Annually, 15–25 million tons of mango peels and stones are generated, posing environmental challenges. Mango peels are rich in fiber, vitamins (C and E), phenolic compounds, and carotenoids. Converting them into powders is promising, with drying methods like infra-red and hot-air-drying preserving bioactive compounds. Incorporating mango peel powders in various foods enhances their nutritional content and antioxidant activity. Mango peel extracts also serve as co-pigments and lipid peroxidation inhibitors (Marcal and Pintado 2021). Imran et al. (2013) elucidated the nutritional profile of mango peel of chaunsa variety that depicted that its mango peel constitute 70% moisture, 2.2% protein, 2.3% fat, 5.0% fiber, 2.5% ash, and 87.8% nitrogen free extract (NFE), while the moisture, protein, fat, fiber, ash, and NFE contents are varied such as in Anwar Ratol (71%, 2.3%, 2.2%, 5.5%, 2.3%, and 87.6% respectively), Langra (69%, 2.2%, 2.2%, 4.8%, 2.2%, and 88.8% respectively), Dusheri (68%, 2.1%, 2.2%, 4.7%, 1.9%, and 89% respectively), and Desi contents (71%, 1.9%, 2.1%, 4.5%, 1.8%, and 89.6% respectively). Another study explored the nutritional and micronutrient composition of different varieties of mango peel and revealed diverse range of vitamin C, fat %, protein %, and total sugar concentrations. Vitamin C contents are highest in chaunsa (27 mg/100 g) followed by langra (24 mg/100 g), desi (13.8 mg/100 g), and sindhri (13.6 mg/100 g). Sindhri has highest fat (0.6%) and total sugar levels (6.9%), while langra has highest protein content that is 0.7% (Leghari et al. 2013). Mango peel also contains micronutrients that provide all the essential vitamins and minerals, and prevents the developing of micronutrient deficiencies and other associated disorders. The micronutrients found in the peel of chaunsa, Anwar Ratol, Langra, Dusheri, and Desi are potassium (K), magnesium (Mg), calcium (Ca), sodium (Na), chromium (Cr), copper (Cu), iron (Fe), and manganese (Mn) that are present in a significant quantity (Imran et al. 2013).

Mango peel has gained significant importance in the pharmaceutical, cosmetic, and nutraceutical industries to provide beneficial mango products as well as to minimize organic wastes. Mango peel is a rich source of dietary fibers, constituting both soluble and insoluble dietary fibers, and polyphenols. The ripen mango peel contains soluble dietary fibers in relatively higher concentrations (28%) than the insoluble dietary fibers (21%). Fresh raw peels of raspuri, badami, and chokanan contain respective 30 g, 42 g, and 38 g per 100 g insoluble fiber and 15 g, 21 g, and 20 g per 100 g soluble dietary fibers. However, their fresh ripe peels constitute 40 g, 50 g, and 45 g per 100 g insoluble fiber and 23 g, 28 g, and 11 g per 100 g soluble dietary fiber respectively (Marcal and Pintado 2021). The consumption of insoluble dietary fibers as well as the polyphenols are beneficial for intestinal health, improving beneficial microbes within the gut that ultimately strengthen immune system against infectious disorders. In contrast, soluble dietary fibers are involved in metabolizing carbohydrates and lipids. Pharmaceutical and nutraceutical industries employ mango peels in the synthesis of numerous drugs that significantly reduce the risk of health challenges and promote human health (Serna-Cock et al. 2016).

Mango peel constitutes pectin that accounts for approximately 13% of all ingredients due to which it finds its application in food processing industries as thickening, emulsifying, sweetening, flavoring, and gelling agent (Wongkaew et al. 2021). The dried powder form of mango peels is involved in baking industries to enhance the nutritional as well as phytochemical composition. The antioxidant rich products are more acceptable by the health-conscious individuals, preventing the deleterious effects on consumer's health (Marcal and Pintado 2021).

6.3 Mango Pulp and Its Nutritional Profile

Mango fruit is termed as indeliquescent drupe, which possess single large seed which is surrounded by fleshy mesocarp wrapped in a leathery skin. On industrial point of view, mango fruit has three parts of interest named as pulp, seed and peel. While conventionally with respect to mango utilization the main focus is on mango both industrially and domestically. Mango pulp is important intermediate for mango leather and beverages production. Mango concentrate and pulp demand is rising because of their utilization as base raw material in industry specially beverage industry, in the formulations of baby foods, and in the dairy industry as flavoring ingredient. The composition of mango pulp differs as it depends on various factors like maturity stage, variety, climate and locality. The pulp on average contains around vitamin C 27.9 mg, 1.85 g fiber, and vitamin A 3895 IU per 100 g of pulp with polyphenolic compounds in significant amounts.

A comprehensive characterization of minor phytochemicals and polyphenols of mango pulp was performed for first time in mango cultivar of 'Keitt', for determining apocarotenoids, hydroxybenzoic acids, flavonoids, and cinnamic acids. 2 monogalloyl glucosides were detected notably, ester-linked form being most predominant among the polyphenolics of non-tannin class. The impact assessment of commercial enzymes responsible for cell-wall degradation during the processing of mango pulp was also assessed, indicating the decrease in ester-linked MGG and ether-linked MGG complete hydrolysis (Krenek et al. 2014).

6.4 Mango Seed Kernal and Its Nutritional Profile

Due to reports that it is a biowaste rich in bioactive chemicals (carotenoids, dietary fibre, phenolic compounds and vitamin C) that are beneficial to human health, mango seeds have piqued special scientific interest. Rich in stearic acids and oleic, it offers an appealing profile of basic amino acid groups and fats (6–16%) and is a decent source of carbs (58–80%) and proteins (6–13%). Because of its strong antioxidant capacity, it has been found to exhibit antibacterial activity against both Gramme positive and Gramme negative bacteria as well as anticancer efficacy against cancers of the breast and colon. Its high tannin concentration has been linked to purported antidiarrheal effects. Furthermore, mango seed fat's physicochemical properties are remarkably similar to those possessed by commercial cocoa butter. Mango seed has a well-documented rational usage as an element in balanced animal feed; nonetheless, the majority of this byproduct is deemed waste and contributes to environmental damage. Mango seeds can be used responsibly to reduce waste and improve health while also lowering the negative effects on the environment and nutritional deficiencies. Mango seeds can also be used as a food ingredient and raw material. Currently, developments aim to produce zero waste, using the wastes produced as basic components for new goods and uses.

After analysis, the seeds of mango kernels from Nigeria revealed the following: 2.0% crude fibre, 2.0% ash, 0.39% tannin, 12.8% fat, 44.4% moisture, and 6.0% protein. The oleic (40.0%) and Stearic (46.3%) acids made up the majority of the fat content, with trace amounts of other fatty acids. Lower quantities of, sodium, iron, zinc, and manganese (less than 11.0 mg/100 g) were observed, while high levels of potassium (365 mg), phosphorus (140 mg), magnesium (100 mg), and calcium (49 mg) were detected per 100 g. Mango seed kernels show potential as a valuable nutrient source for human and animal feed. In another study *Mangifera indica* seed kernel from, Abuja, proximate, phytochemical, along with mineral analysis were performed. Analysis highlighted the moisture content at 5.93%, crude protein at 5.23%, carbohydrate at 76.12%, ash 2.45%, and crude fiber at 0.50%. Analysis of mineral composition revealed the presence of potassium at 116.5 ppm, sodium 23.9 ppm, magnesium 0.57 ppm, and calcium at 0.27 ppm. Phytochemical analysis detected the presence of glycosides, phenols, alkaloids, tannins, volatile oil, steroids, and resins. Presence of such compounds indicates the potential of the fruit in the treatment of various diseases (Joyce et al. 2014).

Mango kernels are produced annually in 123,000 metric tonnes, making them a common waste product of the fruit processing sector. It gives infections and other germs a medium for growth, degrades and spoils, and releases unpleasant odours as a result. Mango peels and seeds, however, are rich in a variety of nutritive and bioactive substances with high industrial potency. The main nutritional and bioactive components found in mango seeds are displayed in Fig. 6.2. These substances antiatherosclerotic, antidiabetic, antimutagenic, anticarcinogenic, and angiogenesis-inhibiting qualities offer enormous potential health benefits. Mango compounds are recognized as health-promoting and extra-nutritional compounds which belongs to a phytochemicals heterogeneous group which include tocopherols, polyphenols, ascorbic acid, phytosterols, and carotenoids. Due to advantageous effects against diseases like cancer and cardiovascular disease, such compounds are of particular interest in consumers and scientific community.

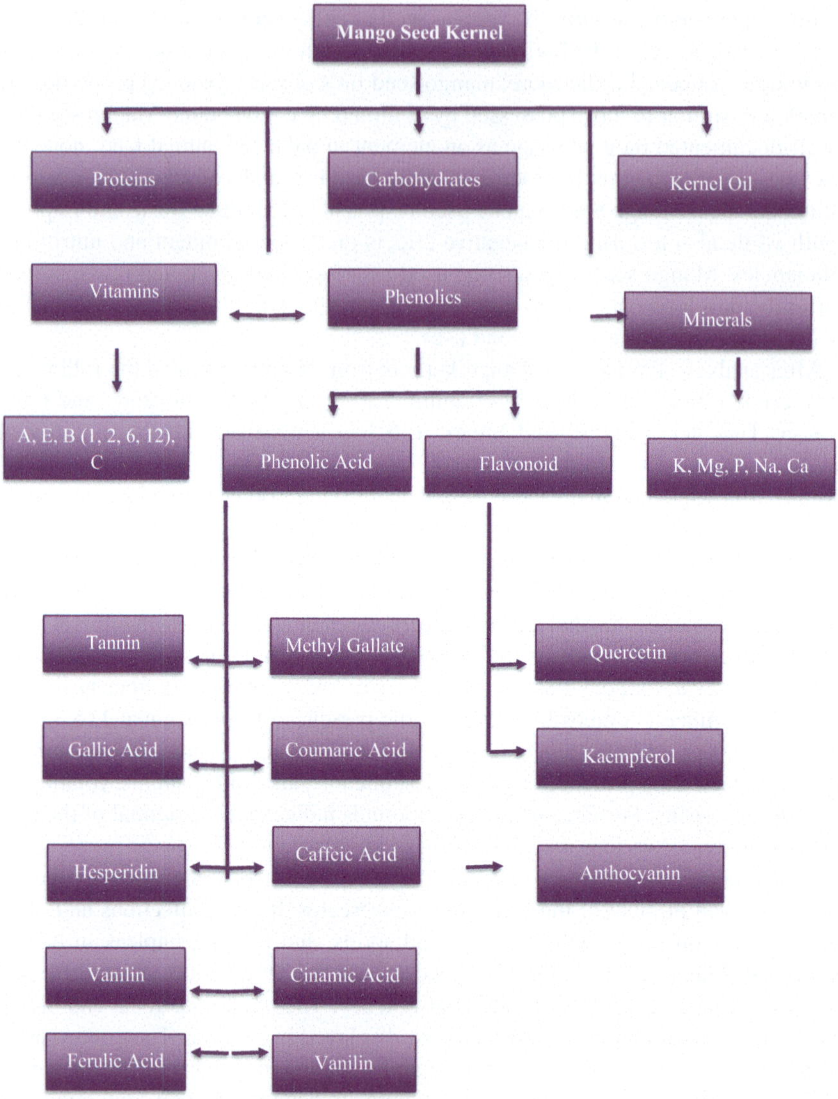

Fig. 6.2 Mango seed kernel nutrient and phytochemical profile

6.5 Few Major Mango Varieties Globally

6.5.1 Julie Mango

Julie mangoes are renowned for their sweet and tangy flavor, coupled with a firm texture. They are abundant in vitamin C, providing a powerful boost to the immune system. Julie mangoes also contain significant levels of beta-carotene, promoting healthy vision and skin. 'Julie' mango, also known as 'Saint Julian', mango cultivar was made popular in Caribbean. It is considered to be one of the best mangoes in the Caribbean (Kirker and Newman 2024).

Few studies aimed to investigate the nutritional compositions and active compounds in Julie varieties seed kernels with medicinal properties. As per GC–MS analysis around 13 different compounds were observed in Julie variety. Compounds like 6-Octadecenoic acid, Octadecanoic acid, n-Hexadecanoic acid and 1, 2, 3-Benzenetriol were present with peak percentage of 5.41%, 16.05% and 39.76%, respectively (Ubwa et al. 2014). The carbohydrates content for the fruit was around 70.22% and fatcontent in the fruit was on the higher side as compared to other mango varieties at 9.23%. The moisture content was observed at 9.72%, ash content 2.96%, content of crudefibre at 1.90% and proteincontent at 6.05%. Mineral composition of Julie fruit was also observed as 4 micro and 5 macro nutrients were detected. Phosphorus content was 1612 ppm, (Ca) at 4596 ppm, (Mg) content at 1985.50 ppm, (K) 1536 ppm, (Zn) 7.65 ppm, (Na)25.85 ppm, (Cu) 4.25 ppm, (Fe) 75.62ppmand (Mn) 57.95ppmwere identified. Julie seed kernels possess bioactive compounds in abundance with medicinal properties (Nwachukwu et al. 2022). Proximate composition of this and the following varieties are given in the tables.

6.5.2 Nam Doc Mai Mango

Nam Doc Mai mangoes are prized for their unique flavor profile cultivated mostly in Thailand, characterized by hints of floral and tropical notes. They are a rich source of vitamin E, an antioxidant that helps protect cells from damage and supports skin health. Nam Doc Mai mangoes also provide a generous amount of vitamin K, essential for blood clotting and bone health. This variety is of particular interest because of its natural fragrance. Since this variety worldwide introduction, Nam Doc Mai cultivar is considered top-quality dessert and is grown commercially for fresh consumption. Growers favor Nam Doc Mai for its productive nature, disease resistance, and consistent fruiting, and multiple variations of original strain is being developed in breeding, especially Golden Flower Water and Nam Dok Mai. 4 mangoes. Despite their prized characteristics, Nam Doc Mai become soft upon ripening, thus creating challenges in commercial cultivation shipping. The variety is notorious for dropping when almost ripe from the tree (Khiatthong 2024).

6.5.3 *Carabao Mango*

Carabao mangoes, also known as Manila mangoes, 3rd major fruit produce in Philippines, are prized for their creamy texture and rich, tropical flavor. They are an excellent source of vitamin A, supporting vision health and immune function. Carabao mangoes also provide a good amount of potassium, aiding in heart health and blood pressure regulation. The variety is rich in β-carotene, which ranges from 820 to 13,500 μg per 100 g, and around 60% of carotenoids present in this specie is β-carotene. It also possess ascorbic acid, which ranges from 7.9 to 175 mg per 100 g of the ripe mango pulp (Guillermo et al. 2019).

6.5.4 *Irwin Mango*

Mangoes from Irwin are renowned for their sweet, fragrant flavour and silky, fiberless texture. They are rich in antioxidants like kaempferol and quercetin, which lessen inflammation and shield cells from oxidative damage. Mangoes from Irwin are also a good choice for folate, which is necessary for DNA synthesis and cell division. The variety's total carotenoid concentration was found to be 3.8 mg per 100 g in one research (Maldonado-Celis et al. 2019).

A few volatile chemicals found in the sap of the Irwin mango cultivated in Pakistan were trans-caryophyllene, α-pinene, γ-terpinene, α-humulene,α-phellandrene, sabinene and 3-carene. The Irwin variety is said to have the highest concentration of 3-carene (Musharraf et al. 2016). While the enzyme functions of sucrose synthase decline at five embryonic stages, the accumulation of sucrose in Irwin is associated with a decrease in sucrose phosphate synthase. This activity suggests that Irwin cultivars be harvested 110 days following the anthesis, or when they achieve physiological maturity. (Wongmetha et al. 2015).

6.5.5 *Palmer Mango*

Palmer mangoes are prized for their firm texture and rich, sweet flavor. They are packed with vitamin C, providing powerful antioxidant protection and immune support. Palmer mangoes also contain folate, promoting healthy cell growth and development. In a study conducted by Akin-Idowu et al. (2020) it was observed that palmer mango has highest length among the ten studied varieties of mango. Its length was observed at 16.72 cm while the 2nd highest was observed at 16 cm for the Kent variety. Palmer variety also boosts highest mineral/ash content which was observed at 2.68 g per 100 g along with the highest fat content noted at 0.93 g per 100 g. Palmer variety moisture content is among the lowest which is recorded at 77.85% and because of its low moisture content as compared to other varieties, palmer also has higher shelf life. In the similar study antioxidant evaluation was also conducted where the highest activity for the total flavonoid (TF) was observed in palmer variety at 48 mg per 100 g, while the second best was way behind at 23 mg per 100 g in Saigon variety.

References

Abbasi AM, Guo X, Fu X, Zhou L, Chen Y, Zhu Y et al (2015) Comparative assessment of phenolic content and in vitro antioxidant capacity in the pulp and peel of mango cultivars. Int J Mol Sci 16:13507–13527. https://doi.org/10.3390/ijms160613507

Abdelaziz SA (2018) Physicochemical characteristics of mango kernel oil and meal. Middle East J Appl Sci 8:01–06

Ajila CM, Bhat SG, Rao UP (2007) Valuable components of raw and ripe peels from two Indian mango varieties. Food Chem 102:1006–1011

Akin-Idowu PE, Adebo UG, Egbekunle KO, Olagunju YO, Aderonmu OI, Aduloju AO (2020) Diversity of mango (*Mangifera indica* L.) cultivars based on physicochemical, nutritional, antioxidant, and phytochemical traits in South West Nigeria. Int J Fruit Sci 20–S376

Choudhary P, Devi TB, Tushir S, Kasana RC, Popatrao DS, K N (2023) Mango seed kernel: a bountiful source of nutritional and bioactive compounds. Food Bioproc Technol 16:289–312

Ediriweera MK, Tennekoon KH, Samarakoon SR (2017) A review on ethnopharmacological applications, pharmacological activities, and bioactive compounds of *Mangifera indica* (mango). Evid Based Complement Alternat Med 2017:6949835

Elegbede JA, Achoba II, Richard H (1995) Nutrient composition of mango (*Mangnifera indica*) seed kernel from Nigeria. J Food Biochem 19:391–398

FAOSTAT (2023) Crops and livestock products

Guillermo MCS, Naciongayo DS, Galela MGC, Dioquino CP, Bayogan EMV, Manliguez CG (2019) Determining 'Carabao' mango ripeness stages using three image processing algorithms. Banwa B 14

Hussain T, Ishttiaq M, Zafar B, Parveen A, Maqbool M, Khan FA (2023) Morphometric characterization of different mango varieties cultivated in tehsil Bernala district Bhimber Azad Kashmir, Pakistan. Pak J Bot 55:1805–1812

Hussain SZ, Naseer B, Qadri T, Fatima T, Bhat TA (2021) Mango (*Mangifera indica*): morphology, taxonomy, composition and health benefits. In: Fruits grown in highland regions of the Himalayas. Springer, pp 245–255

Idris A (2012) Nutritive value of differently processed mango seed kernel

Ifesan BOT (2017) Chemical properties of mango kernel and seed and production of biscuit from wheat-mango kernel flour blends. Int J Food Nutr Res 1–5

Imran M, Butt MS, Anjum FM, Sultan JI (2013) Chemical profiling of different mango peel varieties. Pak J Nutr 12:934

Jamil W, Ahmad S, Ahmad M, Ali S, Abbas MM (2015) Morphophysiological and biochemical profiling of some mango cultivars in Pakistan. J Agric Res 53:3

Joyce OO, Latayo BM, Onyinye AC (2014) Chemical composition and phytochemical properties of mango (*Mangifera indica*) seed kernel. Int J Adv Chem 2:185–187

Kabir Y, Shekhar HU, Sidhu JS (2017) Phytochemical compounds in functional properties of mangoes. In: Handbook of mango fruit. Academic Press, pp 237–254

Kaur A, Brar JK (2017) Use of mango seed kernels for the development of antioxidant-rich biscuits. Int J Sci Res 6:535–538

Khiatthong T (2024) Nam Dok Mai mango product identity study and development in a collaborative farming system in Samut Prakan Province, Thailand. J Southwest Jiaotong Univ 59

Kim H, Castellon-Chicas MJ, Arbizu S, Talcott ST, Drury NL, Smith S, et al. (2021) Mango (*Mangifera indica* L.) polyphenols: anti-inflammatory intestinal microbial health benefits, and associated mechanisms of actions. Molecules 26:2732

Kirker CL, Newman M (2024) Mango: a global history. Reaktion Books

Krenek KA, Barnes RC, Talcott ST (2014) Phytochemical composition and effects of commercial enzymes on the hydrolysis of gallic acid glycosides in mango (*Mangifera indica* L. cv. 'Keitt') pulp. J Agric Food Chem 62:9515–9521

Kumar M, Saurabh V, Tomar M, Hasan M, Changan S, Sasi M, et al. (2021) Mango (*Mangifera indica* L.) leaves: nutritional composition, phytochemical profile, and health-promoting bioactivities. Antioxidants 10:299

Lebaka VR, Wee YJ, Ye W, Korivi M (2021a) Nutritional composition and bioactive compounds in three different parts of mango fruit. Int J Environ Res Public Health 18:741

Leghari MH, Sheikh SA, Soomro AH, Khooharo AA (2013) Quality attributes of immature fruit of different mango varieties. J Basic Appl Sci 9:52–56

Maldonado-Celis ME, Yahia EM, Bedoya R, Landázuri P, Loango N, Aguillón J, et al (2019) Chemical composition of mango (*Mangifera indica* L.) fruit: nutritional and phytochemical compounds. Front Plant Sci 10:1073

Marcal S, Pintado M (2021) Mango peels as food ingredient/additive: nutritional value, processing, safety and applications. Trends Food Sci Technol 114:472–489

Musharraf SG, Uddin J, Siddiqui AJ, Akram MI (2016) Quantification of aroma constituents of mango sap from different Pakistan mango cultivars using gas chromatography triple quadrupole mass spectrometry. Food Chem 196:1355–1360

Mutua JK, Imathiu S, Owino W (2017) Evaluation of the proximate composition, antioxidant potential, and antimicrobial activity of mango seed kernel extracts. Food Sci Nutr 5:349–357

Nwachukwu KC, Ugbogu OC, Nwachukwu NC (2022) Assessment of biologically active components and nutritional contents of seed kernels of Opioro and Julie mango varieties: a right step for drug discovery. Asian J Biol 16:1–10

Nzikou JM, Kimbonguila A, Matos L, Loumouamou B, Pambou-Tobi NPG, Ndangui CB et al (2010) Extraction and characteristics of seed kernel oil from mango (*Mangifera indica*). Res J Environ Earth Sci 2:31–35

Okoth DM, Sila EM, Onyango CA, Owino WO, Musyimi SM, Mathooko FM (2013) Evaluation of chemical and nutritional quality attributes of selected mango varieties at three stages of ripeness, grown in lower Eastern province of Kenya–part 2

Rajasekaran A, Soundarapandian S (2023) Nutritional and medicinal values of *Mangifera indica* L. fruit. Trends Hortic 6:2949

Rather AMU, Hajam MA, Bhat MSA, Malik MI (2023) Horticulture: principles and practices. Academic Guru Publishing House

Ribeiro SMR, Schieber A (2010) Bioactive compounds in mango (*Mangifera indica* L.). In: Bioactive foods promot health. Academic Press, pp 507–523

Serna-Cock L, García-Gonzales E, Torres-León C (2016) Agro-industrial potential of the mango peel based on its nutritional and functional properties. Food Rev Int 32:364–376

Solís-Fuentes JA, del Carmen Durán-de-Bazúa M (2011) Mango (*Mangifera indica* L.) seed and its fats. In: Nuts and seeds in health and disease prevention. Academic Press, pp 741–748

Statista (2022) Mango, mangosteen, and guava production in the Asia-Pacific region in 2022, by country or territory

Tacias-Pascacio VG, Castañeda-Valbuena D, Fernandez-Lafuente R, Berenguer-Murcia Á, Meza-Gordillo R, Gutiérrez LF, et al (2022) Phenolic compounds in mango fruit: a review. J Food Meas Charact 1–18

Ubwa ST, Ishu MO, Offem JO, Tyohemba RL, Igbum GO (2014) Proximate composition and some physical attributes of three mango (*Mangifera indica* L.) fruit varieties. Int J Agron Agric Res 4:21–29

Wongkaew M, Kittiwachana S, Phuangsaijai N, Tinpovong B, Tiyayon C, Pusadee T et al (2021) Fruit characteristics, peel nutritional compositions, and their relationships with mango peel pectin quality. Plants 10:1148

Wongmetha O, Ke LS, Liang YS (2015) The changes in physical, biochemical, physiological characteristics and enzyme activities of mango cv. Jinhwang during fruit growth and development. NJAS Wageningen J Life Sci 72:7–12

Yadav D, Pal AK, Singh SP, Sati K (2022) Phytochemicals in mango (*Mangifera indica*) parts and their bioactivities: a review. Crop Res 57:79–95

Zahoor S, Anwar F, Qadir R, Soufan W, Sakran M (2023) Physicochemical attributes and antioxidant potential of kernel oils from selected mango varieties. ACS Omega 8:22613–22622

Chapter 7
Understanding the Physiology of Mango Trees in the Context of Climate Change

Uzman Khalil, Shah Rukh Khan, Abdullah Jamil, and Ghulam Mustafa

Abstract Climate change significantly affects the physiology of mango trees, influencing their growth, flowering, fruiting, and overall productivity. Rising temperatures, fluctuating rainfall patterns, and increased frequency of extreme weather events disrupt the delicate balance of physiological processes such as photosynthesis, transpiration, and nutrient uptake. Prolonged heat stress can impair pollination and fruit set, while irregular water availability impacts root function and nutrient assimilation. Additionally, shifts in phenology, such as altered flowering and fruiting periods, may result in reduced yield and quality. Understanding the physiological responses of mango trees to these climatic stressors is crucial for developing adaptive management strategies. Approaches such as selecting climate-resilient cultivars, optimizing irrigation practices, and enhancing soil health through organic amendments can mitigate the adverse effects of climate variability. This knowledge provides a foundation for improving mango tree resilience, ensuring sustainable production, and meeting future food security demands amidst a changing climate.

Keywords *Mangifera indica* · Heat stress · Drought · Salinity · Frost

U. Khalil (✉)
Horticultural Sciences Department, University of Florida, Gainesville, FL, USA
e-mail: ukhalil.chaudhry@ufl.edu

S. R. Khan
Plant Molecular and Cellular Biology, University of Florida, Gainesville, FL, USA

A. Jamil
Department of Agronomy, University of Florida, Gainesville, FL, USA

G. Mustafa
Department of Horticulture, MNS-University of Agriculture, Multan, Pakistan

S. A. H. Naqvi et al. (eds.), *Climate Change and Mango Production*,
https://doi.org/10.1007/978-3-031-87035-4_7

7.1 Introduction

Mango (*Mangifera indica* L.) is among the most economically significant and widely cultivated fruit crops in tropical and subtropical regions, valued for its taste, nutritional properties, and cultural importance. With its origins in South Asia, mango cultivation has expanded to over 100 countries, including major producers such as India, China, Indonesia, Pakistan and Mexico. Pakistan ranked fifth among the global mango producers, harvesting approximately 2.3 million tons annually, but productivity remains far below its potential. Despite its resilience to many environmental challenges, mango cultivation faces unprecedented threats from climate change, which alters critical climatic variables such as temperature, rainfall patterns, and atmospheric carbon dioxide (CO_2) concentrations (Antwi-Boasiako et al. 2024).

Mango cultivation thrives on stable environmental factors, including defined temperature ranges, regular rainfall, and distinct dry and wet seasons. Rising atmospheric CO_2 levels, irregular rainfall patterns, and extreme temperature fluctuations disrupt these critical phenological processes, as evidenced by global studies. For instance, higher temperatures are associated with premature flowering, irregular fruit set, and accelerated ripening, ultimately reducing fruit quality and postharvest life (Halder et al. 2024). As indicated by Halder and Hasan (2020), rising CO_2 levels enhance photosynthesis under certain conditions but also lead to respiratory losses, drought stress, and reduced flowering when coupled with high temperatures. Additionally, water stress impacts flowering uniformity, a challenge exacerbated in regions with inadequate irrigation practices (Cavalcante 2022).

The mango tree's physiology thrives under specific climatic conditions characterized by moderate temperatures, consistent rainfall, and distinct dry and wet seasons. Climate change disrupts these conditions, resulting in higher atmospheric CO_2 levels, irregular rainfall, and extreme temperature fluctuations, which affect plant physiological processes, flowering, fruit set, and overall productivity (Cavalcante 2022). Elevated temperatures are associated with premature flowering and irregular fruit development, while prolonged droughts and flooding impact water uptake and nutrient mobilization, impairing photosynthesis and respiration (Ogunbode et al. 2024). Roots can reach depths exceeding 5 m, facilitating drought resistance; however, flowering is contingent on specific environmental cues, such as prolonged water stress or cool temperatures, depending on geographic location (Carr 2014). This indicates a fragile dependency on climatic stability for reproductive success. Furthermore, high temperatures (>30 °C) disrupt floral induction and vegetative development. Conversely, moderate warmth enhances fruit ripening and biochemical quality, reflecting a nuanced interplay between heat and productivity. Moreover, abiotic stresses like salinity and low temperatures further hinder physiological processes essential for reproductive success.

Adaptations such as increased stomatal regulation, changes in enzymatic activity, and altered hormone levels are evident in mango trees' response to these stressors. However, such responses are not always sufficient to sustain high productivity under extreme climatic conditions. For instance, elevated CO_2 levels may initially enhance photosynthesis but often result in nutrient imbalances and lower fruit quality (Halder et al. 2024).

Water management is a critical factor in mitigating climate change impacts. In regions like semi-arid northeast Brazil, precision irrigation techniques, such as micro-sprinklers and drip irrigation, have been shown to maintain optimal water productivity. Mature mango trees require 4–5 mm of water daily, peaking at 6 mm under high transpiration conditions. However, water deficits during flowering and early fruit development can adversely affect fruit retention and quality. Conversely, regulated water stress can enhance sugar concentration and dry matter in fruits, improving quality even at the cost of reduced yield (Carr 2014).

Temperature extremes also influence the vegetative and reproductive cycles of mango trees. While moderate warmth accelerates ripening and enhances fruit quality, excessive heat (>30 °C) reduces inflorescence size, impairs pollen viability, and induces physiological disorders like spongy tissue in certain cultivars. These disruptions underscore the need for adaptive practices, including selective pruning, nutrient optimization, and flowering synchronization to stabilize production amid climatic variability (Halder and Hasan 2020).

This chapter explores the complex interplay between mango tree physiology and climatic factors, emphasizing innovative approaches to enhance resilience and ensure sustainable production despite global climate challenges.

7.2 Physiology of Mango Trees

Mango trees demonstrate a complex and well-adapted physiology under normal environmental conditions, enabling them to thrive across a wide range of tropical and subtropical climates. Their physiological processes are governed by the interaction of structural components such as roots, trunk, leaves, flowers, and fruits, alongside critical processes like photosynthesis, respiration, and transpiration (Figs. 7.1 and 7.2).

7.2.1 Plant Structures (Roots, Trunk, Leaves, Flower and Fruit)

Root physiology plays a crucial role in the overall health and survival of plants, particularly in response to environmental stressors such as desiccation. The roots are responsible for water and nutrient uptake, and their adaptation to varying moisture levels is vital for plant resilience. In the context of mango trees, understanding root physiology is essential for improving their endurance to drought conditions. Research indicates that root depth and spread significantly influence a tree's ability to access water, especially during dry spells. Deeper roots can tap into moisture reserves that

Fig. 7.1 Pictorial view of healthy mango tree, grown under sub-tropical climate

shallower roots cannot reach, thereby enhancing the tree's survival chances in arid environments. Additionally, physiological responses of roots, such as changes in growth patterns and morphology, are critical for optimizing water absorption under stress. The efficiency of water uptake is influenced by root thickness and surface area, which can vary among different plant species. Furthermore, the communication between roots and shoots is vital for coordinating the plant's response to water availability, ensuring that resources are managed effectively (Taria et al. 2022).

The trunk serves as the central axis of the mango tree, facilitating the transport of water, nutrients, and photoassimilates between the roots and aerial parts. Previous models evaluating the diameter of current-year branches suggest that trunk

Fig. 7.2 Pictorial view of a mango tree under severe stress

and branch size are closely associated with xylem conductivity, which constitutes the majority of branch mass. For instance, larger sieve tube elements and more extensive sieve areas at the trunk base enhance transport efficiency, particularly in taller trees. Studies on Quercus species revealed larger sieve tube elements in older trunks, facilitating better transport to basal regions; however, comparative data in mango remains limited. The trunk also displays scaling relationships in vascular elements, with the phloem playing a critical role in transporting photo assimilates to growing tissues such as fruits, flowers, and meristems (Galán Saúco 2009; Olson and Rosell 2013; Jensen et al. 2014; Olson et al. 2020).

Mango leaves are evergreen and monopodial, developing primarily on terminal branches. These leaves play a vital role in photosynthesis and support the production of carbohydrates required for fruiting. The geometry of xylem and phloem vessels within the leaves has a direct impact on leaf mass and photosynthetic efficiency. For example, mango trees grown in subtropical climates with low winter temperatures exhibit reduced leaf mass compared to those in tropical environments. Studies have shown that leaf mass is closely linked to branch vigour, as influenced by the hydraulic traits of the vascular system (Heo et al. 2014; Pace et al. 2015; Barceló-Anguiano et al. 2021).

Mango inflorescences arise on terminal branches, developing on the youngest growth. The vascular tissues, particularly phloem, play a crucial role in providing the necessary nutrients for floral development. The efficiency of phloem transport, determined by sieve tube geometry and scaling, directly impacts flowering and subsequent fruit set. Although the phloem has been studied in forest trees and some fruit crops, there is limited research on its structure and function in mango trees, despite its importance for reproductive success and overall productivity (Normand et al. 2008; Galán Saúco 2009).

The development of mango fruits is closely linked to the phloem's efficiency in transporting photoassimilates, water, and nutrients. The phloem operates under the pressure flow hypothesis, where the scaling of sieve tubes and sieve plate pores directly influences sap flow resistance. Recent findings suggest that smaller, compact mango trees in subtropical climates—often shaped by low temperatures and pruning—exhibit unique phloem-to-xylem ratios. This ratio is correlated with dwarfism, a desirable trait in mango cultivation for its ease of management and potential to maintain high productivity. Moreover, studies on other woody perennials have shown that scaling relationships between sieve tube geometry and branch size significantly affect fruit development and quality (Savage et al. 2017; Boudon et al. 2020).

7.2.2 Photosynthesis, Respiration and Transpiration

The photosynthetic efficiency of mango cultivars plays a pivotal role in identifying suitable varieties for breeding programs aimed at improving productivity, particularly for specific environmental conditions (Lu et al. 2012). High photosynthetic performance during flowering and fruiting has been recognized as a key trait in enhancing mango productivity (Jain and Priyadarshan 2009). However, photosynthesis and related assimilate production are profoundly influenced by interactions between genotype and environmental factors, underscoring the need for a comprehensive understanding of these dynamics in field conditions (Lu et al. 2012).

The mango tree's inconsistent cropping behavior has been attributed to evolutionary adaptations to its natural habitat of hot, dry monsoonal environments, which can impact energy and nutrient dynamics. While fruit production in mangoes is only moderately demanding in terms of energy and minerals, extended periods of heat and moisture stress can elevate respiratory activity, reducing the annual photoassimilate surplus necessary for consistent yields (Lu et al. 2012). Tropical climates, in particular, present environmental challenges to photosynthetic efficiency, such as seasonal variations in light intensity and moisture availability. For example, a study on the polyembryonic cultivar Kensington Pride (KP) in Northern Australia reported significant seasonal variations in net photosynthesis, with notable declines during the wet monsoon compared to the dry season (González and Blaikie 2003). Despite such findings, there remains a lack of extensive field-based research connecting environmental factors, photosynthesis, and the productivity of diverse mango cultivars.

The respiration and transpiration processes in mangoes, especially in the Alphonso variety, are pivotal for maintaining fruit quality and prolonging shelf-life postharvest. After harvest, mangoes remain metabolically active, and their physiological activities—primarily respiration and transpiration—play crucial roles in preserving the integrity of fruit tissues and overall quality. Respiration is a metabolic process where organic compounds are broken down in the presence of oxygen (O_2), producing carbon dioxide (CO_2), water (H_2O), and energy. This process is critical for designing storage systems for mangoes, as it directly affects their physiological state and storage performance (Nakamura et al. 2004). Factors such as temperature, relative humidity, and gas composition significantly influence the respiration rate, determining the fruit's postharvest life and quality. Proper management of these factors in storage conditions can reduce postharvest losses, extend shelf life, and enhance marketability (Peiris et al. 1997). A controlled atmosphere storage system, designed with insights into the rates of CO_2 production and O_2 consumption, is essential for maintaining optimal storage conditions for mangoes (Nakamura et al. 2004). Respiration generates heat, which, if unmanaged, can raise the fruit's temperature, leading to increased transpiration and subsequent moisture loss. This heat generation varies among crops and within mango varieties and is highest at the beginning of the storage period, gradually decreasing under steady-state conditions. Temperature management is particularly critical. Higher storage temperatures lead to elevated respiration rates, increasing physiological weight loss due to moisture loss from the fruit surface. Studies reveal that ambient storage conditions result in significant weight loss compared to cooler conditions, such as storage at 10 °C (Prajapati et al.). Additionally, moisture released during respiration may condense on the inner surfaces of storage chambers, further emphasizing the need for effective temperature and humidity control (Peiris et al. 1997).

7.3 Adaptation to High Temperature

Mango trees exhibit several physiological and biochemical adaptations to high-temperature environments, enabling them to maintain growth and productivity under heat stress. These adaptations include alterations in metabolic pathways, antioxidant defense mechanisms, and morphological changes.

Mango is a tropical fruit that faces significant challenges in Mediterranean climates, particularly due to rising temperatures. The ability of mango trees to tolerate high temperatures is noteworthy, as they can withstand short periods of heat up to 48 °C without severe damage (Rajan 2012). However, temperatures exceeding 45 °C can adversely affect their photosynthetic machinery, leading to concerns about plant health and productivity (Legave et al.). Rising temperatures increase air saturation with water vapor, intensifying water evaporation from fruit surfaces, which results in reduced yields and accelerated ripening (Henson 2011). High temperatures also induce various natural, anatomical, physiological, and biological changes in fruit tissues. For instance, prolonged exposure to high temperatures speeds up fruit ripening, as seen in avocados, by enhancing the activity of cell wall-degrading enzymes such as cellulase and polygalacturonase (Kays 1997). In apples, temperatures above 40 °C exacerbate issues like sunburn, water core, and rapid fruit softening (Woolf et al. 1999). The phenological behavior of mango trees is significantly influenced by temperature, with studies indicating that greenhouse cultivation results in earlier flowering and fruit ripening compared to open-air conditions. This phenomenon is attributed to the elevated temperatures experienced during the initial reproductive stages, rather than a general acceleration of growth throughout the season (Scuderi et al. 2023).

The dynamics of fruit growth under high-temperature conditions reveal that mango fruits exhibit a pattern of shrinking during the hottest parts of the day and expanding at night when temperatures drop. This fluctuation is a direct response to thermal stress, highlighting the physiological adaptations of the mango tree to cope with extreme heat (Legave et al.). Furthermore, while greenhouse cultivation allows for earlier harvests, which can be economically beneficial, it also subjects the plants to potential thermal stress during hot summers. This stress can threaten the overall health and productivity of the mango trees, raising questions about the sustainability of such cultivation practices in the long term (Scuderi et al. 2023).

Mango trees exhibit significant morphological and physiological changes to adapt to high-temperature stress, ensuring survival and productivity. Morphologically, the leaves of mango trees orient themselves to minimize direct exposure to intense sunlight, reducing heat absorption and water loss. This is complemented by structural adaptations such as thicker cuticles and reduced stomatal density, which lower transpirational water loss while maintaining essential gas exchange. Physiologically, high temperatures trigger the activation of the γ-aminobutyric acid (GABA) shunt pathway, a key metabolic response that modulates carbon and nitrogen metabolism to support cellular stability under heat stress. Enhanced proline accumulation, a common osmoprotectant, aids in maintaining cell turgor and stabilizing proteins and membranes under elevated temperatures. Furthermore, the synthesis of heat-shock proteins (HSPs) provides additional protection by preventing the denaturation of vital enzymes and ensuring cellular homeostasis. Together, these morphological and physiological adjustments highlight the remarkable resilience of mango trees in adapting to thermal extremes, a trait that is crucial for sustaining mango production in warming climates (Khanum et al. 2020).

To mitigate the adverse effects of high temperatures, various management techniques can be employed, including shading, roof whitening, and misting systems. However, these methods may compromise the economic viability of greenhouse systems when compared to traditional open-air cultivation. Overall, understanding the growth dynamics and physiological responses of mango trees to high temperatures is crucial for developing effective cultivation strategies in Mediterranean climates, ensuring both plant health and economic sustainability for growers (Scuderi et al. 2023).

7.4 Response to Drought Conditions and Flooding

Mango trees display a dualistic physiological and morphological response to drought and flooding, reflecting their adaptability to extreme water stress conditions. These responses are crucial for sustaining productivity and survival in varying climates, particularly under the influence of climate change.

Drought stress affects mango trees primarily by disrupting their gas exchange processes. Studies reveal that stomatal closure is the first line of defense to conserve water, reducing stomatal conductance (gs) and net photosynthesis (Pn). However, prolonged water deficits lead to downregulation of photosynthetic machinery, impacting maximal net photosynthesis (Amax) and photosynthetic electron transport rates (Jmax). Previous research demonstrated that these reductions in photosynthesis are not due to photodamage but rather the result of sink limitation, as indicated by an increased hexose-to-sucrose ratio. Recovery of these parameters upon rehydration highlights the reversible nature of these changes (Damour et al. 2009).

Biochemical adjustments are critical for drought tolerance. Osmotic regulation through the accumulation of proline, soluble sugars, and other osmolytes maintains cellular turgor and protects against dehydration-induced damage. Proline acts as an Osmo protectant, stabilizing proteins and membranes. Researcher observed that drought-tolerant mango cultivars like 'Choke Anand' accumulate higher proline levels compared to less tolerant cultivars such as 'Khieo Sawoei,' ensuring better recovery post-drought (Elsheery and Cao 2008).

Mango trees' extensive root systems enable them to extract water from deep soil layers, ensuring survival during prolonged dry periods. Researchers highlighted that root can penetrate up to 5 m, providing access to groundwater reserves even in arid conditions. However, severe droughts still negatively impact flowering, fruit set, and fruit size, primarily by reducing the translocation of nutrients and water to developing fruits. Drought also accelerates leaf senescence, further limiting the tree's productive capacity (Carr 2014).

In recent years, the increasing risk of flooding due to climate change driven by global warming has posed significant challenges to mango orchards in tropical regions. As perennial crops, fruit trees require several years to reach fruit-bearing maturity, making it difficult to restore fields affected by flooding to their original productivity (Hara et al. 2014). Consequently, sustainable fruit cultivation becomes challenging in areas frequently experiencing flooding, and even regions currently unaffected may see reduced cultivation due to future flood risks.

Flooding has been shown to reduce the photosynthetic rate in the leaves of many tree species (Schaffer et al. 2006). For example, in passion fruit trees, increased soil moisture content leads to lower photosynthetic rates and a reduction in root starch reserves (Shimada et al. 2018). While similar reductions in photosynthesis have been observed in mango leaves under flooded conditions, there is limited research on how flooding impacts root carbohydrate metabolism in mango trees (Larson et al. 1991). Additionally, submerging the underground parts of trees has been reported to induce anaerobic respiration, contributing to the accumulation of ethanol, acetaldehyde, and ethylene (Larson et al. 1993). Although instances of tree mortality under flooding have been documented, the exact conditions leading to death remain poorly understood.

To adapt, mango trees develop morphological features like adventitious roots and hypertrophied lenticels, which facilitate oxygen diffusion in flooded environments. Ethylene production, triggered by low oxygen availability, plays a key role in lenticel hypertrophy. (Larson et al. 1993) reported that these adaptations significantly enhance the survival of flooded trees by maintaining aerobic respiration in submerged tissues . However, prolonged flooding reduces vegetative growth and fruit production due to sustained nutrient deficiencies and limited gas exchange.

Given that most fruit trees are propagated vegetatively through grafting or cutting, the rootstock plays a significant role in determining flood tolerance (Tamura et al. 1995). For grafted fruit trees, flood resistance largely depends on the rootstock used (Arbona et al. 2008). While certain wild mango species exhibit higher tolerance to flooding (Eiadthong et al. 1999), the precise mechanisms underlying tree mortality in response to submersion are yet to be fully explained.

7.5 Adaptation to Elevated CO_2

Carbon dioxide (CO_2), the most prominent greenhouse gas, is projected to reach atmospheric concentrations of 460–560 ppm by 2050, accompanied by a temperature increase of 2.6–4.8 °C. These changes have significant implications for plant physiology, particularly photosynthesis, which is influenced by CO_2 levels, light, and temperature. Elevated CO_2 enhances photosynthesis by increasing carbon assimilation; however, this effect may be counterbalanced by rising temperatures, which can reduce the efficiency of photosynthetic processes. Elevated temperatures are expected to accelerate plant growth and development, potentially delaying the cessation of growth in perennial plants during early winter. It has been observed that mango trees grown at a day/night temperature of 30/25 °C exhibited a threefold increase in mean leaf mass compared to trees grown at 20/15 °C (Whiley et al. 1989). In contrast there was no significant effect of temperature on leaf mass within the daily temperature range of 20–28 °C. This discrepancy highlights the need for further research to elucidate the interaction between temperature and plant growth. Despite these findings, there is a notable lack of studies on the combined effects of elevated CO_2 and temperature on mango, particularly the cultivar Alphonso. Understanding these dynamics is critical to assessing how mango plants will respond to future climatic conditions and to developing strategies for sustainable cultivation in the face of climate change (Dambreville et al. 2012).

7.6 Adaptation to Salinity

Salinity significantly impacts plant growth by reducing water absorption, leading to rapid declines in growth rate, ion imbalances, and ion toxicity. In India, approximately 4.10 million hectares of land have been reported as saline, with affected regions spanning the Indo-Gangetic plains, arid areas, and coastal zones. This issue continues to grow due to the mismanagement of canal irrigation systems and the use of brackish groundwater for irrigation (Pandey et al. 2014). Scientists identified soil properties as a key factor limiting mango growth, with optimal soil and water conditions being a pH of 7.86 and 7.46, respectively, and an EC of 0.49 mmhos/cm and 0.50 mmhos/cm, respectively. Beyond these thresholds, the establishment of mango orchards becomes challenging (Gupta and Sen 2003).

Salt stress has been shown to significantly hinder seedling growth, reduce chlorophyll content, impair CO_2 assimilation, and disrupt nutrient uptake and ion balance. Mango trees are highly sensitive to salinity, particularly during the seedling stage, which can result in symptoms like scorched leaf tips, leaf curling, and reduced growth, eventually leading to leaf abscission and tree mortality under severe conditions (Dubey et al. 2006; Srivastav et al. 2007). The mechanisms of salt tolerance in mango involve osmotic adjustment through the accumulation of compatible solutes, restricting Na^+ influx, vacuolar sequestration, Na^+ extrusion, and scavenging reactive oxygen species (ROS) to mitigate oxidative stress (Mittler 2002). Despite these mechanisms, the genetic basis of salt tolerance in mango remains largely unexplored.

Salinity-affected soils are projected to increase globally, with India facing an estimated 6.74 million hectares of salt-affected land, and nearly 50% of arable land expected to be impacted by 2050 (Kumar and Sharma 2020). Mango is particularly sensitive to saline conditions, tolerating soil salt concentrations no higher than 0.05%. Changes in climate are expected to exacerbate soil salinization, underscoring the need for systemic research on salt tolerance mechanisms and breeding salt-tolerant rootstocks (Hassani et al. 2021). Salinity stress impairs normal plant development by increasing water stress, ion toxicity (e.g., sodium and chloride), and the generation of ROS, which causes oxidative stress (Munns 2002; Lenis et al. 2011).

Studies indicate that salinity also induces the production of secondary metabolites such as phenolic compounds, flavonoids, and pigments, which protect against oxidative damage. These compounds include hydroxybenzoic and hydroxycinnamic acids like vanillin, protocatechuic acid, and chlorogenic acid, which have been detected in mango (Kim et al. 2009; Palafox-Carlos et al. 2012; Corrales-Bernal et al. 2016). While phenolic acids play a significant role in salt tolerance by scavenging free radicals, further research is needed to understand how these compounds enhance mango's resilience to salinity (Maldonado-Celis et al. 2019).

Advances in transcriptome analysis have provided insights into the molecular responses to salt stress in mango. For example, RNA sequencing of salt-tolerant and salt-sensitive genotypes like 'Turpentine' and 'Mylepelian' revealed differential expression of key genes, including Auxin Response Factor 2 (ARF2), Late

Embryogenesis Abundant (LEA), and CDPK, which were validated through qRT-PCR (Perveen et al. 2023). Recent studies on induced salinity stress in polyembryonic genotypes such as Bappakkai, Kurukkan, and Nekkare showed their ability to tolerate up to 50 mM NaCl, positioning them as promising rootstocks for commercial mango varieties (Nimbolkar et al. 2023). This highlights the potential of using genetic variability and induced mutations to develop salt-tolerant mango cultivars.

7.7 Response to Low Temperature and Frost

Low temperatures are recognized as one of the most significant environmental stressors impacting the growth, productivity, and geographical distribution of crops and horticultural plants (Gadallah et al. 2019). Exposure to cold, non-freezing temperatures triggers genetic, morphological, metabolic, and physiological changes in plants, fostering cold hardiness and freezing tolerance (Nimbolkar et al. 2023). Like plant, mango fruit is also sensitive to low temperature. To mitigate postharvest losses, low-temperature storage slows biological processes and preserves fruit quality. However, mangoes, being tropical and subtropical fruits, are sensitive to chilling injuries (CI) at temperatures below 13 °C (Barman and Asrey 2014). Prolonged cold storage induces metabolic changes that lead to CI, characterized by cellular wall and membrane degradation, reduced ethylene production, and lowered respiration rates (Nair and Singh 2003). Symptoms of CI manifest initially as small patches around lenticels, progressing to large black spots on the fruit's skin due to cellular death (Shewfelt and Del Rosario 2000).

Mango trees thrive in a variety of frost-free climates, performing best in regions with a well-defined, relatively cool dry season and significant heat accumulation during the flowering and fruit development phases (El-Moniem et al. 2010). The optimal temperature range for mango cultivation is 24–26.7 °C, with a minimum threshold of 10–12 °C, below which chilling injuries occur (Nene et al. 1996). The underlying cause of CI is linked to the generation of reactive oxygen species (ROS) within various cell organelles, including mitochondria, plasma membranes, and the cell wall, with mitochondria contributing up to 85% of ROS production (Junmatong et al. 2015). Key ROS such as hydroxyl radicals (OH•) and singlet oxygen ($^1O_2$) attack double bonds in organic molecules, initiating oxidative damage and forming new compounds (Arafat 2005). To alleviate CI, studies have explored treatments like applying organic acids (e.g., oxalic acid at 20 mM), methyl jasmonate, methyl salicylate, or low-temperature conditioning during storage (Wang 2006; Li PeiYan et al. 2015; Lo'ay and Dawood 2019). These strategies aim to reduce oxidative stress and preserve the quality of mangoes during extended storage.

Mangoes exhibit high sensitivity to temperatures ranging between 0and 15 °C, particularly young trees, which are more prone to damage. Although variability in cold tolerance is evident among cultivars following cold spells, detailed information on this variability remains limited, and the capacity of mangoes to acclimate to cold conditions is largely unexplored (McKellar et al.; Farooq and Farooq-E-Azam

2001). Advancing our understanding of the biological responses of mango trees to low temperatures could inform the development of effective cold protection strategies and the breeding of cold-tolerant cultivars, while also providing insights applicable to other tropical and subtropical plants facing similar stress conditions (Hao et al. 2009).

7.8 Conclusion

This chapter provides a comprehensive analysis of mango tree physiology, focusing on the challenges posed by climate change and extreme environmental stressors. Mango trees, known for their economic and cultural significance, exhibit an extraordinary ability for adaptation through mechanisms such as metabolic adjustments, morphological changes, and enhanced stress-response pathways. However, the increasing intensity and frequency of abiotic stresses such as temperature extremes, salinity, flooding, drought, and elevated CO_2 exceed the natural resilience of these trees, threatening productivity and fruit quality. For instance, while elevated CO_2 can temporarily enhance photosynthesis, its long-term benefits are undermined by associated heat stress, nutrient imbalances, and diminished flowering rates. Similarly, adaptations like deep root systems and osmotic adjustments, though critical for drought tolerance, fail to fully mitigate the adverse effects of prolonged water deficits and erratic rainfall.

Precision techniques such as targeted irrigation, canopy management, and controlled stress induction hold promise for optimizing yields under adverse conditions. Furthermore, advances in molecular biology, including the identification of stress-resilient genes and rootstocks, offer transformative potential for breeding programs aimed at creating cultivars tailored to specific stress conditions. Nonetheless, the interplay of multiple stress factors, particularly under natural field conditions, remains poorly understood and warrants extensive multidisciplinary research.

To secure the future of mango cultivation, a holistic approach integrating traditional agronomic practices with cutting-edge scientific innovations is essential. This involves not only addressing immediate challenges but also preparing for the long-term impacts of climate change. Such efforts will be critical in preserving the economic viability of mango farming, ensuring global food security, and sustaining the cultural legacy associated with this iconic fruit (Tables 7.1 and 7.2).

Table 7.1 Impact of biotic factors on mango tree phsiology

Biotic factor	Impact on mango physiology	Examples	Mechanism of impact
Pathogens	Disruption of nutrient and water transport, reduced photosynthesis, and tissue necrosis	*Lasiodiplodia theobromae*, *Colletotrichum* spp., *Phytophthora* spp.	Production of toxins, enzymes, and blockage of xylem vessels causing wilting and dieback
Insect pests	Physical damage to leaves, stems, and fruits; vectoring pathogens	Mango hoppers (*Idioscopus* spp.), fruit flies (*Bactrocera* spp.), thrips (*Scirtothrips dorsalis*)	Feeding on plant tissues, oviposition in fruits, and transmitting fungal or bacterial pathogens
Nematodes	Damage to roots, leading to impaired water and nutrient uptake	Root-knot nematodes (*Meloidogyne* spp.), lesion nematodes (*Pratylenchus* spp.)	Formation of galls or lesions in roots, reducing nutrient translocation to above-ground parts
Weeds	Competition for water, nutrients, and light	*Cyperus rotundus*, *Parthenium hysterophorus*	Outcompeting mango roots for nutrients and water, reducing photosynthesis due to shading
Epiphytes and parasites	Reduction in photosynthesis due to shading or parasitic nutrient extraction	Parasitic plants (*Dendrophthoe falcata*), lichens	Blocking light penetration or directly extracting nutrients and water from host vascular tissues
Endophytic microbes	Can have beneficial or harmful impacts; stimulate growth or increase susceptibility	*Bacillus* spp., *Trichoderma* spp. (beneficial); *Fusarium* spp. (harmful)	Production of growth-promoting compounds or toxins that can alter host growth and defense responses
Symbiotic microorganisms	Positive impacts on nutrient uptake and stress tolerance	Mycorrhizal fungi, nitrogen-fixing bacteria (*Rhizobium* spp.)	Enhancing nutrient absorption (phosphorus, nitrogen), improving drought tolerance, and stress resilience

Table 7.2 Abiotic factors affecting mango tree physiology and their effects on mango health

Abiotic factor	Impact on mango tree physiology	Effects on mango health
Temperature	Alters flowering, fruit set, and ripening processes	• High temperatures cause heat stress, flower drop, and poor fruit quality • Low temperatures can delay flowering and fruiting
Water availability	Affects water uptake, photosynthesis, and nutrient transport	• Drought leads to wilting, reduced growth, and poor fruit size • Waterlogging causes root rot and nutrient deficiencies
Soil salinity	Impacts osmotic balance and ion uptake	• High salinity causes leaf burn, stunted growth, and lower fruit yields
Light intensity	Influences photosynthesis and chlorophyll production	• Insufficient light reduces fruit yield and delays ripening • Excessive light can lead to leaf scorching
Wind	Affects transpiration and structural stability	• Strong winds can cause branch breakage, fruit drop, and increased evaporation, leading to water stress
Nutrient availability	Determines chlorophyll synthesis, growth, and fruit development	• Nutrient deficiencies (e.g., nitrogen, potassium) cause poor growth, fruit drop, and reduced resistance to pests and diseases
pH of soil	Affects nutrient solubility and microbial activity	• Acidic or alkaline soils limit nutrient uptake, leading to deficiencies and weaker tree health
Air pollution	Reduces photosynthetic efficiency and damages leaf tissues	• Pollutants like sulfur dioxide or ozone cause chlorosis, necrosis, and reduced fruit quality
Humidity	Regulates transpiration and disease susceptibility	• High humidity favors fungal diseases like anthracnose • Low humidity leads to increased water loss and stress

References

Antwi-Boasiako A, Amponsah P, Opoku JA, Coulibaly D, Mintah P (2024) Increasing mango production efficiency under the fast-changing climate

Arafat LAET (2005) Chilling injury in mangoes. Wageningen University and Research

Arbona V, Hossain Z, López-Climent MF, Pérez-Clemente RM, Gómez-Cadenas A (2008) Antioxidant enzymatic activity is linked to waterlogging stress tolerance in citrus. Physiol Plant 132:452–466

Barceló-Anguiano M, Hormaza JI, Losada JM (2021) Conductivity of the phloem in mango (*Mangifera indica* L.). Hortic Res 8

Barman K, Asrey R (2014) Salicylic acid pre-treatment alleviates chilling injury, preserves bioactive compounds and enhances shelf life of mango fruit during cold storage

Boudon F, Persello S, Jestin A, Briand A-S, Grechi I, Fernique P, Guédon Y, Léchaudel M, Lauri P-É, Normand F (2020) V-Mango: a functional–structural model of mango tree growth, development and fruit production. Ann Bot 126:745–763

Carr MKV (2014) The water relations and irrigation requirements of mango (*Mangifera indica* L.): a review. Exp Agric 50:1–23

Cavalcante HL (2022) Mango flowering: factors involved in the natural environment and associated management techniques for commercial crops

Corrales-Bernal A, Jaramillo G, Rodríguez B, Kazuz EY, Maldonado-Celis ME (2016) Mango (*Mangifera indica* cv. Azúcar) antiinflammatory and chemopreventive role during colorectal carcinogenesis. Emir J Food Agric (EJFA) 28

Dambreville A, Normand F, Lauri P-É (2012) Plant growth co-ordination in natura: a unique temperature-controlled law among vegetative and reproductive organs in mango. Funct Plant Biol 40:280–291

Damour G, Vandame M, Urban L (2009) Long-term drought results in a reversible decline in photosynthetic capacity in mango leaves, not just a decrease in stomatal conductance. Tree Physiol 29:675–684

Dubey AK, Srivastav M, Singh R, Pandey RN, Deshmukh PS (2006) Response of mango (*Mangifera indica*) genotypes to granded level of salt stress. Indian J Agric Sci 76

Eiadthong W, Yonemori K, Sugiura A, Utsunomiya N, Subhadrabandhu S (1999) Genetic resources of Mangifera species in Thailand. Jpn J Tropic Agric 43:76–83

El-Moniem EAA, Ismail OM, Shaban AEA (2010) Changes in leaves component of some mango cultivars in relation to exposure to low temperature degrees

Elsheery NI, Cao K-F (2008) Gas exchange, chlorophyll fluorescence, and osmotic adjustment in two mango cultivars under drought stress. Acta Physiol Plant 30:769–777

Farooq S, Farooq-E-Azam (2001) Co-existence of salt and drought tolerance in Triticeae. Hereditas 135:205–210

Gadallah FM, El-Yazal S, Abdel-Samad GA (2019) Physiological changes in leaves of some mango cultivars as response to exposure to low temperature degrees. Hortic. Int. J 3:266–273

Galán Saúco V (2009) El cultivo del mango (rev. y amp.). Mundi-Prensa Libros

González A, Blaikie SJ (2003) Seasonal variation of carbon assimilation in mango (cv. Kensington Pride): effect of flowering treatments. Aust J Agric Res 54:309–321

Gupta NK, Sen NL (2003) Studies on initial establishment of mango seedling in saline environment

Halder S, Hasan MA (2020) Climate change and mango production. Chem Sci Rev Lett 9

Halder R, Varma S, Singh M, Dahiya A (2024) From orchard to table: Understanding climate change impacts on mango production in India: a review 472–478

Hao W, Arora R, Yadav AK, Joshee N (2009) Freezing tolerance and cold acclimation in guava (*Psidium guajava* L.). HortScience 44:1258–1266

Hara Y, Yukimatsu H, Thaitakoo D, Tsuchiya K (2014) Agricultural landscape changes and its resilience in response to the 2011 serious floods in the urban fringe of Bangkok

Hassani A, Azapagic A, Shokri N (2021) Global predictions of primary soil salinization under changing climate in the 21st century. Nat Commun 12:6663

Henson R (2011) The rough guide to climate change. Dorling Kindersley Ltd

Jain SM, Priyadarshan PM (2009) Breeding plantation tree crops: tropical species. Springer

Jensen KH, Valente AXCN, Stone HA (2014) Flow rate through microfilters: influence of the pore size distribution, hydrodynamic interactions, wall slip, and inertia. Phys Fluids 26

Jo H, Roszak P, Furuta KM, Helariutta Y (2014) Phloem development: current knowledge and future perspectives. Am J Bot 101:1393–1402

Junmatong C, Faiyue B, Rotarayanont S, Uthaibutra J, Boonyakiat D, Saengnil K (2015) Cold storage in salicylic acid increases enzymatic and non-enzymatic antioxidants of Nam Dok Mai No. 4 mango fruit. Sci. Asia 41:12–21

Kays SJ (1997) Postharvest physiology of perishable plant products. Exon Press, Athens. In, p 532p

Khanum Z, Tiznado-Hernández ME, Ali A, Musharraf SG, Shakeel M, Khan IA (2020) Adaptation mechanism of mango fruit (*Mangifera indica* L. cv. Chaunsa White) to heat suggest modulation in several metabolic pathways. RSC Adv 10:35531–35544

Kim Y, Lounds-Singleton AJ, Talcott ST (2009) Antioxidant phytochemical and quality changes associated with hot water immersion treatment of mangoes (*Mangifera indica* L.). Food Chem 115:989–993

Kumar P, Sharma PK (2020) Soil salinity and food security in India. Front Sustain Food Syst 4:533781

Larson KD, Schaffer B, Davies FS (1991) Flooding, leaf gas exchange, and growth of mango in containers. J Am Soc Hortic Sci 116:156–160

Larson KD, Schaffer B, Davies FS (1993) Floodwater oxygen content, ethylene production and lenticel hypertrophy in flooded mango (*Mangifera indica* L.) trees. J Exp Bot 44:665–671

Legave JM, Normand F, Lauri PE, Climate change and its probable effects on mango production and cultivation, pp 21–31

Lenis JM, Ellersieck M, Blevins DG, Sleper DA, Nguyen HT, Dunn D, Lee JD, Shannon JG (2011) Differences in ion accumulation and salt tolerance among glycine accessions. J Agron Crop Sci 197:302–310

Li P, Yan LP, Zhang X, Chowdhury MGF, Cordasco K, Brecht JK (2015) Prestorage application of oxalic acid to alleviate chilling injury in mango fruit

Lo'ay AA, Dawood HD (2019) Tolerance of 'Baladi'mandarin fruits to cold storage by postharvest pectin/PVA blend with ascorbic acid treatment. Scientia Horticulturae 256:108637

Lu P, Chacko EK, Bithell SL, Schaper H, Wiebel J, Cole S, Müller WJ (2012) Photosynthesis and stomatal conductance of five mango cultivars in the seasonally wet–dry tropics of northern Australia. Sci Hortic 138:108–119

Maldonado-Celis ME, Yahia EM, Bedoya R, Landázuri P, Loango N, Aguillón J, Restrepo B, Guerrero Ospina JC (2019) Chemical composition of mango (*Mangifera indica* L.) fruit: Nutritional and phytochemical compounds. Front Plant Sci 10:1073

McKellar MA, Buchanan DW, Campbell CW, Cold hardiness of two cultivars of avocado and a mango, pp 212–215

Mittler R (2002) Oxidative stress, antioxidants and stress tolerance. Trends Plant Sci 7:405–410

Munns R (2002) Comparative physiology of salt and water stress. Plant, Cell Environ 25:239–250

Nair S, Singh Z (2003) Pre-storage ethrel dip reduces chilling injury, enhances respiration rate, ethylene production and improves fruit quality of 'Kensington'mango. Food Agric Environ 1:93–97

Nakamura N, Dv S, Shiina T, Nawa Y (2004) Respiration properties of tree-ripe mango under CA condition. Jpn Agric Res q: JARQ 38:221–226

Nene YL, Sheila VK, Moss JP (1996) Tapioca: a potential substitute for agar in the tissue culture media. Curr Sci 70:493–494

Nimbolkar PK, Kurian MR, Upreti KK, Laxman RH, Shivashankara KS, Varalakshmi LR (2023) Physio-biochemical responses of polyembryonic mango (*Mangifera indica* L.) genotypes to varying levels of salinity stress. J Hortic Sci 18:150–161

Normand F, Bissery C, Damour G, Lauri PÉ (2008) Hydraulic and mechanical stem properties affect leaf–stem allometry in mango cultivars. New Phytol 178:590–602

Ogunbode TO, Esan VI, Ayegboyin MH, Ogunlaran OM, Sangoyomi ET, Akande JA (2024) Understanding the perception of mango (*Mangifera indica*) farmers on the impact of climate change on mango farming in Nigeria. Int J Agron 2024:6486998

Olson ME, Rosell JA (2013) Vessel diameter–stem diameter scaling across woody angiosperms and the ecological causes of xylem vessel diameter variation. New Phytol 197:1204–1213

Olson ME, Anfodillo T, Rosell JA, Martínez-Méndez N (2020) Across climates and species, higher vapour pressure deficit is associated with wider vessels for plants of the same height. Plant Cell Environ 43:3068–3080

Pace MR, Alcantara S, Lohmann LG, Angyalossy V (2015) Secondary phloem diversity and evolution in Bignonieae (Bignoniaceae). Ann Bot 116:333–358

Palafox-Carlos H, Yahia EM, González-Aguilar GA (2012) Identification and quantification of major phenolic compounds from mango (*Mangifera indica*, cv. Ataulfo) fruit by HPLC–DAD–MS/MS-ESI and their individual contribution to the antioxidant activity during ripening. Food Chem 135:105–111

Pandey P, Singh AK, Dubey AK, Awasthi OP (2014) Effect of salinity stress on growth and nutrient uptake in polyembryonic mango rootstocks. Indian J Hortic 71:28–34

Peiris KHS, Mallon JL, Kays SJ (1997) Respiratory rate and vital heat of some specialty vegetables at various storage temperatures

Perveen N, Dinesh MR, Sankaran M, Ravishankar KV, Krishnajee HG, Hanur VS, Alamri S, Kesawat MS, Irfan M (2023) Comparative transcriptome analysis provides novel insights into molecular response of salt-tolerant and sensitive polyembryonic mango genotypes to salinity stress at seedling stage. Front Plant Sci 14:1152485

Prajapati DM, Jagalan V, Dave KD, Respiration behavior and heat of respiration of mango (cv. Alphonso) under different storage conditions

Rajan S (2012) Phenological responses to temperature and rainfall: a case study of mango. Trop Fruit Tree Species Clim Chang 71

Savage JA, Beecher SD, Clerx L, Gersony JT, Knoblauch J, Losada JM, Jensen KH, Knoblauch M, Holbrook NM (2017) Maintenance of carbohydrate transport in tall trees. Nature Plants 3:965–972

Schaffer B, Davies FS, Crane JH (2006) Responses of subtropical and tropical fruit trees to flooding in calcareous soil. HortScience 41:549–555

Scuderi D, Gianguzzi G, Tinebra I, Gugliuzza G, Farina V (2023) Phenology and fruit growth dynamics of mango (*Mangifera indica* L.) in greenhouse and open air in Mediterranean climate. In: Proceedings of the national academy of sciences, india Section B: biological sciences, vol 93, 731–741

Shewfelt RL, Del Rosario BA (2000) The role of lipid peroxidation in storage disorders of fresh fruits and vegetables. HortScience 35:575–579

Shimada A, Tominaga S, Yamamoto M (2018) Effects of water management on vine growth and fruit quality in passion fruit

Srivastav M, Dubey AK, Pandey RN, Desnmukh PS (2007) Effect of soil salinity on survival, growth and chlorophyll contents of 'Kurukkan' mango (*Mangifera indica*). Indian J Agric Sci 77

Tamura F, Tanabe K, Katayama M (1995) Relationship between water tolerance and cyanide-resistant respiration in pear rootstocks. J Jpn Soc Hortic Sci 64:47–53

Taria S, Alam B, Rane J, Kumar M, Babar R, Singh NP (2022) Deciphering endurance capacity of mango tree (*Mangifera indica* L.) to desiccation stress using modern physiological tools. Scientia Horticulturae 303:111247

Wang CY (2006) Reducing chilling injury and maintaining quality of horticultural crops with natural products and their derivatives. In: IV International conference on managing quality in chains-the integrated view on fruits and vegetables quality, vol 712, pp 285–290

Whiley AW, Rasmussen TS, Saranah JB, Wolstenholme BN (1989) Effect of temperature on growth, dry matter production and starch accumulation in ten mango (*Mangifera indica* L.) cultivars. J Hortic Sci 64:753–765

Woolf AB, Ferguson IB, Requejo-Tapia LC, Boyd L, Laing WA, White A (1999) Impact of sun exposure on harvest quality of 'Hass' avocado fruit. Rev Chapingo Ser Hortic 5:353–358

Chapter 8
Amelioration of Climate Change Impacts in Mango

Moazzam Riaz, Diyan Rashid, Ubaid Ur Rehman, Qurat Ul Ain, Muhammad Akash Athar, Syed Shahzaib, Muhammad Umair, Najma Majeed, Mukhtar Ahmed, Shakeel Ahamd, Muhammad Yousaf Raza, and Ahmad Mahmood

Abstract Climate change has been affecting all spheres of life especially those which depend directly on climate like that of agriculture. In agriculture, the extent of climate change is wide which reaches all the way from vegetables, fruits, and field crops to livestock, and significant losses of produce have been observed. This is especially true for spring and summer crops which cannot withstand high temperatures, erratic rainfall, flooding, and high winds. Mango is one of the most affected crops due to changing climatic conditions. Mango flowering season receiving abrupt rainfall, temperature rise during flowering, fruit quantity and quality loss due to heatwaves, water shortages-led fruit loss, and pollen sterility are a few issues to count. As the changing climate affects the morphology and physiology of the mango tree by altering its growth pattern, the susceptibility to disease and pest attack rises, compromising the quality and quantity of fruit. Therefore, there is an immediate need to ameliorate the effects of climate change on mango through sustainable approaches. Planting trap crops near the mango orchards to attract aphids and mealybugs, using shelterbelts and windbreaks to address the heatwaves, better management of shade, improved irrigation management utilizing drip and sprinkler systems, drought and

---

M. Riaz · D. Rashid · U. U. Rehman · Q. U. Ain · M. A. Athar · S. Shahzaib · N. Majeed
Department of Soil and Environmental Sciences, MNS-University of Agriculture, Multan, Pakistan

M. Umair
Department of Horticulture, MNS-University of Agriculture, Multan, Pakistan

M. Ahmed · M. Y. Raza
Department of Agronomy, Pir Mehr Ali Shah Arid Agriculture University, Rawalpindi, Pakistan

S. Ahamd
Faculty of Agricultural Sciences and Technology, Institute of Agronomy, Bahauddin Zakariya University, Multan, Pakistan

A. Mahmood (✉)
School of Interdisciplinary Engineering & Sciences (SINES), National University of Sciences and Technology (NUST), Islamabad, Pakistan
e-mail: ahmad.mahmood@sines.nust.edu.pk

S. A. H. Naqvi et al. (eds.), *Climate Change and Mango Production*,
https://doi.org/10.1007/978-3-031-87035-4_8

heat tolerant rootstocks, integrated pest management, biological control, intercropping, and cover cropping are a few techniques to consider for amelioration of effects of climate change in mango. Addressing the impacts of climate change on mango production is crucial for sustaining agricultural economy and livelihoods.

Keywords Climate change · Mango production · Climate resilience · Amelioration · Effects

8.1 Introduction

Mango (*Mangifera indica)* ranks among the most significant tropical fruit worldwide in terms of both economy and culture, With India, China, Thailand, Pakistan, Indonesia and Mexico being the major producers. The economic value of Mango can be estimated by considering that the Mango Industry was valued at $50 Billion in 2020 on global scale (https://www.fao.org/faostat/en/#data/QCL). In Pakistan, the total area used for Fruit cultivation is around 853.4 thousand Hectares producing 7178.8 thousand tonnes, while area used for Mango cultivation is around 175 thousand hectares with production of 1784 thousand tonnes of Mango making it the 2nd major fruit of Pakistan after citrus, Pakistan being the 4th largest producer of Mango outputs around 8.5% of total world's Mango production.

Major export markets of Pakistani mango include United Arab Emirates, Saudi Arabia, Oman, United Kingdom, Kuwait, Bahrain, France and Germany, contributing about 41, 16, 11, 10, 3, 2, 1 and 0.90 percent of total mango exports from Pakistan, respectively (Fig. 8.1) (Awan et al. 2010). Mango season is often celebrated as a major cultural event, Mango festivals are widely observed throughout the country as public gatherings and recreational events during the months of June and July.

8.1.1 Climate Change and Mango

Mango is considered one of most sensitive to climate change, for the optimum production of mango requires the specific climatic conditions, typically it requires well distributed, warm and moderately humid climatic conditions, it can affect in following ways: Mango trees thrive in temperatures ranging from 24 °C to 30 °C, but rising temperatures can lead to heat stress, impacting their growth, flowering, and fruiting. Excessive heat often causes flower drop, reduced fruit setting, and ultimately lower yields (Laxman et al. 2016). Additionally, altered rainfall patterns, such as delayed monsoons or unseasonal rains, disrupt the crucial dry period required for flowering and the adequate rainfall necessary for fruit setting. Warmer and more humid conditions also increase the incidence of pests and diseases, with mangoes being particularly vulnerable to fruit flies, aphids, and mealy bugs, as well as diseases like anthracnose, powdery mildew, and bacterial black spots. Moreover, extreme

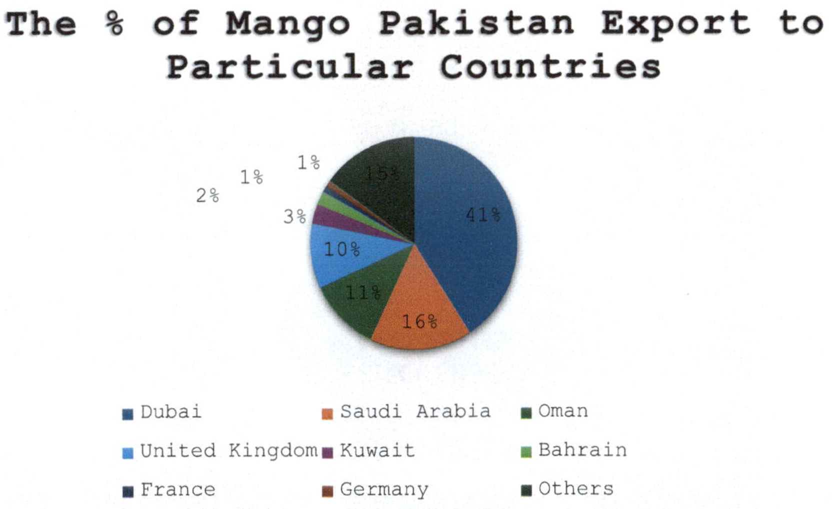

Fig. 8.1 The percentage of Pakistan's mango export to particular countries

weather events such as storms, hail, and cyclones can physically damage mango trees and fruit, further threatening their productivity (Fig. 8.2).

8.2 Mango: A Vulnerable Crop

Mango often addressed as the "king of fruit" is well known for its taste, color, sweetness and cultural significance is also a very vulnerable crop, it being a perishable food product has a little shelf life in raw state and also have a Hugh issue of Post-Harvest Losses. Mangos being a tropical fruit grow in a warm-humid region, provided the said climatic conditions these are most desired for many insect pets and also bacterial and fungal diseases. A 2019 study published in Scientia Horticulture (Khan and Ahmed 2021) found that pests and diseases can cause yield losses exceeding 50%, highlighting the vulnerability of mango production. Beyond biological threats, the ever-changing climate poses another challenge. Research published in Agricultural and Forest Meteorology in 2020 by (Normand et al. 2015) all suggests that rising temperatures and erratic rainfall patterns can negatively impact mango flowering, fruit set, and overall health.

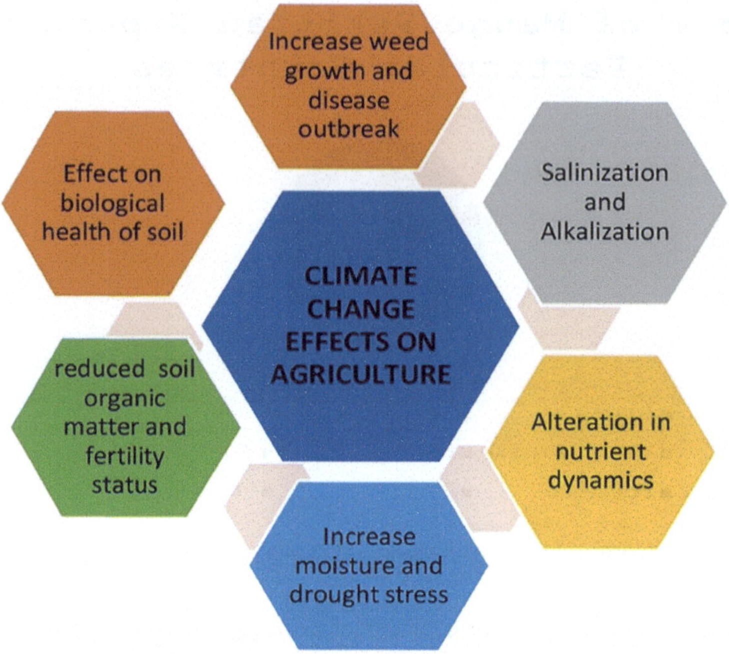

Fig. 8.2 Effect of climate change on tree

8.3 How Climate Change Affects Mango

Climate change is driven by rising anthropogenic greenhouse gas concentration and poses a significant threat to agriculture, including Mango cultivation. Global models predict increasing average temperatures, Higher atmospheric CO_2 levels, greater soil salinity, and irregular rainfall patterns. The variability and frequency of extreme weather events like heatwaves, heavy, droughts, and hurricanes are also expected to rise. Mango is the Fifth most cultivated fruit globally and holds substantial economic and nutritional value, especially in tropical and sub-tropical regions such as Pakistan, China, India, Philippines. Current knowledge indicates that climate variables significantly impact mango tree processes such as photosynthesis, Vegetation and reproduction, and fruit quality. Climate change has a significant impact on mango farming, affecting soil quality, changing blooming and fruit sets, and increasing sensitivity to pests and diseases. This delicate equilibrium which is critical for mango plant is under unprecedented stress from rising temperatures and unpredictable weather patterns. Understanding and treating these effects is critical for maintaining this prized fruit (Choudhary et al. 2022). Assuring sustainable production, and protecting the lives of millions of farmers worldwide.

8.4 Climate Change and Soil Health

Climate change has a negative impact on soil health by generating erosion, lowering moisture content, and changing microbial activity. These changes reduce the nutrient availability which is essential for crop development, risking agricultural productivity and sustainability. Climate change has an influence on soil physical processes including water air and nutrient dynamics, posing concerns such as salinization, decreased water availability, and changes carbon and nitrogen cycle, These changes have an impact on soil health by changing water transport, root penetration, and soil biodiversity, which are all important for agricultural production and sustainability (Girija Veni et al. 2020) (Table 8.1).

The main impacts of climate change on the biological properties of soil are: Alteration of microbial diversity and function, Reduction in Organic matter decomposition, Changes in soil enzyme activity, affect soil pathogens, reduce soil moisture and Causes soil acidification. Climate change has an unexposed effect on the soil. Alters the physical, chemical, and biological properties of the soil and this leads to the processes that lead to the degradation of soil just as Soil erosion issues, Loss of soil moisture, and decreased microbial activity of the soil. Soil erosion affects mango productivity by losing the nutrients from the soil ultimately impacting the vegetation and flowering of the mango, and also reducing the availability of beneficial microorganisms in the soil which have a significant role in nutrient cycling (Pankhurst et al. 1995).

8.5 Heatwave Effect on Mango

Heatwave is another consequence of climate change due to a sufficient increase in temperature by global warming. Heatwaves normally occur between May to August and unfortunately, in the same season their occur the Vegetative growth of Mangoes. It has a severe impact on the production of mangoes. The common Effect is Leaf scorching and sunburn due to prolonged exposure to extreme temperatures leading to premature leaf drop and reducing the photosynthetic capacity of the tree. High temperature during the flowering may cause flower dropping, reducing fruit set and yield. This leads to an increase in fruit drop particularly in young and developing fruits. Fruits exposed to direct sunlight can develop sunburn causing blemishes and reducing marketability.

According to (Bal et al. 2022) Mango orchards of several districts of India were impacted by the heatwave and severe flower and fruit drop, reduction in fruit size was observed in mangoes. The Kesar and Alphanso Mango varieties planted in some regions gave less yield due to unseasonal rainfall, Fog from December to January, and Hot temperatures in March. High temperatures for several days may cause water stress in mangoes. Waster stress limits photosynthesis in response to which plants

Table 8.1 The impacts of climate change and soil health indicators

Soil physical properties		
Name	Uses	References
Soil texture	The percentage of sand, silt and clay in a soil. It has direct influence on climate change. The four climatic scenarios (arid, semi-arid, sub-humid, and humid) significantly affect soil processes, including textural differences in the profile	Brammer and Brinkman (1990)
Soil structure	The arrangement of particles in the soil determines the water and air content. It affects collective stability, organic carbon buildup, water transport, and microbial activity. Decreased organic matter reduces soil stability and raises the danger of erosion. Organic content, farming practices, and natural processes all impact soil structure, which is critical for ecosystem function	Doran and Zeiss (2000)
Soil porosity	The measure of void spaces in the soil, and the pore distribution is very critical for storing root zone water and air necessary for plant growth. It also Elevated CO_2 and temperature, Variable and Extreme Rainfall Events, Reduced Microbial Activity, Reduced Root Growth and Exudates and Increased Rainfall Intensity	Patil and Lamnganbi (n.d.)
Bulk density	Climate change impacts soil bulk density by increasing rainfall intensity, causing compaction, and altering precipitation patterns. This leads to lower soil organic matter, reducing soil structure and aggregation. Prolonged droughts also reduce soil moisture, while changes in root growth patterns and microbial activity affect soil structure and infiltration rates	Al-shammary et al. (2018)
Soil Chemical properties		
Soil pH	It tells about the acidity or alkalinity of the soil. Most soils would not be susceptible to rapid pH changes arising from causes of climate change such as increased temperatures, CO_2 fertilization, variable precipitation, and atmospheric N deposition. Climate change impacts organic matter status, Nitrogen cycling, plant availability, and productivity, ultimately affecting soil pH	Reth et al. (2005)

(continued)

Table 8.1 (continued)

Soil physical properties		
Name	Uses	References
Electrical conductivity	The concentration of soluble salts present in the soil. It is the key indicator to determine the salinity and sodicity in the soil. Climate change can lead to high electrical conductivity due to higher temperatures and low precipitation	Smith et al. (2002)
Sorption and cation exchange capacity	CEC is an important chemical property that determines the sorption and desorption of nutrients and retention of cations and anions between the clay minerals of the soil. Since climate change causes an increase in the temperature this elevated temperature leads to the breakdown of soil Organic matter resulting in a decrease in the CEC of the soils. Heavy rainfall can cause enhanced leaching of base cation, moving alkalinity from soil to streams	Davidson and Janssens (2006)

close their stomata, reducing a tree's ability to produce energy for growth and development, which has an impact on fruit and flower production. According to (Núñez-Elisea and Davenport 1994) Floral morphogenesis in mango trees is triggered by temperatures about 15 °C, but vegetative morphogenesis occurs at 20 °C or above. In the tropics, blooming occurs after dry spells. Mango vegetative shoots flush periodically before developing apical panicles. Dormant buds can undergo vegetative or floral morphogenesis, with apical buds suppressing axillary bud development.

8.6 How Climate Change Affects the Key Processes of Mango Production

8.6.1 Effect on Photosynthesis

Temperature, Light, and CO_2 concentration all have a favorable impact on photosynthesis, but high temperatures (> 45 °C) and strong light due to climate change can impair this process. Increasing CO_2 promotes photosynthesis but also increases respiratory loss. Drought and a large vapor pressure deficit produce fast stomatal closure decreasing the rate of photosynthesis. Flooding although less common in dry places has a detrimental influence on mango photosynthesis by reducing the transpiration and stomatal enclosure (Makhmale et al. 2016).

8.6.2 Effect on Vegetation

In the absence of any limitation by nutrient availability or water, increasing temperatures speed up mango growth cycles and delay winter growth cessation. Temperature increases can result in bigger leaves, although these effects are variety-dependent. Root dry matter allocation is increased by high CO_2. According to (Halder and Abu Hasan 2020) vegetative development is hindered by drought and floods; though a few mango varieties are tolerant of long drought periods, some can be deadly within a few days. Unpredictable cool nights and warm winters interfere with mango flowering, adversely affecting yields. Some of the adaptation strategies adopted for this purpose are growth retardants, moisture stress, post-harvest thinning, trench digging for better drainage, and shoot trimming. Selection and breeding with the use of available genetically diverse resources further reduce the impact of climatic change on mango yield.

8.6.3 Effect on Flowering and Fruit Set

Cool temperatures are mostly responsible for the floral induction of mango trees. As temperature increases it would have a detrimental effect on floral induction. Increasing temperatures during blooming can improve pollen viability and fruit set, especially in chilly locations. Temperature can reduce fluorescence size (Dambreville et al. 2012) and also on the number of flowers per inflorescence (Sukhvibul et al. 1999). As per (Normand et al. 2015) drought and increased VPD have deleterious impacts on fruit set and retention and drought may promote early growth stoppage and vegetative rest, which can aid in floral induction.

8.6.4 Effect on Fruit Quality and Growth

According to (Halder and Hasan 2020) high temperature induces spongy tissue Alphonso mangoes, leading to tissue break down and reduce fruit quality and drought may reduces the size of mango fruits. High temperatures cause physiological changes within the fruit, potentially affecting its overall health and marketability. Thus, Off-season blooming and fruit setting of mango plants are due to the uneven climate change, which provides low-temperature nights and warm winters. High temperatures may halt pollination and fruit set, while severe weather damages the blooms. All these disturbances lead to reduced yields with poor-quality fruits (Patel et al. 2019).

8.7 Climate Change Influence of Pest and Disease Manifestation

Climate change affects disease epidemics on various scales, ranging from microclimates to global ecosystems. The basic plant disease triangle, which includes plant hosts, infections, and the environment, gets more complex with vector-borne diseases, as vectors become an important element. These interactions affect plant development, microclimates, and vector activity, complicating disease dynamics (Jones and Barbetti 2012). The higher temperature allows pests to breed more frequently and extend their active season leading to an increase in population. Similarly, alteration in precipitation creates favorable conditions for the growth and spread of fungal and bacterial diseases. Global climate change will significantly affect agricultural and insect pests through effects on fundamental biology. Effects on pest interactions with the environment and other species will be indirect. As poikilothermic animals, insects are highly responsive to temperature, which is being altered by increased CO_2, increased temperatures, and decreased soil moisture, and could profoundly impact insect population dynamics, leading to increased crop losses. Climate change is creating new ecological niches that allow the development and regional spread of pests. On crop-pest interactions, this intricacy confounds the reality with new challenges to the farmers and continues to threaten global food security by transcending physical and political boundaries (Skendžić et al. 2021).

8.8 Ameliorating Techniques for Mango

8.8.1 Sustainable Approaches for Climate Resilience

Climate change is posing significant challenges to mango cultivation. To mitigate these impacts and enhance orchard sustainability, farmers are increasingly adopting climate-smart practices. Climate-smart fertilizers, such as biofertilizers and controlled-release fertilizers, offer precise nutrient delivery, reducing nutrient leaching and improving soil health. Mulching, a simple yet effective technique, conserves soil moisture, regulates temperature, and suppresses weeds, contributing to overall orchard resilience (Rajan et al. 2020). By integrating these practices, mango growers can optimize resource utilization, enhance productivity, and adapt to changing climatic conditions, ensuring a sustainable future for the mango industry.

8.8.2 Harnessing the Power of Soil to Combat Climate Change

Healthy soils in mango orchards are essential for reducing climate change. The following are some ways that soil health practices lead to a more sustainable future:

8.8.3 Carbon Sequestration

Organic matter in healthy soils stores atmospheric carbon dioxide (CO_2), they function as carbon sinks. Reduced tillage and cover crops are two examples of practices that improve soil health and increase the buildup of organic matter, which efficiently absorbs carbon dioxide from the atmosphere and slows down climate change (Bruce et al. 1999).

8.8.4 Improved Water Management

Water can be stored in soil that is healthy and has high organic matter contents. This corresponds to:

- **Decreased soil erosion and water runoff:** Increase water infiltration reduces soil erosion and surface runoff, safeguarding important topsoil and water resources.
- **Improved drought resistance:** Trees with larger water storage capabilities are able to with stand dry spells more easily, which lowers the demand for irrigation and fresh water supplies.

Mango producers can mitigate climate change by improving water management and sequestering carbon by concentrating on soil health techniques. These methods can improve their orchards long term health and productivity in addition to helping the environment.

8.8.5 Building a Thriving Soil Ecosystem

A robust soil ecology rich in beneficial microorganisms is essential for the ideal growth and fruit yield of mango trees. The following are some ways that you can improve the soil ecology in your orchard (Mason-Jones et al. 2022).

8.8.5.1 Soil Organic Matter (SOM)

A healthy soil ecosystem is built on SOM. It gives plants nutrition, enhances water absorption and retention, and establishes a home for helpful soil organisms. Reduced tillage minimizes soil disturbance, allowing organic materials to break down naturally and integrate into the soil. Adding organic amendments like compost or manure further enriches the soil by enhancing microbial activity and providing a readily available supply of nutrients. Planting cover crops during fallow periods also boosts soil organic matter as these plants decompose, and nitrogen-fixing cover crops can significantly enhance soil fertility. Together, these practices improve soil health, nutrient availability, and overall productivity.

8.8.5.2 Soil Microorganisms

A varied community of soil microorganisms is essential for the cycling of nutrients, the breakdown of organic matter and the stimulation of plant growth. Because they give soil microbes a food source and a home, practices that improve SOM like the ones listed above also help them.

8.8.5.3 Covering Crops

Cover crops actively help soil microorganisms in a number of ways in addition to increasing SOM:

Root Exudates: When cover crops are grown, carbon-rich chemicals known as root exudates are released into the environment. These microorganisms find sustenance in the exudates.

Increased Microbial Diversity: A more resilient and diverse soil ecosystem is fostered by the growth of various microbial populations, which is encouraged by the species of cover crops. Mango may promote a healthy soil ecosystem by putting these principles into practice. This ultimately results in increased fruit output and the sustainability of orchards, as well as healthier trees and better nutritional availability.

8.9 Integrated Pest Management (IPM)

IPM of mango orchards can be done sustainably with the use of IPM. It reduces dependency on chemical pesticides and places a high priority on prevention. IPM reduces dependence on chemicals by focusing on monitoring pest populations and using pesticides sparingly, targeting only the specific pests. This approach not only minimizes the risk of resistance but also protects beneficial insects that contribute to natural pest control. By reducing pesticide usage, IPM lowers the risk of harmful residues on mangoes, resulting in safer, higher-quality fruits for consumers. Additionally, IPM enhances orchard ecology by promoting habitat modifications and fostering natural predators, creating a more balanced ecosystem with increased biodiversity and beneficial insects to aid in pest management (Stenberg 2017).

8.9.1 IPM Practices for Mango

- **Monitoring:** identifying the pest populations and routinely checking trees for identification of pest damage.
- **Cultural practices:** Pest habitats and breeding grounds can be reduced by following sanitation measures such as picking up fallen leaves and fruits trimming appropriately and keeping an orchard clean.
- **Biological Control:** It introducing lacewings and ladybugs, two naturally occurring predators, to manage pest numbers.
- **Judicious use of Pesticides:** The prudent application of pesticides involves choosing those that have the least negative effects on beneficial insects and reserving their usage for dire emergencies (Stenberg 2017).

Mango growers may efficiently control pest and faster a sustainable and healthy orchard environment by putting in place a thorough IPM program.

8.9.2 Improved Water Management

Mango requires water to flourish but in places with scarce water supplies or erratic rainfall, effective management is critical. Your mango trees can benefit from batter water management and rainwater gathering in the following ways:

8.9.2.1 Efficient Irritation System

Drip Irrigation: By limiting evaporation and optimizing water use efficiency, this focused technique uses emitters to provide water directly to the root zones. It permits derivation, which is the practice of applying fertilizers with irrigation water, and is perfect for established trees.

Sprinkler Irrigation: Although less effective than drip irrigation, sprinklers can be used in big orchards or on young trees. To reduce water loss and guarantee even distribution throughout the root zone, choose for low pressure system (Evans and Sadler 2008).

8.9.2.2 Monitoring and Practices

Utilizing soil moisture, we make sure trees receive enough water without being overwhelmed using soil moisture monitors or become knowledgeable about the systems of water stress. This technique, which is especially useful in drier areas is deliberately cutting back on water during particular growth phases like as vegetative growth, in order to increase drought resistance and water use efficiency.

8.9.2.3 Rainwater Harvesting

Rainwater collection is a sustainable way to meet irrigation demands. Building cistern or rain barrels to catch roof runoff. Building ponds or seals to collect and hold rainwater during the rainy season might reduce reliance on groundwater or other external sources by offering a vital water source during dry spells (Pandey et al. 2003).

8.9.3 Drought Tolerant Rootstocks

A long terms advantage is choosing drought-Tolerant rootstocks for grafted mango trees. Because these rootstocks require less frequent irrigation, the orchards total water efficiency can be improved. Look at rootstocks cultivars that are known to withstand drought in your areas such as "Gomera-1". Mango growers can preserve this valuable resource while guaranteeing their trees receive enough water for healthy growth and fruit production by combining these strategies.

8.9.4 Microclimate Management

Mango trees are susceptible to wind and extremely high or low temperatures. With the use of microclimate management strategies, a region can be made more conductive to ideal development and fruit production. A microclimate is a localized set of atmospheric conditions that differ from the surrounding areas and are frequently influenced by elements like wind, temperature, and humidity.

8.9.5 Shelterbelts and Windbreaks

Arranging rows of quickly growing trees or shrubs to form a physical barrier around the edge of the orchard reduce wind speed. It shields trees from harm and minimizes water loss from excessive transpiration. It also claims the orchard's microclimate, which may have an impact on temperature and humidity. Depending on your particular region and the predominant wind pattern l, windbreak design elements like tree species, distance and height should be taken into account. The best windbreak arrangement for your orchard can be determined by speaking with regional agricultural specialists (Staple and Lehane 1955).

8.9.6 Improve Shade Management

While some shades are beneficial for young mango trees, too much shadow in older orchard might hinder flowering and fruit output. The following are some tactics:

- **Selective pruning:** Pruning branches in a waythat optimize light penetration into the canopy helps to form flower buds and improve the quality of fruits.
- **Intercropping with Low-Growing Shade Providers:** Crop that can withstand shade such as shade coffee, without adversely affecting the amount of light that adult trees receive. Intercepts should be properly managed though, to prevent competition for nutrients and water.

Mango producers can improve the growth, fruit quality, and general health of their orchard by using these microclimates management strategies to create a more conductive environment for their plants.

8.9.7 Interrupting and Cover Cropping

Mango trees can be made healthier and more productive by utilizing two helpful techniques which are intercropping and cover crops. Growing two or more than two crops concurrently in the same mango orchard is known as intercropping. This can be especially helpful for young mango trees that gain from the shade that a lower growing intercrop provides. Vegetables. Legumes and peanuts are frequently chosen for intercropping along with mango trees, intercropping is beneficial for Increase the availability of nutrients for mango tree by fixing nitrogen in the soil (for legumes), Suppress weeds by making them compete with them for resources and light. Provide producers with extra revenue.

Planting a crop specially to enhance soil health and inhabit weeds is known as cover crop and are sown in that fallow season in between mango harvest. The cover crops are tilled back into the soil as it reaches maturity which improves soil fertility and water infiltration while also adding organic matter. Mango trees are often covered with grasses, non-leguminous plants and legumes (Snapp et al. 2005). Cover crops can assist with Increase the amount of organic matter and soil fertility. Increase water infiltration and lessen erosion. Control pest and weeds.

8.9.8 Models Application in Mango Yield Estimation

Process based crop models are good helping tools to asses impact of different management and climatic variables on the mango crop yield. AquaCrop model was used in Egypt to simulate mango (*Mangifera indica* L.) yield under different climate, irrigation and soil conditions (Arafat and Maklad 2021). Their results indicated that calibrated AquaCrop model had great potential to simulate mango yield in Egypt. Input data required by the AquaCrop model have been shown in Fig. 8.3. Similarly, soil–plant-atmosphere continuum in AquaCrop have been elaborated in Fig. 8.4. Furthermore, input data required to run the AquaCrop input have been given in Fig. 8.5. Crop yield and water saving potential for AquaCrop model under full and deficit irrigation managements were studied by Abdalhi and Jia. They concluded that AquaCrop model has great potential to simulate full and deficit irrigation managements in crops.

8.9.9 Conclusion

In conclusion, the reduction of the climatic change effects in mango farming is crucial to support the mango agricultural sector to be sustainable and less vulnerable to climatic change impacts. In essence, it is now possible to protect mango production from the effects of climate variability and help improve fruit quality while protecting the sources of food and incomes of millions of producers. Speaking of the advantages of amelioration for the sustainability of agricultural land, it is possible to list several key positive outcomes. Less vulnerability to climate change stress factors like drought, high & low temperatures, and diseases & pests will result in better yield stability and less economic damage.

Further, sustainable practices will strengthen a favorable and more sustainable production environment, thus enhancing the quality of the soil, use of water, and reducing the use of chemicals, enhancing the healthy and sustainable production of the mangoes. Consequently, future adjustments and research are vital indefinitely. Due to the impacts placed on mango productivity, research towards the augmentation of mango varieties that can resist climate change conditions including drought and heat resistance and pests should be a focus of breeding programs. Such knowledge will have to be circulated and the strategies and interventions taken to another level and this can only be done through teamwork especially involving the researchers, policymakers, and the extension agents. Climate change adaptation will lie with the investment in climatology smart agriculture and reassurance of innovation to

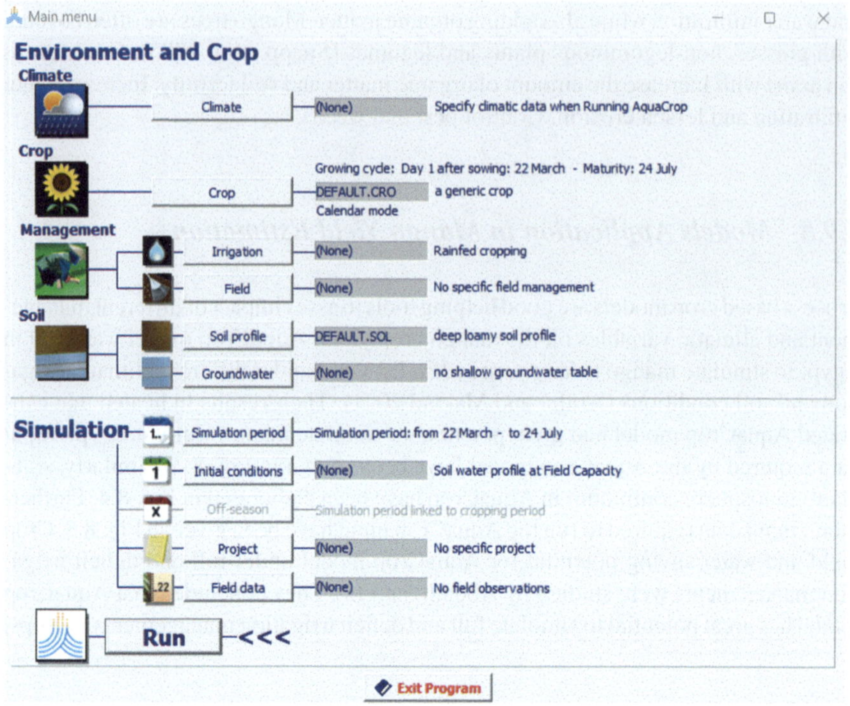

Fig. 8.3 Main interface of AquaCrop model

support mango farming in case of climate-altering and in supporting food security and sustainable expansion of agriculture around the globe.

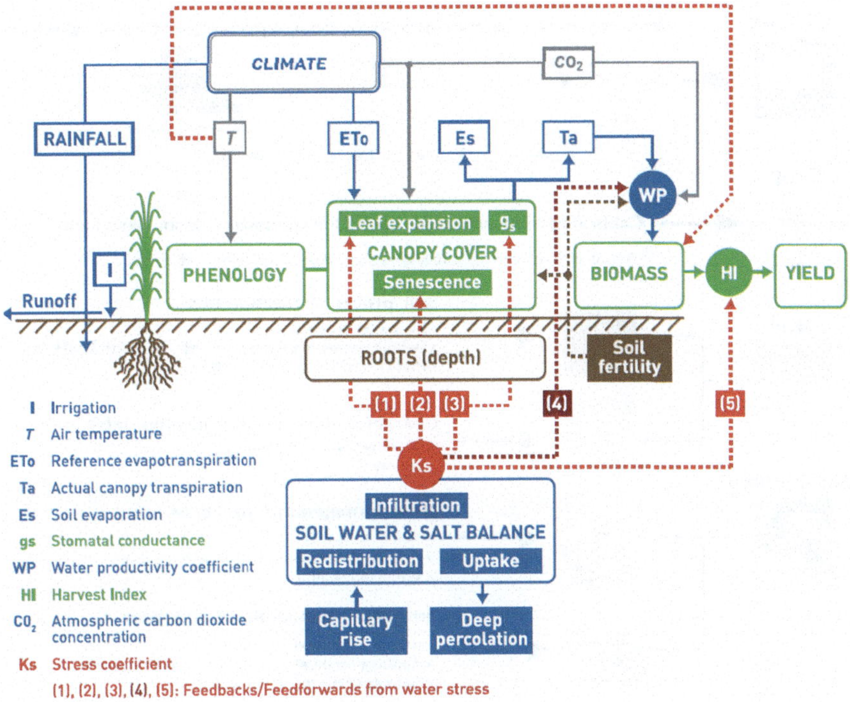

Fig. 8.4 Soil–plant-atmosphere continuum in AquaCrop. *Source* FAO AquaCrop reference manual: Steduto et al. (2008)

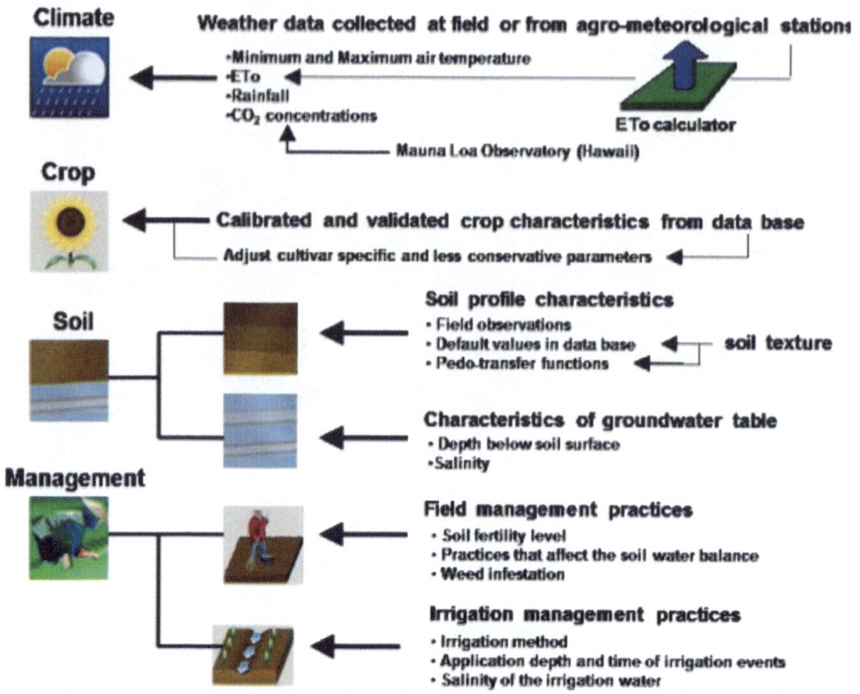

Fig. 8.5 AquaCrop input data requirements. *Source* FAO AquaCrop reference manual: Raes et al. (2009)

References

Al-Shammary AA, Kouzani AZ, Kaynak A, Khoo SY, Norton M, Gates W (2018) Soil bulk density estimation methods: a review. Pedosphere 28:581–596

Arafat IE, Maklad TN (2021) Modeling climate change impact on irrigation water requirement and yield of mango (Mangifera indica L.) in Egypt. Fundam Appl Agric 6(2):155–162

Awan A, Mustafa K, Zafar I, Mushtaq K (2010) Determinants of mango export from Pakistan

Bal S, Prasad JVNS, Singh V (2022) Heat wave 2022: causes, impacts and way forward for Indian agriculture

Brammer H, Brinkman R (1990) Changes in soil resources in response to a gradually rising sea level. In: Scharpenseel HW, Schomaker M, Ayoub A (eds) Developments in soil science, vol 20. Elsevier, pp 145–156

Bruce JP, Frome M, Haites E, Janzen H, Lal R, Paustian K (1999) Carbon sequestration in soils. J Soil Water Conserv 54:382

Choudhary G, Meena KK, Kaswan PK, Yadav S, Didal B, Choudhary S (2022) Impact of climate change on fruit and vegetable crops and its mitigation strategies. J Eco-Friendly Agric 17:205–210

Dambreville A, Normand F, Lauri PÉ (2012) Plant growth coordination in natura: a unique temperature-controlled law among vegetative and reproductive organs in mango. Funct Plant Biol 40:280–291

Davidson EA, Janssens IA (2006) Temperature sensitivity of soil carbon decomposition and feedbacks to climate change. Nature 440:165–173

Doran JW, Zeiss MR (2000) Soil health and sustainability: managing the biotic component of soil quality. Appl Soil Ecol 15:3–11

Evans RG, Sadler EJ (2008) Methods and technologies to improve efficiency of water use. Water Resour Res 44:2007WR006200

FAOSTAT (n.d.) Retrieved 21 July 2024, from https://www.fao.org/faostat/en/#data/QCL

Girija Veni V, Srinivasarao Ch, Sammi Reddy K, Sharma KL, Rai A (2020) Soil health and climate change. In: Prasad MNV, Pietrzykowski M (eds) Climate change and soil interactions. Elsevier, pp 751–767

Halder S, Abu Hasan Md (2020) Climate change and mango production. Chem Sci Rev Lett 9

Jones RAC, Barbetti MJ (2012) Influence of climate change on plant disease infections and epidemics caused by viruses and bacteria. CABI Rev 2012:1–33

Khan M, Ahmed N (2021) Sustainable management of mango nutrition for better yield and quality

Laxman RH, Annapoornamma CJ, Biradar G (2016) Mango. In: Rao NKS, Shivashankara KS, Laxman RH (eds) Abiotic stress physiology of horticultural crops. Springer India, pp 169–181

Makhmale S, Bhutada P, Yadav L, Yadav BK (2016) Impact of climate change on phenology of mango—the case study

Mason-Jones K, Robinson SL, Veen GF, Manzoni S, Van Der Putten WH (2022) Microbial storage and its implications for soil ecology. ISME J 16:617–629

Normand F, Lauri PÉ, Legave JM (2015) Climate change and its probable effects on mango production and cultivation. Acta Hortic 1075:21–31

Núñez-Elisea R, Davenport TL (1994) Flowering of mango trees in containers as influenced by seasonal temperature and water stress. Sci Hortic 58:57–66

Pandey DN, Gupta AK, Anderson DM (2003) Rainwater harvesting as an adaptation to climate change. Curr Sci 46–59

Pankhurst CE, Hawke BG, McDonald HJ, Kirkby CA, Buckerfield JC, Michelsen P, O'Brien KA, Gupta V, Doube BM (1995) Evaluation of soil biological properties as potential bioindicators of soil health. Aust J Exp Agric 35:1015–1028

Patel K, Khan M, Kumar Y, Yadav A (2019) Novel techniques in post-harvest management of mango: an overview. South Asian J Food Technol Environ 5:821–835

Patil A, Lamnganbi M (n.d.) Impact of climate change on soil health: a review. Int J Chem Stud

Raes D, Steduto P, Hsiao TC, Fereres E (2009) Aquacrop reference manual. FAO, Rome, Italy, pp 218

Rajan R, Ahmad M, Pandey K, Aman A, Kumar V (2020) Climate change and resilience in fruit crops, pp 337–3354

Reth S, Reichstein M, Falge E (2005) The effect of soil water content, soil temperature, soil pH-value, and root mass on soil CO2 efflux—a modified model. Plant Soil 268:21–33

Skendžić S, Zovko M, Živković IP, Lešić V, Lemić D (2021) The impact of climate change on agricultural insect pests. Insects 12:440

Smith JL, Halvorson JJ, Bolton H (2002) Soil properties and microbial activity across a 500 m elevation gradient in a semi-arid environment. Soil Biol Biochem 34:1749–1757

Snapp SS, Swinton SM, Labarta R, Mutch D, Black JR, Leep R, Nyiraneza J, O'Neil K (2005) Evaluating cover crops for benefits, costs, and performance within cropping system niches. Agron J 97:322–332

Staple WJ, Lehane JJ (1955) The influence of field shelterbelts on wind velocity, evaporation, soil moisture, and crop yield. Can J Agric Sci 35:440–453

Steduto P, Raes D, Hsiao TC, Fereres E, Heng L, Izzi G, Hoogeveen J (2008) AquaCrop: a new model for crop prediction under water deficit conditions. Drought Manag: Sci Technol Innov 80:285–292

Stenberg JA (2017) A conceptual framework for integrated pest management. Trends Plant Sci 22:759–769

Sukhvibul N, Whiley AW, Smith MK, Hetherington SE, Vithanage V (1999) Effect of temperature on inflorescence development and sex expression of mono- and poly-embryonic mango (Mangifera indica L.) cultivars. J Hortic Sci Biotechnol 74:64–68

Chapter 9
Mango Grafting in the Face of Climate Change

Noor-un-nisa Daudpota

Abstract Grafting plays a significant role in making mango plants more resilient to the challenges imposed by climate change, such as increasing drought, temperatures, irregular rainfall patterns, and soil salinity issues. In the face of climate change, grafting offers an adaptable solution for mango growers to build resilience in their orchards. By choosing suitable rootstocks and scions, farmers can ensure that their mango trees continue to thrive under this sudden face of climate change. In grafting, mango growers combine desirable traits such as heat and drought-tolerant varieties to raise more climate-adapted mango trees. Grafting increases both the quantity and quality of fruit production. In a nutshell, one can say that grafting secures the future of mango production in regions vulnerable to climate shifts.

Keywords Grafting · Climate resilience · Mango production · Vigor rootstocks

9.1 Grafting Methods in Relation to Climate Change

Mango is commercially perpetuated by both sexual (seed) and asexual (via vegetative parts) methods of propagation. The seeds of mango possess both characteristics of mono and poly-embryonic characteristics. Mono-embryonic varieties can produce a single zygotic embryo which is a cross between maternal and parental parents while poly-embroynic seeds can produce more than one embryo and they are zygotic (sexual) and nucellar seedlings which are identical to the parent plant (Bally 2006; Ram and Litz 2009). Mango being high cross-pollinated and heterozygous in nature requires asexual method of propagation for the commercial production of true to type plants. In asexual propagation, grafting is the most commercial method of propagation. Several grafting methods like approach, veneer/side, cleft/top and epicotyl are applied. Most of the grafting methods response best when the bark is slipping and the vegetative buds start pushing in late March and early April with no freezing nights. However, success ratio of grafting is generally based on the expertise

N.-n. Daudpota (✉)
Department of Horticulture, Sindh Agriculture University Tandojam, Sindh, Pakistan
e-mail: nnmemon@sau.edu.pk

S. A. H. Naqvi et al. (eds.), *Climate Change and Mango Production*,
https://doi.org/10.1007/978-3-031-87035-4_9

of the grafting and its specific climatic conditions. The brief of each method is given below.

9.1.1 Epicotyl Grafting

Epicotyl is different from other grafting methods because it works on very young seedlings. Its process ranges from few days to a one-month-old seedlings just above the embryonic leaves. The rootstock seedlings around 6–10 inches in height are used. The top section of the seedlings is cut off and attached with the wedge-shaped scion from desirable tree (Fig. 9.1). The data in Table 9.1 shows that grafting success percentage varies with the age and scion variety as reported by Jamali et al. (2024). They reported more than 70% grafting success from 50 days old rootstock. This method, like other grafting methods, is not suitable under open and harsh conditions of climate since young seedlings are more vulnerable to erratic precipitation, high temperatures, drought, or any other climate-related stressor. More resilient methods such as approach, top, cleft, and veneer work on older established rootstocks that are better equipped to handle stress, which would be more suitable in adapting to the effects of climate change (Ahmed 2020, 2023a; b). This type of grafting is often used in fruit trees specially mangoes and is suited for high density plantation where the canopy of the trees is managed. Epicotyl grafting promotes quicker growth and faster fruit production as compared to other grafting methods. This method is the most suitable method where space is limited. However, the rootstock selection by this method cannot be performed because of very young age of the rootstocks. This might also reduce the threat of tolerance in the face of climate change (Fig. 9.2).

Table 9.1 Effect of rootstock age and various scion varieties on the success (%) of grafting

Rootstock age	Scion varieties			Mean ± SE
	Langra	Sindhri	Chaunsa	
20 days	52.9	52.24	61.24	55.46 ± 5.01c
30 days	54.23	60.45	56.90	57.19 ± 3.12c
40 days	61.88	67.10	61.44	63.47 ± 3.14b
50 days	67.58	71.58	71.58	70.25 ± 2.30a
	62.79 ± 6.87b	54.15 ± 8.41c	68.84 ± 6.22a	

9.1.2 Approach Grafting

Approach grafting is also referred as Inarch grafting in which a rooted rootstock and scion of different plants is grafted with each other. The scion is taken from the

Fig. 9.1 Steps of epicotyl grafting followed at Sindh Agriculture University, Tandojam, Sindh Pakistan

desirable tree and grafted with one year old seedlings. This method is quite expensive, and also a large area is required for the practice and maintenance as compared to other grafting methods in which detached scions from their mother plants are used for grafting. However, this method of grafting provides a significant advantage in the adaptation to climate change since it allows the propagation of resilient varieties. Approach grafting, decreases climatic stress on the grafted plant due to the fact that scion and rootstock have a different root system. This technique has further been associated with assisting the progressive acclimatization of the scion towards

Fig. 9.2 Epicotyl grafted seedlings

the environmental conditions of the rootstock in a way that may lead to enhanced adaptability towards climate stresses.

9.1.3 Veneer/Side Grafting

Veneer/side grafting is another type of asexual propagation in which V-shaped cut is given on one side of the rootstock and wedge-shaped detached scion is placed in the cut of the rootstock. This practice is mostly done on young seedlings from 6 to 12 months old. In some places, it is practiced on over 1-year-old rootstocks also. However, the best time for this grafting is August to September; however it can be implemented also during the months of February and March. This method has a relatively high success rates even under variable conditions make this method effective because the graft union is well-protected. This method is less sensitive to erratic temperatures compared to other more delicate methods of grafting, like the epicotyl grafting process. In the scenario of climate change, veneer grafting is the best method for the proper selection of rootstock and scion. The rootstock and scion are selected on the basis of desirable traits.

9.1.4 *Top/Cleft Grafting*

This is a common method of asexual propagation that involves cutting the top part of an established and mature rootstock and attaching a scion to it by using cleft grafting method. Cleft grafting is an excellent method of asexual propagation that is used for both older as well as young mango plants This method is mainly effective particularly in the sense to rejuvenate older mango trees that are more vulnerable to insect pest and diseases or less productive, to introduce new hardy varieties with improved resistance traits such as temperature and drought tolerant or replacing old age tree plants with new varieties. The erratic precipitation and temperature patterns have highly affected the growth and productivity of the mango trees and there is dire need to adapt or graft mango varieties that can survive in less favorable conditions. There is a need that mango growers should introduce stress tolerant varieties that can bear erratic patterns of global climate change. Cleft grafting method is applied in topworking of the tree. This is an effective method of grafting that is used to replace less productive varieties or more susceptible varieties with more resistant varieties of mango. The grafting is done in winter and the most appropriate technique to induce rapid shoot growth and earlier fruiting. In addition, because the method of grafting is done at the top part of the rootstock branches, it helps in promoting a well-developed, upright tree which may be beneficial in parts of the country exposed to high winds or some other environmental stressors. This grafting is done on two pieces of wood above ground also known as bench grafting and can be done at a workbench or table, as opposed to the other styles which we do in the field. This is a great way of combining the genetic advantages of a scion root (fibrous, vigorous growth, tree size, resistance to diseases) with the genetic advantages of a scion wood that will eventually give the production of fruit.

9.2 Growth Related Issues of Grafting in Relation to Climate Change

Success in grafting largely depends on compatibility and physiological activities of scion and rootstock. Both factors are very responsive to external climatic fluctuations. Climatic variables such as temperature, drought, rainfall, humidity, and wind all play crucial roles in determining graft success. All climatic variables react with the plant processes of callus induction and differentiation, tissue regeneration, and generally development of the grafted tree. The effects of climate change are multifaceted disturbing at each stage of the graft development and growth.

9.2.1 Graft Union

The grafting union a point between the scion and the rootstock is much sensitive to multiple climatic stressors during the grafting process. Cells of the scion and rootstock both proliferate and merge to give rise to a vascular connection for the transport of water, nutrient and organic molecules. High or low temperature extremes can lower the cell activity needed for callus formation. When temperatures are high definitely evaporation rate will be high and at a high evaporation rates, cell turgor reduces that is much essential for cell division. Growers generally graft in the season with temperature ranges between 20 to 30 °C. The temperature above 35 °C, increases cellular respiration along with transpiration. Conversely, the low temperatures below 15 °C also inhibit cellular activity as well as enzymatic functions that ultimately affect graft union.

The low temperature disturbs the sap flow and lower the translocation of nutrients, carbohydrates, and growth hormones of the plants in the grafted union. The cell division of callus and differentiation are primarily temperature-driven functions. Moreover, declining photosynthetic efficiency at low temperatures debilitates the scion further, thus raising the chances for graft failure. In extreme lower temperatures, frost may form ice crystals within plant tissues, causing rupturing of the cells and resulting in complete death of the tissues. The temperature fluctuations also cause thermal stress that affects the major process of grafted plants include expression of genes, auxin transport, and production of the secondary metabolite. The prolonged elevated temperatures interfere with the accumulation of suberin and lignin at the graft union, hence making it more susceptible to the pests.

Graft union is also very sensitive to the temperature fluctuations of day and night. However, this is mainly based on the length and the amount of temperature variations. While differences between daytime and nighttime temperatures may be large, differential growth rates between the scion and the rootstock can create mechanical stress at the graft union. The loss in alignment within the cambial layers results in a decrease in the efficiency of vascular connectivity. Night-time rapid cooling also results in dehydration due to reduced uptake of water by the scion from the rootstock. Such a disequilibrium between supply and demand in water may exacerbate desiccation stress especially for dryland regions wherein diel temperature fluctuations are recurrent.

9.2.2 Disruption of Vascular Connectivity

The most important pathway in the transport of water and nutrients as well as carbohydrates for the plant is through xylem and phloem representing vascular connectivity. In stress condition of heat and water, air pockets are formed in xylem known as embolisms followed by dysfunction. This breaks down the hydraulic conductivity of the plant. Accumulation of reactive oxygen species (ROS) damage cells leading to necrosis or reduces the success of graft union. Extreme temperature and water stresses also affect phloem transport and ultimately photosynthesis. Due to poor establishment of phloem tissues in high temperature and water stresses leading to slow down growth and development of the graft union. Over time, imbalance in source-sink relationship in which scion acts as a source of photosynthates and the rootstock as a sink, results in stunted growth or may even lead to graft rejection.

9.2.3 Oxidative Stress and Accumulation of Reactive Oxygen Species (ROS)

ROS, partially reduced oxygen products, are produced in all aerobic cells of the plants. They are useful as regulators of cell processes and act as signaling molecules in physiological processes, but toxicity and death of cells become more frequent in response to their increased levels. Under climatic stresses, more particularly in drought and high temperatures, more production of ROS such as hydrogen peroxide, superoxide radicals and hydroxyl radicals occur. This causes cellular damage at the graft union, particularly in proteins, nucleic acids, and membrane lipids.

Accumulation of ROS may also disrupt callus formation, vascular connection of the grafted scion and rootstock, transport of ions like calcium and potassium, and signal transduction at the graft site. The presence of antioxidants such as superoxide dismutase (SOD) and catalase is very crucial for detoxifying ROS; however, during extreme conditions, the natural defense system of the plant gets overwhelmed, leading to irreversible graft failure.

9.2.4 Hormonal Signaling

Plant hormones such as auxins, gibberellins and cytokinin have also major role in the success of graft union and its growth and development. These plant hormones are more essential for the proper regulation of shoot and root growth and the establishment of the vascular tissues. The biosynthesis and transport of these plant hormones may be inhibited under harsh weather conditions. For instance, during the drought condition, there would be an increase in abscisic acid (ABA), an inhibiting hormone that prevents cell division and causes the stomata to close. Wilkinson and Davies (2002) reported that under short term drought, ABA synthesizes in roots and transported into guard cells to trigger stomatal closure leading to reduction in transpiration. While long term drought decreased stomatal density in newer leaves. ABA levels in the grafted mango plant can further delay the healing process at the graft union due to inhibiting auxin induced cell proliferation and differentiation. Gibberellin is another crucial plant growth regulator, especially for growth elongation and vascular differentiation. Cytokinin is another major plant growth regulator that is sensitive to heat and drought stress and decreases the activity of cytokinin and finally results in lower uptake of water and root.

9.2.5 Long-Term Graft Incompatibility Problems

Sometimes sudden climatic changes delayed graft compatibility and exert long terms effects on proper growth of the plant. This happens when graft union is successful in its earlier stages but fails later due to physiological or biochemical imbalances between the scion and the rootstock. Climatic changes also worsen compatibility and alter hormonal balance, nutrient partitioning, and vascular functions. In this regard, the trees may initially seem healthy, but they may deteriorate or suffer significant decline in productivity several years after grafting, compromising the longevity of an orchard.

9.2.6 Susceptibility to Pests and Diseases

The grafted or non-grafted plants are becoming more susceptible to the attack of insect pest and diseases in the running irregular climatic patterns. The example of excessive irregular raining causes the multiplication of soilborne pathogens such as Fusarium and Phytophthora, which cause root rot infections. They form colonies that weaken the graft unions causing systemic infections affecting both the scion and rootstock. High temperatures on the other side also render the immunity system of the grafted plants and makes more susceptible to infections from fungal and bacterial pathogens. This is also supplemented by suppressed production of phenolic compounds which a plant uses as its defense mechanism during high temperatures. The altered climate change may also affect the geographical distribution and behavior of the insect pests such as hoppers and aphids that attack mostly the mango trees.

9.3 Adaptive Strategies

As the climate continues to shift, growers must adapt new strategies to produce resilient grafted plants that can face harsh climatic conditions.

9.3.1 Selection of Climate Resilient Rootstocks

In the face of climate change, it becomes very important to select climate-resilient rootstocks for optimizing mango grafting. There must be proper selection on the basis of different environmental stressors such as temperatures, drought, waterlogging and salinity. There is dire need to do research on the identification of rootstocks that are tolerant to the various stressors. The rootstocks that have deep rooting system are more suitable for the areas facing drought. The drought stress induces adverse effects on scion as reduction in leaf area, wilting of leaf, reduced biomass and reduction in photosynthetic products and water contents (Shao et al. 2008). Grafting onto drought tolerant rootstocks can be a good strategy to alleviate these problems (Shehata et al. 2022). *Gomaera* (Spain) and *13/1* (Israil) have been identified as salinity tolerant mango rootstocks. *Pullima* and *Then Ca* are considered the best rootstocks for the dry areas of Sri Lanka and Vietnam, respectively (Saúco 2016). It is necessary that different rootstocks may be explored on the basis of site-specific study in relation to climate change. By selecting such rootstocks growers can enhance the ability of the grafted trees to survive and thrive in better way under climatic stress.

9.3.2 Optimizing Grafting Timings

The timing of grafting operations is becoming increasingly important as climate change disrupts traditional seasonal patterns. To optimize graft success, growers must adjust grafting schedules to align with periods of more stable and favorable climatic conditions. It would be better to plan grafting on the basis of the forecast data of climate and weather. Avoiding of grafting during critical periods of temperature, high humidity and excessive rainfall can reduce the chances of success of the graft union. Early morning or evening of the mild season is better time for grafting.

9.3.3 Proper Selection of Grafting Methods in Relation to Climate Stress

The selection of appropriate grafting method is very much critical for ensuring the successful establishment of mango plants. Each method of grafting has its specific merits and demerits. The selection of grafting method is mainly based on the time of grafting. However, sometimes growers do graft on the right time and get success in the graft union at earlier stage but somehow with sudden erratic changes of seasonal patterns fail the proper growth of the graft. Among the commercial grafting methods, mostly growers follow running traditional method of grafting. Epicotyl grafting method is one of the grafting methods that is quick and less time-consuming method as usually applied on young seedlings just above the cotyledonary leaves. Due to delicate young seedlings, this method cannot run in any climate related stresses. This method of grafting is particularly vulnerable to temperature and drought fluctuations in open field. However, if this method is applied in the shade house that may be better prevented from temperature fluctuation and intrusion of biotic stresses.

However, humidity must be managed in the shade house by using foggers, ventilating system and evaporative cooling systems. This method is beneficial form economical point of view in number of ways. By using this method one can produce number of grafts in very small space i.e. 100 grafts can occupy one square meter space and transport is also easy as compared to other traditional methods of grafting. Besides this method gives quick production within 2 to 3 months, no need to wait for one year to establish first rootstock seedlings. Expenses of the management and care (labour and other related input investments) of the plant are much reduced as compared to the traditional method of grafting. However, it is quite difficult to get large stock of the grafts within short period of time that ranges from 1 to 2 weeks as usually this grafting is done after seed germination within 15 to 30 days.

The top grafting is applied on mature as well as young seedlings. However, this method is more prone to various climate stresses. This is mainly due to the cuts that are done on the upper surface of the trunk or other branches which makes more vulnerable to the risk of desiccation. There is more chance of fungal infections in humid climate especially in heavy rains. Upper surface wounds lose moisture in hot and dry conditions that can impede the union between the scion and rootstock. This is very important for this method that grafting application is done at proper time otherwise success chances are less.

Veneer or side grafting can be more reliable and susceptible to excessive heat or cold fluctuations. This method involves making as sloping cut on the side of the rootstock and inserting a scion that aligns with the cut, which increases the contact area between the scion and rootstock in various climates. However, this method also requires careful timing to avoid climate extremes. This larger surface area enhances cambial alignment and improves sap flow and nutrient exchange, promoting effective integration of the graft union. By keeping the scion partially attached to the rootstock during the healing process, side or veneer grafting reduces the risk of desiccation, allowing the grafted plant to retain moisture and minimizing stress from high temperatures. The ongoing support from the rootstock helps the scion establish itself more efficiently, making this method increasingly popular as a climate adaptation strategy in regions prone to extreme weather.

Approach grafting can be saying more resilient method under climate stressors as both rootstock and scion have separate rooting system and complete established plants. This may reduce the risk of various stressors. This grafting method provides better protection from strong winds, erratic temperatures, heavy rainfall and fluctuations in humidity. This method making it more adaptable to changing environmental conditions compared to other grafting methods. However one big limitation is that you need complete established plants means more time, more space, more labour, more input and care. All these factors put the growers in economical loss along with lengthy time.

It may be concluded that each method has advantages and disadvantages.

9.3.4 *Protective Structures*

This is the best option in relation to grafting methods to face erratic climate changes. The use of protective structures such as shade house, rain shelters, greenhouses, cold frames, plastic tunnels and rain shelters has become a major adaptation strategy in mitigating the sudden climate changes. These structures regulate the temperatures provide protection from both extreme and low temperatures, intrusion of insect pest, maintain moisture levels and promoting faster healing at the graft union. In high temperatures shading structures reduces the evaporative demand on grafted plants, maintain moisture levels and promoting faster healing at the graft union. However, it is observed that humidity levels increase in the structures due to several interrelated factors such as reduces direct sunlight, limited air circulation and decreasing evaporation rates from soil and plant surfaces. This retention of moisture naturally increases humidity levels within microclimate of the structures. Exceeding levels of humidity causes fungal growth. High humidity levels can be reduced by installing vents or fans, foggers, dehumidifiers or you can manage at some level by temperature and water management practices. The use of protectives structures is much beneficial in regions where weather conditions are unpredictable, ensuring more stable grating conditions.

9.4 Insect Pest and Disease Management Under Climate Stress

Selection of disease tolerant rootstock is a valuable adaptative strategy to reduce the risk of infestation at the graft union. Definitely there are number of traditional methods are in use such as use of copper-based sprays, use of antifungal pesticides and other cultural practices that can reduces the chances of diseases attack. The grafting sites should be much away from the diseases affected plants and insect pest. However, selection of proper rootstock is utmost important along with other traditional practices. The rootstock studies in terms of insect pest and diseases at the grafting level is not studied much. There is a hire need to conduct these types of experiments to explore disease resistant rootstocks with reference to climate change. In this perspective there is also a need to explore breeding and biotechnological approaches to develop climate resilient rootstocks and scion varieties. Breeding programs are focusing on rootstocks with deeper root system, which can access water more effectively in drought prone areas. Normand et al. (2015) reported that the genetic improvement of mango plant material with respect to climate change is a promising way of adaptation. Use of advance molecular techniques such as marker-assisted selection (MAS) and genetic engineering are being used to develop mango rootstocks that are better equipped to tolerate extreme of temperatures, salinity and drought. The use of biostimulants would be essential in climate stressed conditions. The grafted seedling quality and growth can be improved with biostimulants and thereby contributing to plant resilience under harsh conditions (Malik et al. 2021) These inventions can complement traditional grafting practices by providing rootstocks with improved traits that can withstand climate induce stresses.

9.5 Education and Capacity Building for Growers

Climate change creates major challenges to the mango growers in all perspectives. It is very important to educate mango growers with the issues related with grafting and their adaptative solutions via trainings and workshops. Brochure and grafting manuals must be distributed by the academia to the mango industry in local as well as national languages.

References

Ahmed M (2020) Introduction to modern climate change. Andrew E. Dessler: Cambridge University Press, 2011, 252 pp, ISBN-10: 0521173159. Sci Total Environ 734:139397. https://doi.org/10.1016/j.scitotenv.2020.139397

Ahmed M (2023a) Climate change and farming system: a review of status, potentials, and further work needs for disaster risk reduction. In: Ahmed M, Ahmad S (eds) Disaster risk reduction in agriculture. Springer Nature Singapore, Singapore, pp 1–19. https://doi.org/10.1007/978-981-99-1763-1_1

Ahmed M (2023b) The science of climate change. In: Brinkmann R (ed) The Palgrave handbook of global sustainability. Springer International Publishing, Cham, pp 195–222. https://doi.org/10.1007/978-3-031-01949-4_22

Bally ISE (2006) Mangifera indica (mango). In: Elevitch CR (ed) Species profiles for Pacific islands agroforestry, Ver. 3.1. Permanent Agriculture Resources (PAR), Hawaii, USA

Jamali MF, Memon NW, Wahocho NA, Talpur A, Ahmed W, Jamali FA, Jamali AR, Mangsi HA, Ali A (2024) Comparative effects of varieties and rootstock age on success and survival rate of epicotyl grafting in mango (Mangifera indica L.). Pak J Biotechnol 21:219–227. https://doi.org/10.34016/pjbt.2024.21.01.891

Malik A, Mor VS, Tokas J, Punia H, Malik S, Malik K, Sangwan S, Tomar S, Singh P, Singh N, Himangini V, Nidhi S, Kumar G, Sandhya V, Karwasra A (2021) Biostimulant-treated seedlings under sustainable agriculture: a global perspective facing climate change. Agronomy 11:1–24

Normand F, Lauri PE, Legave JM (2015) Climate change and its probable effects on mango production and cultivation. Acta Hortic 1075:21–32. https://doi.org/10.17660/ActaHortic.2015.1075.1

Ram S, Litz R (2009) Crop production: propagation. In: Litz RE (ed) The mango, botany, production and uses. CAB International, Wallingford, pp 367–403

Shao HB, Chu LY, Jaleel CA, Zhao CX (2008) Water-deficit stress-induced anatomical changes in higher plants. C R Biol 331:215–225. https://doi.org/10.1016/j.crvi.2008.01.002

Shehata SA, Omar HS, Elfaidy AGS, El-Sayed SSF, Abuarab ME, Abdeldaym EA (2022) Grafting enhances drought tolerance by regulating stress-responsive gene expression and antioxidant enzyme activities in cucumbers. BMC Plant Biol 22:408. https://doi.org/10.1186/s12870-022-03791-7

Saúco VG (2016) Mango rootstocks, literature review and interviews. National Mango Board, Florida, USA. Available at: https://www.mango.org/wpcontent/uploads/2017/10/Mango_rootstock_Final_Report_Eng.pdf. Accessed 4th April 2020

Wilkinson S, Davies WJ (2002) ABA-based chemical signalling: the co-ordination of responses to stress in plants. Plant Cell Environ 25:195–210. https://doi.org/10.1046/j.0016-8025.2001.00824

Chapter 10
Water Management Strategies for Sustainable Mango Cultivation in a Changing Climate

Amar Razzaq, Muhammad Asad Ur Rehman Naseer, Mukhtar Ahmed, Shakeel Ahmad, Muhammad Yousaf Raza, and Syed Atif Hasan Naqvi

Abstract This chapter explores the all-encompassing water management strategies for sustainable mango farming in the face of climate change, with an emphasis on Pakistan's practices. Climate change has greatly affected the mango production through increased temperatures, variability in rainfall patterns, and moreover frequent occurrences of the extreme weather events thereby bringing about a need of the adaptive approaches to efficient water management. The chapter analyzes the conventional irrigation methods and their deficiencies, while also examining the potential of the novel methods of irrigation application, together with rainwater harvesting and recharge of groundwater techniques. One of the key aspects that have been studied in the second part of the chapter is whether or not the soil enhancement strategies such as mulching and cover cropping can improve water availability and efficiency. The incorporation of modern technologies such as remote sensing, GIS, mechanistic modeling and IoT-based smart irrigation systems in water use optimization is presented here. The chapter further considers the economic aspects that hinder the adoption of sustainable practices and the importance of policy support and communal initiatives in water management as well. Analysis of the present-day mango cultivation techniques and application of the emerging technologies e.g. remote sensing and process-based modeling (V-Mango) can help us to develop a

A. Razzaq
Business School, Huanggang Normal University, Huanggang, China

M. A. U. R. Naseer (✉)
Department of Agricultural Business and Marketing, Bahauddin Zakariya University, Multan, Pakistan
e-mail: asadrehman@bzu.edu.pk

M. Ahmed · M. Y. Raza
Department of Agronomy, Pir Mehr Ali Shah Arid Agriculture University, Rawalpindi, Pakistan

S. Ahmad
Institute of Agronomy, Faculty of Agricultural Sciences and Technology, Bahauddin Zakariya University, Multan, Pakistan

S. A. H. Naqvi
Department of Plant Pathology, Bahauddin Zakariya University, Multan, Pakistan

S. A. H. Naqvi et al. (eds.), *Climate Change and Mango Production*,
https://doi.org/10.1007/978-3-031-87035-4_10

framework for climate-resilient water management strategies thereby underlining the importance of continuous innovation and the institutional support for ensuring sustainable production in water-scarce regions.

Keywords Climate-resilient agriculture · Water management · Mango cultivation · Smart irrigation · Soil moisture conservation · Sustainable farming · Water scarcity

10.1 Introduction

10.1.1 Climate Change Impacts on Mango Cultivation

Climate change affects mango cultivation through rising temperatures, changed rainfall patterns, and more frequent heavy weather events that severely affect mango productivity and quality (Ogunbode et al. 2024). High temperatures can thereby result in heat stress that in turn would impair the fruit set and reduction of the yield both in quantity and quality of mangoes, according to climate change studies on mangoes (Legave et al. 2013; Makhmale et al. 2016). Furthermore, the hot climate has eventually an impact on the size of the fruit and hence the market value. The variations in rainfall and thence water supply is particularly critical in the important growth phases inducing severe moisture stress which may further impair the quality and productivity of the fruit (Ahluwalia et al. 2021). These abnormalities are often linked to an uneven water supply, which either results in extended drought intervals or unexpected heavy rain, both of which are detrimental to tree health and growth (Seleiman et al. 2021).

Furthermore, more frequent climate-induced drought events and temperature extremes are likely to increase the stress on mango trees from pests and diseases, thus making sustainable production more of a challenge across the main mango-growing countries like India and Pakistan (Khalifa and Abobatta, 2023). Mango farmers are now adopting adaptive practices such as strategic pesticide applications and modification of crop practices to confront the climate constraints and sustain stable yields (Magesa et al. 2023). Moreover, such adaptive root technologies are fundamental in keeping the steady output and rendering the mango stands resilient to the ongoing climate variability.

Sudden changes in the rainy season or long dry spells might affect the blooming and fruit growth in mangoes which in turn would lead to poor fruit yield and quality. Conversely, southern Pakistan districts are faced with delayed monsoons and decreased rainfall, this is the manifestation of climate change impacts on not only flowering but also stability in mango crops yield (Ahmad et al. 2023). Weather extremes like the cyclones in Mexico's coastal areas and unseasonal rainfall in parts of Brazil have been noted to cause physical damage to mango trees, hence, increase the fruit drop and therefore decrease the yield and economic returns of the

Table 10.1 Impacts of Climate Change on Water Availability for Mango Cultivation

Climate Factor	Impact on Water Resources	Effects on Mango Cultivation
Increased Temperatures	Higher evaporation rates and reduced soil moisture	Water stress leads to poor flowering, fruit drop, and lower yields
Erratic Rainfall Patterns	Uneven water availability (flooding or drought)	Overwatering causes root rot; drought hampers growth and fruit development
Reduced Groundwater Levels	Limited water availability for irrigation	Inadequate water supply affects mango growth, flowering, and fruit quality
Frequent Droughts	Decreased water resources and longer dry spells	Stunted growth, reduced photosynthesis, and higher susceptibility to pests and diseases
Increased Storm Events	Sudden heavy rains and flooding	Waterlogging damages roots, reduces aeration, and increases the risk of fungal diseases

farmers (Lima et al. 2023). These shifting conditions illustrate why it is necessary to develop adaptive strategies to ensure yield stabilization and reduce economic losses in mango-producing regions.

Rising temperatures have shortened the mango phenology, causing earlier flowering and fruit cover periods which can have serious effects on fruit quality due to the imbalance between sweetness and acidity. This type of fast development of mango often brings about unequal flowering and fruiting hence disturbing traditional harvest schedules and also being an operational quandary for the entire mail chain (Makhmale et al. 2016). In Pakistan, temperature-related changes have led to chaotic harvesting times, which in turn affects both local market supply and export stability (Anser et al. 2020). The addition of these challenges calls for an integrated approach using improved water management, point control on pests, and climate-oriented cultivation strategies to increase the resilience of mangoes in the many different areas where mangoes are grown (Antwi-Boasiako et al. 2024) (Table 10.1).

10.1.2 Importance of Water Management in Mango Farming

The water management is of large importance for the cultivation of mango, when climate variability increased. Among mango trees, it is important that the water received actually from the rainy season and other growth stages such as flowering, fruit setting, and fruit development be constant and proper according to the standard of the producer. Poor management of water can cause moisture stress and that, in its turn, can lead to i.e. reduced flowering, poor fruit set, and even poor yields. As for countries like Pakistan, where water scarcity is already a threatening fact made worse by fluctuating rainfall patterns and the rise in temperatures, the project of efficient water-use systems is a massive step to productivity (Durán Zuazo et al. 2019).

The necessity of water management can be seen in both traditional and modern farming systems. Traditional methods of irrigation, such as flood irrigation, when employed can often end up with an excessive waste of water, that may evaporate or fluid will run off the ground (Naseer et al. 2020). On the other hand, the new technique of irrigation known as drip irrigation and micro irrigation have made it possible for the water distribution to be extremely efficient. The water can be directed straight to the root zone area, minimizing the wastage of water, and hence water is always plentiful and available during the crucial phases of the growth of the plant (Razzaq et al. 2023). The use of efficient water management systems significantly improves the water situation in environments such as southern Pakistan where water sources are insufficient, and the demand is high.

Apart from the irrigational technologies, other water management technologies like mulching, rainwater harvesting, and soil moisture maintenance are becoming more common. Mulching allows the soil to retain more moisture by decreasing evaporation, while rainwater harvesting enables the farmers to store water during excess rainfall periods for drought use (Rojano-Cruz et al. 2023). These practices are being increasingly cultivated in Pakistan as farmers try to reduce the effects of unpredictable or irregular rainfall conditions and long-term droughts. When soil moisture management is done effectively, mango yields will be satisfactorily stabilized, moreover, the mango orchards would additionally get a guarded advantage against water stress caused by climate (Saddique et al. 2022).

Additionally, soil management goes hand in hand with water management. Soil, which contains organic substances that are healthier and has a higher water holding capacity, assists in mango cultivation in arid and semi-arid regions. The addition of green matter, for instance, to the soil and the use of agroforestry techniques, also add to the water efficiency of farms in mango-producing belts (Razzaq et al. 2019). In Pakistan, where such soil and water management have been introduced, both the sustainability and productivity of the plantation have been greatly improved and that is, in fact, a good solution for farmers, who still have to deal with the changing climate.

10.2 Climate Change and Water Resources in Mango Regions

10.2.1 Climate Trends Affecting Water Availability

Climate change is the root cause of the transformation of water availability in a wide variety of regions. Therefore, new challenges have emerged in both agricultural planning and irrigation. The increasingly hot weather has caused the evaporation rate to climb up, and as a result, the surface and groundwater levels are shrinking. This situation is happening in the regions that are traditionally dependent on stable water resources, which makes it even more difficult to manage water supplies (Baocheng et al. 2024). In agricultural areas, irregular precipitation further disrupts these temperature-driven changes, and thus the inconsistency of water availability needed for crop production is caused. Consequently, the number of natural disaster events in Pakistan have significantly increased since 1980s as shown in Fig. 10.1.

A further trend in water resources, which is quite significant, is the increased variability of rainfall patterns. A few regions that had previously had the privilege of experiencing seasonal rains have now gone through long dry spells and then been subjected to heavy rainfall. Thus, erratic patterns have given rise to a phenomenon called 'feast or famine'; where water is either in overabundance or in critical shortage. Such droughts not only make it difficult to schedule irrigation but also cause damage to the infrastructure such as irrigation channels and increase the potential for waterlogging and soil erosion (Salma et al. 2012). This adds to the problems for the farmers who are trying to ensure that their crops, including mango orchards, are getting a reliable supply of water.

The frequency of severe weather phenomena, such as prolonged droughts, flash floods, and cyclones, has also risen, thus adding more strain to the water resources. Droughts lead to the depletion of groundwater because farmers are compelled to turn to bore wells and other sources to ensure irrigation during the extended dry periods. On the other hand, excessive rainfall and flooding lead to surface runoff which in turn causes the loss of the most valuable topsoil and nutrients. Both of these extremes are the obstacles that come in the way of sustainable water resource management, thus, the ability to store and use water efficiently throughout the agricultural cycle is compromised (Ahmed et al. 2020).

In Pakistan, these climate results made by the changes in water availability have been remarkable in the worst affected areas, especially the ones with mango cultivation. The monsoon's variability in terms of its late arrival and decreased rainfall has made the irrigation of water in many important mango-producing areas quite challenging (Baocheng et al. 2024). Furthermore, the heavy reliance on groundwater, which results in the degradation of water tables, has made it much tougher

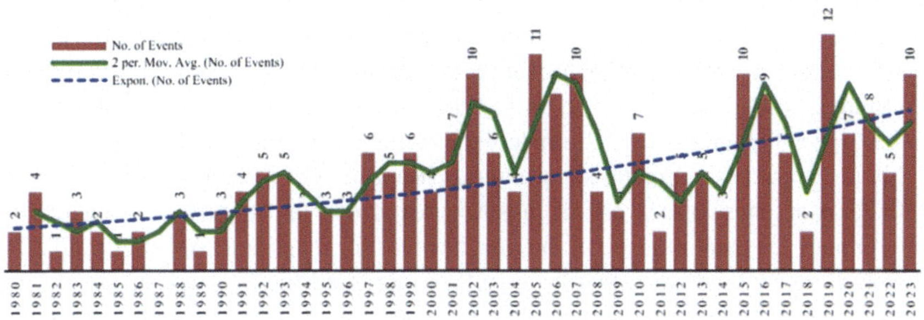

Fig. 10.1 Number of natural disaster events in Pakistan since 1980. *Source* GOP (2024)

for farmers to access the necessary water for their orchards (Dao et al. 2023). Thus, the need for more efficient water use has increased with the implementation of drip irrigation and rainwater harvesting, which is the way to fight the effects of changing water availability and secure the mango production in Pakistan.

10.2.2 Impact of Droughts, Rainfall Variability, and Temperature Extremes

Climate changes, including drought, rainfall variability, and temperature extremes, are some of the factors causing water shortages for agriculture, especially mango cultivation. Farmer's lives are ruined by the continuous lack of water which causes droughts; thus, it is not easy to develop the necessary moisture in the soil for the mango trees (Kumar et al. 2022). Moreover, in these critical times, more and more people resort to groundwater. However, that further exacerbates the situation by reducing the water reserves. Consequently, the increasing use of groundwater becomes a sustainability issue due to the fact that the majority of areas which undergoes droughts more commonly do not have any recharging facilities to compensate the excessive withdrawal.

Rainfall variability, which means the timing and intensity of rainfall are unpredictable, is another major cause. Across many places, traditional rainy seasons have become vibrant, thus bringing both more rainfall and extra dry periods. Unpredictable and sporadic neural rainfall patterns can be caused by both low soil moisture retention and increased surface runoff that decreases the irrigation water. Besides, excess rain at times leads to waterlogging, eventually harming mango tree root health and subsequently productivity (Ayanlade et al. 2018). The rain-fed irrigation systems which are the majority of the mango farmers are still vulnerable during these conditions.

Temperature extremes than ever before are now a further complicating factor in water management. Higher temps intensify the rate of evapotranspiration, which means an outcome pointing to more water being needed by the mango trees. During these periods of high temperatures, the case, the mango orchards need more frequent irrigation to make up for the water loss from both the soil and the plant, because of the intense sun. It takes your capabilities to be at their best to provide this extra demand, especially for areas of Pakistan that have water resources that are almost non-existent. The dual nature of climate change, i.e., the warming trend and the improperly functioning water supply, lead to an increase in moisture stress, which affects the fruit set, growth, and quality. Ultimately the future food supply and consumption is highly prone to these climatic abnormalities (Razzaq et al. 2024a).

In Pakistan, these effects are the most severe because the country is overly dependent on both seasonal monsoons and groundwater for irrigation (Anser et al. 2020). It is short and late seasons that have further increased the pressure on the use of water resources, hence the struggle of mango farmers to secure enough water is. In addition to this, the groundwater levels have dropped further due to both the growing

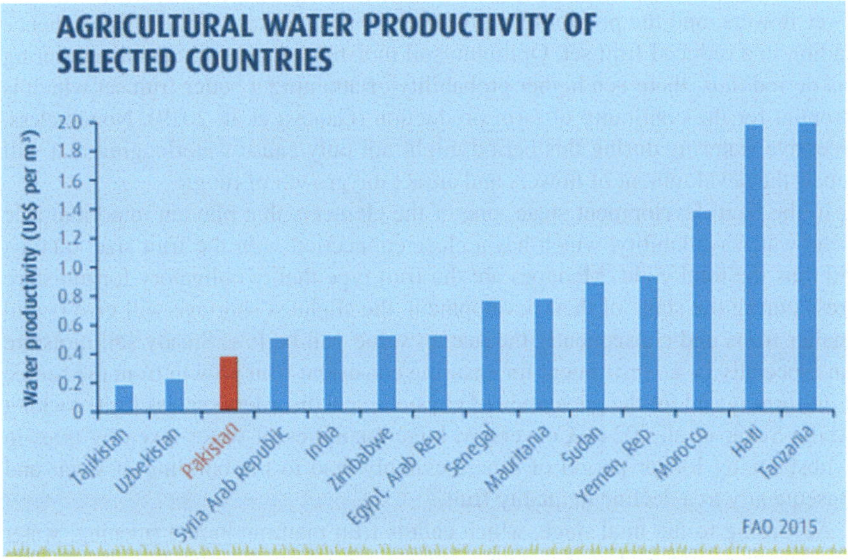

Fig. 10.2 Agricultural water productivity of selected countries. *Source* FAO (2015)

withdrawals and insufficient natural recharge, thus forcing the farmers to apply other effective water management practices (Razzaq et al. 2024b). Drip irrigation and other water-efficiency techniques like rainwater harvesting are becoming more popular to cope with these unexpected extreme conditions and also ensure the success of mango cultivation in the region. As it can be seen in Fig. 10.2, Pakistan has lower agricultural water productivity than many other developing nations.

10.3 Water Needs and Irrigation for Mango Trees

10.3.1 Water Requirements at Different Growth Stages

Mango trees differ in their water needs according to various growth stages, and this poses a huge influence on their productivity and fruit quality. The early phase of vegetative development, the provision of the necessary water is undoubtedly important to the construction of leaf and root components. Sustained water supply should be provided to young trees as through this they form a strong root system which ensures their growth later in life. A low level of water supply at this stage can result in poor tree growth which will limit its capacity to give higher production in the future (Wei et al. 2017).

Another vital phase is flowering where water management should be very precise. It is evident that water Stress during the flowering period may cause the plant to form fewer flowers, and the possibility of the perishing of flowers will be higher, hence leading to a reduced fruit set. Optimum soil moisture content is maintained during this period thus, there is a higher probability of attaining a better fruit set which is essential for the continuity of farm production (Ganeva et al. 2019). Nonetheless, excessive watering during this period might not only cause waterlogging that will impair the development of flowers and attract the growth of fungi.

In the fruit development stage, one of the elements that play an important role is the water availability, which has a close connection with the fruit size, quality, and thus the total yield. Mangoes are the fruit type that is obligatory for moisture stress during the stage of their development, the slightest shortage will give rise to smaller fruits and consequently the market value will be low. Steady soil moisture can especially be an instrument for ensuring consistent fruit growth from the source of uniformity and for the prevention of physiological disorders such as fruit cracking (Abdel-Sattar et al. 2024). Conversely, huge quantities of water taken by trees in a substantially longer period of time inevitably lead to the building of sugar and consequently to a decline in quality fruit.

According to the final stage, which entails fruit maturation and ripening, water application should be handled with caution to conserve fruit quality. The farmer should balance between the adequate supply of the water strength to the crop and the scarcity of the irrigation factor, which is a prerequisite for proper sugar accumulation and lower moisture level before the harvest. Good water resources are scare in areas such as Pakistan and the situation creates the need for the utilization of management practices such as drip irrigation thus enabling the farmers to save water while still providing maximum mango yield even in places where the climate is not stable (Prasad et al. 2018). The water requirements of mango tree at different growth stages are depicted in Fig. 10.3.

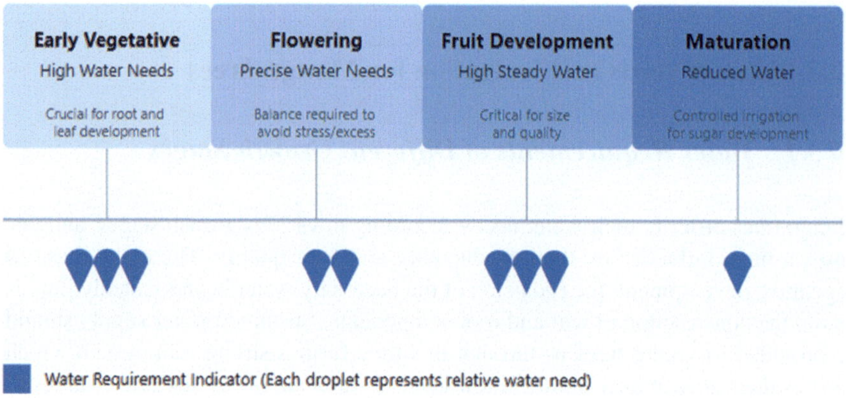

Fig. 10.3 Mango tree water requirements across growth stages. *Source* Author's own illustration

10.3.2 Factors Affecting Irrigation Needs: Soil and Climate Considerations

The water requirements of mango trees depend on several things such as soil and weather characteristics of the area, where soil properties and climatic conditions are the most important factors. Soil texture, structure, and depth along with the root distribution determine how the soil property is affected and the water movement within the soil is presented to the plant water uptake. For example, sandy soils have very low water holding capacities and need more frequent watering while on the other hand, clay soils that have high water holding capacity for a long time may present drainage problems. The regions with predominantly loamy soils have more favorable water retention and drainage balance, thus making irrigation scheduling easier (Mohammad et al. 2015).

The soil condition is also the most important thing for irrigation. Healthy soils with rich organic matter content hold the water and the plants do not need to be irrigated so often. Organic composting or mulching and cultivating cover crops are methods that can move the soil and make water infiltration easy, thus, runoff is reduced (Stevens, 2018). In Pakistan, which has many mango orchards in semi-arid areas, the healthiest soil is the best technique to make irrigation effective and to minimize water shortage (Soomro et al. 2022).

Weather conditions like temperature, humidity, and wind also play a major role in determining plant water requirements. Whenever there is a change in temperature the process of transpiration and evaporation increases the need for more frequent watering to substitute the lost water both from the soil and the plant. Low humidity and strong winds remove water from plants which also increases the water demand (Anser et al. 2020). For example, in southern Pakistan, when the combination of a high temperature with a dry wind is present, the trees girdled this to the mango trees for the irrigation scheduling which will be very difficult. Hence, the careful monitoring of water level will make sure that the mango trees are adequately watered without any scarcity (Ahmad et al. 2023).

Rainfall patterns are the other thing that determines the water demand for irrigation. The regions that experience unpredictable and highly irregular monsoon seasons mean that they will have to engage in supplemental irrigation to make up for the rainless periods. The ways of precipitation, which are becoming more erratic, are the main causes of increasing water stress in Pakistan. Farmers in places along with inconsistent precipitation patterns not only compete in these cases but also use very effective irrigation methods such as drip or sprinkler systems, ensuring their farms remain productive while reducing water consumption.

10.4 Current Water Management Practices

10.4.1 Traditional Irrigation Techniques and Their Limitations

The mango farmers have been using traditional irrigation methods such as flood irrigation and furrow irrigation which have been extensively applied in mango farming for a long period of time. This is the case when large amounts of water are channeled through the field, and although they may look impressive, they are actually a waste of water. Flood irrigation has as its main disadvantage the fact that the water is lost because of evaporation and runoff, which is a typical situation in regions with high temperatures (Razzaq et al. 2018). The first reason relates to the fact that excessive water is applied to the trees beyond the actual needs of the mango trees, which results in water wastage and lower efficiency of the whole system.

Another limitation is the uneven distribution of the water throughout the field. The traditional irrigation systems are not that precise in terms of making sure each tree gets the right amount of water (Razzaq et al. 2019). As a consequence, some of the areas may be waterlogged while others may be under-irrigated. A situation when the soil is excessively moist, due to waterlogging, causes poor aeration, the roots rot, and diseases become highly susceptible. On the contrary, insufficient irrigation results in moisture stress which in turn causes the plant health deterioration, decrease in flowering, and fruit development.

In addition to that, these types of irrigation systems are not only inappropriately used, but they also bring about soil destruction and nutrient leaching. Moreover, in the case when large quantities of water are applied too fast, they can transfer the topsoil and wash out the vital nutrients, hence causing the soil to be less fertile over time (Razzaq et al. 2019). This is primarily a problem in areas where the cultivation of mangoes is done on slopes or where there are soils with high vulnerability to erosion. Besides, the labor-intensive nature of these techniques makes them less feasible in regions with a shortage of workers or increased labor costs.

In the areas, like Pakistan, water resources are already scarce and traditional irrigation practices are adding to the water shortage problems. The farmers mainly are in a difficult position when they try to get the necessary water for their orchards, especially in drought times or when the water is not enough for the irrigation (Razzaq et al. 2022a). This is a reflection of the need for more efficient irrigation systems such as drip or micro irrigation to optimize water use and ensure that mango cultivation is sustainable under the climate changes.

10.4.2 Existing Water Conservation Methods

Currently existing water conservation techniques are of tremendous and first-order importance in implying water efficiency in mango production the water resources are likely to be threatened by the possible climate change. Moreover, covering the soil surface with polyethylene mulch or other organic materials inhibits soil evaporation, thereby enabling moisture storage for longer periods. Mulching is the practice of covering the soil surface with organic materials like straw, leaves, or even plastic sheets to protect the soil from moisture loss. This practice not only prevents evaporation but also limits water-loving weed growth and through the thermal regulation of soil the soil also appears darker (Hahn et al. 2022). However, good water-holding cover mulching has been widely used in Pakistan to alleviate the problems brought about by high temperatures and irregular rainfall thus it can be considered one of the adaptation strategies in water-scarce areas (Ahmad et al. 2023).

Rainwater harvesting is another water-saving one that is practical. Through collecting and storing rainwater, they can guarantee that there is a backup supply during dry spells. In areas where rainfall manifests itself in discrepancies and henceforth it cannot be properly utilized by rainwater harvesting systems, the supply disparity between water availability and crop needs can be bridged (Razzaq et al. 2019). Next to that, water can be employed in direct irrigation and recharging the groundwater which is associated with the occurrence of constant water supply during mango's critical growth phases for both farmers and nature alike. It is a particularly useful method in rural areas of Pakistan, where rainwater dependency sometimes leads to droughts during tree fruit growth phase surpluses.

Soil moisture conservation techniques, including the application of cover crops, are also useful in controlling water use more efficiently. Cover crops promote the natural symbiosis between plants and soil; they maintain soil structure, and reduce erosion, as well as bringing organic substance in the soil which further increases its water holding capacity. Furthermore, no-tillage practices are beneficial to moisture conservation since the soil is preserved under its natural structure and its disturbance is limited (Scavo et al. 2022). These among many others serve their purposes in regions of semiarid climates by ensuring the continued sustenance of the growth of plantations despite limited rainfall and an elevated amount of water loss through Evapotranspiration.

The integration of these water-saving practices along with smart irrigation methods such as drippers and micro-irrigators offers a complete plan to use water resources in a sustainable way (Soomro et al. 2022). The farmers from southern Pakistan resorts to these tactics in combination with each other to counter the scarcity of water because of climate change and climate variability, thus sustaining the mango cultivation (Ahmad et al. 2023). Using these existing water-saving methods increases water use efficiency and also contributes to the long-term resilience of mango orchards in the midst of a changing environment.

10.5 Climate-Adapted Water Management Strategies

10.5.1 Efficient Irrigation Systems: Drip, Sprinkler, and Micro-Irrigation

The progress of water management and resource distribution is the very basic factor for efficient irrigation systems that are characteristic of mango farming, especially when climate change becomes a primary concern for ecosystems worldwide. One of the practices listed water-wise, is drip irrigation that constitutes a daily dose of water to the roots of each tree which positively affect the reduction of water losses due to evaporation and wash-offs. The new watering approach has been so far incredible in making mango trees not only be ensured with constant moisture but also safely handled from the aspect of soil which could easily suffer from issues such as waterlogging and becoming the home of root disease clusters (Baocheng et al. 2024).

Through the judicious reduction of integral amounts of water and maximizing water resources, healthy growth and quality yields can be sustained and increased even in the regions prone to dryness like in Pakistan where the water is in short supply. Another system of irrigation which can effectively deal with loamy soils is given by the introduction of sprinklers that are able to make a perfect imitation of natural showering (Ashraf, 2016). In this way, usage of this system is especially convenient in areas with moderate climate conditions. Sprinkler equipment is more suitable for the soils characterized by good water retention and drainage preliminaries, thus water which regularly penetrates through the whole situation is Sprinkler possibilities are reduced in dry climates because of high rates of evaporation. The majority of water that is intended for the plants would be lost due to evaporation (Ravikumar, 2023). The use of irrigation systems in Pakistan is sometimes feasible in areas with climatic conditions suitable for low evaporation and reliable water supply.

Micro-irrigation technologies, which include both micro-sprinklers and drip irrigation, integrate the advantages of both precise water application and the capability to regulate water delivery according to different plant growth stages. They are important solutions for water-saving with a reduced risk of errors caused by irrigation overproduction, as well as efficiency when higher inputs are required during fruiting and harvesting time (Mohan et al. 2024). The use and the success of micro-irrigation in the southern zone of Pakistan where water shortages have been a main obstacle for the farmers seem to be the best solution in providing the farmers with the chance to overcome the limitations brought about by the unpredictable weather conditions. The pumping of irrigation systems that have been built through such technology by the farmers makes it possible for them to both gain stable and quality crop production as well as boost their mango orchards' resistance to the high-level environmental threats (Ahmad et al. 2023).

10.5.2 Rainwater Harvesting and On-Farm Storage

Water storage and rainwater harvesting are the most effective methods to supplement water for the mango orchards in the areas that face water scarcity and irregular rainfall nowadays. Rainwater harvesting is the practice of collecting rainwater and storing it from rooftops or surface runoff and from the catchment areas especially designed for it. This secondary water source helps the farmers by providing the necessary water for irrigation during the dry seasons, which are caused by the occurrences of natural rainfall being inadequate. Besides, on-farm storage, such as ponds or tanks, can allow captured rainwater to be stored for later use, which is particularly beneficial during the critical periods of flowering and fruit set (Mishra et al. 2020).

Apart from the fact that monsoon has become more and more uncertain in areas such as Pakistan, rainwater harvesting is one way of easing the strain on the depleting groundwater sources. Also, on-farm storage structures can be customized to suit the size of the farm and the specific water needs. The stored reservoir water can be used in the recharge of groundwater during surplus times which contributes to the long-term water balance issues. In semi-arid areas, the method of rainwater harvesting guarantees that there is water for the people even when the rains are late or fail totally.

Another benefit of rainwater harvesting is its ability to reduce soil erosion and nutrient loss by the fact that it cuts down the amount of runoff. Farmers can minimize the erosion of topsoil that provides the nutrients which guarantee soil fertility and structure thus enhancing the productivity of their farmland, by capturing rainwater that would otherwise flee away. The water that is kept can also be useful in the process of irrigation which is by being applied directly to the root zone of mango trees, so that losses through evaporation are minimized and water reaches the places, where it is needed most. The water harvesting if properly managed, can also be a cost-effective solution for resolving the water scarcity problem and making mango orchards more resilient to climate variability (Mishra et al. 2020).

Successful execution of rainwater harvesting and on-farm storage systems needs proper integration of technical know-how and community involvement. The farmers will need training in designing, building, and maintaining rainwater harvesting systems as well as understanding optimal water use practices in order to maximize the benefits. Sustaining these practices in Pakistan would be the people's role and the government's helping hand, and this is what is needed to make sure that farmers have the resources, information, and skills to make water harvesting a part of the water management strategies they have in their farms.

10.5.3 Groundwater Recharge

Groundwater recharge is vital for the sustainability of mango production, especially in regions where over-extraction of groundwater has caused severe depletion of the resource. Mango orchards depend heavily on the groundwater for irrigation, which is mainly the only source of water during the drought periods. Methods such as recharge wells, percolation tanks, and check dams can be employed to prevent water from escaping to the surface and channel it instead to the aquifers that lie beneath the surface. Rechanneling the extra runoff to the recharge constructions causes the farmers to gain in reduced and renewable use of water and the water supply to be secure for future demands (Razzaq et al. 2024b).

Regions such as Pakistan, where aquifers are decreasing, say that groundwater recharge schemes have become crucial in a way. Percolation ponds are some of the more imaginable choices because it is from them that the watershed can be trapped and stored effectively, and then it percolates down through the soil and completes the underneath. In semi-arid zones where the rain falls sporadically but may result in massive runoff that could be wasted, these ponds prove to be extremely important. In the same way, check dams built across small streams retard the surface water thus allowing infiltration and aiding groundwater recharge (Razzaq et al. 2024b).

The production technology based on groundwater's long-term benefits is its ability to stabilize water tables so as to ensure water availability for irrigation in agriculture. The groundwater recharge helps to resolve the immediate water scarcity problem as well as the long-term problem of lowering water tables which may lead to the dwindling of agricultural productivity in the mango-growing areas. The transition from over-extraction of groundwater supplies has recharge efforts as one of the elements that ensure the sustainability of agriculture, the mango orchards are guaranteed access to water, notwithstanding the different growth stages.

Community involvement and government backing up are the terms under which a groundwater recharge project will be implemented successfully. Based on the farmers' proper instructions on the benefits of the techniques and their collective effort, we have the community to bring more effect through the grassroots movement (Ahmad et al. 2020). The Pakistani government's programs of capacity-building in the construction of recharge facilities as well as support for local initiatives have shown some progress, but they do not cover all the water problems. The joint efforts of the authorities, local communities, and agricultural specialists can help to make groundwater recharge become part of the strategies of water management in mango cultivation.

10.6 Soil and Crop Management for Water Efficiency

10.6.1 Mulching and Soil Moisture Conservation

Using mulching in mango orchards is a very effective strategy for saving soil water which is particularly important considering that climate variability and water scarcity are currently here. Mulching, however, when the soil surface is covering the soil surface with organic or inorganic materials such as hay, leaves, wood shavings, or even synthetic films generates a substantially low rate of evaporation from the soil. This is exactly what is needed in conditions of very high temperature, when evaporation is at its peak. Covering the soil with mulch not only promotes water retention but also helps to modulate the temperature thus, creating a more suitable environment for the roots, which is parallel with the statement that roots tend to grow deeper in a cooler environment (El-Beltagi et al. 2022). Apart from being a method of adaptation which has already been cited by many regions that experience limited water supplies, the technique has become a major part of the farming methods, for instance, mulching in cities like Pakistan.

Besides moisture conservation, mulching is a practice that is indeed a more sustainable method of mango cultivation that brings a range of other advantages. The organic mulches composed of materials such as straw or leaves, which get decayed over time add massive amounts of organic matter to the soil have proven to be highly beneficial in this respect. This serves as a basis for promoting good soil structure, allowing for a reduction in water and nutrient deficiencies through a more profound nutrient availability and infiltration among the roots, as well as making the surface of the land a more appealing and effective environment (El-Beltagi et al. 2022). A key link between mulching, which has a beneficial effect on soil health and water efficiency, is good soil that contains a higher organic matter content and thus can store more water and make water distribution smoother for the plants. In desert-like regions such as semi-arid zones, good soil becomes a big challenge because additional water is not available, and therefore mulching is a major means of managing soil fertility (El-Beltagi et al. 2022).

For another reason, using mulch can suppress weed weeds which are great competitors for air, space, water, and nutrients with the main crop which decreases inter-plant competition. Plants like weeds not only consume lots of soil moisture but also reduce the amount of water available to trees, particularly mango trees during the reproductive stages like flowering and fruit set, which is already very crucial for the trees. Weed management through mechanical physical means such as mulching prevents higher underground utilization of water and as a result facilitates more mango trees' production (Iqbal et al. 2020). Whether it's the case of a mango farmer in southern Pakistan where manual weed control and irrigation cost the farmer to a great extent and water rationing is the key to having a productive orchard, this solution is more than relevant.

Altogether, the fact that mulching has been incorporated as an instrument for moisture retention is the key to the success of the mango plantation. Apart from saving water, it also performs the role of improving the soil quality, which then kills weeds or limits their growth and at the same time increases the capability of mango orchards alongside other stress-tolerant plants to withstand climatic effects. Therefore, people living in Pakistan and such regions with high occurrences of drought can cut down their water costs drastically with this tactic thus, simplifying the sustainability program for mango production (Table 10.2).

Table 10.2 Water management strategies for sustainable mango cultivation

Water management strategy	Description	Benefits for mango cultivation
Drip irrigation	Delivers water directly to the root zone with minimal waste	Improves water use efficiency, reduces weed growth, and ensures optimal hydration
Mulching	Covers soil with organic or inorganic material to conserve moisture	Reduces evaporation, improves soil moisture retention, and lowers soil temperature
Rainwater harvesting	Collects and stores rainwater for irrigation	Ensures water availability during dry periods and reduces dependency on groundwater
Deficit irrigation	Applies water below full crop water requirements strategically	Enhances water efficiency and maintains acceptable yields under limited water conditions
Scheduling irrigation	Times watering based on soil moisture and crop needs	Prevents overwatering or underwatering, optimizing growth and fruit production
Soil Moisture monitoring	Uses sensors to measure soil water levels	Guides precise irrigation, avoiding water stress and wastage
Agroforestry practices	Combines mango cultivation with trees and vegetation	Improves soil health, reduces water runoff, and enhances microclimate stability
Use of Treated wastewater	Recycles treated wastewater for irrigation	Reduces freshwater dependency and provides a sustainable water source for mango cultivation

10.6.2 Cover Cropping and Agroforestry for Water Retention

Cover cropping and agroforestry are beneficial systems for the retention of water in mango orchards. Cover crops, like legumes and grasses, are planted between the mango tree rows to prevent soil erosion and reduce evaporation of moisture in the soil. In this way, the crops can maintain a green cover on the soil surface and thus, the soil moisture is preserved, especially during the dry periods. Moreover, cover crops are a source of organic matter as they rot, which results in developing the soil structure and making it a better water retainer (Salceda-Gonzalez, 2023). In dry areas such as Pakistan, the utilization of cover crops has demonstrated potential in augmenting the moisture content of the soil and thus, improving the overall health of the soil, which in turn results in more resilient mango production systems.

Agroforestry, the technique of inserting other trees among mangoes, is as well a key approach to water conservation. In agroforestry systems, by the creation of a multilayered canopy, wind speed is reduced and the soil is shadowed, thus lowering the evaporation rate. Trees such as nitrogen-fixing species can be cultivated with mango trees to improve soil nutrition as well as the addition of leaf litter which adds organic matter. This upgraded soil condition is not only conducive to the retention of water but also gives the mango trees the chance to utilize the deeper soil water during drought periods. Through agroforestry practices it is therefore possible to provide a buffer for mango trees against the effects of varying rainfall and rising temperatures (Fahad et al. 2022).

On the other hand, one of the advantages of cover cropping and agroforestry is that they have a tendency to increase biodiversity in the mango orchards. The existence of diverse plant species is a major contributor to a stable microenvironment, which is, in turn, responsible for improved pest and disease management. The lower pest density translates to more efficient use of water resources, as the healthy trees recover from pest damage without the need for intensive irrigation (Salceda-Gonzalez, 2023). This is of great importance to mango farmers in places like southern Pakistan, where climate change has increased pest prevalence and thus made water management more difficult.

Summarizing the situation, the inclusion of cover crops and agroforestry in mango farming systems is one of the major ways through which water retention is improved and thus climate stressors are avoided. Apart from the fact that water evaporation is minimized, soil enrichment is done as well, and a healthier ecosystem is being created, these practices are offering a sustainable solution to some of the most serious problems mango growers have today.

10.6.3 Enhancing Soil Health for Better Water Retention

The development of soil health is a central method of water retention in mango orchards. Soil that is full of organic matter enables to hold moisture for a longer time, thus, presenting it as the most sustainable form of water management. Organic matter, including compost, manure, and decomposed plant remains, is a factor that assists in the soil structure formation, enhances porosity, and increases the soil's water capacity (Lal, 2020). This is of utmost importance in Pakistan, where water scarcity is a major concern, and the need for water use efficiency becomes a veggie peck. The soil with the highest bacterial activity causes nutrients to be released at the highest rates, which in turn improves soil health and water infiltration, giving the mango tree consistent yields.

Techniques like cover cropping, crop rotation, and the use of organic supplements are also effective in soil health improvement. Cover crops are able to perform such functions as being a natural source of organic matter, a means of weed control, and a means of erosion prevention apart from the main crop. These advantages, in this case, we are talking about better soil moisture retention and the reduced need for frequent irrigation. Crop rotation, on the other hand, acts as a brake on soil depletion by the introduction of a variety of plants to the field. In this way, the nutrient cycle is sustained because depleted nutrients are replenished and soil fertility is increased, leading to the soil's improved capacity to hold water, which is vital for the healthy growth of mango trees (Reeves, 2018).

Soil aeration is the most unimportant but yet the second most important factor to solve the soil health for the increased water holding capacity. Compacted soils tend to have less porosity, thus, restricting the root growth and reducing the soil's capability to store and give water passage. There are good techniques like minimum tillage or controlled aeration that can help farmers to make the soil structure to be more loose, penetrate more water, and also reduce runoffs. Areas of Pakistan that are semi-arid and where soil aeration is introduced to compress areas and where the soil is of poor quality have more root expansion and water availability which contribute to the health and productivity of the mango trees.

In general, the incorporation of organic matter, crop rotation, and aeration are the most effective methods to enhance soil health and in turn, water retention capabilities. These strategies are not only facilitating better water management but also helping mango orchards to be more resistant to climate-induced stress. In regions like Pakistan, where water availability is often unpredictable, farming on well-maintained soil is a sustainable solution that assures efficient water use along with higher mango yields.

10.7 Innovations in Water Management

10.7.1 Remote Sensing and GIS for Monitoring Water and Soil Moisture

The use of remote sensing and GIS technologies in the agricultural sector, especially in mango cultivation, has brought significant changes in water management through the precise monitoring of soil moisture and water distribution. Remote sensing works by employing satellite images and aerial sensors that are able to collect data on the extent of soil moisture, plant health, and the availability of water. This technological advance supports the identification of areas that are characterized by the moisture stress phenomenon, thus enabling the farmers to make sound decisions regarding the right time and place for irrigation (Rocha et al. 2012). In countries like Pakistan, which are suffering from deteriorating water resources, remote sensing helps irrigation to be a smart technology that empowers farmers to use water only where it is most needed, thus eliminating the waste of the resource and enhancing crop productivity.

A geographic information system (GIS) if combined with remote sensing devices, will offer the opportunity for spatial analysis that is highly valuable in dealing with mango orchards efficiently. GIS can be used to map out soil properties, moisture levels, and other vital environmental factors, which will in turn allow the farmers to implement irrigation schedules that are the best for the different zones within their orchards. A zonal approach is quite more productive in heterogeneous terrains where soil structures and water needs differ significantly over a farm setting. In Pakistan, the introduction of GIS-based mapping has made it possible for the authorities to develop better plans and direct their efforts toward the right places, thus, increasing water use efficiency and sustaining the mango production industry.

In addition, these systems also serve as an integral part of long-term water management plans by delivering data about climate developments, rain periods, and water table levels. The inclusion of historical data facilitates gaining insights into the future water needs of farmers, with GIS applications offering the way to plan for drought or heavy rain periods. Apart from that, the remote sensing and GIS methods through data from climate variability are giving growers the power to adjust to the new requirements for the water supply in real-time. This comes as an increase in the adaptability of mango farming to climate changes.

10.7.2 Smart Irrigation and IoT Solutions

Smart irrigation systems and Internet of Things (IoT) solutions bring in great advancements in water-saving techniques in mango cultivation. Smart irrigation, which refers to automatic irrigation systems utilizing real-time data to change water applications as the crops demand, plays a crucial role in this development. Moisture level can be sensed by special sensors located in the soil, while weather stations predict rainfall, temperature, and the rate of evapotranspiration. Using these data resources, such an irrigation system allows mango trees to absorb just the right amount of water they need for survival and growth while, of course, wasting no water the additional part (Obaideen et al. 2022).

IoT solutions are even more useful than the irrigation smart systems because they equip people with one platform that connects a whole array of devices-such as soil moisture sensors, weather stations, and irrigation controllers among others- which communicate with one another seamlessly. Farmers are enabled to supervise these devices through apps on their mobiles, commanding them remotely and adapting irrigation schedules due to this possibility of getting real-time info about their orchards (García et al. 2020). In Pakistan, where labor availability can be a problem, IoT will provide a major assisting mechanism for big mango plantations that are hardly monitored by people only, this means water will be spared and workforce salaries will be lowered.

In the most conducive environment of mango growth, smart irrigation and IoT are the pathways to precision agriculture which promote water and fertilizer delivery to every tree at adaptive levels. Through this precision, production is improved, fruit output is higher, and the environmental impact is lowered. Even though the upfront costs for these technologies might be high, the long-term profits in terms of water review and utilization of the marginal increase in the cost to production make it a good strategy for farmers striving to adopt climate-smart practices in mango cultivation.

10.7.3 Decision Support Tools for Optimized Water Use

Decision support tools (DSTs) are software applications that aid farmers in making informed choices on irrigation and water management. This kind of software uses data from various sources such as climate models, soil sensors, and historic weather records to give suggestions for irrigation scheduling and water usage (Mabhaudhi et al. 2023). The DST can provide farmers with information on the best time to water and the amount of water to be used, which is especially important in areas with fluctuating climate conditions like Pakistan. One well-known DST is the use of crop water stress indices, which help farmers in determining when their mango trees are experiencing water stress and what the most effective way is to respond. These instruments by tracking soil moisture data, temperature, and crop growth stages are able to estimate water requirements and command irrigation strategies. To mango farmers in the least water-supplied areas (such as southern Pakistan), DSTs can help through water use efficiency is taken care of while maintaining good crop health and productivity. As one of the functions of irrigation scheduling, the decision support tools also assist in integrated water resource management by helping farmers foresee their future water needs and make plans for periods of scarcity. If integrated with governmental databases and local weather stations, DSTs are capable of area-wide water management, which means all the water is shared equally among farmers. This joint approach is essential to people with little water in mango growing communities, especially in the face of climate change. A comprehensive overview of traditional vs modern and water conservation technologies in mango farming is shown in Fig. 10.4. A process-based model, Mango-GPP was developed by Tang et al., (2023) to simulate gross primary productivity (GPP) of mangrove ecosystem. They also developed Mango-GPP to simulate GPP under different environments. However, they suggested that there is still need to improve model for LAI prediction and parametrization of different vegetation indices. V-Mango: a functional–structural model of mango tree growth, development and fruit production was developed by Baig et al., (2013). Observed and simulated phenological stages of mango have been shown in Fig. 10.5 as elaborated by Dambreville et al., (2015). Similarly, observed and simulated phenological stages D1 (appearance of the main axis), D2 (bracts begin to fall), E (secondary axis moving away from main axis), F (flowering, from the first to last open flower), G (end of flowering) of the mango inflorescence, and growing fruit (gf) (stage nomenclature and photographs of stages D1–D2–E–F have been elaborated in Fig. 10.6. Furthermore, parameters of the thermal time models for each phenological stage of the growth units and inflorescences for the mango cultivar Cogshall have been given in Table 10.3 while parameters for the sigmoidal growth models of the growth unit axes, leaves and inflorescence axes have been given in Table 10.4 (Baig et al., 2013). Stochastic simulation of the development, fruit production and architecture of a mango tree have been shown in Figs. 10.7 and 10.8 which shows potential of V-Mango to simulate mango tree development and fruit production.

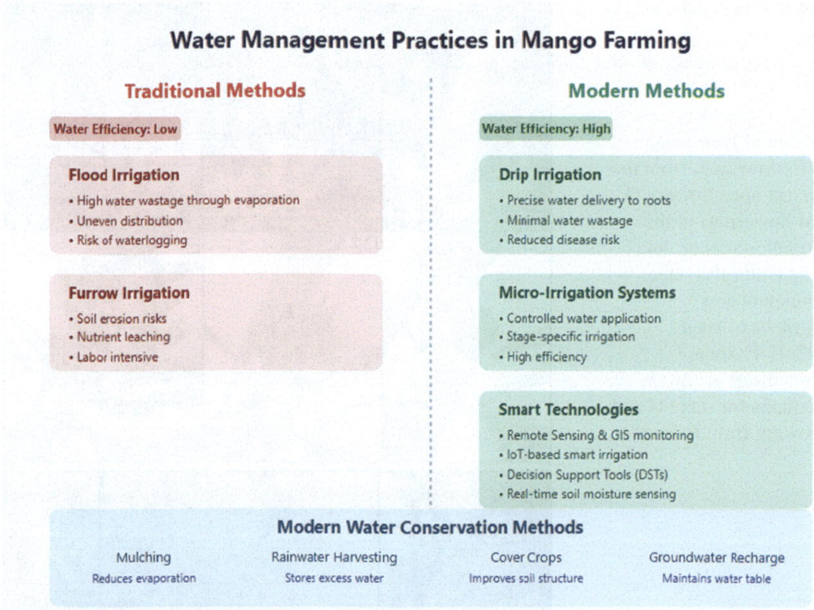

Fig. 10.4 Water management practices in mango farming. *Source* Author's own illustration)

Fig. 10.5 Observed and simulated phenological stages D: appearance of the axis, E: laminas half-opened starting to hang down, F: laminas totally opened hanging limply, G: leaves becoming rigid and moving upward and H: mature growth unit of the mango growth unit. *Source* Dambreville et al., 2015

Fig. 10.6 Observed and simulated phenological stages D1 (appearance of the main axis), D2 (bracts begin to fall), E (secondary axis moving away from main axis), F (flowering, from the first to last open flower), G (end of flowering) of the mango inflorescence, and growing fruit (gf) (stage nomenclature and photographs of stages D1–D2–E–F. *Source* Dambreville et al., 2015; photographs for stage G and the growing fruit: F. Boudon

Table 10.3 Parameters of the thermal time models for each phenological stage of the growth units and inflorescences for the mango cultivar Cogshall. *Source* Baig et al., (2013)

Botanical entity	Parameter	Phenological stages			
		D	E	F	G
Growth unit	Base temperature *T* (°C)	13.4	13.4	13.4	9.8
	Stage duration *TTSS* (°Cd)	38.5	47.6	47.4	316.4
Inflorescence	Base temperature *T* (°C)	11.1	8.7	15.1	–
	Stage duration *TTSS* (°Cd)	70.5	133.3	230.4	–

Table 10.4 Parameters for the sigmoidal growth models of the growth unit axes, leaves and inflorescence axes. *Source* Baig et al., (2013)

Botanical entity	Base temperature *T* (°C)	Growth duration *TTSS* (°Cd)	*tip* (°Cd)	*B* (°Cd)
Growth unit axis	9.2	178.8	89.4	22.4
Leaf	10.7	182.0	91.0	*L*/(0.06 *L*–0.08) (13)
Inflorescence axis	11.1	346.0	136.6	50.9

Where *tip* is the time of maximum growth rate and *B* is a slope parameter. For the leaf growth model, the parameter *B* depends on the final leaf length *L*

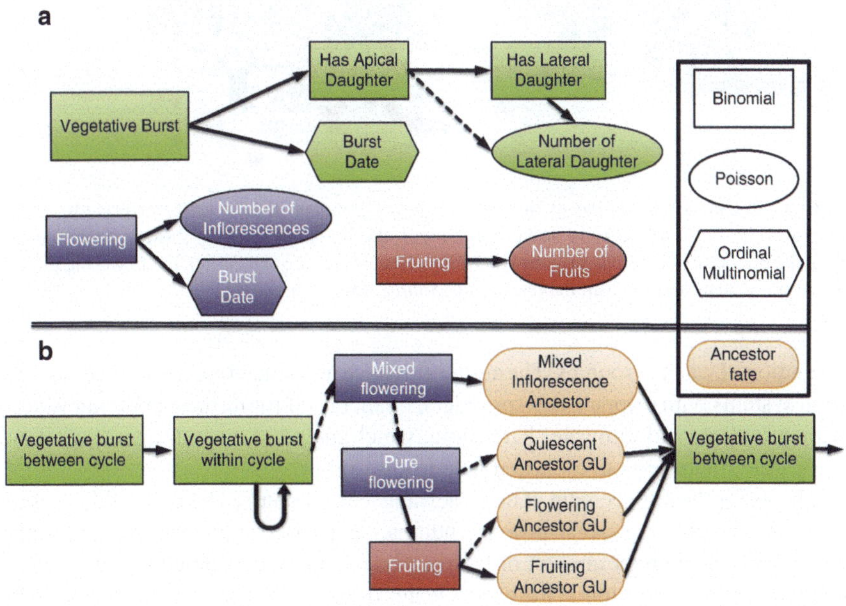

Fig. 10.7 Simulation of mango tree development and fruit production. *Source*: Baig et al., (2013)

10.8 Policy and Institutional Support

10.8.1 Government Policies and Programs for Water Management

Policy measures and programs reflect the government's critical role in the support of sustainable water management in mango cultivation. In Pakistan, some innovative actions have been launched to improve water use efficiency and verify the just

Fig. 10.8 Stochastic simulation of the architecture of a mango tree at different periods of the growing cycle. From left to right and top to bottom, the initial structure, a flush of growth units during the period of vegetative growth (young extending leaves are in yellow), the flowering period with inflorescences and the fruit harvest period. *Source* Baig et al., (2013)

distribution of water resources. The authorities have introduced such initiatives as irrigation systems with drippers and sprinklers, financial aid for farmers applying water-saving practices, and rainwater harvesting, which intend to make farmers switch to proper water management tools that help to save water.

Policies for the replacement of nonrenewable energy sources by renewable ones in the irrigation sector, e.g., solar-powered pumps, can be used as an instrument of water sustainability as a lesser amount of hydrocarbons will be consumed in the process. Such projects not only siphon the excess water from the farmers' fields but also help the farmers to cut on their irrigation expenses, enabling them to use the funds saved to pursue sustainable development. Moreover, government programs, which provide skills and knowledge to farmers, also have an important role in assisting them in realizing the necessity of water conservation and showing them the right later on to learn how to use modern irrigation.

10.8.2 Community-Based Water Initiatives

In a growing number of cases, a community-based water management approach is being acknowledged as the best way to address issues of effective water resources management worldwide. In many rural areas of Pakistan, water user associations (WUAs) have been introduced to the local territories, which contribute to the community-building over the management of irrigation systems and therefore, to the distribution of water that is just. These groups allow farmers to exchange experiences, share the same resources, and use the common infrastructure (for example, pumps and storage facilities) among them, which contributes to the lowering of every farmer's own costs that are connected with the water management process (Ahmad et al. 2023).

In community-level actions, farmers share the responsibility and ownership. This, in turn, leads to the farmers' careful use of water. In the cases where groundwater is overexploited, the community has a say in the monitoring of water extraction rates, which allows water resources to be safeguarded and consequently conserved for upcoming generations (Razzaq et al. 2022b). The community-based approaches by pooling the resources and knowledge facilitate the smallholder farmers to overcome the financial and technical barriers, thus making the sustainable water management to be more accessible.

10.8.3 Extension Services and Capacity Building

Enhancement services and capacity building are extremely significant for the successful adaptation of sustainable water management practices in mango cultivation. Agricultural extension officers, on the other hand, establish the connection between the laboratories and the farmers by enabling the latter to access the latest information as well as research about water conservation, irrigation technologies, and climate adaptation strategies. As a plus, in Pakistan, the role of extension services is not only to inform farmers about the advantages of using the new technology but also to help them to actually implement the technologies in their orchards, such as drip and sprinkler irrigation. Performance and the growth of extension services is crucial to sustainable agriculture development in Pakistan (Ahmad et al. 2020).

Giving farmers the ability and the right knowledge to be able to apply water management in an effective way is done through workshops, field demonstrations, and several other capacity-building programs to support a sustainable agriculture future, a green planet, and a healthier society. This practice, however, is highly effective to fill the knowledge gap and it, in turn, empowers farmers to adopt new methods that increase the resilience of the mango crop under climate circumstances. In turn, the job of the extension services is to make sure that farmers have the skills and the confidence to carry out the water management plan in a sustainable way by constantly giving training and support.

10.9 Challenges to Sustainable Water Management

10.9.1 Socio-Economic and Technical Barriers

The practice of sustainable water management in mango production is often inhibited by both socio-economic and technical barriers. High initial costs related to the installation of efficient irrigation systems such as drip or sprinkler systems can be a major roadblock for smallholder farmers who do not have access to credit or financial support. Even in cases where government subsidies are available, the upfront investment required can discourage farmers from using these technologies, especially in regions with low-income levels.

Technical barriers, like the lack of knowledge about modern irrigation techniques and insufficient access to extension services, additionally, may cause these sustainable practices to be harder to adopt. Quite a number of farmers do not know about the advantages of efficient water use and they may resist changing the traditional irrigation methods that they have used for decades. Overcoming these obstacles entails targeted capacity building, creating better access to finance, and the government actually supporting the process to make water-efficient technologies more affordable and clearer.

10.9.2 Access to Technology and Financial Constraints

In the developing regions, the limited accessibility of technology to mango farmers is a major challenge, including in Pakistan. It is a common reality that the availability of automated irrigation systems, soil moisture sensors, and decision support tools comes in very low numbers to the remote rural regions of a country. Even in such circumstances when the technologies are available, they become quite pricey for small-scale farmers who must not only buy but also operate the technology. Financial constraints, together with the lack of opportunities for formal credit, are the main hindrances for the farmers in their attempts to build the infrastructure that would enable efficient water use.

To overcome these financial obstacles, there is a need for innovative financing models, like microloans and cooperative funding, which can enable farmers to get the resources they need. In addition, public–private partnerships should be involved in reducing the price of advanced technologies and facilitating training on their use. The direct financial support and technical assistance provided to the farmers thus enabling them to implement sustainable water management practices are essential to the development of mango cultivation that is climate adaptive.

10.10 Conclusion

Key strategies for sustainable water management

The sustainable management of water resources is important for maintaining the resilience and productivity of mango orchards in a changing climate. Strategic approaches are key, including the installation of water-saving technologies like drip and micro-irrigation systems that ensure the deep and proper use of water and the prevention of losses due to evaporation and runoff. Rainwater harvesting, groundwater, and soil moisture conservation practices are also important as they provide water during the dry periods. In addition, soil health improvement obtained through organic amendments, cover cropping and mulching will also have a positive effect on the orchard resilience and as a whole is one facet of the advantages of the aforementioned procedures for good water retention.

The integration of advanced technologies, like remote sensing, IoT solutions, and decision, is the key to properly analyzing and making smart decisions. These measures, if accompanied by community-run water management initiatives and appropriate policies, will enable mango growers to deal successfully with climate variability and be able to continue their orchards sustainably.

Future directions for water management in mango cultivation

The future of water management in mango growing will be determined by constant innovation, policy support, and community involvement. The design of cost-effective smart irrigation systems and the extension of extension services will be critical in ensuring all farmers have the know-how and technologies they need to manage water efficiently. Fostering collaboration between the public and private sector is another way to facilitate the implementation of advanced water management solutions by lowering the costs and providing the necessary training and support thus driving the adoption of such solutions.

Research to develop climate-resilient water management practices that are specifically tailored to the challenges faced by mango growers in different regions is required. Clustering the efforts in terms of governments, research bodies, and farming communities will be critical in developing and disseminating these techniques. A move to water-efficient, soil-health, and innovative technologies will aid the mango sector in its fight against climate change thereby making it more sustainable for the coming generations.

References

Abdel-Sattar M, Al-Obeed RS, Makhasha E, Mostafa LY, Abdelzaher RA, Rihan HZ (2024) Improving mangoes' productivity and crop water productivity by 24-epibrassinosteroids and hydrogen peroxide under deficit irrigation. Agric Water Manag 298:108860

Ahluwalia O, Singh PC, Bhatia R (2021) A review on drought stress in plants: implications, mitigation and the role of plant growth-promoting rhizobacteria. Resour Environ Sustain 5:100032

Ahmad B, Pham HD, Ashfaq M, Memon JA, Bano R, Dahri ZH, Mustafa RN, Baig IA, Naseer MAUR (2020) Impact of institutional features on the overall performance assessment of participatory irrigation management: farmers' response from Pakistan. Water 12(2):497

Ahmad B, Mahmood A, Sami A (2023) Haider M (2023) Impact of climate change on fruits and crops production in South Punjab: farmer's perspective. Biol Agric Sci Res J 1:22–22

Ahmed T, Zounemat-Kermani M, Scholz M (2020) Climate change, water quality and water-related challenges: a review with focus on Pakistan. Int J Environ Res Public Health 17(22):8518

Anser MK, Hina T, Hameed S, Nasir MH, Ahmad I, Naseer MAUR (2020) Modeling adaptation strategies against climate change impacts in integrated rice-wheat agricultural production system of Pakistan. Int J Environ Res Public Health 17(7):2522

Antwi-Boasiako A, Amponsah P, Opoku JA, Coulibaly D, Mintah P (2024) Increasing mango production efficiency under the fast-changing climate

Ashraf M (2016) Managing water scarcity in Pakistan: moving beyond rhetoric. In: Proceedings of AASSA-PAS regional workshop on challenges in water security to meet the growing food requirement. Pakistan Academy of Sciences, Islamabad, pp 3–14

Ayanlade A, Radeny M, Morton JF, Muchaba T (2018) Rainfall variability and drought characteristics in two agro-climatic zones: an assessment of climate change challenges in Africa. Sci Total Environ 630:728–737

Baig, MB, Shahid, SA, Straquadine GS (2013) Making rainfed agriculture sustainable through environmental friendly technologies in Pakistan: a review. Int Soil Water Conserv Res 1(2):36–52. https://doi.org/10.1016/S2095-6339(15)30038-1.

Baocheng H, Jamil A, Bellaoulah M, Mukhtar A, Clauvis NK (2024) Impact of climate change on water scarcity in Pakistan. Implications for water management and policy. J Water Clim Change 15(8):3602–3623

Dambreville A, Lauri PE, Normand F, Guédon Y (2015) Analysing growth and development of plants jointly using developmental growth stages. Ann Bot 115(1):93–105

Dao PU, Heuzard AG, Le TXH, Zhao J, Yin R, Shang C, Fan C (2023) The impacts of climate change on groundwater quality: a review. Sci Total Environ:169241

Durán Zuazo VH, Rodríguez Pleguezuelo CR, Gálvez Ruiz B, Gutiérrez Gordillo S, García-Tejero IF (2019) Water use and fruit yield of mango (Mangifera indica L.) grown in a subtropical Mediterranean climate. Int J Fruit Sci 19(2):136–150

El-Beltagi HS, Basit A, Mohamed HI, Ali I, Ullah S, Kamel EA, Shalaby TA, Ramadan KM, Alkhateeb AA, Ghazzawy HS (2022) Mulching as a sustainable water and soil saving practice in agriculture: a review. Agronomy 12(8):1881

Fahad S, Chavan SB, Chichaghare AR, Uthappa AR, Kumar M, Kakade V, Pradhan A, Jinger D, Rawale G, Yadav DK (2022) Agroforestry systems for soil health improvement and maintenance. Sustainability 14(22):14877

FAO (2015) Agricultural water productivity of selected countries. Food and Agriculture Organization, Rome, Italy

Ganeva D, Grozeva S, Pevicharova G (2019) Effect of reduced irrigation on flowering, fruit set and yield of indeterminate tomato. Int J Recent Technol Eng 8(2):932–936

García L, Parra L, Jimenez JM, Lloret J, Lorenz P (2020) IoT-based smart irrigation systems: An overview on the recent trends on sensors and IoT systems for irrigation in precision agriculture. Sensors 20:1042

GOP (2024) National disaster management authority (NDMA), Government of Pakistan, Islamabad, Pakistan

Hahn F, Valle S, Navarro-Gómez C (2022) Pruning and water-saving management effects on mango high-density and mature orchards. Agronomy 12:2623

Iqbal R, Raza MAS, Valipour M, Saleem MF, Zaheer MS, Ahmad S, Toleikiene M, Haider I, Aslam MU, Nazar MA (2020) Potential agricultural and environmental benefits of mulches—a review. Bull Natl Res Cent 44:1–16

Khalifa S, Abobatta W (2023) Climate changes and mango production (temperature)

Kumar L, Chhogyel N, Gopalakrishnan T, Hasan MK, Jayasinghe SL, Kariyawasam CS, Kogo BK, Ratnayake S (2022) Climate change and future of agri-food production. Future Foods. Elsevier, pp 49–79

Lal R (2020) Soil organic matter and water retention. Agron J 112:3265–3277

Legave JM, Normand F, Lauri PE (2013) Climate change and its probable effects on mango production and cultivation. X Int Mango Symposium 1075:21–31

Lima N, Cunha-Lignon M, Martins A, Armani G, Galvani E (2023) Impacts of extreme weather events in Southeast Brazilian mangrove forest. Atmosphere 14:1195

Mabhaudhi T, Dirwai TL, Taguta C, Sikka A, Lautze J (2023) Mapping decision support tools (DSTs) on agricultural water productivity: a global systematic scoping review. Agric Water Manag 290:108590

Magesa BA, Mohan G, Matsuda H, Melts I, Kefi M, Fukushi K (2023) Understanding farmers' choices and adoption of adaptation strategies, and plans to climate change impact in Africa: a systematic review. Climate Serv 30:100362

Makhmale S, Bhutada P, Yadav L, Yadav B (2016) Impact of climate change on phenology of mango: a case study. Ecol Environ Conserv 22

Mishra A, Waghaye A, Chatterjee C, Rautaray S (2020) Rainwater resource management in mango orchards through micro-catchments. Sustain Water Resour Manag 6:45

Mohammad F, Alamoud A, Mahmoud S (2015) Water requirements and water use of mango orchards in Jazan region, Saudi Arabia. J Anim Plant Sci 25

Mohan G, Perarapu LN, Chapagain SK, Reddy AA, Melts I, Mishra R, Avtar R, Fukushi K (2024) Assessing determinants, challenges, and perceptions to adopting water-saving technologies among agricultural households in semi-arid states of India. Curr Res Environ Sustain 7:100255

Naseer MAUR, Ashfaq M, Razzaq A, Ali Q (2020) Comparison of water use efficiency, profitability, and consumer preferences of different rice varieties in Punjab, Pakistan. Paddy Water Environ 18:273–282

Obaideen K, Yousef BA, AlMallahi MN, Tan YC, Mahmoud M, Jaber H, Ramadan M (2022) An overview of smart irrigation systems using IoT. Energy Nexus 7:100124

Ogunbode TO, Esan V, Ayegboyin M, Ogunlaran O, Sangoyomi E, Akande JA (2024) Understanding the perception of mango (Mangifera indica) farmers on the impact of climate change on mango farming in Nigeria. Int J Agron 2024:6486998

Prasad K, Jacob S, Siddiqui MW (2018) Fruit maturity, harvesting, and quality standards. In: Preharvest modulation of postharvest fruit and vegetable quality. Elsevier, pp 41–69

Ravikumar V (2023) Sprinkler and drip irrigation

Razzaq A, Rehman A, Qureshi AH, Javed I, Saqib R, Iqbal MN (2018) An economic analysis of high-efficiency irrigation systems in Punjab, Pakistan. Sarhad J Agric 34:818–826

Razzaq A, Qing P, Abid M, Anwar M, Javed I (2019) Can the informal groundwater markets improve water use efficiency and equity? Evidence from a semi-arid region of Pakistan. Sci Total Environ 666:849–857

Razzaq A, Xiao M, Zhou Y, Liu H, Abbas A, Liang W, Naseer MAUR (2022a) Towards sustainable water use: factors influencing farmers' participation in the informal groundwater markets in Pakistan. Front Environ Sci 10:944156

Razzaq A, Xiao M, Zhou Y, Liu H, Abbas A, Liang W, Naseer MAUR (2022b) Impact of participation in groundwater market on farmland, income, and water access: evidence from Pakistan. Water 14:1832

Razzaq A, Liu H, Xiao M, Mehmood K, Shahzad MA, Zhou Y (2023) Analyzing past and future trends in Pakistan's groundwater irrigation development: implications for environmental sustainability and food security. Environ Sci Pollut Res 30:35413–35429

Razzaq A, Naseer MAUR, Farrukh MU (2024a) Sustainable food consumption and production in the 21st century. Front Media SA 8:1386268

Razzaq A, Zhou Y, Shahzad MA, Wang L, Eliw M (2024b) A tale of two countries: the potential of managed aquifer recharge in Pakistan and Egypt. In: Managed aquifer recharge in MENA countries: Developments, applications, challenges, strategies, and sustainability. Springer, pp 165–183

Reeves D (2018) Cover crops and rotations. In: Crops residue management, pp 125–172

Rocha J, Perdigão A, Melo R, Henriques C (2012) Remote sensing-based crop coefficients for water management in agriculture. In: Sustainable development—authoritative and leading-edge content for environmental management, pp 167–192

Rojano-Cruz R, Martínez-Moreno FJ, Galindo-Zaldívar J, Lamas F, González-Castillo L, Delgado G, Párraga J, Ramírez-González V, Durán-Zuazo VH, Cárceles-Rodríguez B (2023) Impacts of a hydroinfiltrator rainwater harvesting system on soil moisture regime and groundwater distribution for olive groves in semi-arid Mediterranean regions. Geoderma 438:116623–S132

Saddique N, Jehanzaib M, Sarwar A, Ahmed E, Muzammil M, Khan MI, Faheem M, Buttar NA, Ali S, Bernhofer C (2022) A systematic review on farmers' adaptation strategies in Pakistan toward climate change. Atmosphere 13:1280

Salceda-Gonzalez M (2023) Agroforestry and cover crops for improving surface water and groundwater quality. University of Missouri-Columbia

Salma S, Rehman S, Shah M (2012) Rainfall trends in different climate zones of Pakistan. Pak J Meteorol 9

Scavo A, Fontanazza S, Restuccia A, Pesce GR, Abbate C, Mauromicale G (2022) The role of cover crops in improving soil fertility and plant nutritional status in temperate climates: a review. Agron Sustain Dev 42:93

Seleiman MF, Al-Suhaibani N, Ali N, Akmal M, Alotaibi M, Refay Y, Dindaroglu T, Abdul-Wajid HH, Battaglia ML (2021) Drought stress impacts on plants and different approaches to alleviate its adverse effects. Plants 10:259

Soomro SA, Mirjat MS, Mangrio MA, Talpur MA (2022) Soil moisture distribution pattern of surface drip irrigated mango. J Water Land Dev 158:163–163

Stevens AW (2018) The economics of soil health. Food Policy 80:1–9

Tang Y, Li T, Yang X-Q, Chao Q, Wang C, Lai DYF, et al (2023) Mango-GPP: a process-based model for simulating gross primary productivity of mangrove ecosystems. J Adv Model Earth Sys 15:e2023MS003714. https://doi.org/10.1029/2023MS003714

Wei J, Liu G, Liu D, Chen Y (2017) Influence of irrigation during the growth stage on yield and quality in mango (Mangifera indica L.). PLoS One 12

Chapter 11
Socio-economic Impacts of Climate Change on Mango Growers and Communities

Muhammad Asad Ur Rehman Naseer, Amar Razzaq, Shoaib Akhtar, Ahmad Abdul Wahab, Shakeel Ahmad, Ghulam Qadir, Mukhtar Ahmed, Amna Hasnain, and Amjad Malik

Abstract Climate change poses significant threats to mango production and the livelihoods of mango growers and communities worldwide. Through a review of existing literature and case studies from major mango-producing regions, this chapter highlights the vulnerability of mango production to climate-related stresses such as temperature and precipitation changes, extreme weather events, and shifting growing seasons. This chapter also examines the socio-economic impacts of climate change on mango growers and communities, including changes in income, food security, gender roles, and community dynamics. Case studies of socioeconomic impacts from the mango-producing countries such as India, Pakistan, and Brazil are also discussed in the chapter. The findings of this chapter underscore the need for urgent action to address the socio-economic impacts of climate change on mango growers and communities, and to ensure the long-term sustainability of mango production and the livelihoods of the dependent communities.

Keywords Climate change · Food security · Livelihoods · Mango production · Socio-economic impacts

M. A. Ur Rehman Naseer (✉) · S. Akhtar
Department of Agricultural Business and Marketing, Bahauddin Zakariya University, Multan, Pakistan
e-mail: asadrehman@bzu.edu.pk

A. Razzaq
Business School, Huanggang Normal University, Hubei, People's Republic of China

A. A. Wahab · S. Ahmad
Institute of Agronomy, Bahauddin Zakariya University, Multan, Pakistan

G. Qadir · M. Ahmed
Department of Agronomy, Pir Mehr Ali Shah Arid Agriculture University, Rawalpindi, Pakistan

A. Hasnain
Institute of Management Sciences, Bahauddin Zakariya University, Multan, Pakistan

A. Malik
Faculty of Education, University College of the North (UCN), Thompson, MB, Canada

S. A. H. Naqvi et al. (eds.), *Climate Change and Mango Production*,
https://doi.org/10.1007/978-3-031-87035-4_11

11.1 Introduction

11.1.1 Background on Mango Production and Its Importance

Mango farming is an essential component of agriculture in frequent tropical and subtropical areas, with countries such as Pakistan, China, India, Thailand, Mexico, and Indonesia taking the lead in worldwide output (Halder and Hasan 2020). India's mango production reaches over 15 million tons per year, highlighting the fruit's substantial economic importance and its large cultivation in over 100 countries globally (Mitra 2014). Nevertheless, the production of mangoes encounters many issues, such as the negative impact of climate change. These effects include the irregularity of cultivation caused by temperature fluctuations, which have a detrimental impact on blooming and overall yield output (Halder and Hasan 2020). In regions such as Mali, where mango growing is used for export, the susceptibility to fluctuations in rainfall patterns poses a risk to orchards, particularly those that rely on conventional irrigation methods and lack effective moisture conservation strategies (Diarra et al. 2021). These difficulties can be addressed by redesigning methods and strategies to improve mango production, nutritional characteristics, and quality in response to changes in climate and biological pressures (Antwi-Boasiako et al. 2024).

Pakistan holds a leading position in the international mango, ranking as the largest fifth mango grower and largest fourth mango exporter globally. Pakistan has a rich variety of mangoes, with more than 1500 varieties registered. Out of these, for commercial purposes, at least 30 kinds are cultivated, making mango the most significant fruit in the country (Hakim et al. 2024). The mango business in Pakistan encounters difficulties such as the mango hopper insect, etc. which can result in financial losses to mango harvests, prompting research into effective management strategies (Abid et al. 2022). Research and development in postharvest initiatives have greatly increased the mango industry by lowering fruit quality deterioration and developing procedures for better storage, ripening, and disease control. As a result, mango export quality has been dramatically enhanced (Malik et al. 2015). Mango export can be enhanced by managing post-harvest diseases such as stem-end rot and anthracnose by utilizing techniques like hot water treatment. Due to this availability of mangoes in the market enhanced and decreased the occurrence of illnesses, hence improving the export potential (Malik et al. 2018). Climate change took a heavy toll on mango yield for the third consecutive year as the overall production of the major fruit might fall by 600,000 metric tonnes in 2023–2024, exporters and horticulture experts warned on Thursday. Punjab and Sindh are main producers of mango, sharing 70 and 29% of the 1.8 million metric tonnes of national produce. The remaining 1% comes from Khyber-Pakhtunkhwa (K-P). However, poor weather conditions might reduce Punjab's yield by 35–40% and by less than 20% of Sindh. Climate change is negatively impacting mango orchards in the country, resulting in a significant reduction in the production (Fig. 11.1). Due to the non-availability of export quality mangoes, the export target was not achieved in 2023. Land use change is another big concern in Pakistan as most of productive lands have been taken away

Fig. 11.1 Climate change disrupting poor agricultural communities

by housing schemes. Recently larger area of mango farm has been taken away by housing scheme thus leading to reduced mango production as well as destruction of natural ecosystem.

Satellite imagery obtained from Google Earth shows that roughly 6000 acres of mango orchards were cleared between September 2013 and March 2020 to accommodate housing society in Multan (Figs. 11.6 and 11.7). That is, about 500,000 trees were chopped. Entire country, especially Punjab, grapples with pollution and smog, signifying the importance of more green areas in and around the big cities, the monster of private housing schemes continues to devour agricultural land having orchards. Pictures and videos of felled mango trees, some of them showing truckloads of logs, for developing a private housing scheme went viral on social media, drawing fire from those concerned for the environment. What's more alarming is that the said housing scheme is being developed allegedly in violation of the rules and regulations. Multan city in Pakistan Punjab is famous for its mangoes but one fears it won't be long before it loses the title of one of the 'Mango Cities' of Pakistan (Fig. 11.8).

11.1.2 Overview of Climate Change Impacts on Agriculture

The agricultural industry has been significantly affected by climate change worldwide, resulting in several issues like increased/decreased agricultural production, delayed/early planting and harvesting seasons, sensitive heat-related strain, and changed rainfall patterns (Pimpa 2024; Lavudya and Prabhakar 2024; Mahato 2014). Severe weather events such as floods and typhoons cause harm to agricultural infrastructure, forcing farmers to move to a different place. Additionally, unpredictable

Fig. 11.2 Climate change damage to the mango crop in Pakistan. *Source* Pakistani farmer Muhammad Yusuf blames climate change for wreaking havoc on his mango crops in Tando Allahyar village, Picture by Asif HASSAN/AFP

rainfall patterns and increasing temperatures have a negative impact on production and economic stability, declining problems of food scarcity, and insecurity in people's livelihoods (Pimpa 2024). Climate change has widespread consequences for agriculture, including alterations in both the quality and amount of food produced (Ahmed 2020; Ahmed 2023a, b) (Fig. 11.2). This poses significant threats to food security and the overall well-being of human beings (Lavudya and Prabhakar 2024; Semeraro et al. 2023). Furthermore, money is spent in the agricultural sector to minimize climate change effects. This impact varies depending on the region, which highlights the importance of allocating agricultural subsidies more effectively and enhancing investment in research and development. These measures are necessary to improve agriculture's ability to survive and adapt to climate change, while also ensuring its long-term sustainability (Dincă et al. 2023).

11.1.3 Chapter Objectives and Scope

Climate change is now a major concern in relation to agricultural production, and also quite detrimental for the rural communities of many areas on earth that practice monotypic commercial agriculture like cultivators of mango. The objective of the chapter is to address the socio-economic consequences and implications for livelihoods that arise due to climate change impacting mango growers, mapping a host of challenges they are forced to contend with. There are three main goals in this chapter:

- To assess the sensitivity of mango growers and wider communities in which they operate to climate change, including exposure to extreme weather events changes in temperature patterns, or rainfall.
- To investigate the direct and indirect economic impact of climatic changes on mango production in terms of productivity, fluctuations in income generation, marketing capacity as well profitability.
- To look into the impact of climate change on social dynamics in mango growing community—Migration trends, labour availability and transformation in community structure.

11.2 Climate Change Impacts on Mango Production

Climate change has a significant impact on mango production, which causes changes in flowering time, lowers fruit quality, and increases vulnerability to disease and pests. The pollination process is disturbed by increasing temperature, resulting in reduced crop yields. Mango fruiting cycles were also disturbed by fluctuations in precipitation which negatively impact productivity. In addition, mango fruits and trees can be adversely affected by severe conditions such as storms and droughts (Fig. 11.3). These factors collectively threaten the profitability and stability of mango farming and production.

11.2.1 Temperature and Precipitation Changes Affecting Mango Yields

Temperature and precipitation have a significant impact on mango output. Temperature and precipitation patterns have a negative impact on mango growth stages. These patterns influence the flowering process, fruit formation, and overall fruit development (Talib et al. 2020). The rainfall and temperature fluctuations have a prominent effect on yields of cereal crops, such as wheat, maize, rice, sorghum, barley, and pearl millet. The impact of this warming is recorded in many regions (Gadedjisso-Tossou et al. 2021; Wahab et al. 2022). Moreover, the blooming phase of mango is extremely susceptible to cold temperatures, which can trigger flower bud development and also affect fruit (Gill and Jawandha 2012). Moreover, total temperature and rainfall have a strong correlation during the flowering period, development of seedless fruits, and overall yield in mango trees. (Chuan et al. 1996).

Flower bud development is stimulated by cold temperatures, which can impact flowering time and the plant's ability to produce fruit (Gill and Jawandha 2012).

Fig. 11.3 Fruit drop (Jansen 2023)

Elevated temperatures increased the inflorescence duration, while morphological alterations in several components of the mango tree were affected by lower temperatures and its potential production (Gill and Jawandha 2012) (Figs. 11.4 and 11.5). In addition, the growth of fungal infections might be encouraged by the presence of moisture and rain during the stage of flowering and fruiting, leading to the loss of flowers and ultimately decline in overall output (Gill and Jawandha 2012). However, mango production increased because of the changes in the microenvironment, named: Rain Cover Culture, which enhances bisexual flower rates and increases overall yield without lowering fruit quality. Therefore, it presents a promising technique for mango production in some regions with high rainfall (Xu et al. 2017).

11.2.2 Shifts in Growing Seasons and Phenology

Mango phenology is also affected by changing climate and by the changes in growing seasons. Climate change alterations the timing of plant growth activities, resulting in phenology changes by affecting vegetative and reproductive phase duration (Haokip et al. 2020). An increase in temperatures can shorten the time duration between vegetative to flowering that is needed for fruit development, hence affecting the flowering interval (Scuderi et al. 2023). Furthermore, weather factors such as sunlight duration, temperature, and humidity significantly induce the intensity of mango flowering and resulting fruit production (Bajpai et al. 2023). Changing climate, including changes

Fig. 11.4 Heat damage of exposed 'Keitt' mango grown at Pocquereux Fruit Research (la Foa) in New Caledonia 4 days after hot water (HW) treatment of 55 °C for 50 min and 1 day at 20 °C. The sun-exposed red-colored area shows less damage than the shaded green background color of the fruit. Adopted from Kagy et al. (2024)

Fig. 11.5 Image field of the effect of heat stress on mango trees (**a** and **b**). Adopted from Khalifa and Abobatta (2023)

in precipitation and temperature, may change mango flowering time and cause fruit quality and yield loss (Apiratikorn et al. 2012). It also stimulates mango trees to produce flowering during off seasons among different environmental factors, genetic potential, and management strategies (Kumar et al. 2014) (Figs. 11.6, 11.7 and 11.8).

The season of mango growth duration is highly impacted by climate change. Mango trees are impacted by climate change through changes in their flowering patterns and growth as a result of elevated CO_2 levels and rising temperatures. When plants are exposed to high levels of carbon dioxide (CO_2), that alters their vegetative

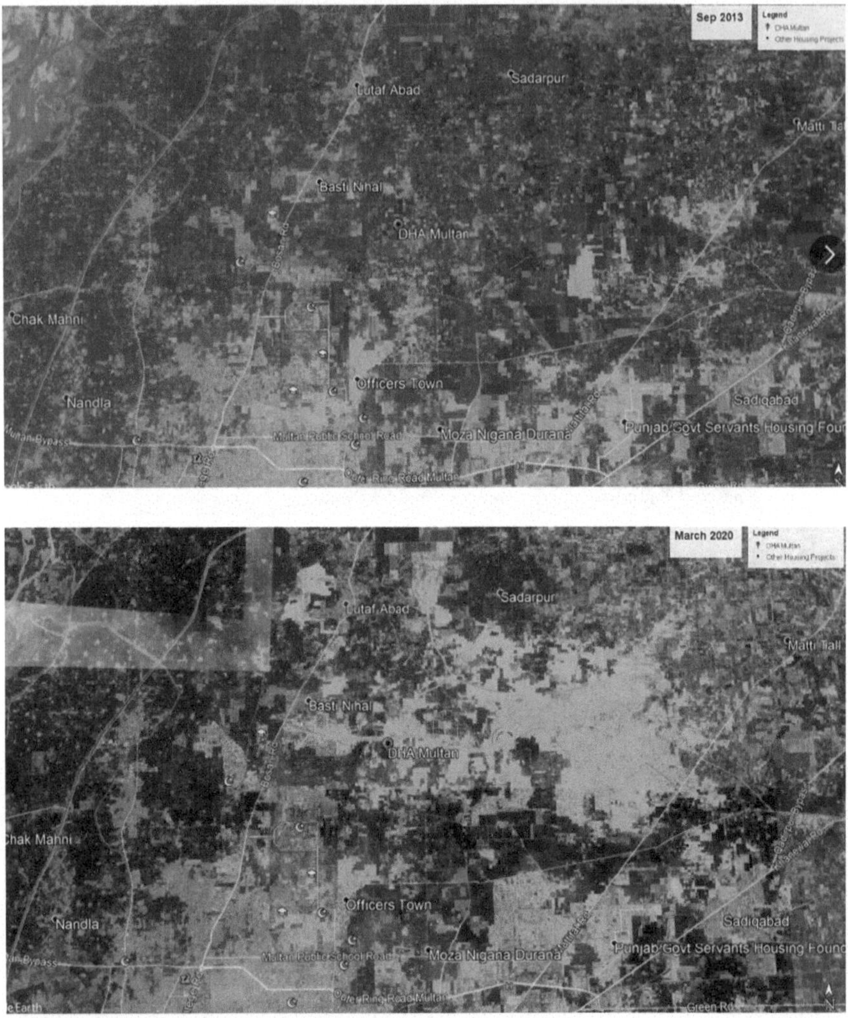

Fig. 11.6 Satellite imagery showing cutting of tress in Multan (Above: September 2013 and below image on March 2020)

and reproductive growth. Specifically, an increase in shoot emergence, but a decrease in the occurrence of blooming (Anwar et al. 2011; Shivashankara et al. 2024). Phenological tests conducted on different mango cultivars demonstrate notable differences in blooming length, panicle development, and overall plant traits. Mango trees are more affected by changes in climate (Kumar et al. 2017; Puche et al. 2012). Climate change increases mango growth duration and whole mango growing seasons will be affected, which could have consequences for mango production and its quality.

Fig. 11.7 Mango tress being transported in Multan city

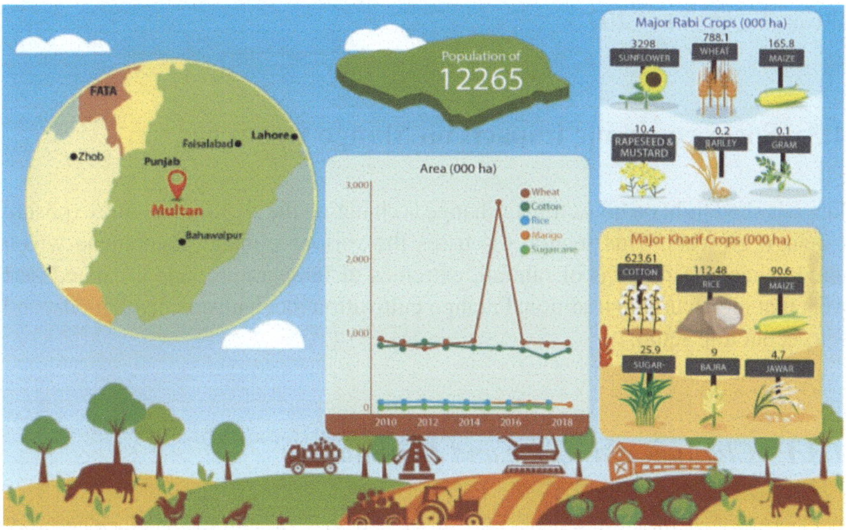

Fig. 11.8 Multan city of mango production transforming to housing societies. *Source* https://www.dawn.com/news/1602000

11.2.3 Increased Frequency of Extreme Weather Events

Floods, tropical cyclones, heat waves, and lightning dramatically affected then mango production worldwide (Ray et al. 2019). Climate change, for example increase of greenhouse gasses by human activities, is leading to irregular rainfall patterns, rising temperatures, and more frequent extreme events, which significantly was challenging to agriculture, including mango cultivation (Legave et al. 2013; Fonseca 2009). Moreover, mango production was decreased up to 37% in Egypt because of severe weather occurrences. This has led to financial losses of about 4.8 billion pounds in 2021 (Khalil and Ali 2023). Rapid taking action to reduce the effect of these severe weather events is essential for retaining mango farming in response to a shifting climate.

Extreme weather conditions have a considerable impact on mango production. Temperatures exceeding 36 °C or falling below 10 °C might cause a delay in fruit growth, which can negatively affect mango trees' capacity to achieve optimal output levels (Goulart and Mertens 2017). Also, in a tropical region with too much rainfall like Costa Riva, postharvest losses may vary between 14 and 84%. Anthracnose, a fungal disease, and rapid loss of moisture due to high temperatures during storage are the major reasons for these losses (Fonseca 2009). In Brazil, when the temperature range is between 0 and − 10 °C, the mortality rate of mango rootstocks reaches about 35–70%, showing a high incidence of this phenomenon in the Carabao mango cultivar (Ribeiro et al. 2002).

11.3 Socio-economic Impacts on Mango Growers

Mangoes shed light on how climate change is changing the socio-economic exposure of mango growers. This section discusses the various climate phenomena which include changing patterns of rainfall, extremes of temperature, and escalated pest levels that are changing traditional mango cultivation mechanisms, livelihoods, and socio-economic setup.

11.3.1 Changes in Income and Livelihoods

Mango yields are directly impacted by climate change, including changes to rainfall patterns and extremes in weather brought about by high temperatures, increased incidences of droughts, and heavy storms. Decreased yields mean smaller incomes for the mango growers who rely on them to earn a living. As follows price fluctuation in the mango market also adds to income instability (Sarker et al. 2014). Smallholder farmers, in particular, can end up chronically indebted from having to borrow money for rental costs when revenue is low—leading to an unbreakable cycle of poverty and financial hardship. Moreover, with the increasing volatility of growing seasons, farmers could be required to invest still more in water conservation practices, irrigation, and climate-resilient crop varieties; this again increases the cost of production and leads to even further margin contraction at the farm-gate level. These shut down grower income also lead to secondary employment or change in crop practice due lack of alternatives bringing a structural shift into traditional livelihoods created around Mango cultivation. Mango has its own specific nutrition, value and taste among all fruits available in summer season in Pakistan. That's why it is not only the national fruit of Pakistan but also the king of fruits. There is a growing demand for mango pulp, not just locally but worldwide for drinks and juices, especially Pakistan's Chaunsa variety. In Pakistan, most mango orchards are owned by small growers who do not possess adequate knowledge of modern techniques for cutting, pruning, insecticides and fertilizer application, and management practices. Given that short-to-medium-term climate predictions have been often erroneous and crop output is a chief source of income generation for growers, any climatic disruption in mango harvest to low draw quality impacts the livelihoods of growers adversely. Shifts in flowering and harvest seasons impact seasonal employment availability, which can lead to increased labor costs (Chay et al. 2019). This part could further look at:

11.3.1.1 Price Volatility

Market prices determined by supply shortages or surpluses make farmers vulnerable to income instability which is made even worse due to the impact on mango yields because of climate change.

11.3.1.2 Increased Input Costs

Shifting pest populations, irrigation requirements, and disease incidence raise costs for pesticides, fertilizers, and water resources.

11.3.1.3 Diversification of Income Sources

Income losses for such a significant number of farmers could prompt some to seek new solutions in the form of crop diversification or off-farm employment.

11.4 Effects on Food Security and Nutrition

Mango is an important fruit crop in many regions and one of the major nutrition fruits that ensure food security. This means an increased risk of climate-induced crop losses, not just in the availability of mangoes but also in nutrition security (Gelaye 2024a, b). Less fruit to pick is directly responsible for the limited mango supply, which has in turn led to lower trade and marketing income. This shift can lead to malnutrition and increased food insecurity in small farming communities which are highly dependent on growing mangoes for their consumption. Moreover, resources targeted at conservation strategies such as irrigation for combatting the impacts of climate, are commonly obtained from other crucial sectors contributing stress to food systems within these localities. But this has a cascade effect on health and well-being; children who depend on mangos to provide the few vitamins they could delight themselves with, will not access as many of these nutrients.

In many communities, it is a staple of the local diet and a good source of nutrients from mangoes. If production is affected by climate change, especially in areas where mango is the main fresh fruit eaten, then this will have implications for food security.

11.4.1 Nutritional Impact

Families who include mangoes in their diet are deprived of essential vitamins and minerals due to less mango production.

11.4.2 Increased Dependency on Imports

Elsewhere, imports can increase where local production is unable to step in to fill the gap—often raising food bills and reducing affordability in the process.

11.5 Impacts on Gender Roles and Responsibilities

Because of climate change, any influence on the lives of people and the course of events is not neutral for gender. In tropical fruit-growing communities, for example, in those villages, several women's activities are tightly integrated into the growing process. This may involve such matters as picking fruit for shipping, selling the harvested crop in markets, and so on. Women, as climate change intensifies, often have even more responsibilities. They have to find water sources, help with driving out mice, feed household food security, and supplement or replace household income. This can lead to apparently traditional gender roles being shaken up, and women stepping into areas of life that were previously male preserves. Decision-making and financial planning are cases in point of this (Mbeye 2016). However, these added responsibilities can also result in increased economic vulnerability and wear and tear on women's bodies, especially for people without many resources.

Measures to reduce these problems must take gender-inclusive approaches into account, helping women to get involved and create appropriate politics in farming, or else focusing on areas like training and providing access to resources where equal opportunities for both sexes can be secured in agriculture practice.

11.5.1 Labor Shifts

With climate impacts necessitating different farming practices, women may take on additional roles in crop management or financial decision-making.

11.5.2 Resource Management

Women often bear responsibility for water and food resources; climate change may increase these burdens, as water becomes scarcer and more difficult to access.

11.5.3 Empowerment and Education

In response to the challenges, some programs may provide education and resources to women, encouraging greater participation in agricultural innovation and management.

11.6 Community-Level Impacts

11.6.1 Impacts on Local Economies and Employment

In producing areas of course mangoes are their staple offering. Mango growing provides work for not only farmers but also those who pick and pack the trees' fruit, as well as transporters and street vendors. This income depends on a good harvest of good quality produce; however, when yields are reduced because there is less to gather and sell then employment is directly affected by climatic regression. And with lower production, many fewer workers have jobs during the agricultural high season. This affects production systems all the way downstream: in transportation and packaging industries, people's ability to find work locally comes under threat. In addition, mangoes are a major export product for many growing areas; reduced output means lower export revenues for the whole region which might otherwise have been channeled into local infrastructure and social welfare projects. The decline in business circles out to agricultural and non-agricultural households alike (Sacramento and Cuizon 2020).

Many regions rely on the annual harvest from their mango orchards to support local families. As a result, when production in these regions decreases due to climate change—and rich countries offer reduced orders or withdraw entirely from buying abroad—it has serious economic repercussions for successive generations of people living there.

11.6.2 Employment Reduction

Lower mango yields can lead to job losses, especially in seasonal harvesting roles, packing facilities, and transportation.

11.6.3 Ripple Effects

Reduced spending by mango farmers affects local businesses that depend on agricultural prosperity, such as equipment suppliers, local markets, and service providers.

11.7 Changes in Social Dynamics and Community Cohesion

Climate impacts disrupt traditional social dynamics within mango-growing communities. Resource scarcity, especially in terms of water and arable land, can lead to competition and conflict among community members. Families that once cooperated in farming and resource-sharing may become increasingly protective of their resources, impacting community cohesion and leading to social fragmentation. Additionally, as economic pressures rise, there may be an increased incidence of social issues such as family disputes, health concerns, and psychological stress. The transformation of traditional livelihoods can also erode cultural and familial ties, as younger generations may leave the community in search of stable employment, reducing the communal support network that once sustained these farming communities (Yobo et al. 2020).

11.8 Effects on Human Migration and Displacement

With escalating climate impacts, migration is becoming a common adaptive strategy itself. As agricultural productivity drops and job opportunities diminish in mango-growing regions, this forces families to move to other areas. It could be the big city or somewhere with more seasons known regularity of conditions. This migration is often hard on family structures, as men may leave their wives and children behind to deal with housekeeping and the farm. Entire communities can become displaced over time, with the resulting loss of traditional agricultural knowledge and skills. The pressure of migration is an added demographic burden to the cities, where resources are finite and so may be crammed into slums housing conditions for millions. All of this urban migration transforms the socio-economic environment of mango-growing areas in an even broader sense (Tables 11.1 and 11.2). Repercussions in the long term could affect both rural and urban communities (Hassan et al. 2024).

Table 11.1 Socio-economic challenges faced by mango growers due to climate change

Climate change factor	Impact on mango cultivation	Socio-economic consequences
Erratic rainfall	Unpredictable flowering and fruiting cycles	Loss of income due to reduced yields and inconsistent mango quality
Increased temperatures	Poor fruit set, smaller fruit size, and lower quality	Decreased market value and export potential of mangoes, leading to financial instability for growers
Frequent droughts	Limited water availability for irrigation	Increased cost of irrigation and reduced profitability due to higher input costs
Pest and disease outbreaks	Higher vulnerability of mango trees to infestations	Increased expenditure on pest control measures, reducing net profits for farmers
Storms and cyclones	Physical damage to mango orchards	Loss of livelihoods for growers and laborers due to destroyed orchards and reduced harvesting opportunities
Declining soil fertility	Decreased productivity of mango farms	Higher costs for fertilizers and soil amendments, straining the economic viability of small-scale growers

Table 11.2 Socio-economic adaptation strategies for mango growers

Adaptation strategy	Description	Benefits for mango growers and communities
Climate-resilient varieties	Cultivation of mango varieties resistant to drought and pests	Ensures stable yields and income despite changing climatic conditions
Irrigation infrastructure	Investment in drip irrigation and rainwater harvesting	Reduces water stress, enhances productivity, and lowers irrigation costs over time
Diversification of income	Engaging in alternative crops or non-farm activities	Reduces dependency on mango cultivation, providing financial security during low-yield years
Access to crop insurance	Insurance schemes to cover climate-related losses	Minimizes financial risks and provides compensation for yield losses caused by extreme weather events
Training and awareness programs	Educating growers on sustainable farming practices	Improves resilience by adopting water-saving techniques, pest management, and climate-smart agriculture
Market support mechanisms	Establishing fair trade policies and stable market linkages	Ensures fair pricing for mangoes, reducing economic vulnerability of farmers
Community-based cooperatives	Groups for sharing resources, knowledge, and market access	Enhances collective resilience, reduces costs, and strengthens the socio-economic fabric of mango-growing regions

11.9 Social and Cultural Impacts on Mango-Growing Communities

11.9.1 Psychological and Health Challenges

The impact of climate change impedes the production of agriculture, but also very severely affects the psychology of mango farmers and their families. With the unpredictability of yields, the need to buy food that is artificially made expensive due to rising debt from investments made in climate-adaptive practices such as irrigation systems that further but at a cost, the cycle of daily habitation is escalating towards exacerbation. With the entire family dependent on the return from mango farming, farmers tend to be more anxious and depressed due to crop loss and drop in income, creating a cycle of intensive stress on the family as a whole which in turn creates subsequent stress on mental health. Farmers are driven to self-harm out of economic despair and the need for mental health support systems (Beyeler et al. 2023). Likewise, temperature increases as a result of climate change may leave workers vulnerable to heat exposure and serious health consequences, such as heatstroke, dehydration, and vector-borne diseases. This vicious cycle is compounded by the lack of access to healthcare in rural regions and shows the need for offerings that would provide both psychological and somatic health assets to agricultural workers in order for them to adapt to climate stressors.

11.9.2 Shifts in Traditional Knowledge and Cultural Practices

Many mango-growing villages have long combined traditional agricultural techniques and seasonal patterns in order to effectively regulate their mango production. Climate change interrupts these patterns, often rendering traditional practice ineffective and therefore undermining centuries of accumulated cultural understanding. The move affects not only how people farm, but also the Each year many celebrations with local flavor in the form of festivals, ceremonies, and special events are tied to mango harvests and other rich foods. For example, if the mango crop is late or fails, festivities may have to be called off or redesigned. This change threatens decades-long traditions that are woven into community life and helped make it survive hardships. The disruptions seen in these sectors may also alienate younger generations from conventional farming practices, which can further compound rural outmigration and reduce intergenerational agricultural knowledge transfer. One solution some communities are turning to is programs that help detail and adapt Indigenous knowledge to changing environmental conditions as they seek to maintain their cultural identity alongside opportunities and challenges for mango farming in a changing climate. Economic Diversification and Resilience-Building Strategies.

11.9.3 Alternative Livelihoods and Income Diversification

Plasticity Buffers the Response of Plant-Microbiome Interactions to Systemic Effects of Environmental Change Alternative livelihoods help mango farmers keep their eggs in multiple baskets. In some of the mango-growing regions, agro-tourism is also becoming a trend where farmers can accommodate visitors to their farms, teach them about mango cultivation and local culture, and earn an income from this. Planting compatible crops such as legumes or short-season vegetables between mango trees can be a source of additional income while enhancing the soil, thus making farms more resilient to climate variability (Marroquín-Agreda et al. 2015; Agreda et al. 2006). Another approach is for mangoes to be processed into products such as chutneys, dried mango slices, or pulp to produce higher-valued, value-added products that would command higher prices on both local and export markets. Apart from providing alternative income streams, these also help in reducing climate change and other mono-cropping-related vulnerabilities for farmers.

11.9.4 Empowering Cooperatives and Value-Addition Initiatives

Smallholder mango production is undertaken by farmers where harvesting is done collectively who can produce, adding value, market reach location is 25 km most often. Mango jams, juices, and dried slices, for example, are less sensitive to market shocks than fresh fruit since they last longer and can be kept or shipped for sale when fresh mango prices fall (Aggarwal et al. 2017). With member farmer cooperatives, they can align to invest in processing facilities where the cost of operations and its resultant benefits are shared among them, lowering costs for individuals. But where cooperatives have emerged, some have developed networks linking smallholders to larger buyers or export markets, ensuring that the smallholders receive higher prices and are less dependent on the services of middlemen. Some cooperatives exchange knowledge between members and provide a platform for farmers to learn about innovations related to climate-resilient practices. Such organizations are an essential component in empowering mango-producing communities to adapt to climate change as they create long-term livelihoods.

11.9.5 Digital and Technological Innovations

As mango farmers adjust to climatic change, technology is an indispensable tool. Farmers also have access to mobile apps and other digital platforms that provide them with real-time insights on weather, pest control, and market prices among others, enabling them to make timely decisions to ensure maximum yield and minimum losses. For example, localized weather forecasts provided through digital platforms enable farmers to adjust the irrigation schedules accordingly and protect the crops from unfavorable weather conditions. E-commerce platforms are also giving mango farmers access to new markets and allowing them to reach consumers directly—in both domestic and international settings—and therefore decreasing reliance on traditional markets, which may be more susceptible to the impacts of climate. Affordable and accessible online training and virtual agricultural extension programs on best practices, climate-smart agriculture, and value addition can be the best possible options for this. Digital inclusion enables smallholder farmers—even in the most remote areas—to tap into resources and information that were previously only available to larger agricultural enterprises, thereby building their resilience to climate-induced economic shocks (Rajan et al. 2020).

11.10 Equity and Justice in Climate Adaptation

11.10.1 Ensuring Equitable Access to Resources

Access to climate adaptation features like loans, grants, and technical support is limited or unequal for many smallholders and marginalized mango farmers, and they find themselves unable to adapt. More well-off farmers and big business agri-firms have more capacity to invest in climate adaptive infrastructure, exacerbating social-economic divides. However, access to equity, technology, and training on how to adapt is critical for sustainable adaptation for these rural mango-growing communities (Zainuri et al. 2019). It would point to the need for programs that reach low-income farmers and other vulnerable populations, to ensure they are not left out of the adaptation process. Resources for inclusive adaptation could include specific microfinance options for smallholder farmers, subsidized climate-smart technology supported by the government, and training programs that are relevant to locals and accessible.

11.10.2 Gender-Inclusive Adaptation Efforts

While women across the value chain—from harvesting and processing to household financial decision-making—are versed in and impacted by climate change, they are left out of relevant adaptation strategies in mango-related international and local policy frameworks. For primary caregivers, and the providers of food, women in farming communities bear even heavier burdens when climate impacts reduce food security or income. Providing women with access to education and information on climate-resilient practices, and involving them in decision-making processes, enhances household resilience and community-level adaptive capacity. For example, gender-sensitive MAPP might include training women on sustainable agricultural practices, financial literacy, and small-scale business skills for mango-focused processing. These initiatives also contribute to social equity by empowering women and giving them more voice and decision-making power in their communities while also building economic resilience.

11.10.3 Advocating for Climate Justice in Agricultural Policies

Mango producers in the developing world have contributed very little to global carbon emissions but are often the first on the front lines facing the impacts of climate on mango production. The principle of climate justice recognizes such imbalance and claims the need for global protection and compensation systems for such vulnerable populations. This might include global approaches, such as carbon credit schemes that reimburse smallholder farmers for adopting sustainable practices or participating in carbon sequestration initiatives. Development aid as well as trade agreements can be used to help mango-growing regions adapt to climate change with financial and technical resources. Pushing for these changes nationally and internationally would not only benefit mango farmers financially but would also work towards a more just paradigm of mitigating the impacts of climate change on agriculture.

11.10.4 Case Studies from Mango-Producing Regions

Case studies of the mango-growing communities affected by climate change from the following mango-producing countries are presented in the next section with examples from India, Pakistan, and Brazil. These case studies illustrate the different experiences and adaptive responses to climate change—including the key socio-economic and climatic factors that determine the climate change impact on mango production and the communities that directly or indirectly depend on mangoes in those regions.

11.10.5 India: Adapting to Erratic Weather Patterns and Monsoon Shifts

India produces mangoes and is home to many different varieties (e.g., Alphonso, Kesar, and Dasheri), making it the world's largest producer of mango. Even so, climate change has posed a problem for mango producers in India, especially with erratic weather patterns and a changing monsoon season (Lal 2003). The monsoon rains have increasingly been late or deficient in recent years, while extreme temperature variations have affected the timing of flowering, fruiting, and harvesting. For example, summer heatwaves that arrive too early can lead to so-called "fruit drop," where mangoes spontaneously fall from their trees, with yields plummeting as a result (Reddy et al. 2017).

The Indian mango grower is not only competing with international growers but is also finding competition for water in Indian states like Maharashtra, Gujarat, and Uttar Pradesh, where agriculture is mainly dependent on monsoon rains. As temperatures increase, these trees use much more water—both for a healthy tree and for quality fruit. Yet, water scarcity became more pronounced, especially within the arid and semi-arid areas, rendering it technically difficult for the smallholders to maintain mango orchards. As a result, followers of this method brought in expensive irrigation systems (drip irrigation and micro-sprinklers) to make better use of the limited water. Although this is positive progress, it is burdened with capital costs as smallholders have to take loans to upgrade, which pulls them into debt and economic insecurity (Nath et al. 2019).

For instance, instead of individual farmers adopting water-efficient technologies, entire communities have come together to form cooperatives—as seen in Gujarat, where growers pool resources to expand access to these technologies and use climate-resilient crop varieties. Even governmental and non-governmental organizations have intervened as well, encouraging climate-smart agricultural practices by intercropping mango trees with legumes and/or cover crops to enrich the soil and minimize water evaporation. Indian research institutions have also contributed by creating mango cultivars with improved tolerance to heat and water stress, but the adoption of these varieties remains to be complimented.

Initiatives, such as the Pradhan Mantri Krishi Sinchayee Yojana (PMKSY) for irrigation and water management projects with working subsidies, have been launched by the government to help farmers adapt to changing climates. Nevertheless, low awareness and its high costs stop many smallholders from adopting it today, which is why more inclusive policies must focus on this gap. Further, mango being an important fruit from both the social as well as economic point of view in India, there are initiatives to promote the consumption of locally grown mangoes among consumers, so that the domestic demand acts as a buffer against climate-induced export variability.

11.10.6 Pakistan: Water Scarcity and Temperature Extremes in South Punjab and Sindh

Mango production is geographically delineated, with its concentration in South Punjab and Sindh provinces of Pakistan, which are also affected by climate change that speeds up the temperature of these regions along with flash floods rainfall due to uncertain and unpredictable weather cycles (Usman et al. 2003). Pakistan is one of the largest producers of mangoes, and the fruit is an integral part of cultural and economic life. But climate change is now seriously threatening this prized harvest. Mangoes face threats to yield and quality from severe heatwaves during flowering season—increasingly frequent in recent years—which lead to less fruit setting, and more erratic rainfall which sets up conditions for pest outbreaks. Additionally, water scarcity creates a further threat to Pakistan's mango sector, especially during critical growth periods. With high demand and little rain in earlier months, rivers and reservoirs in the country are continuously falling short which is further complicating mango orchards to be watered (Mohsin et al. 2021).

Pakistani mango growers are switching to technologies to better deal with water scarcity using techniques like drip irrigation and applying mulch to retain soil moisture. However these techniques are expensive, and many small farmers do not have the capital to implement them on a larger scale. In response, local agricultural extension services, backed by international agencies, are offering training on affordable water conservation techniques including constructing farm ponds to harvest rain and getting mulch of plant residues (Khan and Shaukat 2006). In Sindh too some farmers have started trailing heat and drought-resilient mango varieties but due to limited availability, these varieties are not commonly in use either.

To help mango growers adjust, government-led programs have been implemented like the Agriculture Department initiatives in South Punjab, which encourage climate-smart agriculture. PARC is working on heat-resistant and drought-tolerant mango varieties. In addition, they teach farmers how to manage pests and diseases, as climate change has increased the number and intensity of mango pests like fruit flies, which thrive in warmer conditions and can cause significant damage to mangoes if not managed properly.

Apart from agricultural interventions, the emergence of community adaptation strategies has taken place. Cooperatives have been set up in South Punjab to help farmers pool their resources for outlets, water, and equipment along with pest management controls. By pooling resources, these cooperative models lower the cost on an individual level and create a helpful network to support small-scale growers in implementing practices that would be impossible otherwise. The export market, as well, changed due to climate impact on haphazard natures in yield quality at times, wherein native consumers are now getting more impetus to ensure the loss of exportable is compensated by the local.

11.10.7 Brazil: Drought Resilience in the São Francisco Valley

As one of the top mangoes exporting countries, Brazil has the São Francisco Valley in the northeast region of the country where a large part of the Brazilian mangoes is cultivated. But this part of the world experiences long-term droughts, as well as increasing heat in many places, all of it made worse by climate change and climate disruption. Cycling illustrates the allocation of water resources to crop farming activities by means of irrigation (de Silva et al. 2009). In the São Francisco Valley, farmers have historically received irrigation from the São Francisco River. However, the drop in the incidence of rainfall associated with the rise in evaporation, also driven by global warming, has increased the problem of water scarcity and has become a direct challenge in the life of mango farmers.

Obtaining this state of things and to handle these situations, farmers in Brazil have started using advanced irrigation systems, such as precision drip irrigation that sends water straight to the root zone to minimize waste and optimize water (de Azevedo et al. 2003). Data collected using such practices are now often linked with climate data, enabling the farmers to understand the weather forecast, and soil moisture and schedule watering accordingly. But this sophistication is not usually at the level of smaller landholdings, thereby exposing smallholders to much of the negative impacts of climate.

The Brazilian government is backing mango farmers through the Ministry of Agriculture by supporting research and development. Which involves breeding types of mangoes for adaptation to drought and productivity in water-scarce environments. These drought-tolerant cultivars are now undergoing field trials, but many are unlikely to be widely used until they have been researched and disseminated well (Oki and Quiocho 2020). Farmers in drought-prone areas are also receiving subsidies for water-saving technologies and renewable energy-powered irrigation systems, but these initiatives struggle with funding constraints and high demand.

At the community level, training in organic mango production and agroecological practices to increase soil moisture retention and biodiversity are being replicated in the São Francisco Valley, where the creation of microclimates has protected crops from extreme weather events. The Brazilian Mango Exporters Association (Abmango) has also sought to be a promoter of this movement, connecting growers to the world and indicating the sustainability of Brazilian mango producers.

Additionally, local communities have embraced rainwater harvesting and small dams for affordable water management solutions. Such approaches allow for additional water availability through the dry season, contributing to the management of mango orchards and mitigating the high demand for the São Francisco River. The farmers in the region are generating agro-tourism income inviting tourists to their farms for learning mango cultivation and entering the secondary income which will compensate for their yield loss.

11.11 Summary

The different impacts of climate change on mango production and the adaptation strategies deployed by mango-producing communities are illustrated by individual case studies. Although it is true for the large commercial farmer within whichever region you might be examining, they have the resources to delve into whatever economic sector interests them most often based on how much they can invest into technology and innovative and new irrigation practices; smallholders, on the other hand, continue to carry the burden of the old, have faith in government assistance that is never as speedy or as crucial as the people need it to be, and the cooperation of a community in which they live. We believe that, based on these experiences, a need for targeted financial and technical support arises to promote the resilience of mango growers against climate change threats, through addressing their income stability and food security for their communities.

11.12 Conclusion

Climate change is a complex threat to mango producers and their communities, affecting income, food security, social stability, and migration flows. Mitigating these impacts will require a combination of economic assistance, climate-resilient agricultural strategies, and community-oriented adaptations. By ensuring policies that benefit their women, children, and those with limited resources—the essential vulnerable groups in these communities, for people to be physically and mentally safe within their homes. Finally, sharing knowledge and strategies between the mango-producing countries will help develop robust strategies to preserve growers' livelihoods and the world mango industry from future and ongoing climate change.

References

Abid G, Ahamad F, Khan A, Abdul A, Aziz H, Javed IM, Ahmad S, Raza M, Imran M, Noor M, Ashaq R (2022) Establishment of strategy for mango hopper *Idioscopus clypealis* (Lethierry) management and impact of weather factors on its population in South Punjab, Pakistan. Biol Clin Sci Res J 2022(1):152–160

Aggarwal P, Kaur A, Bhise S (2017) Value-added processing and utilization of mango by-products. In: Handbook of mango fruit: production, postharvest science, processing technology and nutrition, pp 279–293

Agreda FM, Pohlan JÜRGEN, Janssens MJ (2006) Effects of legumes intercropped in mango orchards in the Soconusco, Chiapas, Mexico. In: Conference on International Agricultural Research for Development, University of Bonn, pp 11–13

Ahmed M (2020) Introduction to modern climate change, Andrew E. Dessler: Cambridge University Press, 2011, 252 pp, ISBN-10: 0521173159. Sci Total Environ 734:139397. https://doi.org/10.1016/j.scitotenv.2020.139397

Ahmed M (2023a) Climate change and farming system: a review of status, potentials, and further work needs for disaster risk reduction. In: Ahmed M, Ahmad S (eds) Disaster risk reduction in agriculture. Springer Nature, Singapore, pp 1–19. https://doi.org/10.1007/978-981-99-1763-1_1

Ahmed M (2023b) The science of climate change. In: Brinkmann R (ed) The Palgrave handbook of global sustainability. Springer International Publishing, Cham, pp 195–222. https://doi.org/10.1007/978-3-031-01949-4_22

Antwi-Boasiako A, Amponsah P, Opoku JA, Coulibaly D, Mintah P (2024) Increasing mango production efficiency under the fast-changing climate. In: Hasanuzzaman M, Nahar K (eds) Abiotic stress in crop plants—ecophysiological responses and molecular approaches. IntechOpen, London, pp 121–143

Anwar R, Ahmad S, Rajwana IA, Khan AS, Memon N, Nafees M (2011) Phenological growth patterns and floral malformation of mango (*Mangifera indica* L.) tree under subtropical climate. Pak J Agri Sci 48(2):107–113

Apiratikorn S, Sdoodee S, Lerslerwong L, Rongsawat S (2012) The impact of climatic variability on phenological change, yield and fruit quality of mangosteen in Phatthalung province, Southern Thailand. Agric Nat Resour 46(1):1–9

Bajpai Y, Soni SK, Bajpai A, Tiwari S, Rastogi L, Trivedi M (2023) Patterns of shifting phenostages and flowering time in mango (*Mangifera indica* L.) due to climate change. Preprints 2023, Article 2023061600

Beyeler NS, Nicastro TM, Jawuoro S, Odhiambo G, Whittle HJ, Bukusi EA, Schmidt LA, Weiser SD (2023) Pathways from climate change to emotional wellbeing: a qualitative study of Kenyan smallholder farmers living with HIV. PLOS Global Public Health 3(7):e0002152

Chay KG, Workeneh A, Shifera B (2019) A review on production and marketing of mango fruit. World J Agric Soil Sci 2(2):1–7

Chuan TJ, Yao MingHwi YM, Chi KungShy CK, Liang LienShing LL (1996) Studies on the relationships between climatic elements and the yield of mango in Yuching area. J Agric Res China 45(2):186–194

da Silva VDP, da Cunha Campos JHB, de Azevedo PV (2009) Water-use efficiency and evapotranspiration of mango orchard grown in northeastern region of Brazil. Scientia Hortic 120:467–472

de Azevedo PV, da Silva BB, da Silva VP (2003) Water requirements of irrigated mango orchards in northeast Brazil. Agric Water Manag 58:241–254

Diarra S, Sissoko S, Diawara MO, Traore BM, Sidibe A (2021) Cultivation practices of mango (*Mangifera indica*) varieties Kent and Keitt planters and the origins of orchard vulnerability to rainfall variability in the Koulikoro District. Int J Agric Econ 6:172–180

Dincă G, Netcu IC, El-Naser A (2023) Analyzing EU's agricultural sector and public spending under climate change. Sustainability 16:72

Fonseca JM (2009) Postharvest handling under extreme weather conditions. In: Postharvest handling: a systems approach, Food science and technology series. Academic Press, Elsevier Inc., New York, USA, pp 539–559

Gadedjisso-Tossou A, Adjegan KI, Kablan AKM (2021) Rainfall and temperature trend analysis by Mann-Kendall test and significance for rainfed cereal yields in northern Togo. Sci 3:17

Gelaye Y (2024) A review of storage temperature and relative humidity effects on shelf life and quality of mango (*Mangifera indica* L.) fruit and implications for nutrition insecurity in Ethiopia. Open Agriculture 9(1):20220296

Gelaye Y (2024) A review of storage temperature and relative humidity effects on shelf life and quality of mango (*Mangifera indica* L.) fruit and implications for nutrition insecurity in Ethiopia. Open Agric 9(1):20220296

Gill MS, Jawandha SK (2012) Response of mango (*Mangifera indica* L.) to abiotic stresses: an overview. Int J Agric Environ Biotechnol 5:459–462

Goulart F, Mertens F (2017) The late mangos: Is there any doubt humans are inducing climate change? Biomed J Sci Tech Res 1:1–3

Hakim A, Shafique S, Mehdi M, Baig IA (2024) Farmer to consumer—an online supply chain process system using blockchain technology. Agric Sci J 6(1):1–11

Halder S, Hasan MA (2020) Climate change and mango production. Chem Sci Rev Lett 9:1–8

Haokip SW, Shankar K, Lalrinngheta J (2020) Climate change and its impact on fruit crops. J Pharm Phytochem 9(1):435–438

Hassan Z, Khan FZ, Aldosary AS, Al-Ramadan B, Ahmad A, Manzoor SA, Rahman MT (2024) Roots to roofs: farmers' perceived socio-ecological impacts of converting mango orchards to urban areas in Multan, Pakistan. Environ Challanges 15:100935

Jansen C (2023) Mini tornado rips mangoes off trees. FreshPlaza.com (Published on Wed 13 Dec 2023). Available online: https://www.freshplaza.com/north-america/article/9585267/mini-tornado-rips-mangoes-off-trees/

Kagy V, Duret H, Mangeas M, Bouteiller M, Murcia I, Pineau R, Woolf AB (2024) Sun exposure and high preharvest fruit temperatures increase antioxidants, heat shock proteins and thermotolerance of mango fruit (*Mangifera indica* L.). N Z J Crop Hortic Sci, 1–26

Khalifa SM, Abobatta WF (2023) Climate changes and mango production (Temperature). IgMin Res 1(1):043–046

Khalil A, Ali AA (2023) An econometric analysis for the impact of climatic conditions on the productivity of mango crops in Egypt by using panel data models. Int J Adv Res 11(3):351–360

Khan D, Shaukat SS (2006) The fruits of Pakistan: diversity, distribution, trends of production and use. Int J Biol Biotechnol 3:463–499

Kumar M, Ponnuswami V, Kumar PJ, Saraswathy S (2014) Influence of season affecting flowering and physiological parameters in mango. Sci Res Essays 9(1):1–6

Kumar A, Malik S, Chaudhary P, Kumar N (2017) Studies on the growth and flowering of different mango (*Mangifera indica* L.) cultivars under Western Uttar Pradesh conditions. J Pharm Phytochem 6(6S):439–442

Lal M (2003) Global climate change: India's monsoon and its variability. J Environ Stud Policy 6:1–34

Lavudya S, Prabhakar B (2024) Effect of climate change on agriculture and its management. Environ Conserv J 25:619–627

Legave JM, Normand F, Lauri PE (2013) Climate change and its probable effects on mango production and cultivation. In: X international mango symposium, vol 1075, pp 21–31

Mahato A (2014) Climate change and its impact on agriculture. Int J Sci Res Publ 4:1–6

Malik MT, Tariq T, Khan AH, Ullah H, Imran M, Iqbal J, Zainab A (2018) Outbreak of anthracnose and stem end rot diseases of mango in changing climate and their management through hot water treatment. Pak J Phytopathol 30:91–98

Malik AU, Amin M, Mazhar MS, Johnson PJ, Hofman P, Campbell J, Iqbal M, Iqbal J, Zainab A, Munir M, Collins RJ (2015) Mango value chain improvement through postharvest research and development: a developing country case study. In: Bally I, Tran-Nguyen L, Hall D (eds) XI international mango symposium 1183. Australian Centre for International Agricultural Research, Bruce, Australia, pp 411–420

Marroquín-Agreda FM, Gehrke-Velez MR, Pohlan JA, Lerma-Molina JN, Toledo-Toledo E, Ley-de C (2015) The association of bushy legumes with 'Ataúlfo' mango (*Mangifera indica* L.)+ cv. Ataúlfo affects reproductive biology and enhances productivity in mango plantations in Soconusco, Chiapas, Mexico. Indian Horticult J 5:63–69

Mbeye N (2016) Mapping out evidence-informed decision making landscape for health research in Malawi. AEN EIDM landscape mapping series no 13. Johannesburg: Africa Evidence Network. https://www.africaevidencenetwork.org/resources/landscape-maps/

Mitra SK (2014) Mango production in the world–present situation and future prospect. In: Coates LM, Hofman PJ, Johnson GI (eds) XXIX international horticultural congress on horticulture: sustaining lives, livelihoods and landscapes (IHC2014): IV 1111. International Society for Horticultural Science, Brisbane, Australia, pp 287–296

Mohsin M, Mehmood S, Jamal F, Anwer MM (2021) Modeling and forecasting of declining area and production of mango orchards in district Bahawalpur, Pakistan: a case of tehsil Ahmedpur East. J Agric Res 59:197–212

Nath V, Kumar G, Pandey SD, Pandey S (2019) Impact of climate change on tropical fruit production systems and its mitigation strategies. In: Climate change and agriculture in India: Impact and adaptation, pp 129–146

Oki T, Quiocho RE (2020) Economically challenged and water scarce: identification of global populations most vulnerable to water crises. Int J Water Resour Dev 36:416–428

Pimpa N (2024) Climate change challenges: case of the Thai agriculture business sector. Sustain Econ 2:112

Puche M, Pérez Macias M, Soto E, Figueroa R, Gutierrez M, Avilán Rovira L (2012) Minimum temperature an environmental determinant in the floral initiation of mango. Rev Científica UDO Agrícola 12(1):83–90

Rajan R, Feza M, Pandey K, Aman A, Kumar V (2020) Climate change and resilience in fruit crops. In: Climate change and its effects on Agriculture, pp 337–354

Ray K, Arora K, Srivastav AK (2019) Weather extremes and agriculture. Int Arch Photogramm Remote Sens Spatial Inf Sci 42:493–497

Reddy AGK, Kumar JS, Maruthi V, Venkatasubbaiah K, Rao CS (2017) Fruit production under climate changing scenario in India: a review. Environ Ecol 35(2B):1010–1017

Ribeiro IJA, Soares NB, Pettinelli Júnior ARMANDO, Dudienas C (2002) Comportamento de porta-enxertos de mangueira (*Mangifera indica* L.) submetidos a condições de baixas temperaturas. Rev Bras Frutic 24:249–250

Sacramento NJJE, Cuizon ACBD (2020) The Philippine mango industry governance, prospects, and recommendations: the case of Guimaras Province. Thammasat Rev 23(2):232–260

Sarker FIMGW, Biswas JC, Maniruzzaman M (2014) Climate change adaptation and economic profitability: crop land shifting to mango orchard in Rajshahi region. Bangladesh Rice J 18(1–2):8–17

Scuderi D, Gianguzzi G, Tinebra I, Gugliuzza G, Farina V (2023) Phenology and fruit growth dynamics of mango (*Mangifera indica* L.) in greenhouse and open air in Mediterranean climate. Proc Natl Acad Sci India Sect B Biol Sci 93(3): 731–741

Semeraro T, Scarano A, Leggieri A, Calisi A, De Caroli M (2023) Impact of climate change on agroecosystems and potential adaptation strategies. Land 12:1117

Shivashankara KS, Laxman RH, Geetha GA, Rashmi K, Kannan S (2024) Growth and physiological response of mango (*Mangifera indica* L.) cv. Alphonso under elevated CO_2 conditions. J Hortic Sci 19(1)

Talib SAA, Hassan MHM, Rashid MA, Sabdin ZHM, Rashid MZA, Ibrahim WMW, Mustaffa MAO (2020) Effects of environmental temperature and precipitation pattern on growth stages of *Mangifera indica* cv. Harumanis mango. J Agric Sci 12:26

Usman M, Fatima B, Khan MM, Chaudhry MI (2003) Mango in Pakistan: a chronological review. Pak J Agric Sci 40:151–154

Wahab AA, Javed MA, Khan RMMM, Zakir I, Abbas MT, Ahmed M, Tariq M, Kan M, Dogan H, Ahmad S (2022) Threats to cereal production under climate change. In: Dogan H, Fidan H (eds) Global climate change: agriculture and food science perspective. IKSAD Publishing House, Turkey, pp 3–40

Xu WT, Wang SB, Wu HX, Gao YY, Yao QS, Luo C, Zhan RL, Ma XW (2017) Effects of rain cover cultivation on ecophysiology-factors, yield and fruit quality in mango. In: XII international mango symposium vol 1244, pp 137–142

Yobo CM, Iponga DM, Tieguhong JC, Bengone NN, Ngoye A (2020) Exploring gender dynamics, economics and perceptions of the vulnerability of the bush mango value chain in three provinces of Gabon. Int for Rev 22(3):354–369

Zainuri Z, Sjah T, Nurrachman N, Ayu C (2019) Mango off-season technology (MOST): innovative, applicable, adaptive to climate change, and brings many positive impacts. AIP Conf Proc 2199(1)

Chapter 12
Impact of Soil Health on Mango Cultivation and Production Under Changing Climatic Scenarios

Syed Farhan Haider Bukhari, Muhammad Farhan, Rafia Kiran, Rifat Hayat, and Syed Zeeshan Hassan Bukhari

Abstract Mango production and cultivation heavily rely on soil health factors, especially in light of the strain caused by changing climate conditions. These factors include soil composition, nutrient availability, organic matter, and microbial diversity; all of these appear to be crucial elements that directly affect growth and fruit yield as well as disease resistance. The intricate relationship between soil health and mango output is further complicated by climate change, which is characterized by variations in temperature and precipitation as well as an increase in extreme weather events. This chapter explores how these alterations in soil quality affect pathogen activity, water-holding capacity, and nitrogen cycling. Environmentally friendly methods of soil management, such as crop rotation, cover crops, and organic amendments, are discussed in the paper as a potent way to improve soil resilience and boost mango yields in such delicate areas. It is now known that creative soil management and adaptive farming methods are essential to the long-term viability of mango orchards in the face of climate change.

Keywords Soil health · Climate change · Mango cultivation · Sustainable soil management

S. F. H. Bukhari (✉)
Department of Soil Science, University College of Agriculture and Environmental Sciences, The Islamia University of Bahawalpur, Bahawalpur, Pakistan
e-mail: farhanbukhari001@gmail.com

M. Farhan · R. Kiran
Department of Plant Pathology, Faculty of Agricultural Sciences and Technology, Bahauddin Zakariya University, Multan, Pakistan

R. Hayat
Institute of Soil and Environmental Sciences, Pir Mehr Ali Shah Arid Agriculture University, , Rawalpindi, Pakistan

S. Z. H. Bukhari
Department of Plant Breeding and Genetics, The Islamia University of Bahawalpur, Bahawalpur, Pakistan

S. A. H. Naqvi et al. (eds.), *Climate Change and Mango Production*,
https://doi.org/10.1007/978-3-031-87035-4_12

12.1 Introduction

Despite being somewhat steady for the last 10,000 years or more, the Earth's climate is constantly changing, mostly as a result of natural processes like volcanic activity and human activities (IPCC 2007). Given that crop yields are directly influenced by climate, food production is especially vulnerable to climate change, which might result in a reduction of up to one-third in food yields in the tropics and subtropics (Pender 2008). Even a 2–3 °C increase in temperature might put an extra 30–200 million people at risk of starving, on top of the estimated 12% of the world's population that is now at risk (Stern 2006). Food security will be severely impacted in the coming decades by the drastic changes in agricultural yield and geographical distribution brought about by climate change, especially rising temperatures, changed rainfall patterns, and climate variability. Global warming is predicted to exacerbate the Earth's hydrological cycle and raise the likelihood of droughts and floods, according to climate model simulations (Asadieh and Krakauer 2017). It is estimated that 10% of the world's land area will concurrently see a rise in high extreme streamflow and a fall in low extreme streamflow, suggesting a possible rise in the danger of both drought and flooding. According to models, streamflow will rise in areas close to the Arctic Circle in the future, increasing freshwater supplies to the Arctic Ocean (Lique et al. 2016). At the same time, subtropical drylands will see a further decrease in streamflow (Schewe et al. 2014).

The mango (*Mangifera indica* L.), which originated in South East Asia and is now grown all over the world, is the most significant fruit crop in tropical regions and has socioeconomic value. According to Neguse et al. (2019), mango fruit has emerged as the second most popular tropical crop in terms of both production and farmed acreage. The crop has adapted to grow in a variety of environments with varying soil types, temperatures, and rainfall patterns. Nevertheless, regions with a cool or dry time before flowering, lots of soil moisture, and fairly hot (30–33 °C) temperatures during fruit development are ideal for commercial production (Nel 2024; Zuazo et al. 2021). Numerous processes and life cycles in this environment are influenced by soil health like soil moisture, soil pH, the diffusion of gases and solvents (particularly oxygen), nutrient availability, microbial migration, compound diffusion between organism cells and the environment, and mineralization rate. All of these elements simultaneously impact soil productivity and fertility. The global biogeochemical cycle and ecosystem services, including crop yield in agricultural systems, are impacted by weather phenomena, which also have a major effect on plants and soil biota (Vanbergen and the Insect Pollinators Initiative 2013). Although plant-associated microorganisms are essential for plant growth and health, very little has been discovered about how the plant-microbiome complex reacts to extreme weather conditions like drought and floods (Francioli et al. 2021). Floods and droughts have a wide range of social and economic consequences, and supply chain disruption is one way that they affect agriculture (He et al. 2020). Extreme weather events pose a severe danger to food security due to population growth and the resulting soil deterioration. During the last 12 years, droughts and floods have significantly reduced

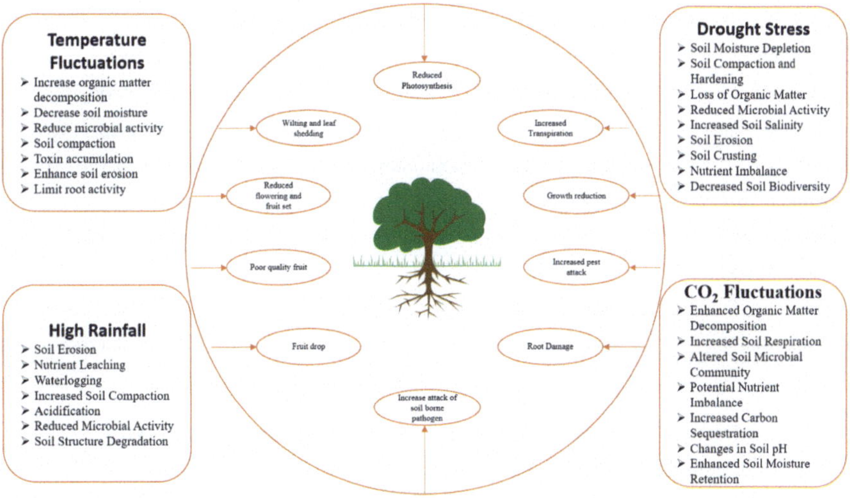

Fig. 12.1 Effect of soil health on mango tree due to different environmental extremes

crop production in the United States alone, contributing to approximately 70% of the 2011 crop yield reduction (Striker 2012). The effects of such phenomena are felt around the globe (Fig. 12.1).

Temperature affects mango flowering frequency, extent, fruit growth, and quality. Rainfall sustains mangos, and commercial mango production depends on equally distributed rainfall. Mangos benefit when water stress occurs during the flowering stage. Water stress not only causes but also speeds up flowering in tropical environments (Satisha et al. 2020). Water stress prolongs the accumulation of floral stimulation and inhibits vegetative flushing. Flooding negatively impacts mango development and gas exchange, however in order to withstand extended flooding stress, the trees grow hypertrophic lenticels on stems just above the waterline. Because the harmful ions lead to nutrient imbalance and hinder growth, fruit bearing, and yield, saline circumstances are not conducive to the successful production of mangoes. Nonetheless, a number of rootstock cultivars have demonstrated varying degrees of resistance to saline stress. In salt-affected soils, the negative impacts on growth and yield could be mitigated by using tolerant rootstocks (Ahmad and Anjum 2020; Sharma and Singh 2021). Given that a variety of abiotic factors influence mango growth, development, and final yields, it is crucial to have a complete understanding of how these factors affect the phenology, physiology, growth, and development of mangos (Laxman et al. 2016). In this chapter we will discuss the impact of soil health on the mango tree physiology, biochemistry, floral biology, biotic stressors and its production under changing climate (Table 12.1).

Table 12.1 Various soil factors for the optimum growth of mango tree

Soil factor	Optimum range	Reference
Soil Ph	5.5–7.5	Bally (2006)
Soil temperature	24–31	Yusof et al. (1969)
Soil texture	Loamy, alluvial, well drained and deep soils	Das et al. (2019)
Organic matter	More than 3%	
Soil moisture	65% to 70% of field water capacity	Wei et al. (2017)
Electrical conductivity	< 2 dS/m	
Nutrient level	Adequate levels of N, P, K	Ram et al. (2020)
Soil aeration	Good	Das et al. (2019)

12.2 Impact of Climate Change on Soil Health Affecting Mango Growth

Numerous aspects of soil health, such as salinity, hydrology, temperature regimes, and the availability of organic matter, could be influenced by climate change. The development and growth of mango trees can be impacted by changes in the soil. The primary impacts of climate change on soil health are listed below.

12.2.1 Soil Physical Parameters

The movement, accumulation, and transformation of materials such water, air, organic matter, soluble salts, carbonates, silicate clay minerals, sesquioxides, and silica are among the main physical processes that occur in soils (Binkley and Fisher 2019). These advantages often include the oxidation and hydration processes that add organic matter, oxygen, and water (Driese et al. 2021). In some places, soluble chemicals deposited by groundwater progressively bring new mineral minerals to the surface. The main losses happen when water carries away dissolved or suspended elements by surface runoff through the porous soil or percolation through the soil profile (Zade et al. 2021). Numerous soil physical processes are expected to be impacted by climate change, including increased temperatures, precipitation extremes, concentration of CO_2 and how they interact. These modifications have the potential to significantly increase the risk of salinization, decrease water availability, modify the nitrogen and carbon cycles, impact soil nutrient retention, and diminish soil biodiversity (Benbi and Kaur 2009). The physical characteristics and procedure of soil impact root development, water retention, and water transport. These physical traits are drastically changed by climate change, which affects the soil's general health.

12.2.2 Soil Texture

The texture of a soil is determined by the ratios of sand, silt, and clay. Climate change directly affects it. The four potential climate scenarios—arid, semi-arid, sub-humid, and humid—have a significant impact on important soil processes, including the textural difference in the soil profile (Gautam et al. 2021). The region faces issues with soil and water conservation due to low rainfall, undulating topography, and light-textured soil that experiences heavy runoff. The main restrictions in this area include insufficient nutrient availability, particularly for phosphate in the uplands, soil erosion and acidity, and a lack of rainfall. The main issues preventing successful double-cropping include irregular rainfall, inadequate irrigation infrastructure, low water retention capacity, and soil permeability (Naik et al. 2015).

The physical characteristics of the soil are greatly influenced by the climate, which has an impact on mango plant cultivation, growth, and development. Accordingly, the texture is one of the most significant soil characteristics impacted (Cardoso et al. 2017). Sand, silt, and clay ratios are changed as a result of increased erosion or sedimentation brought on by weathering-induced changes in the climate. Mango plants, for instance, have soil that encourages root growth but, while having strong water retention qualities, becomes denser or even heavier, which significantly restricts the expansion of those roots (Ganeshamurthy et al. 2018). Healthy root systems depend on both general soil aeration and the availability of vital nutrients, both of which are typically impacted by this changed soil texture.

Because mango trees prefer ideal soil, these plants typically grow stunted and yield less fruit (Tripathy 2021). Drainage and water retention are directly influenced by soil texture. Sand loam is ideal for mango plants because it retains moisture and allows for efficient drainage (Kumar et al. 2017). Drought stress results from a decrease in water retention caused by a shift in soil texture toward coarser materials. This may have an effect on fruit set, flowering, and tree health in general, which could lead to decreased fruit quality and yields (Pinney et al. 1998; Maity et al. 2021). Degraded soils that are less suitable for growing mangoes might result from ongoing erosion and sedimentation over time. The implementation of soil conservation techniques, including contour farming or cover crops, may be necessary for farmers to counteract these changes and restore soil texture.

12.2.3 Soil Structure and Aggregate Stability

Soil structure refers to how primary and secondary particles are arranged and organized within a soil mass. It regulates the soil's water and air content, which is best for microbial life and enzymatic processes, both of which are critical for the growth of mango plants. According to Moebius et al. (2007) soil structure, different chemical and biological characteristics, and management techniques all affect aggregate stability, which is the ability of soil aggregates to withstand outside influences such intense rainfall and plowing. It supports activities like organic carbon accumulation, water infiltration and storage, and root and microbial activity—all of which are critical for plant growth—and is a crucial indication of soil health and ecosystem functionality (Yan et al. 2023).

Therefore, the structure of the soil is crucial to the health of mangoes. Climate change-induced elevation in rainfall which may result in soil compaction, and breaks up soil aggregates and porosity. Because compaction may hinder the roots' ability to reach a considerable depth, it would restrict their ability to penetrate. To be able to absorb water and nutrients, they require deep roots, particularly during dry spells (Cortez et al. 2014). Mango trees struggle to maintain a healthy root system at all times due to poor soil structure. Drought and nutrient deficits are quite dangerous during these periods. Anaerobic conditions caused by poorly organized soil impair root health and overall plant vigor (Phuong 2023).

Mango health depends on root growth and nutrient uptake, which are facilitated by well-structured soil. Anaerobic conditions brought on by poor soil structure might limit the amount of oxygen available to roots. Symptoms include yellowing leaves, poor fruit development, and heightened disease susceptibility may arise from this (Stirling et al. 2016). Mango orchard viability can be lowered by ongoing soil structure degradation, which raises production costs and lowers yields. Enhancing soil structure and climate change resilience can be achieved by putting techniques like crop rotation, decreased tillage, and organic amendments into practice (Chaudhari and Singh 2019; Ganeshamurthy et al. 2019).

The amount and quality of organic matter, the inorganic components in the soil matrix, farming methods, and natural physical processes like freeze–thaw cycles and shrinkage and expansion all have a significant impact on the structure and quality of soil. Decreased soil organic matter makes soil aggregates less stable, infiltration rates lower, and soil aggregates more susceptible to erosion, compaction, and discharge (Bot and Benites 2005; Karmakar et al. 2016) (Table 12.2).

12.2.4 Porosity

Pore size distribution and porosity, which is a measurement of a material's empty spaces as a fraction (volume of empty space to that of total volume), are what allow the soil to retain the root zone air and water required for the growth of plant (Reynolds et al. 2002). Bulk density, microporosity, pore volume functions, and soil physical quality are all closely related to the characteristics of soil pores. Soil porosity and water release characteristics have a direct effect on a number of soil indices, such as soil aeration capacity, plant accessible water capacity, and relative field capacity (Abdallah et al. 2021).

For mango roots to efficiently absorb water and nutrients, high porosity is necessary. Roots may have trouble finding moisture when porosity declines, particularly during dry spells. Because of their close relationship to root growth and soil enzyme activity, soil porosity and pore size distribution might have unexpected effects

Table 12.2 Impact of soil chemical factors on mango tree health

Soil chemical factor	Role in mango tree physiology	Impact on mango tree health
Soil salinity	Influences osmotic potential and water uptake	– High salinity causes osmotic stress, leaf burn, stunted growth, and reduced fruit yield and quality
Nitrogen (N)	Essential for vegetative growth and leaf development	– Excess nitrogen promotes vegetative growth at the expense of flowering – Deficiency leads to yellowing (chlorosis) and stunted growth
Phosphorus (P)	Vital for root development and energy transfer	– Deficiency causes poor root growth, delayed flowering, and smaller fruits
Potassium (K)	Critical for water regulation, fruit quality, and disease resistance	– Deficiency leads to poor fruit color, size, and taste – Increases vulnerability to diseases
Calcium (Ca)	Strengthens cell walls and supports root growth	– Deficiency causes tip burn, poor root growth, and fruit cracking
Magnesium (Mg)	Central component of chlorophyll and photosynthesis	– Deficiency leads to interveinal chlorosis, reduced photosynthesis, and lower fruit production
Iron (Fe)	Required for chlorophyll synthesis	– Deficiency causes chlorosis in young leaves, especially in alkaline soils
Zinc (Zn)	Influences growth hormones and fruit set	– Deficiency results in small leaves (little leaf disorder), poor flowering, and reduced fruit yield
Organic matter (OM)	Improves nutrient cycling and chemical buffering	– Enhances nutrient availability, improves soil structure, and supports overall mango tree health
Toxic elements (e.g., Aluminum, Heavy Metals)	Negatively affect root growth and nutrient uptake	– Presence of toxic elements in acidic soils inhibits growth, reduces yields, and affects fruit quality

(Sanchez Garcia 2024). Because of the impact of raindrops, decreased microbiological activity, decreased root growth and exudates, and decreased aggregate stability, higher rainfall intensities can cause surface sealing in sodic soils. Together, these elements raise the possibility of surface runoff while impeding plant emergence and growth (Singh et al. 2022). If porosity continues to decline due to climate change, mango trees may suffer significant stress, which will result in lower yields and fruit size. Farmers may need to implement conservation practices, such as cover cropping and organic matter incorporation, to improve porosity and soil health.

12.2.5 Infiltration and Plant Available Water

The lower limit of plant available water (so osmotic potential is excluded), porosity, field capacity, micropore flow, and texture are some of the soil characteristics that affect water availability for plant growth and significant soil activities (Jarvis 2007).

Water cannot reach the root zone where mango trees can absorb it due to surface runoff and reduced penetration caused by intense rainfall events. A cycle of drought stress may result from this circumstance, especially during crucial growth times when mango trees need enough moisture to bloom and bear fruit (Nel 2024). In certain situations, inadequate penetration can result in waterlogging, which suffocates roots and encourages root infections. Lower fruit yields and overall tree health are the results, highlighting the necessity of efficient water management techniques in mango farming (Chawla et al. 2011; Singh 2019).

Integrative soil health testing has employed measurements of plant accessible water capacity to evaluate the effects of management. Additionally, the distribution and availability of water in the soil may vary quickly as a result of climate change, particularly in reaction to drought or intense, erratic rainfall events. Extreme drought, intense rainfall, or severe erosion can be lessened by using management techniques including cover crop planting, conservation tillage, and the addition of organic matter, which can assist maintain or improve soil moisture levels and water infiltration (Salvador Sandris et al. 2008).

12.2.6 Bulk Density

In order to describe the condition of soil compactness in relation to land use and management, bulk density is regularly measured. Bulk density generally exhibits an inverse relationship with the amounts of soil organic matter (SOM) or soil organic carbon (SOC), according to Weil and Magdoff (2004). Higher temperatures may expedite decomposition, causing organic carbon to be lost. This might raise the bulk density of the soil, increasing its susceptibility to compaction (Davidson and Janssens 2006). Drought episodes and variable, intense rainfall are stressors associated with climate change and land management practices (Birkas et al. 2009). Mango trees benefit from a lower bulk density since it promotes root growth and improves access to nutrients and water (Kumari et al. 2020).

Increased susceptibility to environmental stresses, decreased fruit output, and stunted growth can result from high bulk density. The passage of water and air, which are necessary for proper root function, is impeded by high bulk density. Compacted soils prevent roots from exploring for moisture and nutrients, which stunts mango tree

Fig. 12.2 Deficit irrigation strategies for subtropical mango farming (Main physiological mechanisms to face water stress in mango tree. ATP, adenosine triphosphate; ABA, abscisic acid; JA, jasmonic acid; ↑, increase; ↓, decrease) (*Source* Zuazo et al. 2021)

growth and lowers fruit quality. High bulk density might prevent mango trees from accessing sufficient nutrients, causing stress symptoms including leaf yellowing and poor fruit development that ultimately reduce output (Habibi et al. 2023; Patil and Lamnganbi 2018). Due to a greater need for fertilizers and irrigation, mango plantations may see decreased output and higher production costs if bulk density stays high. To control bulk density and improve soil health, farmers may need to implement techniques like organic amendments, less tillage, and planned crop rotations (Fig. 12.2).

12.2.7 Rooting Depth

Plant water available capacity, subsurface salinity, SOC content, and other characteristics that reveal significant limitations in the soil profile are likely to be impacted by changes in rooting depth (Birkas et al. 2009; Dalal and Moloney 2000). Subsoil limitations such as salinity and excessive chloride concentrations are anticipated to have a bigger effect on plant availability of water and, consequently, plant productivity under extended drought (Dang et al. 2008; Rengasamy 2010). Mangoes and other deep-rooted trees get advantage from having deeper access to nutrients and water, which increases their resistance to dry spells (Das et al. 2019). Trees may display stress symptoms, including as wilting and nutritional deficits, if rooting depth is limited. These symptoms might impact fruit yield and quality. According to Zaharah and Razi (2009) persistent restrictions on rooting depth may result in lower productivity and heightened vulnerability to drought stress. To promote deeper root growth and improve tree resilience, farmers may need to use techniques including deep tillage, cover crops, and moisture management.

12.2.8 Soil Surface Cover

According to Bruce (1996), soil surface cover is essential for ecological processes such as fixing carbon and nitrogen, promoting the germination of native seeds, and preserving soil surface water and nutrient retention. In the context of climate change, soil structural features like crust and seal formation—especially in sodic soils—are important markers of soil health. Runoff, oxygen diffusion, water infiltration, surface water evaporation, and soil erosion are just a few of the soil processes that these formations might affect (Bolan et al. 2023). Reduced surface cover can cause long-term degradation of mango cultivation areas, leading to lower yields and higher production costs and farmers may need to implement strategies like planting cover crops, maintaining biodiversity, and using mulches to improve surface cover and enhance soil health (Das et al. 2021). Mango orchards may experience higher rates of erosion, nutrient loss, and increased temperatures, all of which can negatively impact tree health and fruit quality (Kundu 2020; Bakshi et al. 2015).

12.2.9 Soil Temperature

The flow of water and gasses, evaporation, heat transmission inside the soil, and the absorption and emission of solar radiation at the soil's surface are some of the elements that affect soil temperature (Karmakar et al. 2016). Like soil moisture, soil temperature is essential for controlling a number of soil functions. Warmer soil temperatures accelerate a variety of processes, including increased nitrification rate, nutrient release, organic matter breakdown, microbiological activity, and general chemical weathering of minerals (Parama 2017). The kind of plant that grows on the soil's surface can affect its temperature, which can change as a result of climatic shifts or adaption techniques. The optimal soil temperature is between 25 and 30 °C for healthy mango root development and nutrient uptake (Fischer et al. 2022; Ullah et al. 2024). High soil temperatures can cause stress to roots, which can result in poor fruit quality, reduced development, and heightened susceptibility to disease. The cycles of flowering and fruiting may also be disturbed by high temperatures. Mango orchards' long-term survival could be in danger if soil temperatures keep rising. Farmers may need to adjust by using soil-healthy measures like applying organic mulches, shade plants, or irrigation methods that assist cool the soil. Maintaining sustainable mango production in a warming environment requires these modifications (Bacelar et al. 2024).

12.3 Soil Chemical Parameters

12.3.1 Soil pH

Parent material, weathering time, vegetation, and climate all affect the pH of soil. It is regarded as a crucial indicator of soil health and has been included in thorough evaluations of soil health to determine the consequences of changes in land use and agricultural practices. Rapid pH changes brought on by climate change-related processes like as CO_2 fertilization, rising temperatures, varying precipitation, and atmospheric nitrogen deposition are not anticipated to cause major issues in the majority of soils. However, the amount of water available to plants, the cycling of carbon and nutrients, the condition of organic matter, and consequently plant productivity will all be impacted by these climate change drivers, which will also affect the pH of the soil (Reth et al. 2005). Increased rainfall can cause alkaline cations to leach from the soil, which can lower pH. Acidic environments can have a detrimental effect on nutrient availability, particularly for vital macronutrients like potassium, phosphorus, and nitrogen (Prakash et al. 2023).

Increased microbial activity brought on by warmer temperatures may hasten the breakdown of organic materials. The resulting organic acids can contribute to soil acidification, further affecting pH levels (Raza et al. 2021). Mango trees prefer a slightly acidic to neutral pH (around 6.0–7.5) for optimal nutrient uptake; a drop in pH can result in nutrient deficiencies, especially of calcium and magnesium, which are necessary for healthy fruit development (Kamarudin et al. 2023). Acidic soils can also increase the solubility of toxic metals, potentially harming root systems and overall tree health. Symptoms may include poor fruit set, stunted growth, and increased susceptibility to diseases. Farmers may need to apply lime or organic amendments to restore pH levels and enhance nutrient availability in order to combat sustained acidification, which can result in chronic nutrient deficiencies and lower mango yields and fruit quality (Ray 2023; Ram et al. 2020). This requires additional management and resources.

12.3.2 Electrical Conductivity

The electrical conductivity (EC) of the soil is used to measure the amount of salt present. It can offer data on crop performance, nutrient cycle, biological activity, and salinity variations. It can be used as a proxy for soil structural deterioration in addition to pH, particularly in sodic soils (Arnold et al. 2005). One chemical indicator that has been used to evaluate how crop management practices affect the biological quality of the soil is electrical conductivity (Gil et al. 2009). Under climate change scenarios, electrical conductivity rises as temperatures rise and precipitation falls (Smith et al. 2002). Rainfall and soluble salt concentrations were found to have a non-linear connection in a study of soils from four climatic regions: Mediterranean, Semi-arid, Mildly Arid, and Arid. Sites that received less than 200 mm of rainfall had significantly higher soluble contents, and vice versa (Gautam et al. 2021). High electrical conductivity indicates higher salinity levels, which can be harmful to mango trees. Elevated salinity can cause osmotic stress, which impairs the tree's ability to absorb water and nutrients; symptoms of salt stress in mango trees include leaf burn, wilting, and reduced fruit quality, which ultimately results in lower yields. Persistently high salinity can make soils unsuitable for mango cultivation, requiring expensive soil remediation techniques (Muhammed et al. 2022; Laxmi et al. 2021). To lessen the effect on mango output, farmers may need to use techniques like choosing salt-tolerant crop cultivars or flushing salts using freshwater irrigation.

12.3.3 Sorption and Cation Exchange Capacity

Important soil characteristics include sorption and cation exchange capacity (CEC), which are crucial for maintaining vital cations like Ca_2+, Mg_2+, and K+ as well as immobilizing potentially hazardous cations like Al_2+ and Mn_2+. Because they show how well the soil can absorb nutrients, pollutants, and pesticides, these traits are useful markers of soil health (Ross et al. 2008). Since the CEC of these soils is connected with that of SOM, the decrease in CEC of low-activity clay and coarse-textured soils may be the result of increased SOM loss and decomposition brought on by high temperatures (Davidson and Janssens 2006). Extreme and prolonged rainfall events may cause increased base cation leaching, which could transport alkalinity from soil to streams.

Increased microbial activity brought on by warmer temperatures can accelerate the breakdown of organic molecules. This reduces the soil's organic matter content, which negatively impacts CEC and the soil's ability to hold onto essential cations like calcium, magnesium, and potassium (Navarro-Pedreño et al. 2021). The stability of clay minerals and their capacity to retain cations can be impacted by changes in precipitation patterns, which can also affect soil moisture levels. Since moisture is essential for preserving exchange sites, extended dry spells may result in decreased cation availability (Mondal 2021). However, sustaining nutrient availability in mango production requires a high CEC. Deficits in certain nutrients, especially potassium and calcium, which are essential for fruit development and general plant health, might result from low CEC (Wahane et al. 2024). Reduced tree vigor, poor fruit set, and fading leaves are signs of insufficient nutrition availability. Mango orchards may have chronic nutrient deficits if CEC keeps dropping, which would lower yields and increasing reliance on synthetic fertilizers (Ram et al. 2020; Ullah et al. 2024).

To increase CEC and guarantee continuous mango production, farmers may need to increase soil organic matter through cover crops or composting.

12.3.4 Plant Available Nutrients

The ability of soil to sustain plant development and the identification of critical or threshold levels that can point to possible environmental risks can both be ascertained by measuring the extractable nutrients in the soil (Dalal and Moloney 2000). Climate change factors like rising temperatures, varying precipitation, and atmospheric nitrogen deposition are predicted to impact nitrogen cycling and possibly the availability of other vital nutrients like sulfur and phosphorus because of the close relationship between soil organic carbon cycling and nutrient cycling, especially nitrogen (N) (Weil and Magdoff 2004). Mango growth, flowering, and fruiting are all impacted by the availability of nutrients. Symptoms including foliage chlorosis, poor fruit growth, and heightened vulnerability to pests and diseases can result from deficiencies in essential nutrients (Kumari et al. 2008). In this case, low potassium levels can lead to greater fruit drop and poor fruit quality. Over time, substantial drops in mango output and quality could result from chronic nutritional deficits brought on by climate change (Bibi et al. 2019). To maintain mango orchards in changing climates and guarantee sufficient nutrient availability, farmers may need to employ integrated nutrient management techniques, such as organic additions and targeted fertilization techniques.

12.4 Soil Biological Parameters

Organisms in the soil are able to adjust to shifting environmental conditions. Biological indicators are essential for evaluating soil quality for mango production and culture in the context of climate change.

12.4.1 Soil Organic Matter

A complex and varied component of soil, soil organic matter (SOM) is made up of both living and non-living materials, each having unique characteristics, roles, and rates of turnover. By supplying vital nutrients and enhancing soil structure, SOM plays a crucial part in preserving soil fertility and promoting plant growth. The stability and well-being of soil ecosystems depend heavily on its dynamic nature (Gurmu 2019). As a source and sink of carbon and nitrogen, as well as a participant in the cycling of phosphorus and sulfur, soil organic matter adds to the soil's charge characteristics. Additionally, it can combine with multivalent ions and organic compounds to create complexes (Rao et al. 2019). In addition to influencing hydraulic qualities, water retention, and aggregate stability, soil organic matter offers habitat and substrates for microbes and wildlife. Loss of soil structure, decreased water retention, increased risk of erosion, increased bulk density, and soil compaction can all result from a decrease in SOM (Cotrufo and Lavallee 2022). In order to mitigate global warming, land use and management methods that result in the accumulation of SOM will aid in the absorption of CO_2 from the atmosphere. By enhancing water storage, SOM can help reduce the effects of flooding after periods of intense rainfall and boost soil resilience by storing water during droughts (Baveye et al. 2020).

High temperatures can hasten the breakdown of organic materials by soil microbes. In addition to lowering SOM, this process contributes to additional climate change by releasing stored carbon into the environment as CO_2. Unbalanced nutrient availability for plants can result from the quick breakdown of organic matter, which can cause a nutrient flush followed by depletion. Temperature and rainfall variations can have an impact on biomass output and plant development. The inflow of organic matter through litterfall and root exudates may decrease if climate change results in less vegetation cover or a change in the species composition, which would lower SOM. Moinet et al. (2018) and Raza et al. (2023).

Rich soils with organic matter assist to maintain moisture and nutrients necessary for the healthy growth of a mango tree. Stunted growth, poor fruit set, and greater susceptibility to pests and diseases are some of the symptoms that might arise from reduced SOM since it can cause poorer water retention and decreased nutrient availability (Nath et al. 2019). Nutrient deficiencies can have a major effect on fruit quality and yield, particularly in the areas of potassium, phosphorus, and nitrogen. Lower agricultural output and weakened soil health can result from SOM's long-term deterioration. To mitigate the impact of climate change on the production of mangoes, farmers may need to implement techniques that improve SOM, such as using organic amendments, cover crops, or no-till farming methods (Swain et al. 2023).

12.4.2 Soil C: N Ratio

Rich soils with organic matter assist to maintain moisture and nutrients necessary for the healthy growth of Microbial activity (mineralization/decomposition) is stimulated by temperature increases and sporadic rainfall. Reduced biomass accumulation, soil carbon depletion, and a drop in the C:N ratio are the results of this (Rosenzweig and Hillel 2000; Lal 2004). Plant water usage efficiency (WUE) rises with atmospheric CO_2. As a result, more biomass will be produced per millimeter of available water (Kimball 2003). The rate of decomposition surpasses the rate of net primary output when there is a greater water shortage. The drier environment produced by this process is perfect for the reduction of organic carbon. Drought-induced biomass losses result in decreased annual and perennial vegetation (Lal 2019). Generally, a C:N ratio of 20:1 to 30:1 is appropriate for the best nutrient availability. A higher C:N ratio may restrict the amount of nitrogen available, which would hinder the growth and fruiting of mango trees. Nitrogen shortages can cause inadequate leaf growth, less flowering, and smaller, worse-quality fruit, all of which can diminish yields. According to Naveena Kumara et al. (2023) mango plantations may eventually suffer from chronic nutrient deficits and decreased soil fertility if the C:N ratio is left unbalanced as a result of climate change. Examples of management strategies include intercropping, green manuring, crop residue management, and conservation tillage techniques. In order to sustain good mango agriculture and balance the nutritional profile of the soil, farmers may need to employ techniques like adding organic amendments or nitrogen-fixing cover crops (Ganeshamurthy et al. 2018).

12.4.3 Soil Microbial Biomass

The biological component of soil organic matter, known as microbial biomass, is thought to be the most reactive and readily broken-down type of carbon in soils. This element is extremely sensitive to environmental changes and is essential to soil health and nutrient cycling. Microbial biomass activity has a major impact on ecosystem processes and soil fertility. It mediates the transfer between various soil organic carbon fractions, is a sensitive indicator of changes in soil processes, and is essential to the dynamics of nutrients and energy in the soil (Saha and Mandal 2009). Nevertheless, it has been demonstrated that soil microbial biomass, like labile C, is sensitive to transient environmental changes (Haynes 2008). Temperature rises have the potential to initially enhance microbial activity, but excessive heat can also stress or even kill microbial communities, which lowers total biomass. This may impair the soil's capacity to sustain plant development and interfere with nitrogen cycle mechanisms. Microbial biomass may be further impacted by fluctuating moisture conditions (Jansson and Hofmockel 2020). While too much moisture can produce anaerobic conditions that are unfavorable for many beneficial bacteria, resulting in a decrease in variety and functional capacity on the other hand drought can cause a decrease in microbial activity and biomass (Schimel et al. 2018).

Healthy microbial biomass is essential for nitrogen cycling, organic matter breakdown, and overall soil fertility. Poor mango tree performance can result from a decrease in microbial biomass, which can also limit nutrient release and lower soil fertility. Fruit productivity and quality may also be impacted by symptoms such as nutrient shortages, slower development rates, and heightened susceptibility to illnesses (Srivastava 2022). Long-term soil deterioration, lower fertility, and decreased agricultural output can result from a persistent drop in microbial biomass. To support sustainable mango farming in the face of climate change, growers may need to implement practices that enhance microbial ecosystems, such as crop rotation, the addition of organic matter, or tillage reduction (Singh and Gupta 2018).

12.4.4 *Enzyme Activity*

Plant-soil system changes may be indicated by soil enzyme activities because they are (1) easily measured, (2) closely related to nutrient cycling and soil biology, (3) incorporate data on the physicochemical and microbiological characteristics of the soil, and (4) react promptly to modifications in soil management (Garcia-Ruiz et al. 2009). Increased CO_2 may alter the amount and quality of carbon taken by plants, thereby raising the activity, abundance, and turnover of microbial enzymes, which can, in turn, influence the functioning of the microbial community in soil. Because of improved microbial metabolism, elevated temperatures can initially increase enzymatic activity. Extreme heat, however, can denature enzymes or prevent their synthesis, which lowers the soil's capacity to efficiently cycle nutrients and decompose organic matter (Lombao et al. 2021). Enzyme activity can be strongly impacted by variations in soil moisture. While waterlogged soils can provide anaerobic conditions that impede enzyme function, reducing nutrient mobilization and availability, drought conditions can result in decreased microbial activity and enzyme production (Li et al. 2024).

In order to ensure healthy plant growth, organic matter must be broken down and nutrients must be released, both of which are made possible by active soil enzymes. These enzymes catalyze metabolic reactions in the soil, which increases the availability of nutrients. They are important markers of soil fertility and health since their activity is impacted by management techniques and soil conditions (Mitra et al. 2022). Mango trees may experience deficits as a result of reduced enzyme activity that impedes nutrient cycling. Farmers may experience reduced yields and financial returns as a result of symptoms such as stunted growth, poor flowering, and worse fruit quality. In order to boost enzymatic activity, farmers may need to implement management techniques including adding organic amendments, keeping moisture levels appropriate, and making sure the plant population is diverse (Elsheery et al. 2020).

12.5 Conclusion

Mango production systems depend heavily on the quality of their soil, particularly in light of the growing difficulties brought on by climate change. The physical and biological characteristics of the soil are changing due to climate change, necessitating the use of creative soil management techniques to preserve mango yields. To maintain soil health, farmers must concentrate on strengthening nutrient cycling, cultivating a diverse microbial population, and improving soil structure. It is advised that agroforestry methods, precision irrigation, and the addition of organic matter be used to lessen the negative consequences of climate change on mango cultivation. Furthermore, using sustainable techniques guarantees that the soil will continue to be productive and fertile for upcoming generations. In addition to encouraging long-term mango farming, these methods reduce possible negative effects on the environment and the economy. In order to stop soil erosion and preserve moisture, farmers should also incorporate conservation strategies like cover crops. Adapting management tactics to particular climatic circumstances can be facilitated by routine soil testing and monitoring. All of these initiatives work together to make mango production systems more resilient and sustainable overall.

References

Abdallah AM, Jat HS, Choudhary M, Abdelaty EF, Sharma PC, Jat ML (2021) Conservation agriculture effects on soil water holding capacity and water-saving varied with management practices and agroecological conditions: a review. Agronomy 11:1681

Ahmad R, Anjum MA (2020) Physiological and molecular basis of salinity tolerance in fruit crops. In: Fruit crops. Elsevier, pp 445–464

Arnold SL, Doran JW, Schepers J, Wienhold B (2005) Portable probes to measure electrical conductivity and soil quality in the field. Commun Soil Sci Plant Anal 36:2271–2287

Asadieh B, Krakauer NY (2017) Global change in streamflow extremes under climate change over the 21st century. Hydrol Earth Syst Sci 21:5863–5874

Bacelar E, Pinto T, Anjos R, Morais MC, Oliveira I, Vilela A, Cosme F (2024) Impacts of climate change and mitigation strategies for some abiotic and biotic constraints influencing fruit growth and quality. Plants 13:1942

Bakshi P, Wali VK, Iqbal M, Jasrotia A, Kour K, Ahmed R, Bakshi M (2015) Sustainable fruit production by soil moisture conservation with different mulches: a review. Afr J Agric Res 10:4718–4729

Bally IS (2006) Mangifera indica (mango). Species profiles for Pacific Island agroforestry, pp 1–25

Baveye PC, Schnee LS, Boivin P, Laba M, Radulovich R (2020) Soil organic matter research and climate change: merely re-storing carbon versus restoring soil functions. Front Environ Sci 8:579904

Benbi DK, Kaur R (2009) Modeling soil processes in relation to climate change. J Indian Soc Soil Sci 57:433–444

Bibi F, Ahmad I, Bakhsh A, Kiran S, Danish S, Ullah H, Rehman AU (2019) Effect of foliar application of boron with calcium and potassium on quality and yield of mango cv. summer bahisht (SB) Chaunsa. Open Agric 4:98–106

Binkley D, Fisher RF (2019) Ecology and management of forest soils. John Wiley & Sons

Birkas M, Dexter A, Szemok A (2009) Tillage induced soil compaction as a climate threat increasing stressor. Cereal Res Commun 37:379–382

Bolan N, Nortje G, Bolan S, Srivastava P, Kumar M, Wang H, Kirkham MB (2023) Soil crust: formation, influence on soil productivity, and management. In: Soil constraints and productivity. CRC Press, pp 179–188

Bot A, Benites J (2005) The importance of soil organic matter: key to drought-resistant soil and sustained food and production. FAO Soils Bull No. 80. Food and Agriculture Organization of the United Nations, Rome, Italy, pp 78

Bruce JG (1996) Morphological characteristics and interpretation of some polygenetic soils in loess in southern South Island, New Zealand. Quaternary International 34:205–211

Cardoso JAF, Lima AMN, Cunha TJF, Rodrigues MS, Hernani LC, Cunha JC, de Oliveira Neto MB (2017) Changing in chemical and physical attributes of a sandy soil under irrigated mango cultivation in semiarid region

Chaudhari SK, Singh A (2019) NRM interventions for enhanced mango productivity and quality. World 5:7–96

Chawla R, Sheokand A, Rai MR, Kumar R (2011) Impact of climate change on fruit production and various approaches to mitigate these impacts. Trop Fruits 26

Cortez JW, Lopes HL, Coelho DS, Arcoverde SN (2014) Spatial variability of mechanical resistance to penetration evaluated with mobile unit on haplic vertisol cultivated with mango. Eng Agric 34:244–253

Cotrufo MF, Lavallee JM (2022) Soil organic matter formation, persistence, and functioning: a synthesis of current understanding to inform its conservation and regeneration. Adv Agron 172:1–66

Dalal RC, Moloney D (2000) Sustainability indicators of soil health and biodiversity. In: Hale P, Petrie A, Moloney D, Sattler P (eds) Management for sustainable ecosystems. Centre for Conservation Biology, Brisbane, pp 101–108

Dang YP, Dalal RC, Mayer D, McDonald M, Routley R, Schwenke GD (2008) High subsoil chloride concentrations reduce soil water extraction and crop yield on Vertisol in northeastern Australia. Aust J Agric Res 59:321–330

Das B, Kandpal BK, Devi HL (2021) Cover crops for orchard soil management. In: Cover crops and sustainable agriculture. CRC Press, pp 147–168

Das SC, Datta M, Ray P, Singh SK, Jena RK, Das B, Ray SK (2019) Mango (Mangifera indica) cultivation in northeastern region of India. Adv Agric Res Technol J 3:54–66

Davidson EA, Janssens IA (2006) Temperature sensitivity of soil carbon decomposition and feedbacks to climate change. Nature 440:165–173

Driese SG, Nordt LC, Stinchcomb GE (2021) Soils, chemical weathering, and climate change in Earth history. Hydrogeol Chem Weather Soil Form 21–65

Elsheery NI, Helaly MN, El-Hoseiny HM, Alam-Eldein SM (2020) Zinc oxide and silicone nanoparticles to improve the resistance mechanism and annual productivity of salt-stressed mango trees. Agronomy 10:558

Fischer G, Cleves-Leguizamo JA, Balaguera-López HE (2022) Impact of soil temperature on fruit species within climate change scenarios. Rev Colomb Cienc Hortic 16

Francioli D, Cid G, Kanukollu S, Ulrich A, Hajirezaei MR, Kolb S (2021) Flooding causes dramatic compositional shifts and depletion of putative beneficial bacteria on the spring wheat microbiota. Front Microbiol 12:773116

Ganeshamurthy AN, Ravindra V, Rupa TR, Radha TK (2019) Long-term conservation horticulture in mango orchards: comparative effects of conventional and conservation management practices on an Alfisol under seasonally dry tropical savanna climate

Ganeshamurthy AN, Rupa TR, Shivananda TN (2018) Enhancing mango productivity through sustainable resource management. J Hortic Sci 13:1–31

Garcia-Ruiz R, Ochoa V, Vinegla B, Hinojosa MB, Pena-Santiago R, Liebanas G et al (2009) Soil enzymes, nematode community and selected physico-chemical properties as soil quality indicators in organic and conventional olive oil farming: influence of seasonality and site features. Appl Soil Ecol 41:305–314

Gautam MK, Maurya A, Singh UP, Kumar R (2021) Impact of climate changes on soil properties

Gil SV, Meriles J, Conforto C, Figoni G, Basanta M, Lovera E et al (2009) Field assessment of soil biological and chemical quality in response to crop management practices. World J Microbiol Biotech 25:439–448

Gurmu G (2019) Soil organic matter and its role in soil health and crop productivity improvement. For Ecol Manage 7:475–483

Habibi F, Liu T, Shahid MA, Schaffer B, Sarkhosh A (2023) Physiological, biochemical, and molecular responses of fruit trees to root zone hypoxia. Environ Exp Bot 206:105179

Haynes RJ (2008) Soil organic matter quality and the size and activity of the microbial biomass: their significance to the quality of agricultural soils. In: Huang Q, Huang PM, Violante A (eds) Soil mineral-microbe-organic interactions: theories and applications. Springer, Berlin, pp 201–230

He X, Pan M, Wei Z, Wood EF, Sheffield J (2020) A global drought and flood catalogue from 1950 to 2016. Bull Am Meteorol Soc 101:E508–E535

IPCC (2007) Climate change—mitigation. Fourth assessment report of the intergovernmental panel on climate change

Jansson JK, Hofmockel KS (2020) Soil microbiomes and climate change. Nat Rev Microbiol 18:35–46

Jarvis NJ (2007) A review of non-equilibrium water flow and solute transport in soil macropores: principles, controlling factors and consequences for water quality. Eur J Soil Sci 58:523–546

Kamarudin KN, Abdul Rahman MH, Mohamad M, Abd Rashid NF, Shahidin NM, Roslan N, Khairun NM (2023) Soil fertility status in relation to insidious fruit rot incidence at Harumanis mango orchard in Perlis: a case study. Malays J Soil Sci 27

Karmakar R, Das I, Dutta D, Rakshit A (2016) Potential effects of climate change on soil properties: a review. Sci Int 4:51–73

Kimball BA (2003) Response of plants to elevated atmospheric CO_2. Indian J Plant Physiol 18–24

Kumar K, Adak T, Singh VK (2017) Green manuring and nutrient management impacting soil properties and sustainability of mango orchard. J Soil Water Conserv 16:72–78

Kumari R, Kundu M, Das A, Rakshit R, Sahay S, Sengupta S, Ahmad MF (2020) Long-term integrated nutrient management improves carbon stock and fruit yield in a subtropical mango (Mangifera indica L.) orchard. J Soil Sci Plant Nutr 20:725–737

Kumari S, Rai R, Kumar R, Prasad B (2008) Critical studies for identification of physiological disorders in mango

Kundu M (2020) Orchard floor management. In: Sustainable agriculture. Apple Academic Press, pp 227–251

Lal R (2004) Soil carbon sequestration to mitigate climate change. Geoderma 123:1–22

Lal R (2019) Carbon cycling in global drylands. Curr Clim Change Rep 5:221–232

Laxman RH, Annapoornamma CJ, Biradar G (2016) Mango. Abiotic stress physiology of horticultural crops 169–181

Laxmi K, Kamal A, Kumar V, Muthukumar M, Bajpai A (2021) Morphological indicators of salinity stress and their relation with osmolyte-associated redox regulation in mango cultivars. J Plant Biochem Biotechnol 30:918–929

Li T, Ge L, Zhao R, Peng C, Zhou X, Li P et al (2024) Phenolic compounds weaken the impact of drought on soil enzyme activity in global wetlands. Front Microbiol 15:1372866

Lique C, Holland MM, Dibike YB, Lawrence DM, Screen JA (2016) Modeling the Arctic freshwater system and its integration in the global system: lessons learned and future challenges. J Geophys Res Biogeosci 121:540–566

Lombao A, Barreiro A, Fontúrbel MT, Martín A, Carballas T, Díaz-Raviña M (2021) Effect of repeated soil heating at different temperatures on microbial activity in two burned soils. Sci Total Environ 799:149440

Maity A, Khayyat M, Azarmi-Atajan F, Agehara S, Sarkhosh A (2021) Soil and nutrition. The pomegranate: botany, production and uses. CAB International, Wallingford, pp 285–319

Mitra D, Mondal R, Khoshru B, Senapati A, Radha TK, Mahakur B et al (2022) Actinobacteria-enhanced plant growth, nutrient acquisition, and crop protection: advances in soil, plant, and microbial multifactorial interactions. Pedosphere 32:149–170

Moebius BN, Van EHM, Schindelbeck RR, Idowu OJ, Clune DJ, Thies JE (2007) Evaluation of laboratory-measured soil properties as indicators of soil physical quality. Soil Sci 172:895–912

Moinet GY, Hunt JE, Kirschbaum MU, Morcom CP, Midwood AJ, Millard P (2018) The temperature sensitivity of soil organic matter decomposition is constrained by microbial access to substrates. Soil Biol Biochem 116:333–339

Mondal S (2021) Impact of climate change on soil fertility. In: Climate change and the microbiome: sustenance of the ecosphere, pp 551–569

Muhammed MA, Mohamed AKS, Qayyum MF, Haider G, Ali HA (2022) Physiological response of mango transplants to phytohormones under salinity stress. Sci Hortic 296:110918

Naik SK, Das B, Kumar S, Bhatt BP (2015) Evaluation of major and micronutrient status of acid soils of different mango orchards. Int J Fruit Sci 15:10–25

Nath V, Kumar G, Pandey SD, Pandey S (2019) Impact of climate change on tropical fruit production systems and its mitigation strategies. In: Climate change and agriculture in India: Impact and adaptation, pp 129–146

Navarro-Pedreño J, Almendro-Candel MB, Zorpas AA (2021) The increase of soil organic matter reduces global warming: myth or reality? Sci 3:18

Naveena Kumara KT, Singh H, Kaur N, Kang BK, Devi I (2023) Uniconazole improves mango flowering and fruit yield by regulating gibberellins and carbon–nitrogen nutrition. Hortic Environ Biotechnol 64:735–752

Neguse TB, Wanzala FK, Ali WM, Owino WO, Mwangi GS (2019) Mango (*Mangifera indica* L.) production practices and constraints in major production regions of Ethiopia. Afr J Agric Res 14:185–196

Nel GP (2024) Establishing the water requirements and crop coefficients of a mature mango orchard under subtropical conditions. Doctoral dissertation, Stellenbosch University

Parama VR (2017) Effect of climate changes on soil properties and crop nutrition

Patil A, Lamnganbi M (2018) Impact of climate change on soil health: a review. Int J Chem Stud 6:2399–2404

Pender JS (2008) What is climate change? And how it will affect Bangladesh. Briefing paper, Dhaka, Bangladesh: Church of Bangladesh Social Development Programme

Phuong NT (2023) Tree age affects soil physico-chemical properties of mango orchard in Dong Thap province. Tap Chi Khoa Hoc Dai Hoc Dong Thap 12:63–70

Pinney K, Labavitch JM, Polito VS (1998) Fruit growth and development. In: Walnut production manual, publication, vol 3373, pp 139–143

Prakash J, Agrawal SB, Agrawal M (2023) Global trends of acidity in rainfall and its impact on plants and soil. J Soil Sci Plant Nutr 23:398–419

Ram RA, Rahim MA, Alam MS (2020) Diagnosis and management of nutrient constraints in mango. In: Fruit crops. Elsevier, pp 629–650

Rao DLN, Aparna K, Mohanty SR (2019) Microbiology and biochemistry of soil organic matter, carbon sequestration, and soil health. Indian J Fert 15:124–138

Ray PK (2023) Nutrition, fertilization, and irrigation. In: Jackfruit: Botany, production and uses. CABI, pp 158–194

Raza S, Zamanian K, Ullah S, Kuzyakov Y, Virto I, Zhou J (2021) Inorganic carbon losses by soil acidification jeopardize global efforts on carbon sequestration and climate change mitigation. J Clean Prod 315:128036

Raza T, Qadir MF, Khan KS, Eash NS, Yousuf M, Chatterjee S, Oetting JN (2023) Unrevealing the potential of microbes in decomposition of organic matter and release of carbon in the ecosystem. J Environ Manage 344:118529

Rengasamy P (2010) Soil processes affecting crop production in salt-affected soils. Funct Plant Biol 37:613–620

Reth S, Reichstein M, Falge E (2005) The effect of soil water content, soil temperature, soil pH-value and root mass on soil CO_2 efflux-A modified model. Plant Soil 268:21–33

Reynolds WD, Bowman BT, Drury CF, Tan CS, Lu X (2002) Indicators of good soil physical quality: density and storage parameters. Geoderma 110:131–146

Rosenzweig C, Hillel D (2000) Soils and global climate change: challenges and opportunities. Soil Sci 165:47–56

Ross DS, Matschonat G, Skyllberg U (2008) Cation exchange in forest soils: the need for a new perspective. Eur J Soil Sci 59:1141–1159

Saha N, Mandal B (2009) Soil health—a precondition for crop production. In: Khan MS, Zaidi A, Musarrat J (eds) Microbial strategies for crop improvement. Springer, Heidelberg, pp 161–168

Salvador Sanchis MP, Torri D, Borselli L, Poesen J (2008) Climate effects on soil erodibility. Earth Surf Process Landf 33:1082–1097

Sanchez Garcia D (2024) Mango industry and the importance of using organic fertilizers for export to the European Union

Satisha J, Laxman RH, Upreti KK, Shivashankara KS, Varalakshmi LR, Sankaran M (2020) Mechanisms of abiotic stress tolerance and their management strategies in fruit crops. In: New frontiers in stress management for durable agriculture, pp 579–607

Schewe J, Heinke J, Gerten D, Haddeland I, Arnell NW, Clark DB, Kabat P (2014) Multimodel assessment of water scarcity under climate change. Proc Natl Acad Sci USA 111:3245–3250

Schimel JP (2018) Life in dry soils: effects of drought on soil microbial communities and processes. Annu Rev Ecol Evol Syst 49:409–432

Sharma PC, Singh A (2021) Fruit cultivation in salt-affected soils: constraints and management options. In: Dryland horticulture. CRC Press, pp 231–253

Singh AK (2019) Commercial cultivation of jackfruit

Singh JS, Gupta VK (2018) Soil microbial biomass: a key soil driver in management of ecosystem functioning. Sci Total Environ 634:497–500

Singh P, Sharma A, Dhankhar J (2022) Climate change and soil fertility. In: Plant stress mitigators: action and application. Springer Nature, Singapore, pp 25–59

Smith JL, Halvorson JJ, Bolton H (2002) Soil properties and microbial activity across a 500 m elevation gradient in a semi-arid environment. Soil Biol Biochem 34:1749–1757

Srivastava AK (2022) Soil fertility management in fruit crops amidst climate change: a review. Int J Innov Hortic 11:259–273

Stern N (2006) The economics of climate change: the Stern review. Cambridge Univ Press, Cambridge, UK

Stirling G, Hayden H, Pattison T, Stirling M (2016) Soil health, soil biology, soilborne diseases and sustainable agriculture: a guide. CSIRO Publishing

Striker GG (2012) Flooding stress on plants: anatomical, morphological and physiological responses. Botany 1:3–28

Swain R, Praveena J, Rout GR (2023) Mitigation strategies of the impact of global climate change on fruit crops. In: Cultivation for climate change resilience, vol 1. CRC Press, pp 28–56

Tripathy MB (2021) Deterioration of soil health and its management. Agric Food e-Newsletter 9

Ullah MA, Khanal A, Joyce P, White N, Macnish A, Joyce D (2024) Internal disorders of mango fruit and their management—physiology, biochemistry, and role of mineral nutrients. Plants 13:2596

Vanbergen AJ, Initiative IP (2013) Threats to an ecosystem service: pressures on pollinators. Front Ecol Environ 11:251–259

Wahane M, Saha D, Pal SK, Bera A, Das TK (2024) Geospatial insights into Alphonso mango cultivation: a comprehensive land suitability study in the coastal belt of Maharashtra, India. Environ Monit Assess 196:708

Wei J, Liu G, Liu D, Chen Y (2017) Influence of irrigation during the growth stage on yield and quality in mango (*Mangifera indica* L). PLoS One 12

Weil RR, Magdoff F (2004) Significance of soil organic matter to soil quality and health. In: Magdoff F, Weil RR (eds) Soil organic matter in sustainable agriculture. CRC Press, Florida, pp 1–43

Yan C, An D, Zhao B, Ma H, Liu Y, Kong R, Su J (2023) Soil aggregates, carbon and nitrogen distribution, and water infiltration as influenced by mulching practices in a laterite mango (*Mangifera indica* L.) orchard. J Soils Sediments 23:3163–3176

Yusof IM, Buchanan DW, Gerber JF (1969) The response of avocado and mango to soil temperature. J Am Soc Hortic Sci 94:619–621

Zade S, Gourkhede PH, Vaidya PH, Singh RS, Sinha SK, Kumar A, Kumar V (2021) Problem soils and their management practices

Zaharah SS, Razi IM (2009) Growth, stomata aperture, biochemical changes, and branch anatomy in mango (*Mangifera indica*) cv. Chokanan in response to root restriction and water stress. Sci Hortic 123:58–67

Zuazo VHD, García-Tejero IF, Rodríguez BC, Tarifa DF, Ruiz BG, Sacristán PC (2021) Deficit irrigation strategies for subtropical mango farming: a review. Agron Sustain Dev 41:13

Chapter 13
Climate-Smart Postharvest Management and Value Addition of Mangoes

Rana Naveed Ur Rehman, Sajid Ali, Safina Naz, Weifeng Chen, Tahira Kalsoom, and Muhammad Azam Khan

Abstract Climate-smart postharvest management and value addition are essential to address the challenges posed by climate change to mango production. Shifting weather patterns, such as rising temperatures and unpredictable rainfall, influence mango ripening rates, shelf life, and market quality. This chapter emphasizes innovative approaches to postharvest handling, including the use of controlled atmosphere storage, optimized cooling systems, and packaging solutions that reduce spoilage and maintain fruit quality. Techniques such as hydro-cooling and the use of edible coatings derived from natural substances are explored for their effectiveness in prolonging freshness and minimizing chemical use. Value addition through mango processing into products like dried slices, purees, and beverages is highlighted as a means to enhance economic returns and reduce postharvest losses. Renewable energy technologies, such as solar drying and energy-efficient cold chains, are also discussed as sustainable solutions for rural and urban stakeholders. The adoption of digital tools, including IoT-based monitoring systems and AI-driven supply chain management, is underscored for improving efficiency and reducing waste. By implementing climate-smart strategies, mango growers and postharvest handlers can ensure that the fruit retains its quality, flavor, and nutritional value while maximizing profitability and contributing to sustainable agricultural practices.

Keywords Climate-smart management · Postharvest practices · Mango value addition · Sustainable processing

R. N. U. Rehman · T. Kalsoom · M. A. Khan
Department of Horticulture, PMAS-Arid Agriculture University, Rawalpindi, Pakistan

S. Ali (✉) · S. Naz
Department of Horticulture, Bahauddin Zakariya University, Multan, Pakistan
e-mail: sajidali@bzu.edu.pk

W. Chen
Hunan Horticultural Research Station, Changsha, China

S. A. H. Naqvi et al. (eds.), *Climate Change and Mango Production*,
https://doi.org/10.1007/978-3-031-87035-4_13

13.1 . Overview of the Mango Industry in the World

Mango (*Mangifera indica* L.) is an important tropical fruit crop which is cultivated in most of the continents. Mango production is concentrated in Asia (77%) followed by the America (including Caribbean) (12%), Africa (10%), and Oceania (0.11%). It is cultivated in more than 100 countries of which more than 65 countries produce each more than 1000 million tons. It is quite interesting fact that since 2000 bulk production of mango is centered in 10 countries which account about 82% of the total production (Evans et al. 2017). The estimated output is about 40 million tons with major contribution by India (15 million tons) followed by China, Thailand, Indonesia, Philippines, Mexico, and Brazil (Mitra 2016). People of Asian countries prefer sweet mango varieties (such as Sindhari, Chounsa, Anwar Ratol etc.); on the other hand, European consumer prefer cosmetically rich Floridian cultivars (such as Haden, Tommy Atkins, Kent etc.) containing abundant anthocyanin (Saúco 2011). Floridian varieties make headways in global export volume with fairly stable prices. In international market, organic mangoes fetch high prices particular in Japan, USA and European markets. Moreover, US, Netherland, England, France, Portugal, Spain, Denmark, Belgium and Sweden are the leading importer of fresh mangoes. In sub-continent (India, Pakistan and Bangladesh), the market availability window exists from May to October. While, Brazil, Kenya, Columbia and Venezuela, harvesting period continues for 9–10 months. The season is also quite long in Costa Rica, Jamaica, Burkina Faso, Puerto Rico Indonesia, Nicaragua, and Mexico. There are many reported varieties of mango cultivated in various countries like India (1500), Thailand (> 100), Pakistan (> 100) and China (> 40).

13.2 . Nutritional Significance

Mango is highly nutritive fruit with rich source of vitamins A and C. Its pulp contains organic acids, sugars, proteins, ascorbic acid, carbohydrates, and minerals (Zafar and Sidhu 2017; Shivran et al. 2023). Mangoes are an excellent source of essential nutrients, vitamins, dietary fiber and minerals delivering significant part (> 50%) of the recommended daily intake (Kabeer et al. 2023). Mangos are a great source of vitamin C, which helps with collagen synthesis and immune system function, improving overall health (Naliyadhara et al. 2023). The rich content of dietary fiber aid in digestion and help in weight management with potential application of obese control (Tariq et al. 2023). Mangoes are packed with antioxidants such as beta-carotene, phenolic compounds and lutein. It has been determined that these antioxidants help shield the body's cells from oxidative damage brought on by free radicals. Therefore, these antioxidants are essential for lowering the chance of developing chronic illnesses (Naliyadhara et al. 2023). Mangoes are known for their distinct flavor and scent, but they are also a great source of antioxidants. There are various kinds of antioxidant enzymes which confer for the up-regulation of antioxidant capacity as well as numerous health benefits (Dumitru et al. 2023). Furthermore, mango peel is a great source of proteins, lipids, cellulose, hemicellulose, pectin, polyphenols, and carotenoids, all of which have strong antioxidant qualities (Evans et al. 2017). Moreover, mango peel flour contains functional ingredients and health benefitting molecules hence it is extensively used in various products such as bread, noodle, cake, bread, biscuit and other bakery products.

13.3 Goals and Principles of Climate-Smart Practices

The fluctuation in atmospheric conditions and consequent environmental patterns due to anthropogenic activity or some natural factors is called 'climate change' (IPCC 2014). The consistent change in temperature and relative levels of greenhouse gases affect the available water and land resources; thus the crop production is greatly influenced in term of quantity and quality. According to the Intergovernmental Panel on Climate Change (IPCC) by the end of this century average air temperature is expected to enhance between 1.4 °C and 5.8 °C (IPCC 2001). IPCC retrieved this data by scientific modeling techniques and relevant information was extracted from atmospheric (air, humidity, temperature) and ocean (levels, movements) behavior. Another study predicted the rise of temperature at the rate of 0.088 °C per decade for this century (Kalnay and Cai 2003). These investigations forecasted the more frequent occurrence of extreme flooding, heat waves and droughts than the past. The turbulence in the weather pattern can seriously damage the several agricultural related operations such as cultivation, harvesting, processing, value chain, industrialization as well as the economic status of developing countries. Climatic change is a kind of threat which threatens the very existence of mankind. Climatic change is induced by many factors (such as bio-diversity, pollution, land-use) which have close relationship with life on earth and their intricate interaction with ocean, troposphere and land.

13.4 Impact of Climate Change on Pre-Harvest Quality

13.4.1 Critical Growth Stages

Mango is tropical fruit crop with distinctive phenological growth stages viz. bud break, emergence of new flushes, flowering, and fruit development. Change in climatic conditions (such as temperature) can significantly affect the anatomical, morphological, biochemical and physiological changes. These changes can delay the consequent phenological growth stage; thereby influencing the drastic reduction in yield and overall production. Rapid change in temperature lead to quick growth rhythms by shortening the time-lag between vegetative flushes and affecting the growth units. It also delays the growth cessations in mango during the start of winter.

A recent study by Dambreville et al. (2013) demonstrated that there is no significant effect of average daily temperature (20–28 °C) on leaf size. However, earlier to this, According to Whiley et al. (1989) a mango tree growing at 30:25 °C (day: night) had an average leaf size that was 300% greater than a tree growing at 20:15 °C (day: night). Similarly, excessive accumulation of carbon dioxide (CO_2) can affect the tree growth by regulating the excessive accumulation of dry matter in the roots (Legave et al. 2013). Elevated levels of CO_2 are reported to influence on the accumulation and composition of mineral nutrients in mango tree. Enhanced concentration of CO_2 (700 μ mol mol^{-1}) lowers the mineral composition (P, N, K, Mg, Ca, Cl, Zn, Fe etc.) in mango leaves as compared to control (350 ppm) (Schaffer et al. 1997). Flooding or drought is another consequent threat of climatic change which hampers the mango tree growth by reducing the vegetative growth.

13.4.2 Effect on Yield and Quality

Fruit yield is greatly influenced by the climatic changes; particularly, after the flower anthesis (period of flower opening) and subsequent events like pollination, fertilization. High temperature or excessive humidity can hamper the bee activity and removal of stigmatic secretions which cause poor pollination, abortion of ovule and ultimately reducing the fruit yield. Although, propensity of fruit set is different for different cultivars however, minute exposure of unfavorable environmental conditions (humidity, rain, temperature, wind, light, drought etc.) during flowering and fruit setting lead to significant cut-off of fruit yield. In the subtropical climate, low (< 10 °C) temperature at night or high (> 30 °C) temperature at day may induce the development of stenospermocarpic fruit. Some cultivars such as Dashehari or Ataulfo are prone to produce the nubbins (small and low weight fruit) due to temperature variations during pollination period (Makhmale et al. 2016; Severiano-Galeana et al. 2024). Temperature plays multiple role in plant growth by regulating the metabolism of photosynthesis and respiration. The ratio of photosynthesis to respiration is crucial for a tree's healthy growth and development, in addition to having a high rate of photosynthesis. At 15 °C, the ratio of photosynthesis to respiration is greater than 10, which explains why plants grow more readily in temperate climates than in tropical ones (Went 1953). High temperature enhances the photosynthesis by modulating the enzymatic activities, biochemical reactions and electron transport chain (Sage and Kubien 2007). However, after certain threshold level, the excessive rise in temperature may lead to dysfunction of enzymes and thus restricting the biochemical processes. Finally, high temperature during growing period lead to early maturation of horticultural crops than the same crops cultivated under normal (optimum) temperature as described by Wurr et al. (1996) and Hall et al. (1996) (Fig. 13.1).

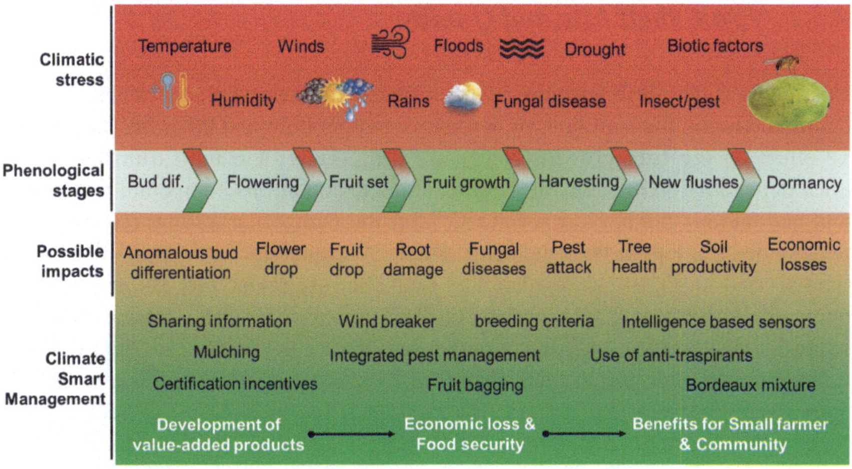

Fig. 13.1 Climate-smart post-harvest management

13.5 Impact of Climate Change on Postharvest Quality

Climate change results in rise of temperature, carbon dioxide, greenhouse gas, and depletion of ozone. The effect of climatic calamities can potentially affect the postharvest quality of mango.

13.5.1 Fruit Harvesting

Fruit harvesting window is greatly influenced by climatic conditions during the fruit growth and development. A rise in rainfall amount and longevity period of high precipitation during the fruit development phase may delay the maturity and downgrade the fruit appearance. In the Philippines, mango fruit is harvested after 100–110 days (after flowering in April-June) due to warm climate. While other picking is made after 120–125 days in January-March due to cool weather (Ram and Rajan 2003). In India, Pakistan and Thailand, mango maturation is delayed in the northern areas than in the central and southern areas (districts); thus mango harvesting window is extended. High temperature causes the significant effect on the physiological changes in mango fruit. For example, prevalence of spongy tissues in Alphonso is due to high temperature of fruit which lead to breakdown of internal-tissues. High temperature enhances the rate of respiration, slows down the rate of transpiration thus influence the normal ripening process of fruit. Fruit exposed to sunlight develop attractive color and aroma, on the other hand, fruit inside the dense canopy are exposed to diseases, insect attack and poor fruit color is developed.

13.5.2 *Fruit Shelf Life and Quality*

Fruit quality is mainly determined by the series of pre-harvest practices and climatic conditions during the fruit growth. Excessive temperature and humidity have negative effect on the fruit growth (color, size etc.). Time of mango fruit harvesting varied depending on the cultivar, regime, climatic conditions, temperature management and various pre-harvest practices. After the mango harvesting, respiration is the main determinant of fruit shelf-life. There are limited methods available to food scientists and biologists to regulate and limit the respiratory process in harvested commodities (Saltveit 2002). In fact, the fruit quality cannot be enhanced during the postharvest period—only the quality loss is delayed. Thus, in order to avoid the quality loss and undesirable changes in fruit, farmers, processors and consumer can adopt several techniques to improve the fruit shelf life. It includes, correct method of harvesting, sorting, packaging, rapid cooling and storage facilities during transporting and marketing. Besides this the fruit temperature is lowered by placing it in the refrigerated conditions to hinder the physiological processes. The physiological processes are also slowed down the storing the fruit in the controlled atmosphere (CA) where low levels of oxygen and elevated CO_2 is provided. These series of strategies are applied to mitigate the challenge of climate change.

13.6 Application of Climate-Smart Agriculture Using Multiple Approaches

Presently, world is facing various global challenges such as population increase, land degradation, urbanization, high demand of agricultural production which has been further exacerbated by the emergence of a new challenge—climate change. To combat this global challenge, national appropriate mitigation actions (NAMA) and national adaptation programs of actions (NAPA) are being designed by the developing countries to work on the sustainable agriculture production system by implementing Paris agreement (UNFCCC 2015). In order to maintain this momentum, the concept of Climate Smart Agriculture (CSA) has been developed to meet the goals of food security, adaptation and mitigation (FAO 2013). The nature of CSA has variety of options for considering the diversified contexts such as social groups (age, gender etc.) and their priorities based on climate regimes, agro-ecological zones, and land uses (Scherr et al. 2012; Neufeldt et al. 2013). CSA is set of all integrated agricultural practices to manage the cropland, forest, livestock and fisheries to address all concerned interlinked challenges of food security. In Kenya, CSA attributed to the fact that: (i) agriculture has great potential for contributing to the achievement of a range of development goals related to food security; and (ii) the recognition of Kenya National Adaptation Plan 2015–2030 on critical CSA practices (GoK 2015).

13.6.1 Pre-Harvest Practices

A collection of integrated farming practices known as Climate Smart Agriculture (CSA) are used to solve the complex and interconnected issues around food security. These practices include the crop rotation, deep ploughing, terracing, mixed cropping, zero tillage, cover crops and mulching.

CSA-certification incentives: A recent study in Ghana concluded that CSA-certification adoptability among famers is influenced by technical, sociodemographic and institutional factors in different ways. The access to high-value-market, credit facilities, membership of mango farmers' association are the promising incentives for farmers to adopt sustainability certification scheme and abandon the use of herbicides, inorganic fertilizer and weedicide (Akrong et al. 2023). It is now challenging to forecast how climate change would affect mango trees due to the lack of a thorough crop model. Hence, these interventions are assessed based on our current knowledge about mango trees. The climatic influence has inevitable impact on the photosynthesis, reproductive and vegetative growth, fruit quality and next-year cropping patterns.

Early crop maturity: Rising temperature can have positive effect on mango by expediting the fruit growth. The estimated time of mango fruit maturity has been decreased by 12–16 days in Australia due to change in winter temperature (1.5 °C) during last 45 years (Olesen 2011). Similarly, good pruning and training practices improves the light exposure which has positive correlation with fruit color development and high photosynthesis. In addition to this, exposure of fruit to high temperature or light can induce the biosynthesis of secondary compounds thus improves nutritional value of fruit (Léchaudel et al. 2013).

Displacement of mango cultivation zones: The suitable way of climatic adoptability is the displacement of mango tree plants to an area where it has appropriate cultivation response. This strategy is good for Caribbean islands where mean temperature and precipitation (annual) is expected to increase by 2.0 °C and decrease by 12% at the end of the century. The island is expected to be drier and warmer thus, the phenological events will be severely affected and consequently the yield may be decreased (Normand et al. 2015).

Revision of breeding criteria: Mango germplasm is need to be improved to combat the climatic change by considering the drought resistant and high temperature tolerant cultivars as breeding criteria. Similarly, ability to induce flower and fruit set at high temperature should be part of the selection criteria. In addition to this, selection of rootstock should be based on the drought tolerance, strong root system, resistance to strong winds, and take-up of high water volume (Normand et al. 2015).

Windbreaks: In addition to this, utilization of windbreak is another sustainable practice to reduce the soil erosion, protect orchard, and provide shelter to pollinating insects. The rate of mortality in fruit plants was $< 30.81\%$ and $> 91.43\%$ in orchards which were either surrounded by windbreaks or without windbreaks, respectively (Rathore et al. 2012).

Fruit bagging: After fruit setting, the fruit should be covered with brown paper at marble stage to get highest fruit weight and free of other physiological defects (such as tissue sponginess). It also reduced the insect attack and fungal infestation and hinder the use of insecticide and fungicide etc. (Haldankar et al. 2015).

Mulching: The covering of soil with plastic sheet or dry grasses to conserve the soil moisture, enhance the microbial activity is called mulching. The mulch has significant role for increasing the yield such as mango (45.23%), guava (25.93%), litchi (12.16%) and pineapple (25.93%) as compared to control (Patil et al. 2013).

Use of anti-traspirants: Certain chemicals (such as chitosane, kaolin) have ability to deflect the solar irradiation. Application of such particles reduce the water loss by hindering the rate of transpiration and lowering the temperature of leaf and fruit surface. The treatment with chitosane (2%) on banana enhanced the bunch weight and maximum finger weight as compare to the control (Ahmed 2014).

Use of Bordeaux mixture: Bordeaux mixture is an excellent substitute of synthetic fungicides. Application of Bordeaux mixture protects the plant from fungus diseases and hinder the frost damage in grapes (Yadollahi 2011).

Application of intelligence based sensors in orchard: Various kinds of intelligence based sensors have been developed to manage the cultural practices in the orchard. These sensors (such as irrigation system) are also connected with other tools (satellite, soil moisture meter, tree size) to retrieve the corresponding data and make a precise decision about the time and amount of irrigation. In case, weather is too hot and soil moisture is lowering abruptly, the automatic system will switch on after certain threshold levels and irrigation will be applied to plant. This robust system of irrigation hampers the plant from drought stress and maintain the plant health by providing the timely input of resources (Chandu and Shukla 2023).

13.6.2 Post-harvest Practices

Development of mango cultivar specific maturity indices: Mango fruits exhibit specific maturity indices upon reaching the harvest. These maturity standards are based on physical and chemical attributes. Physical attributes include the fruit size, weight, shape, specific gravity, external color, firmness and number of days from bloom to harvest. Similarly, biochemical parameters are the total soluble solids, titratable acidity, pH, sugar/acid ratio, tannins, ascorbic acid, and profile of volatile compounds. Fruit quality and postharvest shelf life is greatly influenced by the stage of maturity and time of harvest. Good taste, flavor and juice content of fruit can only be ensured when fruit is reached at maturity phase. Several reports are available suggesting the cultivar specific time of maturity on the basis of physical and biochemical basis (Ahmed and Ahmed 2014; Shamim et al. 2015; Muiruri 2016).

Weather conditions and technique of harvest: After reaching the cultivar specific maturity standards, the mango fruit should be harvested as soon as possible. However, weather condition at the time of harvest play an inevitable role in the postharvest quality retention and shelf life extension of fruit. Scorching heat at noon or afternoon enhances the fruit transpiration rate and thus the fruit size shrinks; hence harvesting must be avoided at noon. The preferable time of mango harvesting should be morning when moisture content is well-retained. Similarly, harvesting should be delayed for certain days when it is raining; because raining soften the fruit. Mango is a fleshy fruit and hence it is very sensitive to mechanical injury (Waskar et al. 1996). It requires special care for mechanical harvesting.

Removal of field heat and sap management: Field heat must be removed after the fruit harvest by placing it in shady pack-house. The heat trap in fruit pulp may expedite the process of ripening; thus restricting the shelf life. In addition to this, mango fruit and stem has an extensive system of ducts for transportation and translocation of photosynthetic assimilates. As soon as the stem is separated from fruit by making a cut, heavy sticky acidic sap exuded which spread on the fruit skin. This sap subsequently affects the cosmetic beauty, shelf life and overall quality of fruit. Hence, de-sapping is performed on the ground, special racks or conveyer machines followed by lime dipping treatment (neutralizing the acidic nature) to reduce the sap burn related issues in mango fruit (Maqbool et al. 2007).

Pre-storage treatments: Mango is climacteric fruit which shows the rapid ripening process after harvesting by enhanced respiration and ethylene production. It is highly perishable fruit and the shelf life is restricted to only 4–8 days at ambient temperature. It is also highly susceptible to various kinds of fungal diseases such as anthracnose (*Colletotrichum gloeosporioides*), and stem end rot (*Lasiodiplodia theobromae*). Latent spores of these fungal diseases transfer on fruit skin during humid weather conditions in the orchard and become active after the fruit harvest. Hence, pre-storage fungicidal treatment is mandatory for quality maintenance during postharvest period. In addition to this, fruit fly is another noxious pest which puncture the fruit skin and lays eggs inside. For this purpose, hot water treatment or vapor heat treatment is mandatory to kill the fruit eggs and hinder the mango quality loss.

Temperature control during transportation and storage: Various kinds of cooling methods have been introduced to remove the heat (field) from fresh produce since 1920s based on the principle that decrease of temperature (10 °C) of pulp temperature extend the shelf life of produce by 2–threefold. Low temperature, reduces the rate of respiration, hinders the enzyme activities, inhibits the pathogenic infestation, minimizes the weight loss and retains the firmness and biochemical attributes by slowing down the physiological processes. Mango fruit can be stored in common cold storage facility having temperature and humidity control. Another method of mango storage is the placing of fruit in modified atmosphere packaging in which respiratory

Table 13.1 Climate-smart postharvest management strategies for mangoes

Postharvest challenge	Climate-smart strategy	Benefits for mango quality and marketability
Heat stress during storage	Use of evaporative coolers or solar-powered cold storage	Maintains mango freshness, reduces spoilage, and extends shelf life
Rapid ripening	Use of ethylene inhibitors or controlled ripening chambers	Slows ripening, enabling better timing for market delivery and reducing postharvest losses
Moisture loss	Application of edible coatings (e.g., aloe vera or wax)	Retains moisture, reduces shrinkage, and preserves fruit appearance
Pathogen attacks	Hot water treatment or UV sterilization	Reduces fungal infections (e.g., anthracnose) and improves food safety
Energy-intensive cooling	Adoption of renewable energy systems for cold storage	Reduces dependency on fossil fuels and ensures eco-friendly storage
Transportation losses	Use of ventilated crates and temperature-controlled trucks	Minimizes bruising and ensures delivery of high-quality mangoes to markets
Packaging waste	Use of biodegradable and climate-friendly packaging	Reduces environmental impact and enhances consumer appeal

quotient are change by lowering the internal oxygen and producing the high carbon dioxide. Most recently, controlled atmosphere is used in which content of oxygen are extremely lowered from 21% to 1–5% and concentration of carbon dioxide is enhanced from 0.03 to 2–5%. The change in gas composition and low temperature can extend the shelf life of mango up to 6 weeks for Kensington Pride and R2E2 (Singh and Zaharah 2010), 4 weeks for Alphonso, and 5 weeks for Banganapalli (Rao and Gopalakrishna 2008).

The climate induced stress under post-harvest system their management strategies and benefits of adopting such strategies are mentioned in Table 13.1.

13.7 Need of Mango Value Addition

Mango is extensively cultivated crop predominantly in the tropical countries. However, only bunch of varieties are cultivated on commercial scale in each country and very minute quantity is exported across the borders. Huge quantity of mango is transported to the domestic market where it is used fresh or various kinds of value added products (such as mango pickles, puree, nectar, leather, chutney, slices in syrup and canned slices) are developed.

Fruit processing waste has attracted a lot of interest lately as a possible source of products with additional value. The mango offers enormous potential for the development of high-value added goods because to its abundance of pectin, adaptable phenolic compounds, and other antioxidants. It is estimated that huge share of food waste (39%) in developing countries come from the processing industry (Mirabella et al. 2014).

Moreover, mango being the tropical fruit, is exposed to hostile weather conditions during different phases of growth and development. Recent climate change has added to its magnitude by unexpected temperature rise, winds, rains and other climate driven events. These hostile events can sweep the 75% of fruits on the mango tree, leading to huge economic losses (Ravani and Joshi 2013). These economic losses can be minimized by collecting the dropped/fallen fruit and developing value added products.

13.8 Packaging and Ripening of Fresh Product

Fresh mango fruit is admired due to its unique taste and juiciness. The demand of fresh mango is on the rise, globally. However, the market potential of fresh commodity is closely associated with fruit quality and accessibility of market. In addition to this, certain pre-harvest and postharvest factors (as discussed above) also contribute to the shelf life and quality of fruit.

Packaging material: In developing countries, mango is sometime packed in wooden crates with newspaper as liner. The surface of wooden crate is coarse and thus it causes physical injuries and bruises on fruit skin during transportation. However, it is banned in international markets considering the quarantine concerns and related disinfestation treatments (FAO 2002). On the other hand, corrugated boxes are commercially used for transportation and trade of fruits and vegetables. These boxes are available with suitable size for presentation at retail and carrying for consumer (Anwar et al. 2008).

Method of fruit ripening: Since mangos are climacteric fruits, they are picked when they reach the green stage and allowed to ripen before being used fresh. However, the ripening process is expedited by the certain chemical exposures according to the market demand. For this purpose, sometime calcium carbide (CaC_2) is used as ripening agent in developing countries. The use of this chemical is discouraged due to deleterious effects on human body as well as the dangers of explosions. The anomalous use of calcium carbide may carryover the toxic material (arsenic and phosphorus) to consumer and make the healthy fruit poisonous. On the other hand, ethanol and ethrel/ethephon (2-chloroethylphosphonic acid) are potential chemicals which are recommended for mango fruit ripening. These gaseous chemicals penetrate into the fruit skin and decompose into the ethylene. Ethanol (with trade name of Ripener-I®) is extensively used for safe mango fruit ripening.

13.9 . Development of Processed Value Added Products

A healthy fresh mango contains edible pulp (33–85%), inedible kernel (9–40%), inedible peel (7–24%) depending on the type of cultivar (Brahmeet et al. 2023). Mango peel and pulp are rich with phenolic, carotenoids, cellulose, pectin, hemi-cellulose, proteins, dietary fiber, lipids and variety of polyphenols. Similarly, mango seed kernel contains significant contents of protein, carbohydrates and considerable concentration of fat and crude fiber.

13.9.1 Unripe Mango Products

Mango is one of the few fruits that can be consumed at every stage of growth i.e. from immature to fully mature or ripened stage. Raw mangoes have tangy taste, rich nutritional value and therapeutic effects.

Mango pickle: Green, horticulturally mature mango fruits have great importance for making pickles and chutneys. Mango pickles can be classified on the basis of method of preservation such as salt pickles, sweet pickles, or oil pickles. Mangoes with or without stones, peeled or unpeeled, are combined with varying amounts of spices and condiments for this purpose.

Mango chutney: A special mixture of sliced, peeled, and semi-ripe mangoes, spices, onion, salt, vinegar, garlic, sugar, or jaggery is called mango chutney. The texture of this product is thick made from brined or fresh mango slices. Some varieties (such as *Fazli* and *Totapuri*) are well known for good color and full of flavor chutneys (Ravani and Joshi 2013).

Green mango powder: Green immature mango slices are dehydrated for making powder. The dehydration process is carried out by sun drying, or cabinet. Cabinet drying is faster than sun drying however, smooth circulation of air may reduce the period of sun drying. The quality of green powder is mainly influenced by the drying temperature, blanching and Sulphur dioxide contents.

Green mango beverage: The green mangoes are sliced and boiled in water followed by straining through muslin cloth. Mango pulp (25%), sugar (40%), salt (2%), citric acid (1.5%) and minute quantity of powdered black pepper and cumin are added according to recipe. Kaur et al. (2006) further added the basil leaves, cardamom and other condiments to improve the recipe.

13.9.2 Ripe Mango Products

Ripe mangoes are admired due to their unique taste and flavor. It is extensively used for fresh table purpose serving. Various kinds of value added products are made using mature and ripe mangoes.

Mango pulp: During peak season, when mango is abundantly available in the market, the pulp can be stored for making other products. The pulp is maintained at 14–18 Brix with addition of sugar syrup and citric acid. Canned mango pulp doesn't require any preservative, however potassium meta-bisulphite is added (as preservative) when pulp is filled in glass jars for longer period storage.

Mango beverages: Mango pulp is highly viscous in nature thus various kinds of beverages can be developed such as nectar, juice and squash. To preserve the soluble solids levels (14–15%) and acidity (0.2–0.3%), pulp (20%) is combined with sugar and citric acid for nectar. In a similar manner, juice is prepared by adding an equal amount of water and juice while keeping the acidity (0.4–0.5%) and soluble solids levels (12–15%) constant. Besides this, mango squash is prepared by juice (25%) and maintaining the soluble solids contents (45%) and acidity (1.2–1.5%) in presence of sodium benzoate (1000 ppm) as preservative. These beverages are than heated (85 °C), filled in cans, sealed, processed and cooled.

Mango leather: The mango pulp having sour taste is mixed with sugar, pectin, potassium metabisulphite, citric acid and is traditionally sun-dried (60–65 °C) to develop unique taste and flavor. Recently pectin is replaced by guava pulp which is natural source of pectin and ascorbic acid (vitamin C). Mango leather is a very popular snake among people of all walk of life.

Mango jam: Mango jam consists of dehydrated slices, equal quantity of sugar, citric acid. The final soluble solid content is maintained as 65–68 Brix and 0.6–0.7% acidity for good taste and appearance.

Mango yoghurt: Mango yoghurt is prepared by gentle mixing of boiled milk and mango poser (10:3) in the presence of starter culture (curd). It is an excellent source of high calorie dessert pudding with unique color, flavor and texture.

13.9.3 Utilization of Other By-products

Mango pulp is extensively used for processing of various kinds of value-added products. However, peel and kernel are discarded during mango processing. The wastage of these byproducts cause pollution in the environment. These byproducts (peel and kernel) can be source of generating additional revenue and extraction of valuable nutrients.

Mango peel: Depending on the cultivar, mango peel makes around 15–25% of the fruit weight. Carotenoids, vitamin C and E, polyphenols, and dietary fiber are all found in good amounts in mango peel. Food technologist suggested that incorporation of mango peel in macaroni improved the antioxidant properties, firmness, cooking and nutraceutical characteristics. Mango jam incorporated with mango peels also showed good sensory properties (Larrauri et al. 1996).

Mango kernel: Depending on the cultivar and fruit size, 20–60% of the total fruit weight is made up of mango seeds or kernels. Mango seed is rich with starch, fatty acid (12.3 g/100 g), and phenolic (112 mg /100 g) and potent source of antioxidants. In addition to this, it contains versatile chemical compounds such as phytosterols (bsitosterol, campesterol etc.), amino acids (lysine, leucine, valine), sesquiterpenoids, phospholipids, and various kinds of triglycerides (Ravani and Joshi 2013). Mango kernel fat blended with other nut fats serves as substitute of cocoa butter.

Further climate smart value addition of mango and their benefits are discussed in Table 13.2.

Table 13.2 Value addition opportunities for mangoes in a climate-smart context

Value addition method	Description	Climate-smart advantage
Mango pulp processing	Extraction and storage of mango pulp for later use	Reduces waste during peak harvest, ensuring year-round availability for products like juices and jams
Drying and dehydration	Solar drying or energy-efficient dehydration techniques	Produces dried mango slices or powders with low energy consumption
Mango-based beverages	Production of juices, nectars, and smoothies	Enhances market value while utilizing lower-quality mangoes that might otherwise be discarded
Pickling and fermentation	Creating mango pickles, sauces, or chutneys	Provides alternative income streams and reduces wastage of surplus mangoes
Freeze-drying	Removes water content without affecting quality	Preserves mango for long-term storage with minimal energy usage in the processing phase
Mango seed products	Utilization of seeds for oil extraction or as animal feed	Adds value to typically discarded parts, reducing overall waste
Organic mango products	Production of organic-certified mango products	Meets the growing demand for sustainable, eco-friendly products while commanding premium market prices

13.9.4 Conclusion

Climate change is a bitter reality of current age which is characterized by the enhanced anthropogenic activities, excessive carbon emissions, depletion of ozone layer, temperature rise, frequent weather twists of excessive rain or drought like conditions and development of poor quality fruits and vegetables. Climate smart agriculture is an integrated approach to deal with the exacerbated challenge of food security with three main aims i.e. a. sustainable production b. developing resilience of in food crops, and c. minimizing the greenhouse gas emission.

References

Ahmed YM (2014) Impact of spraying some antitranspirants on fruiting of Williams banana grown under Aswan region conditions. Stem Cell 5(4):34–39

Ahmed OK, Ahmed SET (2014) Determination of optimum maturity index of mango fruits (Mangifera indica L.) in Darfur. Agric Biol J North Am 5(2):97–103

Akrong R, Akorsu AD, Jha P, Agyenim JB (2023) Towards environmental sustainability: the role of certification in the adoption of climate-smart agricultural practices among Ghanaian mango farmers. Cogent Food Agric 9(1):2174482

Anwar R, Malik AU, Amin M, Jabbar A, Saleem BA (2008) Packaging material and ripening methods affect mango fruit quality. Int J Agri Biol 10:35–41

Chandu E, Shukla MK (2023) IoT-based smart irrigation system for mango orchard. In: Artificial intelligence and machine learning in satellite data processing and services: proceedings of the international conference on small satellites, ICSS 2022. Springer Nature, Singapore, pp 61–69

Dambreville A, Normand F, Lauri PE (2013) Plant growth coordination in natura: a unique temperature-controlled law among vegetative and reproductive organs in mango. Funct Plant Biol 40:280–291

Evans EA, Ballen FH, Siddiq M (2017) Mango production, global trade, consumption trends, and postharvest processing and nutrition. In: Handbook of mango fruit: production, postharvest science, processing technology, and nutrition, pp 1–16

FAO (2002) Food and agriculture organization of the United Nations. Available: http://faostat.fao.org/. Accessed 4 Oct 2020

FAO (2013) Climate smart agriculture sourcebook (Rome)

Galán Saúco V (2011) Current situation and future prospects of worldwide mango production and market. In: Global conference on augmenting production and utilization of mango: biotic and abiotic stresses, vol 1066, pp 69–84

GoK (2015) Kenya national climate change action plan 2015–2030

Haldankar PM, Parulekar YR, Kireeti A, Kad MS, Shinde SM, Lawande KE (2015) Studies on the influence of bagging of fruits at marble stage on quality of mango cv. Alphonso J Plant Stud 4(2):12–20

Hall AJ, McPherson HG, Crawford RA, Seager NG (1996) Using early season measurements to estimate fruit volume at harvest in kiwifruit. N Z J Crop Hortic Sci 24:379–391

IPCC (2001) Climate change: working group II: Impacts, adaptations and vulnerability. Available: http://www.grida.no/climate/ipcc_tar/wg2/005.html. Accessed 13 Mar 2009

IPCC (2014) Summary for policymakers. In: Field CB, Barros VR, Dokken DJ, Mach KJ, Mastrandrea MD, Bilir TE et al (eds) Climate change 2014: impacts, adaptation, and vulnerability. Part A: global and sectoral aspects. Contribution of working group II to the fifth assessment report of the intergovernmental panel on climate change. Cambridge University Press, Cambridge, United Kingdom and New York, NY

Kabeer S, Govindarajan N, Radhakrishnan P, Alharbi HF, Essa MM, Qoronfleh MW (2023) Formulation of fortified instant weaning food from Musa paradisiaca (banana) and Eleusine coracana. Front Nutr 10:1203955. https://doi.org/10.3389/fnut.2023.1203955

Kalnay E, Cai M (2003) Impact of urbanization and land-use change on climate. Nature 423:528–531

Kaur B, Panesar PS, Anal AK, Chu-Ky S (2023) Recent trends in the management of mango by-products. Food Rev Int 39(7):4159–4179. https://doi.org/10.1080/87559129.2021.2021935

Larrauri JA, Rupérez P, Barroto B, Sauracalivta F (1996) Mango peels as new tropical fiber: preparation and characterization. Lebensm Wiss Technol 29(9):729–733

Léchaudel M, Lopez-Lauri F, Vidal V, Sallanon H, Joas J (2013) Response of the physiological parameters of mango fruit (transpiration, water relations and antioxidant system) to its light and temperature environment. J Plant Physiol 170:567–578

Legave JM, Normand F, Lauri PE (2013) Climate change and its probable effects on mango production and cultivation. In: X International mango symposium, vol 1075, pp 21–31

Makhmale S, Bhutada P, Yadav L, Yadav BK (2016) Impact of climate change on phenology of mango—the case study. Ecol Environ Conserv 22(9)–S132

Maqbool M, Malik AU, Jabbar A (2007) Sap dynamics and its management in commercial mango cultivars of Pakistan. Pak J Bot 39(5):1565–1574

Mirabella N, Castellani V, Sala S (2014) Current options for the valorization of food manufacturing waste: a review. J Clean Prod 65:28–41. https://doi.org/10.1016/j.jclepro.2013.10.051

Mitra SK (2016) Mango production in the world—present situation and future prospect. Acta Hortic 1111:287–296. https://doi.org/10.17660/ActaHortic.2016.1111.41

Muiruri J (2016) Evaluation of maturity indices of selected mango varieties and effect on quality attributes. Doctoral dissertation, University of Nairobi

Naliyadhara N, Kumar A, Gangwar SK, Devanarayanan TN, Hegde M, Alqahtani MS, Abbas M, Sethi G, Kunnumakkara A (2023) Interplay of dietary antioxidants and gut microbiome in human health: what has been learnt thus far? J Funct Foods 100:105365. https://doi.org/10.1016/j.jff.2022.105365

Neufeldt H, Jahn M, Campbell BM, Beddington JR, DeClerck F, De Pinto A, Gulledge J, Hellin J, Herrero M, Jarvis A, LeZaks D, Meinke H, Rosenstock T, Scholes M, Scholes R, Vermeulen S, Wollenberg E, Zougmoré R (2013) Beyond climate-smart agriculture: toward safe operating spaces for global food systems. Agric Food Secur 2:12. https://doi.org/10.1186/2048-7010-2-12

Normand F, Lauri PE, Legave JM (2015) Climate change and its probable effects on mango production and cultivation. Acta Hortic 1075:21–31. https://doi.org/10.17660/ActaHortic.2015.1075.1

Olesen T (2011) Late 20th century warming in a coastal horticultural region and its effects on tree phenology. New Zealand J Crop Hort Sci 39(2):119–129

Patil SS, Kelkar TS, Bhalerao SA (2013) Mulching: a soil and water conservation practice. Res J Agric Forestry Sci 1(3):26–29

Rathore AC, Raizada A, Jaya PJ, Sharda VN (2012) Impact of chilling injury on common fruit plants in the doon valley. Curr Sci 102(8):1107–1111

Ravani A, Joshi D (2013) Mango and its by-product utilization—a review. Energy 74(44)

Sage RF, Kubien D (2007) The temperature response of C3 and C4 photosynthesis. Plant Cell Environ 30:1086–1106

Saltveit ME (2002) Respiratory metabolism. Accessed 22 Mar 09

Schaffer B, Whiley AW, Searle C, Nissen RJ (1997) Leaf gas exchange, dry matter partitioning, and mineral element concentrations in mango as influenced by elevated atmospheric carbon dioxide and root restriction. J Am Soc Hortic Sci 122:849–855

Scherr SJ, Shames S, Friedman R (2012) From climate-smart agriculture to climate-smart landscapes. Agric Food Secur 1:12. https://doi.org/10.1186/2048-7010-1-12

Severiano-Galeana F, Rosas-Guerrero V, Alemán-Figueroa L, Lucas-García R, Almazán-Núñez RC, Kuk-Dzul JG (2024) Orchards closer to forest patches produced fewer malformed fruits and more commercial fruits: the importance of legitimate floral visitors. Agric Ecosyst Environ 363:108872

Shamim F, Khan AR, Ali MA (2015) Classification of maturity indices of mango fruit for safe harvesting and storage. Ukr Food J 4(3)

Shivran M, Sharma N, Sharma N, Muthusamy V, Dubey AK, Singh SK, Singh BP, Kumar N, Sevanthi AM, Singh N, Singh NK (2023) Development of ripening gene-specific markers and their association with shelf-life in mango (Mangifera indica L.) varieties. Nat Acad Sci Lett 46(3):179–184. https://doi.org/10.1007/s40009-023-01207-0

Singh Z, Zaharah SS (2010) Controlled atmosphere storage of mango fruit—an overview. In: IX International mango symposium, vol 992, pp 481–492

Sudhakar Rao DV, Gopalakrishna Rao KP (2008) Controlled atmosphere storage of mango cultivars 'Alphonso' and 'Banganapalli' to extend storage-life and maintain quality. J Hortic Sci Biotechnol 83(3):351–359

UNFCCC (2015) UNFCCC conference of the parties: twenty-first session. Available: https://unfccc.int

Waskar DP, Masalkar SD, Gaikwad RS, Damame SV (1996) Effect of method of harvesting on postharvest loss reduction and extension of shelf life of mango. In: V international mango symposium, vol 455, pp 662–668

Went FW (1953) The effect of temperature on plant growth. Annu Rev Plant Physiol 4:347–362

Whiley AW, Rasmussen TS, Saranah JB, Wolstenholme BN (1989) Effect of temperature on growth, dry matter production and starch accumulation in ten mango (Mangifera indica L.) cultivars. J Hort Sci 64:753–765

Wurr DCE, Fellows JR, Phelps K (1996) Investigating trends in vegetable crop response to increasing temperature associated with climate change. Scientia Hortic 66:255–263

Yadollahi A (2011) Evaluation of reduction approaches on frost damages of grapes grown in moderate cold climate. Afr J Agric Res 6(29):6289–6295

Zafar TA, Sidhu JS (2017) Composition and nutritional properties of mangoes. In: Siddiq M (ed) Handbook of mango fruit: production, Postharvest science, processing technology and nutrition. Wiley, pp 217–236 https://doi.org/10.1002/9781119014362.ch11

Chapter 14
Policy Frameworks and Institutional Support for Climate-Resilient Mango Production

Shoaib Akhtar, Muhammad Asad Ur Rehman Naseer, Mukhtar Ahmed, Shakeel Ahmad, Saqib Ali Ateel, Amna Hasnain, and Ahmad Abdul Wahab

Abstract Climate-resilient mango production requires a supportive policy and institutional environments. This chapter examines the global and national policy frameworks, institutional support mechanisms, and financial instruments that can enable climate-resilient mango production. It reviews existing policies and programs in major mango-producing countries like India, China, and Brazil highlighting successes and challenges. The chapter also explores extension services, capacity building, and climate information systems that can support farmers in adopting climate-resilient practices. Case studies from countries with effective policy and institutional support are presented, and emerging opportunities and innovations are discussed. The chapter concludes with recommendations for policymakers and practitioners to enhance policy frameworks and institutional support for climate-resilient mango production.

Keywords Capacity building · Climate-resilient agriculture · Extension services · Financial instruments · Institutional support · Mango production · Policy frameworks

---

S. Akhtar · M. A. U. R. Naseer (✉) · S. A. Ateel
Department of Agricultural Business and Marketing, Bahauddin Zakariya University, Multan, Pakistan
e-mail: asadrehman@bzu.edu.pk

M. Ahmed · A. A. Wahab
Department of Agronomy, Pir Mehr Ali Shah Arid Agriculture University, Rawalpindi, Pakistan

S. Ahmad
Institute of Agronomy, Bahauddin Zakariya University, Multan, Pakistan

S. A. Ateel
Secretary Agriculture and Livestock, Punjab, Government of Punjab, Lahore, Pakistan

A. Hasnain
Institute of Management Sciences, Bahauddin Zakariya University, Multan, Pakistan

S. A. H. Naqvi et al. (eds.), *Climate Change and Mango Production*,
https://doi.org/10.1007/978-3-031-87035-4_14

14.1 Introduction

The agriculture sector of Pakistan which is the backbone of its economy has to face significant challenges in the coming years due to climate variation. With agriculture engaging almost 42% of the workforce and contributing around·19% to GDP, it is the most exposed sector hit by climate variability and extreme weather events. (GoP 2023). Mango production is one of the major activities in many tropical countries, the income source for millions of smallholder farmers and also for national economies (Wangu et al. 2020). Climate-related hazards such as droughts, floods, and heatwaves are becoming increasingly frequent and severe while mango production is at risk due to these climate events affecting yields, quality of fruit, and course farmer incomes. In this regard, policy frameworks and institutional support are required for climate-resilient mango production in order to assist farmers in adapting to climate change by ensuring the overall sustainability of the sector (Imtiaz 2023).

Climate-sensitive breeding programs are needed to accelerate the development of new varieties with a high yield, good consumer acceptability, and resistance/tolerance to economically important diseases. In this context effective policy frameworks can also contribute significantly to promoting climate-resilient mango production by providing incentives for fostering the adoption of climate-smart strategies among farmers; enhancing access to seasonal climate information services and advisory support from extension personnel within the agriculture system; as well as improving the overall resilience in terms of weather sensitivity throughout the value chain (Rhiney and Ajayi 2018). Institutional support, particularly extension services of research and development involving farmer organizations is also vital in building the capacity of farmers to address climate change adaptation through the adoption of appropriate technology for resilient crop productivity practice. Renewed efforts to strengthen policy framework and enabling environment of institutional support can help mango farmers increase their resilience against climate-related events while improving productivity and sustainability; ultimately boosting the overall well-being among them. This has a ripple effect in creating food security, reducing poverty, and spurring economic growth as well as supporting the global goals of climate change mitigation and adaptation.

14.2 Policies for Climate-Resilient Agriculture

14.2.1 National Climate Change Policy (NCCP)

The National Climate Change Policy (NCCP) launched in 2012 “provides a holistic framework for addressing the enormous challenges of climate change across all sectors, including agriculture.” Additional strategies that are needed for adaptation will be considered as well, to further protect the resilience of agricultural systems. Recommendations focused on the development of wind- and drought-resistant

Table 14.1 Key policy interventions for climate-resilient mango production

Policy area	Intervention	Expected outcomes
Climate-smart agriculture	Promotion of climate-resilient mango varieties	Ensures stable yields under varying climatic conditions
Water resource management	Incentives for rainwater harvesting and efficient irrigation systems	Reduces water stress and promotes sustainable water use
Soil health management	Policies encouraging organic farming and soil health monitoring	Improves soil fertility, reduces dependency on chemical inputs, and enhances resilience
Pest and disease control	Subsidized pest management programs for mango growers	Minimizes crop losses and supports sustainable production practices
Research and development	Investment in mango-specific climate adaptation research	Develops innovative solutions for climate-resilient mango farming
Market access	Establishing fair trade agreements and export incentives	Enhances market opportunities and ensures fair prices for mango growers
Financial support	Crop insurance and low-interest loans for farmers	Provides financial security against climate-induced crop failures
Awareness and training	Capacity-building programs on climate-resilient practices	Empowers farmers with knowledge and skills to adapt to climate change

crop varieties, enhancing irrigation practices as well as promoting sustainable land management techniques (Mumtaz 2018) (Table 14.1).

14.2.2 National Agriculture Policy (NAP)

The National Agriculture Policy is the articulation of the government's vision for the agriculture sector to enhance productivity, ensure better returns to farmers, and make farm business sustainable. The policy also factors in climate change, recommending the use of climate-smart agricultural practices. Advocates for efficient irrigation, conservation agriculture, and Integrated pest management in minimizing the effect of climate change on crop yield (Clune 2021; GoP 2023; Akhtar et al. 2018a, b).

14.2.3 The Framework for Implementation of Climate Change Policy (FICCP)

It lays out specifics on agendas and deadlines for various sectors including agriculture. It identifies main adaptation strategies like early warning systems, farmers' capacity building, and climate-intelligent infrastructure (Ali 2022; Akhtar et al. 2018a, b). The framework also emphasizes the importance of research and development in creating innovative solutions for climate-resilient agriculture.

14.3 Institutional Support Mechanisms

14.3.1 Ministry of Climate Change (MoCC)

The MoCC collaborates with other Ministries, Provincial governments, and also international organizations for upscaling successful and sustainable climate-resilient agriculture. Key initiatives include developing national plans of action, funding for climate adaptation projects, and supporting interagency coordination.

14.3.2 Pakistan Agricultural Research Council (PARC)

The Pakistan Agricultural Research Council (PARC) is leading research and development for increased climate-resilient agriculture. PARC is researching climate-resilient crop varieties, sustainable farming practices, and innovative technologies to increase the productivity of agriculture (Akhtar et al. 2019; Faisal et al. 2020). The council also provides extension services to disseminate knowledge and best practices to farmers across the country (PARC 2021).

14.3.3 Provincial Departments of Agriculture

Provincial Departments of Agriculture are at the forefront of ensuring Climate-Smart Agriculture can be implemented on the ground. These departments work in all areas to provide the technical support, training, and material resources needed for a farmer. Province leads the design and delivery of provincial adaptation plans ensuring that local contexts are dealt with properly against challenges (Mensah et al. 2021).

14.3.4 International Development Agencies

Several international development agencies and organizations are actively involved in supporting climate-resilient agriculture in Pakistan. Agencies such as the United Nations Development Program (UNDP), the Food and Agriculture Organization (FAO), and the World Bank collaborate with the government to fund projects, provide technical expertise, and facilitate capacity-building initiatives. These collaborations have led to the successful implementation of various climate adaptation projects across the country (World Bank 2011; Aslam 2016).

14.4 Key Initiatives and Projects

14.4.1 Green Climate Fund (GCF) Projects

Different climate-resilient agriculture projects in the country have received funding from the Green Climate Fund (GCF), Pakistan. The projects mainly focus on improving decentralized water management, promoting climate-smart solutions, and capacity building to strengthen early warning systems. By working with a host of international and local organizations, these GCF-partnered projects address all aspects of climate adaptation (Basak and Karlsson-Vinkhuyzen 2022).

14.4.2 National Disaster Risk Management Fund (NDRMF)

The National Disaster Risk Management Fund (NDRMF) has been established to finance projects that reduce the risk of climate-related disasters and leave no one behind. In the agriculture sector, NDRMF will support projects improving water management, soil conservation, and crop diversification (Thapa 2022). These projects aim to reduce the vulnerability of farming communities to climate-induced shocks and stresses.

14.4.3 Community-Based Adaptation Projects

Community-based adaptation initiatives, often delivered by NGOs and other local groups are crucial for building a more resilient grassroots level. The agro-forestry projects are part of the landscape hotspot implementation efforts that empower local communities to adopt climate-resilient practices like water harvesting, and sustainable land management. The projects are intended to involve communities in the planning and implementation of adaptation measures so as to be context-specific, and sustainable.

14.5 Importance of Policy Frameworks for Mango Production

Policy frameworks are essential tools that have direct implications, as they guide the sector through a set of guidelines and regulations to overcome various uncertainties. This chapter discusses the critical role that policy frameworks can play in mango production, and how well-crafted policies have a proven track record of increasing productivity; ensuring quality; safeguarding stakeholder interests—including farmers—as well as tackling environmental adversities arising from excessive pesticide application.

14.5.1 Enhancing Productivity and Sustainability

14.5.1.1 Research and Development

Understanding the importance of Mango Production, there is a good question: How can we increase this production with improved mango research and development (R&D) technologies? Governments can play a part in developing new strains of high-yielding, disease-resistant mangoes, and more sustainable agricultural practices by funding research institutions and universities. These improvements may result in increased productivity and tolerance for pests, making mango production sustainable.

14.5.1.2 Extension Services

There is also a need to update extension services that provide information on soil health practices and particularly on pest management, irrigation methods, and post-harvest handling which could help farmers adopt the best agriculture-productive sustainable systems.

14.5.2 Ensuring Quality and Standards

14.5.2.1 Quality Control Regulations

Quality control guidelines are maintained by certain regulations put in place by the listed policies for mango production. Regulations are in place to ensure the quality, coloration, sugar, and safety of mangoes. A policy framework then enforces these quality standards to enable consumer trust as well as offer an in-road into domestic and international markets, which otherwise would be unavailable for those without requisite classification.

14.5.2.2 Certification and Labeling

Legalization and certification to the regulation of organic and fair-trade mango production lend a hand to the general development. These certifications give consumers confidence that the mangoes they buy are produced sustainably and ethically. Certification policies ensure that the farmers have access to premium markets and get more value for their products.

14.5.3 Supporting Farmers and Rural Communities

14.5.3.1 Financial Support and Subsidies

Mango farmers, including smallholders, will be able to receive financial support and subsidies. This financial instruments can be in the shape of grants, low-interest loans, and subsidies on inputs (fertilizers/pesticides). Workable financial help they would receive will support them and they can invest this on needed technologies and modern farm practices which automatically reduces their burden on heavy capital sinking towards high productive quality mangoes (Bisen; Jarman et al. 2023).

14.5.3.2 Access to Markets

Market access policies are key for the economic viability of mango production. They mitigate the operations of farms to market infrastructure like roads, storehouses, and transportation systems which shuffle mangoes from nursery to terminal. In addition, we need trade policies that are fair to the farmers who can also help increase their income and gain better market access.

14.5.4 Addressing Environmental and Climate Challenges

14.5.4.1 Climate-Resilient Practices

The promotion of climate-resilient agriculture practices can be encouraged through policy frameworks to mitigate the effects on mango production due to climate change. Methods consist of water-efficient irrigation practices, agroforestry, and conservation agriculture. Adaptation of farmers to changing climatic conditions is enhanced by policies that promote more climate-resilient practices and thus reduce the vulnerability of farmers to extreme weather events (Goswami et al. 2023).

14.5.4.2 Sustainable Land Management

Maintaining land health and reversing further soil degradation requires policies that promote sustainable land management. Adopting the sustainable practice of intercropping crop rotation. Its reduced tillage can enhance fertility, and structure would result in higher mango production leading to long-term sustainability in agriculture (FAO and FAOSTAT 2018).

14.5.5 Facilitating Market Access and Trade

14.5.5.1 Export Promotion

Export promotion enabling policy frameworks can term mangoes as a product of choice for international market scanners. Some of these policies could be trade agreements, export subsidies, and high-value marketing campaigns about the quality and specialty of local mango varieties. Policies that enable export opportunities for mango can help improve the feasibility of producing a profit and hence economic growth (Wardhan et al. 2022).

14.5.5.2 Trade Regulations

Trade regulations must also comply with international standards so that these mangoes can be exported and used without falling prey to an exorbitant import tax. Comprehensive policies that align phytosanitary measures and quality standards with international norms reduce the likelihood of trade disputes and facilitate effective market access. The particular significance of this alignment is felt most closely in exporting countries, or those wanting to realize their latent potential for making such an entry (Lee 2019).

Fig. 14.1 *Source* www.globalvolunteers.org

14.6 Global Policy Frameworks

Guidelines, and strategies by global policy frameworks are being offered to tackle a range of threats such as climate change, food security concerns sustainability issues, or on-trade in agriculture. The frameworks help with global collaboration, and learning best practices and they allow resilient agricultural systems to evolve. This section deals with major global policy frameworks influencing mango production as well as other agricultural practices (Fig. 14.1).

14.7 The United Nations Framework Convention on Climate Change (UNFCCC)

The United Nations Framework Convention on Climate Change (UNFCCC) is an international environmental treaty to which 196 Parties have signed, and it addresses climate change through the reduction of greenhouse gas emissions, adaptation capacity with environment integrity. The UNFCCC was adopted in 1992 and lays the foundation for global responses to changes in climate that can affect agriculture.

14.7.1 Implications for Agriculture

The UNFCCC acknowledges the significant greenhouse gas emissions and related damages from agriculture, as well: The UNFCCC, through instruments such as the Paris Agreement, encourages countries to ready themselves by preparing Nationally Determined Contributions (NDCs) which can contain specific actions that focus on limiting emissions from agriculture and improving practices in resilient farming. This will require the adoption of climate-smart agriculture practices, better water management, and such to assist in the development of different Climate-resilient crop varieties (Rajan et al. 2020).

14.8 The Sustainable Development Goals (SDGs)

The Sustainable Development Goals (SDGs) are a set of 17 global goals adopted by all United Nations Member States in 2015. They provide a comprehensive framework for addressing global challenges, including poverty, inequality, climate change, environmental degradation, peace, and justice, with a target to achieve these goals by 2030.

14.8.1 Implications for Agriculture

Several SDGs are directly relevant to agriculture, particularly Goal 2 (Zero Hunger), Goal 12 (Responsible Consumption and Production), and Goal 13 (Climate Action). These goals emphasize the importance of sustainable agricultural practices, improved food security, and resilience to climate change. For mango production, aligning with the SDGs involves adopting sustainable farming practices, reducing post-harvest losses, and ensuring fair trade and market access for farmers (Bokelmann et al. 2024).

14.9 The Global Framework for Climate Services (GFCS)

The Global Framework for Climate Services (GFCS) is an international initiative coordinated by WMO to enable better access, production, and application of climate information varying from multi-year predictions and seasonal forecasts down to short-term atmospheric tracks. Introduced in 2009, the GFCS aims to better predict and prepare for climate girls with timely forecasts of global warming (Fig. 14.2).

14.9.1 Implications for Agriculture

In agriculture, GFCS will provide farmers with climate services that support decision-making on when to plant or irrigate a crop, and how best to manage pests in their fields. Climate services support mango growers, which can be used to quantify weather

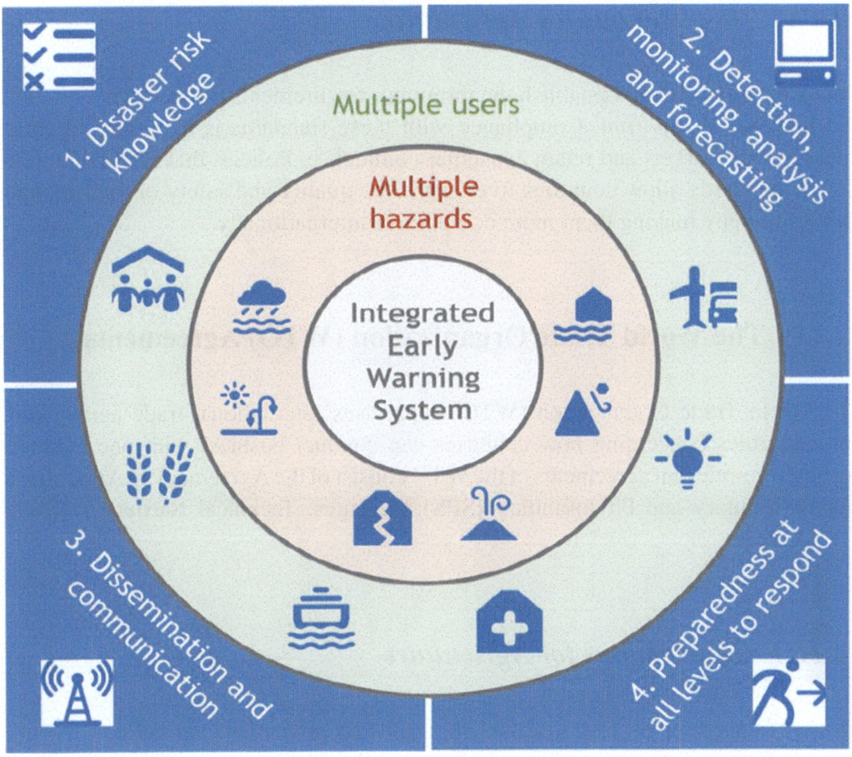

Fig. 14.2 *Source* WMO and the early warnings for all initiative

forecasts with early warning systems and climate risk assessments of good quality (Sivakumar 2011).

14.10 The Codex Alimentarius Commission

Its directives are produced by the Food and Agriculture Organization (FAO) of the United Nations as well as its wagoner agency, which is the World Health Organization (WHO), in order to develop all sorts of international food standards including recommendations on every single stage i.e. from harvesting till distribution within commerce internationally called Codex Alimentarius Commission. They are broad standards meant to help protect consumer health and the integrity of fair-trading practices in the global food trade.

14.10.1 Implications for Agriculture

Codex mango standards establish the minimum requirements for the safety, quality, and labeling of this fruit. Compliance with these standards is necessary to enter international markets and retain consumer confidence. Policies that are in line with Codex standards allow countries to enhance the quality and safety of their mango exports thereby making them more competitive internationally.

14.11 The World Trade Organization (WTO) Agreements

The World Trade Organization (WTO) supervises international trade agreements and sets rules concerning how countries can conduct business with one another. Agriculture-relevant agreements of the WTO consist of the Agreement on Agriculture (AoA), Sanitary and Phytosanitary (SPS) Measures, Technical Barriers to Trade (TBT), etc.

14.11.1 Implications for Agriculture

The WTO agreements seek to establish a fair and predictable agricultural trading system. These arrangements also help in establishing transparency as well as, predictability of trade policies and ensuring that these are based on scientific principles which are important for producing foodstuff such as Mango. Maintaining SPS and TBT standards is a critical step before the exportation of mangoes to overseas markets so as to meet specified health safety requirements.

14.12 The Food and Agriculture Organization (FAO) Policies

The Food and Agriculture Organization (FAO) of the United Nations leads international efforts to alleviate hunger, improve nutrition, and increase global food security. FAO creates, designs, and implements agricultural policy guidelines for sustainable development.

14.12.1 *Implications for Agriculture*

Its policies and programs provide the technical expertise, resources, and training to create sustainable food systems and support biodiversity more efficiently than industrial farming while putting nutrition first place. In the area of mango production, FAO projects are directed toward better cultivation practices, increased market access, and sustainable land management. The aim is to help countries build more resilient and sustainable mango production systems with the support of FAO's wealth of resources and expertise (Canton 2021).

14.13 Paris Agreement and Nationally Determined Contributions (NDCs)

The Paris Agreement is the first-ever universal, legally binding global climate change agreement, adopted at the COP21 conference in December 2015. The target is to achieve the aim of limiting global warming significantly below 2 °C above pre-industrial levels and efforts will be made to limit temperature increase even more, up to a ceiling no higher than 1.5 °C.

Here are key aspects of the Paris Agreement.

14.14 Sustainable Development Goals (SDGs)

The Sustainable Development Goals (SDGs) are a collection of 17 global intentions established by the United Nations in 2015, dealing with poverty among other challenges on earth that each Member State requires acting to end otherwise bad things will continue happening. Many of these goals touch upon agriculture, and in the case of hunger, sustainable development or climate action also rely on agricultural solutions (Halisçelik and Soytas 2019).

Following is the list of SDGs that are related to climate-resilient agriculture:

14.14.1 *SDG 1: No Poverty*

Climate-resilient agriculture contributes to elevating poverty by boosting the productivity and sustainability of farming, which is central to hundreds of millions of smallholder farmers in developing countries.

14.14.2 SDG 2: Zero Hunger

This goal is about getting a grip on hunger and food security. Climate-resilient agriculture seeks to make food systems more secure in the face of climate change and also enhance nutritional outcomes.

14.14.3 SDG 12: Responsible Consumption and Production

One of these is the advancement of sustainable agricultural practices. Climate-resilient agriculture is also about being more efficient and less wasteful, it means using fewer resources far in order to get the same produce produced.

14.14.4 SDG 13: Climate Action

It also aligns with the Adaptation Goal, which Climate-resilient agriculture directly contributes to by implementing practices on the ground that support farmers in adapting and mitigating climate change impacts. Methods such as more modern irrigation systems, drought-resistant crops, and soil conservation regional planning are involved.

14.14.5 SDG 15: Life on Land

This target relates to the areas of forestry, desertification, and biodiversity. Climate-resilient agriculture: This can preserve ecosystems, step up land management, and help restore degraded lands.

14.15 Climate-Resilient Agriculture

14.15.1 Adaptation Strategies

Adopting climate-smart practices to enhance the resilience of agricultural systems. This could mean implementing the latest plant breeding to infuse drought-tolerant genetics in crops, altering when they are planted, and enhancing soil health.

14.15.2 Sustainable Practices

Employing practices that maintain or increase agrarian productivity, while avoiding adverse environmental effects. These include conservation tillage systems, agroforestry practices, and organic farming.

14.15.3 Innovation and Technology

Leveraging technological breakthroughs (i.e., precision agriculture) to promote resource efficiency and productivity, while at the same time minimizing exposure to risks stemming from climate change.

14.15.4 Diversification

These efforts are intended to help agriculture continue in its role of addressing global food needs while also adjusting and contributing effectively to climate change considerations among other sustainability objectives.

14.15.5 Artificial Intelligence and Blockchain

The role of Artificial Intelligence (AI) and Blockchain is critical to predict the climate that will affect agricultural production. Therefore, for climate resilient mango production AI will be helpful in farm practices and farmers information.

14.16 National Policy Frameworks from Major Mango-Producing Countries

14.16.1 India

As is well known India being the largest mango-producing country in the world leading global production too. The country has a detailed policy framework for mango production, targeting productivity improvement, quality, and sustainability. There are following key policies in India for mango production and development.

14.16.2 National Horticulture Mission (NHM)

The NHM, established in 2005, intends to encourage the overall expansion of the horticulture sector, particularly mango production. It supports many activities, including research and development, extension services, infrastructure development, and post-harvest management.

14.16.3 Mission for Integrated Development of Horticulture (MIDH)

NHM aims to promote the general development of the horticulture sector with a special focus on mango production which was established in 2005. It covers various enterprises including research and development, extension services, infrastructure building, and post-harvest management.

14.17 China

Similarly, the second largest mango-producing country China too has formulated a number of policies for their horticultural sector. Crucially, the emphasis is on retrofitting old ways of agriculture to new age quality and best practices that must be environmentally sustainable. There are following key policies in China for mango production and development.

14.17.1 National Modern Agriculture Demonstration Zones

They aim to demonstrate best practices and technologies in agriculture. Again, mango-growing zones include modern irrigation technologies and integrated pest management including high-yielding varieties.

14.17.2 Green Food Certification Program

The project promotes eco-friendly agriculture. Certification: Mango producers are encouraged to get certification over sustainable practices which could provide enhanced market access as well (Paull 2008).

14.18 Brazil

Brazil is a major player of Indian mango in the international market with pro-national productivity, procurement, and export promotion policies envisioned to sustain competitiveness.

14.18.1 National Plan for the Development of Fruit Culture (PNDF)

This plan is oriented towards the development of pear orchards but also includes support for the growth of other fruit sectors such as mangoes. The focus is on R&D, extension, and market development.

14.18.2 Agroecological Transition Program

The program supports the adoption of agroecological practices to ensure sustainable agriculture. Assistance to small mango growers for conversion from conventional to organic farming and adoption of sustainable land management practices.

14.19 Existing Policies Analysis Supporting Climate-Resilient Mango Production

14.19.1 Research and Development

The research and development (R&D) are the apparent show for major mango-producing countries. Such validated knowledge has policy relevance for the development of climate-resilient mango varieties, and improved irrigation practices/ pest management strategies among others. For example, the National Horticulture Mission in India funds research institutions to breed mango varieties that will produce more fruit and be resistant to diseases.

14.19.2 Extension Services

One of the well-used strategies is access to extension services for farmers. These services share the technical knowledge on climate-smart and best practices available in mango production. The Modern Agriculture Demonstration Zones across China are educational headquarters that teach farmers about new farming practices.

14.19.3 Financial Support

Financial incentives—like cost-sharing, grants, and low-interest loans to incentivize the uptake of sustainable farming on the part of farmers are included in many policies as well. Scheme of National Mission for Development Fruit Culture grants infrastructure development and market expansion in Brazil.

14.19.4 Quality Assurance and Certification

This process would be laborious and time-consuming, therefore quality assurance affects the procurement of good markets and consumer trust. China incentivizes its mango producers to become certified and practice sustainability—through initiatives like the Green Food Certification Program (Paull 2008).

14.19.5 Market Access and Infrastructure

Improving market access and infrastructure is a key focus area. Policies support the development of transportation networks, storage facilities, and export promotion strategies. India's Mission for Integrated Development of Horticulture emphasizes the importance of post-harvest management and market linkage. There is a need for ongoing capacity building for farmers and extension workers. Training programs should be regularly updated to reflect the latest advancements in climate-resilient agriculture.

14.20 Institutional Support Mechanisms

14.20.1 Role of Government Agencies

14.20.1.1 Ministry of National Food Security and Research

The Ministry of National Food Security and Research (MNFSR) in numerous countries is the ministry responsible for policy planning, coordination activities operational development projects implementation of federal financial assistance for food security, and agricultural productivity to probe the agriculture sector. It is responsible for supporting mango production in Pakistan through different programs and initiatives. MNFSR has the following major functions.

14.20.1.2 Development of Policy and Implementation

The Ministry of National Food Security and Research (MNFSR) prepares national policies and strategies for different sectors which include eco-green environment-friendly agricultural systems ensuring Mango orchard sustainable production. It doubles as a framework for managing water, controlling pests, and replenishing soil fertility.

14.20.1.3 Research and Development

It also provides funds and coordinates research activities targeting the development of high-yielding, disease-resistant, climate-resilient mango varieties. Central to this function is working in collaboration with agricultural research institutions and universities.

14.20.1.4 Extension Services

One of the principal functions of MNFSR is to provide extension services so that farmers. They provide training and technical assistance on good agricultural practices for mango production and pest management as well as climate change adaptation strategies.

14.20.1.5 Subsidies and Funding

The support under subsidy programs and financial requirements stimulates an efficient up-take of new technologies/piracies for infield mango production which leads to increased resilience and Productivity while also aiding MNFSR in achieving their policy objectives. For mango growers, the National Horticulture Development Program includes certain projects/ schemes that are channeled through MNFSR in Pakistan. These interventions address modern land use approaches, traditional agricultural practices for example in improving means of production from livestock to crop farming at a basic level and access to market approach (GoP 2023).

14.20.2 Ministry of Climate Change and Environmental Coordination

Ministry of Climate Change and Environmental Coordination (MCCEC) Handles climate change issues and manages environmental policies. This work is paramount for mainstreaming climate-resilient practices in agriculture, must be continued at the ministry level, and has to include mango production. It has the following major functions.

14.20.2.1 Climate Policy and Planning

To this end, the MCCEC formulates national climate policies and action plans which also account for preparing the agricultural sector to face uncertainties. These include reaching targets for reducing greenhouse gas emissions, promoting sustainable land use practices, and more.

14.20.2.2 Environmental Regulations and Standards

The ministry prescribes environmental regulations, standards, and practices to support sustainable agriculture. It also has the default recommendations or guidelines for pesticide, water management, and conservation agriculture.

14.20.2.3 Climate Smart Agriculture Practices

MCCEC partners with a host of government agencies, international organizations, and civil society to operationalize climate-smart agriculture programs. Programs in this area of work seek to increase the resilience and adaptive capacity of farmers and promote sustainable practices.

14.20.2.4 Awareness and Capacity Building

The ministry also organizes awareness and capacity-building programs for farmers as well as stakeholders to get them familiar with the changes taking place due to climate change, which is urging us towards practicing reviving outmoded. practice. In India, it has been introduced as the National Action Plan on Climate Change (NAPCC), under which the National Mission for Sustainable Agriculture (NMSA) is operated by the Ministry of Environment, Forest and Climate Change. NMSA encourages various climate-resilient agricultural practices like precision farming, micro-irrigation, and integrated pest management, among others which are best suitable for enhancing productivity in the production of mangoes.

14.21 Role of International Organizations

14.21.1 Food and Agriculture Organization (FAO)

The FAO provides technical assistance, policy advice, and funding to support sustainable agricultural practices worldwide. It plays a crucial role in promoting climate-resilient agriculture, including mango production. FAO has the following function:

- Technical Assistance
- Capacity Building
- Funding and Resources.

The organization provides funding and resources for research and development projects aimed at improving agricultural productivity and resilience (FAO and FAOSTAT 2018).

14.21.2 International Fund for Agricultural Development (IFAD)

IFAD concentrates on rural poverty reduction and agriculture development. It supports smallholder farmers with financial and technical means of adopting sustainable agriculture. IFAD functions include project funding, technical assistance and capacity building. IFAD's training programs provide farmers, and local institutions trained in sustainable agriculture and climate resilience. It promotes climate-resilient practices and market access for mango producers as part of this scheme (Turvey 2017).

14.21.3 The World Bank

The World Bank provides financial and technical support to countries for development projects, including those in the agricultural sector. It focuses on enhancing agricultural productivity, sustainability, and resilience to climate change. The main functions of the World Bank are project financing, policy advice, and research and knowledge sharing.

14.22 Role of CSOs and NGOs

There are many Civil Society Organizations (CSOs) and Non-Governmental Organizations (NGOs) which are the forerunners with experience in facilitating community-based support to climate-resilient agriculture. They make sure the needs and voices of farmers are heard by policymakers. CSOs and NGOs functions include advocacy and awareness; capacity building; implementation of projects; research and innovation; practical action.

14.22.1 Worldwide Fund for Nature (WWF)

Through community and government partnerships, WWF introduces sustainable agriculture while supporting conservation efforts. This includes support for organic agriculture and agricultural value chain to reduce the environmental footprint of cultivation which benefits mango producers (Anyango-van Zwieten et al. 2019).

14.22.2 Oxfam

Oxfam trains smallholder farmers on sustainable agricultural practices and lobbies for fair trade policies. Their works commonly turn around weather smart agriculture and marketing support to farmers (Sen and Majumder 2011).

14.23 Financial Support Mechanisms

14.23.1 Government Funding and Subsidies for Climate-Resilient Agriculture

Government funding and subsidies are important in sustainable agriculture, particularly mango production. These instruments of economic support are intended to lighten the load for farmers and incentivize sustainable farming while building climate resilience in agricultural systems.

Mechanisms.

14.23.2 Direct Subsidies and Grants

Mango farmers are also encouraged by the government through direct subsidies and grants to adopt climate change resilient practices. This can range from subsidies to buying drought-tolerant mango varieties, installing efficient irrigation systems, and using integrated pest management practices.

14.23.3 Research and Development Expenditure

Financial support for research and development (R&D) is vital in moving climate-resilient mango production. They help pay for mango varieties that are resistant to pests, new technologies, and crops adapted to climate change.

14.23.4 Infrastructure Development

Infrastructural investments like water storage facilities, cold storage, and transportation networks also aid in reducing post-harvest losses thereby increasing mango productivity.

14.23.5 Extension and Training Services

By investing in training programs and extension services, farmers can be taught how to adapt the best practices for climate-resilient agriculture. Many subsidy programs are available in India under the National Horticulture Mission (NHM) or The Mission for Integrated Development of Horticulture (MIDH) by the Ministry of Agriculture and Farmers Welfare. These initiatives include assistance in promoting advanced farming practices, establishment of nurseries for quality planting material as well as post-harvesting infrastructure.

14.23.6 International Funding Sources

International finance supports urgently needed climate-resilient agricultural development in developing countries. These resources focus on supporting the resilience of agricultural systems, encouraging sustainable practices, and minimizing climate change vulnerabilities.

Mechanisms.

14.23.7 Green Climate Fund (GCF)

The GCF is a global fund created to channel the money raised by developed countries into efforts of developing nations in dealing with climate change.

14.23.8 Adaptation Fund

The Adaptation Fund finances projects/programs helping vulnerable communities in developing countries to adapt to the negative effects of climate change. This will include activities to improve the resilience of agricultural systems, including mango production.

14.23.9 Global Environment Facilitator

GEF finances projects to address global environmental issues, among them climate change. It seeks to reinforce actions encouraging sustainable land management, conservation of biodiversity, and climate-resilient agriculture. In Pakistan, the Green Climate Fund has approved projects that focus on the resilience of agricultural communities with a view to managing climate change impacts. These projects range from working to better manage water use, encourage new climate-smart agricultural practices, and increase the resiliency of farmers facing more variable climates.

14.24 Private Sector Investment and Financing Options

Private sector investment and financing options are crucial in catalyzing the scale of transformation needed to promote climate-resilient agriculture. This will be facilitated by markets and mechanisms designed to channel hundreds of billions in investment from agribusiness, financial institutions, and impact investors into sustainable agriculture practices that specially build the resilience of farming communities.

Mechanisms.

14.24.1 Agribusiness Investments

Sustainable supply chains such as agribusiness companies investing in climate-resilient practices for farmers. Improving seed, fertilizers, and other inputs in addition to providing technical services or training.

14.24.2 Impact Investing

Generally, impact investors aim to create both social and environmental benefits in addition to financial gains from their investments. Climate change adaptation and investments in climate-resilient agriculture projects can be achieved with a project such as sustainable mango production that provides financial benefits to participating farms whilst addressing one of the heaviest challenges arising from climate impacts.

14.24.3 Agricultural Financing

To encourage farmers to adopt climate-resilient practices, financial institutions provide different kinds of financing—including but not limited to loans and credit lines— in addition to other types such as insurance products. They are financial products that serve to reduce risks related to climate variability and therefore support sustainable agriculture.

Brazil has seen the emergence of alternatives in terms of agricultural finance to support farmers aiming at climate resilience production. Banks will provide farms that practice sustainability (e.g. water-efficient drip irrigation and integrated pest management) with low-interest loans, or credit lines specifically for farmers in many countries.

14.25 Support of Extension Services and Capacity Building Programs

14.25.1 Extension Services

Extension services play a crucial role in connecting research with real-world solutions on the ground. They contribute significantly to improving the climate change resilience of mango production in this zone. Extension services play a key role in boosting productivity, sustainability, and climate change resilience by passing on knowledge to farmers.

Good extension services help to promote climate-resilient mango production in Pakistan by

- Weather forecasts to farmers, climate trends, and advisories inform decisions such as planting/irrigation/pruning dates in agricultural contexts.
- Introduction of grafting, pruning, and IPM (integrated pest management) to strengthen the resistance of a mango tree.
- Highlighting smart irrigation systems, soil conservation practices, and pest management solutions to enhance mango productivity

- Training the farmer on climate-resilient agriculture practices, soil management, and water conservation
- Plant materials (generic and disease-tolerant mango varieties), genuine seeds, fertilizers, and pesticides to the farmers.
- Direct linkages to markets—stable incomes incentivize investment in climate-smart practices.
- Assessing the impact of climate-resilient practices and giving out feedback for course correction.

By playing these roles, extension services can help Pakistani mango farmers adapt to climate change, improve productivity, and ensure a sustainable mango industry.

14.25.2 Capacity Building Programs

Capacity-building programs can support climate-resilient mango production. There should be separate capacity-building programs for farmers, extension agents, and policymakers.

14.25.2.1 Capacity Building Programs for Farmers

- Training on climate-resilient practices
- Soil and water management
- Climate risk management

14.25.2.2 Capacity Building Programs for Extension Agents

- Advanced training in climate-resilient agriculture
- Communication and outreach skills
- Monitoring and evaluation techniques

14.25.2.3 Capacity Building Programs for Policymakers:

- Climate change and agriculture policy
- Climate-resilient agriculture planning
- International best practices

Through capacity building of farmers, extension agents, and policymakers on the one hand these programs can strengthen the resilience of Pakistan's mango production system to climate change; thereby ensuring a sustainable and productive future for its Mango industry. Implemented by the National Mango Board, it offers a capacity development program on mango production for agriculture extension agents and policymakers. The International Mango Training Program provides complete training of mango training to growers in collaboration with international research institutions. It deals with various features of climate-resilient mango cultivation techniques such as best agricultural practices, Integrated Pest and Disease Management (IPDM), value-given market presence, etc. (Mitton et al.2021).

14.26 Climate Information and Early Warning Systems

Climate information is important for planning and strategies to improve the resiliency of the agricultural system as in mango production. Having precise weather data is vital for informing farmers on when to plant, irrigate, and protect crops from pests or trouble during harvesting; thus, making them less vulnerable to erratic climate patterns like floods and cyclones.

The crucial factors that make it essential to have information about the climatic conditions for climate-resilient agriculture are:

- Weather forecasts can be used to determine when is best for farmers to plant and reduce their chances of crop failure.
- With climate data, it is possible to know the exact time for watering crops, which reduces waste and optimizes use.
- Climate information equips farmers to expect, be ready for, or act on upcoming outbreaks of pests and diseases.
- It helps to increase or reduce the level of wastage and environmental pollution due to inaccurate application timing for fertilizer and pesticides.
- Using the information on climate variability, farmers are able to make more informed decisions about their approach to strategies for insuring crops and managing risks.
- Knowing more about the climate can help provide uninformed decisions on these important issues.
- Climate information allows farmers to harvest the rainwater which in turn reduces groundwater dependence.
- Uses climate information to develop advisories for farmers and helps them in taking up weather-resilient practices.

Climate information is essential for decision making and by sharing this climate risk is reduced, resulting in a more resilient agriculture sector of Pakistan.

14.27 Early Warning Systems for Weather-Related Hazards and Climate Risks

Early warning systems for weather-related hazards and climate risks are crucial for climate-resilient agriculture in Pakistan. These systems perform the detection of weather-related hazards, provide timely warnings, and enable preparedness and response. The key components of these systems include weather forecasting, risk assessment, communication networks, and response planning. There are several benefits of these systems such as reduction of crop losses, livelihood security, and help for agricultural planning. In Pakistan, the following are the major warning systems.

1. Pakistan Meteorological Department's Weather Forecasting System
2. National Disaster Management Authority's Early Warning System
3. Agricultural Extension Department's Climate Advisory Services.

By establishing effective early warning systems, Pakistan can reduce the impact of weather-related hazards on agriculture, supporting climate-resilient agricultural practices and enhancing food security. These systems can further be strengthened by:

- Improving weather forecasting accuracy
- Enhancing communication networks
- Developing response plans
- Conducting regular drills and training.

14.28 Challenges and Opportunities

14.28.1 Barriers to Effective Policy and Institutional Support

Despite the critical importance of policy and institutional support for climate-resilient mango production, several barriers hinder the effectiveness of these mechanisms. Addressing these challenges is essential for enhancing the resilience and sustainability of the mango industry. Following are the key barriers to effective policy development.

- The mango farmers do not have access to reliable data, hence rely on it when deciding what trees they should plant.
- An insufficient number of poorly equipped climate-resilient agriculture service providers prevent Mango from accessing new practices.
- Mango Farmer needs the credit but does not have access to them as well insurance which can help in adapting climate resilient practices.
- If governments do not have proper water management practices or infrastructure, this can cause a lack of accessibility to the people and inefficient use.

- Inadequate development of Mango varieties and practices suitable to combat Climate Change.
- This results in confusion and overlapping efforts among government agencies or departments.
- Mango is not heard, and some policies do not address his needs.
- Mango farmers have very little insecurity but among them, the largest is that they do not have access to modern irrigation systems, and greenhouses due to which they could perform climate-resilient practices.
- National policies do not prioritize climate-resilient agriculture limiting govt funds and assistance
- Pakistan's agriculture sector does not benefit from them due to limited international cooperation.

Addressing these barriers can help strengthen policy and institutional support for climate-resilient agriculture in Pakistan, benefiting mango growers and enhancing the sector's overall resilience in Pakistan (Table 14.2), the fragmentation of responsibilities among different ministries and agencies has been identified as a major barrier to effective policy implementation for climate-resilient agriculture. This lack of coordination often results in overlapping efforts and inefficient use of resources.

14.28.2 Emerging Opportunities and Innovations for Climate-Resilient Mango Production

The scene of climate-resilient mango production is now being redefined by new frontiers and innovation opportunities. From precision agriculture and portions of dry climatic smart mango varieties to vertical farming, the work offers a new foothold for farmers wanting their trees better adapted to global warming. Moreover, upcoming technologies such as AI (Artificial Intelligence), ML (Machine Learning), and Blockchain are aiding mango growers in supply chain transparency market access, and decision-making. These innovations not only enhance productivity and sustainability but also enable smallholders to cope with climate change.

Both agroforestry and permaculture practices have been integrated too, which makes the whole system regenerative with a lot of soil conservation that further helps in making mango production resilient. Similarly, climate information services and climate-resilient agriculture insurance offer invaluable assistance to mango growers in their bid for more effective management of risks arising from the changing climatic conditions. While these newfound opportunities and innovations will mature, they pose to be significantly instrumented in the sustainability and longevity of mango production which subsequently benefits both its farmers along with consuming groups. The mango industry has to utilize these innovations if it wants to survive in the changing climate and secure its future.

Table 14.2 Institutional support systems for climate-resilient mango production

Institution/organization	Role in climate-resilient mango production	Examples of support provided
Government agricultural departments	Implementation of climate-smart policies and subsidies	Distribution of drought-resistant mango varieties, support for irrigation systems, and soil health monitoring
Agricultural research institutes	Development of climate-adaptive mango farming techniques	Research on pest-resistant mango varieties and efficient water-use technologies
Non-governmental organizations (NGOs)	Advocacy and field-level support for farmers	Training on sustainable practices, awareness campaigns, and access to funding for small-scale farmers
Farmer cooperatives	Collective decision-making and resource sharing	Shared access to irrigation infrastructure, pest control equipment, and market linkages
Microfinance institutions	Provision of financial assistance to smallholders	Loans for climate-smart technologies, crop insurance, and support for diversification projects
International organizations	Global support for climate adaptation and resilience	Funding and technical assistance from organizations like FAO, World Bank, and UNDP
Extension services	Bridging knowledge gaps between researchers and farmers	Dissemination of climate-resilient practices, pest control measures, and postharvest management strategies
Private sector	Investment in climate-resilient supply chains	Development of sustainable packaging, cold storage solutions, and fair-trade certifications

14.29 Conclusion

Unfortunately, climate change presents major challenges for mango production by reducing yields and quality of the fruit which have negative effects on farmers' livelihoods. Such challenges are complex and need to be addressed by a comprehensive strategy that includes policy frameworks, institutional mechanisms, and financial resources including extension services and capacity-building programs. Introduction Stakeholders can use climate information, early warning systems, and emerging opportunities to build the resilience of mango value chains and increase seeds for solutions.

There is a need for policy frameworks and institutional support in the promotion of climate-resilient mango production. Examples of government-led initiatives to assist farmers in adapting to climate change can be found in the form of national policies implemented by some countries, particularly major mango-producing nations like India, China, and Brazil. By reviewing what is already being done in governance and on the ground, particular areas where more coordination or funding might make sense/where a differentiated approach with mango actors would be useful. Extension services are crucial for sharing knowledge of the best technologies that enhance our resilience to climate change. Farmers, extension workers, and policymakers need to undergo capacity-building programs in order that they could be able to implement climate-resilient mango production practices as well as provide support for them.

At this moment, it is very important for different funding sources including international institutions or their local representatives to look into ways and means of supporting activities at the governmental level that can proliferate climate-resilient practices across the board with regard to mango production. These financial incentives help alleviate the economic pressures on farmers, promote sustainable agriculture, and build greater resilience in our systems. Decisions based on accurate and timely climate information and early warning systems are critical for risk management throughout the development process. Such systems help farmers anticipate unfavorable weather and reduce its impact, reducing the loss of crops every season, and thereby increasing overall farmer resilience. Enhancing the resilience of the mango industry requires addressing barriers to effective policy and institutional support, most notably coordination challenges along with inadequate funding. Green shoots in the form of technology, good agricultural practices for climate-resilient mango, and public–private partnerships also offer viable solutions to ensure sustainable production as well as productivity of mango.

14.30 Future Directions

14.30.1 Enhanced Coordination

Better coordination between government departments, R&D organizations, and other stakeholders is crucial for the successful implementation of policies and programs. Acting together will help make better use of resources and enable farmers to get the necessary support.

14.30.2 Increased Investment in R&D

A vital part is that new mango types and agriculture tactics that can face up to those climate stressors could be advanced through purchasing studies and development. This investment will spur innovation in new, farmer-ready tools to respond to a more variable climate.

14.30.3 Replicating and Scaling Successful Initiatives

Scaling successful climate-resilient initiatives and practices can increase the reach to more farmers while strengthening overall mango industry resilience. To achieve these goals, public-private cooperation and digital platforms will be key.

14.30.4 Enhancing Extension Services

Strengthening extension services could help spread climate-smart practices on a much wider scale. The continuous training and professional development of extension agents will facilitate effective adaptation by farmers to climate change.

References

Akhtar S, Li G, Nazir A, Razzaq A, Ullah R, Faisal M, Naseer MAUR, Raza MH (2019) Maize production under risk: the simultaneous adoption of off-farm income diversification and agricultural credit to manage risk. J Integr Agric 18:460–470

Akhtar S, Li G, Nazir A, Razzaq A, Ullah R, Faisal M, Naseer MAUR, Raza MH (2018a) Maize production under risk: the concurrent adoption of off-farm income diversification and agricultural credit to manage risk. J Integr Agric 18(2):460-470

Akhtar S, Li G, Ullah R, Nazir A, Iqbal MA, Raza MH, Iqbal N, Faisal M (2018b) Factors influencing hybrid maize farmers' risk attitudes and their perceptions in Punjab Province, Pakistan. J Integr Agric 17:1454–1462

Ali M (2022) Diagnosis of the climate smart agriculture and water management (CSAWM) innovation system in Pakistan

Anyango-van Zwieten N, Lamers M, van der Duim R (2019) Funding for nature conservation: a study of public finance networks at World Wide Fund for Nature (WWF). Biodivers Conserv 28:3749–3766

Aslam M (2016) Agricultural productivity current scenario, constraints and future prospects in Pakistan. Sarhad J Agric 32:289–303

Basak R, Karlsson-Vinkhuyzen S (2022) The Green Climate Fund: history, status and legitimacy. In: Handbook of international climate finance. Edward Elgar Publishing

Bisen A (n.d.) An overview on production constraints, strategies, and innovative production technologies of fruit crops

Bokelmann W, Rivera M, González-Puetate IR, Kini J (2024) Open access edited by. In: Transformative food value chains for local development, p 56

Canton H (2021) Food and agriculture organization of the United Nations—FAO. In: The Europa directory of international organizations 2021. Routledge

Clune T (2021) Conceptualising policy for sustainable agriculture development. Aust J Public Admin 80:493–509

Faisal M, Chunping X, Akhtar S, Raza MH, Khan MTI, Ajmal MA (2020) Modeling smallholder livestock herders' intentions to adopt climate-smart practices: an extended theory of planned behavior. Environ Sci Pollut Res 27:39105–39122

FAO and FAOSTAT (2018) Food and agriculture organization of the United Nations. Rome. http://faostat.fao.org. Accessed 15 May 2020

GoP (2023) Economic survey of Pakistan. Economic Advisor's Wing, Finance Division, Islamabad, Pakistan

Goswami M, Gupta AK, Kishan R, Baidya S, Khan YDI, Prakash S, Premkumar A, Nautiyal S (2023) An evaluation of climate-resilient agricultural practices in India: a narrative synthesis of literature. Environ Sustain 6:7–23

Halisçelik E, Soytas MA (2019) Sustainable development from Millennium 2015 to Sustainable Development Goals 2030. Sustain Dev 27:545–572

Imtiaz A (2023) Perceptions of climate change among mango growers in Bangladesh

Jarman A, Thompson J, McGuire E, Reid M, Rubsam S, Becker K, Mitcham E (2023) Postharvest technologies for small-scale farmers in low-and middle-income countries: a call to action. Postharvest Biol Technol 206:112491

Jomo KS, Chowdhury A (2019) World Bank financializing development. Dev 62:147–153

Lee MG (2019) Food classification by the Codex Alimentarius Commission: cereal grain, nuts and seeds, herbs and spices. J Food Hyg Saf 34:212–218

Mensah H, Ahadzie DK, Takyi SA, Amponsah O (2021) Climate change resilience: lessons from local climate-smart agricultural practices in Ghana. Energy Ecol Environ 6:271–284

Mitton RV, Michel M, Ortega L (2021) National mango board's tools for the mango industry

Mumtaz M (2018) The national climate change policy of pakistan: an evaluation of its impact on institutional change. Earth Syst Environ 2:525–535

Paull J (2008) Green food in China. Elementals J Bio-Dyn Tas, 48–53

PARC (2021) Pakistan Agricultural Research Center, Government of Pakistan, Islamabad, Pakistan

Rajan R, Feza M, Pandey K, Aman A, Kumar V (2020) Climate change and resilience in fruit crops. In: Climate change and its effects on agriculture, 337–354

Rhiney K, Ajayi OC (2018) Economic impacts of climate change on priority value chains in the Caribbean: implications for private-sector investment and scaling-up of climate-smart agriculture in the region. CTA

Sen D, Majumder S (2011) Fair trade and fair trade certification of food and agricultural commodities: promises, pitfalls, and possibilities. Environ Soc 2:29–47

Sivakumar MVK (2011) Agricultural drought—WMO perspectives. World Meteorological Organization, Geneva, Switzerland: 24

Thapa R (2022) The preparedness of schools in responding to emerging climate change impacts and risks in Bhutan. RMIT University

Turvey CG (2017) IFAD research series 10—inclusive finance and inclusive rural transformation. IFAD Research Series

Wangu J, Mangnus E, van Westen ACM (2020) Limitations of inclusive agribusiness in contributing to food and nutrition security in a smallholder community: a case of mango initiative in Makueni County, Kenya. Sustainability 12:5521

Wardhan H, Das S, Gulati A (2022) Banana and mango value chains. In: Agricultural value chains in India: ensuring competitiveness, inclusiveness, sustainability, scalability, and improved finance, pp 99–143

World Bank (2011) Vulnerability, risk reduction, and adaptation to climate change: climate change country profile for Pakistan. Global Facility for Disaster Reduction and Recovery

Chapter 15
Significance of Pre- and Post-harvest Fruit Bagging in Mango Cultivation Under Climate Change Scenarios

Ghulam Mustafa, Uzman Khalil, Ahsan Amin, Fahad Hakim, and Aimal Waheed

Abstract Pre- and post-harvest fruit bagging has emerged as a vital strategy in mango cultivation to address challenges posed by climate change. This technique provides a protective barrier against pests, diseases, and environmental stressors such as extreme temperatures, irregular rainfall, and sunburn. Pre-harvest bagging enhances fruit quality by improving skin coloration, reducing blemishes, and minimizing pesticide residues, thereby increasing market value. Post-harvest bagging, on the other hand, maintains fruit quality during transportation and storage by reducing moisture loss, mechanical damage, and contamination. These interventions contribute to sustainable mango production, reducing reliance on chemical inputs while ensuring high-quality yields. The method aligns well with eco-friendly agricultural practices and meets the growing demand for clean and safe produce. Given the increasing impacts of climate variability, integrating bagging into mango farming systems can play a pivotal role in ensuring resilience, profitability, and consumer acceptance.

G. Mustafa (✉)
Department of Horticulture, MNS-University of Agriculture, Multan, Pakistan
e-mail: gmustafabzu@gmail.com

Department of Agriculture Extension, Government of the Punjab, Punjab, Pakistan

U. Khalil
Horticultural Sciences Department, University of Florida, Gainesville, USA

A. Amin
Department of Plant Pathology, Bahauddin Zakariya University, Multan, Pakistan

F. Hakim
Institute of Horticulture, Lithuanian Research Centre for Agriculture and Forestry, Babtai, Kaunas Reg., Lithuania

A. Waheed
Biochemistry Research Group, Faculty of Natural Sciences, Vytautas Magnus University, Kaunas, Lithuania

S. A. H. Naqvi et al. (eds.), *Climate Change and Mango Production*,
https://doi.org/10.1007/978-3-031-87035-4_15

Keywords Mango cultivation · Fruit bagging · Climate resilience · Sustainable agriculture

15.1 Introduction

When human actions influence the atmospheric composition in ways that go beyond the natural fluctuation seen over similar timescales, we say that the climate is changing. Heatwaves, cold snaps, frost, droughts, and floods, among other severe weather occurrences, are likely to become more common as a result of this. Plants are very vulnerable to climate change since these changes affect several processes, such as vegetative growth, blooming, fruit development, and fruit quality. Two important climate change effects, inconsistent rainfall and unexpected high temperatures, are projected to reduce agricultural yield (Ahmed and Ahmad, 2019; Ahmed, 2020, 2023). For example, in crops such as litchi and pomegranate, dryness has been associated with less fruit set and more fruit shattering. Climate change (CC) scenario is already acknowledged as one of those environmental challenges that threaten every facet of life including mango farming.

Elevated temperatures directly affect photosynthesis, leading to changes in sugar levels, organic acids, flavonoid content, fruit firmness, and antioxidant activity. Additionally, rising atmospheric carbon dioxide levels have been shown to impair post-harvest quality, as seen in potatoes, where sugar content decreases and common scab incidence rises. Similarly, changes in temperature and rainfall patterns have disrupted banana cultivation in certain regions. In many fruit crops, high temperatures and moisture stress during flowering reduce pollen and ovule viability, ultimately affecting fruit set and yield. As an example, as temperatures rise, papaya blooms become sterile in stigma and stamen, which causes both male and female flowers to remove their petals and undergo a shift in sex. Mango, guava, and litchi frequently experience flower drops when exposed to low temperatures during flowering (Fig. 15.1).

Research has demonstrated that changing climates significantly impact pollinator populations, such as bees, moths, and butterflies, reducing their abundance, geographic range, and pollination activity. Physiological abnormalities, such as black tip and spongy tissue in mango, fruit breaking and granulation in citrus, are more common when fruits are grown and developed at higher temperatures. In addition, pest dynamics are changing; for example, fruit flies are increasingly attacking citrus, guava, and carambola crops when the weather is hot and humid. These conditions also favor the proliferation of pests like beetles and bugs, as well as diseases like mildew and blight.

To address these challenges, developing crop varieties with higher yield potential and resistance to multiple stresses, such as drought, flooding, and salinity, is crucial. Enhancing the genetic diversity of important tropical fruits is another key strategy. Additionally, climate change is altering the natural vegetation and ecological balance, including changes in the timing of plant growth activities, known as phenology. These shifts lead to an altered balance between vegetative and reproductive phases,

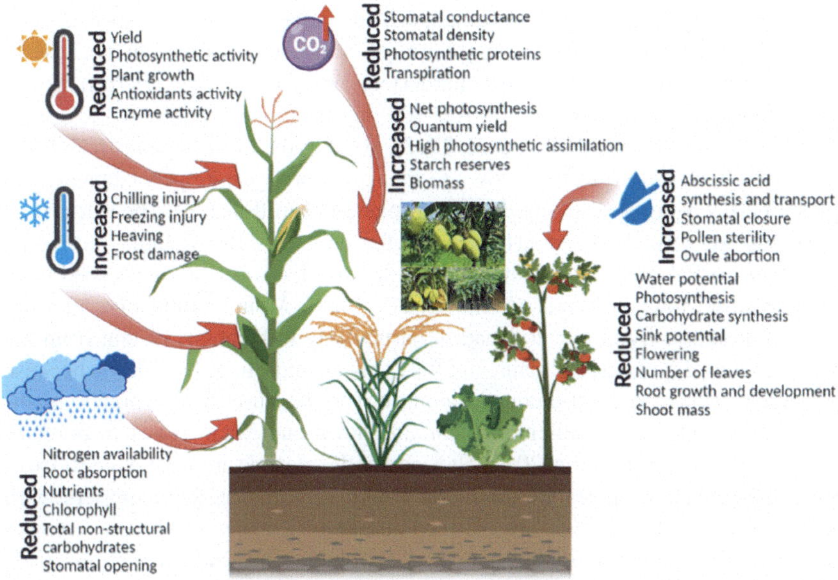

Fig. 15.1 Impact of climate change on crops (modified from Subedi et al. 2023)

with higher temperatures generally shortening the vegetative phase and advancing flowering in many fruit crops.

The exposure of fruits to direct sunlight at elevated temperatures (e.g., 35 °C) results in them being significantly firmer than shaded fruits at lower temperatures (e.g., 20 °C). Higher temperatures during fruit growth also reduce cell wall enzyme activity, delaying ripening and altering fruit chemical composition due to variable precipitation and moisture stress patterns. Temperature increases of 0.7–1.0 °C could shift the areas currently suitable for cultivating premium mango varieties like Dashehari and Alphonso. Similarly, a rise of just 0.2 °C could drastically reduce areas where guava develops its characteristic red color. Changes in flowering times in temperate regions could lead to ecological disruptions, such as the emergence of new insect pests, with tropical insects being particularly vulnerable to warming compared to their temperate counterparts.

Despite the rising levels of atmospheric CO_2, food production remains uncertain due to global warming and altered precipitation patterns. Loss of plant biodiversity and reduction in suitable cultivation areas will exacerbate the challenges for global fruit production. Strategic scientific assessments of these impacts, along with robust adaptation and mitigation plans, are essential to address the looming threats and ensure sustainable fruit production in the face of climate change.

Mango is a flowering plant from the Anacardiaceae family, specifically the genus Mangifera. Many species make up the genus Mangifera; they include the indica, gedebe, minor, and mucronulate varieties (Bally, 2006). *Mangifera indica* produces the majority of commercially traded mango fruit. India is where this species originates. It was sold and grown in neighboring Southeast Asian nations because to its pleasing look, flavor, and scent (Mukherjee, 1953, 1972). *Mangifera indica* is presently produced commercially in over 103 nations throughout the world's tropical and subtropical areas (Dillon et al. 2013). Among the most popular fresh fruits eaten throughout the world right now are mangoes, oranges, grapes, bananas, and apples. In 2019, worldwide mango output surpassed 56 million tons. The majority of the mangoes grown on Earth come from only a few countries: Brazil, India, Indonesia, China, Mexico, Pakistan, and Malawi. India is the world's largest producer, producing over 25 million tons each year (Mehdi et al., 2021). Mango cultivation area has increased dramatically in recent decades, particularly in locations distant from the equator such as Egypt, Israel, Spain, and Italy (Sauco and Lu, 2018). This is because this tropical fruit is in great demand among buyers because to its aromatic qualities, stunning appearance, delicious flavor, and nutritional value as a rich source of minerals and vitamins (Hutton, 2012) . It is one of the most desirable and oldest fruits known to humans. Mangoes are the world's largest producers and exporters, with Pakistan and India leading the way. These two nations' mango production is lower than that of other mango-producing countries. Problems with pests and diseases, ineffective orchard management, a small plant population per hectare, and other factors contribute significantly to mangoes' poor yield. As a result, high density planting and management are required to cope with these adverse scenarios of poor productivity, which provides better yield at lower cost with greater resistance to climatic shocks while maintaining export quality of fruits. High density planting (HDP) is one of the mango farming technologies used across the globe to boost yield while maintaining fruit quality. It has the ability to generate 200% more product than the conventional way (Singh, 2017). Climatic problems like as unexpected temperature increases, humidity, and unseasonal rain have been the primary issues in recent years. Such an unpleasant environment not only affects the exterior look of the fruit, but it also increases the occurrence of pests and illnesses. Consequently, a number of GAP are gaining popularity globally as a means to mitigate losses brought about by both biotic and abiotic causes (Sharma, 2009). Pre-harvest fruit bagging evolved as a successful approach. It was said to have originated in Japan and Korea. Countries like Mexico, Chile, and Argentina will not import fruits unless they are bagged (Sharma et al., 2014). Bagging not only enhances the cosmetic quality of fruits, but it also improves their internal quality by boosting skin coloration and decreasing imperfections via a change in the micro-environment. Thus, an experiment was conducted to investigate the effect of fruit bagging on fruit quality.

Bagging fruits before they are picked is a popular physical protection technique that has been around for a long time and is widespread in Asia. This method has been used to lessen the impact of pests on a variety of fruits, including pitayas, guavas, lychees, pomegranates, and citrus. Mango producers in main producing zones have recently embraced fruit bagging prior to harvest. This approach efficiently protects the fruit against illnesses like anthracnose and stem rot, pest infestations like fruit flies and other insects, and exterior damage including scratches, sunburn, fruit splitting, pesticide residues, and bird damage. Furthermore, it improves the fruit's skin color and general beauty. According to research, fruit bagging increases mango interior quality by creating a micro-environment surrounding the fruit inside the bag. Furthermore, the procedure boosts marketable yield, since bagged mangoes are frequently larger and heavier than unbagged ones.

15.2 Importance of Bagging in Mango Production

Bagging is the process of enclosing fruits with bags to preserve them from different biotic and abiotic influences. According to Fan and Mattheis (1998), this method alters the microclimate inside the bag to promote healthy growth and development, and it also physically protects the fruits, which improves their internal and external quality. Mango fruits, despite their high nutritional value and wonderful flavor, are susceptible to many insects and illnesses due to poor weather circumstances during the ripening process. Mangoes are in short supply despite their great demand in both domestic and international markets, which is mostly attributable to subpar fruit harvesting practices. Mangoes are thus susceptible to a wide range of post-harvest illnesses and disorders that manifest themselves during transit and storage. The overall average mango loss in Bangladesh was reported to be up to 34% (Sarker et al. 2011). To achieve clean, pest-free mangoes, farmers are spraying harmful pesticides on their plantations anywhere from fifteen to sixty-two times (Uddin et al. 2015), which is both unsanitary and harmful to people's health. These synthetic pesticides and fungicides may create serious health concerns for users. As a result, there has been a global push to discover non-chemical ways to decrease fruit illness and disorder occurrence, with the goals of protecting workers, consumers, and the environment (Sharma et al. 2009). Pre-harvest fruit bagging has established as one of the most effective strategies in Australia, Japan, Taiwan, and China (Joyce 1997; Senghor et al., 2007). This method achieves the necessary quality by hanging fruit bunches or individual fruits from the tree for a certain amount of time and then bagging them. Mangoes (Wu et al. 2009; Nagaharshitha et al. 2014; Haldankar et al. 2015; Jakhar and Pathak 2016; Islam et al. 2017a, b), apples (Santos and Wamser 2006; Hao et al. 2011), pear (Feng et al. 2011; Hudina et al. 2012), peaches (Jia et al. 2005; Wang et al. 2010), longan (Yang et al. 2009), and loquat (Xu et al. 2010) are among the many fruits that are physically protected using this technique. Improving fruit color, interior quality, and eliminating problems like splitting, mechanical damage, and sunburn are all ways to increase the fruit's market value (Kim et al., 2010; Zhao et al., 2013; Ding et al., 2003; Amarante et al., 2002a, b; Muchui et al., 2010). It also reduces pesticide residues (Amarante et al. 2002a, b) and enhances insect, disease, and bird damage management (Sarker et al. 2009; Wang et al. 2010). As a consequence, preharvest bagging has emerged as an important strategy for increasing crop commercial value and lowering postharvest losses. It helps produce mangoes of excellent quality that are export-ready by reducing postharvest losses and enhancing fruit appearance and peel color.

One of the most difficult issues in growing high-quality fruit across the world is dealing with severe insect and disease pressure, especially in humid settings. Bagging fruit during its early growth may be an efficient way to physically protect it from pests and illness. Over the course of two years, researchers at an organic farm and a conventional farm conducted a battery of studies. In terms of fruit quality, harvestable yield, and post-harvest disease incidence, the findings showed that bagged fruits were not significantly different from non-bagged fruits. However, in 2015, the early-season cultivar exhibited a 13% increase in marketable output, while in 2016, the mid-season cultivar increased by 11%. It wasn't until 2016 that the commercial farm saw a 100% yield from both the bagged and control treatments due to improper bag application. Although bagging made the mangoes' red blush less noticeable, studies showed that color wasn't a major consideration for buyers. The potential advantages of bagging might be seen in the higher prices that people are prepared to pay for fruit cultivated with less pesticide exposure. Although further economic analysis is needed, the increased price could offset the additional labor and material costs associated with bagging. In a home gardening context, a comparison between bagged and non-bagged fruit, along with different chemical treatments, showed that bagging significantly improved yield, particularly under high disease pressure.

In recent years, various methods have been developed to enhance fruit production and reduce losses caused by biotic and abiotic factors. Methods to enhance fruit quality, look, and resistance to pests and diseases are becoming increasingly important as people become more conscious of the need to reduce pesticide usage for the sake of workers', customers', and the environment's safety (Sharma et al. 2009). According to Fan and Mattheis (1998), pre-harvest fruit bagging has shown to be a successful approach for minimizing both biotic and abiotic stressors. Bobbing for berries, fruits, or bunches of fruits while they're still on the tree is the name of the game with this method. Bagging not only improves the physical appearance and chemical quality of fruits but also protects them from external damages such as cracking, sunburn, and russeting. Amarante et al. (2002a, b), Xu et al. (2010), Joshi et al. (2016a, b), and others have shown that this method is effective for a variety of fruit crops, including improving skin color, decreasing insect and pest infestations, preventing disease outbreaks, minimizing mechanical damage, lowering agrochemical residues, protecting against bird damage, and addressing physiological disorders. Although bags were originally intended for processing and export markets, they are now being used for fruits that are going to be consumed at home. Many different types of fruits are bagged for commercial purposes. These include peaches, guavas, bananas, litchi, grapes, pomegranates, citrus, apples, and guavas.

15.3 Effect of Fruit Bagging

Fruit needs a very precise atmosphere to grow and develop properly. A variety of environmental and biotic variables influence fruit growth and development. For example, temperature fluctuations or a prolonged dry period cause fruit to fracture, reducing its look and marketability (Tables 15.1 and 15.2).

For many fungal illnesses, the optimal circumstances are low temperatures and high humidity. If you bag your fruits or berries, you're really changing the microenvironment inside the bag to make it more conducive to the fruits' growth and development. This approach reduces the prevalence of infections and physiological diseases while also increasing the fruits' postharvest quality. Furthermore, by keeping the fruit at a more consistent temperature inside the bag, bagging prevents frost damage. Bagging boosts the interior temperature by 1–2 °C in winter months and 3–6 °C in hot months (Omar et al. 2014; Santosh et al. 2017).

Table 15.1 Effect of bagging on increasing in fruit size and weight

Fruits/cultivar	Bagging date/ time	Bagging material	Effect	Reference
Mango (Langra and Pant Sinduri)	40 days after fruit set	Brown paper bag	Increase in fruit weight and volume	Joshi et al. (2016b)
Mango (Nam Dok Mai)	For 52 days	2 layer paper bags (black and brown)	Increased fruit weight	Watanawan et al. (2008)
Longan	–	Paper bags	Larger-sized fruit	Yang et al. (2009)
Date palm (Khalas & Sukari)	–	Blue bag	Increase fruit size and bunch weight	Harhash and Al-Obeed (2010)
Litchi	Two months prior to harvest	Brown and butter paper bags	Increase in fruit size and weight	Joshi et al. (2016a)
Guava	20 DAFB	White polyethylene	Increases the fruit weight and size	Meena et al. (2016)
Guava (Bari Peyara-2)	–	White polyethylene	Increases fruit weight and size	Rahman et al. (2017) Rahman et al. (2018)
Papaya	–	Polythene bags	Minimum loss in weight of fruit	Mia (2003)

DAFB days after full bloom

Table 15.2 Effect of bagging on fruit maturity

Fruit/cultivar	Bagging date or time	Bagging materials	Effects	References
Banana	–	Polyethylene bags	Enhance fruit maturity	Johns & Scott (1989)
Litchi	–	Cellophane paper (CP) bags	Delayed maturity	Debnath and Mitra (2008)
Apple (delicious)	–	White paper bag	No effect on fruit maturity	Ju (1998)
Guava	1 month before harvest	Simple news paper	Enhance fruit maturity	Singh et al. (2007)

15.4 Benefits of Fruit Bagging

- Enhances fruit quality and flavor by reducing pesticide residues.
- Improves visual attractiveness, leading to higher market values.
- Helps prevent fruit fly infestations, bird damage, and disease transmission.
- Contributes significantly to the cultivation of organic produce; the bags are biodegradable and recyclable.
- Prevents fruit from breaking and sunburn.
- An environmentally favorable method.

15.5 Steps for Bagging Fruit

- Select fruit trees that are suitable for bagging.
- Thin the fruit according to species and needs.
- Fill each bag with a single fruit or cluster of berries and fasten with thread or coconut midrib.
- Gently raise the bottom of the bag to prevent direct contact with the fruit. For adequate water drainage, make 2–3 tiny holes at the bottom of the bag (Tables 15.1 and 15.2).

15.6 Materials for Bagging

Different types of materials are used for fruit bagging based on fruit variety and location. Common bagging materials include:

- Black and brown paper bags.
- White-coated bags.
- Net bags.
- Plastic Bags

- Banana leaves and cellophane or fabric bags.
- Polyethylene bags might be black or blue.
- Transparent polypropylene micro-perforated bags.

15.7 Impact of Bagging on Fruits

15.7.1 Effect of Bagging on Physiological Factors

15.7.1.1 Fruit Size and Weight

Fruits mature following a long process of sluggish development that begins after fruit set. Fruit size and weight may be influenced by bagging them while they are still in growth. Researchers have found conflicting findings when examining how bagging affects fruit size and weight. These variances might be due to bag type, bagging time, fruit variety, climatic circumstances, and post-harvest storage. As a consequence, bagging may increase, decrease, or have no impact on fruit size and weight (Zhen et al. 2000; Wang et al. 2002; Huang et al. 2007; Chen et al. 2012).

15.7.1.2 Fruit Ripening

Bagging may speed up fruit ripening, as observed in the date palm cultivar "Helali" (Awad 2007). According to Harhash and Al-Obeed (2010), date palm cultivars "Succary" and "Khalas" ripened more quickly when placed in blue polyethylene bags, although both yellow and white bags had beneficial effects as well. Similarly, bagging the black grape variety "Perla" might delay ripening, as pointed out by Signes et al. (2007). An earlier study by Singh et al. (2007) found that ripening guavas were bagged in white plastic bags or newspapers.

15.7.1.3 Fruit Appearance

Fruits are prone to physical flaws and damage during harvesting, processing, packing, and transportation, making them less desirable to customers. People want fruits devoid of blemishes, abrasions, and sores. Pre-harvest fruit bagging helps to lessen or eliminate such mechanical damages while also enhancing fruit colors and so raising its market value (Han et al. 1999).

15.7.1.4 Fruit Color Development

The color of the fruit is a key factor that attracts consumers. An appealing color enhances the fruit's physical appearance, helping to secure higher prices in both domestic and international markets. Numerous studies have shown that pre-harvest fruit bagging can either promote or inhibit fruit coloration, depending on various factors.

15.7.1.5 Color Inhibition

The primary impact of pre-harvest fruit bagging is to limit color development, according to Ju (1998) and Amarante et al. (2002b), albeit this effect is conditional on a number of circumstances. Many factors come into play while bagging fruit, such as the stage of growth of the fruit, the time of bagging, the kind of bag used, the removal of the bag, and the local ambient conditions.

15.7.1.6 Physiological Disorders

Physiological disorders are plant abnormalities that are caused by nonpathogenic causes. These may be caused by nutritional deficit or excess, hormonal imbalance, aberrant growth conditions, and so on (Singh 2002) [93]. Several management measures have been put in place to tackle the many documented problems in various fruit crops. These disorders all affect fruit output and quality. Fruit bagging has been found in research to lower the incidence of some fruit illnesses.

15.7.1.7 Bird Damage

Birds are key pests during fruit growth and ripening, particularly in bananas, mangos, apples, and dates, resulting in large losses. Birds are managed using a variety of tactics, including thumping drums and stretching reflecting ribbons in the field, but the birds rapidly acquire acclimated to these methods (Sharma 2009) [66]. As a consequence, fruit pre-harvest bagging has helped to reduce bird damage to diverse fruit crops.

15.8 The Importance of Fruit Bagging

Fruit bagging is a cultural practice that protects mango fruits during their development. By creating a physical barrier, growers can shield the fruits from.

15.8.1 Pests

Bagging prevents insect access, reducing damage from common mango pests such as fruit flies and beetles (Tables 15.3, 15.4 and 15.5).

Table 15.3 Effect of bagging on fruit appearance

Fruit/cultivar	Bagging material	Effect	Reference
Mango	Paper bag	Reduce incidence of black spots, improve physical quality, light-green skin colour	Sarker et al. (2009)
Litchi	–	Minimum incidence of cracking and sunburn and fruits free from blemishes, superior appearance	Debnath and Mitra (2008)
Banana	–	More attractive fruits, free from skin blemish	Muchui et al. (2010)
Persimmon (Fuyu)	–	Reduces fruit blemishing	Katagiri et al. (2003)
Pear (Doyenne du Comice)	–	Reduces bird damage and skin blemishes, increases marketability	Amarante et al. (2002a)
Papaya	Polyethylene plastic-black bags	Increases the fruit appearance and good firmness	Tran et al. (2015)

Table 15.4 Effect of bagging on fruit colouration

Fruit/cultivar	Bagging material	Effect	Reference
Litchi	Semi-transparent CP bags	Excellent skin colouration on fruit	Hu et al. (2001)
Mango	Two-layer paper bag	Development of greenish-yellow skin colour	Watanawan et al. (2008)
Grape (Perla)	Cellulose bags	Increased uniformity of the fruit colouration	Signes et al. (2007)
Apple (Grany Smith)	–	Enhances development of red colour in green apple	Wang et al. (2010)

Table 15.5 Effect of fruit bagging on insect infestation

Fruit/cultivar	Bagging date or time	Bagging materials	Insect-pest control	References
Mango (Langra)	30 days prior to harvesting	Black polybag, brown paper bags	Fruit fly control (100%)	Sarker et al. (2009)
Litchi	Bagging done after one week of fruit set	Brown and cello phone paper bags, newspaper bags	Minimized the infestation of stone borer and stalk-end borer	Debnath and Mitra (2008)
Guava	42–63 days before harvesting	Biodegradable film. Waxed paper	Control on fruit fly and guava weevil	Bilck et al. (2011)
Apple (Imperial Gala)	Transparent plastic perforated bag	Fruit fly, codling moth, woolly apple aphid	Teixeira et al. (2011)	
Pomegranate (Ganesh, Mridula, Jyothi, Ruby, Jalore Seedless)	60–70 days prior to harvesting	Parchment paper bag	Minimized the infestation of pomegranate butterfly around 90%	Bagle (2011)

15.8.2 Diseases

It also minimizes the risk of fungal and bacterial infections that can thrive in humid environments (Table 15.6).

15.8.3 Environmental Stress

Protects fruits from sunburn, excessive moisture, and physical damage due to wind or rain.

15.9 Techniques for Effective Bagging

Successful fruit bagging involves several key steps.

Table 15.6 Effect of fruit bagging on the incidence of diseases in fruit crops

Fruit/cultivar	Bagging date or time	Bagging materials	Diseases occurrence	References
Mango	During fruit development	White bags	Control anthracnose	Senghor et al. (2007)
Mango (Carabao & Keitt)	60 and 100 days before harvesting respectively	Brown and white paper bags	Reduces occurrence of stem-end rot and anthracnose	Hofman et al. (1997) and Buganic et al. (1997)
Apple (Royal Delicious)	One month before harvesting	Yellow (light) coloured bags	Reduces sooty blotch and fly speck problem	Sharma et al. (2013)
Guava	30 days prior to harvesting	Newspaper bags	Reduces incidence of anthracnose and black spot disease	Martins et al. (2007)
Loquat	During fruit development	Plastic bags	Reduces fruit rot	Ko et al. (2010)
	After fruit setting	White plastic bag	Minimizes the rust	Gong et al. (2002)

15.9.1 Timing

The ideal time to bag mangoes is usually 4–6 weeks after flowering, when the fruits are approximately the size of a marble. Early bagging ensures maximum protection and quality.

15.9.2 Materials

Common materials used for bags include.

15.9.3 Paper Bags

Biodegradable and eco-friendly options, often treated to resist moisture.

15.9.4 Plastic Bags

More durable but raise concerns about environmental sustainability.

15.9.5 Net Bags

Allow for air circulation while providing some level of pest protection.

15.9.6 Bagging Method

Selectively bag the fruits, ensuring they are free from blemishes or signs of pest damage.

- Use soft ties or clips to secure the bags around the fruit stems without causing damage.
- Ensure that the bags are large enough to allow for fruit growth, but not so loose that they fall off.

15.9.7 Monitoring

Regularly inspect the bagged fruits for signs of moisture buildup, pests, or diseases. Remove any damaged bags promptly.

15.9.8 Benefits of Fruit Bagging

The advantages of fruit bagging are manifold, especially in the context of climate change:

Improved fruit quality: Bagged mangoes tend to have fewer blemishes, better color, and increased size, making them more appealing to consumers.
Reduced chemical usage: By minimizing pest and disease incidence, growers can rely less on chemical pesticides, promoting sustainable practices.
Market value: Higher quality fruits can command better prices in the market, increasing profitability for growers.
Environmental protection: By limiting pesticide use and improving fruit quality, bagging contributes to a healthier ecosystem (Table 15.7).

15.9.9 Challenges and Considerations

While fruit bagging offers many benefits, it also presents some challenges:

Labor-Intensiveness: Bagging requires careful labor, which can be a constraint for smaller farms with limited workforce availability.

Table 15.7 Effect of fruit bagging on the physiological disorders of fruit crops

Fruit/cultivar	Bagging date or time	Bagging materials	Disorders	References
Mango (apple)	40–45 days before harvesting	White bags	Reduces lenticels discolouration	Mathooko et al. (2011)
Litchi	Two months prior to harvesting	Brown paper bag	Reduces sun burning and fruit cracking	Joshi et al. (2016a)
Apple (royal delicious)	One month prior to harvesting	Light yellow colour bags	Reduces incidence of brown core, bitter pit and cork pit	Sharma et al. (2013)
Carambola	10–17 days after flowering	Plastic bags	Minimizes fruit dropping	Xu et al. (2008)
Date palm (Zaghloul)	During pollination	Transparent blue polyethylene bags	Reduces fruit cracking	Kassem et al. (2011)
Pear (Conference)	Fruit developmental stage	Plastic bags	Reduces sun-burn	Amarante et al. (2002b)
Pear (Doyenne du Comice)	One month after full flowering	Perforated polyethylene bags	Reduces fruit cracking and russeting	Amarante et al. (2002a)

Cost of Materials: The initial investment in bags and additional labor can be a barrier for some growers, particularly smallholders.

Environmental Concerns: The use of plastic bags raises issues related to waste and pollution. Transitioning to biodegradable options can alleviate some of these concerns.

15.9.10 Best Practices for Implementation

To maximize the benefits of fruit bagging, growers should consider the following best practices:

Education and Training: Providing training sessions for farmers on proper bagging techniques can enhance effectiveness and reduce errors.

Integration with Other Practices: Combine fruit bagging with other sustainable practices such as integrated pest management (IPM) and organic fertilization to create a holistic approach to mango cultivation.

Research and Development: Encourage ongoing research into new materials and methods for bagging, particularly those that reduce environmental impact while maintaining efficacy.

Monitoring and Feedback: Establish a system for growers to share experiences and results, fostering a community of practice that encourages learning and improvement.

15.9.11 The Significance of Fruit Bagging in Mango Fruit Protection

Fruit bagging is a vital agricultural technique used to protect mangoes during their growth stages. This practice forms a protective barrier around the fruit, offering several benefits that enhance both the quality and yield of the crop. The key advantages of fruit bagging include.

15.9.12 Protection from Pests

Bagging shields mangoes from common pests like fruit flies, beetles, and other insects that can cause significant damage, thereby reducing the need for chemical pesticides (Table 15.8).

15.9.13 Prevention of Diseases

By providing a barrier to external environmental factors, bagging helps reduce the risk of fungal and bacterial infections, which thrive in humid conditions that often affect mangoes during growth.

15.9.14 Mitigation of Environmental Stress

The bags protect mangoes from external environmental stressors, including sunburn, rain, and physical damage caused by wind, thus ensuring that the fruits develop without excessive stress, which can affect their quality (Table 15.8).

15.9.15 Critical Practices for Successful Bagging

To effectively protect mangoes using bagging, it's important to follow several critical practices.

Table 15.8 Effect of bagging on fruit quality parameters

Fruit	Quality attributes affected	References
Pear	Reduction in total soluble solids and increase in titratable acidity	Lin et al. (2008)
	Opposite effect on sorbitol and sucrose content	Hudima and Stamper (2011)
Apple	Increases sweetness of the fruits	Bentley and Viveros (1992)
	Reduces TSS content	Chen et al. (2012)
	Increases TSS and vitamin C content	Sharma et al. (2013)
Banana	Not adverse effect on total soluble solids, acidity and fruit firmness	Muchui et al. (2010)
	Improves finger length and finger quality and also provide protection from mechanical damage	Turner (1984)
Mango	Increases TSS, ascorbic acid, titratable acidity	Watanawan et al. (2008)
	Increases internal quality of the fruit with acidity, sugar and carotenoid contents in cultivar Zill	Hongxenia et al. (2009)
	Bagging with newspaper bag and brown paper after 30 days fruit set enhances the fruit quality like TSS, fruit retention, total sugars	Haldanker et al. (2015)
	Bagging with green polyethylene bags showed maximum TSS, sugars, ascorbic acid with minimum acidity	Joshi et al. (2016b)
Litchi	Improves TSS:acidity ratio	Debnath and Mitra (2008)
	Bagging with green polyethylene bags showed maximum TSS, sugars, ascorbic acid with minimum acidity	Joshi et al. (2016a)
Guava	Increase in TSS content	Singh et al. (2007)
Loquat	Increase in TSS with reduced titratable acidity	Liu et al. (2004)
Plum	Reduces soluble solid content	Murray et al. (2005)
Red pitaya	Bagged fruit (7 days after anthesis) shows positive effect on fruit quality	Tuan and Yen (2017)
	Increases TSS, peel thickness and acidity	Costa et al. (2017)

15.9.16 Proper Timing

Bagging should be done 4–6 weeks after flowering, when mangoes are about the size of a marble. Early bagging provides maximum protection and enhances fruit quality.

15.9.17 Choosing the Right Materials

Common materials used for bagging mangoes include:

Paper Bags: Eco-friendly and biodegradable, they often come treated to resist moisture, making them suitable for tropical climates.
Plastic Bags: While durable and effective, plastic bags pose concerns regarding environmental sustainability.
Net Bags: These allow air circulation while offering some protection against pests.

15.9.18 Bagging Method

Carefully select healthy, blemish-free fruits for bagging. Secure the bags with soft ties or clips around the fruit stems to avoid any damage. Ensure the bags are large enough for the fruit to grow but not so loose that they can fall off (Fig. 15.2).

Fig. 15.2 Fruit bagging performed in the orchard to protect mango from various biotic and abiotic factors

15.9.19 Monitoring the Bagged Fruits

It's essential to inspect bagged mangoes regularly for moisture buildup, pests, or any signs of disease. Promptly remove any damaged bags to prevent contamination.

15.9.20 Benefits of Fruit Bagging in Mango Cultivation

The practice of fruit bagging offers multiple benefits that contribute to sustainable and high-quality mango production:

Enhanced Fruit Quality: Bagging results in mangoes with fewer blemishes, improved coloration, and larger sizes, making them more attractive to consumers and increasing marketability.

Reduction in Chemical Pesticide Usage: By preventing pest and disease issues, fruit bagging reduces the need for chemical interventions, supporting more sustainable and environmentally friendly farming practices.

Increased Market Value: Bagging helps produce higher-quality fruits that are in demand, thus enabling growers to fetch better prices, improving their profitability.

Environmental Benefits: The reduced reliance on chemical pesticides, combined with higher fruit quality, leads to healthier ecosystems and a more sustainable agricultural approach (Tables 15.9 and 15.10).

Table 15.9 Effect of bagging on phenolic compound content and anti-oxidant activities

Fruit/cultivar	Effect of bagging in fruits	Reference
Apple (delicious)	Phenolic compound concentration increases by bagging till 60 days then it declined	Ju (1998)
Grape fruit	Increased the concentration of antioxidants	Son and Lee (2008)
Pear (conference)	Increases phenolic compound contents like caffeic acid and epicatechin in the peel	Hudima and Stamper (2011)
Peach (Wanmi)	Bagging did not affect chlorogenic acid and catechol concentrations in fruit skin or flesh	Wang et al. (2010)
Sweet orange	Increases the chemical quality, phenolic compound and antioxidant activity	Xie and Jheng (2013)

Table 15.10 Effect of bag types on appearance, insect-pest attack, disorders and quality of fruits

Fruits	Bagging materials	Best recommendation	Positive influences	References
Peach	Black and White bags	White bags	Improves pulp colour	Takada et al. (2006)
Litchi	Cellophane paper bag, craft and newspaper bags	Fabric and Cellophane bags	Improves fruit colour	Hu et al. (2001)
	Brown and butter paper bags, green polyethylene bags	Brown and butter paper bags	Improves fruit colour and internal quality of the fruits	Joshi et al. (2016a)
Mango	Brown paper and black poly bags	Brown paper bags	Reduces incidence of fruit fly, high TSS and physical quality of fruits	Sarker et al. (2009)
	Newspaper bags, black and brown paper bags	Brown paper bags	Improves skin colour	Ding and Syakirah (2010)
Mango (Kesar)	Newspaper, white paper and brown paper bags	Newspaper and brown paper bags	Enhances peel colouration, fruit length, fruit and pulp weight	Kireeti et al. (2018)
Mango (Alphanso)	Different types of bags	Plastic bags	Enhances the sensory quality and chemical content of fruits	Tendulkar et al. (2018)
Guava	Nylon fabric, Waxed paper and paper bags	Nylon bags	Complete control on fruit fly	Morera-Montoya et al. (2010)
Carambola	Plastic bags, newspaper bags, non-woven cloth bags	Plastic bags	Increases fruit size and TSS	Xu et al. (2008)
Date palm	Black, blue polyethylene bags and white paper bags	Blue and black colour bags	Increases respiration rate	Awad (2007)
Banana	Different coloured bags	Plastic bags	Increases fruit size and enhances fruit maturity	Stover and Simmonds (1987)
Apple	Different coloured bags	Light yellow coloured bags	Improves colour, fruit firmness and reduces storage disorders	Sharma et al. (2013)
		Paper bags	Better calcium absorption by fruit	Dong et al. (2007)

15.9.21 Challenges and Considerations in Fruit Bagging

While fruit bagging offers several advantages, there are some challenges that growers must consider:

Labor-Intensive Process: Bagging requires careful manual labor, which can be a limitation for smaller farms or regions with labor shortages.
Material Costs: The cost of purchasing bags and employing additional labor for bagging can be an obstacle for some growers, especially small-scale producers.
Environmental Impact of Plastic Bags: While plastic bags are effective, they contribute to waste and pollution. Transitioning to biodegradable alternatives, such as paper bags or other eco-friendly materials, can help mitigate these concerns.

15.9.22 Solution for Successful Fruit Bagging Implementation

To maximize the effectiveness of fruit bagging and ensure its long-term sustainability, growers should consider the following best practices:

Farmer Education and Training: Providing proper training on the correct techniques for bagging mangoes will improve the success rate of the practice and reduce common errors.
Integration with Other Sustainable Practices: Combining fruit bagging with other agricultural practices such as integrated pest management (IPM) and organic fertilization can provide a holistic approach that further improves fruit quality and sustainability.
Innovation and Research in Materials: Encouraging ongoing research into alternative, environmentally friendly materials for bagging, such as biodegradable polymers or plant-based materials, will help reduce the environmental footprint of this practice while maintaining its effectiveness.
Monitoring and Feedback Mechanisms: Setting up systems where growers can share their experiences and results will foster a collaborative environment that encourages learning and continuous improvement in bagging techniques.

15.9.23 Future Prospectus

Research on preharvest bagging to improve fruit quality has mainly focused on evaluating fruit quality at harvest, with limited studies on its effects on postharvest quality and storage potential. Therefore, further studies should explore the influence of preharvest bagging on postharvest fruit quality. This review emphasizes that the outcome of preharvest bagging is influenced by several variables, such as the type of bag used, the stage of fruit development when bagged, the duration of bagging, and the exposure to sunlight after removing the bags.

Plastic bags are often recommended for preharvest bagging due to their ability to improve fruit quality and their reusability in various weather conditions. However, they pose disposal challenges as they do not degrade easily, leading to potential environmental issues. While plastic recycling exists, mechanical recycling consumes significant energy compared to producing new items from virgin materials. Moreover, plastic residues can persist in soil for years, hindering plant growth and posing risks to animals, especially when the land is used for grazing.

Paper bags, while also effective at enhancing fruit quality, are susceptible to damage from environmental factors such as wind and rain, which can compromise their protective function and allow pests to reach the fruit. Despite these challenges, preharvest bagging has become a standard practice in the commercial production of fruits like fruits like loquats, peaches, pears, and apples in nations like Japan, China, and the United States. To support the broader adoption of this practice, further cost–benefit analyses are needed.

The development of biodegradable bags is critical for addressing the disposal issues associated with plastic bags. Polyolefin-based biodegradable bags naturally decompose over time with the help of microorganisms, and their degradation can be accelerated with prodegradants like cobalt or manganese stearate, along with exposure to light, heat, and oxygen. Additionally, plant-based biodegradable materials, such as those derived from cassava or corn starch, are being developed to reduce reliance on petrochemicals and lower environmental impact. These starch-based materials decompose within a few months to years, depending on the type of material.

Incorporating agricultural fibers into biodegradable plastics provides added benefits, including low density, recyclability, and biodegradability, while also offering cost-effectiveness. Although research on biodegradable polymers has advanced in other areas, Few studies have investigated their potential for use in fruit bagging prior to harvest. Hence, to determine if biodegradable plastics are practical for this use, more studies are required.

15.10 Conclusion

Fruit bagging has proven to be a valuable practice in mango cultivation, offering numerous benefits such as enhanced fruit quality, reduced pesticide use, and increased profitability. However, challenges such as labor intensity, material costs, and environmental concerns must be addressed. By adopting best practices, investing in research, and utilizing sustainable materials, fruit bagging can continue to play a vital role in ensuring the production of high-quality mangoes while supporting environmental and economic sustainability.

References

Ahmed M (2023) The science of climate change. In: Brinkmann R (ed) The Palgrave handbook of global sustainability. Springer, Cham, pp 195–222

Ahmed M, Ahmad S (2019) Carbon dioxide enrichment and crop productivity. In: Hasanuzzaman M (ed) Agronomic crops, vol 2. management practices. Springer, Singapore, pp 31–46

Ahmed M (2020) Introduction to Modern Climate Change. Andrew E. Dessler: Cambridge University Press, 2011, 252 pp, ISBN-10: 0521173159. Sci Total Environ 734:139397

Amarante C, Banks NH, Max S (2002a) Preharvest bagging improves packout and fruit quality of pears (*Pyrus communis*). New Zeal J Crop Hortic Sci 30(2):93–98

Amarante C, Banks NH, Max S (2002b) Effect of pre-harvest bagging on fruit quality and post-harvest physiology of pears (*Pyrus communis*). New Zeal J Crop Hortic Sci 30:99–107

Awad MA (2007) Increasing the rate of ripening of date palm fruit (*Phoenix dactylifera* L.) cv. 'Helali' by pre-harvest and postharvest treatments. Postharvest Biol Technol 43:121–127

Bagle BG (2011) Studies on varietal reaction, extent of damage and management of anar butterfly, *Deudorix isocrates* in pomegranate. Acta Hortic 890:557–559

Bally IS (2006) Species profiles for Pacific Island agroforestry. In: *Mangifera indica* (Mango). Permanent Agriculture Resources, Hōlualoa, Hawaii, pp 1–25

Bentley WJ, Viveros M (1992) Brown-bagging 'Granny Smith' apples on trees stops codling moth damage. Calif Agric 46:30–32

Bilck PA, Roberto SR, Maria VEG, Yamashita F (2011) Efficacy of some biodegradable films as pre-harvest covering material for guava. Sci Hortic 130:341–343

Buganic Jr RD, Lizada MCC, De-Ramos MB (1997) Disease control in Philippine 'Carabao' mango with pre-harvest bagging and post-harvest hot water treatment. Acta Hortic 455:797–804

Chen CS, Zhang D, Wang YQ, Li PM, Ma FW (2012) Effects of fruit bagging on the contents of phenolic compounds in the skin and flesh of 'Golden Delicious', 'Red Delicious', and 'Royal Gala' apples. Sci Hortic 142:68–73

Costa AC, Ramos JD, Menezes TPD, Laredo RR, Durate MH (2017) Quality of red pitaya fruits submitted to field bagging. Rev Bras Frutic 39:362–377

Debnath S, Mitra SK (2008) Panicle bagging for maturity regulation, quality improvement and fruit borer management in litchi (Litchi chinensis). Acta Hortic 773:201–208

Dillon NL, Bally ISE, Wright CL, Hucks L, Innes DJ, Dietzgen RG (2013) Genetic diversity of the Australian National Mango Genebank. Sci Hortic 150:213–226

Ding P, Syakirah MN (2010) Influence of fruit bagging on postharvest quality of 'Harumanis' mango (*Mangifera indica* L.). Acta Hortic 877:169–174

Ding Q, Han MY, Tian YM (2003) Effect of bagging on nectarine fruit quality and fruit cracking. J Northwest Sci-Technol Univ Agric Nat Sci Ed 31(2):59–61

Dong ZF, Wang YZ, Wang L, Liu CL, Dong XY, Liu GS, Yuan YB (2007) Effects of different bag treatments on the absorption of calcium in 'Red Fuji' apple fruit. Acta Hortic Sinica 34:835–840

Fan X, Mattheis JP (1998) Bagging 'Fuji' apples during fruit development affects colour development and storage quality. HortScience 33:1235–1238

Feng S, Chen X, Zhang Y, Wang Y, Song Y, Chen XL, Liu M (2011) Differential expression of proteins in red pear following fruit bagging treatment. Protein J 30:194–200

Gong JQ, Guan YL, Wang YB, Yu Y, Xie YY (2002) Effect of bagging on fruit quality of loquat. South China Fruits 31(2):30–31

Haldankar PM, Parulekar YR, Kireeti A, Kad MS, Shinde SM, Lawande KE (2015) Studies on influence of bagging of fruits at marble stage on quality of mango cv. Alphonso. J Plant Stud 4(2):12

Han J, Lee H, Jang H (1999) Comparison on skin characteristics between non-bagged and bagged 'Hosui' pear (*Pyrus pyrifolia* Nakai) fruits. J Korean Soc Hortic Sci 40:439–442

Hao Y, Ren H, Guo P (2011) Effects of bagging on the accumulation and transformation of photosynthates in apple fruits. Acta Hortic Sinica 38(2):233–239

Harshash MM, Al-Obeed RS (2010) Effect of bunch bagging color on yield and fruit quality of date palm. Am-Eurasian J Agric Environ Sci 7(3):312–319

Hofman PJ, Smith LG, Joyce DC, Johnson GI, Meiburg GF (1997) Bagging of mango (*Mangifera indica* cv. Keitt) fruit influences fruit quality and mineral composition. Postharvest Biol Technol 12(1):83–91

Hongxenia WU, Song Biao WS, Sheng M, Weihong Z, Gang YI, Lin RUZ (2009) Effect of bagging on fruit quality of Zill mango. J Fruit Sci 26(5):644–648

Hu G, Chen D, Li P, Ouyang R, Gao FW, Ang W, Dong J (2001) Effects of bagging on fruit colouration and phenylalanine ammonia lyase and polyphenol oxidase in 'Feizixiao' litchi. Acta Hort 558:273–278

Huang CH, Chai ML, Pan ZM, Yu B, Jiang ZM, Hu JL, Teng YW (2007) Effects of bagging on fruit skin features and quality of 'Cuiguan' pear cultivar. J Fruit Sci 24:747–751

Hudima M, Stamper F (2011) Bagging of 'Concorde' pears (*Pyrus communis* L.) influences fruit quality. Acta Hort 909:625–630

Hudina M, Stampar F, Orazem P, Petkovsek MM, Veberic R (2012) Phenolic compounds profile, carbohydrates and external fruit quality of the 'Concorde' pear (*Pyrus communis* L.) after bagging. Can J Plant Sci 92(1):67–75

Hutton W (2012) Handy pocket guide to tropical fruits. Tuttle Publishing

Islam MT, Rahman MS, Shamsuzzoha M, Chowdhury AKMMB, Alom R (2017) Influence of pre-harvest bagging on fruit quality of mango (*Mangifera indica* L.) cv. Mishribhog. Int J Biosci 11(3):59–68

Islam MT, Shamsuzzoha M, Rahman MS, Haque MM, Alom R (2017) Influence of pre-harvest bagging on fruit quality of mango (*Mangifera indica* L.) cv. Mollika. J Biosci Agric Res 15(01):1246–1254

Jakhar MS, Pathak S (2016) Effect of preharvest nutrients application and bagging on quality and shelf life of mango (*Mangifera indica* L.) fruits cv. Amrapali. J Agric Sci Technol 18:717–729

Jia HJ, Araki A, Okamoto G (2005) Influence of fruit bagging on aroma volatiles and skin coloration of 'Hakuho' peach (*Prunus persica* Batsch). Postharvest Biol Technol 35(1):61–68

Johns GG, Scott KJ (1989) Delayed harvesting of bananas with 'sealed' covers on bunches. 2. Effect on fruit yield and quality. Aust J Exp Agric 29:727–733

Joshi KK, Singh VP, Saxena D, Mishra DS, Kumar R (2016b) Effect of pre-harvest bagging on fruit quality of litchi (Litchi Chinensis Sonn) Cv. Rose Scented. Ann Hort 9(1):41–44

Joshi KK, Singh VP, Saxena D, Dhami V (2016b) Effect of pre-harvest fruit bagging on quality characteristics of mango cv. Langra and Pant Sinduri. Prog Res Int J 11(3):416–419

Ju Z (1998) Fruit bagging, a useful method for studying anthocyanin synthesis and gene expression in apples. Sci Hortic 77:155–164

Kassem HA, Omar AKH, Ahmed MA (2011) Response of 'Zaghloul' date palm productivity, ripening and quality to different polyethylene bagging treatments. Am-Eurasian J Agric Environ Sci 11:616–621

Katagiri T, Satoh Y, Fukuda T, Kataoka I (2003) Improving marketability of 'Fuyu' persimmon fruit by bagging culture. Acta Hort 601:213–217

Kim Y, Kang S, Cho K, Jeong S (2010) Effects of bagging with different pear paper bags on the color of fruit skin and qualities in 'Manpungbae.' Korean J Hortic Sci Technol 28(1):36–40

Kireeti A, Haldankar PM, Parulekar YR (2018) Studies on effect of type of bag on mango fruit (cv. kesar) at egg stage. Int J Chem Stud 6(6):01–04

Ko Y, Liu CW, Chen SS, Chen CY, Yao KS, Maruthasalam S, Lin CH (2010) First report of fruit rot of loquat caused by an *Alternaria* sp. in Taiwan. Plant Dis 94:481–484

Lin J, Chang Y, Yan Z, Li X (2008) Effects of bagging on the quality of pear fruit and pesticide residues. Acta Hort 772:315–318

Liu YJ, Xu JH, Zhang ZH, Jiang JM, Yu D (2004) Effects of different paper bags on fruit quality of loquat. Acta Agric Univ Jiangxiensis 26:334–337

Martins MC, Lilian A, Lourenço SA, Gutierrez AS, Watanabe HS (2007) Incidence of post-harvest damage in guavas at the wholesale market of São Paulo and its relationship to pre-harvest bagging. Braz Mag Fruit Cult 29:245–248

Meena KR, Maji S, Kumar S, Parihar D, Meena D (2016) Effect of bagging on fruit quality of guava. Int J Bio-Res Stress Manag 7:330–333

Mehdi M, Ahmad B, Penha T, Belik W, Apolinário V, Karatepe I (2021) Economic and social upgrading in global agro-food VCs: Mango in Brazil and Pakistan. In The Phantom of Upgrading in Agricultural Supply Chains (pp. 174–226). Rainer Hampp Verlag

Mia B (2003) Studies on post-harvest behaviour pf papaya. Msc thesis, Department of Horticulture, BAU, Mymen Singh, pp 40–96

Morera-Montoya R, Blanco-Metzler H, Luis-Loria CR (2010) Evaluation of different bagging materials for the control of the fruit fly *Anastrepha* sp. (Diptera, Tephritidae) and fruit pathogens in Taiwanese guava fruits (*Psidium guajava* L.). Acta Horticulturae 849:283–292

Muchui MN, Mathooko FM, Njoroge CK, Kahangi EM, Onyango CA, Kimani EM (2010) Effect of perforated blue polyethylene bunch covers on selected postharvest quality parameters of tissue-cultured bananas (*Musa* spp.) cv. Williams in Central Kenya. J Stored Prod Postharvest Res 1(3):29–41

Mukherjee SK (1953) The mango: its botany, cultivation, uses and future improvement, especially as observed in India. Econ Bot 7:130–162

Mukherjee SK (1972) Origin of mango (*Mangifera indica*). Econ Bot 26:260–264

Murray XJ, Holcroft DM, Cook NC, Wand SJE (2005) Postharvest quality of 'Laetitia' and 'Songold' (*Prunus salicina* L.) plums as affected by pre-harvest shading treatments. Postharvest Biol Technol 37:81–92

Nagaharshitha D, Khopkar RR, Haldankar PM, Haldavanekar PC, Parulekar YR (2014) Effect of bagging on chemical properties of mango (*Mangifera indica* L.) cv. Alphonso. Agrotechnology 3:124

Omar AEDK, Al-Saif AM, Ahmad MAEA (2014) Bagging of bunches with different materials influences yield and quality of Rothana date palm fruit. J Food Agric Environ 12:520–522

Rahman H, Akter A, Rahman J, Riad MI, Rahman MM (2017) Effect of fruit thinning and bagging on the yield and quality of guava. J Agric Sci Technol 6(1):2349–3682

Rahman MM, Hossain MM, Rahim M, Rube MHK, Islam MZ (2018) Effect of pre-harvest bagging on post-harvest quality of guava cv. Swarupkathi. Fundam Appl Agric 3(1):363–371

Santos JPD, Wamser AF (2006) Fruit bagging effect on the damage caused by biotic and abiotic factors in an organic apple orchard. Rev Bras Frutic 28:168–171

Santosh DT, Tiwari KN, Reddy RG (2017) Banana bunch cover for quality banana production—a review. Int J Curr Microbiol Appl Sci 6:1275–1291

Sarkar KM, Alam MM, Rahman A, Bhuiyan MGK (2011) Post harvest losses in mango value chain. Int J Biol Res 10:25–31

Sarker D, Rahman MM, Barman JC (2009) Efficacy of different bagging materials for the control of mango fruit fly, Bangladesh. J Agric Res 34(1):165–168

Sauco VG, Lu P (2018) Mango cultivation practices for the subtropics Víctor Galán Saúco, Instituto Canario de Investigaciones Agrarias, Spain. In: Achieving sustainable cultivation of mangoes (pp. 185–204). Burleigh Dodds Science Publishing

Senghor AL, Liang WJ, Ho WC (2007) Integrated control of *Colletotrichum gloeosporioides* on mango fruit in Taiwan by the combination of *Bacillus subtilis* and fruit bagging. Biocontrol Sci Technol 17:865–870

Sharma RR (2009) Fruit production: problems and solutions. Int Book Distributing Co, Lucknow, India, p 649

Sharma RR, Singh D, Singh R (2009a) Biological control of postharvest diseases of fruits and vegetables by microbial antagonists: a review. Biol Control 50(3):205–221

Sharma RR, Pal RK, Asrey R, Sagar VR, Dhiman MR, Rana MR (2013a) Pre-harvest fruit bagging influences fruit colour and quality of apple cv. delicious. Agric Sci 4:443–448

Sharma RR, Reddy SVR, Jhalegar MJ (2014a) Pre-harvest fruit bagging: a useful approach for plant protection and improved post-harvest fruit quality—a review. J Hortic Sci Biotechnol 89(2):101–113

Sharma RR, Singh D, Pal RK (2013b) Synergistic influence of pre-harvest calcium sprays and postharvest hot water treatment on fruit firmness, decay, bitter pit incidence and postharvest quality of 'Royal Delicious' apples (*Malus domestica* Borkh.). Am J Plant Sci 4:153–159

Signes AJ, Burlo F, Martinez-Sanchez F, Carbonellbarrachina AA (2007) Effects of pre-harvest bagging on quality of black table grapes. World J Agric Sci 3:32–38

Singh J (2002) Basic horticulture. Kalyani Publishers, Hyderabad

Singh H, Verghese A, Stonehouse J, Mumford J, George S, Naik G, Pandey V (2007) Developing bait and lure based integrated pest management module for mango fruit fly (*Bactrocera dorsalis*) management in Orrisa. Indian J Agric Sci 78(7):609–613

Son IC, Lee CH (2008) The effects of bags with different light transmittance on the berry cracking of grape "Kyoho." Int J Agric Environ Biotechnol 49(2):98–103

Stover RH, Simmonds NW (1987) Bananas, 3rd edn. Tropical Agriculture Series, Longman Scientific and Technical, Harlow, UK, p 468

Subedi B, Poudel A, Aryal S (2023) The impact of climate change on insect pest biology and ecology: implications for pest management strategies, crop production, and food security. J Agric Food Res 14:100733

Takada D, Fokuda F, Kubota N (2006) Effects of fruit position in canopy, harvest date and bagging on occurrence of reddish-pulp fruit in peach. Hort Res (Jpn) 5:33–37

Teixeira R, Boff MI, Amarante CV, Steffens CA, Boff P (2011) Effects of fruit bagging on pests and diseases control and on quality and maturity of 'Fuji Suprema' apples. Bragantia 70:688–695

Tendulkar SS, Haldankar PM, Bhuwad AV, Pawasker S, Parulekar YR, Salvi BR (2018) Effect of type bags on chemical property and sensory parameter of mango fruit cv. Alphanso. Int J Chem Stud 6(5):1931–1934

Tran DH, Yen CR, Chen YH (2015) Effect of bagging on fruit characteristics and physical fruit protection in red pitaya (*Hylocerus* spp.). Biol Agric Hortic 31(3):158–166

Tuan NM, Yen CR (2017) Effect of bagging time on fruit yield and quality of red pitaya. Int J Plant Sci 19(6):1–7

Turner DW (1984) Bunch covers for banana. Formers research horticulturist tropical fruit research station Alstonville Division of plant Industries. ORDER, No. H6.3.4

Wang SM, Gao HJ, Zhang XB (2002) Effects of bagging on pigment, sugar and acid development in 'Red Fuji' apple fruit. Acta Hort Sinica 29:263–265

Wang X, Hang B, Liu C (2010) Distribution of calcium in bagged apple fruit and relationship between anti-oxidant enzyme activity and bitter pit. Agric Sci Technol 11:82–85

Watanawan A, Watanawan C, Jarunate J (2008) Bagging 'Nam Dok Mai #4' mango during development affects colour and fruit quality. Acta Hort 787:325–328

Xie RJ, Jheng L (2013) Effect of cultivar and bagging on physico-chemical properties and antioxidant activity of tree sweet orange cultivar (*Citrus sinenesis* L. osbeck). Am Eurasion J Agric Environ Sci 13(2):139–147

Xu CX, Chen HB, Huang RY, He YJ (2008) Effects of bagging on fruit growth and quality of carambola. Acta Hort 773:195–200

Xu HX, Chen JW, Xie M (2010) Effect of different light transmittance paper bags on fruit quality and anti-oxidant capacity in loquat. J Sci Food Agric 90:1783–1788

Yang WH, Zhu XC, Bu JH, Hu GB, Wang HC, Huang XM (2009) Effects of bagging on fruit development and quality in cross-winter off-season longan. Sci Hortic 120:194–200

Zhao J, Wang J, Zhang X, Li H, Gao Z (2013) Effect of bagging on the composition of carbohydrate, organic acid and carotenoid contents in mango fruit. Acta Hortic 992:537–542

Zhen GH, Liao WC, Fan WM (2000) Effects of bag materials and bagging dates on loquat fruits. Fujian Fruits 114:1–4

Chapter 16
Protocols of Clean Pot Media for Containerized Mango Nursery

Sobia Chohan, Muhammad Abid, Shahid Gill, and Muhammad Tariq Malik

Abstract Mango is a significant fruit crop in Pakistan, contributing substantially to both domestic consumption and exports, particularly to the United Arab Emirates and the United Kingdom. Despite an expansion in mango production area, various biotic and abiotic factors hinder the crop's full potential. Notable challenges include quarantine issues such as fruit fly infestations and diseases like anthracnose and Mango Sudden Death Syndrome (MSDS). MSDS remains a major concern, exacerbated by inadequate watering and root damage, with key pathogens including *Ceratocystis fimbriata* and *Lasiodiplodia theobromae*. Effective disease prevention strategies, alongside improved nursery operations and orchard management, are essential for addressing these issues. The recent emergence of MSDS due to contaminated nursery potting mixes underscores the need for stringent sterilisation and pasteurisation protocols in nursery production. By focusing on enhanced disease control and adopting modern production techniques, Pakistan has the potential to increase mango yields to the crop's natural productivity of 25 tonnes per hectare, thus strengthening its position in the global mango market.

Keywords Mango · Disease control · Nursery management · Soil sterilization · Pasteurization · Production techniques · Disease free

16.1 Introduction

Mangos are a popular garden tree in the tropics. Despite their widespread cultivation, mangos need a warm, frost-free region with a dry winter. Rain and excessive humidity during blooming and fruit development decreased fruit output. Mid- to late-winter is when flowering begins, and early to mid-summer is when fruit reaches maturity.

S. Chohan · M. Abid (✉) · S. Gill
Department of Plant Pathology, Faculty of Agricultural Sciences and Technology, Bahauddin Zakariya University, Multan, Pakistan
e-mail: muhammad.abid@bzu.edu.pk

M. T. Malik
Mango Research Institute Multan, Multan, Pakistan

S. A. H. Naqvi et al. (eds.), *Climate Change and Mango Production*,
https://doi.org/10.1007/978-3-031-87035-4_16

Mango trees may grow up to 30 m (100 feet) under some forest conditions, although they typically stand 3–10 m (10–33 feet) tall (Afzal 1987; Ahmed et al. 1995). The canopy has a spreading tendency and is evergreen. Both people and animals may find refuge and shade under the mango's thick canopy. Mango is the second-largest fruit cultivated in Pakistan after citrus having 1844706 hectares area and 1.67 million tons production (2021–2022) (Ministry of National Food Security and Research [MNFSR], Shah et al. 2022). Multan and Bahawalpur Divisions are the main mango-growing areas of Punjab, which accounts for 99013 thousand hectares of mango cultivation with 1.3 million tons (MNFSR, 2021–2022). In Sindh, 59158 hectares (MNFSR, 2021–2022) of the Pakistan's total mango cultivation area is planted to mangos with 0.35 million tons production. The two main mango-growing regions in Sindh are Hyderabad and Mirpurkhas (Shah et al. 2022). A valuable fruit crop for generating foreign cash by exporting to the Middle East and Europe is the mango (Tahir et al. 2003).

Punjab's share of the overall mango output in 2021–22 was 80%, while Sindh's was 20% (Shah et al. 2022). By 2024, pakistan had increased to around 2.8 million tons, positioning Pakistan as the fourth-largest mango producer globally (Worldostats World Data & Statistics). This demonstrates how much effort mango producers have put into improving their farming methods in order to increase fruit output. The United Arab Emirates (UAE) has become the primary export market for Pakistani mangoes, e.g., in the 2024 season, Pakistan exported 13,681 metric tons of mangoes to 42 countries, generating $46.7 million in revenue (Worldostats World Data & Statistics). Recent data indicates that the United Arab Emirates (UAE) has become the largest importer of Pakistani mangoes, with imports totaling $159.74 million over the past five years. This positions the UAE as the primary destination for Pakistani mango exports, surpassing other significant markets such as the United Kingdom, Kazakhstan, and Saudi Arabia (MNFSR, 2021–2022).

Key reasons for the low yield and limited export include the lack of advanced farming technologies, inadequate infrastructure and processing facilities, poor fruit quality, insufficient knowledge of high-quality fruit production techniques, and ineffective management of insect pests, diseases, and disorders (Akem 2006; Al Adawi et al 2006a and b; Ganeshmurthy 2018). Major challenges hindering entry into the global market include diseases such as mango anthracnose and mango sudden death syndrome (MSD), as well as quarantine issues like fruit fly infestations (Soman et al. 2024).

Larson and Crandall 2023) reported soil heating at 60 °C for 30 minutes is an effective way to eliminate plant pathogens, including fungi, bacteria, nematodes, and most viruses. The development of aerated steaming techniques made this approach more feasible. Pathogens are more heat-sensitive than saprophytes, enabling moderate soil steaming to eradicate harmful organisms while preserving beneficial ones, which act as natural antagonists to pathogens (van Loenen et al. 2003). A significant issue with soil sterilization is the risk of recontamination, particularly when intensive methods, such as steaming at 100 °C or using harsh chemicals, create a "biological vacuum." This environment allows pathogens to thrive unchecked upon reintroduction, leading to severe disease outbreaks. Lower-temperature treatments reduce this risk by maintaining beneficial microorganisms that suppress pathogens and support soil health (Narayanasamy 2013).

16.2 Soil and Mango Health

All of the disease-causing microorganisms that target our crops ultimately reside in the soil (Akanmu et al. 2021). The 'root nibblers', or pathogens that covertly keep cutting back root tips or decreasing root efficiency, may be more economically significant than the microorganisms that cause noticeable but infrequent damage to roots, which worried early researchers (Tyler and Triplett 2008). Soil serves as a critical environment for various plant pathogens, including fungi, bacteria, nematodes, and viruses. These organisms can persist in the soil for extended periods, often in dormant forms such as spores or cysts, and can infect plants when conditions are favorable. For example, pathogens like Phytophthora cinnamomi and Stromatinia cepivora (formerly Sclerotium cepivorum) are soilborne fungi that cause diseases such as root rot and white rot in crops like avocados and onions, respectively (Weiland 2021).

Steam and chemical treatments are commonly employed to eliminate disease-causing organisms from soil, each offering distinct advantages. While chemical methods are often preferred for large-scale field applications, steam sterilization is particularly favored in floriculture due to several key benefits (Panth et al. 2020). Steam is generally considered the most effective treatment for controlling soil-borne pathogens, and its penetration and effectiveness can be easily monitored using a thermometer, unlike chemical treatments. Additionally, steam can be safely applied to soil within 30–60 cm of living plants without causing harm, making it a safer option for workers and nearby communities. Moreover, certain chemicals, such as methyl bromide have been found to be ineffective against specific pathogens like Verticillium species and may leave toxic residues harmful to plants like carnations (Panth et al. 2020).

Steam sterilization was first tested for soil treatment in 1888, and florists and nurserymen quickly adopted it, only slightly later than its introduction in medicine. Today, soil steaming is a standard practice in greenhouses and some outdoor settings worldwide. However, the technical and ecological aspects of soil steaming were not extensively studied until the 1950s. This led to the development of aerated steaming in the late 1950s, a method that uses lower temperatures (Larson et al. 2023).

Recent studies show that soil steaming at temperatures of around 65 °C for 30 minutes can effectively deactivate a wide range of pathogens by destroying their cellular structures. This method has gained significant attention in recent years as an alternative to chemical fumigants, offering a safer and more sustainable approach to soil treatment (Gullino et al. 2022). The principle behind soil steaming is based on the fact that parasitic microorganisms, which cause disease in plants, are more susceptible to heat than beneficial saprophytic organisms. Steaming soil at 60–65 °C enables the eradication of these harmful pathogens while preserving a population of beneficial saprophytic microorganisms. These beneficial organisms can act as natural antagonists, supporting plant health and promoting soil fertility. The biological advantage of this method lies in its ability to target specific pathogens while maintaining the ecological balance of the soil (Gullino et al. 2022). With the advent of aerated steaming technology, the process has become more efficient and easier to monitor. Thermometers can be used to track soil temperature, ensuring that the treatment penetrates the soil to a depth of 30–60 cm without causing damage to nearby plant roots. Additionally, unlike chemical treatments, steam sterilization is safe for workers and surrounding communities, posing no harmful residues in the soil after application. One of the major challenges with soil steaming is the potential for reducing populations of beneficial microorganisms along with harmful ones. However, this can be mitigated by integrating organic amendments or introducing beneficial microbes after the steaming process. Practices like these help restore soil health, prevent microbial imbalances, and promote long-term soil sustainability (Gullino et al. 2022; Sahi et al. 2012; Saleem and Akhtar 1989; Shahbaz et al. 2009; Sharma 1993; Sharma and Gupta 1994; Sharma and Gupta 1995).

One of the main challenges in soil treatment is preventing recontamination by pathogens. Overintensive methods, such as chemical treatments or steaming at 100 °C, can create a biological vacuum in the soil. This vacuum allows the first microorganism to recolonize, potentially a pathogen, to proliferate unchecked, leading to significant crop losses. By using lower temperatures for treatment, the risk of a post-steaming outbreak of root disease is reduced, as beneficial antagonistic organisms remain in the soil to suppress pathogens.

16.3 Morphology of the Pathogens Associated with Mango

On the majority of agar plate medium, *Ceratocystis fimbriata* grows well. Initially hyaline, the mycelium eventually becomes dark greenish-brown. Conidiophores that generate chains of hyaline conidia are often numerous within a few days. The endoconidia are round. The appearance of black, ostilate pycnidia with several hyaline, flask-shaped phialides is what distinguishes this coelomycete (*Nattrassia mangiferae*). Initially one-celled and hyaline, phialoconida eventually develop three-celled brown cells with a darker centre than the end cells. Conidia have an ellipsoidal to ovoid form (Azaz 2002; Almeida et al. 2007; Al-Sadi et al. 2010; Marincowitz et al. 2020).

When observed from the front of the petri dish, colonies of *Fusarium solani* display a wide range of colors, including white, cream, tan, salmon, cinnamon, yellow, red, violet, and pink. From the reverse side, the colony colors may appear colorless, tan, crimson, dark purple, or brown. Microscopically, the fungus exhibits hyaline, septate hyphae, along with conidiophores, phialides, macroconidia, and microconidia. The phialides are cylindrical and can be either monophialides or polyphialides (Barbosa et al. 2001; Bernardo and Rene 2012; De Beer et al. 2014).

On potato dextrose agar, *Cladosporium* colonies have a velvety to powdery texture. The colony's front side varies from olivaceous green to black, while the reverse is black. Microscopically, the conidiophores are light brown, branching, and erect (De Beer et al. 2014).

Czaban and Wróblewska (2006) utilized soil dilution plating and soil washing techniques to isolate 1960 microfungi from 105 soil samples, with *Aspergillus* (20 species) and *Penicillium* (24 species) being the most frequently identified genera.

Baruch Rubin et al. (2007) described soil solarization as a method that uses solar energy to heat the upper layers of moist soil covered with transparent polyethylene (PE) sheets, raising the soil temperature to 40–55 °C. When combined with organic amendments, a process known as biofumigation, soil solarization increases soil temperature by an additional 1–3 °C.

Conradie et al. (1992) formulated a semi-selective medium consisting of 20 g Difco Bitek agar, 10 g D-glucose, 1 g L-asparagine, 0.001 g active ingredient (a.i.) chlorothalonil, 0.005 g a.i. dicloran, 0.001 g a.i. vinclozolin, 2 g gallic acid, 0.1665 g streptomycin sulfate, and 1 L distilled water.

Allan et al. (1981) employed various methods to treat infected potting media while aiming to preserve beneficial saprophytic organisms. These methods included (a) chemical sterilization with methyl bromide, (b) steam sterilization at 93 °C, and (c) pasteurization using aerated steam at 63 °C.

In 1994, Bungay introduced three approaches for soil treatment. One of these involved chemical fumigations using methyl bromide and other heavier-than-air toxic gases. Another approach, the dry heat method, required temperatures exceeding 150 °C to achieve complete pathogen eradication. Bungay also utilized the latent heat of water vaporization in steam for sterilization. Additionally, he developed a novel selective medium for monitoring pathogen populations by modifying a standard triphenyl tetrazolium-based basal medium (Figs. 16.1 and 16.2).

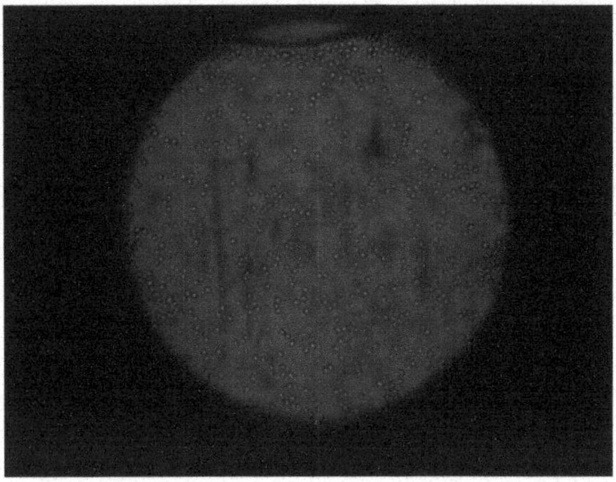

Fig. 16.1 Perithecia of *Natterasia mangiferae*

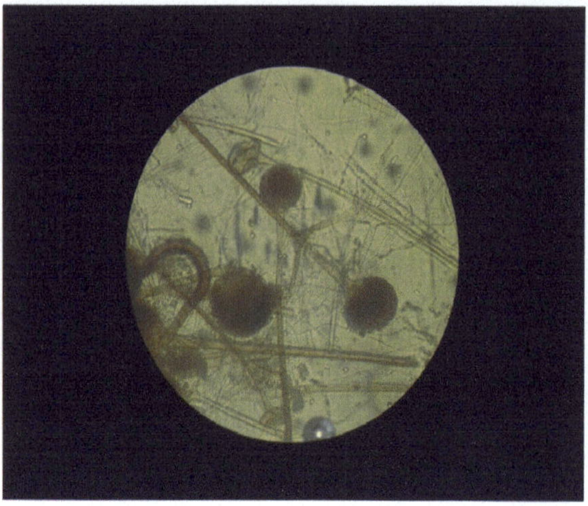

Fig. 16.2 Perithecia of *C. magnicans*

16.4 Description of Mango Sudden Death and Other Diseases

Narayan et al. (2022) identified *C. gloeosporioides* and *C. fimbriata*, either separately or together, as the causes of sudden mortality in mango stone grafts, particularly in the Konkan district of Maharashtra, India. The symptoms were more severe when both pathogens were isolated from infected samples. Mahmood et al. (2002) documented the prevalence of pathogenic fungi in Pakistan by collecting diseased samples from mango tree branches, stems, and roots. These samples revealed the presence of *Ceratocystis fimbriata* (49%), *Colletotrichum gloeosporioides* (30%), *Fusarium oxysporum* (6.6%), *Fusarium solani* (6.6%), and *Fusarium semitectum* (0.66%). In West Bengal, India, *Ceratocystis fimbriata* was linked to guava wilt infections by Pandit and Samajpati (2002), marking the first global report of guava wilt caused by this pathogen. Symptoms included gummosis from the trunk, wilting, and browning of leaves on a single branch, followed by tree death approximately six months after the initial symptoms appeared (Al-Adawi et al. 2003). Mango rapid death was often associated with damage from the bark beetle (*Cryphalus scabercollis*), although *Ceratocystis fimbriata* was also identified as a contributing factor.

Malik et al. (2005a, b, c) conducted a study to identify the primary cause of sudden mortality in mango plants, specifically the Chaunsa cultivar, by collecting diseased samples from the roots, collars, stems, and twigs/branches of plants from six locations in Punjab. The analysis revealed the presence of five fungal species: *Ceratocystis fimbriata*, *Fusarium oxysporum*, *Rhizoctonia solani*, *Macrophomina phaseolina*, and *Phytophthora cinnamomi*. Of these, *Ceratocystis fimbriata* was the most prevalent, found in 82.45% of the cases, followed by *Fusarium oxysporum* at 10.54%. The other fungi were rarely detected in the root samples.

Re-isolation and the emergence of the same symptoms verified that *Ceratocystis fimbriata* was the cause of the mango's rapid demise. *Ceratocystis fimbriata* was identified by Shahbaz et al. (2005) as the cause of the drop in mangoes in Punjab. The inoculated plant known as P2 had the largest diseased area, measuring 4.9 cm. Command The plants stayed healthy and showed no symptoms of illness. Separations from.

Al-Adawi et al. (2006) shown that Mango Sudden Decline Disease in Oman is caused by the vascular wilt pathogen Ceratocystis fimbriata, probably in conjunction with *Ceratocystis fimbriata* and the newly identified Ceratocystis omanensis. These fungi may be recovered from infected seedlings and are isolated from sick trees. According to studies, the bark beetle's piercing behaviour helps it operate as a vector and aid in the quick transmission of illness across Oman. According to Atallah et al. (2024), *Lasiodiplodia theobromae* and *Ceratocystis fimbriata* were linked to an abrupt reduction in Oman. *Ceratocystis omanensis* sp. is another species of Ceratocystis that has been linked to symptoms on damaged trees.

Alizadeh et al. (2024) demonstrated the pathogenicity of *Ceratocystis fimbriata* by artificially inoculating 10-year-old mango trees in Huaral, Peru, leading to symptoms such as dieback, gummosis, and floral necrosis. Meah et al. (1991) found that infection in mangoes only occurred when the fruits began to develop, particularly when temperatures exceeded 25 °C. They also noted that a longer pedicel delayed infection by two to three days compared to fruits without pedicels, suggesting the possibility of an inactive infection. Narasimhudu and Reddy (1992) confirmed the pathogenicity of *Ceratocystis fimbriata* by inoculating immature mango seedlings.

Ploetz et al. (1996) observed signs of decline, such as bud necrosis and tip dieback, following artificial inoculation with various pathogens, including *Ceratocystis fimbriata, Alternaria alternata, Gloeosporium cingulata, Drechslera dominicana*, and *Phomopsis spp.*

In 1998, Lima et al. evaluated the susceptibility of six mango cultivars (*Haden*, *Keit*, *Van Dyke*, *Manguito*, *Rosa*, and *Espada*) to *Ceratocystis fimbriata* isolates (ISO1, ISO2, and ISO3) from northeastern Brazil. They performed inoculation tests on mechanically damaged fruits using PDA discs, measuring lesion areas after 72 h. The largest lesions were produced by the ISO2 isolate (from Petrolina, PE), while *Espada* and *Manguito* were found to be resistant to all three isolates tested. According to Muhae-Ud-Din et al. (2023), the *Lungra* and *Chaunsa* cultivars showed characteristic symptoms of *Ceratocystis fimbriata* 3–4 days after inoculation.

In their 1993 study of the severity of rot produced by *Rhizopusarrhizus*, Patil and Pathak observed that the rot intensity and spore germination % peaked at 35 °C, but for Ceratocystis fimbriata, both peaked at 25 °C. At 100% and at least 30% R.H., rot was the worst and also had an impact on spore germination. Similarly, on Richard's agar at 25 ± 2 C, Shelar et al. (1997a, b) found that Ceratocystis fimbriata, the causative agent of die-back (rapid decline) disease of mango, grew at its maximum. At 0, 5, 10, and 40 C, there was no growth. The effects of temperature, precipitation, and relative humidity on the prevalence of C. gloeosporioides and D. natalensis (Ceratocystis fimbriata), which cause post-harvest rot of mango fruits, were examined by Banik et al. (1998a, b) utilising artificial According to in vitro research, B. theobromae grew best at 320 °C, with excellent mycelial growth occurring at all R.H. ranges between 60 and 100%. Ceratocystis fimbriata was shown to be very prevalent in vivo at temperatures between 33.5 and 36.2 °C and relative humidity levels between 50 and 86.6%. The variability of the Ceratocystis fimbriata spore population was determined by Dhakal (2018). In the plot where the prunings were removed, there were more rain-borne spores. At the start and conclusion of the dry season, when there was significant humidity due to rain, spore collection was higher. Prakash (2004) explored the ecological relationship between two post-harvest infections on mango fruits, *Ceratocystis fimbriata* and *A. niger*. The study found that 30 °C was the optimal temperature for fruit rot development, regardless of whether the pathogen was applied multiple times or just once. At 10 °C, no growth was observed. *Ceratocystis fimbriata* began to cause rot at 15 °C, while *A. niger* showed development at 20 °C. Balasundaran (2001) observed that *Ceratocystis fimbriata* was the dominant fungus in Kerala, India, causing discoloration in all softwood species, especially in the warm and humid climate of the region. The growth of *Ceratocystis fimbriata* on rubber wood was most effective at 30 °C, but impeded at 40 °C, with relative humidity also playing a significant role in fungal development.

Oliveira et al. (2016) assessed the impact of various culture media and temperatures on the growth and sporulation of C. fimbriata from mango. They found that different media influenced mycelial growth and sporulation, with optimal growth observed at 25 °C on 10% V8 agar and malt yeast extract agar (MYEA). Additionally, spore production was most abundant at 25 °C on MYEA, followed by 10% V8, and MEA. These findings suggest that both temperature and culture medium significantly affect the growth and sporulation of C. fimbriata, aligning with the observations regarding the influence of temperature and culture medium on fungal growth. Van Wyk et al. (2007) suggested that the optimal temperature for the development of *Ceratocystis fimbriata* is between 20 and 25 °C (68–77 °F), with no growth occurring at temperatures of 5 °C (40 °F), 10 °C (50 °F), or 35 °C (96 °F).

Leghari (2005) detailed typical disease symptoms caused by *Ceratocystis fimbriata*, including defoliation, blackening, and branch and twig death. Mango rapid decline often presents symptoms such as twig blight, tip dieback, gummosis, and bark splitting (Malik et al. 2005a, b, c), with further signs like root rotting and drying of twigs (Ploetz 2003; Iqbal et al. 2007). Both vegetative and reproductive systems show visible signs of the disease (Naqvi et al. 2014).

Research by Ploetz (2003) noted that the pathogen persisted in soil due to shesham dieback caused by the same fungal species. Masood et al. (2010a, b) observed these symptoms in various Pakistani mango varieties. Al-Adawi et al. (2006) highlighted twig drying and gummosis bark splitting as the most obvious signs of illness. Masood et al. (2008a, b) suggested that poor irrigation and significant plant damage might contribute to the spread of the disease.

In a study conducted by Ploetz et al. (1996), the most frequently isolated fungi from mango trees exhibiting signs of sudden mortality in the U.S. were *Alternaria alternata*, *Cladosporium* sp., *C. gloeosporioides* (G. cingulata), *Dothiorella dominicana*, and *Fusarium* sp. Parkash (2004) linked *Ceratocystis fimbriata* to gummosis, mango stem girdling, and rotting. Mansour et al. (2006) noted that fruit trees affected by root and crown rot might die quickly after heavy rain or spring growth. Collar rot in mangoes can be identified when cankers appear, and if conditions remain favorable, the tree may be girdled and die soon after (Jiskani et al. 2007; Khanzada et al. 2004a, 2004c, 2005, 2009; Khaskheli et al. 2011).

Rodriguez et al. (1999) found that *Ceratocystis fimbriata* was the most prevalent pathogen in 5118 hectares of mango in the Piura region of Peru, causing gummosis and wilting. Symptoms, such as wilting and browning of leaves on a single branch, were described by Al-Adawi et al. (2003), with tree death occurring six months after the appearance of symptoms. The pathogen was isolated from diseased trees, with pycnidia found in the infected bark. *Ceratocystis fimbriata* was first reported as the causal pathogen of mango decline in Pakistan Khanzada et al. (2004b) as the cause of gummosis and decline in Sindh. This suggests that *Ceratocystis fimbriata* is well-established in the soils of Punjabi and Sindhi mango plantations and plays a major role in the sudden decline of mangoes due to collar/stem rot (Mahmood et al. 2007; Mahmood 2008; Malik et al. 1997, 2001, 2003, 2016; DaSilva et al. 2012).

16.5 Production of Aerated Steam

When air is mixed with steam and its temperature is lowered, a controlled environment can be created by adjusting the steam-to-air ratio. For example, a mixture with a 4:1 air-to-steam ratio by volume results in a temperature of 60 °C. This combination can be easily achieved by injecting steam into a larger air flow using a blower with radial blades and a damper to regulate the airflow. The temperature of the mixture is monitored using a precise thermometer.

The air–steam mixture is delivered into a plenum chamber under the soil mix at a pressure of about 15 cm of water. The temperature of the mixture is limited by the amount of air in the system, ensuring that it does not exceed a "ceiling" temperature. As the mixture rises through the medium, it moves as a horizontal front, with the lower side receiving heat, while the upper side remains unheated.

To support the soil mix, a perforated steel plate or expanded metal screen wrapped with hessian is used. The warm air transfers heat to the soil particles, while the steam condenses on cooler particles, effectively heating them. To maintain the balance, the air needs a proper escape route, ensuring the mixture's movement through the soil is efficient.

16.6 Effects on Plant Growth

Over the past five years, several experiments in Pietermaritzburg have compared the effects of pasteurising and sterilising soil mixes using aerated and full steam versus untreated media on plant growth. In earlier studies with vegetable crops, Mukjang et al. (2022) and Li et al. (2022) found that steaming improved plant growth in some cases, but not in all, attributing the variations to differing levels of infection in the soil media. Larson and Crandall (2023) inoculated the media with a bacterial plant pathogen and fresh manure, then compared the effects of steaming at 90 and 60 °C with control treatments for three types of vegetable crops. They observed that steaming generally enhanced the yield of valuable plant parts. However, when fresh manure was added to the soil mix, plant growth and yield were significantly reduced, likely due to toxic substances, with higher temperatures intensifying this effect. In some instances, aerated steam proved superior to full steam, while in other cases it was not, though aerated steam was consistently as effective as full steam in controlling soil-borne diseases and pests (Pernezny and Simone 2000; Phipps 2016; Poland et al. 2006; Ranasingh et al. 2006; Rehman et al. 2011).

16.7 Case Studies

The potting media samples collected from nurseries, such as those from the MRI nursery in Multan, contained several pathogenic fungi, including *Ceratocystis fimbriata*, *Nattrassia mangiferae*, *Fusarium*, and *Cladosporium*, all of which have been linked to mango decline diseases. However, some fungi, such as *Penicillium* and *Aspergillus*, were consistently identified as saprophytes. Microscopic examination also revealed the presence of unidentified nematodes in the colonies, which may be opportunistic organisms, though they seemed to thrive in association with *Ceratocystis fimbriata*. This discovery presents an interesting avenue for further research into potential biological control of *C. fimbriata*.

The most aggressive and virulent pathogen identified in the potting medium from every nursery was *Ceratocystis fimbriata*, which had the highest infection frequency, raising concern about the role of nursery plants in spreading this pathogen across mango plantations. The failure to properly sterilise or pasteurise the potting media is a critical issue, and steam sterilisation at 60 °C for one hour is recommended to address this. Currently, nurseries in Pakistan commonly use solarisation with polythene sheets for 15–20 days in May, which can heat the soil to temperatures between 50 and 53 °C. This method is somewhat effective in killing pathogens, but the temperatures may not be sufficient for complete sterilization.

A compound microscope examination of the contaminated agar colonies revealed the presence of six fungal species: *Fusarium*, *Nattrassia mangiferae*, *Ceratocystis fimbriata*, *Cladosporium*, *Penicillium*, and *Aspergillus*. Among these, *Ceratocystis fimbriata* had the highest infection frequency (77.84%), followed by *Nattrassia mangiferae* (10%), while *Penicillium* had the lowest infection frequency (1.42%). *Cladosporium*, *Fusarium*, and *Aspergillus* were found in the potting medium at 10%, 2.14%, and 5%, respectively (Tables 16.1, 16.2, 16.3, 16.4, 16.5, 16.6, 16.7, 16.8, 16.9, 16.10).

The most effective sterilization method was determined to be T1 (steaming under pressure at 82 °C for 30 min), which eliminated all pathogens. Similarly,

Table 16.1 Extent of different fungi isolated from potting mixes

Nature of bait	Fungi isolated	Infestation (%)
Agar slants	*Ceratocystis fimbriata*	77.84
	Fusarium spp.	2.14
	Cladosporium	10
	Nattrassia mangiferae	4.28
	Penicilium	1.42
	Aspergillus spp.	5

Table 16.2 The extent of different fungi isolated from different nurseries

Name of nursery	Name of fungi	Infestation (%)
MRI (Multan)	1 *Ceratocystis fimbriata*	85.0
	2 *Fusarium*	15.0
	3 Nematodes (*unidentified*)	
MRS (Shujabad)	1 *Nattrassia mangiferae*	30.0
	2 *Ceratocystis fimbriata*	70.0
ARI (Sindh)	*Ceratocystis fimbriata*	100.0
AUS (Sindh)	*Ceratocystis fimbriata*	100.0

Table 16.3 Morphological and cultural features of *ceratocystis fimbriata* isolated from potting mixes

Category	Colony growth rate (%)	Succession of conidia (%)	Presence of conidia (%)	Size of conidia (%)	Presence of clamydospores (%)	Presence of ascospores (%)	Status of perithicia (%)
0	00.0	10.0	00.0	80.0	00.0	00.0	00.0
1	00.0	90.0	20.0	20.0	00.0	00.0	00.0
2	100.0	00.0	80.0	00.0	00.0	00.0	00.0

T1′ (steaming at 82 °C for 30 min with 10 pounds of pressure) was effective in sterilizing the potting medium. In contrast, T2 (steaming without pressure for 30 min) resulted in the highest infection rate of 86% from *Ceratocystis magnicans*, while T2′ (steaming without pressure for 30 min with artificial inoculation of *Ceratocystis magnicans*) had a slightly lower infection rate (85%).

Oven sterilization (T3) at 82 °C for 30 min resulted in a 58% infection rate of *Ceratocystis magnicans*, while T3′ (oven sterilization at 82 °C with artificial inoculation) yielded an infection rate of 53.34%. Solarization (T4) at 71 °C for 6–8 weeks resulted in a 91% infection rate, but artificial inoculation (T4′) reduced the infection rate to a greater extent, confirming the efficacy of sterilization methods in controlling pathogen spread (Tables 16.11, 16.12, 16.13, 16.14, 16.15, 16.16, 16.17, 16.18 and 16.19).

Table 16.4 Isolation and infection frequency of different fungi isolated from silt media

Sources	Total no of agar slants	Total no of petri dished exercised	Total no of bits exercised	Total no of bits colonized	*Ceratocystis magnicans*	Infection frequency (%)	*Natrassia mangiferea*	Infection frequency (%)	Others	Infection frequency (%)
Beaker-1	3	9	9	4	3	75.66 b	1	25.00 a	0	0
Beaker-II	3	9	9	5	5	100.00 a	0	0.00 b	0	0
Beaker-III	3	9	9	4	3	75.00 bc	1	25.00 a	0	0
*LSD	–	–	–	–	–	0.66	–	0.00	–	–

LSD—least significant difference

Table 16.5 Isolation and infection frequency of different fungi isolated from Baggas media

Sources	Total no of agar slants	Total no of petri dished exercised	Total no of bits exercised	Total no of bits colonized	*Ceratocystis magnicans*	Infection frequency (%)	*Natrassia mangiferea*	Infection frequency (%)	Others	Infection frequency (%)
Beaker-1	3	9	9	5	1	20.00 c	4	80.00 a	0	0
Beaker-II	3	9	9	4	2	50.00 a	2	50.00 c	0	0
Beaker-III	3	9	9	3	1	33.00 b	2	77.00 b	0	0
*LSD	–	–	–	–	–	3.11	–	4.99	–	–

LSD—least significant difference

Table 16.6 Isolation and infection frequency of different fungi isolated from Coconut fiber media

Sources	Total no of agar slants	Total no of petri dished exercised	Total no of bits exercised	Total no of bits colonized	*Ceratocystis magnicans*	Infection frequency (%)	*Natrassia mangiferea*	Infection frequency (%)	Others	Infection frequency (%)
Beaker-1	3	9	9	6	5	85.33 a	1	15.00 a	0	0
Beaker-II	3	9	9	5	4	71.66 b	1	06.00 b	0	0
Beaker-III	3	9	9	3	2	41.33 c	1	3.99 c	0	0
*LSD	–	–	–	–	–	4.17	–	2.18	–	–

LSD—least significant difference

Table 16.7 Isolation and infection frequency of different fungi isolated from Mixed pot media

Sources	Total no of agar slants	Total no of petri dished exercised	Total no of bits exercised	Total no of bits colonized	*Ceratocystis magnicans*	Infection frequency (%)	*Natrassia mangiferea*	Infection frequency (%)	Others	Infection frequency (%)
Beaker-1	3	9	9	4	4	100.00 a	0	0.00 c	0	0
Beaker-II	3	9	9	5	4	80.00 a	1	20.00 b	0	0
Beaker-III	3	9	9	4	3	75.00 b	2	25.00 a	0	0
*LSD	–	–	–	–	–	7.17	–	5.79	–	–

LSD—least significant difference

Table 16.8 Isolation and infection frequency of different fungi under the effect of Treatment I (steaming with pressure, 10 lbs, 30 min at 82 °C)

Sources	Total no of agar slants	Total no of petri dished exercised	Total no of bits exercised	Total no of bits colonized	*Ceratocystis magnicans*	Infection frequency (%)	*Natrassia mangiferea*	Infection frequency (%)	Others	Infection frequency (%)
Beaker-1	3	9	9	0	0	0	0	0	0	0
Beaker-II	3	9	9	0	0	0	0	0	0	0
Beaker-III	3	9	9	0	0	0	0	0	0	0
*LSD	–	–	–	–	–	–	–	–	–	–

LSD—least significant difference

Table 16.9 Isolation and infection frequency of different fungi under the effect of Treatment I (steaming with pressure, 10 lbs, 30 min at 82 °C but with artificial inoculation of *Ceratocystis magnicans*)

Sources	Total no of agar slants	Total no of petri dished exercised	Total no of bits exercised	Total no of bits colonized	*Ceratocystis magnicans*	Infection frequency (%)	*Natrassia mangiferea*	Infection frequency (%)	Others	Infection frequency (%)
Beaker-1	3	9	9	0	0	0	0	0	0	0
Beaker-II	3	9	9	0	0	0	0	0	0	0
Beaker-III	3	9	9	0	0	0	0	0	0	0
*LSD	–	–	–	–	–	–	–	–	–	–

LSD—least significant difference

Table 16.10 Isolation and infection frequency of different fungi under the effect of Treatment II (steaming without pressure for 30 min)

Sources	Total no of agar slants	Total no of petri dished exercised	Total no of bits exercised	Total no of bits colonized	*Ceratocystis magnicans*	Infection frequency (%)	*Natrassia mangiferea*	Infection frequency (%)	Others	Infection frequency (%)
Beaker-1	3	9	9	6	5	83 b	1	17 b	0	0
Beaker-II	3	9	9	5	5	100 a	0	0 c	0	0
Beaker-III	3	9	9	4	3	75 c	1	25 a	0	0
*LSD	–	–	–	–	–	11.11	–	–	–	–

LSD—least significant difference

Table 16.11 Isolation and infection frequency of different fungi under the effect of Treatment II (steaming without pressure for 30 min but with artificial inoculation)

Sources	Total no of agar slants	Total no of petri dished exercised	Total no of bits exercised	Total no of bits colonized	*Ceratocystis magnicans*	Infection frequency (%)	*Natrassia mangiferea*	Infection frequency (%)	Others	Infection frequency (%)
Beaker-1	3	9	9	5	4	80 a	1	20 b	0	0
Beaker-II	3	9	9	4	4	100 a	0	0 c	0	0
Beaker-III	3	9	9	4	3	75 b	1	25 a	0	0
*LSD	–	–	–	–	–	7.11	–	13.66	–	–

LSD—least significant difference

Table 16.12 Isolation and infection frequency of different fungi under the effect of Treatment III (heating in oven at 82 °C)

Sources	Total no of agar slants	Total no of petri dished exercised	Total no of bits exercised	Total no of bits colonized	*Ceratocystis magnicans*	Infection frequency (%)	*Natrassia mangiferea*	Infection frequency (%)	Others	Infection frequency (%)
Beaker-1	3	9	9	5	3	60 b	2	40 c	0	0
Beaker-II	3	9	9	4	2	50 c	2	50 a	0	0
Beaker-III	3	9	9	3	2	66 a	1	44 b	0	0
*LSD	–	–	–	–	–	12.33	–	11.66	–	–

LSD—least significant difference

Table 16.13 Isolation and infection frequency of different fungi under the effect of Treatment III (heating in oven at 82 °C but with artificial inoculation)

Sources	Total no of agar slants	Total no of petri dished exercised	Total no of bits exercised	Total no of bits colonized	*Ceratocystis magnicans*	Infection frequency (%)	*Natrassia mangiferea*	Infection frequency (%)	Others	Infection frequency (%)
Beaker-1	3	9	9	5	3	60 a	2	40 b	0	0
Beaker-II	3	9	9	6	3	50 b	3	50 a	0	0
Beaker-III	3	9	9	4	2	50 b	2	50 a	0	0
*LSD	–	–	–	–	–	10.00	–	10.33	–	–

LSD—least significant difference

Table 16.14 Isolation and infection frequency of different fungi under the effect of Treatment IV (solarization at 71 °C for 6–8 weeks)

Sources	Total no of agar slants	Total no of petri dished exercised	Total no of bits exercised	Total no of bits colonized	*Ceratocystis magnicans*	Infection frequency (%)	*Natrassia mangiferea*	Infection frequency (%)	Others	Infection frequency (%)
Beaker-1	3	9	9	5	5	100 a	0	0 a	0	0
Beaker-II	3	9	9	4	3	75ab	1	25 a	0	0
Beaker-III	3	9	9	5	5	100 a	0	0 a	0	0
*LSD	–	–	–	–	–	13.11	–	3.33	–	–

LSD—least significant difference

Table 16.15 Isolation and infection frequency of different fungi under the effect of Treatment IV (solarization at 71 °C for 6–8 weeks but with artificial inoculation)

Sources	Total no of agar slants	Total no of petri dished exercised	Total no of bits exercised	Total no of bits colonized	*Ceratocystis magnicans*	Infection frequency (%)	*Natrassia mangiferea*	Infection frequency (%)	Others	Infection frequency (%)
Beaker-1	3	9	9	5	5	100 a	0	0 a	0	0
Beaker-II	3	9	9	5	5	80 b	1	20 a	0	0
Beaker-III	3	9	9	4	4	100 a	0	0 a	0	0
*LSD	–	–	–	–	–	7.22	–	1.36	–	–

LSD—least significant difference

Table 16.16 Isolation and infection frequency of different fungi under the effect of Treatment V (chemical sterilization with formalin)

Sources	Total no of agar slants	Total no of petri dished exercised	Total no of bits exercised	Total no of bits colonized	*Ceratocystis magnicans*	Infection frequency (%)	*Natrassia mangiferea*	Infection frequency (%)	Others	Infection frequency%
Beaker-1	3	9	9	4	1	25 b	3	75 b	0	0
Beaker-II	3	9	9	5	2	40 a	3	60 a	0	0
Beaker-III	3	9	9	6	1	16 c	5	84 a	0	0
*LSD	–	–	–	–	–	15.11	–	3.36	–	–

LSD—least significant difference

Table 16.17 Isolation and infection frequency of different fungi under the effect of Treatment V (chemical sterilization with formalin but with artificial inoculation)

Sources	Total no of agar slants	Total no of petri dished exercised	Total no of bits exercised	Total no of bits colonized	*Ceratocystis magnicans*	Infection frequency (%)	*Natrassia mangiferea*	Infection frequency (%)	Others	Infection frequency (%)
Beaker-1	3	9	9	5	1	20 a	4	80 a	0	0
Beaker-II	3	9	9	3	1	33 a	2	77 a	0	0
Beaker-III	3	9	9	3	0	0 a	3	100 a	0	0
*LSD	–	–	–	–	–	6.11	–	3.36	–	–

LSD—least significant difference

Table 16.18 Isolation and infection frequency of different fungi under the effect of Treatment VI (chemical sterilization with hydrogen peroxide)

Sources	Total no of agar slants	Total no of petri dished exercised	Total no of bits exercised	Total no of bits colonized	*Ceratocystis magnicans*	Infection frequency (%)	*Natrassia mangiferea*	Infection frequency (%)	Others	Infection frequency (%)
Beaker-1	3	9	9	5	4	80 a	1	20 a	0	0a
Beaker-II	3	9	9	4	2	50 a	1	25 a	1	25a
Beaker-III	3	9	9	3	1	33 a	1	26 a	1	33a
*LSD	–	–	–	–	–	2.33	–	1.22	–	1.11

LSD—least significant difference

Table 16.19 Isolation and infection frequency of different fungi under the effect of Treatment VI (chemical sterilization with hydrogen peroxide but with artificial inoculation)

Sources	Total no of agar slants	Total no of petri dished exercised	Total no of bits exercised	Total no of bits colonized	*Ceratocystis magnicans*	Infection frequency (%)	*Natrassia mangiferea*	Infection frequency (%)	Others	Infection frequency (%)
Beaker-1	3	9	9	5	5	100 a	0	0 a	0	0
Beaker-II	3	9	9	4	3	75 a	1	25 a	0	0
Beaker-III	3	9	9	5	3	60 a	2	40 a	0	0
*LSD	–	–	–	–	–	6.33	–	4.11	–	0

LSD—least significant difference

16.8 Conclusion

In summary, the mango is a vital fruit crop in Pakistan that plays a major role in both domestic consumption and exports, especially to the United Arab Emirates and the United Kingdom. Despite an increase in the area used for mango production, a number of biotic and abiotic factors prevent the output from reaching its full potential. Quarantine problems such fruit fly infestation and diseases like anthracnose and Mango Sudden Death Syndrome (MSDS) provide significant obstacles to increasing exports and production. MSDS is still a major problem, made worse by poor watering and root damage. It is mostly associated with infections like as Ceratocystis fimbriata and Lasiodiplodia theobromae. A thorough strategy to disease prevention, nursery operations, and orchard management is necessary to meet these problems. The need of putting in place efficient sterilisation and pasteurisation procedures in nursery manufacturing is highlighted by the recent outbreak of MSDS via tainted nursery potting mix. Pakistan may be able to increase mango yields to the fruit's natural productivity of 25 tonnes per hectare by concentrating on improving disease control measures and using new production techniques, therefore solidifying its place in the world mango market.

References

Afzal MA (1987) Outbreaks and new records. Pakistan. Black Spot of Mango

Ahmed I, Mahmood A, Majeed K, Saleem A (1995) Evaluation of various fungicides against dieback disease caused by Diplodianatalensis in mango. Pak J Phytopathology 7:208–209

Akanmu AO, Babalola OO, Venturi V, Ayilara MS, Adeleke BS, Amoo AE, Sobowale AA, Fadiji AE, Glick BR (2021) Plant disease management: leveraging on the plant-microbe-soil interface in the biorational use of organic amendments. Front Plant Sci 12:700507

Akem CN (2006) Mango anthracnose disease: present status and future research priorities. Plant Pathol J 5:266–273

Al Adawi AO, Deadman ML, Al Rawahi AK, Khan AJ, Al Maqbali YM (2003) Diplodiatheobromae associated with sudden decline of mango in the Sultanatem of Oman. Plant Pathol 52:419

Al Adawi AO, Deadman ML, Al Rawahi AK, Al Maqbali YM, Al Jahwari AA, Al Saadi BA, Al Amri IS, Wingfield MJ (2006a) Aetiology and causal agents of mango sudden decline disease in the Sultanate of Oman. Eur J Plant Pathol 116:247–254

Al Adawi AO, Deadman ML, Al Rawahi AK, Al Maqbali YM, Al Jahwari AA, Al Saadi BA, Al Amri IS, Wingfield MJ (2006b) Aetiology and causal agents of mango sudden decline disease in the Sultanate of Oman. Eur J Plant Pathol 116(4):247–254

Allan P, Lamb D, Chalton D (1981) Sterilization & Pasteurization of soil mixes. S Afr Avocado Growers' Assoc Yearb 4:124–127

Almeida F, Cerqueira F, Silva R, Uhoa C, Lima A (2007) Mycoparatism studies of Trichodermaharzianum strains against Rhizoctoniasolani: evaluation of coiling and hydrolytic enzyme production. Biotechnol Lett 29(8):1189–1193

Al-Sadi AM, Al-Ouweisi FA, Al-Shariani NK, Al-Adawi AO, Kaplan EJ, Deadman ML (2010) Histological changes in mango seedlings following infection with ceratocystismanginecans, the cause of mango decline. J Phytopathol 158(11–12):738–743

Alizadeh M, Azaryan A, Ghaleh MK, Roshanroo M, Graily-Moradi F (2024) Host and geographical distribution of the plant pathogenic fungus Ceratocystis fimbriata (Ascomycota) worldwide. Nova Hedwigia 118(1–2):41–71

Atallah OO, Hassanin AA, Yassin SM, Aloufi AS, Almanzalawi EA, Abdelkhalek A, Atia MM, Behiry S, Abdelrhim AS, Nehela Y (2024) Pathological characterization and management of lasiodiplodia theobromae, a hemibiotroph with an interkingdom host range. Plant Disease 108(11):3243–3257

Azaz AD (2002) Isolation and identification of soilborne fungi in fields irrigated by GAP in Harran plain using two isolation methods. Turk J Bot 27:83–92

Balasundaran M, Florence EJM, Sharma JK (2001) Increasing eucalypt productivity through high yielding disease resistant clones: method of vegetative multiplication pp 10–11

Banik AK, Kaiser SAK, Dhua RS (1998a) Evaluation of some systemic and non-systemic fungicides against Botryodiplodiatheobromae, the cause of dieback disease of mango (Mangiferaindica L.). J Soils Crops 8:199–222

Banik AK, Kaiser SAKM, Dhua RS (1998b) Evaluation of some systemic and non-systemic fungicides against Botryodiplodiatheobromae, the cause of die-back disease of mango (Mangiferaindica L.). J Soils and Crops 8:119–122

Barbosa M, Rehm K, Menezes M, Mariano RL (2001) Antagonism of Trichoderma species on Cladosporiumherbarum and their enzimatic characterization. Braz J Microbiol 32:98–104

De Beer ZW, Duong TA, Barnes I, Wingfield BD, Wingfield MJ (2014) Redefining ceratocystis and allied genera. Stud Mycol 79(1):187–219

Bernardo AL, Rene T (2012) Prevalence of isolates of Botrytis cinerea resistant to multiple fungicides in Chilean vineyard. Crop Proct. 40:49–52

Conradie E, Swart WJ, Wingfield MJ (1992) Semiselective medium for isolation of Cryphonectria cubensis. Phytopathol 82:1440–1442

Czaban J, Wróblewska B (2006) A simple, direct plating method, alternative to dilution plating, for estimation of the abundance of Penicillium verrucosum on incubated cereal grain. Polish J Microbiol 55(3):237

DaSilva PAV, Martins RB, Michereff SJ, DaSilva MB, Câmara MPS (2012) Sensitivity of Lasiodiplodiatheobromae from Brazilian papaya orchards to MBC and DMI fungicides. Eur J Pl Pathol 13:489–498

Dhakal U (2018) Survey for ceratocystis fimbriata on syngonium species, host range test and molecular characterization of the isolates collected, Master's thesis, University of Hawai'i at Manoa

Documentation of identified and unidentified diseases of mango in Sindh, Pakistan. In: International symposium on prospects of horticultural industry in Pakistan. Institute of Horticultural Science, University of Agriculture Faisalabad, Pakistan, pp 176–190

Ganeshmurthy AN, Rupa TR, Shivananda TN (2018) Enhancing mango productivity through sustainable resource management. J Hortic Sci 13(1):1–31. https://doi.org/10.24154/jhs.v13i1.18

Gullino ML, Garibaldi A, Gamliel A, Katan J (2022) Soil disinfestation: from soil treatment to soil and plant health. Plant Dis 106(4):503–519

Iqbal Z, Valeem EE, Shahbaz M, Ahmad K, Khan ZI, Malik MT, Danish M (2007) Determination of different decline disorders in mango orchards of the Punjab, Pakistan. Pak J Bot 39:1313–1318

Jiskani MM, Pathan MA, Wagan KH, Khaskheli MI (2007) Documentation of identified and unidentified diseases of mango in Sindh, Pakistan. In: International symposium on prospects of horticultural industry in Pakistan. Institute of Horticultural Science, University of Agriculture Faisalabad, Pakistan, pp 176–190

Khanzada MA, Lodhi AM, Saleem S (2004a) Pathogenicity of Lasiodiplodiatheobromae and Fusarium solani on mango. Pak J Bot 36(1):181–189

Khanzada MA, Lodhi AM, Shahzad S (2004c) Pathogenicity of Lasiodiplodiatheobromae and Fusatiumsolani on Mango. Pak J Bot 36:181–189

Khanzada MA, Lodhi AM, Shahzad S (2005) Chemical control of Lasiodiplodiatheobromae, the causal agent of mango decline in Sindh. Pak J Bot 37:1023–1030

Khanzada MA, Lodhi AM, Shahzad S (2009) Effect of soil solarization on mango decline pathogen, *Lasiodiplodia theobromae*. Pak J Bot 41(6):3179–3184

Khanzada MA, Lodhi AM, Shahzad S (2004a) Decline and gummosis diseases of mango in Sindh caused by Lasiodiplodiatheobromae. Online. Plant Health Prog. https://doi.org/10.1094/PHP-204-0302-01-DG

Khaskheli MI, Jiskani MM, Soomro MH, Talpur MA, Poussio GB (2011) Faculty of crop protection, Sindh Agriculture University Tandojam, Pakistan Pak J Agri Agril Engg Vet Sci 27(2):160–167

Larson ER, Crandall SG (2023) Recovery of the soil fungal microbiome after steam disinfection to manage the plant pathogen Fusarium solani. Front Plant Sci 14:14:112851. https://doi.org/10.3389/fpls.2023.1128518

Leghari TN (2005) Epidemiology and yield losses by sudden death syndrome in mango orchards of Sindh and its possible control. M. Sc (Hons). Thesis. Department of Plant Pathology, SAU Tandojam 124 p

Li YB, Zhang ZP, Yuan Y, Huang HC, Mei XY, Du F, Yang M, Liu YX, Zhu SS (2022) Appropriate soil heat treatment promotes growth and disease suppression of panax notoginseng by interfering with the bacterial community. J Microbiol Biotechnol 32(3):294–301. https://doi.org/10.4014/jmb.2112.12005.PMID:35283430;PMCID:PMC9628859

Mahmood A, Saleem AK, Akhtar M (2002) Mango decline in Pakistan and its management. Pak J Phytopathol 14:40–43

Mahmood A, Khan SM, Ali S, Alam S (2007) Methods for the chemical control of quick decline/sudden death of mango. Pak J Phytopathol 19:81–85

Mahmood A (2008) Studies on characterization and management of Lasiodiplodiatheobromae (Pat.)Griff&Maubl.associated with quick decline of mango. Ph.D Thesis, Institute of Plant Pathology, University of the Punjab, Lahore, Pakistan

Malik MT, Haq CA, Khan MI, Javaid MA (1997) Effect of different fungicides against anthracnose/dieback of mango sprayed on nursery and field plants. J Agric Res 35:385–390

Malik MT, Haq CA, Grewal AG, Khan MI, Nisar M (2001) An integrated approach to recover naturally infected mango plants with dieback (Diplodianatalensis). J Agric Res 39:51–58

Malik MT, Haq CA, Grewal AG, Khan MI, Nasir M (2003) Towards the control of a new threat "Quick Decline of Mango." Pak Hort 3:37–40

Malik MT, Dasti AA, Khan SM (2005) Mango decline disorders prevailing in Pakistan. In: Proceedings of international conference on mango and date palm: culture and export. University of Agriculture, Faisalabad, Pakistan. 20–23-June, 25 p

Malik MT, Dasti AA, Khan SM (2005) Mango decline disorders prevailing in Pakistan. In: Malik AU, Pervez MA, Zaif K (eds) Proceedings of international conference on mango and date palm, culture and export. June: 20–23. University of Agriculture, Faisalabad

Malik MT, Dasti AA, Khan SM (2005) Mango decline disorders prevailing in Pakistan. In: Proceedings of international conference on mango and date palm: culture and export. University of Agriculture, Faisalabad, Pakistan, 20–23-June, p 25

Malik MT, Ammar M, Saifullah HU, Mohar TA (2016) Mango diseases and their management. www.aari.punjab.gov.pk/info-desk/publications/brochure.1-9

Mansour FS, Abd-El-Aziz SA, Helal GA (2006) Effect of fruit heat treatment in three mango varieties on incidence of postharvest fungal disease. J Plant Pathol 141–148

Marincowitz S, Barnes I, de Beer ZW, Wingfield MJ (2020) Epitypification of ceratocystis fimbriata. Fungal Syst Evol 6:289–298. https://doi.org/10.3114/fuse.2020.06.14. Epub 2020 Jun 23. PMID: 32904100; PMCID: PMC7451773

Masood A, Saeed S, Sajjad A (2008a) Characterization and damage patterns of different bark beetle species associated with Mango Sudden death Syndrome in Punjab, Pakistan. Pak Entomol 30(2):163–168

Masood A, Saeed S, Sajjad A (2008b) Characterization and damage patterns of different bark beetle species associated with Mango Sudden death Syndrome in Punjab, Pakistan. Pak Entomol 30:163–168

Masood A, Saeed S, Malik MT, Naeem I, Kazmi MR (2010a) Methodology for the evaluation of symptoms severity of Mango Sudden Death Syndrome in Pakistan. Pak J Bot 42:1289–1299

Masood A, Saeed S, Malik MT, NaeemIqbal, Kazmi MR (2010b) Methodology for the evaluation of symptoms severity of Mango Sudden Death Syndrome in Pakistan. Pak J Bot 42(2):1289–1299

Meah MB, Plumbley RA, Jeger MJ (1991) Growth and infectivity of Botryodiplodia theobromae causing stem-end rot of mango. Mycolog Res 95(4):405–408

Muhae-Ud-Din G, Tarafder E, Hameed A, Alam MW, Yong W (2023) Ceratocystis fimbriata causing postharvest fruit rot of pear in Punjab, Pakistan. Plant Dis 2023. https://doi.org/10.1094/PDIS-07-23-1448-PDN. Epub ahead of print. PMID: 37700477

Mukjang N, Mombrikotb SB, Bell T (2022) Microbial community succession in steam-sterilized greenhouses infected with fusarium oxysporum. Environ Microbiol Rep 14(4):577–583. https://doi.org/10.1111/1758-2229.13072. Epub 2022 Apr 20. PMID: 35445561; PMCID: PMC9544407

Narasimhudu Y, Reddy PSN (1992) A note on gummosis of mango

Narayanasamy P (2013) Physical techniques for biological crop disease management. Springer, Dordrecht , pp 57–105. https://doi.org/10.1007/978-94-007-6377-7_3

Narayan H, Srivasatava P, Chandra Bhatt S, Joshi D, Soni R (2022) Plant pathogenesis and disease control. Plant Prot: From Chem Biol 95–114

Naqvi SAH, Perveen R, Manzoor SA, Abbas Z (2014) Occurrence of mango anthracnose on inflorescence in Multan (Punjab) and its in-vitro control through various fungicides. Int J Biosci 4:184–191

Oliveira LS, Damacena MB, Guimarães LM, Siqueira DL, Alfenas AC (2016) Ceratocystis fimbriata isolates on Mangifera indica have different levels of aggressiveness. Eur J Plant Pathol 145:847–856

Pandit PK, Samajpati N (2002) Wilt disease of guava (*Psidium guajava* L.) by Botryodiplodia theobromae Pat pp 71–73

Panth M, Hassler SC, Baysal-Gurel F (2020) Methods for management of soilborne diseases in crop production. Agriculture 10(1):16

Pernezny K, Simone GW (2000) Diseases of mango (Mangiferaindica L.). www.apsnet.org/online/common/names/mango.asp

Phipps N (2016) Tips to sterilize potting soil, garden soil and soil for seeds. Gardening how to› Soil & Fertilizers.http://www.gardeningknowhow.com/garden-how-to/soil-fertilizers/sterilizing-soil.htm

Poland TM, Haack RA, Petrice TR, Miller DL, Bauer LS (2006) Laboratory evaluation of the toxicity of systemic insecticides for control of *Anoplophoraglabripennis*and *Plectroderascalator*(Coleoptera: Cerambycidae). J Econ Entomol 99:85–93

Ploetz RC, Benscher D, Vazquez A, Colls A, Nagel J, Schaffer B (1996) A re-evaluation of mango decline in Florida. Plant Dis 80:664–668

Ploetz RC ed (2003) Diseases of tropical fruit crops. Cabi Publishing

Prakash OM (2004) Diseases and disorders of mango and their management. In: Diseases of fruits and vegetables volume I: diagnosis and management. Springer, Netherlands, Dordrecht, pp 511–619

Ranasingh N, Saturabh A, Nedunchezhiyan M (2006) Use of Trichoderma in disease managemant, Orissa Review, September-October, pp 68–70

Rehman A, Saleem M, Mehboob S, Bokhari AA (2011) Fungi associated with rhizosphere soil in mango decline orchards and their in vitro control. Pak J Phytopathol 23:112–117

Rodríguez H, Fraga R (1999) Phosphate solubilizing bacteria and their role in plant growth promotion. Biotechnol Adv 17(4–5):319–339

Rubin B, Cohen O, Gamliel A (2007) Soil solarization_ an environmentally-friendly alternative. UNEP 71–136

Sahi ST, Habib A, Ghazanfar MU, Badar A (2012) In vitro evaluation of different fungicides and plant extracts against Botryodipldiatheobromae, the causal agent of quick decline of mango. Pak J Phytopathol 24:137–142

Saleem A, Akhtar KM (1989) Mango diseases and their control. In: Choudhry AG, Syed MA (eds) Proceedings of the international mango conference workshop Feb. 27 to Mar. 01, 1989. Agri Ext Multan, pp 143–154

Shahbaz M, Iqbal Z, Saleem A, Anjum MA (2009) Association of lasiodiplodia theobromae with different decline disorders in mango (*Mangifera indica* L.). Pak J Bot 41(1):359–368

Shah MH, Rafique R, Almas M, Usman M, Yasin S, Bibi S (2022) A review: fruit production industry of Pakistan trends, issues and way forward. Pakistan J Agric Res 35(3)

Sharma I (1993) A note on population dynamics and etiology of dieback of mango in Himachal Pradesh. New Agric. 2:229–230

Sharma IM, Gupta VK (1994) Evaluation of different fungicides on Botryodiplodiatheobromae, casual organism of canker and dieback of mango. Indian J Mycol Pathol 24:146–147

Sharma IM, Gupta VK (1995) Evaluation of different fungicides on B. Theobromae Pat. causal organism of canker and dieback of mango. Ind J Mycol Pl Pathol 24:146–147

Shelar SA, Padule DN, Sawant DM, Konde BK (1997a) In vitro evaluation of fungicides against Botryodiplodiatheobromae Pat. The cause of Dieback disease of mango (Mangiferaindica L.). Ind J Pl Proct 25:118–120

Shelar SA, Padule DN, Sawant DM, Konde BK (1997b) In vitro evaluation of fungicides against *Botryodiplodiatheobromae*, the cause of die-back disease of mango. Indian J Plant Prot 25:118–120

Soman P, Chaudhari AU, Krishna B, Balasubramanyam VN (2024) Intensive cultivation of mango using ultra-high density planting (UHDP), and drip and fertigation technologies for higher productivity, pp 88–102. https://doi.org/10.9734/bpi/racas/v1/7229c

Tahir FM, Ibrahim M, Hamid K (2003) Effect of drought stress on vegetative and reproductive growth behaviour of mango (Mangiferaindica L.). Asian J Plant Sci 2(1):116–118

Van Wyk M, Al Adawi AO, Khan IA, Deadman ML, Al Jahwari AA, Wingfield BD, Ploetz R, Wingfield MJ (2007) Ceratocystismanginecans sp. nov., causal agent of a destructive mango wilt disease in Oman and Pakistan. Fungal Diversity 27:213–230

Tyler HL, Triplett EW (2008) Plants as a habitat for beneficial and/or human pathogenic bacteria. Annu Rev Phytopathol 46:53–73. https://doi.org/10.1146/annurev.phyto.011708.103102. PMID: 18680423

van Loenen MCA, Turbett Y, Mullins C, Feilden NEH, Wilson MT, Leifert C, Seel WE (2003) Low temperature-short duration steaming of soil kills soil-borne pathogens, nematode pests and weeds. Eur J Plant Pathol 109(9):993–1002. https://doi.org/10.1023/B:EJPP.0000003830.49949.34

Weiland JE (2021) The challenges of managing Phytophthora root rot in the nursery industry. Plant Health Prog. https://doi.org/10.1094/PHP-02-21-0036-FI

Chapter 17
Climate Variability and Sustainable Management of Mango Diseases: Emphasizing Modern Tools and Forecasting Models

Rafia Kiran and Muhammad Farhan

Abstract The mango, a fruit with significant global economic value, is a member of the *Mangifera indica* L. species and, as such, is susceptible to a number of diseases, the severity of which has been linked to variations in climate. Changes in temperature, precipitation patterns, and humidity in many situations promote the growth of pathogens, which exacerbates the symptoms of diseases including bacterial canker, powdery mildew, anthracnose, and mango sudden death syndrome. Mango yields and quality are impacted by these climate-driven changes, which presents challenges for both farmers and researchers. The chapter examines the complex relationships between climate change and mango diseases, how these relationships affect conventional management approaches, and the need for adaptable and sustainable solutions. To be more precise, it emphasizes how eco-friendly methods can be integrated with conventional ones by using cultural practices that are appropriate for the area, biological control agents, and resistant cultivars. We also discuss the latest methods of detection and surveillance that employ technologies for climate forecasting to anticipate disease outbreaks. This chapter offers an overview of the existing frameworks for managing diseases sustainably by presenting integrated disease management techniques that combine the newest technologies with agronomic approaches. We present case studies of effective initiatives that lessen the adverse effects of climatic variability on the production of mangoes in mango-producing areas. With a stronger emphasis on sustainable practices, this chapter aims to give stakeholders practical knowledge on how the mango industry is adapting to climate change.

Keywords Mango diseases · Forecasting models · Disease detection · IDM · Remote sensing · Emerging diseases

R. Kiran (✉) · M. Farhan
Department of Plant Pathology, Faculty of Agricultural Sciences and Technology, Bahauddin Zakariya University, Multan, Pakistan
e-mail: rafiakiran@student.bzu.edu.pk

S. A. H. Naqvi et al. (eds.), *Climate Change and Mango Production*,
https://doi.org/10.1007/978-3-031-87035-4_17

17.1 Introduction

Climate variability combined with plant disease concerns is becoming a bigger problem for the world's agriculture sector (Singh et al. 2023). The phrase "climate change" describes modifications to long-term patterns of meteorological variables like temperature and precipitation. Either natural variability or human action may be to blame for these shifts (Makhmale et al. 2016). Diseases and climate change are directly and significantly correlated (Shah et al. 2018). Mangos are susceptible to various diseases throughout their life cycle, from young plants in nurseries to mature fruits being stored or transported. Almost every component of a plant, including the stem, branch, twig, root, leaf, petiole, flower, and fruit, is susceptible to several infections, and very few plant organs are immune (Prakash 2004). Important diseases include Mango malformation (Freeman et al. 2014), Fruit Anthracnose (Awa et al. 2012), bacterial black spot (*Xanthomonas campestris pv. mangiferaeindicae*), powdery mildew (*Oidium mangiferae*), root rot (Khaskheli 2020), Mango dieback (Qi et al. 2013), canker of mango (Quimbita-Reyes et al. 2023), infloresence blight (Serrato-Diaz et al. 2013), stem end rot (Zhang et al. 2018), mango sudden death syndrome (Kazmi et al. 2005) and *Alternaria* rot (*Alternaria alternata*) (Ploetz 2003). Diseases such as anthracnose, bacterial blight, dieback, and scab are most effected by climate change (Yeo et al. 2022). Theoretically, there are a number of ways in which climate change could promote plant infection, such as by modifying the evolution of pathogens, affecting host–pathogen interactions and vector physiology, and encouraging the emergence of new pathogen strains that could undermine host-plant resistance (Cohen and Leach 2020).

In this ever-changing world's climate, some diseases may find some surroundings to be too hot, dry, or moist, while other previously unsuitable ecosystems may change and become more favorable for the spread of diseases (Salkeld et al. 2023). Diseases caused by pathogens *Colletotrichum gloeosporioides* (Sharma and Kulshrestha 2015) and *Oidium mangiferae* are favoured by high temperature and high humidity (Pérez-Rodríguez et al. 2017). Understanding how these various pathogens interact with and react to various disease drivers (such as other pathogens, host/vectors, commensal microorganisms, and environment) as well as how they collectively react to climate change is a critical challenge in predicting plant diseases in space and time, and their management (Velásquez et al. 2018). Climate change adaptation technologies can increase overall production and profitability by modifying crop management practices and controlling diseases, such as planting time and density, as well as applying nitrogen and irrigation (Habib-ur-Rahman et al. 2022).

Integrated disease management is the advised strategy because there isn't a single alternative management strategy that can offer the same degree of consistency and control as chemical control, and because there is increasing awareness of the long-term consequences of chemical control on non-target organisms and the environment (Alemu 2014). In order to combat mango diseases in the face of climatic change, integrated disease management (IDM) incorporates genetic techniques, biological and chemical controls, cultural practices, and monitoring and surveillance (Dugyala

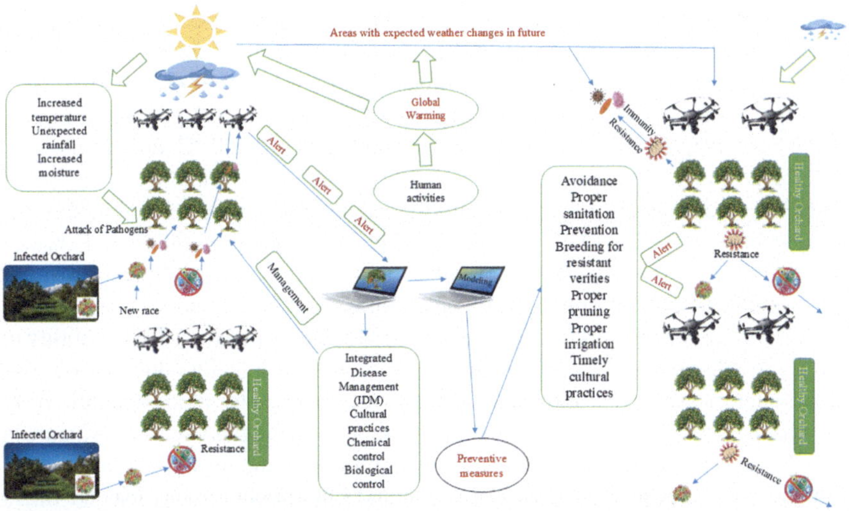

Fig. 17.1 Integrated disease management

2024). Comprehensive management of mango diseases requires knowledge of the biology of the pathosystem and the various control measures, including their viability from an ecological standpoint (Arauz 2000). Drones are essential for tracking the spread of plant pathogens and for their identification, diagnosis, and maintenance of crops' health (Abbas et al. 2023). According to recent research, CRISPR-Cas9 has the ability to modify important genes linked to disease resistance in mangos, with the goal of strengthening these plants' tolerance to several diseases (Ahmad et al. 2021a, b). This chapter discus the increasing severity and incidence of mango diseases in face of climatic changes. This also discuss the sustainable management approaches to control mango diseases in changing climate (Fig. 17.1).

17.2 Impact of Climate Variability on Mango Diseases

Currently, one of the major issues is global climate change. Since the beginning of the industrial revolution at the turn of the millennium, it has caused a progressive increase in the average annual temperature of the planet. Discussions over the rise in Earth's temperature have heated up in response to the recent wave of extreme weather events. High levels of industrial and commercial activity, which include emissions of the primary greenhouse gases—carbon dioxide, methane, etc.—explain the temperature shift (Albergel et al. 2010). According to Lipczynska-Kochany (2018). Climate change is described using long-range patterns in temperature and precipitation as well as other environmental variables like pressure and humidity. Moreover, the most well-known local and worldwide repercussions of climate change include the erratic weather patterns, melting of the world's ice sheets, and the resulting rise in sea level (Lipczynska-Kochany 2018). The best estimate of the total increase in global surface temperature caused by humans between 1850–1900 and 2010–2019 is 1.07 °C, with a probability range of 0.8–1.3 °C. Well-mixed greenhouse gases (GHGs) are likely to have contributed to a warming of 1.0–2.0 °C during this time, while other human drivers—primarily aerosols—were responsible for a cooling of 0.0–0.8 °C. Additionally, internal variability and natural drivers—namely, solar and volcanic activity—were likely to have contributed a change in the global surface temperature of − 0.2 to + 0.2 °C (Lindsey and Dahlman 2020). Specifically, the study concluded that "seasonally variable regimes are becoming more variable" and that "future climate projections indicate increased seasonal precipitation variability in highly variable regimes." A shift in the pattern of water availability due to variations in precipitation and evaporation is indicated by the fact that locations with relatively low seasonality of precipitation would get wetter during the rainy seasons (Konapala et al. 2020). The study revealed that the variability of rainfall in the Sahel (which spans from Senegal in West Africa to Sudan and Ethiopia in East Africa and varies latitudinally from just north of the tropical forests to just south of the Sahara desert) is influenced by both oceanic and regional land changes. However, the past few decades have seen a shift toward more intense and sporadic rainfall that occurs later in the season. The drought of the 1970s and 1980s was mostly caused by anomalies in sea surface temperature, and current projections of rising greenhouse gas concentrations suggest that rainfall patterns may still be altered. Improved rainfall projections, enhanced intraseasonal trends, and increased chances of extreme occurrences necessitate improved observational and modeling efforts (Biasutti 2019). One of the study in Sri Lanka's highlights changing rainfall patterns, including longer dry periods and increased rainfall intensity (Ratnayake and Herath 2005). According to the experts, Southeast Asia's monsoon rainfall and seasonal air flow are becoming more variable due to global warming. Since the 1970s, there has been evidence of a notable increase in worldwide precipitation anomalies, which is closely linked to rising temperatures. The Indian summer monsoon moved westward, and it is predicted that the monsoon onsets will occur 15 days later than usual. Decadal variability in rainfall and the possibility of sporadic flooding are the results of these changes, which have an effect on the region's infrastructure, food security, finances, and people (Loo et al. 2015). As per a recent study, variations in weather patterns' dynamics can be held responsible for about 60% of the variations in seasonal rainfall patterns and only around 20% of the temperature rise recorded over Europe since 1990. While temperature trends are less affected by dynamical component changes in weather type and frequency, long-term rainfall variability is greatly affected (Hoffmann and Spekat 2021). This changing climate is expected to increase plant disease incidence, shift disease patterns to new regions, alter host–pathogen interactions, and extend disease seasons due to warmer temperatures and changing precipitation (Fig. 17.2).

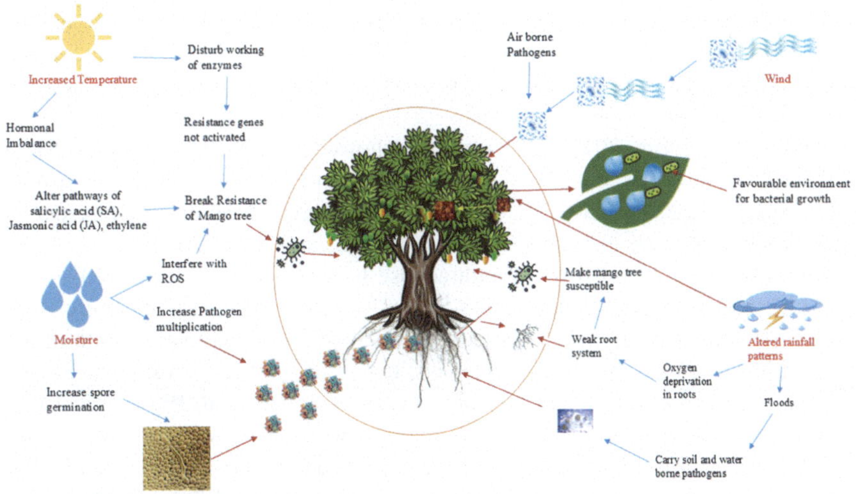

Fig. 17.2 Role of changing climatic factors in susceptibility of mango against diseases

17.3 Disease Dynamics Under Climate Variability

Numerous elements of disease development are impacted by climate change at varying degrees of pathosystem integration, and these effects can be complicated and unpredictable (Jeger and Pautasso 2008). Some important diseases of mango are given below in scenario of climate change.

17.3.1 Anthracnose

Mango anthracnose disease, or MAD, is a devastating disease that can cause up to 100% crop losses in poorly managed plantations (Dofuor et al. 2023). It is caused by *Colletotrichum gloeosporioides*, which propagates via tree canopies as water-borne conidia during rainfall. Young flush leaves, flowers, and fruit are affected by this disease. Climatic factors have a significant impact on the development and severity of anthracnose. Warm temperatures and high humidity are often the environmental factors that promote *C. gloeosporioides* spore germination and dispersion. In the range of 95–100% relative humidity (RH), conidia germinate and develop appressoria (Dodd et al. 1991). The altered rainfall pattern has the potential to significantly increase humidity, which increases the frequency of mango anthracnose since the extended wet periods on the fruit's surface are perfect for the germination of fungal spores (Arauz 2000). Mango trees may become more vulnerable to anthracnose infection during dry spells that are followed by unexpectedly high rains because stressed plants are less able to coordinate their defense mechanisms (Litz 2009). Rainy seasons facilitate the easy dispersal of pathogen spores by splashing water, hence amplifying the disease's spread within an orchard (Freeman et al. 1998). The incidence and severity of mango anthracnose were studied in Dashehari mango orchards of Lucknow during three consecutive years, 2013–2015. Maximum disease incidence attained to 40% during 2013. Cooler months showed higher Humid Thermal Ratio (HTR) values, positively correlated to enhanced severity of anthracnose. The results of this study suggested that HTR could be used to predict the disease in order to guide timely protective measures to farmers (Shukla and Adak 2017).

Similarly, rapid progression of anthracnose lesions occurs in warm (20–30 °C for 12 h) (Kamle et al. 2013), whereas the anthracnose ambient temperature is 28 ± 2 °C (Diedhiou et al. 2007). Temperatures above 28 °C may have a role in the spread of the *C. gloeosporioides* epidemic (Jenny et al. 2019). As a result, the fungus is more common in tropical and subtropical regions (Ploetz et al. 1996). According to further studies, plant diseases are moving into areas where they were previously absent due to temperature increases brought on by numerous human activities. In a similar vein, environmental variables can affect the host plant's potential for a number of functions, including protein, nitrogen, starch, and sugar concentrations, all of which can increase or decrease the plant's resistance to disease (Elad and Pertot 2014). A study on the impact of temperature on diseases that affect mangos after harvest found that, among other things, the infections *Lasiodiplodia theobromae* and *Colletotrichum gloeosporioides*, which cause stem-end rot and anthracnose, reached their maximum mycelial growth at 25–30 °C. In the examined regions, the highest disease incidence observed was 28.76% during periods of maximum temperature (Parthasarathy et al. 2015). The average annual temperature has increased by 0.0067 degrees Celsius over the course of 172 years, starting in 1850. For 2050, extrapolate a 1.5556 ºC global temperature variation, which is far higher than the UNFCCC's permitted maximum increase (Wang et al. 2023). This increase in global temperature is probably going to encourage mango anthracnose's spread and severity. As a result of global warming, the environment is becoming even more favorable for the disease to spread, increasing the risk of infection and making control issues more challenging.

17.3.2 Powdery Mildew

The fungus *Pseudoidium anacardii*, formerly known as *Oidium mangifera*, is the causal agent of powdery mildew, which can cause major yield losses to mango crops globally. The ideal temperature range for the development of disease is normally between 11 and 14 °C at minimum and 27 and 31 °C at maximum, along with relative humidity levels between 64 and 72%. Even with the broad reach and impact of climate change, losses are steadily rising (Iqbal et al. 2024). The pathogen thrives in environments with high relative humidity, often 60–90%. Fungus growth and spore germination are permitted under these conditions (Nelson 2008). Increased intensity of rainfall raises humidity levels, which in turn promotes powdery mildew breakouts. However, even if they improve the disease's severity, showers or mists, particularly during flowering, significantly reduce production due to flower and fruit drop (Shukla et al. 2016a, b). A seven-year study on powdery mildew in South Gujarat revealed a negative correlation between the disease's occurrence and severity on mangos and morning and average relative humidity, but a positive correlation with maximum temperature and evaporation. In terms of variability, the average RH accounted for 59% and 62%, respectively, making it the best predictor of disease incidence and severity (Bana et al. 2020). The incidence of powder mildew in mangoes from Maharashtra's Konkan region was positively and significantly associated with the minimum temperature. In March of both the 2013–14 and 2014–15 seasons, the highest illness intensity—85.60 and 73.40%, respectively was developed. Only the minimum temperature revealed a significant positive link among the climatic elements; the maximum temperature and the evening humidity showed a positive but non-significant correlation. The morning humidity exhibited a non-significant negative connection (Raut et al. 2017). In the Hyderabad study, 28.2 °C and 49% relative humidity (RH) had the highest frequency of mango powdery mildew, especially in Siroli varieties. In the Sindhri variety, the lowest incidence happened at a higher temperature of 40 °C and 42% RH. As a result, the correlation between temperature, humidity, and disease incidence changed, with certain varieties showing higher susceptibility at intermediate temperatures (Abro et al. 2019). In places with prolonged wetting periods, increasing humidity under shifting rainfall patterns might result in a more frequent and severe infection of mango powdery mildew. It is anticipated that the shifting environment would enhance the conditions for the disease's emergence and growth, making containment of the epidemic more difficult.

17.3.3 Mango Dieback

Mangoes are among the many crops that have been known to experience dieback, cankers, and fruit rot due to *Lasiodiplodia theobromae*. This infection can lead to severe fruit rot and dieback in mango crops, resulting in a reduction in fruit quantity and quality. Per reports, these diseases can result in up to 100% loss of seed germination from afflicted crops, which can have a severe negative impact on productivity (Akhtar and Dubey 2021). The ideal temperature range for *L. theobromae* is between 27 and 33 °C, although it can thrive at ranges as wide as 9–39 °C. The fungus may probably become ready to expand to new areas of infection in hosts at higher latitudes that are renowned for being unsuited as global warming proceeds through the drive of climate change (Mehl et al. 2017). The pathogen *L. theobromae* is more active during times of increased temperature and humidity, which may increase the degree of infection and severity of the disease in mangos (Legave et al. 2013). Because of the favorable growth and development conditions provided by high temperatures and humidity, *L. theobromae* is most commonly found in countries located in the world's tropics and subtropics. This covers significant mango and other sensitive crop-growing regions, such as portions of Africa, Asia, the Americas, and Australia (Correia et al. 2016). There have already been reports of *L. theobromae* cases in Tamil

Nadu, Karnataka, and Andhra Pradesh, India. This is explained by the warm weather that permeates the aforementioned regions and the agricultural practices that, for the most part, involve high moisture content, which is ideal for the pathogens (Gnanesh et al. 2022).

It is known that abiotic stressors such as heat, drought, or water stress cause the pathogen *Lasiodiplodia theobromae* to become more aggressive. The latter effects of climate change can cause these deficiencies, which can lead to illness and eventual collapse of the mango trees. If the attack continues, infected trees may initially exhibit moderate symptoms like wilting and leaf curling, which could eventually result in complete defoliation (Saeed et al. 2017). As global temperatures rise, this will expand the range of geographical areas with acceptable climates for *L. theobromae* to flourish. The fungus is mostly restricted to latitudes between 40° N and 40° S; however, warming may allow it to expand, increasing the pool of possible hosts (Félix et al. 2016).

17.3.4 Mango Sudden Death Syndrome

The occurrence of mango sudden death syndrome (MSDS), a dangerous disease of mango trees brought on by infections such as *Ceratocystis fimbriata* and *Lasiodiplodia theobromae*, is certain to be significantly impacted by climate change. Conditions that encourage disease propagation, increasing severity and incidence, can be brought about by temperature increases and changes in precipitation patterns. For example, elevated temperatures can cause stress to mango trees, making them more vulnerable to infection (Khaskheli et al. 2011). Elevated precipitation may also equate to increased humidity, which would facilitate the spread of fungal spores and the effectiveness of vectors like as the mango bark beetle, *Hypocryphalus mangiferae*, hence augmenting the likelihood of MSDS outbreaks (Galdino et al. 2016). Furthermore, the interaction between temperature and humidity can impact the severity of the disease; excessive humidity at optimal temperature can lead to rapid disease progression, as seen in areas with high rainfall (Kazmi et al. 2005). Modified meteorological circumstances could increase the danger of MSDS in regions of South Asia, Brazil, and Oman—already the most susceptible to mango diseases—thus jeopardizing mango output and farmers' livelihoods (Asif et al. 2011).

17.3.5 Mango Malformation

Mango malformation disease (MMD) was initially identified in India in 1891 and is seen in Asia, Africa, and the Americas. *Fusarium moniliforme* was used to replicate the vegetative form of MMD in 1966, while isolates of *F. moniliforme var. subglutinans* from both vegetative shoots and floral tissue were used to replicate the floral form. Later on, the fungus were identified as *F. subglutinans*. Based on nuclear and mitochondrial DNA sequences, a novel species, *F. mangiferae*, was identified in 2002. This species contained strains of *F. subglutinans* from South Africa, Israel, Florida, Egypt, Malaysia, and Singapore, some of which had been demonstrated to cause MMD through artificial inoculation (Marasas et al. 2006). It is anticipated that unfavorable factors related to climate change, such as rising temperatures and unpredictable precipitation patterns, may encourage the emergence and spread of mango malformation. Higher humidity and temperatures at these times may intensify the pathogen's aggressiveness, which would increase infection rates and the severity of the disease in mango plants (Ansari et al. 2015).

Thus, the regular flowering and fruiting cycles of mango trees may be disrupted by temperature variations associated with climate change, providing windows of opportunity for infections by pathogens (Legave et al. 2013). Furthermore, the combination of ethylene with additional stresses like flooding or extremely high temperatures might exacerbate the severity of malformation symptoms. Endogenous ethylene content in the deformed tissues was found to be many times more than in the healthy tissues, indicating that ethylene functions as a stress signal that encourages deformity (Ansari et al. 2013). An other intriguing discovery pertains to the infection *Fusarium mangiferae* and its association with this illness. It is probable that this pathogen plays a more significant role in the ethylene generation, further complicating the dynamics of the sickness (Ansari et al. 2015). Apart from the mango bud mite, other insects that feed on sap, like aphids and thrips, can also serve as carriers for the mango malformation disease. These insects help the illness spread throughout orchards and maybe to other areas by spreading fungal spores from diseased plants to healthy ones (Gamliel-Atinsky et al. 2009). Thus, variations in temperature would have an impact on the mango bud mite's life cycle and population dynamics. Raising temperatures can lead to an increase in mite populations, which can cause more harm to mango plants and more deformity (Legave et al. 2013). The geographic range of locations with climates conducive to mango malformation pathogens will increase if global temperatures rise further. As a result, the disease will spread to previously inappropriate hosts at higher latitudes, invading new areas (Misra 2012).

17.3.6 Bacterial Canker

The most crucial fact is that *Erwinia billingiae* and *Xanthomonas axonopodis pv. mangiferaeindicae* are the bacteria that cause bacterial canker of mangos. It's one of the most serious diseases that mango trees (*Mangifera indica*) are susceptible to. It lowers fruit quality and yield, which results in major economic losses in mango production. The growing occurrence of extreme weather events, such as drought and intense rain, can make managing bacterial canker even more difficult (Gagnevin and Pruvost 2001). Changes in the climate can result in higher humidity and erratic rainfall patterns, which create an environment that is favorable to the growth of bacteria. Elevated relative humidity facilitates the spread of bacterial spores. However, moisture causes the water-soaked lesions that are typical of bacterial canker on fruits and plants (Zombré et al. 2017). This could change the bacterial canker's geographic distribution. Stated differently, shifting climate and precipitation patterns may render previously unsuitable regions for bacterial canker conducive to its growth, hence resulting in the spread of epidemics to previously unaffected areas (Midha et al. 2012). It makes managing bacterial canker more difficult. The altered climate may make traditional cultural and chemical control approaches ineffective. This necessitates the development of fresh approaches to adaptive management that take into consideration how the dynamics of illness are impacted by climate change (Ploetz 2003).

17.4 Role of Climate Variability in Emerging Diseases and Pathogen Evolution

The spread of current infections and the introduction of new ones could be accelerated by climate change. For example, variations in humidity and temperature may allow *Fusarium mangiferae*, a fungus that causes deformity in mangoes, to expand its habitats. According to certain research, climate fluctuation may increase the degree of mango floral malformation, especially in India where environmental conditions are more conducive to the pathogen's proliferation (Usha et al. 2022). With shifting climatic circumstances, the prevalence of already-existing diseases like mango sudden death syndrome (*Ceratocystis fimbriata*) and bacterial canker (*Xanthomonas axonopodis pv. mangiferaeindicae*) is expected to rise. For example, increased humidity and warmth might strengthen the bacterial canker vector and cause a more significant epidemic (Leng et al. 2023). Some important diseases of mango that are newly reported and some that are shifted into new regions due to the changing climate are discussed in Table 17.1.

17.5 Sustainable Management Practices

17.5.1 Breeding and Developing Resistant Varieties

Significant strides have been made in the cloning of R genes over the last thirty years. From the cloning of R genes in model species into agricultural plants and their wild counterparts, recombinant DNA technology and genome sequencing, together with bioinformatic tools and plant transformation and husbandry, have come a long way, aiding translational research. Up to date, scientists have identified around 450 R genes that play significant roles in plant immunity (Chen et al. 2024). In order to successfully breed disease resistance, traditional breeding takes time. Furthermore, the resistance genes may be broken because these races are very new to exist. This makes it extremely difficult to engineer disease resistance for future fast tracking

Table 17.1 Emergence of new diseases and existing mango diseases in new regions during past half century (1974–2024) due to climate change

S. No.	Disease	Pathogen	Year	Geographic region	Type	References
1	Mango malformation disease (MMD)	*Fusarium pseudocircinatum*	2014	Mexico	New region	Freeman et al. (2014)
2	Mango malformation disease (MMD)	*Fusarium tupiense*	2012	Senegal	New region	Senghor et al. (2012)
3	Mango malformation disease (MMD)	*Fusarium mangiferae*	2006	Spain	New region	Crespo et al. (2012)
4	Mango malformation disease (MMD)	*Fusarium mangiferae*	2013	Sri Lanka	New region	Sinniah et al. (2013)
5	Mango malformation disease (MMD)	*Fusarium mangiferae*	2012	China	New region	Zhan et al. (2012)
6	MSDS	*Ceratocystis fimbriata, Lasiodiplodia theobromae, Ceratocystis omanensis*	1998	Oman	New region	Al Adawi et al. (2006)
7	Mango malformation disease (MMD)	*Fusarium nivale* (fr.) Ces.	2008	Pakistan	New region	Khaskheli et al. (2008)
8	Mango leaf anthracnose	*Colletotrichum asianum*	2020	Vietnam	New region	Li et al. (2020)
9	Mango anthracnose (post-harvest)	*Colletotrichum karstii*	2013	Brazil	New region	Lima et al. (2013)
10	Mango anthracnose (post-harvest)	*Colletotrichum siamense*	2017	China	New region	Liu et al. (2017)
11	Mango leaf anthracnose	*Colletotrichum grossum*	2018	Cuba	New region	Manzano León et al. (2018)
12	Mango anthracnose	*Colletotrichum scovillei*	2019	China	New region	Qin et al. (2019)
13	Rachis necrosis and inflorescence blight of mango	*Neofusicoccum mangiferae*	2014	Puerto Rico	New region	Serrato-Diaz et al. (2014)

(continued)

Table 17.1 (continued)

S. No.	Disease	Pathogen	Year	Geographic region	Type	References
14	Mango anthracnose	*Colletotrichum asianum*	2023	Peru	New region	Vilcarromero-Ramos et al. (2023)
15	Fruit anthracnose in Mango	*Colletotrichum gloeosporioides*	2012	Southwestern Nigeria	New region	Awa et al. (2012)
16	Mango anthracnose	*C. gloeosporioides, C. kahawaw* subsp. *ciggaro, C. karstii*	2015	Italy	New region	Ismail et al. (2015)
17	Mango leaf anthracnose	*C. asianum, C. fructicola,* and *C. siamense C. gloeosporioides*	2018	Guangxi province, China	New region	Mo et al. (2018)
18	Postharvest anthracnose disease of mango	*Colletotrichum alienum*	2021	China	New region	Ahmad et al. (2021a, b)
19	Apical necrosis of mango	*Pseudomonas syringae* pv. syringae	1998	Spain	New disease	Cazorla et al. (1998)
20	Wilt disease of mango	*Ceratocystis fimbriata*	2018	Uttar Pradesh, India	New region	Shukla et al.(2018)
21	Gray leaf spot of mango	*Pestalotiopsis mangiferae*	2007	Taiwan	New disease	Ko et al. (2007)
22	Mango leaf blotch	*Pseudoplagiostoma mangiferae*	2022	Taiwan	New region	Zhou et al. (2022)
23	Mango bacterial canker	*Xanthomonas citri* pv. *mangiferae indicae*	2015	Benin	New region	Zombré et al. (2015)
24	Bacterial black spot of mango	*Xanthomonas citri* pv. *Mangiferae indicae*	2013	Hawai'i	New region	Yasuhara-Bell et al. (2013)
25	Bacterial black spot of mango	*Xanthomonas campestris pv. Mangiferae-indicae*	1981	Darwin area	New disease	Pitkethley (2006)

(continued)

Table 17.1 (continued)

S. No.	Disease	Pathogen	Year	Geographic region	Type	References
26	Bacterial black spot of mango	*Xanthomonas citri* pv. *Mangiferae indicae*	2014	Iran	New region	Najafipour et al. (2014)
27	Bacterial necrosis	*Pantoeavagans*	2023	China	New region	Chen et al. (2023)
28	Bacterial black spot of mango	*Xanthomonas campestris pv. manqiferae indicae*	1989	Malaysia	New region	Lim et al. (1989)
29	Mango bacterial canker	*Xanthomonas citri* pv. *Mangiferae indicae*	2011	Ghana	New region	Pruvost et al. (2011)
30	Mango bacterial canker	*Xanthomonas citri* pv. *Mangiferae indica*	2012	Mali	New region	Pruvost et al. (2012)
31	MSDS	*Ceratocystis ssp.,*	1995	Muzafargargh in Pakistan	New region	Asma Rashid et al. (2013)
32	Red rust disease of mango	Cephaleurosvirescens	2020	Cameroon	New region	Patrice et al. (2020)
33	Powdery mildew of mango	*Erysiphequercicola*	2020	Japan	New region	Ajitomi et al. (2020)
34	Powdery mildew of mango	*Pseudoidiumanacardii*	2013	Mexico	New region	Félix-Gastélum et al. (2013)
35	Powdery mildew of mango	*Oidium mangiferae*	1983	Hawai'i	New region	Nelson (2008)
36	Powdery mildew of mango	*Erysiphequercicola* and *Erysiphealphitoides*	2017	Mainland Europe	New region	Desprez-Loustau et al. (2017)
37	Mango wilt	*Ceratocystis fimbriata*	2017	China	New region	Zhang et al. (2017)
38	Verticillium wilt of mango	*Verticillium dahliae*	2010	Southern Spain	New region	Baeza-Montañez et al. (2010)
39	Inflorescence blight of Mango	*Lasiodiplodia theobromae*	2013	Puerto Rico	New region	Serrato-Diaz et al. (2013)
40	Dieback of mango	*Fusarium decemcellulare*	2013	China	New region	Qi et al. (2013)
41	Dieback disease of mango	*Lasiodiplodia theobromae*	2023	Saudi Arabia	New region	El-Komy et al. (2023)
42	New wilt disease on mango	*Ceratocystis fimbriata*	2023	Indonesia	New region	Pratama et al. (2023)

(continued)

Table 17.1 (continued)

S. No.	Disease	Pathogen	Year	Geographic region	Type	References
43	Stem end rot of mango fruit	*Lasiodiplodia pseudotheobromae*	2021	Pakistan	New region	Alam et al. (2021)
44	Stem end rot of mango	*Phomopsis mangiferae*	2009	Taiwan	New region	Ko et al. (2009)
45	Postharvest stem end rot of mango	*Diaporthe pseudomangiferae*	2023	China	New region	Li et al. (2023)
46	Bacterial necrosis of mango	*Pseudomonas syringae* pv. syringae	2008	Australia	New region	Golzar and Cother (2008)
47	Fruit rot disease of mango	*Botryosphaeria dothidea* and *Neofusicoccum mangiferae*	2008	Taiwan	New region	Ni et al. (2010)
48	Mango anthracnose	*Colletotrichum asianum*	2015	South Africa	New region	Sharma et al. (2015)
49	Leaf blight of mango	*Drechslera australiensis*	2010	India	New Disease	Sangeetha Chinnusamy et al. (2010)
50	*Cladosporium* infection of mango inflorescence	*Cladopsorium tenuissimum*	2013	Iran	New region	Baharvandi and Zafar (2015)

and introgression into adaptable cultivars; hence, molecular technologies should be used to supplement traditional breeding methods (Garoma et al. 2024). The fruit industry must develop improved varieties since fruit cultivation is constantly under strain from changing climates and invasive pathogens that adapt quickly. In light of this, innovative breeding methods have surfaced as a viable means of hastening the creation of more suited cultivars in order to feed the world's expanding population. In certain plant species, cisgenesis, CRISPR/Cas editing, and accelerated breeding have already shown to be effective methods for improving agricultural traits. For the purpose of producing exogenous DNA-free fruit tree species, advancements in protoplast regeneration and distribution strategies—such as those involving nanoparticles and virally derived replicon-based techniques—are presented (Campa et al. 2024). The invention of speed breeding (SB) technology allows for the acceleration of crop improvement and the compression of the breeding cycle. All of these are done in an effort to change the key regulating elements that affect plant growth and development in a way that promotes flowering and seed set, such as photoperiod, light intensity and quality, temperature, relative humidity, planting density, and plant nutrition. Each of these is done in a controlled environment. Additionally, the SB method can be used in conjunction with marker-assisted selection and sophisticated genomics to maximize genetic gain in crop breeding initiatives (Pasala et al. 2024). Tommy Atkins and Ataulfo varieties of mango showed resistance against mango Anthracnose in Mexico (Sanchez-Arizpe et al. 2021). In a research it was found that cultivars 'Hindi Besennara', 'Sherokerzam', 'Mulgoa', 'Baneshan', 'Rose' and 'Alumpur Baneshan', are considered as relatively resistant against *C. manginecans* (Al Adawi et al. 2010). Mango breeding programs have been started with clear goals in many nations, and a huge number of hybrids are undergoing rigorous screening. It is therefore anticipated that a large number of additional hybrid cultivars will be made available soon. Understanding the phenology, inheritance patterns, and sophisticated hybridization procedures of mangoes has proven to be quite beneficial in overcoming issues like as irregular bearing, pest and disease susceptibility, and poor quality for eating and storing. The creation of genetic markers improves the management of mango hybrid populations by lowering breeding uncertainty. Clonal selection produces high-yielding, valued outcomes in many cultivars, making it worthwhile to pursue in many nations where the cultivation of particular cultivars may persist for an extended period of time (Salvi et al. 2019).

17.5.2 Integrated Disease Management (IDM)

The use of integrated Disease management (IDM) is one tactic that can promote environmentally friendly agricultural practices and lower the chance of new pests and/or diseases spreading. The goal of the integrated disease management method is to protect plants in a way that minimizes negative environmental effects while increasing production through preventative or indirect measures (Skendžić et al. 2021). The application of antagonistic microorganisms, field sanitation, bagging at the early fruit formation stage, scheduled chemical sprays, pruning and removal of diseased twigs and leaves, and covering orchard ground with vinyl sheets are some of the methods for integrated disease management (IDM), for controlling mango diseases (Ann et al. 2013). For example, mango diseases like anthracnose can be managed by the application of micronutrients such as boron (B), copper (Cu), and zinc (Zn) along with proper amounts of nitrogen, phosphorus, and potassium (NPK) and FYM (farmyard manure) (Nazir et al. 2024). Moreover, a study revealed that the post-harvest pathogens of mango were completely suppressed by the essential oils of oregano, rosemary pepper, cinnamon bark, and clove, with MIC values ranging from 250 to 2000 $\mu L\ L^{-1}$. Carvacrol (69.1%) was the primary chemical in oregano; thymol (77.2%) was present in rosemary and pepper oil; the primary component of cinnamon bark was discovered to be cinnamondehyde (85.1%), while cloves contained eugenol (84.84%). When these substances were tested for their antifungal properties, thymol and carvacrol shown stronger inhibitory effect against fungus (Vilela et al. 2024). Furthermore one of the study's findings demonstrated that treating post-harvest diseases of mango with a fresh formulation containing 108 spores/mL of *Talaromyces tratensis* KUFA 0091, applied three times throughout the flowering, fruit-setting, and fruit-developing stages, reduced the diseases by 80–87%. 75–78% fewer illnesses were caused by the dry formulation. At modest doses of 106 spores/mL, both the fresh and dry formulations demonstrated strong control against mango diseases, ranging from 70 to 79%. The potency of this fungus was equivalent to that of the fungicides mancozeb 1600 ppm, azoxystrobin 250 ppm, and propineb 1750 ppm. It may be employed in integrated mango disease management as a biological control agent (Dethoup et al. 2024). Among the different biocontrol agents tested against *Lasiodiplodia theobromae*, *Trichoderma viride* showed the maximum inhibition of growth, i.e., 71.47%. *Trichoderma harzianum* showed 67.15% inhibition. Rhizospheric LMA showed 47.84% inhibition. *Pseudomonas fluorescens* showed minimum inhibition, i.e., 3.46%. The above results indicated that potential use in management of mango dieback (Parmar et al. 2024).

17.5.3 Early Detection and Monitoring

Effective management strategies and proactive research depend on the accurate, fast, and early diagnosis and detection of plant diseases. The pathogen is discovered long after the disease first manifests, and symptoms-based disease diagnosis techniques are laborious and imprecise. Plant disease detection has been modernized by molecular biology techniques that use essential biomolecules like DNA (quantitative polymerase chain reaction (PCR), microarray, reverse transcriptase PCR, RNA-seq-based next-generation sequencing), protein (Western blot, enzyme-linked immunosorbent assay), and RNA (reverse transcriptase PCR) (Sapre et al. 2021). The methods used for pathogen identification include microarray, PCR-based techniques, dot blot hybridization, PCR–RFLP, reverse transcription-PCR, nested PCR, multiplex PCR, and more. The application of cutting-edge technology such as proteomics and bioinformatics to the study of plant pathogenic organisms has made it possible to detect precise and targeted DNA sequences, which will ultimately improve the accuracy of cutting-edge methods for the detection of plant illnesses (Kumar 2024). Because of their low cost and high throughput potential, serological methods, mostly based on enzyme-linked immunosorbent assay (ELISA), are the most widely used approach for pathogen identification. This method is not very sensitive or dependable for detecting many infections when an illness is still asymptomatic (Trippa et al. 2024). However, these procedures are very inconvenient for widespread on-site use because they frequently call for expensive, specialized instruments, aseptic environments, and skilled workers. These issues can be resolved with the use of the innovative CRISPR technology in illness diagnostics, which offers excellent specificity, selectivity, and a low detection limit. Furthermore, by combining several signal-sensing approaches with CRISPR-Cas-based detection, the ability to recognize targets in situ can be further improved. The resulting system can then be integrated into smart devices (Kasfy et al. 2024).

Alternative methods have been proposed recently that use electromagnetic radiation to interact with objects and living things in a proximity or remote sensing manner. Improvements in satellite and aerial detection techniques, telecommunications systems, and optical sensors enable the application of these analysis techniques to images obtained by satellites (remote sensing by satellite), small manned planes or helicopters, or aerial platforms with a remote pilot (unmanned aerial vehicles; UAVs), commonly referred to as "drones," in a variety of fields (Casagli et al. 2017). Depending on the kind of observation required, UAV platforms can be further divided into fixed-wing and rotary-wing categories. Whereas rotary-wing platforms can scan smaller areas from medium–low altitudes, fixed-wing platforms can monitor large areas from medium–high altitudes. Precision agriculture and many other fields have been using UAVs more and more in recent years, necessitating the need for test systems that can ensure and validate the mechanical and electrical performance components of their propulsion subsystems (Adamo et al. 2017). The nondestructive, quick, and high throughput features of spectroscopy have been used in the detection of plant diseases. Based on the spectrometer's carrying platform, spectral detection devices are classified into three categories: large laboratory spectrometer, aerial vehicle-mounted spectrometer, and portable hand-held spectrometer (Fang et al. 2023). Optical methods like chlorophyll fluorescence, RGB imaging, multi- and hyperspectral sensors, and thermography have shown promise in automated, objective, and repeatable detection systems for the early detection and quantification of plant diseases. As an optical examination that provides more information on agricultural plant viability, 3D scanning has been incorporated recently. There are several systems available for multiscale monitoring of individual crop organs or entire fields, ranging from local to remote sensing. Highly advanced and novel methods of data analysis that produce fresh insights from sensor data for intricate plant-pathogen systems enable accurate and dependable disease diagnosis (Mahlein 2016). Since new types of gas sensor devices and disease detection methods have been developed that use sensor arrays specifically tailored for different chemical classes of volatile organic compounds (VOCs) that are closely linked to particular diseases, the use of e-nose devices for disease diagnostic applications has expanded to plant and animal hosts for a range of analyses. Non-invasive early detection of plant and animal diseases is one of these analyses (Wilson 2018). There are some techniques and tools mentioned in Table 17.2 that are used for the early detection of plant tree diseases and can be employ to detect the diseases of mango tree.

17.5.4 Forecasting

Problems with current plant disease management are brought on by chemical residues, fungicide resistance, climate change, and declining biodiversity. Plant diseases can be anticipated in advance by using forecasting models for plant diseases (Senthilraja et al. 2024). Plant disease warning systems are recognized techniques for estimating the likelihood of a disease developing and its severity, which can help with management choices. Daily decisions like when to apply a fungicide, the risk of seasonal diseases, or even longer time horizons that could influence policy decisions can be influenced by these systems. These systems are similar in that they predict the onset of disease by using information about one or more disease triangle components. Plant disease warning systems, however, do not follow a uniform framework; instead, several strategies are employed (Gent and Del Ponte 2024). The availability of more automatic sensors for measuring weather and soil variables (air and soil temperature, rain, atmospheric humidity, wetness duration, wind speed and direction, and soil moisture) that influence the emergence of disease epidemics has significantly improved the ability to measure environmental factors and, consequently, investigate their impact on epidemiological components. Therefore, this was essential for creating environmental-influenced plant disease models (Kranz and Rotem 2012). Over time, numerous forecasting models have been created and applied to predict a range of diseases worldwide. For the tomato-infecting fungus Alternaria solani, which develops early blight when environmental circumstances are suitable, a computerized forecasting system known as FAST was created in 1978. There are computerized forecast models for potato late blight, such as BLITECAST, JHULSACAST, etc. (Nath et al. 2020). Using a rule-based paradigm, an expert system has been built for the diagnosis of plant diseases in the Barracuda mango (Nam-Dok Mai). After the system was put to the test, 129 photos of leaves were categorized into three groups: Normal, Algal Spot, and Anthracnose. Its accuracy of 89.92% in this area suggests that it will likely be useful in aiding plant disease detection and effective disease management by agriculturists. Results suggest that by offering a precise diagnosis and assistance with decision-making, this approach may be helpful in lowering treatment errors (Trongtorkid and Pramokchon 2018). There are different models built to predict the mango diseases shown in Table 17.3.

Table 17.2 Modern tools for disease detection in forest trees

S. No.	Detection tool/ technique	Working	Disease	Tree	References
1	Vis–NIR hyperspectral imaging	In particular, visible to near-infrared wavelengths are captured by Vis–NIR hyperspectral imaging. The measurements of 460, 520, 690, 760, and 950 nm may be significant in characterizing the symptoms of anthracnose	Anthracnose	Mango	Velásquez et al. (2024)
2	Hyperspectral imaging (HSI) integrated with machine learning (ML) techniques	Hyperspectral imaging (HSI) captures spectral and spatial data across a wide range of wavelengths	Anthracnose	Mango	Siripatrawan and Makino (2024)
3	Multirotor unmanned aerial vehicle (UAV) equipped with high-resolution visible and multispectral imaging systems	The detection of the disease is based on a Linear Discriminant Analysis (LDA) classifier trained on segmented trees from five-band multispectral image stacks	Olive quick decline syndrome (OQDS)	Olive	Di Nisio et al. (2020)
4	Visible-near-infrared (Vis–NIR) Spectrometry	Reflectance spectra of leaves in the visible and near-infrared spectrum bands are used in Vis–NIR spectroscopy to identify Fire Blight in pear plants	Fire blight	Pear	Bagheri and Mohamadi-Monavar (2020)

(continued)

Table 17.2 (continued)

S. No.	Detection tool/ technique	Working	Disease	Tree	References
5	Unmanned aerial vehicle (UAV) combined with a multispectral radiometer	After mounting the multispectral radiometer on a UAV, data on the various wavelengths of light reflected by olive trees is obtained by flying over olive orchards to capture high-resolution multispectral photos	Olive quick decline syndrome (*Xylella fastidiosa* subsp. pauca (*Xfp*))	Olive	Castrignanò et al. (2020)
6	Electronic nose	Electronic nose detects volatile organic chemicals released from affected plant tissues by specific pathogens using a collection of chemical sensors	Ash shoot dieback	Ash	Borowik et al. (2021)
7	Ground penetrating radar (GPR)	Using electromagnetic waves, ground-penetrating radar is a non-invasive imaging method that can detect and map subterranean structures	Ash dieback, acute Oak decline (AOD)	Ash and Oak	Giannakis et al. (2019)
8	Drone remote sensing	Drone photos are analyzed using the SVM and ANN algorithms' image processing techniques to identify pine trees impacted by pine wilt	Pine wilt disease	Pine	Syifa et al. (2020)

(continued)

Table 17.2 (continued)

S. No.	Detection tool/ technique	Working	Disease	Tree	References
9	Multispectral imaging	Using filters and neural networks (MLP and RBF), multispectral pictures captured with an MCA-6 Tetracam camera are utilized to distinguish between different phases of avocado tree laurel wilt	Laurel wilt	Avocado	De Castro et al. (2015)
10	Remote sensing with thermal and hyperspectral imagery	To identify and categorize the severity of Verticillium wilt in olive trees, high-resolution thermal and hyperspectral pictures was subjected to the techniques of support vector machine classification and linear discriminant analysis	Verticillium wilt	Olive	Calderón et al. (2015)
11	Hyperspectral imaging (HSI) combined with artificial neural network (ANN)	Oil palm trees are captured by UAVs in hyperspectral photos, which are subsequently pre-processed to improve spectral properties. Utilizing the ground truth data, train an MLP model to discover the spectral characteristics linked to various BSR stages	Basal stem rot	Palm oil	Lee et al. (2022)

(continued)

Table 17.2 (continued)

S. No.	Detection tool/ technique	Working	Disease	Tree	References
12	Hyperspectral imaging with unmanned aerial vehicle (UAV) combined with random forest algorithm	Based on a random forest algorithm, oil palm trees are then classified as healthy, early-stage, late-stage, or dead based on detailed spectral data obtained by utilizing a hyperspectral imaging UAV	Basal stem rot	Palm oil	Kurihara et al. (2022)
13	YOLO v4 with SENet and feature enhancement module	This model processes drone-derived images to accurately identify and locate infected trees	Pine wilt	Pine	Zhang et al. (2024)
14	Linear discriminant analysis (LDA) for hyperspectral data analysis	Red-edge bands and first derivative reflectance at 714 nm displayed the strongest separability for effective early detection, according to the findings of an LDA analysis conducted to compare separability between healthy trees and pine trees in the early stages of wilt disease	Pine wilt	Pine	Li et al. (2024)

(continued)

Table 17.2 (continued)

S. No.	Detection tool/technique	Working	Disease	Tree	References
15	Multi-scale selective attention convolutional neural network (MSA-CNN)	The MSA-CNN model achieves up to 98.44% OA in the identification of powdery mildew in rubber plants by utilizing important features that are learned from ideal wavelengths, vegetation indices, and texture features. This could guarantee a noticeable enhancement in detection at various spatial resolutions	Powdery mildew	Rubber	Zeng et al. (2024)
16	Digital image analysis	Spruce needles with yellow discolouration from Chrysomyxa rhododendri are identified using digital RGB pictures. A very precise and effective method of disease monitoring is provided by correlating image data with manual needle counts to determine the severity of the disease	Needle bladder rust	Spruce	Ganthaler et al. (2018)
17	UAV-based canopy height models and spectral analysis	–	Forest diseases	Forest trees	Balková et al. (2020)

(continued)

Table 17.2 (continued)

S. No.	Detection tool/ technique	Working	Disease	Tree	References
18	VIS–NIR spectroscopy with machine learning classification	The collection of canopy reflectance spectra from citrus trees in the visible to near-infrared bands for the classification of healthy to HLB-infected trees	HLB	Citrus	Mishra et al. (2012)
19	Hyperspectral imaging with UAV	During the procedure, hyperspectral imaging collects reflectance data on fruit and leaves from different phases of citrus canker. The RBF and KNN classification algorithms then process the data	Citrus canker	Citrus	Abdulridha et al. (2019)
20	Hyperspectral imaging with spectral information divergence (SID)	Hyperspectral imaging captures reflectance data from citrus fruit in the 450–930 nm range. The SID classification method, using a reference spectrum for canker, quantifies spectral similarities to differentiate canker lesions from normal fruit peels and other diseases	Citrus canker	Citrus	Qin et al. (2009)

17.6 Conclusion

The sustainability of mango production is seriously threatened by the dynamic interplay between climate variability and mango diseases. In light of the changing climate, traditional methods of managing diseases are increasingly becoming inefficient, necessitating the development of new approaches. The critical requirement of incorporating climate-adaptive techniques into the management of mango diseases was emphasized in this chapter. Therefore, adopting resistant cultivars, improving

Table 17.3 Different forecasting models built to predict the mango diseases

S. No.	Disease	Model name	Purpose	Efficiency/ correlation	References
1	Anthracnose	Modified VGGNet model named V2IncepNet	Detection and classification of anthracnose disease	92% efficiency	Nagaraju et al. (2020)
2		Logistic model	Disease severity	Minimum temperature and rainfall	Singh and Deepankar (2020)
3		Convolutional neural networks (CNN) and support vector machines (SVM)	Disease severity	97% accuracy and over 90% precision	Mir et al. (2024)
4		ARIMA	Disease severity	Influenced by relative humidity	Pérez-Quintanilla et al. (2013)
5		NIR-HSI/ 2B-CNN algorithmic scheme	Detects anthracnose from day 0 to day 3 of spore inoculation	94% accuracy for calibration and 71% for prediction	Phanomsophon et al. (2024)
6	Powdery mildew	Linear regression models	Disease incidence and severity	77% of the variation in disease incidence and severity, with high R^2 values of 0.77 for incidence and 0.76 for severity	Shukla et al. (2016a, b)
7		Linear regression models	Disease incidence and severity	Relative humidity as the strongest predictor, explaining 59% of the variability in disease incidence and 62% of the variability in disease severity	Bana et al. (2020)

(continued)

Table 17.3 (continued)

S. No.	Disease	Model name	Purpose	Efficiency/ correlation	References
8	Malformation	MLR prediction models	Intensity of disease	Wind speed, minimum and maximum temperature	Reddy et al. (2019)
9		Multiple linear regression (MLR) and logistic regression	Disease intensity	97% accuracy (favourable weather: maximum temperature and relative humidity)	IV (2007)
10	Mango diseases	Random forest regression model	Disease attack		Jawade et al. (2020)

biological control techniques, and using climate models' predictability may help the stakeholders better prepare for diseases such mango sudden death syndrome, powdery mildew, and anthracnose. IDM adoption offers sustainable practices in a comprehensive approach to disease and health (DAH) by fusing cutting-edge technologies with environmentally friendly methods. As a result, there is a need for advanced precision farming techniques that make use of real-time data on disease outbreaks and climatic trends. Artificial intelligence and remote sensing technology combined with spontaneous modeling may help provide early warning signs of illness dangers so that appropriate action can be taken. Furthermore, cooperation between disciplines and geographical areas will advance global causes of disease management strategies in the future. Therefore, improved farmer awareness and training programs will suffice, supported by supporting policies. The mango industry will be able to endure and prosper in the face of inescapable future problems posed by climate change if research, policy, and field practices are connected with these concepts of climate resilience.

References

Abbas A, Zhang Z, Zheng H, Alami MM, Alrefaei AF, Abbas Q, Naqvi SAH, Rao MJ, Mosa WF, Abbas Q, Hussain A (2023) Drones in plant disease assessment, efficient monitoring, and detection: a way forward to smart agriculture. Agronomy 13:1524

Abdulridha J, Batuman O, Ampatzidis Y (2019) UAV-based remote sensing technique to detect citrus canker disease utilizing hyperspectral imaging and machine learning. Remote Sens 11:1373

Abro SA, Jatoi GH, Gadhi MA, Memon KA, Keerio AU, Abro MA, Koondhar IA, Nizamani ZA, Soomro AH (2019) Prevalence of powdery mildew of mango in Hyderabad and its correlation with temperature and humidity. Biocell 43:323–335

Adamo F, Andria G, Di Nisio A, Lay-Ekuakille A, Mattencini G, Spadavecchia M (2017) Designing and prototyping a sensors head for test and certification of UAV components. Int J Smart Sens Intell Syst 10:1–27

Ahmad S, Tang L, Shahzad R, Mawia AM, Rao GS, Jamil S, Wei C, Sheng Z, Shao G, Wei X, Hu P (2021a) CRISPR-based crop improvements: a way forward to achieve zero hunger. J Agric Food Chem 69:8307–8323

Ahmad T, Wang J, Zheng Y, Mugizi AE, Moosa A, Nie C, Liu Y (2021b) First record of *Colletotrichum alienum* causing postharvest anthracnose disease of mango fruit in China. Plant Dis 105:1852

Ajitomi A, Takushi T, Sato Y, Arasaki C, Ooshiro A (2020) First report of powdery mildew of mango caused by *Erysiphe quercicola* in Japan. J Gen Plant Pathol 86:316–321

Akhtar J, Dubey SC (2021) *Lasiodiplodia theobromae*, a potential post-harvest threat to agri-horticultural crops and its morpho-molecular diversity. Indian J Plant Prot 49:125–130

Al Adawi AO, Deadman ML, Al Rawahi AK, Al Maqbali YM, Al Jahwari AA, Al Sadi BA, Al Amri IS, Wingfield MJ (2006) Aetiology and causal agents of mango sudden decline disease in the Sultanate of Oman. Eur J Plant Pathol 116:247–254

Al Adawi AO, Al Sadi BA, Al Jabri MH, Deadman ML, Barnes I, Wingfield BD, Wingfield MJ (2010) Evaluation of mango cultivars for resistance to infection by *Ceratocystis manginecans*. Acta Hortic 992:393–406

Alam MW, Malik A, Rehman A, Sarwar M, Ahmad S, Shafeeq T, Hameed A, Rajput NA, Atiq M (2021) First report of Lasiodiplodia pseudotheobromae causing stem end rot of mango fruit in Pakistan. Plant Dis 105:2249

Albergel C, Calvet JC, Gibelin AL, Lafont S, Roujean JL, Berne C, Traullé O, Fritz N (2010) Observed and modelled ecosystem respiration and gross primary production of a grassland in southwestern France. Biogeosci 7:1657–1668

Alemu K (2014) Dynamics and management of major postharvest fungal diseases of mango fruits. J Biol Agric Healthc 4

Ann PJ, Tsai JN, Ni HF, Yang HR (2013) Current status on occurrence and management of major diseases of mango in Taiwan

Ansari MW, Bains G, Shukla A, Pant RC, Tuteja N (2013) Low temperature stress ethylene and not Fusarium, might be responsible for mango malformation. Plant Physiol Biochem 69:34–38

Ansari MW, Rani V, Shukla A, Bains G, Pant RC, Tuteja N (2015) Mango (*Mangifera indica* L.) malformation: a malady of stress ethylene origin. Physiol Mol Biol Plants 21:1–8

Arauz LF (2000) Mango anthracnose: economic impact and current options for integrated management. Plant Dis 84:600–611

Asif I, Faisal FS, Munawar KR, Chrys AN, Gul Bhar P, Nazim LH (2011) Trend of mango sudden death syndrome (MSDS) in relation to fungal microflora and nematodes fauna in Punjab, Pakistan. Pak J Nematol 29:1

Asma Rashid AR, Iram S, Ahmad IA (2013) Study on incidence, molecular characterization and pathogenesis of different fungi associated with sudden death of mango

Awa OC, Samuel O, Oworu OO, Sosanya O (2012) First report of fruit anthracnose in mango caused by *Colletotrichum gloeosporioides* in Southwestern Nigeria. Int J Sci Technol Res 1:30–34

Baeza-Montañez L, Gómez-Cabrera R, García-Pedrajas MD (2010) First report of Verticillium wilt caused by Verticillium dahliae on mango trees (*Mangifera indica*) in southern Spain. Plant Dis 94:380

Bagheri N, Mohamadi-Monavar H (2020) Early detection of fire blight disease of pome fruit trees using visible-NIR spectrometry and dimensionality reduction methods. J Agric Mach 10:37–48

Baharvandi HA, Zafari D (2015) First report of *Cladosporium* infection of mango inflorescence in Iran. J Plant Pathol 97

Balková M, Bajer A, Patočka Z, Mikita T (2020) Visual exposure of rock outcrops in the context of a forest disease outbreak simulation based on a canopy height model and spectral information acquired by an unmanned aerial vehicle. ISPRS Int J Geo-Inf 9:325

Bana JK, Choudhary JS, Ghoghari PD, Sharma H, Kumar S, Patil SJ (2020) Influence of weather parameters on powdery mildew of mango inflorescence in humid tropics of South Gujarat. J Agrometeorol 22:488–493

Biasutti M (2019) Rainfall trends in the African Sahel: characteristics, processes, and causes. Wiley Interdiscip Rev Clim Change 10

Borowik P, Oszako T, Malewski T, Zwierzyńska Z, Adamowicz L, Tarakowski R, Ślusarski S, Nowakowska JA (2021) Advances in the detection of emerging tree diseases by measurements of VOCs and HSPs gene expression, application to ash dieback caused by *Hymenoscyphus fraxineus*. Pathogens 10:1359

Calderón R, Navas-Cortés JA, Zarco-Tejada PJ (2015) Early detection and quantification of Verticillium wilt in olive using hyperspectral and thermal imagery over large areas. Remote Sens 7:5584–5610

Campa M, Miranda S, Licciardello C, Lashbrooke JG, Dalla Costa L, Guan Q, Spök A, Malnoy M (2024) Application of new breeding techniques in fruit trees. Plant Physiol 194:1304–1322

Casagli N, Frodella W, Morelli S, Tofani V, Ciampalini A, Intrieri E, Raspini F, Rossi G, Tanteri L, Lu P (2017) Spaceborne, UAV and ground-based remote sensing techniques for landslide mapping, monitoring, and early warning. Geoenviron Disasters 4:1–23

Castrignanò A, Belmonte A, Antelmi I, Quarto R, Quarto F, Shaddad S, Sion V, Muolo MR, Ranieri NA, Gadaleta G, Bartoccetti E (2020) Semi-automatic method for early detection of *Xylella fastidiosa* in olive trees using UAV multispectral imagery and geostatistical-discriminant analysis. Remote Sens 13:14

Cazorla FM, Torés JA, Olalla L, Pérez-García A, Farré JM, de Vicente A (1998) Bacterial apical necrosis of mango in southern Spain: a disease caused by *Pseudomonas syringae* pv. syringae. Phytopathology 88:614–620

Chen X, Sun Q, Tang L, Guo T, Huang S, Mo J, Li Q (2023) First report of bacterial necrosis caused by *Pantoea vagans* in mango in China. Plant Dis 107:1935

Chen R, Gajendiran K, Wulff BB (2024) R we there yet? Advances in cloning resistance genes for engineering immunity in crop plants. Curr Opin Plant Biol 77:102489

Cohen SP, Leach JE (2020) High temperature-induced plant disease susceptibility: more than the sum of its parts. Curr Opin Plant Biol 56:235–241

Correia KC, Silva MA, de Morais Jr MA, Armengol J, Phillips AJ, Câmara MP, Michereff SJ (2016) Phylogeny, distribution and pathogenicity of *Lasiodiplodia* species associated with dieback of table grape in the main Brazilian exporting region. Plant Pathol 65:92–103

Crespo M, Cazorla FM, Hermoso JM, Guirado E, Maymon M, Torés JA, Freeman S, de Vicente A (2012) First report of mango malformation disease caused by *Fusarium mangiferae* in Spain. Plant Dis 96:286

De Castro AI, Ehsani R, Ploetz R, Crane JH, Abdulridha J (2015) Optimum spectral and geometric parameters for early detection of laurel wilt disease in avocado. Remote Sens Environ 171:33–44

Desprez-Loustau ML, Massot M, Feau N, Fort T, de Vicente A, Torés JA, Ortuño DF (2017) Further support of conspecificity of oak and mango powdery mildew and first report of *Erysiphe quercicola* and *Erysiphe alphitoides* on mango in mainland Europe. Plant Dis 101:1086–1093

Dethoup T, Kaewkrajay C, Nujthet Y (2024) Field efficacy in controlling mango diseases of antagonist Talaromyces tratensis KUFA 0091 in fresh and dry formulations. Eur J Plant Pathol 1–12

Di Nisio A, Adamo F, Acciani G, Attivissimo F (2020) Fast detection of olive trees affected by *Xylella fastidiosa* from UAVs using multispectral imaging. Sensors 20:4915

Diedhiou PM, Mbaye N, Drame A, Samb PI (2007) Alteration of post-harvest diseases of mango (*Mangifera indica*) through production practices and climatic factors. Afr J Biotechnol 6:1087–1094

Dodd JC, Estrada AB, Matcham J, Jeffries P, Jeger MJ (1991) The effect of climatic factors on *Colletotrichum gloeosporioides*, causal agent of mango anthracnose, in the Philippines. Plant Pathol 40:568–575

Dofuor AK, Quartey NKA, Osabutey AF, Antwi-Agyakwa AK, Asante K, Boateng BO, Ablormeti FK, Lutuf H, Osei-Owusu J, Osei JHN, Ekloh W (2023) Mango anthracnose disease: the current situation and direction for future research. Front Microbiol 14:1168203

Dugyala S (2024) Preserving mango quality: integrated approaches to post-harvest disease management in Telangana

Elad Y, Pertot I (2014) Climate change impacts on plant pathogens and plant diseases. Crop Improv 28:99–139. https://doi.org/10.1080/15427528.2014.865412

El-Komy MH, Ibrahim YE, Al-Saleh MA (2023) First report of *Lasiodiplodia theobromae* causing dieback, a destructive disease on mango trees, in Saudi Arabia. Plant Dis 107:563

Fang S, Cui R, Wang Y, Zhao Y, Yu K, Jiang A (2023) Application of multiple spectral systems for the tree disease detection: a review. Appl Spectrosc Rev 58:83–109

Félix C, Duarte AS, Vitorino R, Guerreiro AC, Domingues P, Correia AC, Alves A, Esteves AC (2016) Temperature modulates the secretome of the phytopathogenic fungus *Lasiodiplodia theobromae*. Front Plant Sci 7:1096

Félix-Gastélum R, Herrera-Rodríguez G, Martínez-Valenzuela C, Longoria-Espinoza RM, Maldonado-Mendoza IE, Quiroz-Figueroa FR, Martínez-Álvarez JC, García-Pérez LM, Espinosa-Matías S (2013) First report of powdery mildew (*Pseudoidium anacardii*) of mango trees in Sinaloa, Mexico. Plant Dis 97:994

Freeman S, Katan T, Shabi E (1998) Characterization of *Colletotrichum* species responsible for anthracnose diseases of various fruits. Plant Dis 82:596–605

Freeman S, Otero-Colina G, Rodríguez-Alvarado G, Fernández-Pavía S, Maymon M, Ploetz RC, Aoki T, O'Donnell K (2014) First report of mango malformation disease caused by *Fusarium pseudocircinatum* in Mexico. Plant Dis 98:1583

Gagnevin L, Pruvost O (2001) Epidemiology and control of mango bacterial black spot. Plant Dis 85:928–935

Galdino TVS, Kumar S, Oliveira LS, Alfenas AC, Neven LG, Al-Sadi AM, Picanco MC (2016) Mapping global potential risk of mango sudden decline disease caused by *Ceratocystis fimbriata*. PLoS ONE 11

Gamliel-Atinsky E, Freeman S, Sztejnberg A, Maymon M, Ochoa R, Belausov E, Palevsky E (2009) Interaction of the mite Aceria mangiferae with *Fusarium mangiferae*, the causal agent of mango malformation disease. Phytopathology 99:152–159

Ganthaler A, Losso A, Mayr S (2018) Using image analysis for quantitative assessment of needle bladder rust disease of Norway spruce. Plant Pathol 67:1122–1130

Garoma B, Yadesa L, Tilahun B, Asefa G (2024) Breeding for resistance to biotic stresses in crops: conventional to genome editing

Gent DH, Del Ponte EM (2024) Plant disease warning systems. In: Agrios' plant pathology. Academic Press, pp 247–257

Giannakis I, Tosti F, Lantini L, Alani AM (2019) Diagnosing emerging infectious diseases of trees using ground-penetrating radar. IEEE Trans Geosci Remote Sens 58:1146–1155

Gnanesh BN, Arunakumar GS, Tejaswi A, Supriya M, Manojkumar HB, Devi SS (2022) Characterization and pathogenicity of *Lasiodiplodia theobromae* causing black root rot and identification of novel sources of resistance in mulberry collections. Plant Pathol J 38:272

Golzar H, Cother EJ (2008) First report of bacterial necrosis of mango caused by *Pseudomonas syringae* pv. syringae in Australia. Australas Plant Dis Notes 3:107–109

Habib-ur-Rahman M, Ahmad A, Raza A, Hasnain MU, Alharby HF, Alzahrani YM, Bamagoos AA, Hakeem KR, Ahmad S, Nasim W, Ali S (2022) Impact of climate change on agricultural production; Issues, challenges, and opportunities in Asia. Front Plant Sci 13:925548

Hoffmann P, Spekat A (2021) Identification of possible dynamical drivers for long-term changes in temperature and rainfall patterns over Europe. Theor Appl Climatol 143:177–191

Iqbal S, Atiq M, Fayyaz M, Zakria M, Rajput NA, Kachelo GA, Ahmad I, Usman M, Mehmood A (2024) Powdery mildew of mango: current status, perspective and emerging tools for management. Agric Sci J 6:92–101

Ismail AM, Cirvilleri G, Yaseen T, Epifani F, Perrone G, Polizzi G (2015) Characterisation of *Colletotrichum* species causing anthracnose disease of mango in Italy. J Plant Pathol 167–171
IV SR (2007) Modelling epidemiology of *Fusarium mangiferae* in relation to mango malformation. Doctoral dissertation. Indian Agricultural Research Institute, New Delhi, India
Jawade PB, Chaugule D, Patil D, Shinde H (2020) Disease prediction of mango crop using machine learning and IoT. In: Advances in decision sciences, image processing, security and computer vision: international conference on emerging trends in engineering (ICETE), vol 1. Springer International Publishing, pp 254–260
Jeger MJ, Pautasso M (2008) Plant disease and global change: the importance of long-term data sets. New Phytol 8–11
Jenny F, Sultana N, Islam M, Khandaker MM, Bhuiyan MAB (2019) A review on anthracnose of mango caused by *Colletotrichum gloeosporioides*. Bangladesh J Plant Phytopathol 35:65–74
Kamle M, Kumar P, Gupta VK, Tiwari AK, Misra AK, Pandey BK (2013) Identification and phylogenetic correlation among *Colletotrichum gloeosporioides* pathogen of anthracnose for mango. Biocatal Agric Biotechnol 2:285–287
Kasfy SH, Hia FT, Islam T (2024) Do CRISPR-based disease diagnosis methods qualify as point-of-care diagnostics for plant diseases? Nucleus 1–14
Kazmi MR, Fateh FS, Majeed K, Muhammad A, Kashkhely IH, Ahmad I, Jabeen A (2005) Incidence and etiology of mango sudden death phenomenon. Pak J Phytopathol 17:154–158
Khaskheli MI (2020) Mango diseases: impact of fungicides. Horticultural Crops 143
Khaskheli MI, Pathan MA, Jiskani MM, Wagan KH, Soomro MH, Poussio GB (2008) First record of *Fusarium nivale* Fr. Ces. associated with Mango Malformation Disease (MMD) in Pakistan. Pak J Bot 40:2641–2644
Khaskheli MI, Jiskani MM, Soomro MH, Talpur MA, Poussio GB (2011) Prevalence of mango sudden decline/death syndrome (MSDS) on various varieties at the orchards of different age in the vicinity of Tando Qaiser, Hyderabad, Sindh. Pak J Agric Agril Engg Vet Sci 27:160–167
Ko Y, Yao KS, Chen CY, Lin CH (2007) First report of gray leaf spot of mango (*Mangifera indica*) caused by *Pestalotiopsis mangiferae* in Taiwan. Plant Dis 91:1684–1684
Ko Y, Liu CW, Chen CY, Maruthasalam S, Lin CH (2009) First report of stem-end rot of mango caused by *Phomopsis mangiferae* in Taiwan. Plant Dis 93:764–764
Konapala G, Mishra AK, Wada Y, Mann ME (2020) Climate change will affect global water availability through compounding changes in seasonal precipitation and evaporation. Nat Commun 11:3044
Kranz J, Rotem J (eds) (2012) Experimental techniques in plant disease epidemiology. Springer Science & Business Media
Kumar S (2024) Recent advances in early detection of plant diseases using molecular tools. New Adv Crop Dis Manag 25
Kurihara J, Koo VC, Guey CW, Lee YP, Abidin H (2022) Early detection of basal stem rot disease in oil palm tree using unmanned aerial vehicle-based hyperspectral imaging. Remote Sens 14:799
Lee CC, Koo VC, Lim TS, Lee YP, Abidin H (2022) A multi-layer perceptron-based approach for early detection of BSR disease in oil palm trees using hyperspectral images. Heliyon 8
Legave JM, Normand F, Lauri PE (2013) Climate change and its probable effects on mango production and cultivation. In: X international mango symposium, vol 1075, pp 21–31
Leng LY, Ahmed OH, Jalloh MB, Awang A, Razak NA, Musah AA, Shahlehi S (2023) Climate change and its impact on mango pests and diseases. J Agric Crops 9:391–399
Li Q, Shu J, Tang L, Huang S, Guo T, Mo J, Ning P, Hsiang T (2020) First report of mango leaf anthracnose caused by *Colletotrichum asianum* in Vietnam. Plant Dis 104:10–1094
Li S, Xu L, Zhang W (2023) First report of postharvest stem-end rot of mango fruit (*Mangifera indica*) caused by *Diaporthe pseudomangiferae* in China. Plant Dis 107:582
Li N, Huo L, Zhang X (2024) Using only the red-edge bands is sufficient to detect tree stress: a case study on the early detection of PWD using hyperspectral drone images. Comput Electron Agric 217:108665

Lim TK, Sijam K, Ong ET (1989) Bacterial black spot of mango—a new disease in Malaysia. In: III international mango symposium, vol 291, pp 317–323
Lima NB, Marques MW, Michereff SJ, Morais MA Jr, Barbosa MAG, Camara MPS (2013) First report of mango anthracnose caused by Colletotrichum karstii in Brazil. Plant Dis 97:1248
Lindsey R, Dahlman L (2020) Climate change: global temperature. Climate.gov, p 16
Lipczynska-Kochany E (2018) Effect of climate change on humic substances and associated impacts on the quality of surface water and groundwater: a review. Sci Total Environ 640:1548–1565
Litz RE (ed) (2009) The mango: botany, production, and uses. CABI
Liu LP, Shu J, Zhang L, Hu R, Chen CQ, Yang LN, Lu BH, Liu YN, Yu L, Wang X, Li Y (2017) First report of postharvest anthracnose on mango (*Mangifera indica*) caused by *Colletotrichum siamense* in China. Plant Dis 101:833
Loo YY, Billa L, Singh A (2015) Effect of climate change on seasonal monsoon in Asia and its impact on the variability of monsoon rainfall in Southeast Asia. Geosci Front 6:817–823
Mahlein AK (2016) Plant disease detection by imaging sensors—parallels and specific demands for precision agriculture and plant phenotyping. Plant Dis 100:241–251
Makhmale S, Bhutada P, Yadav L, Yadav BK (2016) Impact of climate change on the phenology of mango—the case study. Ecol Environ Conserv 22:S132
Manzano León AM, Serra Hernández W, García Pérez L, Crespo K, Guarnaccia V (2018) First report of leaf anthracnose caused by *Colletotrichum grossum* on mango (*Mangifera indica*) in Cuba. J Plant Pathol 100:329
Marasas WFO, Ploetz RC, Wingfield MJ, Wingfield BD, Steenkamp ET (2006) Mango malformation disease and the associated *Fusarium* species. Phytopathology 96:667–672
Mehl J, Wingfield MJ, Roux J, Slippers B (2017) Invasive everywhere? Phylogeographic analysis of the globally distributed tree pathogen *Lasiodiplodia theobromae*. Forests 8:145
Midha S, Ranjan M, Sharma V, Pinnaka AK, Patil PB (2012) Genome sequence of *Xanthomonas citri* pv. mangiferaeindicae strain LMG 941
Mir TA, Banerjee D, Chauhan R, Singh M, Yamsani N (2024) Forecasting anthracnose severity levels in mango leaf using hybrid models. In: 2024 IEEE 9th international conference for convergence in technology (I2CT). IEEE, pp 1–6
Mishra AR, Karimi D, Ehsani R, Lee WS (2012) Identification of citrus greening (HLB) using a VIS-NIR spectroscopy technique. Trans ASABE 55:711–720
Misra AK (2012) Mango diseases in relation to climate change. In: National conference on threatening diseases of horticultural, medicinal, aeromatic and field crops in relation to changing climatic situation & zonal meeting, 1st edn. Indian Phytopathological Society (MEZ). CISH, Lucknow
Mo J, Zhao G, Li Q, Solangi GS, Tang L, Guo T, Huang S, Hsiang T (2018) Identification and characterization of *Colletotrichum* species associated with mango anthracnose in Guangxi, China. Plant Dis 102:1283–1289
Nagaraju Y, Sahana TS, Swetha S, Hegde SU (2020) Transfer learning-based convolutional neural network model for classification of mango leaves infected by anthracnose. In: 2020 IEEE international conference for innovation in technology (INOCON). IEEE, pp 1–7
Najafipour G, Boroumand M, Ayazpour K (2014) First report of bacterial black spot of mango in Iran. Indian J Fundam Appl Life Sci 4:190–196
Nath RK, Begum KH, Choudhury MR (2020) Plant disease forecasting models. Indian J Pure Appl Biosci 8:454–461
Nazir W, Naqvi SAH, Ahmed N, Alrefaei AF, Zulfiqar MA (2024) Optimizing micronutrient supplementation in mango orchards for the suppression of mango anthracnose (*Colletotrichum gloeosporioides*). Appl Fruit Sci 1–14
Nelson SC (2008) Mango powdery mildew
Ni HF, Liou RF, Hung TH, Chen RS, Yang HR (2010) First report of fruit rot disease of mango caused by *Botryosphaeria dothidea* and *Neofusicoccum mangiferae* in Taiwan. Plant Dis 94:128–128

Parmar KA, Rakholiya KB, Patel PR, Singh N (2024) In vitro evaluation of antagonistic effect of biocontrol agents (BCA) against mango dieback incited by *Lasiodiplodia theobromae* (Pat.) Griffon & Maubl. Int J Environ Clim Change 14:479–482
Parthasarathy S, Prabakar K, Thiribhuvanamala G (2015) Influence of temperature in the incidence and management of major post-harvest diseases of mango (*Mangifera indica* L.). Impact Clim Risks Agric Hortic Prod 48–57
Pasala R, Chennamsetti M, Patil B, Kadirvel P, Geethanjali S, Nagaram S, Sajja S, Vennapusa AR, Prasad PV, Mathur RK (2024) Revolutionizing crop production: the imperative of speed breeding technology in modern crop improvement. Crop Breed Genet Genomics 6:1–23
Patrice ND, Alain H, Bertrand MS, Norbert KT, Nourou KN, Brice TT, Honorine AG, Dolaris JA, Bouba D, Zachee A (2020) First report of red rust disease caused by *Cephaleuros virescens* on mango (*Mangifera indica*) trees in Cameroon. Int J Phytopathol 9:187–193
Pérez-Quintanilla JN, Pinson-Rincón EP, Díaz-Vicente VM, Guillén-Sánchez D, Garibay-Gálvez CC, Martínez-Guerrero MA, Hernández-Pérez R (2013) Prediction of anthracnose (*Colletotrichum gloeosporioides* Penz.) on mango (*Mangifera indica* L.) in Soconusco, Chiapas. J Food Agric Environ 11:1336–1340
Pérez-Rodríguez A, Monteón-Ojeda A, Mora-Aguilera JA, Hernández-Castro E (2017) Epidemiology and strategies for chemical management of powdery mildew in mango. Pesqui Agropecu Bras 52:715–723
Phanomsophon T, Sirisomboon P, Jongyingcharoen JS, Junto A, Pokhrel DR, Sharma S, Posom J, Sirisomboon CD, Rerngsamran P, Pornchaloempong P, Hashim N (2024) NIR-HSI/2B-CNN algorithmic scheme for early detection of anthracnose on mango fruits
Pitkethley R (2006) Bacterial black spot of mangoes
Ploetz RC (2003) Diseases of mango. In: Diseases of tropical fruit crops, pp 327–363
Ploetz RC, Benscher D, Vázquez A, Colls A, Nagel J, Schaffer B (1996) Mango decline: research in Florida on an apparently widespread disease complex. In: V international mango symposium, vol 455, pp 547–557
Prakash OM (2004) Diseases and disorders of mango and their management. In: Diseases of fruits and vegetables volume I: diagnosis and management. Springer, Netherlands, pp 511–619
Pratama R, Muslim A, Suwandi S (2023) First report of new wilt disease on *Mangifera indica* caused by *Ceratocystis fimbriata* in Indonesia. Biovalentia Biol Res J 9:36–40
Pruvost O, Boyer C, Vital K, Gagnevin L, De Bruno Austin L, Rey JY (2011) First report in Ghana of *Xanthomonas citri* pv. mangiferaeindicae causing mango bacterial canker on *Mangifera indica*
Pruvost O, Boyer C, Vital K, Vernière C, Gagnevin L, Traoré YN (2012) First report in Mali of *Xanthomonas citri* pv. mangiferaeindicae causing mango bacterial canker on *Mangifera indica*. Plant Dis 96:581–581
Qi YX, Pu JJ, Zhang X, Zhang H, Lu Y, Yu QF, Zhang HQ, Xie YX (2013) First report of dieback of mango caused by *Fusarium decemcellulare* in China. J Phytopathol 161(10):735–738
Qin J, Burks TF, Ritenour MA, Bonn WG (2009) Detection of citrus canker using hyperspectral reflectance imaging with spectral information divergence. J Food Eng 93:183–191
Qin LP, Zhang Y, Su Q, Chen YL, Nong Q, Xie L, Yu GM, Huang SL (2019) First report of anthracnose of *Mangifera indica* caused by *Colletotrichum scovillei* in China. Plant Dis 103(5):1043
Quimbita-Reyes AM, Cabrera-Asencio I, Serrato-Diaz LM, Rivera-Vargas LI (2023) First report of Lasiodiplodia mahajangana causing canker of mango (*Mangifera indica*) in Puerto Rico. Plant Dis 107(2):554
Ratnayake U, Herath S (2005) Changing rainfall and its impact on landslides in Sri Lanka. J Mt Sci 2:218–224
Raut RA, Patil PD, Haldavnekar PC, Munj AY, Baviskar SB, Sanas MP (2017) Epidemiological studies on mango powdery mildew (*Oidium mangiferae* Berthet) cv. Alphonso. IJCS 5(3):525–526
Reddy IS, Usha K, Sreevastv M, Singh B (2019) Forecasting model for mango (*Mangifera indica*) malformation in New Delhi and Lucknow. Ind J Pure Appl Biosci 7(5):557–562

Saeed EE, Sham A, AbuZarqa A, Al Shurafa K, Al Naqbi T, Iratni R, El-Tarabily K, AbuQamar SF (2017) Detection and management of mango dieback disease in the United Arab Emirates. Int J Mol Sci 18(10):2086

Salkeld D, Hopkins S, Hayman D (2023) Climate change and emerging infectious diseases. In: Emerging zoonotic and wildlife pathogens: disease ecology, epidemiology, and conservation. Oxford University Press

Salvi BR, Varadkar RS, Dalvi NV (2019) Recent developments in conventional mango breeding. Adv Agricult Res Technol J 3:30–42

Sanchez-Arizpe A, Galindo-Cepeda ME, Arispe-Vazquez JL, Genis-Velazquez R, Vazquez-Badillo ME, Antonio-Bautista A (2021) Natural resistance of two mango (*Mangifera indica* L.) commercial cultivars to anthracnose caused by *Colletotrichum gloeosporioides* Penz. Penz. and Sacc. Aust J Crop Sci 15(8):1198–1203

Sangeetha Chinnusamy SC, Hubballi MH, Sornakili AS, Saralemma BS (2010) First report of leaf blight of mango caused by Drechslera australiensis in India

Sapre S, Gontia-Mishra I, Thakur VV, Sikdar S, Tiwari S (2021) Molecular techniques used in plant disease diagnosis. In: Food security and plant disease management. Woodhead Publishing, pp 405–421

Senghor AL, Sharma K, Kumar PL, Bandyopadhyay R (2012) First report of mango malformation disease caused by *Fusarium tupiense* in Senegal. Plant Dis 96(10):1582. https://doi.org/10.1094/PDIS-07-12-0623-PDN

Senthilraja N, Gangwar RK, Borad CK (2024) Challenges and opportunities in plant disease management: a brief review. Madras Agric J 111(1–3):1

Serrato-Diaz LM, Perez-Cuevas M, Rivera-Vargas LI, French-Monar RD (2013) First report of *Lasiodiplodia theobromae* causing inflorescence blight of mango. Plant Dis 97(10):1380

Serrato-Diaz LM, Rivera-Vargas LI, French-Monar RD (2014) First report of *Neofusicoccum mangiferae* causing rachis necrosis and inflorescence blight of mango (*Mangifera indica*) in Puerto Rico. Plant Dis 98(4):570

Shah L, Ali A, Yahya M, Zhu Y, Wang S, Si H, Rahman H, Ma C (2018) Integrated control of fusarium head blight and deoxynivalenol mycotoxin in wheat. Plant Pathol 67(3):532–548

Sharma M, Kulshrestha S (2015) *Colletotrichum gloeosporioides*: an anthracnose causing pathogen of fruits and vegetables. Biosci Biotech Res Asia 12(2):1233–1246

Sharma G, Gryzenhout M, Hyde KD, Pinnaka AK, Shenoy BD (2015) First report of *Colletotrichum asianum* causing mango anthracnose in South Africa. Plant Dis 99(5):725

Shukla PK, Adak T (2017) Anthracnose disease dynamics of mango orchards in relation to humid thermal ratio under subtropical climatic condition. J Agrometeorol 19(1):56–61

Shukla PK, Adak T, Misra AK (2016a) Weather based models in predicting powdery mildew of mango for Malihabad mango belt. J Ecofriendly Agric 11(2):142–148

Shukla PK, Tarun Adak TA, Gundappa G (2016b) Seasonal dynamics of powdery mildew of mango and its management under subtropics

Shukla PK, Varma S, Fatima T, Bajpai A, Mishra R, Misra AK, Gundappa MM (2018) First report on wilt disease of mango caused by *Ceratocystis fimbriata* in Uttar Pradesh, India. Indian Phytopathol 71(1):135–142

Singh R, Deepankar (2020) Modeling for anthracnose development in mango in relation to weather parameters. Australas Plant Pathol 49:285–294

Singh BK, Delgado-Baquerizo M, Egidi E, Guirado E, Leach JE, Liu H, Trivedi P (2023) Climate change impacts on plant pathogens, food security and paths forward. Nat Rev Microbiol 21(10):640–656

Sinniah GD, Adikaram NKB, Vithanage ISK, Abayasekara CL, Maymon M, Freeman S (2013) First report of mango malformation disease caused by *Fusarium mangiferae* in Sri Lanka. Plant Dis 97(3):427

Siripatrawan U, Makino Y (2024) Hyperspectral imaging coupled with machine learning for classification of anthracnose infection on mango fruit. Spectrochim Acta A Mol Biomol Spectrosc 309:123825

Skendžić S, Zovko M, Živković IP, Lešić V, Lemić D (2021) The impact of climate change on agricultural insect pests. Insects 12(5):440

Syifa M, Park SJ, Lee CW (2020) Detection of the pine wilt disease tree candidates for drone remote sensing using artificial intelligence techniques. Engineering 6(8):919–926

Trippa D, Scalenghe R, Basso MF, Panno S, Davino S, Morone C, Giovino A, Oufensou S, Luchi N, Yousefi S, Martinelli F (2024) Next-generation methods for early disease detection in crops. Pest Manag Sci 80:245–261

Trongtorkid C, Pramokchon P (2018) Expert system for diagnosis mango diseases using leaf symptoms analysis. In: 2018 international conference on digital arts, media and technology (ICDAMT). IEEE, pp 59–64

Usha K, Nagaraja A, Sahoo RN, Singh B, Kumar N (2022) Predicting the probability of occurrence of floral malformation in mango (*Mangifera indica* L.) under climate change scenario. Int J Phytol Res 2:6–13

Velásquez AC, Castroverde CDM, He SY (2018) Plant–pathogen warfare under changing climate conditions. Curr Biol 28:R634

Velásquez C, Prieto F, Palou L, Cubero S, Blasco J, Aleixos N (2024) New model for the automatic detection of anthracnose in mango fruits based on vis/NIR hyperspectral imaging and discriminant analysis. J Food Meas Charact 18:560–570

Vilcarromero-Ramos RL, Díaz-Valderrama JR, Caetano AC, Huamán-Pilco J, Mansilla-Córdova PJ (2023) First report of anthracnose caused by *Colletotrichum asianum* on mango (*Mangifera indica*) in Peru. Plant Dis 107:2245

Vilela ESD, Terao D, do Nascimento de Queiroz SC, da Silva AM, de Holanda Nunes Maia A, Fracarolli JA, Dorta C, da Silva dos Santos L (2024) Essential oils on the control of fungi causing postharvest diseases in mango. Braz J Microbiol 55:689–698

Wang L, Wang L, Li Y, Wang J (2023) A century-long analysis of global warming and earth temperature using a random walk with drift approach. Decis Anal J 7:100237

Wilson AD (2018) Applications of electronic-nose technologies for noninvasive early detection of plant, animal and human diseases. Chemosensors 6:45

Yasuhara-Bell J, de Silva AS, Alvarez AM, Shimabuku R, Ko M (2013) First report in Hawai'i of *Xanthomonas citri* pv. mangiferaeindicae causing bacterial black spot on *Mangifera indica*. Plant Dis 97:1244–1244

Yeo YS, Kone Y, Dembele DD, Rey JY, Kone D (2022) Mango farmer's perception of climate change and its impacts on mango pests and diseases occurrences in Côte D'Ivoire. In: Alpha, Rey JY, Kone D (eds) Mango farmer's perception of climate change and its impacts on mango pests and diseases occurrences in Côte D'Ivoire

Zeng T, Wang Y, Yang Y, Liang Q, Fang J, Li Y, Zhang H, Fu W, Wang J, Zhang X (2024) Early detection of rubber tree powdery mildew using UAV-based hyperspectral imagery and deep learning. Comput Electron Agric 220:108909

Zhan RL, Yang SJ, Liu F, Zhao YL, Chang JM, He YB (2012) First report of *Fusarium mangiferae* causing mango malformation in China. Plant Dis 96:762. https://doi.org/10.1094/PDIS-12-11-1030

Zhang ZP, Li Q, Luo LX, Li JQ, Hao JJ (2017) First report of mango wilt caused by *Ceratocystis fimbriata* in *Mangifera indica* in China. Plant Dis 101:1042–1042

Zhang H, Wei YX, Qi YX, Pu JJ, Liu XM (2018) First report of Lasiodiplodia brasiliense associated with stem-end rot of mango in China. Plant Dis 102:679–679

Zhang Z, Han C, Wang X, Li H, Li J, Zeng J, Sun S, Wu W (2024) Large field-of-view pine wilt disease tree detection based on improved YOLO v4 model with UAV images. Front Plant Sci 15:1381367

Zhou ZY, Tsao WC, Chung WH, Wang CL (2022) First report of mango leaf blotch caused by *Pseudoplagiostoma mangiferae* in Taiwan. Plant Dis 106:2749

Zombré C, Sankara P, Ouédraogo SL, Wonni I, Pruvost O, Boyer C, Vernière C, Adandonon A, Vayssières JF, Ahohuendo BC (2015) First report of *Xanthomonas citri* pv. mangiferaeindicae causing mango bacterial canker on *Mangifera indica* L. in Benin. Plant Dis 99:1854–1854

Zombré C, Wonni I, Ouédraogo SL, Kpemoua KE, Assignon K, Sankara P, Vernière C, Boyer C, Boyer K, Javegny S, Pruvost O (2017) First report of *Xanthomonas citri* pv. mangiferaeindicae causing mango bacterial canker on *Mangifera indica* in Togo. Plant Dis 101:503–503

Chapter 18
Polyembryonic Mango Germplasm's Reaction to Symptoms Development and Resistance Dynamics Against Notorious Soil-Borne Fungi

Ateeq Ur Rehman, Prince Sajjad, and Muhammad Tariq Malik

Abstract Polyembryonic mango varieties, known for producing multiple seedlings from a single seed, present valuable potential in breeding programs and orchard management. This study examines the response of polyembryonic mango germplasm to the development of symptoms and resistance dynamics against critical soil-borne fungal pathogens, including *Fusarium* spp., *Rhizoctonia solani*, *Lasiodiplodia theobromae*, and *Ceratocystis fimbriata.* A controlled evaluation was conducted to assess disease incidence, symptom progression, and host resistance traits across diverse polyembryonic lines. Findings revealed significant variability in resistance levels, with certain germplasm demonstrating delayed symptom onset, reduced lesion severity, and enhanced plant survival compared to monoembryonic varieties. Resistance mechanisms were associated with increased lignification, phenolic accumulation, and robust root development, which inhibited pathogen invasion and colonization. Notably, polyembryonic rootstocks exhibited superior resilience under pathogen pressure, highlighting their potential as a foundational tool for sustainable mango production. The study emphasizes the strategic importance of polyembryonic germplasm in breeding programs aimed at enhancing resistance to soil-borne diseases, including those caused by *Ceratocystis*, which has emerged as a significant threat. These findings contribute to the development of integrated disease management strategies and climate-resilient mango cultivation systems.

Keywords Polyembryonic mango · Soilborne pathogens · *Ceratocystis fimbriata* · Resistance mechanisms · Mango disease management

A. U. Rehman (✉) · P. Sajjad
Department of Plant Pathology, Bahauddin Zakariya University, Multan, Pakistan
e-mail: ateequrrehman@bzu.edu.pk

M. T. Malik
Mango Research Institute, Multan, Pakistan

S. A. H. Naqvi et al. (eds.), *Climate Change and Mango Production*,
https://doi.org/10.1007/978-3-031-87035-4_18

18.1 Introduction

In most tropical and subtropical parts of the globe, mangos are the most significant fruit that is farmed. Over 90 nations have produced and farmed mangoes, and within the last 30 years, their output has doubled worldwide. Asia, the world's top producer of mangos, is where they originated. India and Pakistan are the world's top two mango-producing nations (FAO 2013). A total of 172,000 ha are used for mango orchard cultivation. With 1.65 million tonnes produced annually, the mango ranks second among Pakistan's fruits and accounts for 24.44% of the country's total fruit output (MNFSR 2015). Although the area planted with mango orchards is growing at the present, productivity is still very modest. A number of biotic and abiotic variables in mango plantations are responsible for this decline in yield. In Pakistan, the potential output is 25 tonnes per hectare, while the average yield is now 9.68 tonnes per hectare. Lack of updated production technologies, poor infrastructure and processing facilities, low-quality fruit, and poor control of insect pests, illnesses, and disorders are the main causes of this low yield and export rate. Mangos are susceptible to a variety of diseases, including decline disorders, anthracnose, malformation powdery mildew,, apical necrosis, and leaf infections. In addition to these illnesses, a dangerous condition known as SUDDEN DEATH OF MANGO (*Ceratosystis fimbriata*) is becoming worse every day. It first occurred in Pakistan between 1998 and 1999. Plants that seem healthy expire in a matter of days, losing the valuable riches. The same ailment, commonly referred to as Secca or Murcha, has been documented in Brazil since 1960.

In Pakistan, some progress has been made in managing this illness. The creation of resistant of this disease is the only way to give the Pakistani mango industry a solid foundation against this hazard in the long run, even though various fungicides and chemical management techniques have been assessed as short-term strategies. However, since the mango is a cross-pollinated crop, it is highly challenging to get the homogeneous true to type and resistant/tolerant roots stock for the soil-born disease in order to fulfil the aforementioned requirement. Tissue culture, air layering, and polyembryonic mango cultivars might all help accomplish this goal. For the creation of resistant/tolerant roots against MSD induced by *Ceratosystis* (soil-born fungus), the first two approaches seem to be laborious. It can be completely controlled by using polyembryonic mango varieties (Fig. 18.1).

Therefore, it is essential to start a thorough investigation into this area of MSD care. The creation of tolerant or resistant rootstock to combat this disease would undoubtedly signal a significant milestone for Pakistan's mango sector. The standardised techniques of inoculation for screening scion wood and the resistant/tolerant roots against MSD among the various polyembryonic types are covered in this book chapter.

Fig. 18.1 Under controlled nursery conditions grafting of polyembryonic germplasm was exercised upon local desi root stock

18.2 Description About *Ceratocystis fimbriata*

18.2.1 Taxonomy

Kingdom: Fungi
Phylum: Ascomycota
Class: Pyrenomycetes
Family: Ceratocystiaceae
Order: Ophiostomatales
Genus: Ceratocystis
Species: *Ceratocystis fimbriata*

18.2.2 Morphological Characters

On the majority of agar plate medium, *Ceratocystis fimbriata* grows well. Initially hyaline, the mycelium eventually becomes dark greenish-brown. Conidiophores that generate chains of hyaline conidia are often numerous within a few days. The endoconidia are round.

18.3 Description About *Nattrassia mangiferae*

18.3.1 Taxonomy

Kingdom: Fungi
Phylum: Ascomycota
Class: Pyrenomycetes
Order: Dothideales
Family: Botryosphaeriaceae
Genus: *Nattrassia*
Species: *Nattrassia mangiferae*

18.3.2 Morphological Characters

The occurrence of black, ostilate pycnidia with several hyaline, flask-shaped phialides is what distinguishes this coelomycete. Initially one-celled and hyaline, phialoconida eventually develop three-celled brown cells with a darker centre than the end cells. Conidia have an ellipsoidal to ovoid form.

Among the tropics and subtropics, the fungus is a widespread soil-borne saprophyte or wounds parasite. These moulds often need damaged plant tissue in order to parasitise it. This fungus is characterised by the production of pycnidia, which are where the spores are generated. Rarely and gradually, pycnidia grew on artificial medium. However, if they are created, they appear as pinhead-sized black balls to the unaided eye. The elliptical spores are quite large in comparison to other spores. Mature spores are dark, distichous, and have thick walls, while young, underdeveloped spores are colourless and unicellular. The fungus's rapidly expanding mycelium is initially snow-white before changing colour in three to four day.

18.4 Symptoms

The disease is typified by the drying of limbs and branches, which is followed by total defoliation, giving the tree a fire-scorched look. The bark darkens and gets discoloured, indicating the Die-Back. As the black areas grow, young green branches are withering, first in the core and then moving outward across its veins of the leaf edges. The impacted leaf becomes dark, and its edges turn up. The limb or branch dies, shrivels, and falls at this point. The main trunk, stem, and branches of the diseased plants all exhibit copious gum production. The gum first emerges as a little droplet. But as the illness worsens, it spreads and engulfs the majority of the trunk and branches. In extreme circumstances, a branch's exterior wood breaks and fractures and releases a gum-like, yellowish to brown fluid. When the cambium and phloem of ancient branches and twigs are cut longitudinally, the vascular tissues exhibit internal discolouration and brown streaks. When compared to plants that get normal watering, plants developing under water stress exhibit more severe symptoms, suggesting that water stress seems to predispose the plant and exacerbate the severity of illness.

18.5 Etiological History

In Oman, sudden mango death was first noted in 1998 (Al Adawi et al. 2003). *Ceratocystis fimbriata,* a virulent fungal infection, has been identified as the cause of this illness (Al Adawi et al. 2006; van Wyk et al. 2007). Initially, a number of reasons, including bark beetle attacks, were blamed for the abrupt fall in mango production. *Hypocryphalus mangiferae*, a bark beetle species that was identified from damaged mango trees, is now a vector of *C. fimbriata* (Al Adawi et al. 2013).

Al-Subhi et al. (2006) Al Adawi et al. (2006) and cite *Ceratocystis fimbriata* Ellis and Halst, *C. omanensis* Al-Subhi, M. J. Wingfield, and *Lasiodiplodia theobromae* (Pat.) M. van Wyk and Deadman, Griffon and Maubl as the three fungal infections linked to mango sudden death disease in Oman. Additional research conducted in Oman and Pakistan by van Wyk et al. (2007) shows that the pathogen causing the abrupt mortality of mangos is *Ceratocystis fimbriata* M. van Wyk, A. Al Adawi, and M. J. Wingf. sp. nov. According to van Wyk et al. (2007), *Ceratocystis fimbriata* and *C. fimbriata* are closely related. In Oman and Pakistan, it has also been noted that the other three related fungi on mango trees exhibit comparable symptoms (Malik et al. 2005a, b; van Wyk et al. 2005, 2007) (Malik et al. 2005a, b) Initially documented in 1996 in the Muzafer Garh region, this illness spread to other parts of Pakistan in 2005 and resulted in losses of up to 60% (Al-Subhi et al. 2006; Saeed and Masood 2008). This rapid mortality disease has been affecting mango plantations in Sindh and Southern Punjab (Masood et al. 2008).This disease is now inflicting significant financial losses in all of Pakistan's orchards.

18.6 Pathogen Biology

The fungus *C. fimbriata* is thought to constitute a species complex due to its broad geographic distribution and diverse host range (Webster and Butler 1967). According to Tarigan et al. (2011), *C. cacaofunesta* and *C. fimbriata* are closely related. *Ceratocystis fimbriata* was formerly thought to be most closely related to *C. acaciivora*, which produces a debilitating canker and wilt disease of *Acacia mangium* that is grown on Indonesian plantations (Tarigan et al. 2011). But according to a DNA research, *C. manginecans* and *C. acaciivora* are synonyms (Fourie et al. 2015).

According to Fateh et al. (2006); Al Adawi et al. (2006) Numerous fungus, such as *Botryodiplodia theobromae*, *Ceratocystis fimbriata*, *Nattrassia mangiferae*, and *Fusarium* sp., have been identified in mango orchards that have declined. Decker et al. (1983) noted that of the 85 species of plant parasitic nematodes documented in association with the root system of various crop plants throughout the Sudan, approximately 55 occurred in association with the roots of crop plants. This suggests that some nematodes were also implicated in MSD in Pakistan.

According to van Wyk et al. (2007) On the media, a greyish olive colony appears. Hyphae are segmented and smooth. Ascomatal bases are dark, globose, and range in diameter from 192 to 254(− 281) μm. Dark brown ascomatal necks that lighten towards the apices 557–635 (− 673) μm in length, (25–)32 42 (− 48) μm in width at the base, and (14–)16–22(− 26) μm in width at the tip. Hyaline, divergent, ostial hyphae measure (42–)45–59(− 69) μm in length. Evanescent asci, invisible. Hyaline, hat-shaped ascospores measure 3–4 μm in length, 4–5 μm in width (without sheath), and 7–8 μm in width (with sheath). Conidiophore appears in two morphological types. Phialidic, lageniform, and hyaline are the main conidiophores. Secondary conidiophores are short, hyaline, tube-shaped, and flaring at the mouth.

Mango isolates of *Ceratocystis fimbriata* were shown to be more virulent than isolates from other hosts by Sangchote (1989). Okrong was the most prone to infection among the six mango cultivars that were evaluated. Compared to fruits with shorter pedicels, those with longer pedicels exhibited disease signs more slowly (Ribeiro 1980; Al Adawi et al. 2006; Masood et al. 2009). Research indicates that the bark beetle *Hypocryphalus mangiferae* Stebbing (Coleoptera: Curculionidae: Scolytinae) contributes to the spread of illness to healthy plants. *H. mangiferae* colonizes sick or dead parts of infected trees. Additionally, beetle galleries were found in the bark of trees that were affected (Masood et al. 2008). However, further research is still required to fully understand how the illness and the beetle interact. The interaction between the beetle and the fungus is mutualistic. The biology and host tree of *H. mangiferae* are poorly understood (Beeson 1941; Masood et al. 2008). According to some reports, it is indigenous to south Asia, which includes Pakistan and India (Butani 1993; Atkinson and Peck 1994; Mukherjee and Litz 2009). Usually, this beetle targets dead or decaying trees.

After emerging from pupae, young bark beetles, a monogamous species, develop by feeding on the twig crotches for healthy mango trees. When young beetles emerge from sick trees, they feed on the bark and infect healthy host trees because they are largely polluted with fungal spores (Webber and Gibbs 1989). The average adult lifespan is 23 days. The larval stage mines into the inner bark tissues after hatching in the outermost part of maternal galleries, and it completes its development in the pupal cells that form at the end of each feeding gallery.

According to reports, bark beetles prey on mango trees that are under stress from diseases, drought, wounds on the stems or roots, or nutritional deficiencies that might contribute to the development of mango sudden death disease (Wood 1982; Ploetz et al. 1996; Masood et al. 2009). Due to stressed conditions, many plants release monoterpenes, which draw in bark insects (Kelsey and Joseph 2001).

According to Lim and Khoo (1985), black root rot is the reason why mangoes die suddenly. They also noticed that the roots of affected trees had blackened decay and an unpleasant, foul stench. Gum droplets leak out of the stem in severe infection instances, and the bark becomes dark brown and cracked longitudinally. Parkash and Srivastava stated that bark rots entirely and mango tree.

Macrophoma mangiferae, which causes blight on mango seedlings and early grafted plants, was isolated by Verma and Singh, whereas *Ceratocystis fimbriata* was identified as an agent that causes both dieback and bark canker. *Ceratocystis fimbriata, Pestalotia mangifera*, Phoma, *Colletotrichum gloeosporioides* sp., *Sclerotium* (Corticium) *rolfsii, Rhizoctonia solani, Diplodia* sp., and *Fusarium solani* were among the fungus that Sharma (1991) recovered from 225 sick samples of mango dieback. While other necrotic infections might facilitate *B. theobromae* invasion, *Ceratocystis fimbriata* was the most prevalent isolate and appeared to be the primary cause of dieback (sudden death) in this region. According to Narasimhudu and Reddy's (1992) 1988 trials in Andhra Pradesh, India, gummosis severely damaged the trees of the Panchadarakalasa X Willard mango hybrid, but the other 9 hybrids were found to be disease-free. Similarly, this disease only affected this hybrid, which was the only accession stored at the Regional Fruit Research Station in Andhra Pradesh's Anantharajupet.

In the study by Ploetz et al. (1996) on mango trees exhibiting sudden death symptoms in the United States, the most commonly isolated fungi were *Alternaria alternata, C. gloeosporioides* (*G. cingulata*), *Cladosporium* sp., *Penicillum* sp., *Pestalotiopsis* sp. *Dothiorella dominicana, Fusarium* sp., *Phomopsis* spp. and *Ceratocystis fimbriata.*

According to Parkash, *Ceratocystis fimbriata* is linked to gummosis, mango stem girdling, and rotting. Halsted 1890 noted that ten fruit trees suddenly collapsed because of root and crown rot, and that plants may die quickly after growing again in the spring and after a period of heavy rain. Fruit trees with collar rot are positively detected when canker appears and a fungus is isolated. If the circumstances for disease continue to be favorable, these cankers may fully girdle the tree.

Ceratocystis fimbriata was found to be the most prevalent pathogen in all assessed locations, inflicting harm to the mango trees, according to Rodriguez et al. evaluation on the incidence of the illness on 5118 ha of mango in the Piura area of Peru. Al Adawi et al. (2003) documented trees that had gummosis from their trunks and leaves that wilted and finally turned brown on a single branch. As the virus progressed, tree death occurred around six months following the onset of symptoms. Pycnidia were seen in the lesions on the infected bark, and the pathogen B. theobromae was isolated from the affected trees. Mahmood and Gill (2002) carried out the first research in Pakistan, reporting this illness for the first time and naming.

Ragab et al. (1971) characterised *Ceratocystis fimbriata* as a robust and aggressive pathogen that causes a variety of disease signs, including twig blight and tip dieback. *Ceratocystis fimbriata, Fusarium oxysporum*, and *Fusarium solani* were among the fungal species that Prakash (1984) collected from dwindling mango trees; nevertheless, they were believed to be secondary colonisers. A variety of fungal species, including *Ceratocystis fimbriata, Fusarium oxysporum*, and *F. solani*, were identified by Lim and Khoo (1985) from dwindling mango trees. Numerous fungal species, such as *Ceratocystis fimbriata, F. oxysporum, F. solani*, and *Phytophthora palmivora*, were recovered by Kueprakone et al. (1986) from declining mango trees in various parts of Thailand. These fungi cause a variety of disease symptoms, such as wilt, root rot, crown rot, collar rot, and ultimately the death of the mango plant. In 1987, Saengkong isolated *Dothiorella*, *B. theobromae*, and *C. gloeosporioides* (*Glomerella cingulata*).

Verma and Singh Isolated *M. mangifera*, *Ceratocystis fimbriata*, and *P. mangifera* are responsible for the following diseases: dieback, bark canker, mango seedling and young grafted plant blight, and grey blight colour of leaves. Several fungi were associated with ill samples, including *Oidium* spp., *Diplodia* sp., *Curvularia* sp., *F. equisetti*, and *Botryosphaeria* ribis, according to Ramos et al. (1997).

According to Parkash, *Ceratocystis fimbriata* causes mango stem girdling, gummosis, and rotting. Ploetz and Pegg, (1997) looked at how many fungi combined to cause a decrease in Florida, USA. The most common attacking fungi were *L. theobromae, Fusarium species, Dothiorella dominicana, Cladosporium species, C. gloeosporioides* (*G. cingulata*), *A. alternata*, *Penicillium* species, *Pestalotiopsis* species, and *Phomopsis* species. The number of species that were separated was influenced by the sample's date and origin. Mango stone grafts abruptly die due to *Ceratocystis fimbriata*, *B. theobromae* and *C. gloeosporioides*, either alone or in combination, according to Savant and Raunt. This significant illness phenomenon was common in the Thekonkan district of Maharashtra, India. When both viruses were isolated from contaminated samples, the symptoms were more noticeable.

Mahmood et al. (2002) the amount of pathogenic fungus in Pakistan was determined by collecting sick samples of mango tree branches, stems, and roots. These samples included *Colletotrichum gloeosporioides* (30%), *Ceratocystis fimbriata* (49%), *F. oxysporum* (6.6%), *F. semitectum* (0.66%) and *F. solani* (6.6%). *Ceratocystis fimbriata* was linked by Pandit and Samajpati (2002) to guava orchard wilt infections in West Bengal, India. A short description was given of the illness symptoms and the tests used to determine the pathogen. This was the world's first report of guava wilt caused by *Ceratocystis fimbriata.* Trees with gummosis from the trunk and wilting and eventually browning of leaves on a single branches were described by Al Adawi et al. (2003). Tree death happened as the virus worsened, some six months after the symptoms first appeared. Mango rapid death was often linked to damage caused by the bark beetle (*Cryphalus scabercollis*), although fungus, *Ceratocystis fimbriata*, were also mentioned.

In order to determine the true cause of abrupt mortality, Malik et al. (2005a, b) gathered sick samples of the root, collar, stem, and twig/branch of mango plants of the cultivar Chaunsa from six districts in Punjab. Five fungus were found in the isolates: *Phytophthora cinnamomi*, *Macrophomina phaseolina*, *Rhizoctonia solani*, *Fusarium oxysporum*, and *Botryodiplodia theobromae.* The prevalence of infection with *Ceratocystis fimbriata* mae was 82.45%, that of *F. oxysporum* was 10.54%, and that of other fungi was extremely low and only discovered in roots. Re-isolation and the emergence of the same symptoms verified that *Ceratocystis fimbriata* was the cause of the mango's rapid demise.

Shahbaz et al. (2009) identified *Ceratocystis fimbriata* as the cause of Punjab's declining mango population. The inoculated plant known as P2 had the largest diseased area, measuring 4.9 cm. Command The plants stayed healthy and showed no symptoms of illness. The identity of the infected isolate, BT-3, of the fungus *Ceratocystis fimbriata*, was verified by isolates obtained from the inoculated plant sections.

Mango abrupt decline disease in Oman is caused by the vascular wilt pathogen *Ceratocystis fimbriata*, probably in conjunction with *Lasiodiplodia theobromae* and the newly identified *Ceratocystis omanensis*, according to Al Adawi et al. (2006). These fungi may be recovered from infected seedlings and are isolated from sick trees. According to studies, the bark beetle's piercing behaviour helps it operate as a vector and aid in the quick transmission of illness across Oman.

Ali et al. (2020) revealed that *Lasiodiplodia theobromae* and *Ceratocystis fimbriata* were linked to an abrupt fall in Oman. *Ceratocystis omanensis* sp. is another species of *Ceratocystis* that has been linked to symptoms on damaged trees.

By artificially inoculating 10-year-old mango trees in Huaral, Peru, with the fungus *Ceratocystis fimbriata*, Rodríguez and Mattos (1988) demonstrated pathogenicity by causing symptoms such as dieback, gummosis, and floral necrosis. According to Meah et al. (1991), inoculating mangoes with *Ceratocystis fimbriata* only resulted in infections as the fruits started to develop, particularly above 25 °C. Additionally, evidence of a potential inactive infection was gathered. When compared to fruits without pedicel, the presence of a lengthy pedicel postponed infection by two to three days.

By inoculating immature mango seedlings, Narasimhudu and Reddy (1992) verified the pathogenicity of *Ceratocystis fimbriata*. Ploetz et al. (1996) noted that all or part of the decline signs, such as gummosis, bud necrosis, vascular discolouration, and tip dieback, were induced following artificial inoculation with *A. alternata*, *G. cingulata*, *D. dominicana*, *Ceratocystis fimbriata*, and *Phomopsis* spp. The loss of mangoes in Florida was shown to be a disease complex.

Lima et al. (1998) sp., *Phomopsis* spp. *Ceratocystis* fimbriata Six mango (*Mangifera indica*) cultivars—Haden, Keit, Manguito, Rosa, Van Dyke, and Espada—were tested for their reactions to *Ceratocystis fimbriata* isolates from different regions of northeastern Brazil. After 72 h, They measured the lesion's area (2 mm) after printing it out on transparent plastic film. By applying a PDA disc and the pathogen's structures on mechanically damaged fruits, they conducted the inoculation test.

Wagan and associates (2001) in contrast to cultivars Neelam, Sindhri, and Desi, *Ceratocystis* fimbriata was found to produce usual symptoms on varieties Lungra and Chaunsa 3–4 days after inoculation. Pure cultures of *Ceratocystis fimbriata* were injected onto mango seedlings by Mahmood et al. (2002), which confirmed the reason of the fast At 100% and at least 30% R.H., rot was the worst and also had an impact on spore germination. In a similar vein, Shelar et al. (1997) found that the causative agent of die-back (rapid decline) illness in mangos, *Ceratocystis fimbriata*, reached its maximal growth on Richard's agar at 25.

Banik et al. (1998) investigated the impact of temperature, precipitation, and relative humidity on the prevalence of *D. natalensis* (*Ceratocystis fimbriata*) and *C. gloeosporioides*, which cause post-harvest rot of mango fruits by artificial inoculation. According to in vitro research, B. theobromae grew best at 320 C, with strong mycelial development across the 60–100% R.H. range. *Ceratocystis fimbriata* was shown to be very prevalent in vivo at temperatures between 33.5 and 36.2 °C and relative humidity levels between 50 and 86.6%. The variability of the *Ceratocystis fimbriata* spore population was determined by Felipe and Luis. In the plot where the prunings were removed, there were more rain-borne spores. At first, spore collection was higher during rainy spells that produced high humidity.

Jadeja (2000) investigated the ecological relationships between two post-harvest infections on mango fruits: *Ceratocystis fimbriata* and *A. niger*. It was discovered that 30 °C was the ideal temperature for the development of fruit rot, regardless of whether the pathogen was implanted many times or all at once. At 10 °C, no development was seen in either scenario. However, in the cases of *Ceratocystis fimbriata* and *A. niger*, the rot development began at 15 °C and 20 °C, respectively.

According to Florence and Sharma, *Ceratocystis fimbriata* was the predominant fungus in Kerala, India, producing discolouration in all soft wood species. The usual warm and muggy weather that prevailed in that state made the issue worse. *Ceratocystis fimbriata* grew best on rubber wood at a temperature of 30 °C. At the higher temperature of 40 °C, growth was impeded. *Ceratocystis fimbriata* development was also shown to be significantly impacted by relative humidity.

Eng et al. (2003) examined how temperature and starting pH affected the radial speed and growth density of *Ceratocystis fimbriata* RC1 in order to demonstrate the fungus's growth features. Temperature and culture medium had an impact on growth density, but only temperatures over 40 °C had an effect on the radial velocity of growth. Additionally, neither metric was impacted by the culture media's original pH.20–25 °C (68–77 °F) is the ideal temperature for *C. fimbriata* development, according van Wyk et al. (2007). No growth occurs above 5 °C (40 °F), 10 °C (50 °F), and 35 °C (96 °F).

18.7 Disease Cycle of Pathogen

According to Saeed and Masood (2008), *C. fimbriata* is spread by the mango bark beetle *Hypocryphalus mangiferae* (Stebbing) on mango trees. The bark beetle (Scolytidae: Coleoptera) is regarded as a significant forest pest. The world's tropical and sub-tropical areas are home to around 6 thousand species of bark beetles (Lieutier et al. 2004).

According to Saeed and Masood (2008), *H. mangifera* needs sick or dead tree stems and creates tiny, sawdust-emitting pores that are around 2 mm in diameter. When the bark is removed from the wooden stem, irregular tunnels that look black because of a fungus inside them are seen.

The bug *H. mangiferae* feeds on phloem. These are usually yellowish and soft-bodied, but as they emerge, they become darker brown and rigid, producing galleries that are black in colour. On trees that were weakened by stressors like disease, salt, water, or nutrition, their population grew. found that, in Punjab, Pakistan's subtropical climate, *H. mangiferae* reached its peak activity in May and continued to be active to varying degrees from February to November (Masood et al. 2009).

The adults of *H. mangiferae* are cylindrical, measuring 1.99 + 0.01 mm in length and 0.089 + 0.007 mm in width, according to Lieutier et al. (2004). When they first emerge, they are light brown; subsequently, they become deeper brown. Larval galleries with branches emerge from the maternal galleries, which have an uneven pattern. The patterns of galleries vary across species. *H. mangifeare* and *Dendroctonus* sp. exhibit the same pattern.

According to Bright and Skidmore (2002), the antennal club of *H. mangiferae* is distinguished by three procurved sutures and five segments. The club was longer than the scape and semi-circular. They had Emerginate eyes. The pronotum also has tiny pores in it. The head is mostly or completely hidden in the thorax, and the dorsal portion of the pronotum becomes declivous.

The majority of *H. mangiferae*'s presence was in phloem (Phloeophagous species). It attacked the main stem's base, which had thicker bark, more so than the branches, which had thinner bark. Uneven galleries with a dark look were seen when the woody stem's bark peeled off. Additionally, pupae were found near exit holes, and larvae were seen tunnelling separate tunnels. Tiny holes made by bark beetles were found in most sick mango trees, and frass exuded from the holes was dispersed throughout the main stem (Figs. 18.2 and 18.3).

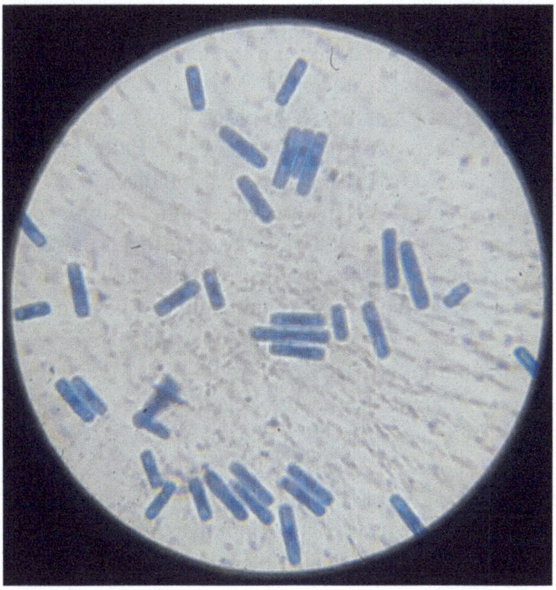

Fig. 18.2 Spores of Ceratocystis fimbriata

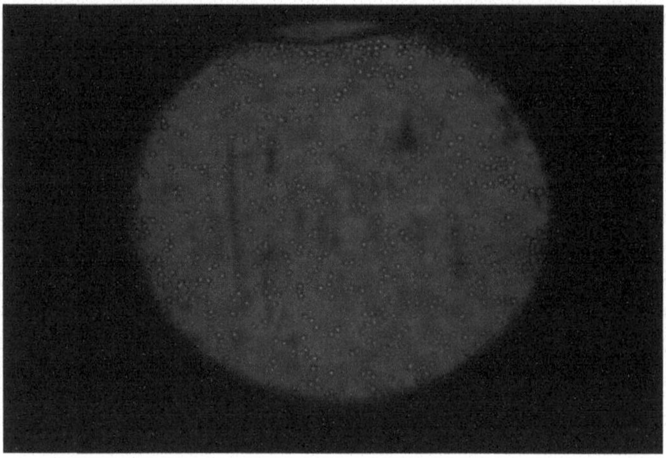

Fig. 18.3 *Natrassia mangiferae*

18.8 Symptomatology

Khuhro et al. (2005) according to reports, the disease's first outward signs include bark splitting, streaking, and vascular discolouration under the gummosis. As the tree dies, its leaves wilt yet still hang there.

When damaged trunks are scratched, rotten cankers that sometimes release foul-smelling fluids are revealed (Masood et al. 2010). In various mango orchards in Brazil or Pakistan, these symptoms may manifest alone or in combination with two or more other symptoms (Ploetz et al. 1996; Iqbal et al. 2007). Improper nutrient flow is caused by blocked vascular bundles.

Al-Sadi et al. (2010) observe the signs of abrupt death in mangos. First, the bark of the sick tree will leak a sticky substance that is the result of cell degeneration. Vascular discolouration will then appear. Additionally, a diseased tree exhibits withering on one side. The tree will die completely within six months after the onset of symptoms. The signs of a *C. fimbiata* infection in mango seedlings include vascular system discolouration, necrosis, and gummosis. Mycelium was also found in the vascular system of the infected mango seedlings. Beetle-induced damage to the stem was also seen. There have also been reports of xylem vessel tylosis.

According to Leghari (2005), the disease's usual symptoms include defoliation, blackening, and the death of twigs and branches from top to bottom. Twig blight, partial or whole tip dieback, gummosis, and bark splitting are indicators of a mango's rapid decline (Malik et al. 2005a, b). The disease's common symptoms, which might be seen alone or in combination, were gummosis, root rotting, bark splitting, drying of branches, and drying of twigs (Ploetz 2003; Iqbal et al. 2007). Both physiological and reproductive organs exhibit the visible signs (Naqvi et al. 2014).

Prior studies have shown that the pathogen persisted in the soil as a result of shesham dieback, which is brought on by the same fungal species (Ploetz 2003).

Masood et al. (2010) noted these typical disease signs on a variety of Pakistani mango varieties. According to Al Adawi et al. (2006), the most obvious signs of illness on different mango cultivars were twig drying and gummosis bark breaking. According to Masood et al. (2008), poor irrigation and serious plant damage may increase the likelihood that the disease will spread.

Al Adawi et. al (2006) mango sudden death is clearly caused by the vascular pathogen *Ceratosystis fimbriata*, and the bark beetle (*Hypocryphalus mangiferae*) contributed to the disease's quick spread across Oman.

According to Muhammad Tariq Malik et al. (2010), *Lasiodiplodia theobromae* and *Nattrassia mangiferae* may function as secondary colonisers or weaker pathogens in mango sudden death when compared to *Ceratosystis fimbriata*. They also displayed the symptoms of sudden death disease, such as vascular discolouration, wilting, and cankers.

Araujo (2014) plants from the cultivar Tommy and Ubá were somewhat resistant, whereas Espada, Haden, and Palmer were more vulnerable, according to a histopathological analysis of the mango-*Ceratocystis fimbriata* interaction.

In order to screen blueberries for anthracnose root rot resistance, Miles et al. (2012) examined several inoculation methods. Using conidial suspension, the inoculation techniques of spray, droplet, and injection inoculation of entire fruit were assessed on mature fruit of a resistant and susceptible cultivar.

The studies were carried out by Susheela (2012) to assess the current techniques for screening and inoculated for immunity against chilli anthracnose. It was discovered that the best way to find the resistant hybrids in the field was to use spray inoculation. Araujo (2014), The stem tissues of mango trees from the resistant Ubá and susceptible Palmer cultivars were examined for alkaloids and phenolics. It was also looked at how these secondary metabolites affected the in vitro development of *C. fimbriata*. Six phenolic chemicals and two alkaloids were found to be more abundant in the infected plants from cv. Ubá than in the inoculated plants from cv. Palmer.

According to Florence, fungus such Alternaria, Fusarium, Ceratocystis, and *Botrydiplodia theobromae* discoloured the mango's deeper wood as well as its surface. Sharma (1991) noted that of the 219 isolates isolated from 225 diseased samples from mango orchards, the following were found to be pathogenic: *Pestalotia mangiferae*, *Phoma* sp., Scleratium (Corticium) rolfsii, *Rhizoctonia solani*, *Diplodia* sp., *Fusarium solani*, and *Botrydiplodia theobromae*, Colletotrichum gloeosporiodies, and *Glomerella cingulata*. Infections of mixed types were prevalent. Although other infections that induce necrosis may make it easier for *Botrydiplodia theobromae* to invade, this organism was the most common isolate and seems to be the main cause of die back decline condition in this area.

Mangos are infected by *Fusarium solani*, and Kore and Mane (1992) reported that disease development increased as soil moisture decreased, reaching its peak at 22% soil moisture. According to Narasimhudu and Reddy (1992), gummosis had a significant negative impact on mango plants. The virulence of *Botryodiplodia theobromae* was proven when it was isolated from sick trees. The most significant preharvest illnesses and disorders of mangos, according to Colyn and Schaffer (1993), are floral malformation, blossom blight, powdery mildew, (*Dothiorella dominicana*), and bacterial black spot (*Xanthomonas campestris* PV. Mangiferaeindicae).

Dothiorella dominicana has been discovered as a significant mango pathogen in South Africa, according to Darvas (1993) research. It was linked to fruit rot, fruit stem-end rot, bloom blight, and branch die-back decrease. Blossom blight is the term used to describe disease signs on the flowers. The infection may induce a drop in branch die-back by spreading into the wood beneath the dead flower axis. Additionally, it may cause die-back decrease by directly infecting branch tips. The pathogen overwinters in dead wood to develop pycnidia. In this investigation, no reproductive stage of the fungus was discovered. In the Philippines, *Phytophthora palmivora* was identified by Tsao et al. from infected mango crown and root tissues exhibiting signs of crown and/or root rot and wilt.

Ploetz et al. (1996a) found that in Florida, USA, mangoes, *Glomerella cingulata, Dothiorella dominicana, Alternaria alternata, Botrydiplodia theobromae*, and Phomopsis species caused all or some of the decline symptoms at artificial inoculation. These symptoms included gummosis, bud necrosis, tip dieback, and vascular discolouration. Ploetz et al. (1996b) identified Fusarium decemcellulare as a pathogen associated with giant mango galls in the United States. Mango Die-Back decrease condition is a complicated disease that involves several fungus, according to the study conducted by the aforementioned experts.

Twig discolouration and darkening, gum leaking, leaf withering, die back, browning of vascular bundles, and plant death are all indicators of die back (Narasimhudu and Reddy 1992; Khanzada et al. 2004). Prior studies have shown that the cause of mango dieback is *Lasiodiplodia theobromae* (Pat.) Griff. and Maubl. (syn. *Botryodiplodia theobronae* Pat.) (Sutton 1980; Khanzada et al. 2004).

Many researches (Das Gupta and Zachariah 1945; McSorley et al. 1981; Narasimhudu and Reddy 1992; Sharma 1993; Schaffer 1994; Simone 1999; Gonztilez et al. 1999) have shown similar correlations between *L. theobromae* and mango decline. *Anacardium occidentale* gummosis in Brazil (Cardoso et al. 2004), collar rot of *Arachis hypogaea* (Phipps and Porter 1998), leaf necrosis and stem canker on *Protea cynaroides* (Denman et al. 2003), and gummosis of *Prunus armeniaca* and *Prunus persica* trees (Li et al. 1995) are some of the similar symptoms that the fungus causes in other economically important plants. According to research, *L. theobromae* is linked to over 50 plant species in Pakistan (Ahmed et al. 1995). Other examples include canker in *Thuja occidentas* (Sandrock et al. 1999), die back in *Citrus japonica* (Ko et al. 2024), storage rot of *Colocasia esculenta*, crown rot of *Musa acuminata*, and Anthony et al. 2004; Alam et al. 2001). die back in *Citrus limon* (Mortuza and Ilag 1999).

Table 18.1 Different fungi isolated from disease samples

Nature of bait	Fungi isolated	Infestation % age
Leave	*Ceratocystis fimbriata, Nattrassia mangiferae*	77.84
		61.66
Stem	*Ceratocystis fimbriata, Nattrassia mangiferae* *Penicilium*	84.33
		61.11
		11.33
Blossoms	*Ceratocystis fimbriata, Nattrassia mangiferae* *Cladosporium*	71.23
		43.44
		23.33

18.9 Case Study of Grafting of Seedling Mango with Polyembryonic Scion Wood

The available seedlings were successfully grafted with the scion woods of the desired polyembryonic variety through the wedge grafting.

18.9.1 Isolation of the Virulent Strain of Ceratocystis fimbriata *and* Natrassia mangiferae

Four fungus were found to be associated with the isolated colonies when examined under a compound microscope: *Nattrassia mangiferae, Ceratocystis fimbriata, Fusarium*, Cladosporium, and *Penicilium. Ceratocystis fimbriata* was found to be much more common than *Nattrassia mangiferae* in all of the infected petri dishes. while the least frequent species found were *Cladosporium* and *Penicilium* (Table 18.1).

18.9.2 Screening of Polyembryonic Germplasm of Mango Against Different Treatments (Inoculation) with Ceratocystis fimbriata *and* Natrassia mangiferae *with Flap and Slit Methods*

The polyembryonic germplasm's reaction to the development of symptoms and its resistance or susceptibility varied, according to the results of the fifteen varieties' pathogenicity tests against *Ceratocystis fimbriata* and *Natrassia mangiferae.* It was discovered that the resistant cultivars, Rockdale Siagon and Bullock Heart, exhibited the least amount of MSD symptoms during the appropriate period of time. However, every other variety exhibited a vulnerable reaction to *Natrassia mangiferae* and *Ceratocystis fimbriata*. All of the polyembryonic types showed some degree of necrosis, withering, yellowing, drying of the leaves, leathering, gummosis, canker, dieback, wilting, and drying. On various types, the most prevalent disease signs were wilting, leathering, and yellowing of the leaves (Tables 18.2, 18.3, 18.4, 18.5, 18.6, 18.7 and 18.8).

18.9.3 Re-isolation of the Fungi from the Inoculated Mango Germplasm After Symptoms Development via Both Methods i.e., Flap and Slit

The same fungi were found to be re-isolated from the affected plants with variable frequency, confirming the connection between the fungus and symptoms of the mango sudden death disease on different mango polyembryonic germplasm. By using these methods, it was shown that Bullock Heart and Rockdale Siagon were immune to the MSD illnesses. However, the strength of the separate mycoflora varied across the remaining species (Tables 18.9 and 18.10).

18.10 Conclusion

The polyembryonic germplasm's reaction to the development of symptoms and its resistance or susceptibility varied, according to the results of the fifteen varieties' pathogenicity tests against *Ceratocystis fimbriata* and *Natrassia mangiferae*. It was discovered that the resistant cultivars, Rockdale Siagon and Bullock Heart, exhibited the least amount of MSD symptoms during the appropriate period of time. However, every other variety exhibited a vulnerable reaction to *Natrassia mangiferae* and *Ceratocystis fimbriata*. All of the polyembryonic types showed some degree of necrosis, drying of the leaves, withering, yellowing, gummosis, leathering, canker, dieback, and drying, wilting. The most frequent illness signs seen in various types were wilting, leathering, and yellowing of the leaves.

Table 18.2 Evaluation of mango germplasm towards the mango sudden death symptoms expression in response to treatment-I (*Ceratocystis fimbriata*) with flip method

Variety	Symptoms characterization % age										
	Necrosis	Withering	Yellowing	Drying	Leathering	Gummosis	Canker	Dieback	Wilting	Lesions	Browning
Bullock heart	0	0	0	0	0	0	0	0	0	0	0
Elephant tusk	1	0.6	0.2	1.2	0.6	0	1	1	1	0	1
Rosa	0.6	0.8	0.4	1.2	1.6	0	0	0	1	0	1
Kao Sack	1.4	1	2.1	0.8	2.0	0	0	0	1	1	0
Rock dale Siagone	0	0	1	0.8	1.4	0	0	0	0	0	0
Vasturi	1.6	0.3	1	0.4	1.6	0	0	0	0	0	0
Banana long	1.2	0.1	1	1.3	0.6	0	0	0	0	1	0
31-1	1.2	1	1.2	1.3	0.4	1	0	0	0	1	0
Palmar	0.6	1.2	1.4	1	0.2	1	1	0	0	1	0
Xooi Tony	0.8	1.4	0.4	1	0.2	0	0	0	1	0	0
Brown seedling	0.6	0	0.6	1	0.2	0	0	0	1	0	0
R2E2	1	0	0	0	1	0	0	0	0	0	0
Olor	1	0.4	0	0	1	0	0	0	0	1	0
Carabao super	1.2	0.4	0	1.2	1.2	0	0	0	0	1	0
Sapa	0.8	0.4	1.4	0.4	0.6	1	0	0	0	0	0

*% is the mean of three plants worked as replications showed in digits

Table 18.3 Evaluation of mango germplasm towards the mango sudden death symptoms expression in response to treatment-II (*Ceratocystis fimbriata*) with slit method

Variety	Symptoms characterization % age										
	Necrosis	Withering	Yellowing	Drying	Leathering	Gummosis	Canker	Dieback	Wilting	Lesions	Browning
Bullock heart	0	0	0	0	0	0	0	0	0	0	0
Elephant tusk	1	1.6	0.8	1	0.6	0	1	1	0	0	0
Rosa	0.6	0.8	0.4	1.2	1.6	0	0	0	1	0	0
Kao Sack	1.2	0	0	0.6	1.2	0	0	0	1.2	0.4	1
Rock dale Siagone	0.4	0	1.2	0.4	1.6	0.4	1	1	0.2	1	1
Vasturi	1.2	0.4	0	0.6	1.2	1	0	0	0	1	0
Banana long	1	0.4	0	1.4	0.2	0	1	1	0	1.2	0
31-1	1.4	0	1.4	0	0.6	0	1	1	1	1	0
Palmar	0.8	1.4	1.6	0	0.4	0	1.2	1	0	0	0
Xooi Tony	0.4	1.2	0.6	0	0.2	0	0	0	0	0	0
Brown seedling	0.4	1	0.4	0	0	1	1	0	0	0	0
R2E2	0	0	0	0	0.4	0	0	0	0	0	0
Olor	0	0.6	1	1.2	0	1	1	0	0	1.2	0
Carabao super	1	0	0	1.4	0	0	0	0	0	1	0
Sapa	0	0	1	0.2	0.8	1.2	0	0	0	0	0

Table 18.4 Evaluation of mango germplasm towards the mango sudden death symptoms expression in response to treatment-III (*Natrassia mangiferae*) with flap method

Variety	Symptoms characterization % age										
	Necrosis	Withering	Yellowing	Drying	Leathering	Gummosis	Canker	Dieback	Wilting	Lesions	Browning
Bullock heart	0	0	0	0	0	0	0	0	0	0	0
Elephant tusk	0	1.2	0.4	0	0.2	0	0	0	1	1	0
Rosa	0.4	0.2	0.2	1	1	0	1	1	1	0	0
Kao Sack	1.4	0	0	0.4	1.2	0	0	0	1.2	0.4	1
Rock dale Siagone	0.4	0	1.2	0.8	1.6	0.4	1	1	0.2	1	1
Vasturi	1.2	0.4	0	0.6	1.2	1	0	0	0	1	0
Banana long	1	0.4	0	1.2	0.1	0	0	0	0	1	0
31-1	1.4	0	1.4	0	0.6	0	1	1	1	0	0
Palmar	0.6	1.4	1.6	0	0.4	0	1.2	1	0	0	0
Xooi Tony	0.4	1.2	0.6	0	0.8	0	0	0	0	0	0
Brown seedling	0.4	1.2	0.8	0	0	1	1	0	0	0	0
R2E2	0	0.8	0	0	0.4	0	0	0	0	0	0
Olor	0	0.8	1.2	1.2	0	1	1	0	0	1	0
Carabao super	1	0	0	1.4	0	0	0	0	0	1	0
Sapa	0	0	1.2	0.8	0.8	1	0	1	0	0	0

Table 18.5 Evaluation of mango germplasm towards the mango sudden death symptoms expression in response to treatment-IV (*Natrassia mangiferae*) with slit method

Variety	Symptoms characterization % age									
	Necrosis	Withering	Yellowing	Drying	Leathering	Gummosis	Canker	Dieback	Wilting	Lesions	Browning
Bullock heart	0	0	0	0	0	0	0	0	0	0	0
Elephant tusk	0	1	0.2	0	0.2	0	0	0	1	1	0
Rosa	0.8	0.4	0.2	1	1	0	1	1	1	0	0
Kao Sack	0.8	0	0	0.4	1.2	0	0	0	1.2	0.4	1
Rock dale Siagone	0.4	0	0.4	0.4	1	0.2	0	0	0.1	1	0
Vasturi	1	0.2	0	0.4	1	1	0	0	0	1	0
Banana long	1	0.4	0	1.2	0.1	0	0	0	0	1	0
31-1	1.4	0	1.4	0	0.4	0	1	1	1	0	0
Palmar	0.4	1.2	1.6	0	0.2	0	1.2	1	0	0	0
Xooi Tony	0.2	1.2	0.6	0	0.6	0	0	0	0	0	0
Brown seedling	0.2	1	0.8	0	0	1	1	0	0	0	0
R2E2	0	0.4	0	0	0.2	0	0	0	0	0	0
Olor	0	0.4	1.2	1.2	0	1	1	0	0	1	0
Carabao super	1	0	0	1.4	0	0	0	0	0	1	0
Sapa	0	0	1.2	0.6	0.4	1	0	1	0	0	0

Table 18.6 Evaluation of mango germplasm towards the mango sudden death symptoms expression in response to treatment-V (*Ceratocystis fimbriata* + *Natrassia mangiferae*) with flap method

Variety	Symptoms characterization % age									
	Necrosis	Withering	Yellowing	Drying	Leathering	Gummosis	Canker	Dieback	Wilting	Lesions	Browning
Bullock heart	0	0	0	0	0	0	0	0	0	0	0
Elephant tusk	0.2	1	0.4	0	0.4	0.2	0.2	0.2	1.2	1.2	0
Rosa	0.8	0.4	0.2	1	1	0	1	1	1	0	0
Kao Sack	0.8	0	0	0.4	1.2	0	0	0	1.2	0.4	1
Rock dale Siagone	0.4	0	0.4	0.4	1	0.2	0	0	0.1	1	0
Vasturi	1.2	0.2	0	0.4	1	1	0	0	0	1	0
Banana long	1	0.4	0	1.2	0.1	0	0	0	0	1	0
31-1	1.4	0	1.4	0	0.4	0	1	1	1	0	0
Palmar	0.4	1.2	1.6	0	0.2	0	1.2	1	0	0	0
Xooi Tony	0.2	1.2	0.8	0	0.6	0	0	0	0	0	0
Brown seedling	0.2	1	0.6	0	0	1	1	0	0	0	0
R2E2	0	0.4	0	0	0.2	0	0	0	0	0	0
Olor	0	0.4	1.4	1.4	0	1	1	0	0	1	0
Carabao super	1	0	0	1.4	0	0	0	0	0	1	0
Sapa	0	0	1.2	0.6	0.4	1	0	1	0	0	0

Table 18.7 Evaluation of mango germplasm towards the mango sudden death symptoms expression in response to treatment-VI (*Ceratocystis fimbriata* + *Natrassia mangiferae*) with slit method

Variety	Symptoms characterization % age										
	Necrosis	Withering	Yellowing	Drying	Leathering	Gummosis	Canker	Dieback	Wilting	Lesions	Browning
Bullock heart	0	0	0	0	0	0	0	0	0	0	0
Elephant tusk	0.2	1.2	0.4	0	0.4	0.6	0.8	0.8	1.4	1.4	0
Rosa	0.8	0.4	0.2	1	1	0	1	1	1	0	0
Kao Sack	0.6	0	0	0.4	1.2	0	0	0	1.2	0.4	1
Rock dale Siagone	0.4	0	0.4	0.4	1	0.2	0	0	0.1	1	0
Vasturi	1.2	0.2	0	0.4	1	1	0	0	0	1	0
Banana long	1.2	0.4	0	1.2	0.1	0	0	0	0	1	0
31-1	1.6	0	1.4	0	0.4	0	1	1	1	0	0
Palmar	0.6	1.2	1.6	0	0.2	0	1.2	1	0	0	0
Xooi Tony	0.4	1.2	0.8	0	0.6	0	0	0	0	0	0
Brown seedling	0.4	1	0.6	0	0	1	1	0	0	0	0
R2E2	0	0.4	0	0	0.2	0	0	0	0	0	0
Olor	0	0.4	1.4	1.4	0	1	1	0	0	1	0
Carabao super	1.2	0	0	1.4	0	0	0	0	0	1	0
Sapa	0	0	1.2	0.6	0.4	1	0	1	0	0	0

Table 18.8 Evaluation of mango germplasm towards the mango sudden death symptoms expression in response to treatment-VII (control) untreated

Variety	Symptoms characterization % age										
	Necrosis	Withering	Yellowing	Drying	Leathering	Gummosis	Canker	Dieback	Wilting	Lesions	Browning
Bullock heart	0	0	0	0	0	0	0	0	0	0	0
Elephant tusk	0	0	0	0	0	0	0	0	0	0	0
Rosa	0	0	0	0	0	0	0	0	0	0	0
Kao Sack	0	0	0	0	0	0	0	0	0	0	0
Rock dale Siagone	0	0	0	0	0	0	0	0	0	0	0
Vasturi	0	0	0	0	0	0	0	0	0	0	0
Banana long	0	0	0	0	0	0	0	0	0	0	0
31-1	0	0	0	0	0	0	0	0	0	0	0
Palmar	0	0	0	0	0	0	0	0	0	0	0
Xooi Tony	0	0	0	0	0	0	0	0	0	0	0
Brown seedling	0	0	0	0	0	0	0	0	0	0	0
R2E2	0	0	0	0	0	0	0	0	0	0	0
Olor	0	0	0	0	0	0	0	0	0	0	0
Carabao super	0	0	0	0	0	0	0	0	0	0	0
Sapa	0	0	0	0	0	0	0	0	0	0	0

Table 18.9 Infestation percentage of re-isolated fungi from various varieties through flap method

Varieties	Fungi isolated	Colonies of fungi	Total colonies	Infection %	Response
Bullock heart	*Ceratocystis fimbriata*	0	0	0	Resistant
	Natrassia mangiferae	0		0	
	Others	0		0	
Elephant tusk	*Ceratocystis fimbriata*	2	5	40	Susceptible
	Natrassia mangiferae	1		20	
	Others	2		40	
Rosa	*Ceratocystis fimbriata*	3	5	60	Susceptible
	Natrassia mangiferae	1		20	
	Others	1		20	
Kao Sack	*Ceratocystis fimbriata*	3	5	60	Susceptible
	Natrassia mangiferae	2		40	
	Others	0		0	
Rock dale Siagone	*Ceratocystis fimbriata*	0	0	0	Resistant
	Natrassia mangiferae	0		0	
	Others	0		0	
Vasturi	*Ceratocystis fimbriata*	3	5	60	Susceptible
	Natrassia mangiferae	2		40	
	Others	0		0	
Banana long	*Ceratocystis fimbriata*	2	4	50	Susceptible
	Natrassia mangiferae	2		50	
	Others	0		0	
31–1	*Ceratocystis fimbriata*	2	4	50	Susceptible
	Natrassia mangiferae	1		25	
	Others	1		25	

(continued)

Table 18.9 (continued)

Varieties	Fungi isolated	Colonies of fungi	Total colonies	Infection %	Response
Palmar	*Ceratocystis fimbriata*	2	4	50	Susceptible
	Natrassia mangiferae	2		50	
	Others	0		0	
Xooi Tony	*Ceratocystis fimbriata*	1	3	25	Susceptible
	Natrassia mangiferae	1		25	
	Others	1		25	
Brown seedling	*Ceratocystis fimbriata*	3	5	60	Susceptible
	Natrassia mangiferae	2		40	
	Others	0		0	
R2E2	*Ceratocystis fimbriata*	2	4	50	Susceptible
	Natrassia mangiferae	2		50	
	Others	0		0	
Olor	*Ceratocystis fimbriata*	3	5	60	Susceptible
	Natrassia mangiferae	2		40	
	Others	0		0	
Carabao super	*Ceratocystis fimbriata*	3	5	60	Susceptible
	Natrassia mangiferae	2		40	
	Others	0		0	
Sapa	*Ceratocystis fimbriata*	4	6	66	Susceptible
	Natrassia mangiferae	2		33	
	Others	0		0	

Table 18.10 Infestation percentage of re-isolated fungi from various varieties through slit method

Varieties	Fungi isolated	Colonies of fungi	Total colonies	Infection %	Response
Bullock heart	*Ceratocystis fimbriata*	0	0	0	Resistant
	Natrassia mangiferae	0		0	
	Others	0		0	
Elephant tusk	*Ceratocystis fimbriata*	3	4	75	Susceptible
	Natrassia mangiferae	1		25	
	Others	0		0	
Rosa	*Ceratocystis fimbriata*	2	5	40	Susceptible
	Natrassia mangiferae	1		20	
	Others	2		40	
Kao Sack	*Ceratocystis fimbriata*	3	5	60	Susceptible
	Natrassia mangiferae	1		20	
	Others	1		20	
Rock dale Siagone	*Ceratocystis fimbriata*	0	0	0	Resistant
	Natrassia mangiferae	0		0	
	Others	0		0	
Vasturi	*Ceratocystis fimbriata*	1	4	25	Susceptible
	Natrassia mangiferae	1		25	
	Others	2		50	
Banana long	*Ceratocystis fimbriata*	3	5	60	Susceptible
	Natrassia mangiferae	2		40	
	Others	0		0	
31-1	*Ceratocystis fimbriata*	2	4	50	Susceptible
	Natrassia mangiferae	2		50	
	Others	0		25	

(continued)

Table 18.10 (continued)

Varieties	Fungi isolated	Colonies of fungi	Total colonies	Infection %	Response
Palmar	*Ceratocystis fimbriata*	2	4	0	Susceptible
	Natrassia mangiferae	2		50	
	Others	0		0	
Xooi Tony	*Ceratocystis fimbriata*	1	3	25	Susceptible
	Natrassia mangiferae	1		25	
	Others	1		25	
Brown seedling	*Ceratocystis fimbriata*	3	5	60	Susceptible
	Natrassia mangiferae	2		40	
	Others	0		0	
R2E2	*Ceratocystis fimbriata*	2	4	50	Susceptible
	Natrassia mangiferae	2		50	
	Others	0		0	
Olor	*Ceratocystis fimbriata*	3	5	60	Susceptible
	Natrassia mangiferae	2		40	
	Others	0		0	
Carabao super	*Ceratocystis fimbriata*	3	5	60	Susceptible
	Natrassia mangiferae	2		40	
	Others	0		0	
Sapa	*Ceratocystis fimbriata*	4	6	66	Susceptible
	Natrassia mangiferae	2		33	
	Others	0		0	

References

Ahmed I, Mahmood A, Majeed K, Saleem A (1995) Evaluation of various fungicides against dieback disease caused by *Diplodianatalensis* in mango. Pak J Phytopathol 7:208–209

Al Adawi AO, Deadman ML, Al Rawahi AK, Khan AJ, Al Maqbali YM (2003) Diplodia theobromae associated with sudden decline of mango in the Sultanate of Oman. Plant Pathol 52(3)

Al Adawi AO, Deadman ML, Al Rawahi AK, Al Maqbali YM, Al Jahwari AA, Al Saadi BA, Al Amri IS, Wingfield MJ (2006) Etiology and causal agents of mango sudden decline disease in the Sultanate of Oman. Eur J Plant Pathol 116:247–254

Al Adawi AO, Al Jabri RM, Deadman ML, Barnes I, Wingfield B, Wingfield MJ (2013) The mango sudden decline pathogen, Ceratocystis manginecans, is vectored by Hypocryphalus mangiferae (Coleoptera: Scolytinae) in Oman. Eur J Plant Pathol 135:243–251

Alam MG, Tokunaga S, Maekawa T (2001) Extraction of arsenic in a synthetic arsenic-contaminated soil using phosphate. Chemosphere 43(8):1035–1041

Ali SS, Asman A, Shao J, Balidion JF, Strem MD, Puig AS, Meinhardt LW, Bailey BA (2020) Genome and transcriptome analysis of the latent pathogen Lasiodiplodia theobromae, an emerging threat to the cacao industry. Genome 63(1):37–52

Al-Subhi AM, Al Adawi AO, Wyk MV, Deadman ML, Wingfield MJ (2006) *Ceratocystis omanensis*, a new species from diseased mango trees in Oman. Mycol Res 110:237–245

Al-Sadi R, Ye D, Said HM, Ma TY (2010) IL-1β-induced increase in intestinal epithelial tight junction permeability is mediated by MEKK-1 activation of canonical NF-κB pathway. Am J Pathol 177(5):2310–2322

Anthony S, Abeywickrama K, Dayananda R, Wijeratnam S, Arambewela L (2004) Fungal pathogens associated with banana fruit in Sri Lanka, and their treatment with essential oils. Mycopathologia 157:91–97

Araujo L, Bispo WM, Cacique IS, Moreira WR, Rodrigues FÁ (2014) Resistance in mango against infection by Ceratocystis fimbriata. Phytopathology 104(8):820–833

Atkinson TH, Peck SB (1994) Annotated checklist of the bark and ambrosia beetles (Coleoptera: Platypodidae and Scolytidae) of tropical southern Florida. Fla Entomol 313–329

Beeson KC (1941) The mineral composition of crops with particular reference to the soils in which they were grown: a review and compilation

Bright DE, Skidmore RE (2002) A Catalog of Scolytidae and Platypodidae (Coleoptera).: Supplement 2 (1995–1999). NRC Research Press

Butani DK (1993) Mango: pest problems

Cardoso JE, Santos AA, Rossetti AG, Vidal JC (2004) Relationship between incidence and severity of cashew gummosis in semiarid north-eastern Brazil. Plant Pathol 53(3):363–367

Colyn J, Schaffer B (1993) The South African mango industry. Fourth International Mango Symposium, Miami, Florida, USA, 5-10 July 1992. Acta Horticulturae 341:60–68

Darvas JM (1993) Dothiorella dominicana, an important mango pathogen in South Africa. Acta Hortic 341:321–328

Das Gupta SN, Zachariah AT (1945) Studies in the diseases of Mangifera indica. Part V. On the die-back disease of the Mango tree

Decker H, Yassin AM, El Amin EM (1983) Incidence of plant parasitic nematodes on horticultural crops in the Sudan. In: VIII African Symposium on Horticultural Crops, vol 143. pp 397–406

Denman S, Crous PW, Groenewald JZ, Slippers B, Wingfield BD, Wingfield MJ (2003) Circumscription of Botryosphaeria species associated with Proteaceae based on morphology and DNA sequence data. Mycologia 95(2):294–307

Fateh FS, Kazmi MR, Ahmad I, Ashraf M (2006) *Ceratocystis fimbriata* isolated from vascular bundles of declining mango trees in Sindh, Pakistan. Pak J Bot 38:1257–1259

Food and Agriculture Organization of the United Nations (FAO) (2013) The state of food and agriculture 2013: food systems for better nutrition. FAO, Rome

Fourie A, Wingfield MJ, Wingfield BD, Barnes I (2015) Molecular markers delimit cryptic species in Ceratocystis sensu stricto. Mycol Prog 14:1–8

Gonztilez E, Umana G, Arauz LF (1999) FLUCTUACION POBLACIONAL DE Botryodiplodia theobromae Pat. EN MANGO

Halsted BD (1890) Some fungous diseases of sweet potato. New Jersey Agric Coll Exp Stn Bull 76:7–14

Iqbal Z, Valeem EE, Shahbaz M, Ahmad K, Khan ZI, Malik MT, Danish M (2007) Determination of different decline disorders in mango orchards of the Punjab, Pakistan. Pak J Bot 39:1313–1318

Jadeja KB, Mayani NG, Patel VA, Ghodasara MT (2000) Chemical control of canker and gummosis of citrus in Gujarat

Kelsey RG, Joseph G (2001) Attraction of Scolytus unispinosus bark beetles to ethanol in water-stressed Douglasfir branches. For Ecol Manage 144(1–3):229–238

Khanzada MA, Lodhi AM, Saleem S (2004) Pathogenicity of *Lasiodiplodia theobromae* and *Fusarium solani* on mango. Pak J Bot 36:181–189

Khaskheli MI, Pathan MA, Jiskani MM, Wagan KH, Soomro MH, Poussio GB (2008) First record of *Fusarium nivale* (Fr.) ces. associated with mango malformation disease (MMD) in Pakistan. Pak J Bot 40:2641–2644

Khuhro SN, Lohar MK, Arain A, Bukero MA (2005) Population of thrips, Thrips tabaci (Lind) and its natural enemies on onion crop. In: Proceedings of Pakistan Congress of Zoology, vol 25. pp 41–48

Ko JG, Eun CH, Kim IJ (2024) Development of Yein-Early, a Unique Fruit-Color and Leaf-Shape Mutant of Citrus unshiu, and Its Specific Selection Marker. Curr Issues Mol Biol 46(9):10606–10617

Kore SS, Mane AV (1992) Dry root-rot of kagzilime seedlings caused by Fusarium solani

Kueprakone U, Branch MD, Bunnaep L (1986) Occurence of mulberry root rot and pathogenicity tests of the disease by Phytophthora spp

Leghari TN (2005) Epidemiology and yield losses by sudden death syndrome in mango orchards of Sindh and its possible control. M.Sc. thesis. Department of Plant Pathology, SAU Tandojam, p 124

Li WW, Liu LQ, Zhang QP, Zhou WQ, Fan GQ, Liao K (2021) Phylogeography of Prunus armeniaca L. revealed by chloroplast DNA and nuclear ribosomal sequences. Sci Rep 11(1):13623

Lieutier F, Day KR, Battisti A, Grégoire JC, Evans HF (eds) (2004) Bark and wood boring insects in living trees in Europe: a synthesis. Dordrecht. Kluwer Academic Publishers, The Netherlands

Lim TK, Khoo KC (1985) Diseases and disorders of mango in Malaysia

Mahmood A, Gill MA (2002) Quick decline of mango and in vitro response of fungicides against the disease. Int J Agri Biol 1:39–40

Mahmood A, Saleem A, Akhtar KM (2002) Mango decline in Pakistan and its management. Pak J Phytopathol 14:40–43

Masood A, Saeed S, Sajjad A, Ali M (2009) Life cycle and biology of mango bark beetle, Hypocryphalus mangiferae (Stebbing), a possible vector of mango sudden death disease in Pakistan. Pak J Zool 41(4)

Malik MT, Dasti AA, Khan SM (2005a) Mango decline disorders prevailing in Pakistan. In: Proceedings of international conference on mango and date palm: culture and export, University of Agriculture, Faisalabad, Pakistan, 20–23-June, p 25

Malik MT, Khan SM, Dasti AA, Kazmi MR (2005b) First record of *Ceratocystis manginecans* causal organism of mango sudden death in Pakistan. Pak J. Phytopathol. 17:187–191

Malik MT, Akem C, Khan SM, Khan MA, Dasti AA, Kazmi GA, Awan MZ (2010) Confirmation of the capability of Ceratocystis fimbriata to cause mango sudden death syndrome in Pakistan. Pak J Phytopathol 22:120–125

Masood A, Saeed S, Sajjad A (2008) Characterization and damage patterns of different bark beetle species associated with mango sudden death syndrome in Punjab, Pakistan. Pak Entomol 30:163–168

Masood A, Saeed S, Malik MT, Iqbal N, Kazmi MR (2010) Methodology for the evaluation of symptoms severity of mango sudden death syndrome in Pakistan. Pak J Bot 42:1289–1299

McSorley R, Campbell CW Goldweber S (1981) Observations of mango decline in South Florida. Proc Fla Hort 93:132–133
Meah MB, Plumbley RA, Jeger MJ (1991) Growth and infectivity of Botryodiplodia theobromae causing stem-end rot of mango. Mycol Res 95(4):405–408
Miles TD, Hancock JF, Callow P, Schilder AM (2012) Evaluation of screening methods and fruit composition in relation to anthracnose fruit rot resistance in blueberries. Plant Pathol 61(3):555–566
Mirkova E (1983) Prilozenie na Trichoderma harzianum Rifai za borba s fuzariinoto uvrhvane (*Fusarium oxysporum* sch. f. dianthi (Prill. Et Del.) Snyd. Et Hans.) pri oranzerinia karamfil," Selskostopanska akademia, Gradinarska i lozarska nauka, god XX(1):65–69
Mortuza MG, Ilag LL (1999) Potential for biocontrol of Lasiodiplodia theobromae (Pat.) Griff. & Maubl. in banana fruits by Trichoderma species. Biol Control 15(3):235–240
MNFSR (2015) Ministry of National Food Security and Research, Islamabad Mortuza MG, Ilag LL (1999) Potential for biocontrol of Lasiodiplodia theobromae (Pat.) Griff. & Maubl. in banana fruits by Trichoderma species. Biol Control 15(3):235–240
Mukherjee SK, Litz RE (2009) Introduction: botany and importance. In: The mango: botany, production and uses Wallingford, CABI, UK, pp 1–18
Narasimhudu Y, Reddy PS (1992) A note on gummosis of mango
Naqvi SAH, Perveen R, Manzoor SA, Abbas Z (2014) Occurrence of mango anthracnose on inflorescence in Multan (Punjab) and its in-vitro control through various fungicides. Int J Biosci 4:184–191
Pandit PK, Samajpati N (2002) Wilt disease of guava (Psidium guajava L.) by Botryodiplodia theobromae Pat
Phipps PM, Porter DM (1998) Collar rot of peanut caused by Lasiodiplodia theobromae. Plant Dis 82(11):1205–1209
Ploetz RC, Benscher D, Vázquez A, Colls A, Nagel J, Schaffer B (1996) Mango Decline: Research in Florida on an apparently wide-spread disease complex. In: V International Mango Symposium, vol 455. pp 547–557
Ploetz R, Pegg K (1997) Fusarium wilt of banana and Wallace's line: was the disease originally restricted to his Indo-Malayan region? Australas Plant Pathol 26:239–249
Ploetz RC (2003) Diseases of tropical fruit crops. In: Ploetz RC (ed) Diseases of mango. CABI Publisher, pp 327–363
Ploetz RC, Benscher D, Vazquez A, Colls A, Nagel J, Schaffer B (1996) A re-evaluation of mango decline in Florida. Plant Dis 80:664–668
Prakash S, Ram S (1984) Naturally occurring auxins and inhibitor and their role in fruit growth and drop of mango 'Dashehari'. Sci Hortic 22(3):241–248
Ragab MM, Sabet KA, Dawood NA (1971) Botryodiplodia theobromae Pat., the cause of fruit rot and die-back of mango in ARE
Ramos LJ, Davenport TL, McMillan RT Jr, Lara SP (1997) The resistance of mango (Mangifera indica) cultivars to tip dieback disease in Florida. Plant Dis 81(5):509–514
Ribeiro IJA (1980) Seca de manguera. Agentes causais e studio da molesta. In: Anais do I Simposio Brasiliero Sobre a Cultura de Mangeura. Sociedad Brasileira de Fruticultura, Jacoticobal, pp 123–130
Rodríguez C, Mattos L (1988) Dieback of mango (Mangifera indica L.) and behaviour of five varieties with respect to the causal agent
Saeed S, Masood A (2008) Association of Bark beetle *Hypocrphalus mangiferae* Stebbing (Coleptera: Scolytidae) with pathogens *Ceratocystis fimbriata* and *Phompsis* sp., in relation to mango sudden death in Pakistan. In: 93rd annual meeting of ESA, Milwaukee, Wisconsin, USA
Schaffer B (1994) Mango decline. In: Ploetz RC, Zentmyer GA, Nishijima W, Rohrbach HD (eds) Compendium of tropical fruit diseases, APS Press St. Paul. pp 43
Sangchote S (1989) Botryodiplodia stem end rot of mango and its control. In: III International Mango Symposium, vol 291. pp 296–303

Sandrock DR, Woodward JL, Dirr MA (1999) Susceptibility of atlantic white cedar cultivars to Botryospheria and Seiridium cankers. In: SNA Research Conference, vol 44. pp 204–206

Shahbaz M, Iqbal Z, Saleem A, Anjum MA (2009) Association of *Lasiodiplodia theobromae* with different decline disorders in mango (*Mangifera indica* L.). Pak J Bot 41:359–368

Sharma IM (1991) A note on population dynamics and etiology of dieback of mango in Himachal Pradesh

Sharma I (1993) A note on population dynamics and etiology of dieback of mango in Himachal Pradesh. New Agric 2:229–230

Sharma IM, Gupta VK (1995) Evaluation of different fungicides on *B. Theobromae* Pat. causal organism of canker and dieback of mango. Ind J Mycol Plant Pathol 24:146–147

Shelar SA, Padule DN, Sawant DM, Konde BK (1997) In vitro evaluation of fungicides against *Botryodiplodia theobromae* Pat. The cause of Dieback disease of mango (*Mangiferaindica* L.). Ind J Plant Proct 25:118–120

Simone GW (1999) Disease Control in Mango (Mangifera indica). Plant Pathology Department Document PDMGV3-22. Florida Cooperative Extension Service, Institute of Food and Agricultural Sciences, University of Florida

Susheela K (2012) Evaluation of screening methods for anthracnose disease in chilli. Pest Manag Hortic Ecosyst 18(2):188–193

Sutton BC (1980) The Coelomycetes, Fungi Imperfecti with Pycnidia. Acervuli and Stromata, pp 470–472

Tarigan M, Roux J, Van Wyk M, Tjahjono B, Wingfield MJ (2011) A new wilt and die-back disease of Acacia mangium associated with Ceratocystis manginecans and C. acaciivora sp. nov. in Indonesia. S Afr J Bot 77(2):292–304

Van Wyk M, Al-Adawi AO, Wingfield BD, Al-Subhi AM, Deadman ML, Wingfield MJ (2005) DNA based characterization of Ceratocystis fimbriata isolates associated with mango decline in Oman. Australas Plant Pathol 34:587–590

Van Wyk M, Al Adawi AO, Khan IA, Deadman ML, Al Jahwari AA, Wingfield BD, Ploetz R, Wingfield MJ (2007) Ceratocystis manginecans sp. nov., causal agent of a destructive mango wilt disease in Oman and Pakistan. Fungal Divers 27:213–230

Webber JF, Gibbs JN (1989) Insect dissemination of fungal pathogens of trees

Webster RK, Butler EE (1967) A morphological and biological concept of the species Ceratocystis fimbriata. Can J Bot 45(9):1457–1468

Wood SL (1982) The bark and ambrosia beetles of North and Central America (Coleoptera: Scolytidae), a taxonomic monograph

Chapter 19
Climate-Shift-Driven Mango Tree Scarification and Floral Diseases Mitigation Through Exogenous Application of Synthetic Chemicals: A Case Study on Cultivar Chaunsa Summer Bahisht

Muhammad Abid, Sobia Chohan, and Ahmed Raza Alvi

Abstract Climate change has exacerbated mango cultivation challenges, particularly in tree scarification and floral diseases, significantly impacting fruit yield and quality. This study investigates the effectiveness of exogenous applications of synthetic chemicals in mitigating these climate-driven stresses, focusing on the commercially important mango cultivar 'Chaunsa Summer Bahisht'. Field trials were conducted to assess the efficacy of different chemical treatments, including fungicides and stress-mitigating compounds, on scarification prevention and floral diseases control. Results demonstrated a significant reduction in tree scarification rates and floral disease incidence in treated trees compared to untreated controls. Improved floral retention, reduced necrosis, and enhanced fruit set were observed, suggesting the potential of these interventions to counteract adverse climatic effects. Moreover, the treatments enhanced physiological parameters such as chlorophyll content and photosynthetic efficiency, contributing to overall tree health. The study underscores the role of targeted chemical applications in safeguarding mango production against climate variability. However, careful management of chemical use is recommended to ensure environmental sustainability and compliance with food safety standards. This case study highlights the need for integrated approaches to manage climate-induced stresses in mango orchards, ensuring long-term productivity and quality of this economically significant crop.

Keywords Mango cultivation · Chaunsa summer · Climate resilience · Synthetic chemicals · Floral disease

M. Abid (✉) · S. Chohan · A. R. Alvi
Department of Plant Pathology, Faculty of Agricultural Sciences and Technology, Bahauddin Zakariya University, Multan, Pakistan
e-mail: muhammad.abid@bzu.edu.pk

S. A. H. Naqvi et al. (eds.), *Climate Change and Mango Production*,
https://doi.org/10.1007/978-3-031-87035-4_19

19.1 Introduction

This chapter is adapted from a previously published article by the authors (Alvi et al. 2022). Mango (*Mangifera indica* L.) trees are susceptible to a number of diseases throughout their developmental phases, but those that occur during the flowering period result in greater losses and have a direct impact on production (Sao-Jose et al. 2000; Cavalcante et al. 2022). Mango flowers are largely susceptible to apical necrosis, blossom blight, and powdery mildew, which may be readily controlled using an integrated disease management strategy that includes fungicide spraying throughout the blooming season (Reuveni and Reuveni 1995; González et al. 2021). However, in many cultivations, such as Chaunsa 'summer bahisht' (S.B.), the ecological cues for flower initiation are generally vague, inconsistent, or poorly defined, making it unpredictable for mangoes to blossom from one season to the next (Katuromunda et al. 2022). Certain substances, like as calcium nitrate ($CaNO_3$) and potassium nitrate (KNO_3), are utilised to promote blooming as well as break dormancy, particularly in the (Bondad and Linsangan 1979; Bueno and Valmayor 1974). Rainfall during the flowering season is another environmental factor that facilitates the spread diseases like powdery mildew, blossom blight, apical necrosis, and others.

Since 2012, apical necrosis has become a common threat in Pakistani mango orchards, and the symptoms it causes are very similar to those of a frost bite. According to English et al. (1980) trees' vegetative growth slows as disease frequency rises. Every year, those trees switch roles as carriers. Its pathogenicity test has not yet been established conclusively, but given its high occurrence, a thorough investigation of this danger was necessary. According to reports, mango apical necrosis is most common during the winter dormant season and is distinguished by the rapid emergence of necrotic scratches on the leaves and sprouts (Carzola et al. 1998; García et al. 2019). Bud failure and necrosis of vegetative and floral buds are among the disease's symptoms. Necrotic lesions are the primary source of the buds, which may sometimes extend from the petiole of the leaf to the stem. Additionally, the flower panicles are impacted, which results in a reduction in fruit set and significant financial losses. Intervenial, pointy, water-soaked patches that may combine, become black, and become somewhat elevated are the result of leaf lesions. When the rate of disease is high, the tree's vegetative development is halted, and necrotic fruit lesions are not seen.

Mango flowers go through eleven different phases of blossoming before developing into fruit. These phases, known as dormant bud to fruit development, range from "A" to "K." The apical necrosis pathogen has been shown to persist on flowers until the "D" stage, or mouse ear stage. It is challenging to determine if the blackening of these blooms is indeed caused by frost or by another biotic invasion, such as *Alternaria, Colletotrichum,* or *Pseudomonas.* Mango blossom blight is caused by Alternaria and *Colletotrichum gloeosporioides.* Mangoes in Australia's New South Wales are particularly vulnerable to this disease, which often causes the defoliation of fresh growth and breaks early leaves. Anthracnose produces severe blossom blight during blooming, which destroys inflorescences (flower panicles) and inhibits fruit set. The fruit may be infected until it is time to harvest, although the visible symptoms often do not appear until the fruit begins to mature. Fruits may get infected until it is time to harvest them, although visible signs often do not show up until they starts to mature. Lesions on leaves, dead twigs, mummified inflorescences, and flower sepals all produce fungal spores. However, the primary cause of the inoculum for bloom blight is diseased young leaves. Because of the wide variety of ecological conditions, spores are readily formed on leaves and are kept apart by rain everywhere.

Surface wetting and temperature conditions are critical factors in inflorescence infection. Compared to higher temperatures (22–25°C), wetness takes a much longer period at lower temperatures (10–13°C). Given this dire circumstance, research must be done to concurrently address biotic and abiotic stressors throughout the mango blooming phases. When inflorescence becomes black and uneven patches develop and merge, the panicles are lost. Blackening of the blooms and the growth and merging of irregular patches are signs of floral panicle death. According to Ploetz and Gregory (1993) substantial yield loss ranging from 80–100% is caused by mango malformation disease. Every year, there were reports of severe deaths from this ancient, troublesome disease. According to Singh et al. (1961) plants between the ages of 4 and 8 experience significant loss (90.9%) due to vegetative deformation. Every tree may suffer damage ranging from 50 to 80 percent, and in extreme situations, the misfortune may be almost total, as the twisted inflorescence fails to produce fruiting bodies (Summanwar 1967). Additionally, it was noted that there are two types of malformation that have the same aetiology: vegetative and floral (Kumar and Beniwal 1987a). According to Ploetz (1994) this disease results in significant injury rates in the plant marketplace and early destruction to the young plants' vegetative shoots. Malformations may also impair bloomy organs and shoots. The mature plant's inflorescence is attacked by deformity. According to Sao-Jose et al. (2000) the percentage of malformed flowers was 54%, with a deformity severity of 17%. Previous claims of aetiologies such as acarological (Singh et al. 1961) and physiologically (Sattar 1946) were disproved since they lacked etiological associations. New taxa have been discovered by testing of Fusarium strains isolated from deformed inflorescences of various foreign sources; *F. mangiferae* is related to strains that were previously described as *F. subglutinans* and are thought to be the cause of MMD globally. The isolation of deformed pieces verified the ascendancy of *F. maniferae* Britz. The mango tissues of different ancestries showed up to 97.0% infection, with a wide dispersion of *F. maniferae* Britz (Iqbal et al. 2011). Freeman et al. (2014) shown via artificial inoculations that *F. mangiferae* Britz is the cause of mango malformation infection.

Powdery mildew is caused by the fungus *Oidium mangiferae* Bert. a biotrophic organism that thrives and develops in host tissue. This condition, which ranges from 10 to 25%, causes significant disasters. In order for this disease to grow and thrive among plants, favourable environmental conditions are crucial. In the crucial circumstances when conidia are generated, which infect other leaves, blooms, and fruits, the leaves of plants or the ground are thought to be a primary source for the pathogen's survival. White, shallow, fine mycelium on flowers, leaves, and young organic products are signs of the infection. The pathogen's main impact is on the leaves, which eventually causes the blooms to drop before reaching pea size. In February and March, disease signs begin to appear on the bodies that produce fruit, flower scales, flower heads, and new leaf stem. It was evident from an epidemiological standpoint that the fungus generates a ludicrous infection of a host plant at 100% relative wetness and 26 °C. However, if the temperature increases to 15–32°C and the moisture content reaches 16–100%, disease may also occur. The whole fruiting body is damaged by severe floral infection, which also causes the bloom to stay closed and collapse after blossoming.

19.2 Use of Synthetic Chemicals in Promoting Flowering

Farmers must provide the plants the best growing circumstances possible in the current field conditions and shield them from pests like weeds, insects, and fungus and pathogenic microorganisms in order to produce mangoes with high-quality fruits and optimum yields. One of the primary causes of the sharp rise in productivity in contemporary agriculture is the widespread use of synthetic fertilizer, micronutrients, fungicidal chemicals, and pesticides, whether in the field or in an orchard. But according to a number of papers in internationally recognised journals, using synthetic compounds like salicylic acid, BTH, methyl jasmonate, etc., may increase a plant's resistance to disease and support its regular physiological processes (Patel and Gupta 2022). In a similar vein, the experiment was carried out to reduce the risks of inflorescence disease under field settings and to employ some documented substances to break the dormancy or stimulate blooming in mango trees. A brief historical summary of the use of several synthetic compounds for the aforementioned topic on a particular mango varaties known as Chaunsa Summer Bahisht, which is renowned worldwide for its flavour and taste.

19.2.1 Potassium Nitrate and Urea Spray

Potassium nitrate (KNO_3) is reported to promote the flowering in mango cultivars. The effects of shoot decapitated and potassium nitrate spray (80 g/liter) alone and in combination on the blossoming and fruit development of the monoembryonic "Haden" and polyembryonic "Manila" mango trees were assessed via experiments. During flowering, the produced vegetative branches of these two cultivars were cured. When these two types were simply sprayed, the number of flowering shoots was at its maximum; in "Haden" and "Manila," the number of shoots compared to unsprayed plants was three times and nearly twice, respectively. It was evident that "Manila" had a better fruit arrangement than "Haden" in every way. The amount of fruit/shoot or per panicle quadrupled when the potassium nitrate spray was sprayed to the Manila alone"' (Gonzales et al. 1992). The widespread use of synthetic fertilizers, micronutrients, fungicides, and pesticides in field and orchard management is a significant contributor to increased productivity in modern agriculture (Hassan et al. 2020).

Potassium nitrate and varying amounts of paclobutrazol (PBZ) were used in order to assess the impact on fruit quality and inflorescence in both mango replicates. The Cotaxtla Experimental Station of the INIFAP in Veracruz, Mexico, served as the site of the experiment. Manila Cotaxtla 1 and Manila Cotaxtla 2 clones, which were 14 years old, were planted in 8 × 8 m plots without irrigation. f) 4% KNO_3 without PBZ, g 1.5 g PBZ, without KNO_3, h control with neither product applied, and a 0.0, b 0.5, c 1.0, d 1.5, and e 2.0 g PBZ per metre of canopy diameter, all in combination with the application of 2% KNO_3, was the experimental design known as Randomised Complete Block Design, or RCBD. The soil was treated with paclobutrazol, and the leaves were treated with the two potassium nitrate treatments. Firmness, titratable acidity (TA), total soluble solids (TSS), weight of fruit, fruit loss in weight, and the number of panicles were all considered. Compared to the control, more mature panicles were produced by the administration of potassium nitrate and paclobutrazol, with greater values noted as the PBZ dosage rose. Regarding the grade of fruit of both varieties there was no negative impact of paclobutrazol. Greater weight loss, less TA, less stiffness, and a higher TSS content were seen in treatments with higher PBZ dosages. The mature fruits' TA varied between the two varieties (Martínez 2008).

The leaves of the "Tommy Atkins" mango were evaluated in Ethiopia using a single application of potassium nitrate (KNO_3) as well as urea at different stages. Unpredictable blooming, constant, intense vegetative growth, and uneven bearing were the trees' defining characteristics. With three replications and three trees per plot, the trees chosen for the RCBD experiment were uniform. Because the leaves of the underdeveloped flushes were dark green, spraying was done on them both at the maturity stage and while they were young. The outcomes of using KNO_3, particularly in the combination with urea (5-L solution of 4% KNO_3 + O), were noted. Better outcomes were obtained with 5 g urea tree-1 and 5 L of 4% KNO_3 + 1 g urea tree for The quality features of plants that were sprayed and those that were not showed a non-significant change. It was determined that the reason for the more notable flowering and yield results seen in comparison to the unsprayed trees was the administration of nitrogen by the splashing of urea and potassium nitrate (KNO_3) (Yeshitela et al. 2004).

Alternaria leaf blight is renowned as the worst cotton disease in northern Australia. At the Katherine Research Station in Australia, a trail was set up to eliminate the Alternaria leaf blight of cotton and learn more about how potassium nitrate may assist treat the disease. The disease's outward manifestation is the dropping of leaves at one-week intervals on the bottom, middle, and top of the plant. When comparing the middle canopy level to the upper and lower canopies, the measures that were given to the canopies revealed more shedding. The foliar spray of KNO_3 significantly decreased the amount of leaf shedding that was computed based on the experiment's duration. Additionally, the middle canopy leaves were affected in the firstIn comparison to the untreated plots, the experimental plants' middle canopy KNO_3 treatment decreased leaf loss throughout the second and third weeks of July and

August. According to the experiment, using potassium nitrate (KNO_3) may help reduce the effects of Alternaria leaf blight on cotton in Australia (Bhuiyan et al. 2007).

Nunez-Elisea (1985) found that when potassium nitrate (KNO_3) was sprayed on "Haden" and "Manila" mango trees, inflorescences appeared on 60 to 76% of the shoots, respectively, but only 32 and 20% of the equivalent controls did the same. Although potassium nitrate and ethephon stimulate mango blooming in the tropics, Davenport and Núñez-Elisea (1990) found that they had little effect on flowering growth in South Florida. The ongoing difficulty between ethylene production and blooming teaches that increased ethylene amalgamation in leaves or buds may not be enough to change botanical acceptability of mango trees. Tongumpai et al. (1991) found that paclobutrazol application in the soil caused a significant increase in flowering and slightly earlier blooming, but a combination of paclobutrazol application in the soil (6 g/tree) and potassium nitrate foliar spraying eight weeks following paclobutrazol treatment triggered early blossoming. When compared to the untreated plants, Sergentet al. (1996) observed that a high concentration of potassium nitrate (KNO_3) (3.6 and 4.6%) induced early flowering and reaping (30–45 days before). According to Oosthuyse (1996) KNO_3 treatment, especially at 4% level, was somewhat phytotoxic to the leaves and the inflorescence and caused some inflorescence branches to become necrotic. Blooming improvement by KNO_3 was not accounted for in subtropical districts, where winter conditions are sufficiently inductive for mango blossoming.

In South Africa, KNO_3 is currently applied commercially to the varieties in issue via inflorescence. The average fruit weight was unaffected by the potassium nitrate (KNO_3) treatment. When potassium nitrate was applied, fruit production per tree rose along with the favourable cost (Barba 1974). It was shown that every tree's quantity of inflorescence rose significantly after potassium nitrate (KNO_3) was applied. The genetic potential of the varieties may be the cause of the variation in the quantity of inflorescence treated with different chemicals (Quijada et al. 2000).Every tree in every experimental block that received potassium nitrate (KNO_3) treatment produced more fruit, and there was no discernible difference in the average weight of the fruit at harvest in September 1996. The use of potassium nitrate was shown to boost relative benefit and production in all trees (Barba 1974). Akhter et al. (2004) revealed that among the seven mango kinds, Hybrid-10 treated with KNO_3 (80 g/L) had the most inflorescences per plant, whereas the untreated Mixed Special plant had the fewest. Amrapali produced the most fruits after being treated with NAA (40 ppm).Potassium nitrate (KNO_3) was shown to aid in the opening of sprouts for vegetative growth in non-inductive conditions and reproductive development in inductive conditions by Yeshitela et al. (2004). For "Tommy Atkins" and "Keitt" mango trees, the inductive condition was determined to be 10/150C for 12 h of light and 12 h of dark.via an experiment, Yeshitela et al. (2004) demonstrated that the primary cause of the increased flowering in the "Tommy Atkins" mango cultivar was thought to be nitrogen supplementation via the application of 4% KNO_3 and urea (0.5 g/tree).

19.2.2 Calcium Nitrate:

This study examined the effects of a 0.8% $Ca(NO_3)_2$ late spray on the calcium content and common mineral-to-mineral correlations in the fruits of a five-year-old apple tree named "Elise." Trees got treatment three times, 10 days apart, prior to fruit harvest. The fruit's mineral content was assessed ten days after the completion of all treatments and at harvest. The effects of the calcium nitrate spray on the foliage were modified by weather conditions in certain experimental years, however in general, the calcium content of fruits assessed both after treatment and after harvest increased. Only in 2005 did calcium nitrate spraying cause a drop in the nitrogen/calcium and potassium/calcium ratios assessed after treatments and fresh-picked fruits (Domagala-Swiatkiewicz and Błaszczyk 2009).

Sarrwy et al. (2012) examined the effect of spray date palm with 250 and 500 ppm of boric acid and 1% and 2% of calcium nitrate, either alone or when combined with treatment that included both quantities of boric acid and calcium nitrate. The use of calcium nitrate and/or boric acid spray to the date palm inflorescence had a substantial effect on fruit set, production, and the physical and chemical properties of the Amhat date palm. The best treatment for fruit quality and yield throughout the two trial seasons was a 500 parts per million boric acid and 2% calcium nitrate spray. Only in 2004 did calcium nitrate treatments result in a considerable rise in the standard weight of fruits and a reduction in their dry matter.

Researchers looked at how late spraying with 0.8% $Ca(NO_3)_2$ affected the calcium level and the relationships between the mineral elements in the fruit of a five-year-old apple tree called "Elise." Three spray applications were made at 10-day intervals prior to fruit harvest. Ten days after each treatment and upon harvest, the mineral content of the fruit was evaluated. Weather variables in specific experimental years altered the impact of the calcium nitrate foliar spray, which generally raised the content of calcium in fruits measured both after treatments and after harvest. According to research by Karuppaiah et al. (2001) foliar spraying of calcium nitrate boosted cucumber yield by up to 14.5 tonnes per hectare and increased the level of NPK in plants. Calcium is essential for the development of young tissues, roots, and fruits. Nitrate may also be directly absorbed by plants, increasing their intake of calcium. Furthermore, Itagi and Hiki (1956) said that urea fertiliser, because of its 46% N content, reduces improper and malformed fruits when sprayed.

Borkowski et al. (2006) investigated how various substances affected Chinese cabbage. This plant's rapid development makes it very vulnerable to tip burn injuries, which are physiological conditions linked to a lack of calcium and water. The primary factor damaging the whole commercial yield of Chinese cabbage is tip burn. Other vegetables and leaves are also affected by the condition. The susceptibility of Chinese cabbage cultivars to tip burn varies. Since cultivars did not differ in the amount of calcium in their leaves, it seems to be related to the vulnerability to stress brought on by high temperatures, which encouraged fast growth, transpiration, and water deficit. As an elicitor, chitosan made plants more resistant to many diseases. To find out whether chitosan or different methods may similarly improve

Chinese cabbage's resistance to tip burn, experiments were conducted. The items formulated with 1.5% calcium nitrate and 2.5% Biochikol 020 PC, calcium nitrate and Tytanit, or calcium nitrate alone demonstrated the best results in the tip burn control, according Borkowski and Kowalczyk (2003) experiment. It's interesting to see that plants treated with calcium nitrate, Tytanit, and Biochikol 020 PC all experienced tipburn at rates similar to the control. Since calcium makes cell membranes more stiff, Saure (1998) discovered that burn injuries were less severe and that the amount of calcium in leaves rose after applying a calcium nitrate spray. Supplemental calcium nitrate may counteract the negative effects of excessive salinity on cucumber plant fruit output and whole plant biomass, according to Kaya and David (2002). Domagala-Swiatkiewicz and Błaszczyk (2009) found that applying a calcium nitrate foliar spray raised the amount of calcium in fruits measured both after treatment and after harvest; however, the weather in the specific experimental years affected this impact. Only in 2005 did the nitrogen/calcium and potassium/calcium ratios in newly harvested fruits and fruits analysed after treatments decrease as a consequence of calcium nitrate spraying.

Yildirim et al. (2007) Although calcium nitrate is one of the main sources of nitrogen when combined with other fertilizers to promote the development of trees and vegetables, it has been noted that it dissolves slowly in soil solutions and cannot act quickly. As a consequence, calcium nitrate applied topically increases fruit production and yield in both young and developing fruits. According to Sarrwyet al. (2012) spraying blooms with boric acid and calcium nitrate affects their physiological and chemical characteristics as well as their fruit production and setting. The best treatment for fruit quality and yield throughout the two trial seasons was a 500 ppm boric acid and 2% calcium nitrate spray. The effects of urea (3 g/liter), boric acid (0.5 g/liter), and calcium nitrate (10 g/liter) on a greenhouse cucumber in both quantity and quality were examined by Nasrollahzadeh et al. (2015). Plant in length, percent of leaf dray matter, leaf weight ratio (LWR), fruit production grade 1 fruit yield, quantity of fruit, and percentage of grade 1 fruit were all included. The findings indicated that, among the principal effects, calcium nitrate treatment had the most effect on the variety of characteristics (Tables 19.1, 19.2 and 19.3)

19.3 Use of Synthetic Chemicals in Controlling Mango Diseases

19.3.1 Apical Necrosis

A three-stage spray plan has been suggested by Prakash and Raoof (1985) (emerging of panicle, opening of flowers, and fruit set). It was found that using a 0.1% bavistin spray 20 days apart helped to eradicate the disease. According to Bancal and Huet (2000) lesions in plants infected with foliar biotrophic diseases are potent nitrogen sinks, which accelerates nitrogen demand and causes apical leaf necrosis. Several

Table 19.1 Dynamics of mango inflorescence attributes and inflorescences diseases to the exposure of different treatments alone and in combinations in the field on mango trees (Chaunsa Summer Bahisht)

Treatments	F.E	A.N	B.B	M.M	P.M	P.T	F.S/ No of fruits
KNO_3 (8.34)*	8.67 ± 0.33 a	4.67 ± 0.33	5.67 ± 0.33	1.33 ± 0.33	0.00 ± 0.00	1.00 ± 0.00	119.33 ± 0.88
$CaNO_3$(7.69)	7.33 ± 1.09 b	5.33 ± 1.09	6.66 ± 0.82	2.33 ± 0.27	0.00 ± 0.00	1.00 ± 0.00	11.67 ± 0.72
Cu $(OH)_2$(8.01)	5.33 ± 0.67 d	4.33 ± 0.67	3.33. ± 0.33	1.33 ± 0.23	0.00 ± 0.00	1.00 ± 0.00	111.33 ± 0.58
Cabrio top (7.68)	4.00 ± 2.31 e	2.67 ± 0.33	1.67 ± 1.33	1.33 ± 0.33	0.00 ± 0.00	1.00 ± 0.00	109.33 ± 0.88
Contaf Plus (8.13)	3.33 ± 0.67 f	1.33 ± 0.13	0.00 ± 0.00	0.00 ± 0.00	0.00 ± 0.00	1.00 ± 0.00	103.00 ± 1.53
KNO_3 + Cu $(OH)_2$(8.24)	6.66 ± 0.58 c	4.66 ± 0.14	4.67 ± 0.58	0.00 ± 0.00	0.00 ± 0.00	1.00 ± 0.00	111.33 ± 0.88
$CaNO_3$ + Cu $(OH)_2$(8.11)	5.33 ± 0.33 d	2.67 ± 0.13	2.99 ± 0.00	0.33 ± 0.13	1.00 ± 0.33	1.00 ± 0.00	114.33 ± 0.31
KNO_3 + Cabrio top (8.10)	5.33 ± 0.33 d	4.67 ± 0.33	3.00 ± 0.38	1.00 ± 0.00	1.00 ± 0.00	1.00 ± 0.00	107.33 ± 1.15
$CaNO_3$ + Cabrio top (7.90)	5.99 ± 0.33 d	3.67 ± 0.33	2.00 ± 0.00	1.00 ± 0.0	0.00 ± 0.00	1.00 ± 0.00	112.33 ± 0.31
KNO_3 + Contaf Plus (8.30)	4.00 ± 1.33 e	4.67 ± 0.33	4.66 ± 0.00	0.33 ± 0.11	1.66 ± 0.00	1.01 ± 0.00	101.33 ± 0.88
$CaNO_3$ + Contaf Plus (8.12)	5.00 ± 1.15 d	4.67 ± 1.03	6.66 ± 0.00	0.00 ± 0.00	1.00 ± 0.00	1.00 ± 0.00	110.33 ± 1.76
Control (unsprayed plants)	1.33 ± 0.33 g	11.33 ± 0.88	10.33 ± 1.86	9.66 ± 1.11	3.36 ± 0.13	3.00 ± 1.13	77.09 ± 1.76

* Values in the parenthesis in the treatment column are the negative log of hydrogen ion concentration (pH) of the solutions prepared in the laboratory and sprayed in the field, F.E = Fruit emergence, A.N = Apical necrosis, B.B = blossom blight, M.M = Mango malformation, P.M = Powdery mildew, P.T = Phytotoxicity, F.S = Fruit setting on trees

Table 19.2 Disease decrease percentage of mango inflorescence diseases after exposure of different treatments alone and in combinations in the field on mango trees (Chaunsa Summer Bahisht)

Treatments	Apical necrosis	Blossom blight	Mango malformation	Powdery mildew
KNO_3	35.00	82.86	91.99	96.33
$CaNO_3$	56.25	71.43	87.32	93.11
Cu $(OH)_2$	87.50	88.57	98.34	94.33
Cabrio top	75.00	85.70	97.34	99.66
Contaf Plus	62.50	97.10	88.23	99.99
KNO_3 + Cu $(OH)_2$	37.50	57.10	81.12	99.99
$CaNO_3$ + Cu $(OH)_2$	92.50	71.40	96.66	98.99
KNO_3 + Cabrio top	75.00	57.11	92.34	89.11
$CaNO_3$ + Cabrio top	87.50	94.28	99.11	98.37
KNO_3 + Contaf Plus	87.50	88.57	93.45	96.78
$CaNO_3$ + Contaf Plus	25.00	73.42	81.11	79.32
Control	—	—	—	—

Table 19.3 Effect of different chemicals and fungicides treatments to induce inflorescence/flower setting in mango trees (Chaunsa Summer Bahisht)

Treatments	Flower emergence/ Fruit setting	Increase in flower emergence (%)	Increase in flower emergence (times)
KNO_3	8.67	551.88	5.51
$CaNO_3$	7.33	451.14	4.51
Cu $(OH)_2$	5.33	300.76	3.00
Cabrio top	4.00	200.75	2.00
Contaf Plus	3.67	175.94	1.75
KNO_3 + Cu $(OH)_2$	6.66	400.76	4.00
$CaNO_3$ + Cu $(OH)_2$	4.00	200.75	2.00
KNO_3 + Cabrio top	5.99	350.38	3.50
$CaNO_3$ + Cabrio top	5.33	300.76	3.00
KNO_3 + Contaf Plus	5.33	300.76	3.00
$CaNO_3$ + Contaf Plus	5.00	275.94	2.75
Control	1.33	—	—

of the evaluated compounds have shown encouraging results against bacterial apical necrosis, suggesting that they may be intriguing substitutes for conventional chemical disease management. The source of apical necrosis, *Pseudomonas syringae*, causes necrotic damage to mango buds and leaves, which results in severe production losses because Cazorla et al. (2006) As a consequence, fosetyl-Al, acibenzolar-S-methyl, Bordeaux mix, gibberelic corrosive, silicon gel, a combination of acibenzolar-S-methyl and Bordeaux blend, and combined applications of fosetyl-Al with either silicon gel or Bordeaux blend were tested. The phosphonate subsidiary fosetyl Al and the plant defence activator acibenzolar-S-methyl, either by themselves or in combination with other mixes, were the drugs that consistently decreased the adverse effects of bacterial apical putrefaction to levels equivalent to the usual Bordeaux blend treatment. Some of these drugs may be elective (Paulin et al. 1990).

The quantity of necrotic buds and leaves was significantly decreased by these medications; nevertheless, some fluctuations in *P. syringae*-like level of population were noted. Commercial copper-based treatments with Bordeaux mixture, which are used to give mango harvests, do not seem to have bactericidal activity, according to an examination of the antibacterial and inhibitory convergence of cupric mixes against *P. syringae* strains obtained from mango tissues. Due to the Bordeaux blend treatment, fosetyl-al was effective but ineffective in controlling bacterial apical rot of mangoes. It also had conflicting effects on other bacterial diseases, such as pear fire scourge, citrus infection, and certain bacterial disease of ornamental plants (Chase 1993).Hattingh et al. (1989) regarded P. syringaepv. syringae as a weak pathogen that only infects the host when it is anxious, such as during tree sluggishness and bud break. According to Cazorla et al. (1998), pathogenicity studies on mango trees showed that P. syringaepv. Syringa induced apical corruption, but that the beginning of infection improvement was influenced by environmental circumstances. Over the course of three years, populations of P. syringae and aggregate bacteria as well as the number of dynamic ice cores were tracked. While P. syringae was sometimes found on solid trees, its largest populations were linked to chilly, rainy times that correlated with the most notable disease severity. Infectivity titration assays were used to assess the middle viable measurement. Throughout the three exploratory seasons, the infection frequencies in buds was consistently detected in February, with August showing the lowest level. Since it has been shown that the severity of the apical putrefaction infection is associated with cool and wet winters after P. syringaepv. syringae colonises mango tissues, more efficient handling of bacterial apical rot of mango is required during the cool and highly wet seasons (Cazorla et al. 1998).

19.3.2 Blossom Blight

Two splashing methods to combat the Glomerella ingulata-caused mango bloom curse were tested in Dominica, according to Prior and Ryder (1987). Before flowering started, a single dose of 4% copper oxychloride (18 g per tree) was applied to the leaves in one treatment. In the other, 2% was applied every two weeks, starting after blooming. Although the repeated evaluations showed no significant effects, the first single dose resulted in a significant increase in the natural product set. The expense of disease control may be reduced by early application followed by a low volume sprinkling of protectant fungicides. At one recording date, the expansion of the 0.5% polyisobutene sticker, which was attempted in the single measures experiment, produced a consistently improved natural product set when compared to the non-sticker treatment; the overall evidence was empowering. According to in vitro testing, the parasite was more sensitive to chlorothalonil and some systemic fungicides than it was to copper. The best way to prevent anthracnose is to avoid growing mangoes in areas where the disease is likely to emerge. In addition to basic copper sulphate, fungicides that are effective against anthracnose and approved for use in Hawaii include sulphur, benomyl, captan, and essential copper sulphate. Splashes and cleans with sulphur have been shown to be quite effective and are not included in resilience. In any event, sulphur has to be used carefully because of the (Palti et al. 1974).

According to Nelson (2008a, b) sporadic fungicide rains are necessary to reduce bloom plague on very vulnerable mango cultivars and under extremely favourable conditions. For adequate disease management, utilisation planning and recurrence are very crucial. When panicles initially appear, splashes should begin. They should continue at the recommended intervals until organic products reach a length of about 11/2–2 inches. A variety of foliar fungicides are used in Hawai ‘i to combat the mango bloom curse. These include products that contain neem oil that has been cleaned up, copper hydroxide, essential cupric sulphate, copper salts of phosphorous corrosive, harpin protein, copper salts of greasy and rosin acids, and chlorothalonil. The adequacy and mode of action of these things vary. For specific recommendations and current item names, see the CTAHR Cooperative Extension Service. According to McMillan (1973) benomyl (8 oz per 100 ladies) significantly reduced disease and increased total yields. Consolidating benomyl with oil–water emulsion and Nu-Film-17 has been shown to have remarkable disease management effects. Ruehle (1953) showed that using zineb, maneb, or captan (1V2 lbs/100 lady) on a weekly basis during sprouting was followed by using copper (3 lbs/100 lady) on a monthly basis after a natural product set provided excellent disease control. In experiments conducted over the last four years, the new fungicide benomyl has shown significant promise as a fungicide for avocados and mango trees (McMillan 1971). It has proved very effective against mango powdery mildew, avocado scab, and mango and avocado anthracnose.

19.3.3 Powdery Mildew:

When fungicides are used properly, the disease may be efficiently managed. Poor scheduling of treatments may lead to needless crop losses, thus spray applications must be managed carefully. The first application should be made 50% of the way through flowering or as soon as disease symptoms appear. According to Nelson

(2008a, b) three spray treatments are often used. Wetable sulphur or sulphur dust sprays are widely used and provide a decent level of control, but if they are sprayed in bright circumstances, they may be phytotoxic to blossoms and young fruit (Nelson 2008a, b). Based on the relationship between temperature and relative humidity and the development of the disease, predictive models have been created to enhance the management of *O. mangiferae* (Sinha et al., 2001). Predictive management may be used to improve the effectiveness of chemical control. Managing the powdery mildew may be achieved by spraying the inflorescence at intervals of 10–14 days. Fresh soft tissue of every flower, leaf, and fruit segment exhibits the telltale symptoms of disease. The whole fruit is harmed if there is significant damage, and the blooms fall off the inflorescence and stay immature (Naqvi et al. 2014).The mango types may be divided into eight categories based on how they performed in respect to the occurrence of powdery mildew. Dusehri, Chaunsa Samar Bahisht, Malda, and Ratole No. 12 had the highest illness incidence (33.33%, 26.66%, and 26.66%, 26.66%), while Almas and Sensation had the lowest disease incidence (3.66% and 3.66%). These two cultivars demonstrated disease tolerance. We may infer that the fruit production of the mango cultivars was unaffected by the presence or lack of powdery mildew symptoms (Naqvi et al. 2014).

Burning, dried-up springs with droplets at night are conducive to hazardous outbreaks in South Africa, according to Wager (1937).When blooms are severely damaged, they may not open and may even fall off the inflorescence, harming the whole fruit. Conidia germinate best between 20–250C and somewhat high relative humidity, according to Uppel et al. (1941). Furthermore, rain during blossoming promotes disease, according to Bose (1953). According to Wagle (1928) wind-borne conidia generate mycelium to finish their life cycle on shoots that grow in 9 days after causing infection 5–7 h after germination. The prominent symptom is a velvet-like, powdery coating with a dark to smoky grey backdrop (Paltiet al. 1974). Adult fruit that has white velvet-like powder on it will have corky surfaces or superficial irregular reddish brown spots (Joubert et al. 1993; Akhtar and Alam 2000).Powdery mildew may be cured, according to Khalid et al. (2000). To achieve maximal disease control, it may be managed by using fungicides at the right time and with the right tools. According to Paltiet al. (1974) powdery mildew disease may be controlled by spraying throughout the blooming stage at intervals of 10–14 days. Sulphur.

19.3.4 Mango Malformation:

Severe mango malformation disease has been identified as the main obstacle to mango output in India and other mango-producing nations worldwide (Crane and Campbell 1994). According to (Hiffnyet al. 1978) mango malformation is a major factor in destroying mango output and resulting in significant financial losses. According to Summanwar (1967) mango deformity causes a larger loss of trees because inflorescences do not produce fruits. In some situations, the damage ranges from 50 to 60%, and in extreme situations, the loss might reach 100%. Uttar Pradesh has a higher rate of disease than the eastern areas. Prasad et al. (1965). Malformation was generally categorised as either floral or vegetative by Kumar and Beniwal (1987a, b). Varma et al. (1969) It was shown that the excess deformity on a single panicle might range from mild to moderate to severe. Chakrabarti and Kumar (1997) noted that the ecological component and mangiferin content cause periodic differences in disease occurrence. The most seriously impacted panicles in Egypt are those that emerge on spring shoots (Shawky et al. 1980). According to reports, the disease is more prevalent in Florida under foggy conditions (Campbell and Marlatt, 1986). Each cultivar and each cycle has a different level of deformity (Azzouset al. 1978). The majority of Indian and Pakistani varieties have irregular fruit sets and alternating bearers. Depending on the cultivar type, degree of hypertrophy, and infection level, the indications in the deformed shoots may present differently.

According to Ploetz and Gregory (1993) substantial yield loss ranging from 80–100% is caused by mango malformation disease. Every year, there were reports of severe deaths from this ancient, troublesome disease. According to Singh et al. (1961) plants between the ages of 4 and 8 experience significant loss (90.9%) due to vegetative deformation. Every tree may suffer damage ranging from 50 to 80 percent, and in extreme situations, the misfortune may be almost total, as the twisted inflorescence fails to produce fruiting bodies (Summanwar 1967). Additionally, it was revealed that there are two types of malformation that have the same aetiology: vegetative and floral. (Kumar and Beniwal 1987a). Ploetz (1994) According to reports, this disease results in significant injury rates in the plant market and early destruction to the young plants' vegetative shoots. Malformations may also impair bloomy organs and shoots. The mature plant's inflorescence is attacked by deformity. According to Sao-Jose et al. (2000) the percentage of malformed flowers was 54%, with a deformity severity of 17%. Acarological (Singh et al. 1961) viral (Das et al. 1989) and physiological (Sattar 1946) aetiologies were earlier proposed but disproved because they lacked etiological correlation. New taxa have been discovered by testing of Fusarium strains isolated from deformed inflorescences of various foreign sources; F. mangiferae is related to strains that were previously described as F. subglutinans and are thought to be the cause of MMD globally. F. maniferae Britz's ascendency was verified by.

19.3.5 Findings of One of the Research

One research aimed to identify the most exciting synthesised, chemical fungicide, chemicals of laboratory grade or their possible combination that not only promotes floral growth but also triggers blooming and combats inflorescence diseases, which are the main reason of reduced production. A field experiment with a range of treatments was carried out for this investigation. The treatment options were, Cantof Plus, KNO_3 and Champion, KNO_3, $CaNO_3$, Champion, Cabrio-Top KNO_3 and Cabrio-top, KNO_3 and Cantof Plus, $CaNO_3$ and Champion, $CaNO_3$ Cabrio-top, and $CaNO_3$ andcontaf-plus.

In comparison to the control, the findings indicated that the treatment KNO_3 (8.67) produced the highest flowering/flower emergence, $CaNO_3$ (7.33), KNO_3 with a mix of champion (6.66), and KNO_3 with the combination of cabrio top (5.33). The plants treated with the fungicides alone, i.e., Contaf plus, showed the smallest amount of flowering, 3.33, compared to the control, which only showed 1.33. When compared to the control, which registered 11.33, Contaf plus supplied the greatest control on mango apical necrosis, scoring 1.33, followed by cabrio top (2.67). When comparing Contaf plus to the control, the former had the best control, with a 0.00 mean for blossom blight and a 1.67 mean for cabrio top was completely controlled by KNO_3 with champion and $CaNO_3$ with cabrio top, as opposed to 9.66 for the control. In contrast to the other disorders, KNO3 and $CaNO_3$ also did the best for controlling mango malformation, with respective scores of 1.33 and 2.33.

Powdery mildew was found to be minimal on the inflorescence because all treatments were effective in controlling mango powdery mildew, with KNO_3 and $CaNO_3$ showing the highest results against the disease. When chemicals or any other treatments were administered, either alone or in combination with other treatments, phytotoxicity was seen on every plant. The highest level of toxicity was seen on inflorescence when fungicides alone were used, followed by fungicides combined with synthetic compounds.

While fungicides alone did not seem to set maximum flowering, treatment KNO_3 and $CaNO_3$ showed the highest fruit setting, followed by a mixture of chemicals and fungicides.

19.4 Incidence Percentage of Different Diseases Appeared on Mango Inflorescence on Chaunsa Cultivar

The disease frequency of each inflorescence disease was evaluated and analysed using the disease incidence formula. It was found that the blossom blight disease had the highest number of inflorescence infections (309), followed by apical cell death of the mango (281 inflorescences), mango malformation (127 inflorescences), and powdery mildew infection (Fig. 19.1). The highest documented disease frequency was 42.90% for BB, followed by 39.00% for AN, 17.70% for MM, and a minimum of 0.40% for powdery mildew (Figs. 19.2 and 19.3).

19.5 Case Studies of Synthetic Fungicides Applications

In the "Haden" and "Manila" mango trees, Nunez-Elisea (1985) discovered that the (KNO_3) spray produced inflorescences on 60 and 76% of the shoots, respectively, in contrast to 32 and 20% of the similar controls. Although fungicides by themselves did not seem to cause any plant to blossom to their full potential, we did discover

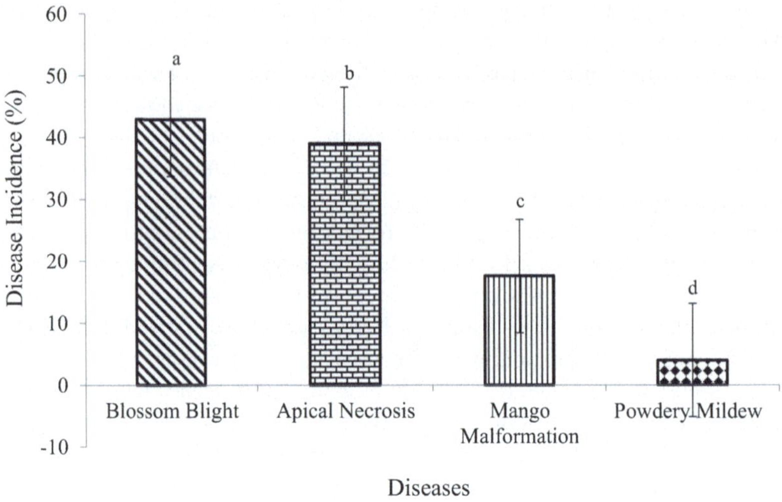

Fig. 19.1 Summary of disease incidence percentage of various pathosystems recorded on mango plant inflorescence

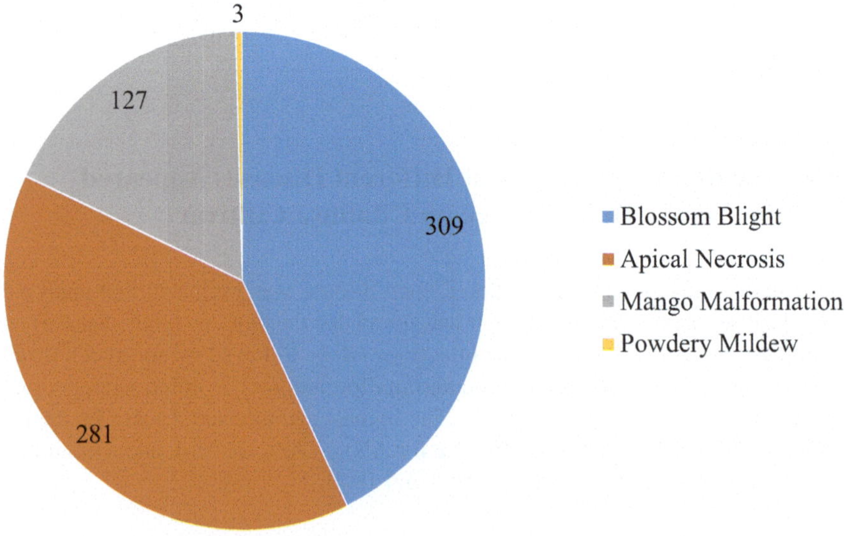

Fig. 19.2 Summary of inflorescences infected by various pathosystems on Chaunsa as indicated in number against the total inflorescence observed during an experiment

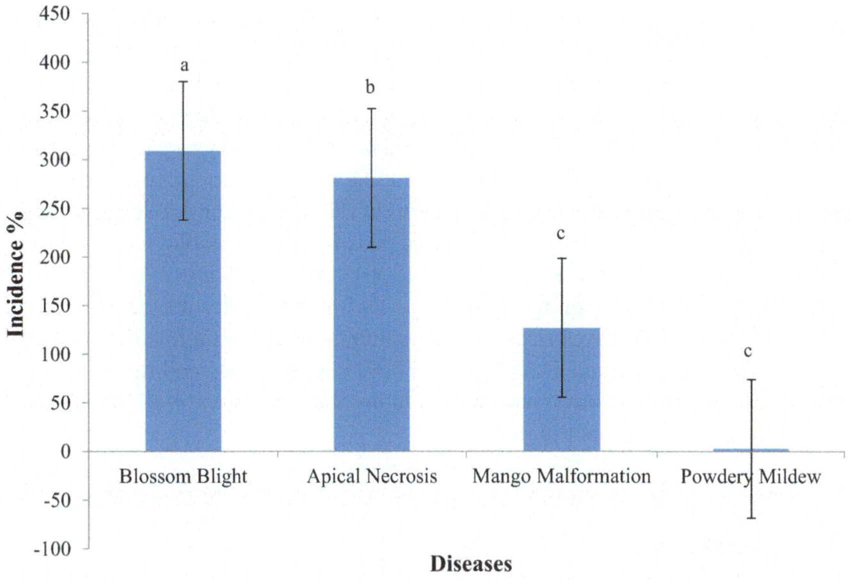

Fig. 19.3 Incidence percentage of different diseases appeared on mango plant inflorescence

that treatment KNO_3 and $CaNO_3$, which were followed by a combination of chemicals and fungicides, produced the highest fruit setting. According to Davenport and Núñez-Elisea (1990) ethephon and KNO_3 are supposed to encourage mango flowering in the tropics, however they had no impact in South Florida during the floral promotive and non-promotive stages.

Contaf plus had the best control over mango apical necrosis, scoring 1.33 compared to the control's 11.33, and cabrio top came in second with 2.67. With a mean of 0.00 for blossom blight and a mean of 1.67 for cabrio top, the former offered the best control when compared to the control, which had a mean of 10.33. Mango malformation disease, believed to be a century-old mango ailment, was completely controlled by KNO_3 with champion and $CaNO_3$ with cabrio top, as opposed to 9.66 for the control. When compared to other disorders, KNO_3 and $CaNO_3$ likewise performed well in controlling mango malformation (1.33 and 2.33, respectively). Powdery mildew was found to be minimal on the inflorescence because all treatments were effective in controlling mango powdery mildew, with KNO_3 and $CaNO_3$ showing the highest results against the disease. When chemicals or any other treatments were administered, either alone or in combination with other treatments, phytotoxicity was seen on every plant. The highest level of toxicity was seen on inflorescence when fungicides alone were used, followed by fungicides combined with synthetic compounds. Similar results were obtained by Tongumpai et al. (1991) Researchers found that applying $Cu(OH)_2$ to the soil caused blooming to rise significantly and to occur somewhat sooner. They also found that applying potassium nitrate foliar spray eight weeks after the treatment dramatically decreased inflorescence disease. When Malik et al. (1997) studied the effectiveness of foliar spraying fungicides against several inflorescence fungal diseases on field plants throughout the blooming season, they came to similar conclusions. According to Quijada et al. (2000) plants treated with KNO_3 produced the most inflorescences per plant. When plants are treated with different chemicals, the number of inflorescences may vary depending on the genetic potential of the different varieties. In non-inductive settings, potassium nitrate promoted the emergence of buds for both physiological and reproductive growth, according to Yeshitela et al. (2004). After a calcium nitrate spraying, Saure (1998) found that damage from burns and phytotoxicity are decreased and the calcium content in leaves rises because calcium produces more stiff cell walls. However, Cengiz Kaya and David Higgs (2002) suggest that additional calcium nitrate might mitigate the adverse effects of high salinity on the overall biomass and fruit production of cucumber plants.

19.6 Conclusion

This study highlights the potential of targeted chemical treatments, including fungicides and stress-mitigating compounds, to mitigate climate-induced challenges in mango cultivation, particularly in addressing tree scarification and floral diseases. The significant reduction in scarification rates, disease incidence, and floral necrosis, along with improvements in floral retention, fruit set, and physiological parameters such as chlorophyll content and photosynthetic efficiency, demonstrates the efficacy of these interventions. These findings emphasize the importance of integrating chemical applications into mango orchard management to counteract the adverse effects of climate variability. However, sustainable and judicious use of chemicals is crucial to balance enhanced productivity with environmental safety and food quality standards. This study underscores the urgent need for integrated and sustainable strategies to ensure the resilience and productivity of mango orchards under changing climatic conditions.

References

Akhtar KP, Alam SS (2000) Powdery mildew disease of mango. A Review. Pak J Biol Sci 3:1119–1122

Akhter MA, Rahim M, Alam MS (2004) Effect of spraying of different chemicals on the flowering and fruiting of seven mango varieties

Alvi AR, Chohan S, Abid M, Malik MT, Riaz HM, Mudassar S (2022) Flower induction and control of inflorescence diseases in mangoes using selected fungicides and chemical fertilizers. Int J Phytopathol 11(1):01–08

Azzous S, Hamdy ZM, Dahshan IM (1978) Studies on malformed inflorescence of mango: the degree of susceptibility among different varieties. Agric Res Rev 56:17–27

Bancal MO, Huet P (2000) Fonctionnement d'un peuplement de blé d'hiver soumis à une attaque tardive de rouille brune (*Puccinia recondita*). Colloques-INRA 515–530

Barba RC (1974) Induction of flowering of the mango by chemical spray. Proc Crop Sci Soc Philipp 5:154–160

Bhuiyan SA, Boyd MC, Dougall AJ, Martin C, Hearnden M (2007) Effects of foliar application of potassium nitrate on suppression of Alternaria leaf blight of cotton (Gossypium hirsutum) in northern Australia. Australas Plant Pathol 36(5):462–465

Bondad ND, Linsangan E (1979) Flowering in mango induced with potassium nitrate. HortScience 14:527–528

Borkowski J, Felczyńska A, Stępowski J, Niekraszewicz A (2006) Effect of different compounds Biochikol 020 PC, calcium nitrate, Tytanit and Pomonit on the healthiness and the yield of Chinese cabbage. Polish Chitin Soc, Monograph XI:201–207

Borkowski J, Kowalczyk W (2003) Relation between tipburn appearance of Chinese cabbage and calcium content and location (English summary). Acta Agrobot 56:53–60

Bose SK (1953) Powdery mildew of mango in Kumaon and its control. Proc Indian Sci Congr 78

Bueno PB, Valmayor RV (1974) Potassium nitrate, key to mango flowering. Agric Los Banos

Campbell CW, Marlatt R (1986) Current status of mango malformation disease in Florida. Proc Int Am Soc Trop Hort 30:223–226

Cavalcante ÍHL, Martins JDL, Lima Neto AJ (2022) Integrated management strategies for mango diseases: a review. Crop Prot 150:105789

Cazorla FM, Arrebola E, Olea F, Velasco L, Hermoso JM, Pérez-García A, Torés JA, Farre JM, De-Vicente A (2006) Eur J Plant Pathol 116:279–288

Cazorla FM, Torés JA, Olalla L, Pérez-García A, Farré JM, De-Vicente A (1998) Bacterial apical necrosis of mango in southern Spain: a disease caused by *Pseudomonas syringae* pv. *syringae*. Phytopathology 88:614–620

Chakrabarti DK, Kumar R (1997) Probability of break out of mango malformation epidemic in West Bengal, India. In V Int Mango Symp 455:609–611

Chase AR (1993) Efficacy of fosetyl-Al for control of some bacterial diseases on ornamentals. Plant Dis 77:771–776

Crane JH, Campbell CW (1994) The Mango. The Horticultural Science Department, Florida Cooperative Extension Service, Institute of Food and Agricultural Sciences, University of Florida

Das GC, Rao DP, Lenka PC (1989) Studies on mango malformation in the mango clone, Chiratpudi. In II Int Symp Mango 231:866–872

Davenport TL, Núñez-Elisea R (1990) Ethylene and other endogenous factors possibly involved in mango flowering. In Int Symp Cult Subtrop Trop Fruits Crop 275:441–448

Domagala-Swiatkiewicz I, Blaszczyk J (2009) Effect of calcium nitrate spraying on mineral contents and storability of 'Elise' Apples. Polish J Environ Stud 18(5):971–976

English H, DeVay JE, Ogawa JM, Lownsbery BF (1980) Bacterial canker and blast of deciduous fruits. Leaflet-University of California, Cooperative Extension Service

Freeman S, Otero-Colina G, Rodríguez-Alvarado G, Fernández-Pavía S, Maymon M, Ploetz RC, Aoki T, O'Donnell K (2014) First report of mango malformation disease caused by *Fusarium pseudocircinatum* in Mexico. Plant Dis

García D, Pérez JA, Téliz D (2019) Mango anthracnose: economic impact and current options for integrated management. J Plant Pathol 101(3):553–565

González M, Restrepo S, Castaño M (2021) Advances in the management of mango diseases: a review. Trop Plant Pathol 46(1):1–12

Hattingh MJ, Roos IMM, Mansvelt EL (1989) Infection and systemic invasion of deciduous fruit trees by *Pseudomonas syringae* in South Africa. Plant Dis 73:784–789

Hiffny HAA, Barkouki M, El-Banna GS (1978) Morphological and physiological aspects of the floral malformation in mangoes. Egypt J Hort 5:43–53

Iqbal Z, Akhtar N, Ghazanfar MU, Shehzad SM, Ahmad S, Asif M, Yasin M, Pervez MA (2011) Management of mango malformation through physical alteration and chemical spray. Afr J Agric Res 6:1897–1901

Itagi T, Hiki M (1956) On the occurrence of deformed fruits in cucumber. I. The influence of some cultural practices on the occurrence of deformities. Hort Abstr 27:1465

Karuppaiah P, Manivannan K, Sriramachandrasekaran MV, Kuppusamy G (2001) Responses of cucumber to foliar application of nutrients on lignite mine spoil. J Indian Soc Soil Sci 49:150–153

Kaya C, David H (2002) Effect of silicon on plant growth and mineral nutrition of maize grown under water-stress conditions. J Plant Nutr 25:861–871

Khalid P, Akhtar K, Alam SS (2000) Powdery mildew of mango. A Review Pak J Biol Sci 3:1119–1122

Kumar J, Beniwal SPS (1987a) A method of estimating cultivar susceptibility against mango malformation. Int J Pest Manag 33:208–210

Kumar J, Beniwal SPS (1987b) A method of estimating cultivar susceptibility against mango malformation. Trop Pest Manag 33:208–211

Martínez AR (2008) Effects of paclobutrazol and KNO_3 over flowering and fruit quality in two cultivars of mango Manila. Interciencia 33:518–522

McMillan Jr RT (1971) Control of anthracnose and powdery mildew of mango with systemic and non-systemic fungicides. Trop Agric (in press)

McMillan RT Jr (1973) Control of anthracnose and powdery mildew of mango with systemic and non-systemic fungicides. Trop Agric 50:245–248

Naqvi SAH, Perveen R, Manzoor SA, Umar HMI, Iqbal MT, Liaquat F, Majid T, Irshad A (2014) Evaluation of various mango varieties against the infection dynamics of powdery mildew (Oidium mangiferae Bert.). Am J Plant Sci 5:2372–2377

Nasrollahzadeh-asl N, Delshad M, Kashi AK (2015) The effects of foliar application of urea, calcium nitrate, and boric acid on growth and yield of greenhouse cucumber (cv. Khassib). Biol Forum 7:712

Nelson SC (2008) Mango anthracnose (*Colletotrichum gloeosporioides*). Univ Hawaii Manoa, Coll Trop Agric Hum Res, Coop Ext Serv. Plant Dis Aug. 2008 PD-48

Nelson SC (2008) Mango powdery mildew. Univ Hawaii Manoa, Coll Trop Agric Hum Res, Coop Ext Serv. UH–CTAHR-PD-46, Aug. 2008

Nunez-Elisea R (1985) Flowering and fruit set of a monoembryonic and polyembryonic mango as influenced by potassium nitrate sprays and shoot decapitation. Proc Fl Hort Soc 98:179–183

Oosthuyse SA (1996) Effect of KNO3 sprays to flowering mango trees on fruit retention, fruit size, tree yield, and fruit quality. In V Int Mango Symp 455:359–366

Palti J, Pinkas Y, Chorin M (1974) Powdery Mildew Mango Plant Rep 58:45–49

Patel N, Gupta V (2022) Role of plant growth regulators in enhancing disease resistance and physiological functions in crops: a review. J Plant Growth Regul 41(3):1234–1245

Paulin JP, Chartier R, Lecomte P, Brisset MN, Lachaad G, Larue P (1990) Experiments with Aliette (Phosetyl-aluminium) in fire blight control. Acta Hortic 273:383–390

Ploetz RC (1994) Distribution and prevalence of *Fusarium subglutinans* in mango trees affected by malformation. Can J Bot 72:7–9

Ploetz RC, Gregory N (1993) Mango malformation in Florida: distribution of *Fusarium subglutinans* in affected trees, and relationship among strains within and among different orchards. Acta Hortic 341:388

Prakash O, Raoof MA (1985) Spray schedule for the control of powdery mildew disease of mango. Indian J Plant Pathol 3:51–52

Prasad A, Singh H, Shukla TN (1965) Present status of mango malformation disease. Indian J Hortic 22:54–265

Prior C, Ryder K (1987) Effect of low volume copper sprays with polyisobutene sticker on mango blossom blight (Glomerella cingulata) in Dominica. Int J Pest Manag 33(4):350–352

Quijada Q, Camacho R, Fonseca Y, Rivas J (2000) Effect of potassium nitrate on flowering of mango Haden. FONAIAP-Divulga 66:2–4

Reuveni M, Reuveni R (1995) Efficacy of foliar sprays of phosphates in controlling powdery mildews in field-grown nectarine, mango trees, and grapevines. Crop Prot 14:311–314

Ruehle GD (1953) Organic fungicides for control of anthracnose of mango. In: Proc Fla Mango Forum

Sao-J AR, Souza SE, Pina AV, Ataide EM (2000) Incidence and severity of mango malformation in Bahia State, Brazil. Acta Hortic 509:765–771

Sarrwy SMA, Gadalla EG, Mostafa EAM (2012) Effect of calcium nitrate and boric acid sprays on fruit set, yield, and fruit quality of cv. Amhat date palm. World J Agric Sci 8:506–515

Sattar A (1946) Diseases of mango in the Punjab. Punjab Fruit J 10

Saure MC (1998) Causes of the tipburn disorder in leaves of vegetables. Sci Hortic 76:131–147

Sergent E, Ferrari D, Leal F (1996) Effects of potassium nitrate and paclobutrazol on flowering induction and yield of mango (*Mangifera indica* L.) cv. Haden. in V Int Mango Symp 455:180–187

Shawky I, Zidan Z, El-Tomi A, Dahshan D (1980) Flowering malformation in relation to vegetative growth of 'Taimour' mangos. Egypt J Hortic 7:1–8

Singh LB, Singh SM, Nirvan RS (1961) Studies on mango malformation: review, symptoms, extent, intensity, and cause. Hortic Adv 5:197–207

Sinha P, Prajneshu, Varma A (2001) Studies on determining favourable factors for the germination of conidia of Oidium mangiferae. Indian Phytopathol 54(2):197–200

Summanwar AS (1967) Mango malformation, serious economic consequences. Indian Hortic 11:12–16

Tongumpai P, Jutamanee K, Subhadrabandhu S (1991) Effect of paclobutrazol on flowering of mango cv. Khiew Sawoey. In III Int Mango Symp 291:67–70

Varma A, Roychoudhuri SP, Lele VC, Ram A (1969) Towards the understanding of the problem of mango malformation. In Symp Mango Mango Cult 24:237–237

Wager UA (1937) Mango diseases in South Africa. Farming S Afr 12:321–324

Wagle PV (1928) Studies in the shedding of mango flowers and fruits. Part I. Mem Dep Agric India Bot Sci 8:219–249

Yeshitela T, Robbertse PJ, Stassen PJC (2004) Effects of various inductive periods and chemicals on flowering and vegetative growth of 'Tommy Atkins' and 'Keitt' mango (*Mangifera indica*) cultivars. N Z J Crop Hortic Sci 32:209–215

Yildirim E, Guvenc I, Turan M, Karatas A (2007) Effect of foliar urea application on quality, growth, mineral uptake, and yield of broccoli (*Brassica oleracea* L., var. italica). Plant Soil Environ 53:120

Chapter 20
Impact of Micronutrients for Mitigating Mango Malformation Disease Under Changing Climate

Awais Balban, Wajid Nazir, Ummad Ud Din Umar, Niaz Ahmed, Ateeq Ur Rehman, and Ghulam Mustafa

Abstract In tropical and subtropical areas, the mango (*Mangifera indica* L.) is an important fruit crop in terms of both nutrition and economics. The high demand for mango fruits and processed products, driven by their taste and associated health benefits, exists in both domestic and international markets. To meet these demands, it is essential to enhance mango production by leveraging various available tools to address environmental stresses, particularly in the context of climate change. This chapter aims to illustrate the effects of climate change on mango growth, quality, and yield, highlighting the interventions implemented to mitigate these impacts. It examines the role of climate variability in influencing mango tree development and yield parameters, as well as the improvement objectives and strategies employed to enhance production and quality. The chapter also reviews the progress made in overcoming biotic stresses that impede mango production and the interventions undertaken to improve the nutritional and quality traits of mangoes, providing directions for future research on mango cultivation. Furthermore, the study emphasises how effective nutrition management plans may be equally important driving for maximising mango production results. These solutions have the potential to improve overall agricultural efficiency by carefully finding a balance between the demands of fruit quality and quantity.

A. Balban · W. Nazir (✉) · U. U. D. Umar · A. U. Rehman
Department of Plant Pathology, Bahauddin Zakariya University, Multan, Pakistan

U. U. D. Umar
e-mail: ummad.umar@bzu.edu.pk

N. Ahmed
Department of Soil Science, Bahauddin Zakariya University, Multan, Pakistan
e-mail: niaz.ahmad@bzu.edu.pk

G. Mustafa
Department of Horticulture, MNS-University of Agriculture, Multan, Pakistan

Department of Agriculture Extension, Government of the Punjab, Punjab, Pakistan

S. A. H. Naqvi et al. (eds.), *Climate Change and Mango Production*,
https://doi.org/10.1007/978-3-031-87035-4_20

Keywords Mangifera indica · Soil health · Micronutrients · Fungal infection · Vegetative and floral malformation

20.1 Introduction

Farmers, including smallholders, are increasingly interested in diversifying their farm produce to include fruit crops like mangoes, driven by the goal of making a profit. This diversification plays a crucial role in achieving food security in an era of rapidly changing climate. Mangoes (*Mangifera indica* L.) are particularly important in reducing hunger and ensuring food security. Their popularity continues to rise due to evolving consumer preferences (Altendorf 2017). This can be partly attributed to its unique aroma, vibrant color, juicy pulp, exceptional taste, and nutritional value (Ntsoane et al. 2019; Tharanathan et al. 2006). Mangoes contain essential nutrients including calcium, iron, phosphorus, and vitamins C and A (Lebaka et al. 2021). It is recommended that fruit nutrients are essential for both physical and mental health, particularly in children (Carrillo et al. 2019; Lamprecht et al. 2013). These attributes crown the mango as the king of fruits. Among tropical fruit crops, mangoes rank fifth globally in terms of production among major fruit crops (FAO 2022; Okino-Delgado et al. 2020). In 2021, the global production of mangoes totaled 58.3 million metric tons. This volume is projected to increase by 1.9%, reaching 65 million metric tons by 2026 (MMO 2023). Mangoes are cultivated in over 90 countries worldwide, with India being the largest producer (FAO 2023). Asia is the leading region in mango production, accounting for over 70% of the global annual yield. Africa contributes 10% to the total annual mango production (FAO 2021). After Mexico and the Philippines, Pakistan comes in third place in terms of mango exports, with a 7.60% market share (Nazir et al. 2024b). Mango production presents a valuable opportunity as an alternative livelihood, particularly in rural areas, and offers substantial profit potential (Testa et al. 2018; Yiran et al. 2023). The rising demand for high-quality mangoes to satisfy local, national, and international market standards has led to the development of new production strategies and techniques. These are crucial for adapting to the rapidly changing climate. For example, in many mango-producing regions, rainfall is insufficient to meet the water needs of mango trees. This necessitates the use of irrigation systems, biostimulants, and paclobutrazol to enhance flowering, fruit yield, and quality (Ferreira et al. 2020; Lobo et al. 2019; Simões et al. 2020).

Micronutrient deficiencies impact various processes in plants, such as impaired plant productivity, plant growth, and quality (Berdanier and Berdanier 2015; Gurjar et al. 2015; Souri and Aslani 2018; Souri and Bakhtiarizade, 2019). Micronutrient deficiencies are also necessary for various physiological and metabolic processes in plants. According to Ahmed et al. (2009), micronutrients are essential for respiration and photosynthesis. The deficiency of iron has an impact on certain fruit trees' output, fruit quality, chlorophyll content, and mineral nutrients (Souri et al. 2018; Tagliavini et al. 2000). A micronutrient termed as iron is necessary for the growth and development of plants. It is essential for several physiological functions, including as respiration, DNA synthesis, and photosynthesis. Numerous enzymes in plant cells that are involved in redox reactions and electron transport include iron (Sharma and Dietz 2006). According to Marschner (2011), zinc (Zn) is an essential element for plants because of its vital involvement in critical factors such fruit set, retention, yield, and quality. Insufficient zinc may significantly hinder plant development, resulting in a significant reduction in total output (Ojeda-Barrios et al. 2014). This highlights how important it is to keep zinc levels in check in order to maximize agricultural yield and guarantee healthy plant development. One essential micronutrient needed by plants for the creation of proteins and enzymes is zinc. It participates in a number of enzymatic processes, such as those that are connected to protein synthesis, auxin production, and glucose metabolism. Zinc is necessary for healthy root growth and is involved in the creation of DNA and RNA (Broadley et al. 2007; Yruela 2005). Plant cell wall development and stability depend on boron. It is essential for the movement of glucose and nutrients throughout the plant, as well as for the production of proteins and nucleic acids. According to Goldbach and Wimmer (2007) and Gupta (1993), boron is necessary for healthy seed and fruit development and is also involved in the creation of pollen tubes.

Unfortunately, mango production faces several bottlenecks that affect yield. Key factors contributing to the yield gap include pests, diseases, insects, weeds, changes in rainfall patterns, seed quality, soil fertility, and irregular flowering, among others. These issues represent significant constraints on mango production and productivity. Gaining a deeper understanding of these challenges can help the mango industry enhance output by developing more effective and efficient management systems. Additionally, various efforts are focused on nutritional management, employing diverse fertilization strategies to mitigate physiological disorders and improve fruit quality (Simões et al. 2022; Taiz et al. 2017; Zhang et al. 2019). Given the extensive research on mango cultivation, this chapter zeroes in on the impacts of climate change on mango production, particularly focusing on how it affects tree growth, development, yield, and fruit quality. It also examines farmers' responses and adaptation strategies for managing climate change. Additionally, the chapter covers general objectives for mango improvement, the challenges faced, and the methods employed. Finally, it discusses strategies for post-harvest management and value addition to enhance mango quality.

20.2 Climate Change and Mango Production

Human activity has led to an increase in greenhouse gas emissions, which has resulted in climate change and global warming. Depending on the region, the consequences of climate change might be neutral, detrimental, or beneficial (Janjua et al. 2014). Climate is still a major factor in determining a place's farming ability and allocating specific regions for growing particular crops (Drebenstedt et al. 2023; Fain et al. 2018; Janjua et al. 2014). Changes in temperature, precipitation amounts, atmospheric CO_2 concentration, drought, and soil salinity are only a few of the effects of climate variability. Therefore, the current environmental circumstances have an impact on the growth, development, flowering, and yields of mango trees. Mangoes' fruit set, quality, and yields may be adversely affected by a change in the necessary climatic conditions (Shahraki et al. 2023).

20.2.1 Effects of Climatic Change on the Development and Growth of Mango Trees

Temperature and rainfall play a major role in determining the mango's vegetative and phenological phases. For example, a mango tree thrives best at temperatures between 20 and 29 °C, and below 15 °C to encourage flowering (Whiley et al. 1989). Temperatures between 20 and 28 °C have a noticeable impact on the growth of leaves (Dambreville et al. 2012), likely due to dry materials moving to the root when CO_2 concentrations are high (Whiley et al. 1989). Compared to annual crops, mangoes respond to temperature increases differently (Litz 2009). Because of this, it may survive desiccating circumstances, which is a beneficial characteristic for yielding in subsequent growing seasons. On the other hand, stress can cause a plant's development to be delayed and result in a significant loss of output (Morison et al. 2006). Mango trees are negatively impacted by flood and drought stress by growing less. Crop variety matching to environment is a necessary step for profitable mango production in both established and emerging regions. Mango blossoming depends on both the wet and dry seasons, with the wet seasons favouring vegetative flow emission (Ramírez and Davenport 2010; Spreer et al. 2009; Sukhvibul et al. 1999). Rain usually happens during mango blossoming, which can be detrimental and result in a complete crop failure (Makhmale et al. 2016). Mango trees are unable to blossom because of their constant growth in the face of high temperatures, heavy rains, and humidity. However, severe dryness and a vapour pressure deficit have a detrimental effect on photosynthesis because they cause the mango plant's stomata to close quickly owing to changes in the weather (Legave et al. 2013). In most varieties, flower buds emerge in the shoot where growth terminates. Mango tree growth chamber experiments verify that low overnight temperatures of 8–15 °C combined with daytime temperatures below 20 °C cause blossoming as soon as the shoot initiation occurs (Figs 20.1, 20.2 and 20.3) (Shu and Sheen 1987).

20.2.2 Effects of Climatic Change on the Quality and Production of Mango Trees

Mango tree development, yield, and quality are impacted by biochemical, anatomical, physiological, and morphological changes brought on by stress imposed by climate variations. For example, unseasonal rains combined with sharp temperature

Fig. 20.1 (A-D) Vegetative malformation on mango tree inflorescence where it can be observed that the inflorescence turned into scale like pesudo leaves

Fig. 20.2 (A-D) Floaral malformation on mango tree inflorescence where it can be observed that the flowers remain unopened and show swelled buds

changes cause a decrease in mango fruit output and quality. For example, changes in the plant's morphological characteristics (vigour along with canopy), reproductive qualities (fruiting potential as well as fruit size), and quality attributes (shelf life, juice content, insect and disease assault) result in a decrease in both the quantity and quality of mangoes. Mango plant exposure to high temperatures along with low humidity

Fig. 20.3 (A-B) This image show mix vegetative and floral malformation symptoms on the same inflorescence

reduces the effectiveness of photosynthesis, transpiration, and leaf water potential (Faria et al. 2016; Khanum et al. 2020). The mango plant lowers its canopy area in an effort to counteract the detrimental impacts of these pressures, which inhibits stomatal conductance therefore may ultimately have an impact on production (Shanthala et al. 2021). Rainfall during the pre-flowering and blooming stages of the mango fruit reduces yields. Additionally, it prevents pollination and fruit set. In a similar vein, storms significantly reduce fruit yield in the latter stages of fruit growth. The amount of water in the soil affects the fruit's size, yield, and quality criteria (starch, sugar, soluble solids, and vitamin C content) in mangoes (Wei et al. 2017). Numerous studies demonstrate to the critical role that irrigation plays in augmenting rainfed agriculture in a rapidly changing environment to achieve better yield and higher-quality mango fruits (Durán Zuazo et al. 2019; Hahn and García 2022; Wei et al. 2017). Climate variation affects both the quality of the fruit and how ripe mango fruits look. This is partially explained by the high prevalence of diseases and pests like anthracnose in areas with high humidity and rainfall (Makhmale et al. 2016; Ploetz 2003). Mango fruits may become less visually appealing during these times because of many factors, including the peel turning black (Makhmale et al. 2016). Conversely, low humidity and precipitation encourage fruit set, harvesting, and the mitigation of biotic stresses (Zuazo et al. 2021). Mango genetic resources have a great deal of diversity, which helps collecting and breeding strategies adapt to changing climates. Fortunately, there are genotype-specific differences in mango productivity under both wet and dry environments. Mango yields and quality will therefore be less impacted by climatic variability when these differences are harnessed. Strategies for adaptation are essential in the fight against climate change with a positive impact on mango yields and quality. For example, crop types that are matched to the climate produce more mangoes in traditional as well as modern producing locations.

20.2.3 Mango Farmers' Responses to Climate Change and Methods for Adaptation

Farmers have implemented various strategies to mitigate the effects of climate change on mango seedling production. These strategies include the development of drainage channels, the use of agrochemicals, grafting, the use of better-quality varieties, the establishment of trees as shade, irrigation, and improved soil nutrition (Asare-Nuamah et al. 2022). The action taken of mango growers in response to climate change emerges from the need to preserve both their way of life and their agricultural land (Esperanza et al. 2018). The production of mangos depends critically on the availability of resources and information on reducing the adverse effects of climate change. By using them, farmers may lessen these consequences by providing water and modifying their schedules in accordance with weather forecasts. Strategies like post-harvest pruning of sizable branches of vigorous as well as late varieties grown in heavy soils to increase soil moisture loss during winter periods, and substitution as well as enhancement of low temperature through the use of growth retardant combined with the imposition of water stress for approximately 4 months prior to blossoming (Halder and Hasan 2020). Comparably, digging trenches alongside the drip line improves drainage, avoiding instances where moisture stress occurs during rainy spells. In the medium as well as long run, traditional methods of controlling water shortfalls are impractical. It is recommended that mango growers utilize deficit irrigation techniques in response to drought situations in order to conserve water, enhance water usage effectiveness, and preserve production and quality (Zuazo et al. 2021). Strategies for acclimating to climate change that other experts have suggested include cover belts, dense plant populations, and drip watering techniques. The adaptation of mango growers to the changing environment is hindered by many issues that might stem from institutional, economical, and technical limitations (Singh et al. 2017).

20.3 Mango Enhancement in a Rapidly Changing Climate

20.3.1 Overall Goals for Improving Mango Trees

The majority of fruit crops, including mangos, have inadequate genetic advancements. Natural selection over many generations has produced the majority of the advancements in mangoes. Mangos with a high yield and excellent quality are required because to the increase in market and customer expectations as well as the rapidly changing climate. Unfortunately, the process of breeding mangoes takes several years—sometimes up to 20 years. This is attributed to its protracted juvenile stage, high heterozygosity, one seed per fruit, refractory seeds, broad area for assessing hybrids, and incompatibility between cultivars, polyembryony, and high frequency of naturally occurring fruit shedding. To overcome these challenges, mangoes must be improved in an innovative manner. However, in addition to its asexual development, mango germplasm has a wide range of genetic variability, making selection easier at any stage of the breeding cycle (Iyer and Dinesh 1997).

The requirements of society in connection to agriculture are at the core of efforts to enhance mangoes. Consequently, this results in three primary pillars: high-quality fruit production, a decrease in environmental impact, and excellent production returns. To prevent trade-offs amongst them, these pillars must be an essential part of any program for improving mango production. Nonetheless, a lot of studies also concentrate on the use of value-addition development and post-harvest techniques to improve the safety and quality of collected fruits for their ultimate customer while lowering mango fruit waste (Arinloye et al. 2017; Esguerra et al. 2018). As a result, hybrid mangoes have been marketed globally (Jha et al. 2010; Rajwana et al. 2010). This is accomplished by developing, launching, and carrying out well-defined breeding objectives. It is anticipated that new cultivars will be introduced that have become well-adapted to the stresses both biotic and abiotic brought about by climate change. Mango cultivar preferences differ across growers, buyers, and sectors. Nonetheless, a variety of characteristics, including hardness, resistance to diseases and pests, early and frequent bearing, dwarfism, appealing look, decent size, absence of physiological problems, and extended shelf life, are desirable characteristics of mango cultivars. The biennial bearing ability of a great number of commercial mango cultivars on the market has been enhanced. The combined use of many genomics approaches has improved mango quality while reducing uncertainty (Gómez-Ollé et al. 2023; Srivastav et al. 2023).

20.3.2 The Problems of Producing Mango Trees in a Rapidly Changing Climate

Natural selection over many generations is mostly responsible for the majority of mango advancements. the increase in market and customer expectations combined with a rapidly changing environment. Unfortunately, the process of breeding mangoes takes several years—sometimes up to 20 years. Its generation duration, lengthy juvenile stage, great heterozygosity, single seed in fruit, resistant seeds, cultivar incompatibility, polyembryony, broad region for assessing hybrids, and high frequency of naturally occurring fruit shedding are blamed for this. Progress in conventional fruit tree breeding is hindered by the long juvenile period, extended generation time, and high field area and people resource needs for sustaining and assessing hybrid populations. It is challenging to match the needs of processors and customers for quality features using traditional breeding methods (Kuhn et al. 2017). Mango breeders also have to deal with high heterozygosity, poor crossing rates (0.1%), polyembryony, a high frequency of fruitlet fall, and a single seed per fruit in addition to these constraints (Bally et al. 2009). Long juvenile periods pose a difficulty to mango breeding programs as they impact the assessment, choice, and distribution of hybrid cultivars. In addition, the expense of keeping the hybrid seedlings must be considered, along with the requirement for broad areas.

20.4 Impact of Micronutrients on Different Mango Tree Parameters

Micronutrients, also referred to as microscopic nutrients, are important for affecting many aspects of mango tree development and growth. Here, we examine the various impacts of these nutrients on several facets of the life cycle of mango trees.

20.4.1 Vegetative Growth Enhancement

Micronutrients are essential for the mango (*Mangifera indica*) plant's vegetative growth and general health, vigour, and development. The mango tree is an economically significant fruit tree (Das et al. 2019; Vittal et al. 2023). Mango trees' numerous physiological functions depend on a wide variety of nutrients in trace levels. Micronutrients, which include elements like iron, copper, zinc, boron, manganese, and molybdenum, are just as important for healthy growth and development as macronutrients like nitrogen, phosphorus, and potassium, which are widely known for their impact on plant growth (Milošević and Milošević 2023; Ryan et al. 2013). There has been a lot of study done on improving vegetative development in plants. Notably, a study on mango trees by Singh and Rajput (1976) showed that different $ZnSO_4$ concentrations had favorable impacts on vegetative development metrics. Plant height, trunk girth, and young plant spread were all affected by the administration of zinc and boron, which enhanced terminal shoot length, leaf count, and leaf area per shoot. Similarly, Masroor et al. (2016) emphasized the need of using zinc, namely 1% $ZnSO_4$ as a foliar spray, to raise the zinc content of mango leaves. This novel method revealed the possibility of applying certain micronutrients to promote vegetative growth. Furthermore, Elsheery et al. (2020) examined the impact of nano zinc oxide (nZnO) on vegetative development and found that applying 100 units of nZnO significantly increased leaf area. The effectiveness of nano-based formulations in promoting plant development was highlighted in this investigation. Furthermore, zinc's critical role in starch formation—a process that leads to higher plant weight and total vegetative growth—was clarified by Kacha et al. (2021). Together, our study results highlight the many ways that zinc and its compounds support vigorous vegetative development in a range of plant species.

20.4.2 Mineral Composition of Leaves and Zinc Foliar Spray

The precise impact of taking micro nutrients on the mineral content of the leaves of mango have been clarified by a number of research (Ahmad et al. 2018). This resulted in a more complete nutritional profile in the leaves by favourably influencing the level of other minerals in addition to improving the overall micronutrient balance (Nazir et al. 2024a; Razzaq et al. 2013). Recent studies have focused a lot of emphasis on leaf mineral content and how it relates to Zn foliar spray. Elsheery et al. (2020) carried out additional research to examine how zinc foliar spray affects the mineral content of mango leaves. Their results demonstrated how this application strategy might improve the general health of plants by enhancing the mineral content of leaves. In a similar vein, Ahmad et al. (2018) highlighted the significant impact of zinc foliar spray on the mineral content of leaves. The correlation between zinc treatment and the mineral composition of leaves has encouraging implications for regulating mineral concentrations and perhaps affecting other vegetative parameters in the process. The combination of these study findings emphasizes the value of zinc foliar spray as an effective method of adjusting leaf mineral content, which in turn affects plant health.

20.4.3 Flowering and Fruit Set

By significantly affecting the blossoming and fruit set processes of *Mangifera indica* trees, micronutrients have a crucial role in determining the output and quality of mangos, an economically significant fruit crop (Suman et al. 2017). The importance of micronutrients like iron, copper, zinc, boron, manganese, and molybdenum in promoting ideal flowering and fruit development cannot be overstated, even though macronutrients like nitrogen, potassium, and phosphorus are well known for their functions in plant growth. Research on the complex link between micronutrients and plant blooming processes has been concentrated. A research by Elsheery et al. (2020) shown the possible advantages of spraying mango plants with ZnO and Si NPs. This method significantly decreased floral deformity, which in turn enhanced the flowering process. The importance of micronutrient supplementation in maximizing flower growth is shown by this discovery. Furthermore, Masroor et al. (2016) and Kacha et al. (2021) shed light on the critical function that zinc plays in affecting blooming. Both researchers highlighted the significant effect of $ZnSO_4$ administration on increasing the quantity of flowers per panicle through their separate investigations. All of these findings point to the critical role that micronutrients—especially those based on zinc—play in influencing the way that plants blossom. The synergy of study results emphasizes how certain nutrient management techniques may have a beneficial impact on fruit sets and flowering, ultimately increasing crop output.

20.4.4 Weight and Fruit Retention

A comprehensive understanding of the precise impacts of applying micronutrients on mango fruit preservation and weight is provided by empirical research. Notably, studies by Merwad et al. (2016) highlighted the roles that calcium, zinc, boron, and NAA play in bolstering fruit retention. An increased percentage of retained fruits was seen in mango trees treated with calcium, zinc, boron, and NAA. This suggests that sufficient amounts of boron, zinc, calcium, and NAA, are essential for enhancing fruit-holding resilience.

Research on the complex interactions between micronutrients and fruit growth has been very popular. Using certain micronutrient treatments, Ahmad et al.'s study (2018) produced impressive results in terms of high fruit retention percentages. This research raises the possibility that micronutrients might help fruits stay on trees longer. Furthermore, Masroor et al. (2016) and Kacha et al. (2021) provided evidence of the impact of zinc on fruit weight. Their own investigations clarified the ways in which zinc treatment techniques—foliar or soil-based—could enhance fruit weight. All of these research highlight how important it is for targeted zinc therapies to affect fruit growth and quality.

20.4.5 Sugar Content, Vitamin C Concentration, and Total Soluble Solids

A great deal of research has been done on the relationship between the biochemical makeup of fruits and the uses of micronutrients. Dheware et al. (2020) revealed a rise in Total Soluble Solids (TSS) content and shed light on how certain fertilizer treatments affect fruit quality. This research emphasizes how specific nutrition management may affect fruit composition.

Accordingly, Patel et al. (2018) conducted research on innovative organic liquid fertilizer treatments, leading to fruit quality gains that were multifarious. Interestingly, these treatments resulted in higher sugar and vitamin C content as well as greater TSS levels. This finding demonstrates how well novel fertilization techniques may improve the nutritious qualities of fruits.

20.4.6 Fruit Quality Improvement

Mango trees' fruit quality is greatly improved by micronutrients. Although they are needed in lesser amounts than macronutrients, they are nonetheless necessary for a number of physiological and biochemical processes that affect the growth, flavour, texture, colour, and overall quality of fruit (Barrett et al. 2010). This is how certain micronutrients might affect mango fruit quality enhancement.

20.4.6.1 Iron (Fe)

Chlorophyll, which gives leaves their green hue and indirectly affects the development of fruit color, requires iron for synthesis (Sun et al. 2022). A sufficient amount of iron guarantees effective photosynthesis which results in the synthesis of carbohydrates that add to the sweetness and flavor of fruit (Ribeiro and Schieber 2010).

20.4.6.2 Zinc (Zn)

Auxins are plant hormones that are important in fruit growth and ripening; zinc is involved in their production (Kazemi 2014). Masroor et al. (2016) found that adequate zinc levels can encourage consistent ripening, optimal fruit size, and improved flavor development.

20.4.6.3 Boron (B)

For the development and stability of cell walls, boron is essential (Liu et al. 2019). It affects the fruit's cell wall strength, which affects the fruit's texture (Huang et al. 2023). Sufficient quantities of boron can reduce fruit breaking and enhance fruit firmness (Shivashankar 2014).

20.4.6.4 Manganese (Mn)

For many of the enzymes involved in the production of secondary metabolites, including those that produce the chemicals that give fruits their flavor and fragrance, manganese is a cofactor (Schwab 2003). Mangoes with higher manganese levels have been shown to have better flavor and fragrance (Appiah et al. 2011).

20.4.6.5 Copper (Cu)

Several enzymatic processes, such as those pertaining to the production of lignin and the creation of color, require copper (Lin et al. 2005). It can affect the fruit's skin color and enhance mangoes' aesthetic attractiveness (Lebaka et al. 2021).

20.4.6.6 Molybdenum (Mo)

Molybdenum is needed for two processes: fixing nitrogen and turning nitrates into amino acids, which are the fundamental blocks of proteins and flavorings. Mangoes with optimal molybdenum levels had higher protein content and a more flavorful texture (Gericó et al. 2020).

20.4.6.7 Magnesium (Mg)

According to Moretti et al. (2021), magnesium is necessary for the creation of enzymes needed for the metabolism of carbohydrates and chlorophyll. It ensures effective photosynthesis and sugar production, which has an indirect impact on the colour development and sweetness of fruit (Ngereza and Pawelzik 2016).

20.4.6.8 Calcium (Ca)

Calcium is essential for the integrity and structure of cell walls (Khaliq et al. 2016). A healthy amount of calcium can help avoid conditions like bitter pit, which degrade the texture and freshness of fruit (Sharma et al. 2013). Additionally, calcium keeps fruit solid throughout shipping and storage (Wang et al. 2014).

20.5 Approaches Used in Mango Improvement

The development of novel varieties and assessment, selection (clones, monoembryomic as well as polyembryomic seedlings), aided and controlled pollination, in vitro selection, embryonic rescue, molecular markers, and multi-omics approaches have all contributed to the enhancement of the mango tree. Utilizing omics technologies for reverse and forward breeding, such as transcriptomics, proteomics, and metabolomics, makes it easier to identify and develop phenotypes controlling important genes. It has been revealed that *Agrobacterium tumefaciens* genetically modifies mangos by using genes that control fruit ripening. At the seedling stage, markers like SNPs are effective in exposing the selfing rate and pollinator behaviour of mangoes (Michael et al. 2023). Each of the two primary methods (conventional and traditional) has advantages and disadvantages. In a similar vein, rootstocks have been employed to enhance mango tree cultivation and mitigate pressures resulting from various causes such as salt, infections, and drought. In particular, by increasing salt tolerance, nanoparticles (nano-silicon and nano-zinc oxide) increased mango productivity (Elsheery et al. 2020). Researchers have employed a range of innovative strategies that integrate the most recent methods with traditional breeding procedures to increase biotic stress tolerance for mango breeding initiatives.

20.5.1 Traditional Breeding Techniques for Mango Improvement

Mango development against biotic stresses has always been based on traditional breeding techniques (Table 20.1). In these methods, several mango varieties are cross-pollinated in order to combine advantageous traits like pest and disease tolerance. Mango cultivars with higher resistance to biotic stress can be found and improved by breeders through recurrent selection and assessment of progenies (Bally et al. 2009). Similar to this, by combining traditional breeding techniques with rigorous screening and assessment protocols, breeders can produce mango cultivars that are more resilient to biotic stress. In order to expedite the process of breeding stress-tolerant mango varieties and enhance overall efficacy, it is imperative to integrate these traditional techniques with cutting-edge instruments such as molecular markers and biotechnological advancements (Tongyoo et al. 2021). However, traditional breeding is costly and may not provide results for several generations.

20.5.2 Current Approaches

Mangoes can only be enhanced with creative approaches. Mango germplasm has a broad range of genetic variety and is asexually propagated, which makes selection easier at any stage of the breeding cycle (Iyer and Dinesh 1997). To increase mango production to improve yield and quality, a variety of techniques including transgenic methods, genome editing technologies, and marker-assisted selection (MAS) are used.

Marker-assisted selection (MAS): In addition to traditional breeding techniques, MAS is a powerful tool for improving mangoes. The aim of MAS is to identify molecular markers associated with features of relevance (disease or insect resistance). Because these markers enable breeders to choose plant with desirable traits at the initial seedling stage, phenotypic assessments take less time and money. Using MAS, mango cultivars with increased resistance to infections like powdery mildew and anthracnose have been successfully created (Parmar et al. 2017). For example, Kuhn et al. (2017) employed MAS to identify progeny that have the necessary alleles for desired traits in mangos; the process involves genotyping the progeny. Random approaches: a direct and focused genetic engineering technique that involves inserting specific genes in mango plants to confer resistance to biotic stress. Transgenic methods can be used to introduce genes from other species, such as those from plants with resistance to diseases or insecticidal proteins, into the genome of mangos. The researchers created a set of trait-associated SNP markers that are that may be used to find these variations, reducing the amount of time, and money required for

Table 20.1 Traditional mango improvement technique

Technique	Description	Examples	References
Germplasm selection	Accessions that are biotic stressor-tolerant or resistant move on to the next phase	Anthracnose and powdery mildew	Dangana et al. (2018)
Cross-pollination and recurrent selection	Mango cultivars are deliberately crossed-pollinated in order to combine genes that are beneficial for resistance to biotic stress. Selecting the progeny with the highest success rate for successive breeding cycles	Hybrid mango varieties have higher tolerance to pests and diseases	Kumar (2006)
Backcrossing	Transferring certain genes related to biotic stress tolerance from one donor parent to another	Mango variety created via backcrossing with increased resistance to a certain disease	Dijkman and Soule (1951)
Mass selection	Selecting mango trees with desirable traits, such as resistance to biotic stresses, from a large population	Local landraces that are naturally resistant to particular pests or diseases	Dinesh et al. (2011)
Line breeding	Selecting and breeding mango trees with advantageous traits over many generations' results in the development of genetically homogeneous lineages	Mango lines with improved biotic stress	Parmar et al. (2017)
Inbreeding and selection	Regulated inbreeding to focus on features that are important for biotic stress resistance	Mango types that are more resilient to specific diseases	Litz and Gomez-Lim (2002)
Open pollination and selection	Enabling natural pollination to occur on mango trees to promote genetic variation	Mango trees that are openly pollinated have special biotic stress tolerance characteristics	Shamili et al. (2012)

field application, upkeep, and long-term material assessment. In conclusion, trait-associated SNP markers used in the study can help breeders more successfully identify these progenies, and MAS is a useful tool for selecting mango progenies having desirable traits.

Transgenic approaches: A clear and focused genetic engineering technique for inserting specific genes into mango plants to confer tolerance to biotic stress. Transgenic methods can be used to introduce genes from other species, such as disease-resistant plants as well as insecticidal proteins, into the genome of mangos (Singh et al. 2013). This method has showed promise in the production of mango cultivars with increased resistance to pests such as fruit flies. However, concerns regarding potential environmental and regulatory issues are also raised by transgenic technologies.

Transgenic technology is the agricultural innovation that is expanding the fastest (Penna and Jain 2023). It describes a set of techniques for utilizing unusual methods to transfer desired genes from any source—including microorganisms, plants, animals, and even intentionally generated genes—into a specific plant. The capacity to modify plants using the genes controlling a range of important agronomically relevant traits—features that may be derived from any creature, including bacteria, plants, and other animals—is the main advantage of transgenic technology. The target plant may so effortlessly take on distinctive traits from any background.

It was revealed that transgenic mango embryos (*Mangifera indica*) cv. "Ataulfo" were produced using the defensin J1 gene. Defensins are small peptides rich in cysteines that have antibacterial action against a range of pathogens, including as viruses, fungi, and bacteria. *Colletotrichum gloeosporioides* is the fungal pathogen that causes anthracnose, and the transgenic mango plants showed increased resistance to it (Rivera et al. 2011). The study suggests that genetic engineering could make it easier to produce mango varieties that are resistant to disease.

The emergence of genome editing technologies, especially CRISPR-Cas9, has fundamentally altered the way in which mangoes may be made more resilient to biotic stress. Genome editing allows for the exact modification of certain genes inside the mango genome without the need to introduce foreign DNA. With the use of this technique, breeders may now enhance the expression of genes linked to disease resistance or pest tolerance (Dongariyal et al. 2023). The use of genome editing holds great promise for speeding up the generation of stress-tolerant mango varietals while also addressing the ethical concerns associated with transgenic techniques (Kaur et al. 2020). Using CRISPR-Cas9, important genes associated with disease resistance in mangos have been altered. It has been shown that by focussing on genes involved in plant-pathogen interactions, such as those responsible for resistance to anthracnose and powdery mildew, it is possible to create mango cultivars that are resistant to disease (Ahmad et al. 2021). Scientists may be able to increase mangos' tolerance to environmental stresses like salt and drought by using CRISPR-Cas9. The objective is to focus on genes associated with stress response pathways in order to produce mango types that are more resilient to adverse environmental circumstances. The genes in charge of fruit ripening, size, and flavour may be altered using CRISPR-Cas9. It is possible to generate mango cultivars with improved flavours and longer shelf lives by concentrating on certain genes related to fruit texture, sugar content, and fragrance (Mathiazhagan et al. 2021). Furthermore, research is being done on CRISPR-Cas9 as a means of enhancing resistance to viral infections that harm mangoes and result in illnesses like mango malformation sickness and distorted mango leaves. By introducing specific mutations in genes that are susceptible to viruses, researchers hope to create mango cultivars with increased viral resistance (Goulin et al. 2019).

Multi-omics approaches: By improving our understanding of the interactions between mango genes and stressors, multi-omics methods (genomics, transcriptomics, proteomics, and metabolomics) have made it much easier to understand the molecular pathways underlying the fruit's ability to withstand biotic stress (Yang et al. 2021). By understanding the connections between important genes and the metabolic processes they play in stress reactions, breeders may create specialized breeding plans for biotic stress resistance. The research of mangoes across the transcriptomics, comparative genomics, and genotyping levels is made easier by recent genome sequencing on mango cultivars (Tommy Atkins, Carabao, Irwin and Alphonso), which offer trustworthy gene sets and references (Bally et al. 2021; Cortaga et al. 2022; Wang et al. 2020). In order to investigate genetic diversity, create genetic linkage maps, and find markers for important features, a variety of molecular markers are also employed (pulp colour, taste, fruit colour, and flavour) (Arogundade et al. 2022; Jena and Chand 2021; Kuhn et al. 2017; Song et al. 2023). These studies help identify the genes or proteins involved in different phases of fruit growth, early and late blooming, and biotic stress conditions. Through metabolomic study on the mango, which provides insight into the metabolites, hormones, and pathways responsible for fruit softness and flower induction, it is feasible to regulate post-harvest and biennial blooming (Mathiazhagan et al. 2022).

Public–private cooperation: Public research organisations and commercial companies must work together closely to strengthen mangos for biotic stress resistance (Jain 2005). Public institutions often own large germplasm collections, the capacity to carry out basic research, and expertise in breeding techniques. Conversely, private companies can invest in state-of-the-art biotechnological machinery and sell the developed mango varieties that are resistant to stress. Working together, these sectors can accelerate the development and dissemination of improved mango cultivars that are resilient to a range of biotic stresses. It is imperative that the vast collection of germplasm used in mango breeding be evaluated (Dinesh et al. 2011). In the context of breeding mangos for biotic stress, germplasm is used to identify sources of resistance to various biotic pressures. Since polyembryonic cultivars are less variable and are usually found in places with high rainfall, it is crucial to screen them for these types of stresses. It's also essential to provide variation to these cultivars in order to raise their quality. The identification of nucellar and zygotic embryony using markers may help in the selection of novel recombinants.

Mango development for biotic stress resistance has evolved significantly as a result of the integration of traditional breeding procedures with state-of-the-art biotechnological methodologies. Marker-assisted selection (MAS), genome editing technologies, transgenic techniques, and omics technologies have revolutionised the way mango breeders develop stress-tolerant cultivars. Working together, the public and private sectors have increased the efficacy and efficiency of mango improvement initiatives. As research into biotic stress tolerance in mango breeding progresses, there is hope that this treasured tropical fruit may be preserved for future generations through sustainable mango production and availability.

20.6 Enhancing Mangoes' Tolerance to Biotic Stresses

Mango growing has a lot of challenges due to biotic stresses. An extensive array of ailments resulting from fungus, bacteria, viruses, and nematodes provide a significant risk to agricultural yield (Suzuki et al. 2014). Furthermore, a range of pests, such as mites and insects, can cause significant damage to mango trees and mangoes. Mango farmers worldwide still confront persistent problems with common diseases, pests like fruit flies and mango hoppers, and diseases including bacterial black spot, powdery mildew, and anthracnose (Pena et al. 1998). For example, fruit flies (Diptera: Tephritidae) are responsible for billions of dollars' worth of yield loss each year worldwide; losses in Brazil alone amount to over USD 242 million, while Mali's whole mango crop suffers losses of 50% (Oliveira et al. 2013; Vayssières et al. 2004). According to recent research, sunlight entering mango plantations triggers the fruits' defensive mechanisms. Mango fruit, for example, becomes resistant to the fungal disease *Colletotrichum gloeosporioides* by triggering the manufacture of phenylpropanoid through defense-related pathways and hormones (brassinosteroid, ethylene, and jasmonic acid) (Sudheeran et al. 2021). Mango trees' resistance to fruit fly infestations and mango wilt disease is increased by using rootstocks derived from polyembryonic plants (Ribeiro et al. 1993; Rossetto et al. 1996). Mango fruit trap detection has been developed by current research to track their population in the field and determine an affordable threshold for management (Hahn et al. 2023). Researchers are eager to strengthen the mango cultivars' genetic foundation in order to increase their tolerance to a variety of biotic stressors (Table 20.2). However, integrated management techniques are mostly used in the contemporary management of these pressures.

Table 20.2 The main biotic stressors in the development of mangoes

Stress	Key management approach	References
Anthracnose	Increased resistance to *Colletotrichum gloeosporioides* due to the MiPR1A gene	Li et al. (2023)
Powdery mildew	Resistant cultivars and sulphur-based fungicides application	Rajan et al. (2021)
Bacterial black spot	Copper-based insecticides and trimmed tissues	Pruvost et al. (2005) and Jean-martial et al. (2021)
Fruit fly	Fruit fly integrated pest control strategies include protein bait sprays, trap-based monitoring	Muriithi et al. (2020) and Kibira et al. (2015)
Anthracnose	Fungicide use combined with cultural activities (pruning and cleanliness)	Zainab and Shinkafi (2016)
Mango hopper	Biological control (parasitoids) and neem-based insecticides	Bokonon-Ganta et al. (2002)

20.7 Strategies for Improving Value Addition and Post-harvest

20.7.1 Mango Post-harvest Losses

Poverty development and reduction were facilitated by smallholder mango producers' encouragement of mango production. But value addition difficulties and post-harvest losses have prevented it from reaching its full market potential (Amin et al. 2021). Mango post-harvest losses are caused by a variety of causes, including bacterial and fungal infections, mechanical damage, physiological changes, and insufficient storage and transit (Liu et al. 2023). Unpredictable rainfall patterns and severe temperatures are two additional elements that contribute to losses brought on by climate change. The primary issues facing the mango sector, however, are the short shelf life of mango fruit, the challenge of controlling post-harvest quality, and the absence of preservation technologies. Everything that happens after a crop is harvested up until it is sold or consumed is referred to as post-harvest procedures. Mango harvesting, packaging, storage, shipping, processing, and marketing are some of these crucial processes. Adopting efficient post-harvest procedures is essential since it minimizes the amount of degradation while simultaneously increasing the mango's overall quality and worth. Understanding the best handling conditions for your particular product can help you minimize losses and maximize the result.

20.7.2 Post-harvest Intervention and Management

Because mangos are climacteric fruits with a short shelf life, post-harvest handling is essential to preserving quality and minimizing losses. Climate change has made post-harvest issues more severe in recent years, requiring creative solutions to guarantee both food security and economic sustainability. The methodical handling of agricultural goods to maintain their freshness and quality and extend their shelf life prior to selling in order to boost market value is known as post-harvest management. Making sure the mangoes are gathered at the optimal maturity level and then ripened to get the ideal flavor and texture is the first step in post-harvest treatment (Normand et al. 2015; Amin et al. 2021). Mangoes that are collected too late, for example, are more likely to rot, while those that are harvested too early will not have fully developed their taste and scent. Mangoes should be handled carefully once they have been picked (Fig. 20.4). Mangoes are extremely vulnerable to injury both before and after harvesting. Once the fruit is injured, all biological processes, such as the formation of ethylene and respiration, start up quickly and cause the fruit to deteriorate quickly. Sorting, pre-cooling/washing, grading, packing, storing, shipping, and marketing are a few of the steps that are considered essential once mangos are harvested (Liu et al. 2023). One extremely significant stage in increasing the shelf life of fruits has been reported to be spreading them out for almost 24 h on a cushioned floor made of paddy

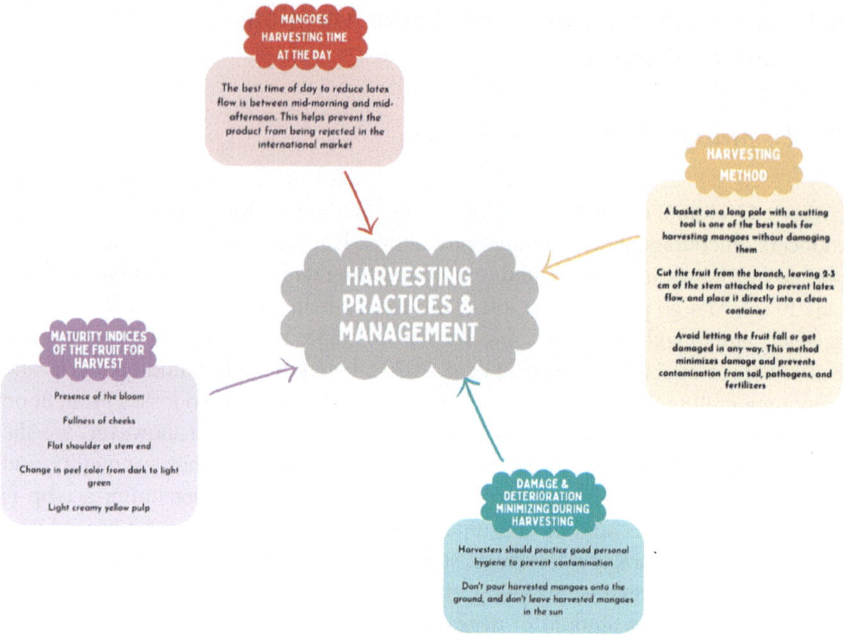

Fig. 20.4 Harvest management and technical procedures

straw in the orchard's yard, then washing them to eliminate dirt. They must to be kept cold and packaged in hygienic, well-ventilated containers. Mangoes should be stored between 10 and 12 °C (Khan 2011).

20.7.3 Techniques to Extend the Shelf Life of Mangos

Mango shelf life may be extended using a variety of techniques, including waxing, modified atmosphere packaging (MAP), controlled-environment storage, and hot water treatment.

- **Waxing**: Mangos can have their shelf life and marketable duration extended by using waxing or coatings (Amin et al. 2021). Mangoes have been coated with beeswax-emulsified chitosan-Aloe vera to increase their shelf life.
- **Modified atmosphere packaging (MAP)**: By changing the atmospheric composition surrounding the fruit, modified atmosphere packaging (MAP) lowers the amount of ethylene produced and slows down respiration. The effectiveness of MAP in prolonging mango shelf life and lowering post-harvest losses has been demonstrated in a number of studies.
- **Controlled atmosphere (CA) storage**: This entails preserving particular temperature ranges and gas concentrations, such as carbon dioxide and oxygen. Studies

have shown that CA storage may successfully postpone mango ripening and increase the fruit's shelf life (Khan 2011).

- **Hot water treatment**: Another quarantine technique that works well for managing fruit fly infection is hot water treatment. Research has indicated that the application of hot water treatment can effectively decrease fruit fly infestation while maintaining fruit quality (Narang and Kataria 2014). The bacteria and fungus that might cause rot are killed when the mangoes are briefly submerged in hot water.

20.7.4 Value Addition Improvement Strategies

The process of improving the quality, quantity, and their marketability of agricultural goods is known as value addition. It comprises converting unvalued raw resources into more desirable forms. Value addition improves food security by lowering post-harvest losses, increasing opportunities for employment, and increasing revenue for farmers and agribusinesses. Value addition may help maintain a sustainable environment by cutting down on waste and greenhouse gas emissions. Mangoes may be processed into a wide range of goods, which can be exported to other nations and increase their shelf life. These products include pulp, juice, pickle, chutney, puree, jam, and concentrates. Mangoes become more marketable when these items are employed as components in different cooking goods and drinks (Malik et al. 2015). Strategies for improving value addition include:

- **Drying and dehydration**: These are conventional preservation techniques that lower moisture content to prevent enzymatic reactions and microbiological development. Vacuum drying and sun drying are two popular methods for preserving mangoes (Sharma and Thakur 2012).
- **Mango by-product utilisation**: Mango peels and seeds are byproducts of the processing of mangos that can be used to make value-added goods including mango kernel oil, pectin, and animal feed (Sudheeran et al. 2019).

Post-harvest and value-addition enhancement measures are critical to increasing mango output and decreasing losses in the era of climate change. The implementation of novel technology and processing methods has the potential to result in amplified market prospects, noteworthy revenue creation, and comprehensive sustainability in the mango industry.

20.8 Conclusion

Mangoes have a well-established export market and offer significant opportunities for international trade, both in their fresh and processed forms. The mango industry also plays a crucial role in providing livelihoods to growers and those involved in its marketing chain. There is a growing need for the development of climate-smart mango varieties with enhanced qualities. Optimizing essential nutrients is a vital strategy for mitigating the adverse impacts of climate change on mango production. Through the precise management of nutrients such as nitrogen, phosphorus, potassium, and micronutrients, mango growers can enhance the resilience of mango trees to environmental stressors, including increased temperatures, extreme weather events, and altered precipitation patterns. Trees can sustain strong growth, effective water usage, and enhanced resilience to diseases and pests—which are frequently made worse by climate change—by ensuring a balanced availability of nutrients. The development of pest (like fruit flies) and disease (like black spot disease) resistance, early maturation, wide adaptability, and discovery of critical features regulating yield, pulp colour, and fruit form should also be the focus of current mango improvement programs.

The adaptation of nutrient management strategies in response to climate variability ensures that mango production can remain viable and productive. Collaborative efforts between researchers, farmers, and policymakers are essential to develop and disseminate best practices that incorporate the latest scientific advancements. By prioritizing the optimization of essential nutrients, the mango industry can safeguard its future against the challenges posed by climate change, securing both economic stability for growers and the availability of high-quality mangoes for consumers worldwide.

References

Ahmad I, Bibi F, Ullah H, Munir TM (2018) Mango fruit yield and critical quality parameters respond to foliar and soil applications of zinc and boron. Plants 7(4):97

Ahmad S, Tang L, Shahzad R, Mawia AM, Rao GS, Jamil S, Tang S (2021) CRISPR-based crop improvements: a way forward to achieve zero hunger. J Agric Food Chem 69(30):8307–8323

Ahmed N, Ahmad F, Abid M, Ullah MA (2009) Impact of zinc fertilization on gas exchange characteristics and water use efficiency of cotton crop under arid environment. Pak J Bot 41(5):2189–2197

Altendorf S (2017) Global prospects for major tropical fruits. In: The food outlook: biannual report on global food markets. FAO, Rome, pp 68–81

Amin U, Khan MKI, Khan MU, Akram ME, Pateiro M, Lorenzo JM, Maan AA (2021) Improvement of the performance of chitosan—aloe vera coatings by adding beeswax on postharvest quality of mango fruit. Foods 10(10):2240

Appiah F, Kumah P, Idun I (2011) Effect of ripening stage on composition, sensory qualities and acceptability of Keitt mango (*Mangifera indica* L.) chips. Afr J Food Agric Nutr Dev 11(5):5096–5109

Arinloye ADD, Degrande A, Fassinou Hotegni VN, Asaah E, Bockarie R, Nyemeck JB, Kalinganire A (2017) Value chain development for mango (*Mangifera indica*) around Outamba Kilimi National Park in Sierra Leone: constraints and opportunities for smallholders. Agric Food Secur 6:1–13

Arogundade O, Matthew JO, Akinyoola OI, Akin-Idowu PE, Akinyemi SOS (2022) Phenotypic and molecular characterization of mango cultivars in Southwest Nigeria. Int J Fruit Sci 22(1):151–159

Asare-Nuamah P, Antwi-Agyei P, Dick-Sagoe C (2022) Mitigating the risks of climate variability and change on mango seedlings in Ghana: evidence from mango seedlings producers in the Yilo Krobo Municipality. Environ Challenges 8:100594

Bally IS, Lu P, Johnson PR (2009) Mango breeding. In: Jain SM, Priyadarshan PM (eds) Breeding plantation tree crops: tropical species. Springer, New York, NY, pp 51–82

Barrett DM, Beaulieu JC, Shewfelt R (2010) Color, flavor, texture, and nutritional quality of fresh-cut fruits and vegetables: desirable levels, instrumental and sensory measurement, and the effects of processing. Crit Rev Food Sci Nutr 50(5):369–389

Berdanier CD, Berdanier LA (2015) Advanced nutrition: macronutrients, micronutrients, and metabolism, 2nd edn. CRC Press, Boca Raton, FL

Bokonon-Ganta AH, de Groote H, Neuenschwander P (2002) Socio-economic impact of biological control of mango mealybug in Benin. Agric Ecosyst Environ 93(1–3):367–378

Broadley MR, White PJ, Hammond JP, Zelko I, Lux A (2007) Zinc in plants. New Phytol 173(4):677–702

Carrillo JÁ, Zafrilla MP, Marhuenda J (2019) Cognitive function and consumption of fruit and vegetable polyphenols in a young population: is there a relationship? Foods 8(10):507

Cortaga CQ, Lachica JAP, Lantican DV, Ocampo ETM (2022) Genome-wide SNP and InDel analysis of three Philippine mango species inferred from whole-genome sequencing. J Genet Eng Biotechnol 20(1):46

Dambreville A, Normand F, Lauri PÉ (2012) Plant growth coordination in natura: a unique temperature-controlled law among vegetative and reproductive organs in mango. Funct Plant Biol 40(3):280–291

Dangana MC, Auta Y, Mailafiya SC, Moses BE, Oluwafemi OKJ, Ebyna KC (2018) Determination of heavy metals in four mango fruit varieties sold in Minna modern market, Niger State, Nigeria

Das SC, Datta M, Ray P, Singh SK, Jena RK, Das B, Ray SK (2019) Mango (*Mangifera indica*) cultivation in north-eastern region of India. Adv Agric Res Technol J 3:54–66

Dheware RM, Nalage NA, Sawant BN, Haldavanekar PC, Raut RA, Munj AY, Sawant SN (2020) Response of micronutrient on the quality yield of mango cv. Alphonso under Konkan agro-climatic conditions. J Pharmacogn Phytochem 9(3):2004–2006

Dijkman MJ, Soule MJ Jr (1951) A tentative method of mango selection. Proc Fla State Hortic Soc 64:257–261

Dinesh MR, Vasugi C, Ravishankar KV (2011) Morphological, molecular characterization and breeding for biotic and abiotic stress in mango (*Mangifera indica* L.). In: Global conference on augmenting production and utilization of mango: biotic and abiotic stresses, vol 1066, pp 37–46

Dongariyal A, Chandra AK, Dongriyal A, Kumar A, Sharma P (2023) Tending genome editing via CRISPR/Cas9-induced mutagenesis: opportunity and challenges for yield, quality and nutritional improvement of fruit crops. Sci Hortic 311:111790

Drebenstedt I, Marhan S, Poll C, Kandeler E, Högy P (2023) Annual cumulative ambient precipitation determines the effects of climate change on biomass and yield of three important field crops. Field Crops Res 290:108766

Durán Zuazo VH, Rodríguez Pleguezuelo CR, Gálvez Ruiz B, Gutiérrez Gordillo S, García-Tejero IF (2019) Water use and fruit yield of mango (*Mangifera indica* L.) grown in a subtropical Mediterranean climate. Int J Fruit Sci 19(2):136–150

Elsheery NI, Helaly MN, El-Hoseiny HM, Alam-Eldein SM (2020) Zinc oxide and silicone nanoparticles to improve the resistance mechanism and annual productivity of salt-stressed mango trees. Agronomy 10(4):558

Esguerra EB, Rolle R, Rahman M (2018) Post-harvest management of mango for quality and safety assurance. In: Guidance for horticultural supply chain stakeholders. Food and Agriculture Organization of the United Nations, Rome, Italy

Esperanza D, Putri G, Rasmikayati E, Qanti S, Yuwariah Y (2018) Adaptation behavior of mango farmers to climate change. J Manajemen Agribisnis 15:268–268

Fain SJ, Quiñones M, Álvarez-Berríos NL, Parés-Ramos IK, Gould WA (2018) Climate change and coffee: assessing vulnerability by modeling future climate suitability in the Caribbean Island of Puerto Rico. Clim Change 146:175–186

FAO (2021) FAOSTAT crop production [Online]. Available from: http://www.fao.org/faostat/en/#home

FAO (2022) Major tropical fruits: preliminary results 2021. Rome

FAO (2023) FAOSTAT crop production [Online]. Available from: http://www.fao.org/faostat/en/#home

Faria LN, Donato SL, Santos MRD, Castro LG (2016) Manejo da irrigação na indução floral da mangueira 'Tommy Atkins' no semiárido baiano. Eng Agríc 36:387–398

Ferreira KM, Simões WL, Mouco MAC, Silva JL, Silva JS, Mesquita AC (2020) Efficient management of the application of paclobutrazol for the production and quality of 'Tommy Atkins' mango. Res Soc Dev 9:e348984894

Gericó TG, Tavanti RFR, Lima JP, Ribeiro RP, Santos LCC, Silva MS, Reis AR (2020) Cobalt and molybdenum stimulate compounds of primary metabolism, nitrogen forms, and photosynthetic pigments in peanut plants (*Arachis hypogaea* L.). J Plant Nutr 43:1907–1922

Goldbach HE, Wimmer MA (2007) Boron in plants and animals: is there a role beyond cell-wall structure? J Plant Nutr Soil Sci 170:39–48

Gómez-Ollé A, Bullones A, Hormaza JI, Mueller LA, Fernandez-Pozo N (2023) MangoBase: a genomics portal and gene expression atlas for *Mangifera indica*. Plants 12(6):1273

Goulin EH, Galdeano DM, Granato LM, Matsumura EE, Dalio RJD, Machado MA (2019) RNA interference and CRISPR: promising approaches to better understand and control citrus pathogens. Microbiol Res 226:1–9

Gupta UC (1993) Boron and its role in crop production. CRC Press

Gurjar TD, Patel NL, Panchal B, Chaudhari D (2015) Effect of foliar spray of micronutrients on flowering and fruiting of Alphonso mango (*Mangifera indica* L.). The Bioscan 10:1053–1056

Hahn F, García JA (2022) Mango stem response under different irrigation regimes. Int J Fruit Sci 22:35–56

Hahn F, Valle S, Rendón R, Oyorzabal O, Astudillo A (2023) Mango fruit fly trap detection using different wireless communications. Agronomy 13:1736

Halder S, Hasan MA (2020) Climate change and mango production. Chem Sci Rev Lett 9:55–57

Huang W, Shi Y, Yan H, Wang H, Wu D, Grierson D, Chen K (2023) The calcium-mediated homogalacturonan pectin complexation in cell walls contributes to the firmness increase in loquat fruit during postharvest storage. J Adv Res 49:47–62

Iyer CPA, Dinesh MR (1997) Advances in classical breeding and genetics in mango. Acta Hortic 455:252–267

Jain SM (2005) Major mutation-assisted plant breeding programs supported by FAO/IAEA. Plant Cell Tissue Organ Cult 82:113–123

Janjua PZ, Samad G, Khan N (2014) Climate change and wheat production in Pakistan: an autoregressive distributed lag approach. NJAS Wageningen J Life Sci 68:13–19

Jean-Martial KKF, N'Goran N, Gaston KK, Seydou T, Dramane DD, Brahima C, Daouda K (2021) Current situation of mango bacterial black spot caused by *Xanthomonas campestris* pv *mangiferaeindicae* in the Poro and Tchologo regions in Côte d'Ivoire. Am J Food Nutr 9:117–121

Jena RC, Chand PK (2021) Multiple DNA marker-assisted diversity analysis of Indian mango (*Mangifera indica* L.) populations. Sci Rep 11:10345

Jha SK, Sethi S, Srivastav M, Dubey AK, Sharma RR, Samuel DVK, Singh AK (2010) Firmness characteristics of mango hybrids under ambient storage. J Food Eng 97:208–212

Kacha HL, Patel HC, Paradava DR (2021) Effect of soil and foliar applications of micronutrients on flowering and yield of mango: performance of mango influenced by soil and foliar applications of micronutrients. J AgriSearch 8:40–44

Kaur N, Awasthi P, Tiwari S (2020) Fruit crops improvement using CRISPR/Cas9 system. In: Genome engineering via CRISPR-Cas9 system. Academic Press, pp 131–145

Kazemi M (2014) Influence of foliar application of iron, calcium and zinc sulfate on vegetative growth and reproductive characteristics of strawberry cv. 'Pajaro'. Trakia J Sci 12(1):21–26
Khaliq G, Mohamed MTM, Ghazali HM, Ding P, Ali A (2016) Influence of gum arabic coating enriched with calcium chloride on physiological, biochemical and quality responses of mango (*Mangifera indica* L.) fruit stored under low temperature stress. Postharvest Biol Technol 111:362–369
Khan MA (2011) Postharvest management of mango: a review. Sci Hortic 129:1–12
Khanum Z, Tiznado-Hernández ME, Ali A, Musharraf SG, Shakeel M, Khan IA (2020) Adaptation mechanism of mango fruit (*Mangifera indica* L. cv. Chaunsa White) to heat suggest modulation in several metabolic pathways. RSC Adv 10:35531–35544
Kibira M, Affognon H, Njehia B, Muriithi B, Mohamed S, Ekesi S (2015) Economic evaluation of integrated management of fruit fly in mango production in Embu County, Kenya. Afr J Agric Res Econ 10:343–353
Kuhn DN, Bally IS, Dillon NL, Innes D, Groh AM, Rahaman J et al (2017) Genetic map of mango: a tool for mango breeding. Front Plant Sci 8:577
Kumar N (2006) Breeding of horticultural crops: principles and practices. New India Publishing
Lamprecht M, Obermayer G, Steinbauer K, Cvirn G, Hofmann L, Ledinski G, Greilberger JF, Hallstroem S (2013) Supplementation with a juice powder concentrate and exercise decrease oxidation and inflammation, and improve the microcirculation in obese women: randomised controlled trial data. Br J Nutr 110:1685–1695
Lebaka VR, Wee YJ, Ye W, Korivi M (2021) Nutritional composition and bioactive compounds in three different parts of mango fruit. Int J Environ Res Public Health 18(2):741
Legave JM, Normand F, Lauri PE (2013) Climate change and its probable effects on mango production and cultivation. In: X international mango symposium, vol 1075, pp 21–31
Li JJ, Luo C, Yang XZ, Peng LH, Lu TT, Yang JH, He XH (2023) Genome-wide identification of the mango pathogenesis-related 1 (PR1) gene family and functional analysis of *MiPR1A* genes in transgenic *Arabidopsis*. Sci Hortic 321:112254
Lin CC, Chen LM, Liu ZH (2005) Rapid effect of copper on lignin biosynthesis in soybean roots. Plant Sci 168:855–861
Litz RE (ed) (2009) The mango: botany, production and uses. Cabi
Litz RE, Gomez-Lim MA (2002) Genetic transformation of mango. In: Khachatourians G, McHughern A, Scorza R, Nip WK, Hui YH (eds) Transgenic plants and crops. Marcel Dekker, New York, pp 412–436
Liu Y, Riaz M, Yan L, Zeng Y, Cuncang J (2019) Boron and calcium deficiency disturbing the growth of trifoliate rootstock seedlings (*Poncirus trifoliate* L.) by changing root architecture and cell wall. Plant Physiol Biochem 144:345–354
Liu B, Xin Q, Zhang M, Chen J, Lu Q, Zhou X, Li X, Zhang W, Feng W, Pei H, Sun J (2023) Research progress on mango post-harvest ripening physiology and the regulatory technologies. Foods 12(1):173
Lobo JT, De Sousa KDSM, Neto VBDP, Pereira RN, Silva LDS, Cavalcante IHL (2019) Biostimulants on fruit yield and quality of mango cv. Kent grown in semiarid. J Am Pomol Soc 73(3):152–160
Makhmale S, Bhutada P, Yadav L, Yadav BK (2016) Impact of climate change on phenology of mango—the case study. Ecol Environ Conserv 22(9):S127–S132
Malik AU, Amin M, Mazhar MS, Johnson PJ, Hofman P, Campbell J, Holmes R, Rehman A, Dunne AJ, Mehdi M, Adeel A, Ayyaz S, Sun T, Collins RJ (2015) Mango value chain improvement through postharvest research and development: a developing country case study. In: XI international mango symposium, Darwin, NT, pp 411–420
Mango Genome Consortium, Bally ISE, Bombarely A, Chambers AH, Cohen Y, Dillon NL, Innes DJ, Islas-Osuna MA, Kuhn DN, Mueller LA, Ophir R, Rambani A, Sherman A, Yan H (2021) The 'Tommy Atkins' mango genome reveals candidate genes for fruit quality. BMC Plant Biol 21:1–18

Mango Market Outlook (2023). Accessed 30 May 2023. Available from: https://www.reportlinker.com/clp/global/2974#introduction

Marschner H (ed) (2011) Marschner's mineral nutrition of higher plants. Academic Press

Masroor HM, Anjum MA, Hussain S, Ejaz S, Ahmad S, Ercisli S, Zia-Ul-Haq M (2016) Zinc ameliorates fruit yield and quality of mangoes cultivated in calcareous soils. Erwerbs-Obstbau 58(1)

Mathiazhagan M, Chidambara B, Hunashikatti LR, Ravishankar KV (2021) Genomic approaches for improvement of tropical fruits: fruit quality, shelf life and nutrient content. Genes 12(12):1881

Mathiazhagan M, Padala S, Doddahejjaji SGC, Murugan S, Makki DR, Kundapura RV (2022) Omics of mango: a tropical fruit tree. In: Rout GR, Peter KV (eds) Omics in horticultural crops. Academic Press, pp 427–448

Merwad MA, Eisa RA, Saleh MMS (2016) The beneficial effect of NAA, Zn, Ca and B on fruiting, yield and fruit quality of Alphonso mango trees. Int J Chemtech Res 9(3):147–157

Michael VN, Crane J, Freeman B, Kuhn D, Chambers AH (2023) Mango seedling genotyping reveals potential self-incompatibility and pollinator behavior. Sci Hortic 308:111599

Milošević T, Milošević N (2023) Role of nutrients in modifications of fruit quality and antioxidant activity. In: Singh VP, Siddiqui MH (eds) Plant ionomics: sensing, signaling, and regulation. Wiley, UK, pp 148–170

Moretti LG, Crusciol CAC, Bossolani JW, Garcia A, Rossi R, Moreira A (2021) Thermomagnesium as a fertilizer for soybean: carbohydrate metabolism, silicon–magnesium fertilizer, and grain yield. J Plant Nutr 44:2108–2122

Morison IL, Growth MMP, Change C (2006) Blackwell Publishing Ltd.

Muriithi BW, Gathogo NG, Diiro GM, Mohamed SA, Ekesi S (2020) Potential adoption of integrated pest management strategy for suppression of mango fruit flies in East Africa: an ex ante and ex post analysis in Ethiopia and Kenya. Agriculture 10:278

Narang RS, Kataria AK (2014) Postharvest management of mango: a global perspective. Postharvest Biol Technol 84:1–16

Nazir W, Naqvi SAH, Ahmed N, Alrefaei AF, Zulfiqar MA (2024a) Optimizing micronutrient supplementation in mango orchards for the suppression of mango anthracnose (*Colletotrichum gloeosporioides*). Appl Fruit Sci 1–14

Nazir W, Naqvi SAH, Ahmed N, ur Rehman A, Alrefaei AF, Zulfiqar MA, ud Din Umar U (2024b) Optimizing micronutrient supplementation in mango orchards for the suppression of mango anthracnose (*Colletotrichum gloeosporioides*). Appl Fruit Sci 66(4):1417–1430

Ngereza JA, Pawelzik E (2016) Nutritional characterization of organically and conventionally grown mango (*Mangifera indica* L.) and pineapple (*Ananas comosus*) of different origins. J Crop Sci Agron 1:1–17

Normand F, Lauri PE, Legave JM (2015) Climate change and its probable effects on mango production and cultivation. Acta Hortic 1075:21–31

Ntsoane ML, Zude-Sasse M, Mahajan P, Sivakumar D (2019) Quality assessment and postharvest technology of mango: a review of its current status and future perspectives. Sci Hortic 249:77–85

Ojeda-Barrios DL, Perea-Portillo E, Hernández-Rodríguez OA, Avila-Quezada G, Abadía J, Lombardini L (2014) Foliar fertilization with zinc in pecan trees. HortScience 49:562–566

Okino-Delgado CH, Prado DZ, Pereira MS, Camargo DA, Koike MA, Fleuri LF (2020) Mango. In: Galanakis CM (eds) Valorization of fruit processing by-products. Academic Press, Elsevier, Amsterdam, The Netherlands, pp 167–181

Oliveira CM, Auad AM, Mendes SM, Frizzas MR (2013) Economic impact of exotic insect pests in Brazilian agriculture. J Appl Entomol 137:1–15

Parmar N, Singh KH, Sharma D, Singh L, Kumar P, Nanjundan J, Saha P, Jatav HS, Srivastav M, Singh NK, Thakur AK (2017) Genetic engineering strategies for biotic and abiotic stress tolerance and quality enhancement in horticultural crops: a comprehensive review. 3 Biotech 7:1–35

Patel RJ, Patil SJ, Tandel BM, Ahlawat TR, Amarcholi JJ (2018) Effect of micronutrients and banana pseudostem sap at different pH levels of foliar spray solution on fruit quality of mango (*Mangifera indica* L.) cv. Kesar. Int J Curr Sci 6:852–854

Pena JE, Mohyuddin AI, Wysoki M (1998) A review of the pest management situation in mango agroecosystems. Phytoparasitica 26:129–148

Penna S, Jain SM (2023) Fruit crop improvement with genome editing, *in vitro*, and transgenic approaches. Horticulturae 9:58

Ploetz R (2003) Diseases of tropical fruit crops. CABI Publishing, Wallingford, pp 327–363

Pruvost O, Roumagnac P, Gaube C, Chiroleu F, Gagnevin L (2005) New media for the semiselective isolation and enumeration of *Xanthomonas campestris* pv. *mangiferaeindicae*, the causal agent of mango bacterial black spot. J Appl Microbiol 99:803–815

Rajan S, Srivastav M, Rymbai H (2021) Genetic resources in mango. In: Kole C (ed) The mango genome. Compendium of plant genomes. Springer, pp 45–73

Rajput CB, Singh NP (1976) Leaf analysis and phosphorus fertilization of guava (*Psidium guajava* L.). J Jpn Soc Hortic Sci 44(4):355–359

Rajwana IA, Malik AU, Khan AS, Saleem BA, Malik SA (2010) A new mango hybrid shows better shelf life and fruit quality. Pak J Bot 42:2503–2512

Ramírez F, Davenport TL (2010) Mango (*Mangifera indica* L.) flowering physiology. Sci Hortic 126:65–72

Razzaq K, Khan AS, Malik AU, Shahid M, Ullah S (2013) Foliar application of zinc influences the leaf mineral status, vegetative and reproductive growth, yield and fruit quality of 'Kinnow' mandarin. J Plant Nutr 36:1479–1495

Ribeiro SMR, Schieber A (2010) Bioactive compounds in mango (*Mangifera indica* L.). In: Watson RR, Preedy VR (eds) Bioactive foods in promoting health. Academic Press, pp 507–523

Ribeiro IJA, Rossetto CJ, Donadio LC, Sabino JC, Martins AIM, Gallo PB (1993) Mango wilt. XIV selection of mango (*Mangifera indica* L.) rootstocks resistant to the mango wilt fungus *Ceratocystis fimbriata* Ell & Halst. Acta Hortic 370:159–166

Rivera Domínguez M, Astorga Cienfuegos KR, Vallejo Cohen S, Vargas Arispuro I, Sánchez Sánchez E (2011) Transgenic mango embryos (*Mangifera indica* cv. 'Ataulfo') with the defensin J1 gene. Rev Mex Fitopatol 29(1):78–80

Rossetto CJ, Ribeiro IJA, Gallo PB, Soares NB, Sabino JC, Martins ALM et al (1996) Mango breeding for resistance to diseases and pests. Acta Hortic 455:299–304

Ryan J, Rashid A, Torrent J, Yau SK, Ibrikci H, Sommer R, Erenoglu EB (2013) Micronutrient constraints to crop production in the Middle East–West Asia region: significance, research, and management. Adv Agron 122:1–84

Schwab W (2003) Metabolome diversity: too few genes, too many metabolites? Phytochemistry 62:837–849

Shahraki AS, Caloiero T, Bazrafshan O (2023) Influence of climatic factors on yields of pistachio, mango, and bananas in Iran. Sustainability 15:8962

Shamili M, Fatahi R, Hormaza J (2012) Characterization and evaluation of genetic diversity of Iranian mango (*Mangifera indica* L., Anacardiaceae) genotypes using microsatellites. Sci Hortic 148:230–234

Shanthala L, Ramachandra RK, Mohan KS, Swamy GSK, Honnabyraiah MK, Pallavi HM (2021) Evaluation of mango (*Mangifera indica* L.) rootstocks under induced moisture stress condition for various morpho-physiological traits. Int J Curr Microbiol Appl Sci 10:143–148

Sharma SS, Dietz KJ (2006) The significance of amino acids and amino acid-derived molecules in plant responses and adaptation to heavy metal stress. J Exp Bot 57(4):711–726

Sharma SK, Thakur NK (2012) Postharvest management of mango: a review. Indian J Hortic 69:1–16

Sharma RR, Singh D, Pal RK (2013) Synergistic influence of pre-harvest calcium sprays and postharvest hot water treatment on fruit firmness, decay, bitter pit incidence and postharvest quality of 'royal delicious' apples (*Malus x domestica* Borkh). J Plant Sci 4:153

Shivashankar S (2014) Physiological disorders of mango fruit. In: Janick J (ed) Horticultural reviews: volume 42. Wiley, pp 313–348

Shu ZH, Sheen TF (1987) Floral induction in axillary buds of mango (*Mangifera indica* L.) as affected by temperature. Sci Hortic 31:81–87

Simões WL, do Carmo Mouco MA, de Andrade VPM, Bezerra PP, Coelho EF (2020) Fruit yield and quality of Palmer mango trees under different irrigation systems. Comun Sci 11:e325

Simões WL, Silva JSD, Mouco MADC, de Oliveira CP, Silva DJ, Oliveira FFD (2022) Marine calcium application on 'Palmer' mango production. Rev Bras Eng Agríc Ambient 26:618–623

Singh Z, Singh RK, Sane VA, Nath P (2013) Mango—postharvest biology and biotechnology. Crit Rev Plant Sci 32(4):217–236

Singh NP, Ashok A, Pavithra S, Balaji SJ, Anand B, Khan MA (2017) Mainstreaming climate change adaptation into development planning. In: Policy paper 32. ICAR-National Institute of Agricultural Economics and Policy Research, New Delhi

Song M, Wang H, Fan Z, Huang H, Ma H (2023) Advances in sequencing and key character analysis of mango (*Mangifera indica* L.). Hortic Res 10(2):259

Souri MK, Aslani M (2018) Beneficial effects of foliar application of organic chelate fertilizers on French bean production under field conditions in a calcareous soil. Adv Hortic Sci 32(2):265–272

Souri MK, Bakhtiarizade M (2019) Biostimulation effects of rosemary essential oil on growth and nutrient uptake of tomato seedlings. Sci Hortic 243:472–476

Souri MK, Naiji M, Aslani M (2018) Effect of Fe-glycine aminochelate on pod quality and iron concentrations of bean (*Phaseolus vulgaris* L.) under lime soil conditions. Commun Soil Sci Plant Anal 49(2):215–224

Spreer W, Ongprasert S, Hegele M, Wünsche JN, Müller J (2009) Yield and fruit development in mango (*Mangifera indica* L. cv. Chok Anan) under different irrigation regimes. Agric Water Manag 96(4):574–584. https://doi.org/10.1016/j.agwat.2008.09.020

Srivastav M, Radadiya N, Ramachandra S, Jayaswal PK, Singh N, Singh S (2023) High-resolution mapping of QTLs for fruit color and firmness in Amrapali/Sensation mango hybrids. Front Plant Sci 14:1135285. https://doi.org/10.3389/fpls.2023.1135285

Sudheeran PK, Love C, Feygenberg O, Maurer D, Ovadia R, Oren-Shamir M et al (2019) Induction of red skin and improvement of fruit quality in 'Kent', 'Shelly' and 'Maya' mangoes by preharvest spraying of prohydrojasmon at the orchard. Postharvest Biol Technol 149:8–26

Sudheeran PK, Sela N, Carmeli-Weissberg M, Ovadia R, Panda S, Feygenberg O et al (2021) Induced defense response in red mango fruit against *Colletotrichum gloeosporioides*. Hortic Res 8

Sukhvibul N, Whiley AW, Smith MK, Hetherington SE, Vithanage V (1999) Effect of temperature on inflorescence and floral development in four mango (*Mangifera indica* L.) cultivars. Sci Hortic 82(1–2):67–84

Suman M, Sangma PD, Singh D (2017) Role of micronutrients (Fe, Zn, B, Cu, Mg, Mn, and Mo) in fruit crops. Int J Curr Microbiol Appl Sci 6:3240–3250

Sun T, Rao S, Zhou X, Li L (2022) Plant carotenoids: recent advances and future perspectives. Mol Hortic 2(1):3

Suzuki N, Rivero RM, Shulaev V, Blumwald E, Mittler R (2014) Abiotic and biotic stress combinations. New Phytol 203(1):32–43

Tagliavini M, Abadía J, Rombolà AD, Abadía A, Tsipouridis C, Marangoni B (2000) Agronomic means for the control of iron deficiency chlorosis in deciduous fruit trees. J Plant Nutr 23:2007–2022

Taiz L, Zeiger E, Møller IM, Murphy A (2017) Fisiologia e desenvolvimento vegetal, 6th edn. ArtMed, Porto Alegre, p 858

Testa R, Tudisca S, Schifani G, Di Trapani AM, Migliore G (2018) Tropical fruits as an opportunity for sustainable development in rural areas: the case of mango in small-sized Sicilian farms. Sustainability 10(5):1436

Tharanathan RN, Yashoda HM, Prabha TN (2006) Mango (*Mangifera indica* L.) "the king of fruits"—an overview. Food Rev Int 22(2):95–123

Tongyoo P, Duangjit J, Sharma N, Chunwongse J (2021) Molecular mapping and breeding in mango. In: The mango genome, pp 153–163

Vayssières JF, Fatogoma S, Moussa N (2004) Inventory of species of fruit flies (Diptera: Tephritidae) linked to the mango tree in Mali and testing of rational control. Fruits 59:1–14

Vittal H, Sharma N, Dubey AK, Shivran M, Singh SK, Meena MC et al (2023) Rootstock-mediated carbohydrate metabolism, nutrient contents, and physiological modifications in regular and alternate mango (*Mangifera indica* L.) scion varieties. PLoS One 18:e0284910

Wang Y, Xie X, Long LE (2014) The effect of postharvest calcium application in hydro-cooling water on tissue calcium content, biochemical changes, and quality attributes of sweet cherry fruit. Food Chem 160:22–30

Wang P, Luo Y, Huang J, Gao S, Zhu G, Dang Z (2020) The genome evolution and domestication of tropical fruit mango. Genome Biol 21:1–17

Wei J, Liu G, Liu D, Chen Y (2017) Influence of irrigation during the growth stage on yield and quality in mango (*Mangifera indica* L). PLoS One 12

Whiley AW, Rasmussen TS, Saranah JB, Wolstenholme BN (1989) Effect of temperature on growth, dry matter production, and starch accumulation in ten mango (*Mangifera indica* L.) cultivars. J Hortic Sci 64:753–765

Yang Y, Saand MA, Huang L, Abdelaal WB, Zhang J, Wu Y (2021) Applications of multi-omics technologies for crop improvement. Front Plant Sci 12:563953

Yiran GA, Ablo AD, Asem FE (2023) Is mango farming sustainable? An integrated analysis of remote sensing techniques and smallholder farmers' perception in mango farming communities in Ghana. Ghana Soc Sci J 20:71–96

Yruela I (2005) Copper in plants. Braz J Plant Physiol 17:145–156

Zainab M, Shinkafi S (2016) Isolation and identification of fungi responsible for leaf spots disease of mango (*Mangifera indica* L.) in Sokoto state, Nigeria. Bayero J Pure Appl Sci 9:166–173

Zhang D, Chong W, Li XL (2019) Yield gap and production constraints of mango (*Mangifera indica*) cropping systems in Tianyang County, China. J Integr Agric 18:1726–1736

Zuazo VHD, García-Tejero IF, Rodríguez BC, Tarifa DF, Ruiz BG, Sacristán PC (2021) Deficit irrigation strategies for subtropical mango farming. A review. Agron Sustain Dev 41:1–13

Chapter 21
Management of Mango Diseases in Climate Change Scenarios Focusing on Nanoparticles

Muhammad Shahbaz, Amjad Ali, Noor Fatima, Syed Atif Hasan Naqvi, and Jaya Seelan Sathiya Seelan

Abstract Climate change exacerbates the prevalence and severity of mango diseases due to rising temperatures, erratic rainfall, and humidity fluctuations, creating conducive environments for pathogenic outbreaks. Traditional disease management strategies, such as chemical fungicides, face limitations due to environmental concerns and resistance development. Nanotechnology offers a promising solution for managing mango diseases effectively and sustainably under these changing conditions. Nanoparticles, such as silver, copper, and zinc-based formulations, exhibit antimicrobial properties that target a wide range of pathogens while minimizing the ecological footprint. Their small size allows for better penetration and controlled release, enhancing their efficacy and reducing the quantity required. Recent studies have demonstrated the potential of nanoparticles in suppressing fungal pathogens like *Colletotrichum gloeosporioides* and *Lasiodiplodia theobromae*, which cause devastating diseases like anthracnose and stem-end rot. Additionally, nanoparticle applications can be integrated with existing disease management practices, such as resistant cultivars and biocontrol agents, to provide a holistic approach. This chapter explores the role of nanoparticles in mitigating mango diseases in the context of climate change, highlighting their mechanisms, efficacy, and environmental safety. By embracing nanotechnology, mango cultivation can achieve enhanced disease resistance and sustainable productivity amidst the challenges posed by global climate shifts.

M. Shahbaz (✉) · N. Fatima · J. S. S. Seelan (✉)
Institute for Tropical Biology and Conservation, Universiti Malaysia Sabah, Jalan UMS, 88400 Kota Kinabalu, Malaysia
e-mail: MUHAMMAD_SHABAZ_DX22@iluv.ums.edu.my

A. Ali
Department of Plant Protection, Faculty of Agricultural Sciences and Technologies, Sivas University of Science and Technology, 58140 Sivas, Türkiye

S. A. H. Naqvi
Department of Plant Pathology, Bahauddin Zakariya University, 60000 Multan, Pakistan

S. A. H. Naqvi et al. (eds.), *Climate Change and Mango Production*,
https://doi.org/10.1007/978-3-031-87035-4_21

Keywords Mango diseases · Nanoparticles · Climate change · Sustainable management

21.1 Introduction

Mango (*Mangifera indica* L.) is among the most cherished tropical fruits globally, renowned for its nutritional value, flavor, and cultural significance (Gelaye 2024). It is cultivated in over 100 countries, with India, China, Thailand, and Mexico being the top producers. Mango plays a vital role in the livelihoods of millions of farmers, contributing significantly to global fruit trade and food security (Ranjan et al. 2022). Mango trees have ecological benefits, such as carbon sequestration and soil stabilization, making them crucial for sustainable agricultural systems. Mango holds immense economic importance as a staple fruit crop and a key contributor to the economies of tropical and subtropical regions. Its global production exceeds 55 million metric tons annually, with a steady rise driven by increasing demand (Vuppalapati 2023). This fruit is a major source of foreign exchange earnings in producing countries, with exports valued at billions of dollars each year (Huang et al. 2022). The mango industry supports millions of smallholder farmers, providing employment in production, processing, and export chains. For instance, in India, which accounts for over 40% of global mango production, the industry directly supports approximately 2.5 million farming households (Wardhan et al. 2022). The United States and Europe are major importers, with consumers valuing mango not only for its taste but also for its health benefits, including being a rich source of vitamins A and C, fiber, and antioxidants. The mango industry also has cultural significance, celebrated in festivals and culinary traditions worldwide (Kiloes et al. 2022).

Despite its importance, mango production faces significant challenges due to various diseases that reduce yield and fruit quality (Rahaman et al. 2023). Key fungal diseases include anthracnose, caused by *Colletotrichum* spp., and powdery mildew, caused by *Oidium mangiferae*, which lead to pre- and post-harvest losses (Samwel et al. 2020). Bacterial diseases such as bacterial canker (*Xanthomonas campestris pv. mangiferae indicae*) also pose a serious threat, causing defoliation, fruit blemishes, and yield reduction (Sossah et al. 2024) mango malformation, disrupt flower and fruit formation, further compounding economic losses. These diseases not only impact yields but also incur high management costs, which strain farmers' finances (Shahbaz et al. 2023; Shahid et al. 2024). Post-harvest diseases alone can result in losses of up to 30% in some regions, reducing export potential and profitability. Furthermore, quarantine regulations in international trade due to disease presence restrict market access, impacting export revenues. As mango cultivation expands into new areas due to increasing global demand, emerging pathogens and favorable conditions for disease outbreaks exacerbate these challenges (Lengai et al. 2022).

The impact of climate change on agriculture is undeniable, with far-reaching implications for plant pathology (Rubayet and Hossain 2024). Rising temperatures, altered precipitation patterns, and increased frequency of extreme weather events significantly affect the incidence and severity of crop diseases (Prasad et al. 2021). These changes influence the distribution, life cycles, and virulence of pathogens, presenting a growing threat to global food security. Mango cultivation is particularly vulnerable, given its sensitivity to climatic variability and its reliance on predictable seasonal patterns (Gullino et al. 2022). Climate change has created a dynamic shift in pathogen ecology and disease epidemiology. Higher temperatures accelerate the life cycle of fungal pathogens such as *Colletotrichum* and *Oidium*, leading to increased disease incidence. Changes in rainfall patterns, such as prolonged wet seasons, create favorable conditions for spore germination and pathogen spread (Islam et al. 2024). Drought conditions, on the other hand, stress mango trees, weakening their natural defense mechanisms and increasing susceptibility to bacterial and viral infections (Dofuor et al. 2023). Similarly, climate change has disrupted the synchrony between host phenology and pathogen activity, intensifying pre-harvest losses. Altered wind patterns have also been implicated in the dispersal of spores over long distances, introducing pathogens to new areas (Ren et al. 2020). The interplay between climate change and disease prevalence is multifaceted. Elevated CO_2 levels and temperature shifts alter the physiological responses of mango plants, such as stomatal conductance, potentially increasing susceptibility to certain pathogens (Shah et al. 2024). For example, bacterial canker severity has been linked to increased rainfall and high humidity, conditions becoming more frequent due to climate change (Singh et al. 2023). Extreme weather events, such as cyclones and hailstorms, create physical damage to mango trees, providing entry points for pathogens (Rana et al. 2022). The cumulative effect of these changes has been a noticeable rise in the frequency and intensity of mango disease outbreaks, causing widespread losses and threatening the sustainability of mango cultivation (Yuen et al. 2021). Climate-induced stress on mango crops has made traditional disease management practices less effective, necessitating innovative solutions (Tengsetasak et al. 2024).

Addressing the complex challenges of mango disease management in a changing climate requires the integration of advanced technologies (Ranganna et al. 2024). Among these, nanotechnology has emerged as a promising tool, offering innovative solutions for sustainable agriculture. Nanoparticles, with their unique physicochemical properties, have shown potential in enhancing disease management by targeting pathogens directly and boosting plant defense mechanisms (Malik et al. 2023). Nanoparticles are microscopic particles with dimensions measured in nanometers, giving them unique properties such as high surface area, enhanced reactivity, and precise delivery mechanisms. In agriculture, metallic nanoparticles (e.g., silver, copper), polymeric nanoparticles, and bio-nanocomposites have gained attention for their antimicrobial properties and potential to improve plant health (Joudeh and Linke 2022). When applied to mango disease management, nanoparticles act through mechanisms such as disrupting microbial cell membranes, generating reactive oxygen species (ROS), and inhibiting pathogen reproduction. Nanoparticles can also serve as carriers for active ingredients, ensuring controlled release and reducing chemical use, thereby minimizing environmental impact (Tariq et al. 2022). For instance, nano-silver has shown efficacy against anthracnose pathogens, while copper nanoparticles have demonstrated potential in controlling bacterial canker.

Furthermore, nanoparticles are being integrated into coatings for post-harvest disease management, significantly reducing losses during transport and storage (Rameez et al. 2024).

This chapter aims to explore the potential of nanotechnology in managing mango diseases within the context of climate change. It focuses on the implications of climate-induced changes in disease dynamics, evaluates the effectiveness of nanoparticle-based interventions, and discusses their integration into sustainable disease management strategies. By addressing the current challenges and future prospects, this chapter provides a comprehensive overview of how nanotechnology can revolutionize mango disease management in a changing world.

21.2 Mango Diseases: A Comprehensive Overview

Mango cultivation is significantly hindered by various diseases that reduce yield and compromise fruit quality (Ullah et al. 2024). These diseases are caused by a diverse range of pathogens, including fungi, bacteria, and viruses, which affect different parts of the plant, such as leaves, flowers, fruits, and stems (Table 21.1). Effective management of these diseases is critical for sustaining mango production and meeting the growing global demand (Dofuor et al. 2023). This section provides an overview of the most common mango diseases, emerging challenges due to climate change, and the limitations of current management practices.

Table 21.1 Key mango diseases, causal pathogens, and favorable conditions

Disease	Causal pathogen	Symptoms	Favorable conditions
Anthracnose	*Colletotrichum gloeosporioides*	Black, sunken lesions on fruits, flowers, and leaves	Warm, humid, rainy weather
Powdery mildew	*Oidium mangiferae*	White powdery growth on leaves and flowers, reduced fruit set	Dry climate with high humidity
Bacterial canker	*Xanthomonas campestris pv. mangiferaeindicae*	Water-soaked lesions, necrosis, defoliation	Wet, humid conditions with rain splash
Mango malformation	*Fusarium* spp.	Deformed flower and vegetative growth	High humidity and fluctuating temperatures
Stem-end rot	*Lasiodiplodia theobromae*	Black rot at stem-end of fruits	Warm, moist post-harvest conditions

21.2.1 Common Mango Diseases and Pathogens

21.2.1.1 Anthracnose (*Colletotrichum* spp.)

Anthracnose is one of the most destructive diseases affecting mango, primarily caused by *Colletotrichum gloeosporioides* (Jeevanantham et al. 2024). This fungal disease affects all parts of the plant, including leaves, flowers, and fruits, leading to significant pre- and post-harvest losses. Symptoms include black, sunken lesions on leaves and fruits, often accompanied by a pinkish spore mass. In flowers, anthracnose causes blossom blight, which reduces fruit set and yields. The disease thrives in warm and humid conditions, making it particularly problematic in tropical and subtropical regions. Rainy weather exacerbates the spread of spores, resulting in widespread infections during flowering and fruiting seasons. The fungus also survives as latent infections, which manifest post-harvest during storage and transport (Singh and Afaque 2021). Control measures include the application of fungicides, but their effectiveness is often limited due to environmental concerns and resistance development. Integrated disease management strategies, including cultural practices such as pruning and maintaining orchard hygiene, are essential for reducing anthracnose prevalence (Jagtap et al. 2020).

21.2.1.2 Powdery Mildew (*Oidium mangiferae*)

Powdery mildew, caused by the obligate fungal pathogen *Oidium mangiferae*, is another significant disease of mango. It primarily affects young leaves, flowers, and developing fruits, covering them with a white, powdery fungal growth. Infected flowers often fail to set fruits, leading to yield losses of up to 90% in severe cases. The disease is most prevalent in dry climates with high humidity during flowering, as these conditions favor spore germination and disease progression. Early symptoms include the appearance of white spots on leaves, which coalesce to form larger patches (Iqbal et al. 2024). If left untreated, the fungus weakens the tree, reducing its overall productivity. Effective control relies on timely fungicide application, but the repeated use of chemicals has raised concerns about environmental toxicity. In addition, powdery mildew tends to reoccur seasonally, requiring continuous monitoring and preventive measures such as removing infected plant parts and enhancing airflow within orchards (Carmona et al. 2020).

21.2.1.3 Bacterial Canker (*Xanthomonas* spp.)

Bacterial canker, caused by *Xanthomonas campestris pv. mangiferae indicae*, is a severe bacterial disease affecting mango trees. It is characterized by water-soaked lesions on leaves, stems, and fruits, which eventually turn necrotic. The disease causes significant defoliation, dieback of branches, and blemished fruits, reducing both yield and market value. The bacteria are disseminated through rain splash, wind, and contaminated tools, spreading rapidly under wet and humid conditions (Adikaram 2024). Wounds or natural openings in plant tissues serve as entry points for the pathogen. Control is challenging due to the limited effectiveness of copper-based bactericides, which are the primary chemical control option. Moreover, excessive reliance on these chemicals has led to resistance development in bacterial populations. Cultural practices, including sanitation, pruning infected parts, and avoiding overhead irrigation, are critical for managing bacterial canker. Research into biological control agents, such as beneficial microbes, offers promise for sustainable management (Scortichini 2022).

21.2.1.4 Other Emerging Diseases Due to Climate Variability

Climate change has introduced new challenges to mango disease management by altering the dynamics of existing pathogens and facilitating the emergence of new ones. Diseases such as mango malformation, black spot, and stem-end rot are becoming more prominent in regions experiencing erratic rainfall and temperature fluctuations (Abdurrehman et al. 2024). Mango malformation, caused by *Fusarium* spp., leads to abnormal flower and vegetative growth, significantly reducing yields. Black spot disease, caused by *Pseudocercospora mangiferae*, results in circular, black lesions on leaves and fruits, affecting photosynthesis and fruit quality. (Khaskheli 2020). Stem-end rot, caused by pathogens such as *Lasiodiplodia theobromae*, affects fruits post-harvest, causing significant losses during storage and transport. These emerging diseases highlight the need for adaptive management strategies that account for the changing climate and its impact on pathogen behavior (Huda-Shakirah et al. 2022).

21.2.2 Challenges in Current Disease Management Practices

Chemical fungicides and bactericides are the cornerstone of current mango disease management practices. However, their effectiveness is often limited by improper application methods, timing, and dosage. Over-reliance on chemical controls has led to the development of resistant strains of pathogens, reducing the efficacy of widely used products. Additionally, the frequent use of broad-spectrum chemicals disrupts the ecological balance, affecting beneficial organisms in the orchard ecosystem. Smallholder farmers, who constitute the majority of mango producers, often lack access to affordable and effective chemicals, resulting in suboptimal disease control (Khursheed et al. 2022). Moreover, the high cost of chemical inputs strains the finances of resource-poor farmers, particularly in developing countries. Pathogen resistance to chemical controls is a growing concern in mango disease management (Shanka 2020). Repeated applications of fungicides with the same mode of action have selected for resistant populations of *Colletotrichum*, *Oidium*, and *Xanthomonas*

(Li et al. 2020). This resistance renders conventional treatments ineffective, necessitating higher dosages or the use of newer, more expensive chemicals. The environmental impact of chemical residues in soil and water is another significant issue, contributing to biodiversity loss and human health risks. Many countries have implemented stringent regulations on chemical pesticide use, further complicating disease management. The need for sustainable and eco-friendly alternatives, such as biocontrol agents and advanced technologies like nanoparticles, has become increasingly evident in the face of these challenges (Khan et al. 2023).

21.3 Climate Change Scenarios and Their Implications

Climate change is reshaping agricultural landscapes, with significant implications for plant pathology. Rising temperatures, increased humidity, and altered precipitation patterns are transforming the dynamics of plant-pathogen interactions, particularly in mango cultivation. These changes are leading to shifts in disease prevalence, severity, and distribution, necessitating a deeper understanding of their underlying mechanisms (Angelotti et al. 2024).

21.3.1 Effects of Increased Temperature and Humidity

Climate change significantly impacts the delicate balance between mango plants and their associated pathogens. Elevated temperatures accelerate the physiological processes of both host plants and pathogens, altering their interactions. For instance, warmer climates enhance the virulence of certain fungal pathogens, such as *Colletotrichum* spp., which thrive in higher temperatures and humid conditions. This results in earlier onset and prolonged periods of infection. Moreover, stressed mango plants under heat or water scarcity often exhibit weakened immune responses, making them more susceptible to infections (Yadav et al. 2023). High humidity, a common consequence of climate change, creates a conducive environment for spore germination and pathogen proliferation. Studies indicate that under elevated CO_2 and temperature conditions, pathogens can adapt to overcome host resistance, leading to more severe infections. These dynamics highlight the complex interplay of abiotic and biotic stressors in a changing climate (Misiou and Koutsoumanis 2022).

Climate change is altering the phenology of mango diseases, with significant implications for disease management. Higher temperatures shorten the latency periods of many fungal pathogens, allowing for multiple infection cycles within a single growing season (Gullino et al. 2022). For example, *Oidium mangiferae* spores germinate faster under warm and humid conditions, leading to rapid disease outbreaks. Similarly, *Xanthomonas campestris pv. mangiferae indicae*, responsible for bacterial canker, exhibits enhanced survival and spread during periods of excessive rainfall. Changes in wind patterns driven by climate variability contribute to the wider dispersal of airborne pathogens, introducing diseases to previously unaffected regions (Umar et al. 2022). Prolonged wet seasons create conditions for spore survival and dissemination, while erratic rainfall can delay flowering and fruiting in mango trees, disrupting the synchrony between plant development and disease susceptibility (Strickland 2023).

21.3.2 Predicted Trends in Disease Prevalence and Severity

Climate change is driving significant shifts in the geographic distribution of mango diseases. As temperatures rise, pathogens once confined to tropical and subtropical regions are expanding into higher altitudes and latitudes (Ameen et al. 2023). For instance, anthracnose, previously a problem primarily in lowland orchards, is now emerging in cooler regions as warming trends make these areas more conducive to fungal growth. Such geographic shifts pose challenges for farmers who lack experience managing these diseases and for regulatory bodies tasked with monitoring disease spread. Additionally, changes in rainfall patterns and humidity levels influence the regional suitability for pathogen survival, creating new hotspots for disease outbreaks (Palanivel and Shah 2021) This redistribution of diseases disrupts traditional mango-growing zones, potentially reducing production in historically favorable areas while increasing pressure in newly affected regions. Predictive modeling tools are essential for mapping these shifts and guiding proactive management strategies (Hendrichs et al. 2021).

21.4 Nanoparticles in Disease Management

Nanotechnology has emerged as a transformative approach to managing plant diseases, including those affecting mango, by providing innovative tools to address challenges in pathogen control (Table 21.2). Nanoparticles (NPs), due to their unique physicochemical properties, offer enhanced efficiency in combating pathogens while reducing environmental risks compared to conventional treatments. These advanced materials have diverse applications in agriculture, such as antimicrobial agents, carriers for agrochemicals, and enhancers of plant immunity, making them valuable components in sustainable disease management strategies (Rana et al. 2021).

Nanoparticles are defined as materials with dimensions ranging from 1 to 100 nm, a scale that imparts them with distinct properties beneficial for agricultural use. Based on their composition, nanoparticles are categorized into metallic, polymeric, lipid-based, carbon-based, and bio-nanoparticles (Shahbaz et al. 2024). Metallic nanoparticles, such as silver (Ag), copper (Cu), and zinc oxide (ZnO), are particularly valued for their potent antimicrobial activities. Polymeric nanoparticles, synthesized from natural or synthetic polymers, serve as effective carriers for controlled release of pesticides or fungicides. Lipid-based nanoparticles encapsulate bioactive compounds,

Table 21.2 Properties and applications of common nanoparticles in mango disease management

Nanoparticle type	Properties	Applications in mango disease management
Silver (AgNPs)	Strong antimicrobial activity, ROS generation	Effective against anthracnose and powdery mildew
Copper (CuNPs)	Enzyme inhibition, membrane disruption	Effective against bacterial canker and fungal pathogens
Zinc Oxide (ZnO NPs)	Enhanced nutrient bioavailability	Soil treatments to suppress root-borne pathogens
Chitosan NPs	Biodegradable, boosts plant immunity	Controls fungal infections and strengthens plant defenses
Carbon-Based NPs	High stability, antimicrobial activity	Pathogen suppression and soil health improvement

enhancing their stability and bioavailability (Alavi and Kennedy 2021). Carbon-based nanoparticles, including graphene oxide and carbon nanotubes, are increasingly studied for their role in improving soil health and pathogen management. Bio-nanoparticles, derived from biological sources, represent an eco-friendly option for sustainable agriculture. These various types of nanoparticles cater to specific needs, providing versatility in their applications for mango disease management (Eivazzadeh-Keihan et al. 2022).

The unique properties of nanoparticles make them particularly suitable for pathogen control. Their high surface area-to-volume ratio increases their interaction with pathogens, enhancing their antimicrobial efficacy. Nanoparticles exhibit superior solubility, which improves the bioavailability of active ingredients, ensuring better disease control. Metallic nanoparticles such as silver and copper demonstrate strong antimicrobial activity through mechanisms like membrane disruption and reactive oxygen species (ROS) generation (Sayed et al. 2022). Polymeric and lipid-based nanoparticles enable the controlled release of bioactive agents, reducing the frequency of applications and lowering environmental risks. Nanocarriers are capable of delivering active compounds directly to infection sites, minimizing off-target effects and increasing treatment precision. These properties, combined with their potential for environmental safety, make nanoparticles a compelling alternative to conventional chemical treatments (Farasati Far et al. 2023).

Nanoparticles combat pathogens through both direct and indirect mechanisms of action. Directly, nanoparticles disrupt microbial cells by generating ROS, which cause oxidative damage to cellular components like DNA, proteins, and lipids, ultimately leading to cell death. Metallic nanoparticles, such as silver and copper, are particularly effective at creating oxidative stress (Umair Raza et al. 2023). Additionally, nanoparticles disrupt microbial membranes, leading to structural damage and leakage of cellular contents. They also interfere with enzymatic processes critical for pathogen survival, effectively halting their metabolic functions. Indirectly, nanoparticles strengthen plant defenses by activating systemic resistance. For example, they stimulate the production of phytohormones such as salicylic acid and jasmonic acid, which regulate plant immune responses (Hassan et al. 2022). Nanoparticles also enhance the activity of antioxidant enzymes, such as superoxide dismutase and peroxidase, which neutralize ROS generated during pathogen attacks. Furthermore, nanoparticles induce the deposition of lignin and callose in plant cell walls, fortifying tissues against pathogen invasion. These dual mechanisms of action—direct pathogen targeting and indirect defense activation—provide comprehensive protection against mango diseases (Jomova et al. 2024).

Nanoparticles can be applied in various ways to manage mango diseases effectively. Foliar sprays are a common method, where nanoparticle-based solutions are applied directly to leaves and fruits, ensuring rapid action against surface pathogens (Mahmoud and Gaber 2024). Metallic nanoparticles such as silver and copper are frequently used in foliar sprays to combat fungal infections like anthracnose and powdery mildew. Soil treatments with nanoparticles offer another avenue for disease control, targeting soil-borne pathogens and improving root health. For instance, zinc oxide nanoparticles can enhance nutrient uptake while suppressing soil-borne diseases (Rajwade et al. 2020).

21.5 Case Studies: Nanoparticles in Managing Mango Diseases

21.5.1 Studies on Nano-Silver and Nano-Copper

Nano-silver (AgNPs) and nano-copper (CuNPs) have been extensively studied for their effectiveness in managing fungal and bacterial diseases in mango (Table 21.3). Nano-silver exhibits broad-spectrum antimicrobial properties, making it effective against common mango pathogens such as *Colletotrichum* spp., the causative agent of anthracnose. Its mode of action involves the generation of reactive oxygen species (ROS), which disrupts microbial cell membranes and causes intracellular damage, leading to cell death (Odetayo et al. 2022). Similarly, nano-copper has demonstrated potent activity against both fungal and bacterial pathogens, including *Oidium mangiferae* (powdery mildew) and *Xanthomonas campestris pv. mangiferae indicae* (bacterial canker). Studies reveal that nano-copper disrupts cellular processes by binding to proteins and DNA, interfering with essential enzymatic functions in pathogens (Gupta et al. 2020). Research has shown that foliar sprays of nano-silver and nano-copper reduce disease severity in mango orchards, particularly under humid conditions that favor pathogen proliferation. These nanoparticles not only inhibit the growth of pathogens but also prevent spore germination, thereby controlling the spread of diseases. Field trials in mango orchards have highlighted their potential to replace or reduce chemical fungicides and bactericides, offering a more sustainable approach to disease management (Tiwari et al. 2022).

Table 21.3 Case studies on the application of nanoparticles in mango disease management

Study	Nanoparticle used	Pathogen targeted	Outcome
Field trials in humid tropical orchards	Nano-silver (AgNPs)	*Colletotrichum* spp. (anthracnose)	70% reduction in disease severity
Laboratory experiments on infected fruits	Nano-copper (CuNPs)	*Xanthomonas campestris* (bacterial canker)	Significant inhibition of bacterial growth
Post-harvest storage evaluation	Chitosan nanoparticles	*Lasiodiplodia theobromae* (stem-end rot)	Extended shelf life and reduced rot incidence

21.5.2 Nano-Based Pesticides for Mango Orchards

Nano-based pesticides represent a significant advancement in disease management for mango orchards. These formulations utilize nanoparticles as carriers for active ingredients, improving their solubility, stability, and bioavailability (Sarmah et al. 2023). Field studies have demonstrated that nano-pesticides are highly effective under diverse climatic conditions, including regions prone to high humidity and temperature fluctuations. For example, nano-formulated copper-based pesticides have shown enhanced efficacy against bacterial canker compared to conventional copper fungicides. Their ability to adhere to leaf surfaces for extended periods ensures prolonged protection, even during rainy seasons. Similarly, nano-silver-based pesticides have proven effective in controlling anthracnose, with significantly reduced spore germination rates observed in treated orchards. Field trials have indicated that mango trees treated with nano-pesticides exhibit higher fruit quality and yield compared to those managed with traditional chemicals. Moreover, nano-pesticides reduce the frequency of applications, lowering labor and input costs. By addressing challenges associated with conventional pesticides, such as low persistence and environmental contamination, nano-based solutions have emerged as a promising tool for sustainable mango cultivation (Abdollahdokht et al. 2022).

21.5.3 Limitations and Risks of Using Nanoparticles

21.5.3.1 Potential Toxicity to Non-target Organisms and the Environment

Despite their effectiveness, nanoparticles pose potential risks to non-target organisms and the environment. Studies have shown that high concentrations of metallic nanoparticles, such as silver and copper, can be toxic to beneficial soil microbes, disrupting the microbial ecosystem essential for soil fertility. Similarly, aquatic ecosystems are at risk when nanoparticles leach into water bodies, where they can accumulate in fish and other aquatic organisms. Nano-silver, for instance, has been observed to cause oxidative stress in algae and inhibit photosynthesis, raising concerns about its ecological impact. The long-term persistence of nanoparticles in the environment is another area of concern, as their small size and high reactivity can lead to bioaccumulation in food chains. Furthermore, the effects of nanoparticles on human health, particularly through occupational exposure during application or consumption of treated fruits, require thorough evaluation. These risks necessitate the development of safe application guidelines and environmentally friendly nano-formulations to mitigate potential hazards.

21.6 Regulatory Challenges and Public Perception

The use of nanoparticles in agriculture faces significant regulatory challenges, as most countries lack comprehensive guidelines for their approval and use. The absence of standardized testing protocols for evaluating the safety and efficacy of nano-based agrochemicals complicates their adoption. Regulatory bodies are often cautious due to the potential risks associated with nanoparticles, delaying their approval for commercial use. Additionally, public perception plays a critical role in the acceptance of nanotechnology in agriculture. Concerns about the safety of nano-treated crops and potential environmental impacts have led to skepticism among consumers and farmers. Educational initiatives and transparent communication about the benefits and risks of nanoparticles are essential to build trust and promote their adoption. Addressing regulatory challenges and public concerns is crucial for realizing the full potential of nanoparticles in sustainable mango disease management.

21.7 Conclusions

The integration of nanoparticles in mango disease management presents a transformative approach to addressing the challenges posed by climate change. With their unique properties, such as enhanced antimicrobial activity, targeted delivery, and environmental safety, nanoparticles offer a sustainable alternative to conventional chemical treatments. Their application has shown significant potential in controlling major mango diseases, including anthracnose, powdery mildew, and bacterial canker, while reducing the reliance on harmful pesticides. Moreover, the ability of nanoparticles to strengthen plant defenses and adapt to diverse climatic conditions makes them a crucial tool for climate-resilient agriculture. By complementing traditional practices and biological control methods, nanotechnology ensures a holistic approach to managing mango diseases. However, realizing these benefits requires careful implementation, supported by robust scientific research and tailored application strategies. The future of mango cultivation depends on leveraging these advancements to sustain productivity while protecting ecosystems.

While nanoparticles offer numerous benefits, challenges remain in ensuring their safe and effective use. The potential risks to non-target organisms, environmental accumulation, and long-term ecological impacts must be thoroughly addressed. Additionally, the high costs of nanoparticle development and the lack of awareness among farmers may hinder widespread adoption. Regulatory frameworks that establish clear guidelines for the safe use of nanomaterials are critical to promoting their responsible integration into agriculture. Farmer training programs are equally essential to enhance understanding and foster trust in these innovations. A balanced approach, combining the advantages of nanoparticles with eco-friendly practices and regulatory oversight, is key to achieving sustainability. By addressing these challenges and investing in research and education, nanotechnology can become a cornerstone of sustainable mango disease management in the face of climate change (Figs. 21.1 and 21.2).

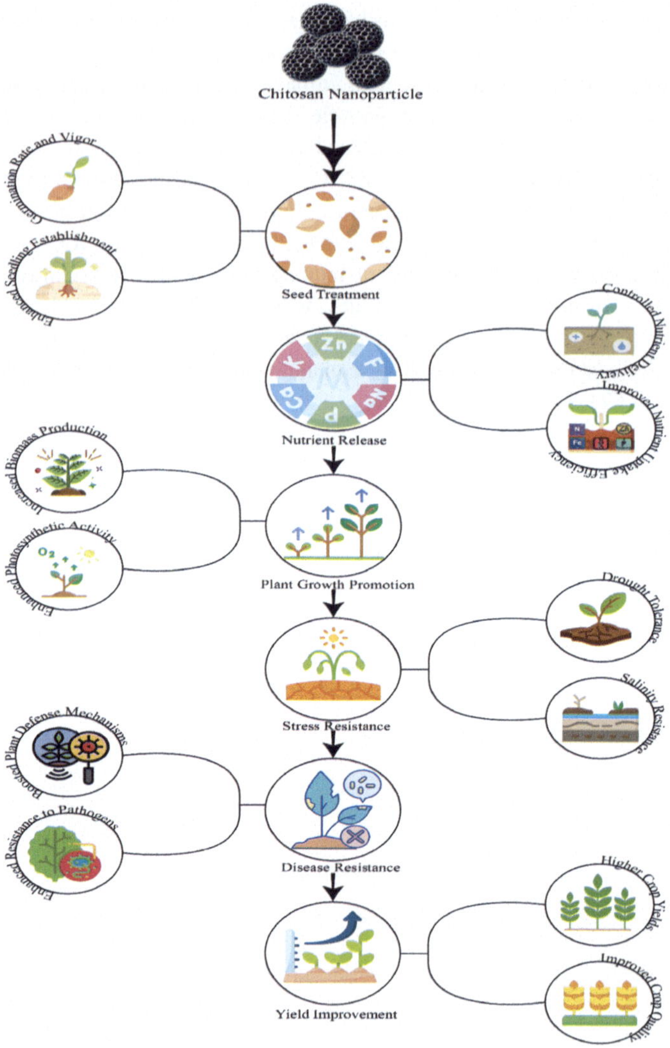

Fig. 21.1 Chitosan based nanoparticles application can impact yield of the crop and promote defence against diseases

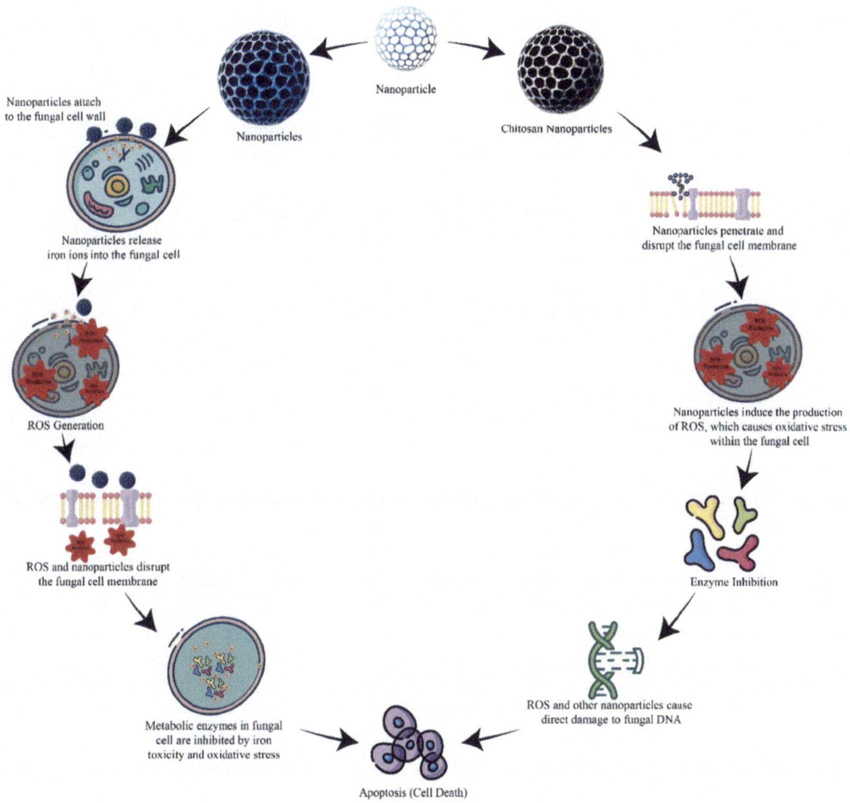

Fig. 21.2 Antifungal mode of action of chitosan-based nanoparticles wherein NPs induce the production of ROS which causes the oxidative stress within the fungal cell

References

Abdollahdokht D, Gao Y, Faramarz S, Poustforoosh A, Abbasi M, Asadikaram G, Nematollahi MH (2022) Conventional agrochemicals towards nano-biopesticides: an overview on recent advances. Chem Biol Technol Agric 9(1):13

Abdurrehman TM, Subhani MN, Malik MT, Hasnain A, Abbas A, Naqvi SAH (2024) Unveiling intricacies of mango malformation dynamics caused by *Fusarium solani* through extensive field surveys and laboratory, epidemiology, and model-based insights. Appl Fruit Sci 66(2):677–688

Adikaram N (2024) Diseases of major tropical and subtropical fruit plants. In: Agrios' plant pathology. Academic Press, pp 807–821

Alavi M, Kennedy JF (2021) Recent advances of fabricated and modified Ag, Cu, CuO and ZnO nanoparticles by herbal secondary metabolites, cellulose and pectin polymers for antimicrobial applications. Cellulose 28:3297–3310

Ameen M, Mahmood A, Ahmad M, Mansoor Javaid M, Nadeem MA, Asif M, Khan BA (2023) Impacts of climate change on fruit physiology and quality. In: Climate-resilient agriculture: crop responses and agroecological perspectives, vol 1. Springer International Publishing, Cham, pp 93–124

Angelotti F, Hamada E, Bettiol W (2024) A comprehensive review of climate change and plant diseases in Brazil. Plants 13(17):2447

Carmona M, Sautua F, Pérez-Hérnandez O, Reis EM (2020) Role of fungicide applications on the integrated management of wheat stripe rust. Front Plant Sci 11:733

Dofuor AK, Quartey NKA, Osabutey AF, Antwi-Agyakwa AK, Asante K, Boateng BO, Ninsin KD (2023) Mango anthracnose disease: the current situation and direction for future research. Front Microbiol 14:1168203

Eivazzadeh-Keihan R, Noruzi EB, Chidar E, Jafari M, Davoodi F, Kashtiaray A, Mahdavi M (2022) Applications of carbon-based conductive nanomaterials in biosensors. Chem Eng J 442:136183

Farasati Far B, Naimi-Jamal MR, Sedaghat M, Hoseini A, Mohammadi N, Bodaghi M (2023) Combinational system of lipid-based nanocarriers and biodegradable polymers for wound healing: an updated review. J Funct Biomater 14(2):115

Gelaye Y (2024) A review of storage temperature and relative humidity effects on shelf life and quality of mango (*Mangifera indica* L.) fruit and implications for nutrition insecurity in Ethiopia. Open Agric 9(1):20220296

Gullino ML, Albajes R, Al-Jboory I, Angelotti F, Chakraborty S, Garrett KA, Stephenson T (2022) Climate change and pathways used by pests as challenges to plant health in agriculture and forestry. Sustainability 14(19):12421

Gupta S, Rautela P, Azad CS, Singh KP (2020) Major diseases of mangoes and their management. In: Diseases of fruits and vegetable crops. Apple Academic Press, pp 125–144

Hassan HU, Raja NI, Abasi F, Mehmood A, Qureshi R, Manzoor Z, Proćków J (2022) Comparative study of antimicrobial and antioxidant potential of *Olea ferruginea* fruit extract and its mediated selenium nanoparticles. Molecules 27(16):5194

Hendrichs J, Vreysen MJB, Enkerlin WR, Cayol JP (2021) Strategic options in using sterile insects for area-wide integrated pest management. In: Sterile insect technique. CRC press, pp 841–884

Huang KM, Guan Z, Hammami A (2022) The US fresh fruit and vegetable industry: an overview of production and trade. Agriculture 12(10):1719

Huda-Shakirah AR, Mohamed Nor NMI, Zakaria L, Leong YH, Mohd MH (2022) Lasiodiplodia theobromae as a causal pathogen of leaf blight, stem canker, and pod rot of *Theobroma cacao* in Malaysia. Sci Rep 12(1):8966

Iqbal S, Atiq M, Fayyaz M, Zakria M, Rajput N A, Kachelo GA, Mehmood A (2024) Powdery mildew of mango: current status, perspective and emerging tools for management. Agric Sci J 6(1):92–101

Islam T, Danishuddin, Tamanna NT, Matin MN, Barai HR, Haque MA (2024) Resistance mechanisms of plant pathogenic fungi to fungicide, environmental impacts of fungicides, and sustainable solutions. Plants 13(19):2737

Jagtap GP, Talukdar D, Dey U (2020) Major seed-borne diseases in important vegetables: symptomatology, aetiology and its economic importance. Seed-Borne Dis Agric Crop: Detect Diagn Manag 543–575

Jeevanantham S, Praveen A, Livitha R, Balamurugan K (2024) Post harvest anthracnose of mango caused by *Colletotrichum gloeosporioides*: a review. Arch Curr Res Int 24(2):106–115

Jomova K, Alomar SY, Alwasel SH, Nepovimova E, Kuca K, Valko M (2024) Several lines of antioxidant defense against oxidative stress: antioxidant enzymes, nanomaterials with multiple enzyme-mimicking activities, and low-molecular-weight antioxidants. Arch Toxicol 98(5):1323–1367

Joudeh N, Linke D (2022) Nanoparticle classification, physicochemical properties, characterization, and applications: a comprehensive review for biologists. J Nanobiotechnology 20(1):262

Khan NA, López-Maldonado EA, Majumder A, Singh S, Varshney R, López JR, Aljundi IH (2023) A state-of-art-review on emerging contaminants: environmental chemistry, health effect, and modern treatment methods. Chemosphere 344:140264

Khaskheli MI (2020) Mango diseases: impact of fungicides. Horticultural Crops 143

Khursheed A, Rather MA, Jain V, Rasool S, Nazir R, Malik NA, Majid SA (2022) Plant based natural products as potential ecofriendly and safer biopesticides: a comprehensive overview of

their advantages over conventional pesticides, limitations and regulatory aspects. Microb Pathog 173:105854
Kiloes AM, Azizan FA, Checco J, Joyce D, Abdul Aziz A (2022) What do consumers want in fresh mangoes? A systematic literature review. Int J Food Sci Technol 57(3):1473–1492
Lengai GM, Fulano AM, Muthomi JW (2022) Improving access to export market for fresh vegetables through reduction of phytosanitary and pesticide residue constraints. Sustainability 14(13):8183
Li W, Deng Y, Ning Y, He Z, Wang GL (2020) Exploiting broad-spectrum disease resistance in crops: from molecular dissection to breeding. Annu Rev Plant Biol 71(1):575–603
Mahmoud GAE, Gaber SH (2024) Nanotechnology applications in postharvest disease management. In: Recent advances in postharvest technologies: postharvest applications, vol 2. Springer Nature Switzerland, Cham, pp 225–245
Malik S, Muhammad K, Waheed Y (2023) Emerging applications of nanotechnology in healthcare and medicine. Molecules 28(18):6624
Misiou O, Koutsoumanis K (2022) Climate change and its implications for food safety and spoilage. Trends Food Sci Technol 126:142–152
Odetayo T, Tesfay S, Ngobese NZ (2022) Nanotechnology-enhanced edible coating application on climacteric fruits. Food Sci Nutr 10(7):2149–2167
Palanivel H, Shah S (2021) Unlocking the inherent potential of plant genetic resources: food security and climate adaptation strategy in Fiji and the Pacific. Environ Dev Sustain 23(10):14264–14323
Prasad P, Bhardwaj SC, Thakur RK, Adhikari S, Gangwar OP, Lata C, Kumar S (2021) Prospects of climate change effects on crop diseases with particular reference to wheat. J Cereal
Rahaman M, Chowdhury M, Rahman MA, Ahmed H, Hossain M, Rahman MH, Biswas M (2023) A deep learning based smartphone application for detecting mango diseases and pesticide suggestions. Int J Comput Digit Syst 13(1):1–1
Rajwade JM, Chikte RG, Paknikar KM (2020) Nanomaterials: new weapons in a crusade against phytopathogens. Appl Microbiol Biotechnol 104(4):1437–1461
Rameez M, Khan N, Ahmad S, Ahmad MM (2024) Bionanocomposites: a new approach for fungal disease management. Biocatal Agric Biotechnol 103115
Rana RA, Siddiqui MN, Skalicky M, Brestic M, Hossain A, Kayesh E, Islam T (2021) Prospects of nanotechnology in improving the productivity and quality of horticultural crops. Horticulturae 7(10):332
Rana VS, Sharma S, Rana N, Sharma U, Patiyal V, Banita, Prasad H (2022) Management of hailstorms under a changing climate in agriculture: a review. Environ Chem Lett 20(6):3971–3991
Ranganna G, Gavhale KN, Singh A, Mali VS, Shingade DM, Khatri R (2024) The role of biotechnology in the future of fruit crop production: a review. J Adv Biol Biotechnol 27(9):967–987
Ranjan P, Brahmi P, Tyagi V, Ranjan JK, Srivastava V, Yadav SK, Singh K (2022) Global interdependence for fruit genetic resources: status and challenges in India. Food Security 14(3):591–619
Ren P, Néron V, Rossi S, Liang E, Bouchard M, Deslauriers A (2020) Warming counteracts defoliation-induced mismatch by increasing herbivore-plant phenological synchrony. Glob Change Biol 26(4):2072–2080
Rubayet MT, Hossain MM (2024) Climate change and its impacts on disease dynamics in major cereal crops. In: Climate change and soil-water-plant nexus: agriculture and environment. Springer Nature Singapore, Singapore, pp 265–295
Samwel J, Msogoya T, Kudra A, Mtui HD, Baltazari A, Sullivan JA, Mwatawala MW (2020) Pre-harvest field application effects of hexanal formulation on fruit losses in mango grown in eastern Tanzania. J Crop Improv 34(3):366–377
Sarmah K, Borah R, Boruah S, Sarmah D (2023) Nanopesticides: revolutionizing pest management with nanotechnology. Biol Forum Int J 15(7):210–218
Sayed FAZ, Eissa NG, Shen Y, Hunstad DA, Wooley KL, Elsabahy M (2022) Morphologic design of nanostructures for enhanced antimicrobial activity. J Nanobiotechnology 20(1):536

Scortichini M (2022) Sustainable management of diseases in horticulture: conventional and new options. Horticulturae 8(6):517

Shah IH, Manzoor MA, Jinhui W, Li X, Hameed MK, Rehaman A, Chang L (2024) Comprehensive review: effects of climate change and greenhouse gases emission relevance to environmental stress on horticultural crops and management. J Environ Manage 351:119978

Shahbaz M, Akram A, Raja NI, Mukhtar T, Mehak A, Fatima N, Abasi F (2023) Antifungal activity of green synthesized selenium nanoparticles and their effect on physiological, biochemical, and antioxidant defense system of mango under mango malformation disease. PLoS ONE 18(2):e0274679

Shahbaz M, Seelan JSS, Abasi F, Fatima N, Mehak A, Raza MU, Proćków J (2024) Nanotechnology for controlling mango malformation: a promising approach. J Biomol Struct Dyn 1–21

Shahid A, Ikhlaq M, Noreen A, Amjad MA, Bashir MA, Rasool F, Latif A (2024) Characterization of different mango (*Mangifera indica*) varieties of South Punjab

Shanka D (2020) Roles of eco-friendly low input technologies in crop production in sub-Saharan Africa. Cogent Food Agric 6(1):1843882

Singh BK, Delgado-Baquerizo M, Egidi E, Guirado E, Leach JE, Liu H, Trivedi P (2023) Climate change impacts on plant pathogens, food security and paths forward. Nat Rev Microbiol 21(10):640–656

Singh D, Afaque M (2021) Postharvest diseases of cucurbitaceous fruits and their management. In: Postharvest handling and diseases of horticultural produce. CRC Press, pp 363–380

Sossah FL, Aidoo OF, Dofuor AK, Osabutey AF, Obeng J, Abormeti FK, Borgemeister C (2024) A critical review on bacterial black spot of mango caused by *Xanthomonas citri pv. mangiferae indicae*: current status and direction for future research. Forest Pathol 54(3):e12860

Strickland DA (2023) Refining chemical management of Apple Powdery Mildew in New York State (Doctoral dissertation, Cornell University)

Tariq M, Mohammad KN, Ahmed B, Siddiqui MA, Lee J (2022) Biological synthesis of silver nanoparticles and prospects in plant disease management. Molecules 27(15):4754

Tengsetasak P, Tongkoom K, Yomkerd J, Susawaengsup C, Khongdee N, Chatsungnoen T, Bhuyar P (2024) Sustainable strategies for fresh mangosteen: adapting to climate challenges. Earth Syst Environ 1–19

Tiwari AK, Gupta MK, Pandey G, Tilak R, Narayan RJ, Pandey PC (2022) Size and zeta potential clicked germination attenuation and anti-sporangiospores activity of PEI-functionalized silver nanoparticles against COVID-19 associated Mucorales (*Rhizopus arrhizus*). Nanomaterials 12(13):2235

Ullah MA, Kiloes AM, Aziz AA, Joyce DC (2024) Impact of factors contributing to internal disorders of mango (*Mangifera indica* L.) fruit—a systematic literature review. Scientia Horticulturae 331:113150

Umair Raza M, Abasi F, Shahbaz M, Ehsan M, Seerat W, Akram A, Proćków J (2023) Phytomediated silver nanoparticles (AgNPs) embellish antioxidant defense system, ameliorating HLB-diseased 'Kinnow' Mandarin plants. Molecules 28(5):2044

Umar UD, Ahmed N, Zafar MZ, Rehman A, Naqvi SAH, Zulfiqar MA, Marc RA (2022) Micronutrients foliar and drench application mitigate mango sudden decline disorder and impact fruit yield. Agronomy 12(10):2449

Vuppalapati C (2023) Specialty crop mangoes. In: Specialty crops for climate change adaptation: strategies for enhanced food security by using machine learning and artificial intelligence. Springer Nature Switzerland, Cham, pp 419–514

Wardhan H, Das S, Gulati A (2022) Banana and mango value chains. In: Agricultural value chains in India: ensuring competitiveness, inclusiveness, sustainability, scalability, and improved finance, pp 99–143

Yadav S, Korat JR, Yadav S, Mondal K, Kumar A, Kumar S (2023) Impacts of climate change on fruit crops: a comprehensive review of physiological, phenological, and pest-related responses. Int J Environ Clim Chang 13(11):363–371

Yuen KW, Hanh TT, Quynh VD, Switzer AD, Teng P, Lee JSH (2021) Interacting effects of land-use change and natural hazards on rice agriculture in the Mekong and Red River deltas in Vietnam. Nat Hazard 21(5):1473–1493

Chapter 22
Ecology of *Ceratocystis* Induced Infections with an Emphasis on MSDS

Muhammad Farhan and Rafia Kiran

Abstract Mango Sudden Death Syndrome (MSDS) is one of the most devastating disease brought on by *Ceratocystis fimbriata*. A vast variety of plant species, mostly woody and agricultural ones, are host to *C. fimbriata*. It is known that 74 species belonging to 55 genera and 35 plant families are infected by *C. fimbriata*. The pathogen enters the tree by wounds from insect feeding, storm damage, or pruning. It avoids the tree's natural defense mechanism as a result. *C. fimbriata*. Because the pathogen obstructs the tree's vascular tissues and interferes with water transport, it is extremely aggressive and causes the tree to wilt, dieback, and eventually die. The ecology and management of *C. fimbriata* are complicated since it can infect crops other than mangos. Therefore, management necessitates a thorough understanding of the pathogen's ecology, including spore dispersion processes, environmental conditions, and the part insect vectors play in the infection process. The three factors have the potential to lessen the impact and spread of MSDS, improving mango production outcomes.

Keywords Mango Sudden Death Syndrome · *Ceratocystis fimbriata* · Mango Wilt · *Ceratocystis manginecans* · *Ceratocystis omanensis* · *Lasiodiplodia theobromae*

22.1 Introduction

Ceratocystis fimbriata is a highly pathogenic fungus that negatively impacts a wide range of economically significant trees and crops, including eucalyptus, *Colocasia esculenta*, coffee, cacao, sweet potatoes, mangoes, and rubber tree. Mango Sudden Death Syndrome is one of the most devastating disease brought on by *C. fimbriata*. This disease suffered severally substantial losses in mango output on plantations located within the tropical and subtropical areas of the world. It has been observed that

M. Farhan (✉) · R. Kiran
Department of Plant Pathology, Faculty of Agricultural Sciences and Technology, Bahauddin Zakariya University, Multan, Pakistan
e-mail: muhammadfarhan@student.bzu.edu.pk

S. A. H. Naqvi et al. (eds.), *Climate Change and Mango Production*,
https://doi.org/10.1007/978-3-031-87035-4_22

C. fimbriata caused devastating losses in Pakistani mango plantations, particularly in the Punjab and Sindh regions. According to Guimarães et al. (2021) Mango Sudden Death Syndrome was the primary cause of tree losses on numerous farms in Pakistan and Brazil, with losing as much as 60% of their mango trees. Similarly, Oman had a sharp drop in mango production in 1999. More than 60% of mango trees in the Al Batinah region of northern Oman were impacted by this disease after it first appeared (Al Subhi et al. 2006). Knowing the ecology of *C. fimbriata* creating effective control strategies to prevent its spread requires knowledge of the interactions that *C. fimbriata* has with its surroundings, host plants, and vectors.

In the case of MSDS, bark beetles, among other insect vectors, play a fundamental role in transporting *C. fimbriata* spores to healthy mango trees where entry of the pathogen occurs through wounds or natural openings. The pathogen's life cycle is closely linked to its survival in plant debris, decaying wood, and the soil, where it remains viable as spores or mycelium for a very long time. Spores are also dispersed by wind, water, and human activities, making *C. fimbriata* extremely effective in its transmission within and between orchards. Mangos thrive in warm, humid conditions that are also ideal for the growth and spread of this pathogen. Rainfall, temperature, and insect activity are some of the environmental factors that contribute to the disease's rapid spread. After entering the host, *C. fimbriata* colonizes the vascular system, which blocks water transport and causes rapid tree decline (Harrington 2013). Since characteristics like mean annual temperature (bio1), mean diurnal temperature range (bio2), and precipitation of the driest month (bio14) play a significant impact in the potential spread of MSDS, it poses a threat to mango production, particularly in northern Oman. Disease outbreaks are a real possibility due to the predicted future climate. Therefore, in order to minimize damage, it is imperative to develop strategic measures for the protection of mango production regions that are at risk (Al-Ruheili et al. 2021). Because the pathogen is so adaptable and has such a wide host range, this chapter provides an overview of some of the ecological factors that influence the spread of the disease, and places special emphasis on the importance of MSDS in managing this disease (Tables 22.1 and 22.2).

22.2 Taxonomy and Description of Mango Sudden Death Syndrome

C. fimbriata was the species for which Ellis and Halsted first created the genus *Ceratocystis* in 1890 after it was discovered to be associated to black rot of sweet potatoes in the USA (Halsted 1890). *Ophiostoma, Sphaeronema, Enddoconidiophora,* and *Ceratocystiopsis* were among the numerous taxa in which *C. fimbriata* was analyzed in the years that followed (Upadhyay 1981). The comparable physical characteristics of *Ceratocystis* and related taxa contributed significantly to the taxonomic problems surrounding these fungi. The aforementioned genera of fungi are characterized by lengthy, beaked ascomata that produce sticky masses of spores at their apices.

Table 22.1 Frist reports of different species of *Ceratocystis* causing Mango Sudden Death Syndrome/mango blight/mango wilt in different geographic regions

S. No.	Disease name	*Ceratocystis* species	Country	Year	References
1	Mango Sudden Death Syndrome	*Ceratocystis fimbriata*	Brazil	1960	Viégas (1960)
2	Mango Sudden Death Syndrome	*Ceratocystis fimbriata*	Pakistan	1996	Malik et al. (2004)
3	Mango Sudden Death Syndrome	*Ceratocystis fimbriata*	India	1997	Ploetz and Prakash (1997)
4	Mango Sudden Death Syndrome	*Ceratocystis fimbriata*	Oman	1998	Al Adawi (2002)
5	Mango sudden decline syndrome	*Ceratocystis omanensis*	Oman	2006	Al Subhi et al. (2006)
6	Mango wilt disease syndrome	*Ceratocystis manginecans*	Oman and Pakistan	2007	Van Wyk et al. (2007)
7	Mango blight syndrome	*Ceratocystis fimbriata*		2008	Batista et al. (2008)
8	Mango blight syndrome	*C. mangicola* and *C. mangivora*	Brazil	2011	Van Wyk et al. (2011)
9	*Ceratocystis* wilt syndrome	*Ceratocystis acaciivora*	Pakistan, Oman and Brazil	2015	Oliveira et al. (2015)
10	Mango wilt syndrome	*Ceratocystis fimbriata*	China	2017	Zhang et al. (2017)
11	Mango wilt syndrome	*Ceratocystis fimbriata*	India	2018	Shukla et al. (2018)
12	Mango wilt syndrome	*Ceratocystis fimbriata*	Indonesia	2023	Pratama et al. (2023)

Typically, evanescent asci form ascospores (Elliott 1923). The taxonomy enigma surrounding *Ceratocystis* and *Ophiostoma* was ultimately cleared up with the availability of DNA sequence data that offered phylogenetic insights into their relationship (Hausner et al. 1993). These taxa are unrelated, as demonstrated by the first phylogenetic trees developed by Spatafora and Blackwell (1994). The current knowledge that *Ophiostoma* is found in the Ophiostomatales in the Sordariomycetidae and *Ceratocystis* is found in the Ceratocystidaceae (Microascales) in the Hypocreomycetidae is based on a substantial body of evidence (Réblová et al. 2011; De Beer et al. 2013). In 1960s, *C. fimbriata* was detected to cause wilt disease of mango in Brazil. At that time disease was named as *Ceratocystis* wilt of mango (Viégas 1960).

Initial reports during 1970s from Pakistan and India linked the disease to widespread rots of the roots and stems, which were attributed to the fungal diseases *Rhizoctonia solani* and *Fusarium oxysporum*. Those investigations did not, however, specifically identify the primary causative organism, indicating the complexity of the problem (Ploetz 2003). Despite this, a general reduction in mango trees that

Table 22.2 Morphological and microscopic characteristics of three species of *Ceratocystis* involved in Mango Sudden Death

Morphological characteristics	*Ceratocystis manginecans*	*Ceratocystis fimbriata*	*Ceratocystis omanensis*
Colony color	Grayish olive in colour on 2% MEA	Initially white but turned to olive-grey with white to grey at the margin after culturing for 5–7 days	White to cream when it first grows on culture media. As the colony matures, it may turn light brown to dark brown
Hyphae	Smooth and segmented	Septate and hyaline	Smooth-edged hyphae with septa
Ostiolar hyphae	Hyaline, divergent (42–45), (59–69) μm long	Divergent hyaline and aseptate ostiolar hyphae, ranging from 15.5 to 53.5 μm in length, were diverse	Divergent two types of hyphae were present, one with smooth walls and another with a granular appearance
Odour	Banana like	Banana like	Banana like
Ascomatal bases	Globose, black (153–192), (254–281 μm in diameter)	Globose to sub-globose in shape, and they are often darkly pigmented (100–200 μm in diameter)	Globose, black and covered with short spines
Ascospores	Hyaline, hat-shaped, 3–4 μm in length, 4–5 μm wide excluding sheath, 7–8 μm wide including sheath	Hyaline hat-shaped ascospores, with sizes ranging from 4.4–5.2 × 3.5–3.7 μm	Hat-shaped
Conidiophores	**Primary conidiophores** Phialidic, lageniform, hyaline (72–81), (109–144) μm **Secondary conidiophores** Tube-like, flaring at mouths, short, hyaline (59–65), (77–84) μm long, (5–8) μm wide at bases	Simple or branched (20–80 μm in height)	Hyaline (transparent) and can appear simple or branched (30–200 μm)

(continued)

Table 22.2 (continued)

Morphological characteristics	*Ceratocystis manginecans*	*Ceratocystis fimbriata*	*Ceratocystis omanensis*
Conidia	**Primary conidia** Hyaline, cylindrical (15–)23–29(–33) μm in length, 3–6 μm wide **Secondary conidia** Hyaline, barrel-shaped (8–)9–11(–12) μm in length, 5–7(–8) μm wide	Hyaline cylindrical and barrel-shaped conidia The cylindrically shaped conidia were approximately 18.3–28.7 × 4.2–4.5 μm, while the barrel-shaped conidia ranged from 8.6–12.7 × 4.2–4.6 μm	Cylindrical conidia barrel-shaped conidia in a chain
Aleuroconidia	–	Aleuroconidia is arranged singly or basipetal in chains, thick-walled, pale brown to brown in colour, ovoid to obpyriform in shape, and has a smooth surface	–
Chlamydospores	Brown, thick-walled, globose to sub-globose	Formed by thickening of the walls of endoconidia	Globose or oval-shaped
Perithecial necks	–	Slender perithecial necks, measuring from around 282.7 to 776.3 μm in length, 12.8–28.7 μm wide at the tip, and 26.5–44.4 μm at the base, were also similar in colour	–
Perithecia	–	Long-necked perithecia, measuring from 133.2 to 281.5 μm in length and between 128.2 and 283.0 μm in width, were dark brown to black in colour and globose shaped	All isolates produce perithecia
References	Van Wyk et al. (2007)	Al Subhi et al. (2006)	Syazwan et al. (2021)

included abrupt wilting and tree death was widely documented in Oman in the 1990s. Consistent isolation of *C. fimbriata* from affected trees marks a significant advancement in our knowledge of the disease's etiology. This fungus originated as a wilt disease on other fruit trees and later emerged as the main causative agent of MSDS. As a secondary pathogen, *Lasiodiplodia theobromae* is best regarded as such (Al Adawi et al. 2006). In the tropics and subtropics, *L. theobromae* is one of the most prevalent opportunistic tree diseases (Mehl 2017). As with many other species of *Botryosphaeria*, the fungus is the asexual form of *Botryosphaeria rhodina* (Sutton 1980) and is found in asymptomatic tree tissue as an endophyte (Johnson and Kotze 1994). Additionally, another species of *Ceratocystis* was discovered in a study using DNA sequences and morphology. Analogous morphological analyses revealed a resemblance between *C. moniliformis* and the fungus found on dying mango trees in Oman. Ascospores in the shape of hats, disc-shaped plates at the base of ascomatal necks, and spines on ascomatal bases are characteristics shared by both fungi. It was confirmed that the sequences of ITS1-2, 5.8S RNA, b-tubulin, and Transcription Elongation Factor (EF1-a) were distinct for the fungus from Oman and for *C. moniliformis* or any other close related. It was established that the fungus belonged to one of four sub-clades that were phylogenetically well-supported. Isolates of *C. bhutanensis*, *C. moniliformis*, and *C. moniliformopsis* made up the other subclades. The *Ceratocystis* sp., linked to mango wounds in Oman was determined to represent a novel species, which has been named *Ceratocystis omanensis* sp. nov. based on morphological characteristics and differences in DNA sequences for three gene regions (Al Subhi et al. 2006). A unique species within the *C. fimbriata* sensu lato species complex has been identified from mango isolates from Pakistan and Oman. *Ceratocystis manginecans* sp. nov. is the name of the fungus, which is different even though it resembles existing isolates from mangos in Brazil (Van Wyk et al. 2007). A study was conducted with the objective of analyzing *Ceratocystis* isolates from infected mango trees in Brazil causing mango blight. The identification of the isolates was based on the sequencing data from ITS1 + 5.8S + ITS2 rDNA, a portion of the Beta-tubulin 1 gene, and a portion of TEF1-alpha gene. Within *C. fimbriata* s.l., the Brazilian isolates comprised two clearly characterized and separate clades. These were also different from *C. manginecans*, which in Oman and Pakistan causes the same sickness linked to *H. mangiferae*. Combining morphological traits with sequence comparisons enables the description of isolates that belong to both clades as *C. mangicola* sp. nov. and *C. mangivora* sp. nov. (Van Wyk et al. 2011).

Various countries have identified differences in the morphology of *Ceratocystis* isolates linked to Mango Sudden Death Syndrome. Gum exudation, necrotic lesions in the roots and stems, and dark streaks in the vascular tissue are some of the visible symptoms of some of the infected mango trees (Van Wyk et al 2013). Hyaline ascospores, cylindrical conidia, and globose to subglobose perithecia were all present in the same characteristics in isolates from Pakistan and Oman (Van Wyk et al. 2007). However, spore shape and colony growth varied slightly for the isolates from Brazil and Egypt, probably due to host-specific and environmental variables (Gomdola et al. 2022). Because they show the adaptability and pathogenicity of various strains of *Ceratocystis* infecting mango trees worldwide, these morphological distinctions are important for identifying and describing the geographic spread of MSDS (Al Subhi et al. 2006). Additional studies concentrated on environmental variables that had exacerbated the disease, such as high temperatures, dampness, or heavy rain. Stress factors like dryness and low soil nourishment made the tree more vulnerable. Soils with compromised drainage were more conducive to the pathogen's growth. By 2013, researchers had focused on the intricacy of MSDS, characterizing it as the result of a complex interaction between multiple fungi, environmental stressors, and management practices, as opposed to a single pathogen acting alone. This represented a move toward integrated disease management strategies (Al Adawi et al. 2013).

22.3 Host Range of *Ceratocystis fimbriata*

A vast variety of plant species, mostly woody and agricultural ones, are host to *C. fimbriata*. It is known that 74 species belonging to 55 genera and 35 plant families are infected by *C. fimbriata*. These data imply that the pathogen is extremely widespread. *C. fimbriata*, which affects trees and crops, has been detected in 46 different countries (Alizadeh et al. 2024). A number of plant species are affected, including the mango *Ipomea batatas* (Weber and Stahmann 1964), *Acacia mangium* (Tarigan et al. 2011), *Ficus carica* (Harrington et al. 2011), *Tectona grandis* (Firmino et al. 2012), *Metrosideros polymorpha* (Fortini et al. 2019), *Mangifera indica*, sweet potato (*I. batatas*), cocoa (*Theobroma cacao*), coffee (*Coffea* spp.), taro (*C. esculenta*), rubber tree (*Hevea brasiliensis*), and plane tree (*Platanus* spp.). This fungus has considerably led to a lot of damage in agriculture and forestry because of its vast capability to infect various types of woody plants and crops (Harrington and Wingfield 1998). Due to its genetic diversity and plasticity, which allow it to infect diverse plant species from different families, including crops, ornamental plants, and forest trees, it is incredibly adaptable, resistant, and has a wide host range. Its pathogenic strategies, which include latent infections and the synthesis of enzymes, enable it to elude host defenses and, as a result, adapt well to a variety of situations. Insect vectors and human activity also contribute to the proliferation of species, including mango, sweet potato, coffee, cocoa, and eucalyptus (Marincowitz et al. 2020).

(a) **Mango**: Mango Sudden Death Syndrome is a deadly disease that is brought on by *C. fimbriata*, which is found in mangos (*Mangifera indica*), that is the main concern of this chapter. As soon as the pathogen enters the tree's vascular tissues, afflicted branches and entire trees quickly die. Shoots at their tips are the first obvious indicators when yellowing of the leaves progresses to wilting and browning. The bark develops dark streaks and occasionally a sticky discharge is seen. It causes a complete obstruction of the vascular system, which in a matter of weeks results in a serious deterioration and death. Since a complete mango orchard is destroyed, especially in the areas where the disease has established a stronghold, the losses are extremely noteworthy. Mango farmers, particularly those in tropical and subtropical locations, face a significant threat from the disease due to its rapid spread and stubborn nature (Naqvi and Perveen 2015).

(b) **Sweet potato**: "Black rot" on sweet potatoes (*I. batatas*) is one of the most pathogenic diseases caused by *C. fimbriata*. Although it may also infect the stem and leaf sections, it usually targets the roots. Early signs like withering or yellowing leaves and reduced growth could be confused with other physiological stressors that the plant is experiencing. Lesions on the stems that are dark and sunken, and—more significantly—blackened and necrotic root tissue. The necrosis may lead to the entire degeneration of root structure, and infected roots may have a foul smell, indicating tissue degradation. These diseases might result in lower-quality sweet potato crops as well, which is economically detrimental to the growers. Moisture and temperature are two environmental elements that impact how severe black rot is. Because of the damp surroundings, high humidity fosters a setting that makes the agent more likely to spread. Consequently, crop rotation, irrigation control, and cultivars resistant to disease are among the management measures used to lessen the effects of *C. fimbriata* in sweet potatoes (Tyson et al. 2020).

(c) **Cacao**: *C. fimbriata* is one of these infections that causes "cacao vascular wilt" or "cacao black pod disease," which are serious diseases that affect cacao. The pathogen mostly causes wilting and mortality in infected trees by attacking the vascular tissues. The first signs are drooping and yellowing of the leaves, which are followed by browning and withering of the stems and pods. As the disease progresses, dark streaks may show up in the vascular tissue, and lesions on the stems and pods cause them to deteriorate. Not only does this deterioration lead to lower cacao bean yields and quality, but it also presents a serious financial risk to cocoa growers, particularly in areas where the disease is prevalent. Conditions like as temperature and humidity interact with disease severity because of the manner by which *C. fimbriata* spreads and affects the plant. Using resistant cacao cultivars, adhering to hygiene regulations, and keeping an eye out for infection symptoms early on can all help reduce the amount of cacao disease that is contracted (Engelbrecht et al. 2007).

(d) **Coffee**: One of the most significant pathogens for coffee (*Coffea* spp.) is *C. fimbriata*, which is responsible for a condition known as "Coffee trunk canker". The pathogen causes drooping, yellowing of the leaves, and eventual death of the afflicted plants by infecting the vascular tissues of the coffee. The initial symptoms can be withered leaves and slowed development, which can be mistaken for other environmental stressors. Later in the disease's course, vascular tissue may start to show streaks, and blackened necrotic lesions are commonly discovered in the roots. The plant may be severely unable to absorb water and nutrients as a result of this degradation, which would significantly reduce yields and coffee bean quality. Climate has an impact on how severe it is. Increased humidity encourages the bacterium to proliferate. Because diseased litter in the soil might act as a source of inoculum, the issue keeps getting worse. Using disease-resistant coffee cultivars, keeping plantations clean, and removing sick plant parts to stop further spread are additional efficient ways to manage the disease-causing pathogen *C. fimbriata* (Bhat et al. 2002).

(e) **Eucalyptus**: The well-known eucalyptus (*Eucalyptus* spp.) pathogen *C. fimbriata* is responsible for "eucalyptus canker" or "*Ceratocystis* wilt" (Roux et al. 2000). Early indications of the disease, which usually affects the vascular tissues of eucalyptus trees, include yellowing leaves, wilting, and branch dieback. Early in the disease, water-soaked lesions are often the first signs of infection on the bark; however, they can grow and eventually encircle parts of the stems with dark cankers. Due to the disease, branches may perish, and in acute situations, the infection might fully harm the whole tree. Temperature and humidity levels can affect how severe the illness is, and the pathogen often grows more readily in moist environments. An inoculum source also occurs if contaminated plant detritus is present in the soil. Proper field sanitation, removal of diseased plant materials, and early emergence of infection indicators are effective strategies for the management of *C. fimbriata* on eucalyptus. Another way to control it is to plant resistant eucalyptus cultivars (Roux and Wingfield 2009).

(f) **Rubber tree**: One of the most significant diseases of rubber trees (*H. brasiliensis*) is thought to be caused by *C. fimbriata*, which is frequently referred as "rubber tree wilt" or "*Ceratocystis* wilt." The disease mostly affects the rubber tree's avascular tissues, which causes severe wilting, leaf yellowing, and ultimately the death of affected trees. The early symptoms, which include abrupt leaf drooping and reduced growth, are usually comparable to those of nutrient deficits or water stress. Dark streaks may appear in the vascular tissue as the disease progresses, and lesions in the bark may cause necrosis and branch dieback. Given that the fungus grows more active in moist environments, environmental factors, particularly temperature and humidity, may have an impact on infectiousness. The fungus can cause a sharp decline in latex yields from infected rubber trees, posing a serious financial danger to rubber farms. Techniques of Management Sanitation, planting resistant rubber tree varieties, and defoliation and eradication of infected plant components are examples of control measures (Valdetaro et al. 2015; Fadli et al. 2022).

22.4 Rise of *Ceratocystis* Diseases

The emergence and spread of *C. fimbriata* infections can be attributed to a number of interconnected variables, including changes in the environment, agricultural practices, and the pathogen's biology. Large-scale plantations and monoculture farming are common features of modern agricultural systems, which have produced the perfect environment for the development and growth of this kind of disease. In these farming techniques, this also frequently reduces plant resistance, increasing the susceptibility of crops to infection. Furthermore, once the disease is introduced to previously uninfected areas, the free movement of agricultural products and materials between regions facilitates its easy spread (Alizadeh et al. 2024).

22.4.1 Agricultural Practices

Large-scale monoculture farming of genetically similar crops, like eucalyptus, mango, or rubber, frequently increases the crops' vulnerability to *C. fimbriata.* When a pathogen such as *C. fimbriata* infects a single plant in monoculture, it can readily infect additional plants of the same species that have fewer genetic variants to fend off the infection. Usually, monocultures lack the natural diversity that would prevent the disease from spreading. Trade and transportation of diseased plant material, such as seeds, cuttings, wood, or soil from affected areas, is another way that it spreads. When environmental conditions improve, the disease agent that was previously dormant in plant or soil tissues will eventually infect the newly deviated areas. Pruning shears, saws, and even tractors can propagate fungal spores from infected plants to healthy ones. Without proper sanitation procedures, the disease can easily move from one tree or plantation to another during pruning, harvesting, or planting operations. The pathogen survives in any infected material that is left in the field, such as diseased trunks, fallen leaves, or trimmed branches. Spores of infected material can spread to healthy plants through the wind, water, and even insect activity. The disease may continue to spread if infected material is improperly disposed of by burning or combining it without removing the spores. *C. fimbriata* persists in soil due to crop rotation not being practiced and the continuous planting of the same crop in the same fields. The pathogen reinfects newly planted crops after surviving the planting season by either surviving in soil or plant debris. The disease cycle can be broken by rotating crops with non-host crops; in the absence of crop rotation, the pathogen persists. Through runoff or shared water sources, any pathogen-associated spores could be transported by water and spread infection to uninfected plants. It has been discovered that the failure to implement significant controls over the transportation of plant materials across regions is the cause of *C. fimbriata*'s global spread. Transporting plant materials across regions without adequate treatment and monitoring creates opportunities for the infection to spread to new places, where it will subsequently settle in the vulnerable crops. Similarly, the disease can proliferate in nurseries, greenhouses, and farms when proper sanitation practices are not followed (Nasution et al. 2019).

22.4.2 Buildup of Inoculum of **Ceratocystis**

When fungal spores or other infectious forms of *C. fimbriata* accumulate in a particular habitat, including soil, plant detritus, or infected plant tissues, this is referred to as inoculum buildup. It can effectively limit the disease's severity and spread; generally speaking, the more inoculum load that accumulates in the environment, the more likely it is that the infection will infect healthy plants.

Plant waste such as fallen leaves, branches, and sick trunks can provide *C. fimbriata* with a long-term habitat. Spores are released into the immediate vicinity by the infection, which persists in the infected debris that is left behind in the field. This encourages the growth of inoculum, particularly in plantations or orchards with inadequate sanitation. Infected plant roots or decaying organic materials can harbor *C. fimbriata* for years. Once in the soil, it can withstand a variety of environmental factors. This lengthens the interval between inoculum application and infection as long as sensitive crops are planted in that soil. This offers a sustained source of inoculum, particularly for fields of vulnerable crops that are continually cultivated (Flood 2021). Large quantities of spores can be produced by the fungus, particularly under ideal circumstances like high humidity, which aids in the buildup of inoculum and the quick spread of disease. The ongoing existence of vulnerable hosts creates an environment in which the inoculum can gradually build up and result in recurrent infections. It grows and flourishes in damp, humid environments. The pathogen's spores can readily spread and survive in areas with high rainfall or poor management (such as excessive irrigation or poor drainage), building up enormous inoculum levels in soil and debris. The chance of inoculum buildup increases with the likelihood that *C. fimbriata* have already established themselves in the soil or plant material. In fields that have already been contaminated, new plantings might become infected very quickly, resulting in a more devastating outbreak (Erakyathappa et al. 2022).

22.4.3 One Pathogen on Multiple Hosts and Vice Versa

The genus *Ceratocystis*, and in particular *C. fimbriata*, provides a good illustration of both phenomena: one pathogen infecting multiple hosts and multiple strains or species of the pathogen infecting a single host. Its multidimensional nature is characterized by this dichotomy of adaptation, making it a significant ecological and agricultural danger. Numerous plant species from different genera are infected by this fungus, which has a very broad host range. This genus includes some valuable trees and crops. Its wide host range can be attributed to the following factors. The fungus *C. fimbriata* has a high degree of genetic diversity, which increases its ability to adapt to a variety of hosts and environmental factors. It causes the tissues of the plants it infects to deteriorate by producing poisons and enzymes. As a result, it can infect plants from a variety of taxa. *C. fimbriata* produces both asexual conidia and sexual ascospores, which are easily dispersed among different plant species by insects, water, tools, and human activity (Harrington et al. 2011).

However, it is known that distinct species or strains of the *Ceratocystis* genus can infect the same host with varying degrees of aggression and virulence. For instance, different strains of *Ceratocystis* infect mangoes in different locations, and in Pakistan and Brazil, some of these strains produce Mango Sudden Death Syndrome. Although they all infect the same host species, these strains may have differed slightly from one location to another due to local conditions that may have shaped their evolution. The wilt disease in eucalyptus trees that occur in different places of the world is caused by different species or strains of *Ceratocystis*. Aggression and resistance to environmental stimuli may vary throughout the strains. The genus *Ceratocystis* may experience hybridization between species or strains, which could result in novel pathogens with distinct traits or degrees of pathogenicity that can infect the same host (Rashid et al. 2013).

22.5 Disease Cycle

C. fimbriata can persist for extended periods of time as mycelium, conidia, and chlamydospores in soil, dead wood, and plant detritus. The pathogen typically survives throughout the dry season, which runs from November to February, when the conditions for infection are less ideal (Accordi 1989). The pathogen may spread through a variety of dispersion routes, with bark beetle insect vectors playing a significant role. Because they transmit fungal spores and eat on afflicted trees, these insects serve as spore vectors. They are most active during the warmer months of March–June (Iton 1959; Moller and DeVay 1968). During the monsoon season, which normally lasts from June to September, the spores are also spread by wind and water. Wounds or natural openings are the source of infection, which is typically brought on by insect activity or mechanical harm. Because it promotes the pathogen's proliferation, this stage is mainly prevalent from March to June. Following inoculation, the fungal pathogen *C. fimbriata* colonizes the vascular tissues of mango trees, causing wilting, leaf yellowing, vascular discolouration, and branch dieback—all of which are typical symptoms that are typically seen between June and September. This pathogen contributes to sporulation by producing both asexual conidia and sexual ascospores. Sporulation happens during the monsoon season, when high humidity and the accompanying moisture provide ideal circumstances for spore formation and release. In addition to human operations like trimming and harvesting, the cycle continues as spores spread to infect new hosts. In order to reduce the risk of infection and restrict the spread of this pathogen that is harmful to mangos, good management of *C. fimbriata* entails a set of procedures that address the removal of infected plant material, adequate sanitation, and control of insect vectors.

22.6 Pathogen Survival

On crops: Through the development of resistant structures, such as asexual structures called conidiophores and sexual fruiting bodies called peritheciums, *C. fimbriata* endures in soil, plant waste, and afflicted tissues. The spores are shielded by thick melanized tissues, which enables the pathogen to withstand adverse conditions like drought or dry seasons. In addition to other woody tissues of the hosts, such as mango, cacao, eucalyptus, and rubber trees, the pathogen can live for months or even years in cranked branches and decomposing roots. The disease spreads to healthy plants because it continues to sporulate dead or dying plant tissues. Later, when the rainy season presents appropriate conditions, the spores produced will be flushed out. If plant waste is not removed, spores produced by *C. fimbriata* from rotting wood or the roots of badly infected mangoes or coffee trees may be added to the inoculum build-up. Although the majority of wood-infecting fungus are similarly subject to competition from saprophytic bacteria, *C. fimbriata* remains comparatively resilient. The infection cannot survive or spread if infected material is removed and proper cleanliness is maintained. In woody tissues, where spores can be released under the right circumstances, the pathogen may live for several years (Saranya 2021).

Alternate hosts: When healthy mango trees are planted in orchards, *C. fimbriata* spores can spread from sick trees or other hosts in the same general vicinity and result in new infections. The majority of new infections in orchards where *C. fimbriata* has already taken over come from internal inoculum sources, such as dead wood, infected plant debris, or already infected trees. There are several ways that *C. fimbriata* spores might spread across an orchard or plantation, including wind, water, insects, and contaminated equipment. Instead of living directly in the soil, *C. fimbriata* mostly lives in the tissues of afflicted plants. The pathogen obtains its inoculum from infected wood and plant detritus. The virus lives in decaying plant tissues, dead roots, and cranked branches. It continuously generates spores that, in the right circumstances, could lead to more infections (Fourie 2014).

22.7 Sources of Inoculum and Spore Dispersal

The pathogen develops into the vascular system after infecting trees or crops, producing various spore fruiting structures. Tissue produced by infections serves as a reservoir for inoculum, which can subsequently infect healthy plants. Even after infected trees die, the disease can persist in wood from dead trees or pruning and continue to spawn for a considerable amount of time. Studies have shown that the disease may persist and produce viable spores under ideal conditions in wood samples from mango trees that had been infected for a number of years. On the trunks and branches of diseased trees, *C. fimbriata* produces cankers that girdle and kill substantial portions of the plant. These cankered tissues frequently contain high amounts of the pathogen's spores, which are discharged in damp or humid environments. This indicates that rubber and eucalyptus tree cankers and necrotic tissues serve as long-term sources of inoculum, sustaining the illness on the farm. A viable spore from infected plant debris left on the ground will cause the disease, even though *C. fimbriata* does not live in soil for very long compared to many other soil-borne diseases. New infections in the field are caused by spores in rotting roots, leaves, or trimmed limbs that are discarded without a remedy. When certain conditions are met, the diseased plant tissues cause spores to be released through various dispersal methods (Billings et al. 2008).

The pathogen *C. fimbriata* is now much more successful because to its spore dispersal mechanism. The virus spreads swiftly across an orchard or plantation because it generates both sexual ascospores and asexual conidia, which are both discharged from fruiting structures it creates on infected tissues. Spores can be disseminated through a variety of ways.

22.7.1 Vector Dispersal

The spread of *C. fimbriata* by beetles is the most important way that the infection spreads among the different hosts through insect vectors. The bark beetles and wood-boring beetles are the most important bugs linked to helping *C. fimbriata* to migrate. Fimbriata spores are essential in the epidemiology of diseases like Mango Sudden Death Syndrome, eucalyptus canker, and rubber tree wilt because they can distinguish between diseased and healthy trees. Because beetles can bore into wood and consume tree tissues, they become very effective vectors for *C. fimbriata.* Beetles came into direct contact with the *C. fimbriata* spores when they dug into infested trees. These spores adhere to the beetle bodies, particularly the exoskeleton, legs, and mouthparts. They include sexual ascospores and frequently asexual conidia. The infected wood frequently produces enormous amounts of wood fimbriata. As they dig into the trees for lodging, the beetles move the spores to the healthy trees, so transmitting them from tree to tree. The tree's vascular system is directly sliced by these new wounds, and *C. fimbriata* infects the vessel. Aside from these insect actions, the fungus would have trouble entering the tree's internal tissues through the outer protective layers (Cassol 2023).

Seasonal factors have a significant impact on beetle vectors. *C. fimbriata* should be sporulating during warm-humid portions of the year, when beetles would normally be active. In more tropical and subtropical areas, it frequently falls around the start of the rainy season, which makes it the perfect period for disease epidemics because beetle activity rises and the environment is conducive to spore dispersal. Mango Sudden Death Syndrome, for instance, has been linked to increased activity in bark beetles during the growth season. This is mainly because the beetles have a tendency to disseminate the disease to plants that haven't been infected before (Harrington et al. 2011) (Figs. 22.1, 22.2, 22.3 and 22.4).

22.7.2 Wind

Ceratocystis conidia and ascospores are small and light, making them likely to float on air currents and be easily carried short distances by wind or inclement weather. Spores on the tissues of diseased plants detach from the plant and fly into the air during severe winds or storms. New infections are started when wind-borne spores strike exposed surfaces of nearby healthy plants, such as open wounds or natural openings. In contrast, the infection may spread across a greater distance by an insect-mediated dispersal process. On the other hand, wind dispersal is typically far more confined across shorter distances. The same orchard or plantation is often traversed by wind-blown spores, which infect nearby plants. Because the spores are easily transferred from infected to healthy plants, this short-distance transmission can result in extremely dangerous epidemics in intense plantings. If infected plant debris, like as falling leaves, rotting wood, or clipped branches, is left in an orchard or plantation, the

Fig. 22.1 Symptoms of Mango Sudden Death Syndrome **a** gumosis, **b** stem cracking, **c** stem discoloration, **d** stem pitting by beetles, **e** wilting and **f** complete death of mango tree

likelihood of wind spread increases. Wind activity may affect the pathogen's fruiting bodies, such as perithecia or conidiophores, when these infected materials dry up and decompose. When spores are released from these fruiting structures in windy conditions, they are carried by the air currents and eventually land on new hosts. Wind dispersal is a bigger problem in places where plant material is left unmanaged, like after pruning or during the dry season (Billings et al. 2008).

22.7.3 Water

Water is a key dissemination mechanism for *C. fimbriata*, particularly in situations with inadequate drainage or excessive rainfall. Water-borne dispersal occurs during floods, through irrigation systems, and through surface runoff. Spores of the fungus from decomposing wood, soil, or infected plant waste infect healthy trees when precipitation or irrigation enters orchards and plantations. *C. fimbriata* spores. Specifically, water readily removes Fimbriata's conidia, which are asexual spores. Even contaminated tissues or soil can spread these spores, particularly if the soil does not cover the infected branches or roots. In low-lying locations where runoff builds up, water dispersal is very effective, putting healthy plants at risk for infection. It is

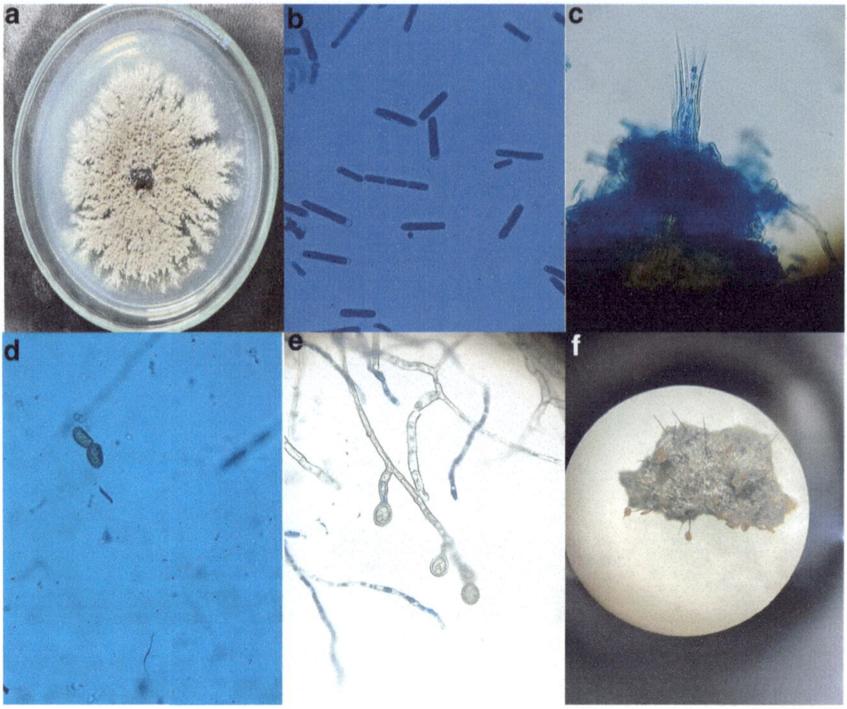

Fig. 22.2 **a** Colony of *Ceratocystis*, **b** endoconidia, **c** ostiolar hyphae with ascospores, **d** ascospores, **e** aleurioconidia and **f** ascomata

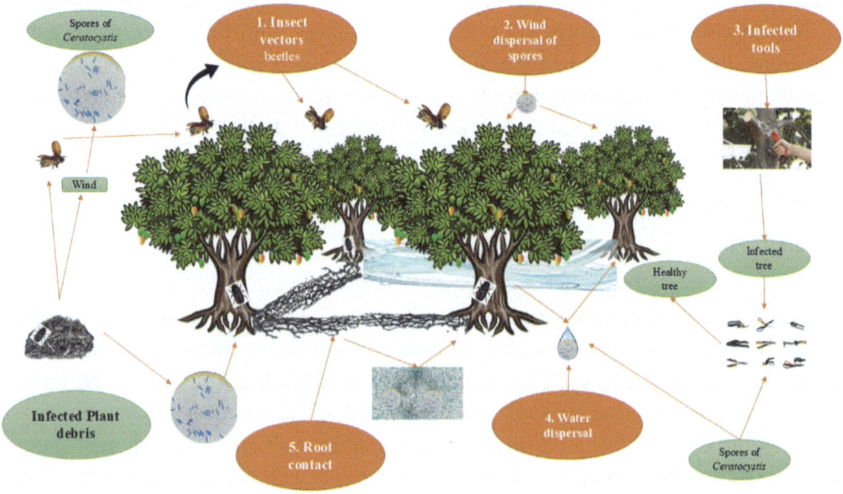

Fig. 22.3 Disease cycle of *Ceratocystis fimbriata* on mango

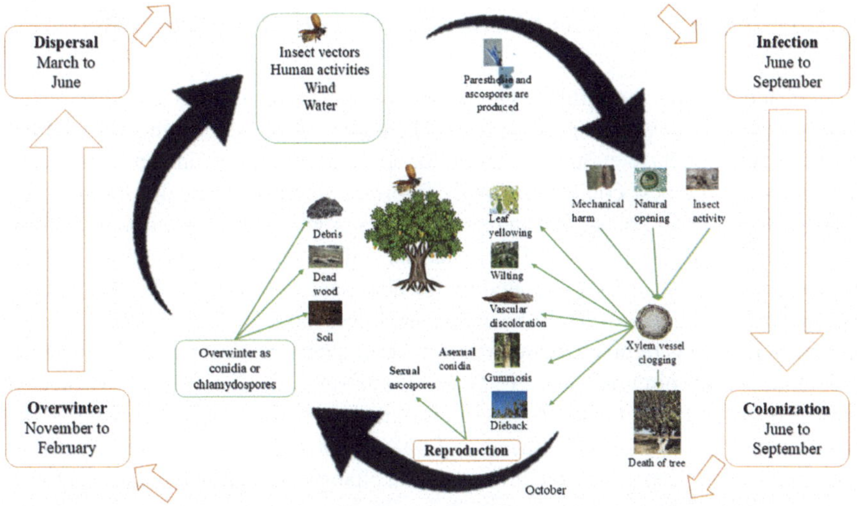

Fig. 22.4 Dispersal of *Ceratocystis fimbriata*

possible for the infection to enter new hosts through open wounds or natural openings. If the area is prone to flooding and has inadequate drainage, water-borne *C. fimbriata* spores can spread farther. Removing contaminated plant material, ensuring appropriate drainage, and avoiding wet circumstances that encourage the spread of the infection are all examples of good management practices (Billings et al. 2008).

22.7.4 Human Activities

C. fimbriata spreads mostly through human-mediated dissemination. Diseased planting material, equipment, or tools are frequently moved from one field to another as a result of pruning, harvesting, and other farming operations. Unintentional transfer from one field to another is frequently the result of this. If the equipment used to prune diseased plants are not properly cleaned and sterilized, spores will remain on the healthy plants. Secondary infections are the main result of this (Harrington et al. 2011). Additionally, infected planting materials—such as seeds, cuttings, or rootstocks—are transported from one area to another, which has raised serious concerns as it makes it easier for *C. fimbriata* to spread over large distances (De Beer et al. 2014).

22.7.5 Impact of Environment on Dispersal of Ceratocystis

Rainfall, temperature, humidity, and other environmental factors all have a big impact on how much *C. fimbriata* spores spread. The pathogen grows best in warm, humid environments that have enough moisture for spore development and spread. Generally speaking, the best conditions for *C. fimbriata*'s spread occur during the rainy season when coffee, cocoa, and mangoes are being grown in tropical and subtropical areas (Araujo et al. 2014). The rapid distribution of spores within an orchard or plantation can be easily facilitated by the abundance of insect and water activity during these months. Conversely, because there is less moisture during the dry season, pathogen activity is reduced, which restricts spore generation and spread. However, under unfavorable circumstances, *C. fimbriata* can survive in dead tissues or infected plant detritus and wait for the right circumstances to resume spore production and dissemination (Li et al. 2016). Because of this seasonal variation in disease activity, it is necessary to keep an eye on environmental changes and implement preventative measures when the spores are most likely to spread.

22.8 Pre-infection, Infection and Colonization

Pre-infection, infection, and colonization are the three main phases of the *C. fimbriata* disease process on mangos (*Mangifera indica*), which results in Mango Sudden Death Syndrome. A clear image of how a disease infects a mango tree and spreads within the host can be created by comprehending each of the latter.

22.8.1 Pre-infection

Conidia, which are asexual spores, and ascospores, which are sexual spores, are two types of *C. fimbriata* spores that can form on infected plant waste, cankered tissues, or rotted wood during the pre-infection stage. The bark beetle, water runoff, wind, and human activity all contribute to the further spread of these spores. When suitable environmental conditions are present, the virus can wait for extended periods of time in the environment. The spores typically enter healthy trees through wounds or natural openings brought on by insect feeding, mechanical damage, or pruning (Araujo et al. 2014).

22.8.2 *Infection*

The spores enter the wounds, germinate when the temperature and moisture levels are right, and begin to produce hyphae that gradually infiltrate the host tissues. The pathogen enters the mango tree through open wounds and breaks past the outer bark defense to reach the vascular tissues, primarily the xylem. Wilting and yellowing leaves are also possible early indicators of plant infection, albeit they might not be readily apparent in the early stages (Engelbrecht et al. 2007).

22.8.3 *Colonization*

During the colonization stage, *C. fimbriata* grows inside the xylem, blocking the tree's water-conducting tissues. The fungus breaks down the components of the cellular wall both toxically and enzymatically. As it spreads throughout the vascular system, it also begins to interfere with the flow of water. Wilting is the first symptom, which is followed by dieback, yellowing leaves, and eventually the tree's death. The pathogen's fruiting bodies, or perithecia, release spores to perpetuate the infection cycle (Arriel et al. 2016).

22.9 Conclusion

The ecological relations of *C. fimbriata* and the infections it causes, especially Mango Sudden Death Syndrome (MSDS), are complex and multifaceted, being simultaneously influenced by environmental factors, pathogen characteristics, and host susceptibility. Such devastating diseases in mango (*Mangifera indica*) indicate the adaptability and resilience of *C. fimbriata* as a pathogen; therefore, an understanding of the ecological interactions favoring its survival and spread is critical for managing and controlling the disease. *C. fimbriata* thrives in warm, humid tropical and subtropical climates, which facilitate the pathogen's growth and spread. The pathogen seems to mostly survive in decaying wood, infected plant waste, and cankered tissues that serve as sources of inoculum. While insect vectors including bark beetles, wind, water, and humans aid in the spread of *C. fimbriata*, high humidity and rainy conditions promote the formation and shedding of spores. Insect activity and MSDS disease cycles are tightly associated. It generally happens during warm, humid months when the beetles are active. The latter turns becomes a host to transfer the *C. fimbriata* spores from the diseased mango to a healthy one. The pathogen enters the tree by wounds from insect feeding, storm damage, or pruning. It avoids the tree's natural defense mechanism as a result. *C. fimbriata*. Because the pathogen obstructs the tree's vascular tissues and interferes with water transport, it is extremely aggressive and causes the tree to wilt, dieback, and eventually die. The ecology and management of *C. fimbriata* are complicated since it can infect crops other than mangos. Therefore, management necessitates a thorough understanding of the pathogen's ecology, including spore dispersion processes, environmental conditions, and the part insect vectors play in the infection process. The three factors have the potential to lessen the impact and spread of MSDS, improving mango production outcomes.

References

Accordi SM (1989) The survival of *Ceratocystis fimbriata* f. sp. platani in the soil

Al Adawi AO (2002) The biology of *Diplodia theobromae* associated with mango sudden decline disease in Oman. Dissertation, Sultan Qaboos University

Al Adawi AO, Deadman ML, Al Rawahi AK, Al Maqbali YM, Al Jahwari AA, Al Saadi BA, Al Amri IS, Wingfield MJ (2006) Aetiology and causal agents of mango sudden decline disease in the Sultanate of Oman. Eur J Plant Pathol 116:247–254

Al Adawi AO, Al Jabri RM, Deadman ML, Barnes I, Wingfield B, Wingfield MJ (2013) The mango sudden decline pathogen, *Ceratocystis manginecans*, is vectored by Hypocryphalus mangiferae (Coleoptera: Scolytinae) in Oman. Eur J Plant Pathol 135:243–251

Alizadeh M, Azaryan A, Ghaleh MK, Roshanroo M, Graily-Moradi F (2024) Host and geographical distribution of the plant pathogenic fungus *Ceratocystis fimbriata* (Ascomycota) worldwide. Nova Hedwigia 118(1–2):41–71

Al-Ruheili AM, Boluwade A, Al-Subhi AM (2021) Predicting mango sudden decline due to *Ceratocystis fimbriata* under a changing climate. Arab J Plant Prot 39(3)

Al Subhi AM, Al-Adawi AO, Van Wyk M, Deadman ML, Wingfield MJ (2006) *Ceratocystis omanensis*, a new species from diseased mango trees in Oman. Mycol Res 110(2):237–245

Araujo L, Bispo WMS, Cacique IS, Moreira WR, Rodrigues FA (2014) Resistance in mango against infection by *Ceratocystis fimbriata*. Phytopathology 104(8):820–833

Arriel DAA, da Silva Guimarães LM, de Resende MDV, Neto FPL, Silva DFSHS, de Siqueira DL, Alfenas AC (2016) Genetic control of resistance on *Mangifera indica* to *Ceratocystis* wilt. Sci Hortic 211:312–318

Batista DDC, Terao D, Barbosa MAG, Barbosa FR (2008) Seca-da-mangueira: detecção, sintomatologia e controle

Bhat SS, Rajamani R, Nandagopal N (2002) Screening of *Coffea arabica* L. cultivars against coffee trunk canker pathogen, *Ceratocystis fimbriata*

Billings RF, Appel DN, Center LBJ (2008) The proceedings of the 2nd national oak wilt symposium

Cassol SA (2023) *Ceratocystis fimbriata*–Actinidia spp. pathosystem: mechanical dispersal, disease control and pathogenicity of isolates from different hosts

de Beer MC, Wroblewski JM, Noffsinger VP, Ji A, Meyer JM, van der Westhuyzen DR, de Beer FC, Webb NR (2013) The impairment of macrophage-to-feces reverse cholesterol transport during inflammation does not depend on serum amyloid A. J Lipids 1:283486

De Beer ZW, Duong TA, Barnes I, Wingfield BD, Wingfield MJ (2014) Redefining *Ceratocystis* and allied genera. Stud Mycol 79(1):187–219

Elliott JA (1923) The ascigerous stage of the sweet potato black-rot fungus. Phytopathology 13:56

Engelbrecht CJ, Harrington TC, Alfenas A (2007) *Ceratocystis* wilt of cacao—a disease of increasing importance. Phytopathology 97(12):1648–1649

Erakyathappa NS, Venkatappa D, Gopal SC, Sridhara S, Kalaiah HM, Ugalat J (2022) Molecular characterization and genetic variation in *Ceratocystis fimbriata* Ell. and Halst. on pomegranate. J Fungi 8(12):1276

Fadli NN, Dharma A, Syafrizayanti N, Asmarayani R, Efdi M (2022) Antifungal activity of three Piper species against *Ceratocystis fimbriata* in rubber trees. Arch Phytopathol Plant Protect 55(12):1379–1393

Firmino AC, Tozze HJ Jr, Furtado EL (2012) First report of *Ceratocystis fimbriata* causing wilt in *Tectona grandis* in Brazil. New Dis Rep 25(24):2044–0588
Flood J (2021) Coffee wilt disease
Fortini LB, Kaiser LR, Keith LM, Price J, Hughes RF, Jacobi JD, Friday JB (2019) The evolving threat of rapid 'Ōhi'a death (ROD) to Hawai'i's native ecosystems and rare plant species. For Ecol Manag 448:376–385
Fourie A (2014) Distinguishing between cryptic species in the *Ceratocystis fimbriata* sensu lato species complex. University of Pretoria (South Africa)
Gomdola D, Bhunjun CS, Hyde KD, Jeewon R, Pem D, Jayawardena RS (2022) Ten important forest fungal pathogens: a review on their emergence and biology. Mycosphere 13(1):612–671
Guimarães LM, Nunes AS, Santos SA, Resende MD, Damacena MB, Siqueira DL, Alves RS, Alfenas AC (2021) Resistance of mango cultivar Ubá to *Ceratocystis fimbriata* depends on the pathogen's physiological variability. Crop Prot 143:105560
Halsted BD (1890) Some fungous diseases of sweet potato. New Jersey Agric Coll Exp Stn Bull 76:7–14
Harrington TC (2013) *Ceratocystis* diseases. In: Infectious forest diseases. CABI, Wallingford, UK, pp 230–255
Harrington TC, Wingfield MJ (1998) The *Ceratocystis* species on conifers. Can J Bot 76(8):1446–1457
Harrington TC, Thorpe DJ, Alfenas AC (2011) Genetic variation and variation in aggressiveness to native and exotic hosts among Brazilian populations of *Ceratocystis fimbriata*. Phytopathology 101(5):555–566
Hausner G, Reid J, Klassen GR (1993) On the subdivision of *Ceratocystis* s.l., based on partial ribosomal DNA sequences. Can J Bot 71(1):52–63
Iton EF (1959) Studies on a wilt disease of cacao at River Estate
Johnson GI, Kotze JM (1994) Stem-end rot. In: Ploetz RC, Zentmyer GA, Nishijima WT, Rohrbach KG, Ohr HD (eds) Compendium of tropical fruit disease. APS Press, USA, pp 39–41
Li Q, Harrington TC, McNew D, Li J, Huang Q, Somasekhara YM, Alfenas AC (2016) Genetic bottlenecks for two populations of *Ceratocystis fimbriata* on sweet potato and pomegranate in China. Plant Dis 100:2266–2274
Malik MT, Altaf AD, Sultan MK (2004) Some manageable predisposing factors of collar/stem rot of mango. Pak J Phytopathol 16:37–42
Mehl J, Wingfield MJ, Roux J, Slippers B (2017) Invasive everywhere? Phylogeographic analysis of the globally distributed tree pathogen Lasiodiplodia theobromae. Forests 8(5):145
Marincowitz S, Barnes I, De Beer ZW, Wingfield MJ (2020) Epitypification of *Ceratocystis fimbriata*. Fungal Syst Evol 6:289–298
Moller WJ, DeVay JE (1968) Insect transmission of *Ceratocystis fimbriata* in deciduous fruit orchards
Naqvi SA, Perveen R (2015) Mango quick decline manifestation on various cultivars at plants of particular age in the VI. Pak J Phytopathol 27:31–39
Nasution A, Glen M, Beadle C, Mohammed C (2019) *Ceratocystis* wilt and canker—a disease that compromises the growing of commercial *Acacia*-based plantations in the tropics. Aust For 82:80–93
Oliveira LS, Harrington TC, Ferreira MA, Damacena MB, Al-Sadi AM, Al-Mahmooli IH, Alfenas AC (2015) Species or genotypes? Reassessment of four recently described species of the *Ceratocystis* wilt pathogen, *C. fimbriata*, on *Mangifera indica*
Ploetz RC (2003) Diseases of mango. In: Diseases of tropical fruit crops, pp 327–363
Ploetz RE, Prakash O (1997) Foliar, floral, and soil borne diseases. In: Litz RE (ed) The mango: botany, production and uses. CAB International, Oxon, UK, pp 281–225
Pratama R, Muslim A, Suwandi S (2023) First report of new wilt disease on *Mangifera indica* caused by *Ceratocystis fimbriata* in Indonesia. BIOVALENTIA Biol Res J 9:36–40
Rashid AR, Shazia Iram SI, Ahmad IA (2013) Study on incidence, molecular characterization and pathogenesis of different fungi associated with sudden death of mango

Réblová M, Gams W, Seifert KA (2011) *Monilochaetes* and allied genera of the *Glomerellales*, and a reconsideration of families in the *Microascales*. Stud Mycol 68:163–191

Roux J, Wingfield MJ (2009) *Ceratocystis* species: emerging pathogens of non-native plantation *Eucalyptus* and *Acacia* species. South For J For Sci 71:115–120

Roux J, Wingfield MJ, Bouillet JP, Wingfield BD, Alfenas AC (2000) A serious new wilt disease of *Eucalyptus* caused by *Ceratocystis fimbriata* in Central Africa. For Pathol 30:175–184

Saranya R (2021) *Ceratocystis fimbriata*—a threatening pathogen

Shukla PK, Varma S, Fatima T, Bajpai A, Mishra R, Misra AK, Gundappa MM (2018) First report on wilt disease of mango caused by *Ceratocystis fimbriata* in Uttar Pradesh, India. Indian Phytopathol 71:135–142

Spatafora JW, Blackwell M (1994) The polyphyletic origins of ophiostomatoid fungi. Mycol Res 98:1–9

Sutton BC (1980) The Coelomycetes. Commonwealth Mycological Institute, Kew, Surrey, England, p 643

Syazwan SA, Mohd-Farid A, Wan-Muhd-Azrul WA, Syahmi HM, Zaki AM, Ong SP, Mohamed R (2021) Survey, identification, and pathogenicity of *Ceratocystis fimbriata* complex associated with wilt disease on *Acacia mangium* in Malaysia. Forests 12:1782

Tarigan M, Roux J, Van Wyk M, Tjahjono B, Wingfield MJ (2011) A new wilt and die-back disease of *Acacia mangium* associated with *Ceratocystis manginecans* and *C. acaciivora* sp. nov. in Indonesia. S Afr J Bot 77:292–304

Tyson JL, Manning MA, Wright PJ (2020) Pathogenicity of *Ceratocystis fimbriata* from New Zealand kumara on kiwifruit cultivars. N Z Plant Prot 73:6–12

Upadhyay HP (1981) A monograph of *Ceratocystis* and *Ceratocystiopsis*, 176 pp

Valdetaro DC, Oliveira LS, Guimarães LM, Harrington TC, Ferreira MA, Freitas RG, Alfenas AC (2015) Genetic variation, morphology and pathogenicity of *Ceratocystis fimbriata* on *Hevea brasiliensis* in Brazil. Trop Plant Pathol 40:184–192

Van Wyk M, Al Adawi AO, Khan IA, Deadman ML, Al Jahwari AA, Wingfield BD, Ploetz R, Wingfield MJ (2007) *Ceratocystis manginecans* sp. nov., causal agent of a destructive mango wilt disease in Oman and Pakistan. Fungal Divers 27:213–230

Van Wyk M, Wingfield BD, Al-Adawi AO, Rossetto CJ, Ito MF, Wingfield MJ (2011) Two new *Ceratocystis* species associated with mango disease in Brazil. Mycotaxon 117:381–404

Van Wyk M et al (2013) Phylogenetic relationships and taxonomy of species in *Ceratocystis* sensu lato. University of Pretoria, South Africa

Viégas AP (1960) Seca da mangueira. Bragantia 19:162–182

Weber DJ, Stahmann MA (1964) *Ceratocystis* infection in sweet potato: its effect on proteins, isozymes, and acquired immunity. Science 146:929–931

Zhang ZP, Li Q, Luo LX, Li JQ, Hao JJ (2017) First report of mango wilt caused by *Ceratocystis fimbriata* in *Mangifera indica* in China. Plant Dis 101:1042–1042

Chapter 23
Mitigation for Fungal Disease of Mango in Changing Climate Scenarios

Rabia Tahir Bajwa, Mustansar Mubeen, Qaiser Shakeel, and Ifrah Rashid

Abstract The farming industry is being adversely affected by climate change. Specifically examines the impact of climate change on fungal diseases that affect mangoes. It also explores the areas of uncertainty surrounding this issue and proposes potential strategies to alleviate its consequences. Furthermore, mango is vulnerable to several diseases during its entire lifespan. The primary abiotic elements linked to climate change that impact mango diseases encompass alterations in precipitation patterns, fluctuations in wind, rising temperatures, elevated atmospheric CO_2 levels, and shifts in light intensity. These factors influence the occurrence and spread of fungal diseases, reproduction, dispersion, and migration. Abiotic variables also impact the growth, and reproduction of plants. The interaction of these elements is intricate, and additional research is required to gather pertinent data for comprehending the correlations between these parameters and the occurrence of diseases. Creating forecasting models using this data and implementing IDM strategies can be beneficial diseases management on mango trees and food security. Moreover, early disease management is crucial to reduce mango production loss.

Keywords IDM · Climate change · Sustainable agriculture · Disease forecasting

23.1 Introduction

One of the most significant problems the world is now experiencing is climate change. It is characterised by notable variations from the long-term averages of climatic variables, such temperature and precipitation (World Meteorological Organization 1992; Malhi et al. 2021). The world's climate has changed significantly in recent decades due to increasing human activity that has changed the natural composition of the

R. T. Bajwa · Q. Shakeel (✉) · I. Rashid
Cholistan Institute of Desert Studies, The Islamia University of Bahawalpur, Bahawalpur, Pakistan
e-mail: qaiser.shakeel@iub.edu.pk

M. Mubeen
Department of Plant Pathology, College of Agriculture, University of Sargodha, Sargodha, Pakistan

S. A. H. Naqvi et al. (eds.), *Climate Change and Mango Production*,
https://doi.org/10.1007/978-3-031-87035-4_23

atmosphere (IPCC 2007; Malhi et al. 2021). Greenhouse gas concentrations have sharply grown since 1750, with carbon dioxide accounting for 40%, methane for 150%, and nitrogen oxides for 20%, respectively (IPCC 2014) (Sathaye et al. 2006). increased from 22.15 billion metric tonnes in 1990 to 36.14 billions metric tonnes in 2014 (Abeydeera et al. 2019). Since 1975, the average global temperature has increased by 0.15 to 0.20 degrees Celsius per ten years (Arora et al. 2005; NASA 2020), with an expected increase of 1.4 to 5.8 degrees Celsius by 2021. Greenhouse gas emissions from combustion of fossil fuels, especially carbon dioxide (CO_2), as well as non-CO_2 GHGs which includes CH4, nitrous oxide (N_2O), and chlorofluorocarbons, or CFCs, aggravate global warming. The biggest contributor to greenhouse emissions in our atmosphere is the gas carbon dioxide, which is produced by 65% of fossil fuels and industrial processes, 11% by forestry, and 21% by other land uses such as the production of CH4 (16%), N_2O (6%), and fluorinated emissions (2%). Prior to 1750, the contribution of fossil fuel emissions to CO_2 emissions was negligible; but, as industrialisation advanced, it did so swiftly. Since 1751, over 1.5 trillion metric tonnes of CO_2 have been emitted into the atmosphere globally. Regional differences exist in the emission, however. With more than 514 billion metric tonnes of CO_2 emissions, Europe is the world's biggest emitter. With total emissions of carbon dioxide of 457 billion metric tonnes apiece, North America and Asia set records. Compared to China, which emits 200 billion metric tonnes of CO_2, the United States produces 399 billion metric tonnes, or 25% of all historic releases since 1751. Twenty-two percent of all history has been made up of the twenty-eight nations that comprise the European Union (EU-28) and share common goals. Due to its low per-capita emissions, Africa only contributes 3% of the world's CO_2 emissions. Conversely, countries like Brazil and India who have previously produced fewer emissions now significantly raise global emissions (CDIAC 2020). As atmosphere CO_2 levels rise, crop fertilisation increases and warming-related decreases in energy requirements take place (Malhi et al. 2021). It's expected that in the foreseeable future, climate change will worsen. In Pakistan's Punjab province, forecasts for the Kharif and Rabi seasons include higher both minimum and maximum temperatures. The mean hottest temperatures and average lowest temperature during the Kharif season are predicted to rise by 1–3 °C and 2–4 °C, respectively, based on predictions made for the forthcoming mid-century (2040–2069). On the other hand, it is anticipated that Rabi's average temperature will rise by 2.1–3.5 °C and 2–4 °C, correspondingly. Furthermore, estimations of the changes in rainfall between the regions have been made, with a particular emphasis on the Kharif month (25–35%) and minor variations in the period of rabi (Bokhari et al. 2017). Due to its extensive reach and weather-related unpredictability, agriculture is the industry most at risk from climate change, which will have a significant financial impact (Mendelsohn 2009; Malhi et al. 2021). Crop output is significantly impacted by changes in weather phenomena like temperature and rainfall. The crop, the area, and the extent of parameter change all affect the effects of rising temperatures, unpredictable precipitation, and CO_2 fertilisation. It has been shown that rising temperatures lower yields. (Adams et al. 1998). Climate, crop adaptability ability, crop type, and the impact of CO_2 fertilisation are some of the factors influencing Iranian agricultural productivity (Karimi et al. 2018). It has been observed that Cameroon's farmers' net income is greatly reduced when precipitation declines or temperatures rise. The country's revenue has fluctuated because to bad governance and the unpopularity of Cameroon's agricultural exports (Molua and Lambi 2007; Malhi et al. 2021). Depending on the area and irrigation technique, climate change has varying effects on agricultural output. Extending irrigated areas may boost crop production, but doing so may have negative ecological effects. Malhi et al. (2021) and Kang et al. (2009). Many crops will probably produce less due to their shorter growing seasons caused by higher temperatures (Mahato 2014). In both the warm and tropical regions, it is anticipated that a 2 °C increase in temperature would result in a drop in the overall grain production (Challinor et al. 2014). Because tropical crops remain closer to their warmer optimal points and hence experience severe stress from increasing temperatures, climate change often has a greater impact on tropical locations.

23.2 Climate Change and Agriculture

It has been known for more than 2,000 years that plant diseases are influenced by the climate. It is anticipated that the climate change will have a significant impact on agricultural markets, resulting in a 0.26% decrease in world GDP (Costinot et al. 2016). An evaluation of the effects of climate change on food cropping systems is not simple due to the intricate nonlinear associations between a plant and the environment (Luck et al. 2011). The host may undergo significant changes as a result of climate change, and these changes may have indirect impacts on the pathogen. These changes are overlooked when analyzing the response of pathogen to climate change in isolation. Diseases pose a danger to food security since they cause losses of minimum 10% of the world's food output (Strange and Scott 2005; Yáñez-López et al. 2012). Agrios (2004) calculated that disease-related losses cost US$ 220 billion annually. In addition to causing direct losses, disease control techniques, particularly those involving chemicals, can also have negative social and economic effects, contaminate the environment, and leave persistent chemicals in food. Given the intimate connection between diseases and the environment, changes in the present phytosanitary situation are likely to result from climate change. Because there may be a reduction, a rise, or no influence on the various pathosystems in each region, the impacts may be positive, negative, or neutral. In order to prevent more significant losses, it is crucial to analyze the possible effects of climate change on plant diseases and adopt adaptation measures, such as developing resistant cultivars, novel control strategies, or adapted techniques (Nazir et al. 2018).Because of their large numbers, simple reproduction and dissemination, and short generation times, plant diseases are common in both natural and controlled systems and are among the first to show the influence of climate change. They therefore represent a core class of biological indicators that, in addition to causing losses in productivity and possibly posing a threat to the sustainability of agroecosystems, require to be assessed in relation to the effects of climate change (Yáñez-López et al. 2012; Nazir et al. 2018). All of the major crops, including fruit crops, are impacted by climate change. *Mangifera indica* L., the scientific name for the mango, is a member of the family Anacardiaceae and is regarded as one of the most important tropical fruits in the world. The phrase "king of fruits" is frequently used to describe the mango (Purseglove 1972). Southeast Asia, or Indo-Burma, is where it first appeared (Popenoe 1932; Mukherjee 1951). Mango varieties are known for their vivid colours, sweet scent, rich taste, and high nutritional value. Fruit often contains a sugar content ranging from 10–20%, making it a significant source of vitamin A, C, and abit of vitamin B as well (Usman et al. 2003; Javaid et al. 2012).The cultivation of mango in India dates back thousands of years (Mukherjee 1953; Kostermans and Bompard 1993). Cultivation of this crop dates back to the ancient Indian civilization and has been practiced in India for the past 4-6thousand years (DeCandolle 1884). The development and culture of the sub-continent were primarily influenced by the Mughal Emperors, particularly Akbar. Akbar played a significant role in the establishment of Lakh Bagh, as well as in promoting the practices of nursery men, amateur gardeners, and growers through the process of selection following cloning (Usman et al. 2003; Javaid et al. 2012). It has now become an indispensable component of our cultural and historical heritage. Pakistan ranks as the fifth greatest producer of mangoes globally. Mango is the 2nd most abundant fruit grown in Pakistan, following citrus fruits. Punjab and Sindh are the primary regions in the country where mangoes are cultivated on a large scale. Mango is cultivated either in February or in the first week of March. June and July are the months when mango harvesting is at its highest. Pakistan cultivates around 200 different varieties of mango. Pakistan, a prominent global supplier of mangoes, exports approximately 7–10% of its entire mango production, which is valued at over US$ 20 million annually. Pakistan mostly exports mangoes to Dubai, China, Iran and Europe (Javaid et al. 2012). Mangoes can thrive in various soil types, although they yield optimal outcomes when cultivated deeply in well-drained sandy loam soil. Avoid soil with hardpan, adhesive clay, and waterlogged situations. Mangoes thrive in a broad pH range, specifically between 5.5 and 8.7. Increased precipitation in rainy places leads to a higher frequency of flushes, resulting in abundant vegetative growth. Flower induction requires exposure to low temperature stress, which is a significant characteristic of the sub-tropical regions, or water stress in tropical regions. Optimal ripening necessitates hot and arid environments. Winter frost poses a significant risk and has a harmful impact on large trees. The environmental conditions in Punjab and Sindh are highly suitable and favorable for the cultivation of mangoes (Usman et al. 2003; Javaid et al. 2012).In the 1970s, we had the position of the second largest global producer of mangoes. However, our current standing has significantly declined to the 5th position. The decrease in mango output is a significant issue and has been attributed to a combination of difficult problems. Possible significant causes contributing to low yield, in addition to grower neglect, may include complex issues such as poor agronomic practices, insect pest and major mango diseases including anthracnose, powdery mildew, mango deformity, and die back diseases (Javaid et al. 2012).The absence of proactive measures and subsequent persistence in mango research incentives for export (Usman et al. 2003). The low production can be attributed to the extensive variability in climatic conditions, environmental circumstances, and the multitude of disease and abiotic disorder issues that are present. More than 140 diseases are recognized for their ability to inflict harm on crops. To effectively mitigate the risks posed by diseases, it is crucial to precisely ascertain and diagnose the issues at hand. The anticipated alterations in the tropical environment are projected to intensify the infestation of diseases on mango trees. There is not enough study on climate change effect on mango diseases, despite the fact that climate change poses a threat to plant preservation in the near future.

23.3 Major Fungal Diseases of Mango

Mango is susceptible to several diseases throughout its whole growth cycle, starting from the seedling stage in the nursery to its mature fruit during storage or transportation. Field diseases lead to crop loss, however post-harvest diseases are significantly associated with dead in both the export and local markets. Nearly each plant organ is susceptible to diseases, including the branch, stem, root, petiole, leaf, twig, flower, and fruit. These diseases present themselves in various forms such as rot, scab, malformation, mildew, die back, sooty mold, necrosis, stem bleeding, canker, spots, wilt, unclear etiology, and problems. Certain diseases have become a constraining element in the cultivation of mangoes. In general, fungal infections affecting leaves and fruits are more prevalent in Pakistan. Some of major fungal diseases of mangoes are discussed below (Fig. 23.1).

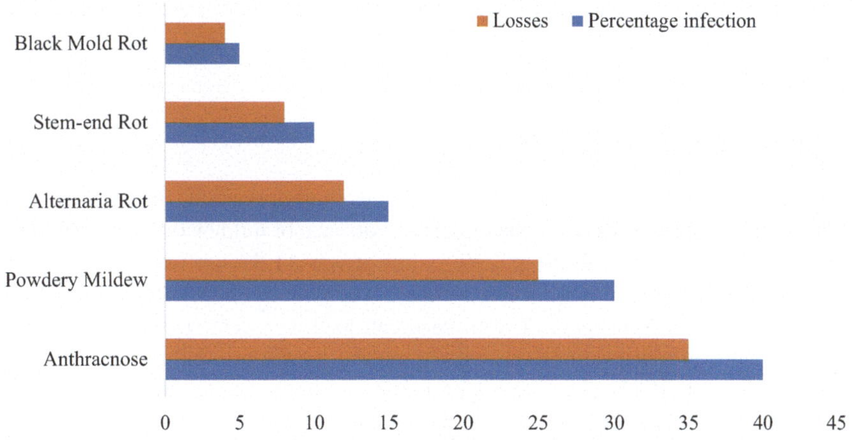

Fig. 23.1 Infection percentage and losses caused by fungal diseases in mango tree orchards

23.3.1 Powdery Mildew

In 1914, Berthet was the first to document the occurrence of powdery mildew on mango trees in Brazil. He subsequently termed this disease *Oidium mangiferae* Berthet. Previously, it was regarded as a disease of minimal significance, but it is now gaining significance in most of mango-producing countries worldwide, impacting nearly all varieties, whether severely or mildly. The powdery mildew is quite prevalent in Pakistan and poses a substantial risk to mango production (Javaid 2005). The disease severity is primarily influenced by the prevailing climate scenarios. Additionally, India, Iran, and Mexico have also seen significant declines in mango harvests due to powdery mildew. While infection can develop on both foliage and blossoms, the majority of losses are primarily caused by infection of the blossoms.

23.3.1.1 Geographical Distribution

Powdery mildew has been found to cause significant harm primarily in regions located upto latitudes 40° Northern-Southern of the Equator. The reported locations of this disease include Pakistan, Myanmar, Israel, Nepal, New South Wales (Zaire),Bangladesh, India, Sri Lanka, Lebanon (Asia), Congo, Zimbabwe, Egypt, Mozambique, Ethiopia, Malawi, Mauritius, Costa Rica, Reunion, Zambia, Jamaica, Tanzania, South Africa (Africa), Kenya, USA (California and Florida), Queensland and New Caledonia (Australia), Mexico, and Guatemala (Central America), as well as Brazil. Colombia, Venezuela and Peru are countries located in South America (Prakash et al. 1996). Additionally, there have been reports of this infection occurring in the Canary Islands and Cuba (Hernandez et al. 1955; Padron 1983). The illness is widespread across all over Pakistan.

23.3.1.2 Economic Significance

Significant losses occur as a result of the shedding of unimpregnated diseased blooms and underdeveloped fruits. In Australia, the occurrence of mildew on untreated trees was found toreacha maximum of 23%, while it was 11.5% on trees treated with Mancozeb (Peterson et al. 1991). According to Anonymous (1968), there was a 20% decline in yield in Venezuela. The disease can lead to a decrease in crop output of up to 90%, primarily due to its impact on the formation and growth of fruits (Schoeman et al. 1995). This pathogen causes an approximate 20% decrease in mango yield in India during some years (Anonymous 1930). One calculation placed the loss between 22.35 - 90.41% (Prakash and Raoof 1994). In Punjab's major mango-growing regions, the infection's prevalence varied from 0 to 70% (Javaid 2005).

23.3.1.3 Symptoms

The symptoms are evident on the inflorescence, flowering stalk, leaves, and immature fruits. Inflorescences are vulnerable from the moment as the main axis undergoes a change in color until fruit set occurs. The hallmark of the illness is the colorless, powdery fungal growth on affected areas, which is accompanied by many conidia organized in chain on conidiophores. Flowers that are infected with the mildew fungus shed their petals. The sepals are more vulnerable compared to the petals. The impacted flowers exhibit a lack of ability to bloom and may shed early. The immature fruits are completely enveloped by the fungal epidermis of the affected fruit, resulting in the formation of splits and corky tissues. Older fruits exhibit purplish brown mottled patches on their skin. Before they are completely ripe, the fruits may stay on the tree until they reach the size of peas, at which point they may fall off. Usually, between 20 and 40 percent of blooms and fruit are lost due to illness. On the other hand, the harvest of mangoes could completely disappear during an outbreak (Prakash et al. 1996). When immature leaves turn from brown to a pale green colour, infection is frequently seen. Attacks can occur on both surfaces of the young leaves, which show up as tiny, uneven, greyish spots. Usually, though, the symptoms are more apparent on the inner surface of the leaves. Frequently, these patches merge together and cover larger surfaces, resulting in a purplish brown hue. As time progresses, the patches undergo a change in color, becoming darker. When the conditions are right, a dense, white growth covers the regions that have been invaded. This growth consists of a conidia and mycelium mat. Usually, the infection is limited to the area that includes the affected leaves' main and secondary veins. As such, these leaves frequently display deformation, twisting, and curling.

23.3.1.4 Impact of Climate Change on Disease Development

Elevations in CO_2 and temperature are anticipated to elicit intricate impacts on plant pathogens. An elevation in both the CO_2 concentration and temperature resulted in a noticeable escalation of powdery mildew on plants. The concentration of CO_2 and temperature have been shown to have an impact on the severity of diseases and the formation of mycotoxins. The growth of powdery mildew was more significantly influenced by the combination of CO_2 and temperature rather than by CO_2 levels alone (Gullino et al. 2018). Nevertheless, an elevation in carbon dioxide levels can trigger physiological alterations that have the potential to impact the progression of diseases (Table 23.1).

23.3.1.5 Mitigation Strategies for Powdery Mildew

Select arid regions with high temperatures for mango growth; if feasible, refrain from selecting regions that have regular rainfall throughout the flowering season. Prune plants to ensure that blossoms are within the reach of spraying tools and to enhance

Table 23.1 Impact of climate change on the severity of mango diseases

Disease	Pathogen	Impact of climate change	Expected change in incidence and severity
Mango dieback	*Lasiodiplodia theobromae*	Increased temperature and drought stress create favorable conditions for the pathogen, while tree stress reduces resistance	Higher incidence and severity in arid and semi-arid regions
Anthracnose	*Colletotrichum gloeosporioides*	Elevated humidity and warmer temperatures increase the pathogen's sporulation and spread	Increased incidence and severity, particularly in regions with higher rainfall or humidity
Powdery mildew	*Oidium mangiferae*	Warmer and drier conditions can favor the pathogen's development, especially during flowering	Extended outbreaks during flowering seasons in dry regions with warm temperatures
Sooty mold	*Capnodium spp.*	Climate change-induced increases in pest populations (e.g., hoppers, aphids) lead to more honeydew excretion, promoting mold growth	Higher severity due to increased pest prevalence
Fruit rot	*Phomopsis mangiferae, Botryosphaeria spp.*	Prolonged rainfall and higher humidity during fruiting seasons can enhance fruit rot pathogen survival and infection rates	Higher prevalence during wetter seasons, with increased post-harvest losses
Bacterial canker	*Xanthomonas campestris pv. mangiferaeindicae*	Increased temperatures and humidity stress mango trees, making them more susceptible to bacterial infections	Increased severity in warm and humid regions
Stem-end rot	*Lasiodiplodia spp., Phomopsis spp.*	Higher temperatures and moisture can enhance pathogen activity and stress mango trees, making them more prone to infection during post-harvest handling	Higher post-harvest losses in regions with increasing temperatures and fluctuating humidity levels

(continued)

Table 23.1 (continued)

Disease	Pathogen	Impact of climate change	Expected change in incidence and severity
Black spot	*Asperisporium caricae*	Warmer and wetter conditions can favor the spread of the pathogen and spore germination, especially in tropical regions	Increased incidence in humid and tropical zones
Malformation	*Fusarium mangiferae*, hormonal imbalance	Altered temperature patterns and higher CO_2 levels affect flowering patterns, potentially exacerbating the physiological and fungal components of the disease	More unpredictable and severe cases due to climate-induced flowering stress
Root rot	*Phytophthora spp.*	Prolonged waterlogging due to erratic rainfall increases the survival and infectivity of the pathogen	Higher incidence in areas experiencing increased flooding and poorly drained soils
Alternaria leaf spot	*Alternaria alternata*	Fluctuations in temperature and humidity favor spore production and infection cycles	Increased outbreaks during alternating wet and dry spells

the flow of air in the canopy. Collect and dispose of mango leaves that have fallen to the ground, and eliminate panicles that are heavily infested. Implement intercropping techniques by cultivating mango trees alongside other kinds of fruit trees or forest species. Ensure that tall weeds are kept at a distance from mango trees. Israel and South Africa have indicated that applying phosphate fertilizer solutions, such as monopotassium phosphate (KH2PO4) at a concentration of 0.5%, individually or in combination with standard powdery mildew fungicides, is an economically and successful method. To effectively manage mango powdery mildew on highly susceptible cultivars, it is recommended to apply fungicide treatments at the appropriate times. The majority of the fungicides approved for the management of powdery mildew on mangoes have the potential to be effective. The initial spray should be applied when the flowers are at 50% of their full bloom, and subsequent sprays should be done every 7–14 days until the fruits have formed. Sulfur dusts or wettable sulfurscan effectively reduce powdery mildew, but they may also cause phytotoxicity, or "burning," to flowers and early fruit if applied during bright and warm weather.

23.3.2 *Anthracnose*

Anthracnose is a global disease that causes severe damage to mangoes both prior to and after they are harvested. Mainly, *C. gloeosporides* is the pathogen of this disease. The disease adversely affects affected mango trees, resulting in reduced productivity and diminished fruit quality. Anthracnose has the potential to result in a complete loss of crop production in farms that are not adequately managed and provide favorable conditions for the pathogen to propagate. Since the disease is common in practically every area where the crop is grown and causes large economic losses, a great deal of research has been done on a number of topics, including postharvest losses, destruction, characterization, evaluation, and classification, as well as resistance management and genomics control.

23.3.2.1 Geographical Distribution

The first documented report of Colletotrichum species on a global scale was made by Corda in 1831. Subsequently, the disease has been spreading to several trees and crops, including mangoes, on a global scale (Ciofini et al. 2022). MAD is prevalent in multiple nations across different continents. In Africa, it is seen in Ghana, Ethiopia, Ethiopia, Côte d'Ivoire, South Africa and Nigeria. In Oceania, it is present in Australia. In Asia, it is observed in Pakistan, Bangladesh, India, China, Taiwan and Indonesia. In the Americas, it is seen in Peru, Mexico and Colombia.

23.3.2.2 Economic Importance

Mango anthracnose commonly results in reduced mango fruit yield due to a decrease in either the quality or the quantity of the harvested fruits. According to local sources, Mango Anthracnose Disease (MAD) poses a serious threat to Ghana's mango crop, causing a 30% decrease in output in one particular district (Honger et al. 2014). Roughly 39% of India's decline in mango yield can be attributed to the anthracnose disease (Prakash and Srivastava 1987). Mango Anthracnose Disease causes 30–60% of mango harvests to be lost each year, and under ideal conditions, it can even result in a 100% loss (Kamle and Kumar 2016). Anthracnose disease was responsible with a 50.28% yield loss in Gondunglegi, Indonesia (Kumari et al. 2017; Arauz 2000). Between 1990 and 1992, a 29.6% postharvest loss was noted in Himachal Pradesh, India (Sharma and Verma 2007). In Hyderabad, a significant proportion of mango fruits, ranging from 20 to 30%, experienced decay as a result of the presence of *C. gloeosporioides*. It was reported that mango productivity in Bangladesh is reduced by 25–30% due to anthracnose and stem end rot infections (Hossain and Ahmed 1994). According to studies conducted by Sardsud et al. (2003), mangoes in Thailand observed a loss rate of 62.8% during mango harvesting and 63.2% postharvest losses due to this disease. Anthracnose is responsible for a significant reduction of 30–60%

in China's mango yield every year (Li et al. 2019). In Pakistan, the losses can reach 10–90% and even 100% losses can be observed under unmanaged plantations where conditions become favourable for pathogen dispersal.

23.3.2.3 Symptoms

The disease manifests as elliptical or erratic brown to dark brown depressed lesions of different sizes, scattered across the entire leaf surface, particularly on fresh leaves but even on mature leaves. As the leaves age, these spots grow into bigger lesions that are irregular or rectangular in form and have a crimson halo around them (Qin et al. 2019).The fungus may cause blossom blight on blooming plants. Long lesions ranging in hue from dark grey to black may develop in the stem. Little black spots are also forming on the open blooms; these spots will eventually grow larger and cause the plant to die. The procedure that causes the blighted blossoms to turn from brown to black and become brittle (Alemu et al. 2014). Occasionally, the fungus penetrates the afflicted tree's branches, twigs, and stems. Twigs exhibit symptoms in the form of tiny, enlarged oblong and necrotic lesions that gradually merge and spread. Severe infections can lead to the invasion of twigs by the fungus, resulting in dieback (Tovar-Pedraza et al. 2020).

23.3.2.4 Impact of Climate Change on Disease Development

Anthracnose is a significant disease that impacts all stages of mango growth and leads to a substantial decrease in yield (Uddin et al. 2018). Elevated levels of both temperature and relative humidity contribute to the occurrence of anthracnose disease (Benatar et al. 2021). Furthermore, the spores of anthracnose are distributed through the splashes of rain and irrigation water (Uddin et al. 2018). Wind impacts the development of diseases by positively interacting with the distribution of resources and the density of pests. The conidia of anthracnose, which are found on mature plants and are spread by wind and rain, might act as the main source of infection to contaminate fruit. They can also serve as secondary inoculum source on plants that already have anthracnose lesions (Gao et al. 2021) (Table 23.1).

23.3.2.5 Mitigation Strategies for Anthracnose

Management strategies must be economically viable, highly effective, and environmentally safe for both consumers and famers. An effective way to manage anthracnose is to apply treatments either before or after harvest, or some combination of the two. There is a diverse selection of strategies available for managing mango anthracnose in practical situations. The key components include fungicide treatment, adherence to suitable cultural practices, and the selection of cultivars. The diverse reactions of different cultivars to the infection in different places have hindered the continuous use of resistance as a means of controlling anthracnose. By selecting the right mango variety, farmers may produce a crop with fewer cases of anthracnose, leading to higher yields and higher-quality fruit. Furthermore, resistance improves sustainability and cost-effectiveness by lowering the requirement for fungicides to preserve the crop (Leadbeater 2014). A high degree of relative humidity or wet circumstances are required for the illness to thrive (Božič and Kanduč 2021). Thus, it is advantageous to establish farms in regions characterized by a pronounced dry season, as this facilitates the cultivation of fruit without the risk of diseases. Furthermore, it has been proposed that field sanitation methods should include the collection and incineration of debris from decaying trees and fruit (Sosnowski et al. 2009).The primary emphasis and scrutiny in combating mango anthracnose has been on the application of fungicides. The application of fungicides minimizes harm to inflorescence and fruit. Depending on the destination of the exported fruit, several fungicides may be applied. Non-systemic fungicides such as captan, zineb, and maneb were shown to be successful against the fungus when administered once a week during flowering and once a month during fruit development (Sardrood and Goltapeh 2018). In South Africa, it is advised to apply copper oxychloride either alone or in combination with zineb prior to harvesting in order to prevent anthracnose after 14 days in wet conditions and 28 days in dry conditions (Akem 2006). Additionally, copper oxychloride and mancozeb were administered alternately every month until harvest time after the fruit had begun to mature. While copper fungicides are suggested, they are less efficient than dithiocarbamates in managing high disease pressure, which raises concerns about their potential harm to mango blooms (Thind and Hollomon 2018). Fungicides that have the ability to respond after development of an infection, such as imidazole prochloraz and benzimidazoles are successful in combating anthracnose. Regular applications of benomyl, occasionally combined with protectant fungicides, were employed to prevent the emergence of resistance in pathogen population. Prochloraz has been applied both as a preventive measure and as a treatment (Chiangsin et al. 2016). One of the most important aspects of postharvest management strategies for mango anthracnose infection may involve focussing on the mitigation of dormant infections on mangoes. Postharvest treatments have traditionally been used to combat anthracnose in ripe fruit. Research conducted by Mirshekari et al. (2012) has demonstrated the effectiveness of hot water dips in managing anthracnose. From a commercial standpoint, latent infection is eradicated through the application of chemicals, heat, or a combination of both methods. Researchers have proposed subjecting individuals to temperatures ranging from 50 to 55 °C for periods of time ranging from 3 to 15 min (Mirshekari et al. 2012). Various fungicides have been employed post-harvest to manage anthracnose. Thiabendazole and Prochlorazcan be used. Nevertheless, the efficacy of these treatments fluctuates depending on the disease severity (Shi et al. 2020). Postharvest procedures, such as utilizing a controlled atmosphere and cold storage, preserve the quality of produce by delaying the ripening process and preventing damage. This technique has various limitations that reduce its potential benefits. Mangoes are vulnerable to chilling and can be damaged when exposed to temperatures under 10–13 °C (Patel et al. 2016). When fruit is left to mature in its natural habitat, disease develops. The use of postharvest biological control to prevent mango anthracnose disease has been sparse and inconsistent. Anthracnose pathogen *C. gloeosporioides* has been found to have antagonistic relationships with a number of microorganisms (Kankam et al. 2022). According to Zin and Badaluddin (2020), *Trichoderma* spp., Pseudomonas spp., and Bacillus spp. are recognized for their antagonistic qualities against plant diseases. They have been shown to be more effective than *C. gloeosporioides* in controlling the spread of anthracnose disease by outcompeting it for resource and habitat. However, despite its seeming potential, there is currently no evidence of this discovery's actual use in practice. Recent research has also demonstrated the efficacy of utilizing botanical substances for the treatment of anthracnose (Alemu et al. 2014). Research has shown that the growth of *C. gloeosporioides* in a controlled laboratory environment can be inhibited by water-based extracts derived from the leaves of *Azadirachtaindica* and *Eucalyptus camaldulensis*. Even in field settings, the use of foliar extracts reduced the frequency and impact of anthracnose disease on mango trees (Haider et al. 2020). The deficiencies and irregular effectiveness of postharvest treatments highlight the necessity for a comprehensive approach to tackle mango anthracnose disease.

23.3.3 Mango Malformation

The primary etiological agent of malformations diseases in mangos is Fusarium mangiferae. Recent research has shown that the infection may have a fungal cause. The occurrence of malformation poses a significant risk to the mango cultivation regions in Pakistan, resulting in crop losses of up to 70%. The intricate nature of the disease is evident from the varied assertions made by researchers from several countries regarding its origin, such as physiological, fungal, viral, nutritional and acarological.

23.3.3.1 Geographical Distribution

The occurrence of mango malformation was initially identified in 1891 by Mariesin Darbhanga, Bihar (Watt 1891). Subsequently, the disease was documented in several regions of the country, however significant focus was only directed towards it in the mid-1950s when it reached a critical level. The disease primarily affects the trees in northern India, with more than 50 percent of them being impacted. Many nations where mangoes are grown have confirmed the existence of this phenomenon: Egypt (Hassan 1944); South Africa (Schwartz 1968); Central America, USA (Florida); Mexico; Israel (Malo and Mc Millan 1972); Pakistan, Cuba, Bangladesh; Sudan; Australia; Brazil; and the United Arab Emirates (Kumar and Beniwal 1992; Kumar et al. 1993).

23.3.3.2 Symptoms

The workers pointed out two clearly different types of symptoms: floral malformation (MF) and vegetative malformation (MV).The occurrence of vegetative malformation is more prominent in early seedlings and seedling trees compared to grafted plants. The impacted seedlings exhibited an overabundance of vegetative branches, characterized by restricted development, enlargement, and unusually short spaces between nodes. Floral malformation (MF) is a condition characterized by abnormal development of the inflorescence. The primary symptoms of MFare the decreased length and compression of internodes, resulting in a distorted, broom-like appearance. Abnormal panicles typically do not produce fruits. Even the fruits that ripen throughout their respective seasons do not exceed the pea size. These have been discovered to produce regular fruit throughout non-seasonal periods (Jawanda 1963).

23.3.3.3 Impact of Climate Change on Disease Development

Multiple environmental factors play a role in the development and intensity malformation disease. Fluctuations in temperature and excessive rainfall not only affect the reproductive patterns of mango trees, but also make them more susceptible to a range of diseases, with the most significant being mango malformation disease (Ramírez and Kallarackal 2018). The interaction between temperature and humidity is crucial in establishing optimal circumstances for the proliferation and onset of pathogen growth and disease. During the flowering period, elevated temperatures and relative humidity facilitate the spread of the infection, resulting in a higher occurrence of the disease. The onset of mango malformation infection is greatly influenced by environmental factors (Noriega-Cantú et al. 1999) (Table 23.1).

23.3.3.4 Mitigation Strategies for Mango Malformation

The reviewed research clearly indicates that there are currently no specific control methods that can be recommended for mango malformation. Nevertheless, the subsequent actions can potentially decrease the occurrence of malformation infection:

- It is recommended to refrain from using scion sticks from trees that have deformed flowering for propagation purposes. The indexing healthy trees of mango is performed to obtain material for propagation purposes.
- Only authorized samples should be utilized for reproduction.
- Once a symptom is observed, it is recommended to prune the afflicted terminals, including the basal 15–20 cm of seemingly in good physical shape area and bunt.
- To prevent the disease from recurring, it is advisable to apply fungicides as a prophylactic precaution on orchards that are situated in areas prone to disease.

23.4 General Mitigation Strategies to Address Climate Change Impact on Mango Diseases

The host plants, diseases, and relationships between these creatures are all impacted by the substantial changes that climate change brings about in the mango ecosystem. Global agricultural systems are expected to experience both positive and negative consequences from climate change, with the negative effects outweighing the positive ones. A number of measures have been implemented in an attempt to lessen the impact of global warming on agricultural productivity, such as the introduction of superior seedlings and types with increased disease resistance. Enhancing the genetic traits of mango planting materials in response to climate change shows great potential for adaptation. When selecting a cultivar or rootstock, the goal should be to build tolerance to adverse weather conditions and to have a robust root system that efficiently absorbs water. Furthermore, mango seedling growers employ adaptation measures to offset the impacts of climate change, such as implementing irrigated farming, constructing gutters, providing shelter, and incorporating mulch into the soil. An effective approach to mitigate the hazards of spreading new infections is the implementation of integrated disease management (IDM), which is an environmentally friendly strategy. The integrated pest control approach focuses on employing preventive or indirect methods in plant protection to minimize environmental harm while maximizing crop productivity. While IDM tactics have been implemented for certain diseases, pesticides continue to be the favored choice for farmers in managing mango diseases. Successful implementation of disease management systems with low pesticide input necessitates a thorough understanding of the biology of pathogen and their interactions with the crop and the environment. Gaining knowledge about the variables that influence crop infection, such as the interplay between the host and the pathogen, forms the foundation for effective disease control (Table 23.2).

Table 23.2 Improved Management of fungal diseases under climate change

Fungal disease	Pathogen	Challenges under climate change	Improved mitigation strategies
Mango dieback	*Lasiodiplodia theobromae*	Higher temperatures and drought stress increase tree susceptibility and pathogen virulence	• Develop drought-tolerant and dieback-resistant cultivars • Use of endophytic biocontrol agents like *Trichoderma spp.* • Improved irrigation management to reduce tree stress
Anthracnose	*Colletotrichum gloeosporioides*	Elevated humidity and rainfall increase sporulation and disease spread	• Use of resistant mango varieties • Application of biostimulants (e.g., seaweed extracts) to boost plant immunity • Smart spraying technology for targeted fungicide application
Powdery mildew	*Oidium mangiferae*	Warmer and drier conditions favor pathogen growth during flowering	• Early detection using weather-based disease forecasting systems • Use of sulfur-based organic fungicides • Improved canopy ventilation through strategic pruning
Sooty mold	*Capnodium spp.*	Increased pest populations lead to more honeydew deposits, enhancing mold growth	• Strengthen integrated pest management (IPM) programs to control pests • Application of biopesticides like neem oil • Regular washing of honeydew deposits using water sprays

(continued)

Table 23.2 (continued)

Fungal disease	Pathogen	Challenges under climate change	Improved mitigation strategies
Fruit rot	*Phomopsis mangiferae, Botryosphaeria spp.*	Prolonged rainfall and humidity increase pathogen survival and infection rates	• Use of pre-harvest fungicide sprays during critical periods • Develop post-harvest treatments like hot water dips or UV irradiation • Improved orchard sanitation and drainage
Stem-end rot	*Lasiodiplodia spp., Phomopsis spp.*	Increased temperature and humidity promote infection during post-harvest handling	• Post-harvest treatments using natural antifungal compounds (e.g., essential oils) • Adoption of cold chain logistics to minimize pathogen activity • Improved packaging materials to reduce moisture accumulation
Black spot	*Asperisporium caricae*	Warmer and wetter conditions favor disease outbreaks	• Implementation of weather-based disease warning systems • Use of resistant cultivars • Enhanced pruning to reduce leaf wetness
Alternaria leaf spot	*Alternaria alternata*	Fluctuating temperatures and humidity favor spore germination and infection cycles	• Foliar sprays with systemic fungicides during high-risk periods • Use of bio-based resistance inducers like salicylic acid • Timely removal of infected leaves

(continued)

Table 23.2 (continued)

Fungal disease	Pathogen	Challenges under climate change	Improved mitigation strategies
Root rot	*Phytophthora spp.*	Erratic rainfall and waterlogging increase pathogen survival and infectivity	• Installation of efficient drainage systems to prevent waterlogging • Application of soil amendments like biochar to improve soil health • Use of antagonistic microbes such as *Pseudomonas spp.*
Malformation	*Fusarium mangiferae*, hormonal imbalance	Stress caused by erratic climate patterns disrupts flowering, increasing disease severity	• Use of plant growth regulators (e.g., GA3) to reduce hormonal imbalance • Timely fungicide sprays to control fungal infections • Resistant cultivars to minimize malformation susceptibility

References

Abeydeera LHUW, Mesthrige JW, Samarasinghalage TI (2019) Global research on carbon emissions: a scientometric review. Sustainability 11:3972

Adams RM, Hurd BH, Lenhart S, Leary N (1998) Effects of global climate change on agriculture: an interpretative review. Clim Res 11:19–30

Agrios GN (2004) Plant pathology, 5th edn. Elsevier, London, p 922

Akem CN (2006) Mango anthracnose disease: present status and future research priorities. Plant Pathol J 5:266–273. https://doi.org/10.3923/ppj.2006.266.273

Alemu K, Ayalew A, Woldetsadik K (2014) Antifungal activity of plant extracts and their applicability in extending the shelf-life of mango fruits. Food Sci Qual Manag 33:47–53

Arauz LF (2000) Mango anthracnose: economic impact and current options for integrated managaement. Plant Dis 84:600–611. https://doi.org/10.1094/PDIS.2000.84.6.600

Arora M, Goel NK, Singh P (2005) Evaluation of temperature trends over India/Evaluation de tendances de temperature en Inde. Hydrol Sci J 50:81–93

Benatar GV, Wibowo A, Suryanti (2021) First report of *Colletotrichum asianum* associated with mango fruit anthracnose in Indonesia. Crop Prot 141:105432

Bokhari SAA, Rasul G, Ruane AC, Hoogenboom G, Ahmad A (2017) The past and future changes in climate of the rice-wheat cropping zone in Punjab, Pakistan. Pak J Meteorol 13:26

Božič A, Kanduč M (2021) Relative humidity in droplet and airborne transmission of disease. J Biol Phys 47:1–29. https://doi.org/10.1007/s10867-020-09562-5

CDIAC (2020) Carbon dioxide information analysis center. www.cdiac.ess-dive.lbl.gov. Accessed 13 November 2020

Challinor AJ, Watson J, Lobell DB, Howden SM, Smith DR, Chhetri N (2014) A meta-analysis of crop yield under climate change and adaptation. Nat Clim Chang 4:287–291

Chiangsin R, Wanichkul K, Guest DI, Sangchote S (2016) Reduction of anthracnose on ripened mango fruits by chemicals, fruit bagging, and postharvest treatments. Australas Plant Pathol 45:629–635. https://doi.org/10.1007/s13313-016-0456-x

Ciofini A, Negrini F, Baroncelli R, Baraldi E (2022) Management of post-harvest anthracnose: current approaches and future perspectives. Plants 11:1856. https://doi.org/10.3390/plants11141856

Costinot A, Donaldson D, Smith C (2016) Evolving comparative advantage and the impact of climate change in agricultural markets: evidence from 1.7 million fields around the world. J Pol Econ 124:20–25

DeCandolle AP (1884) Origin of cultivated plants. London

Gao Y, Ren C, Liu Y, Zhu J, Li B, Mu W, Liu F (2021) Pepper-maize intercropping affects the occurrence of anthracnose in hot pepper. Crop Prot 148:105750

Gullino ML, Pugliese M, Gilardi G, Garibaldi A (2018) Effect of increased CO_2 and temperature on plant diseases: a critical appraisal of results obtained in studies carried out under controlled environment facilities. J Plant Pathol 100:371–389

Haider E, Khan MA, Atiq M, Shahbaz M, Yaseen S (2020) Phytoextracts as management tool against fungal diseases of vegetables. Int J Biosci 16:303–314. https://doi.org/10.12692/ijb/16.3.303-314

Hassan AS (1944) Notes on eriophes mangiferae. S N Bull dela Soc Fouad IER.Ent 28:179

Hernandez J, Gallo L, Lobet L, Jaizmevega MC (1955) A preliminary study of the fungal species associated with pine apple, papaya and mango in Canary Islands. An Inst Nac Invest Agrar Agric 28:171–180

Honger JO, Offei SK, Oduro KA, Odamtten GT, Nyaku ST (2014) Identification and species status of the mango biotype of *Colletotrichum gloeosporioides* in Ghana. Eur J Plant Pathol 140:455–467. https://doi.org/10.1007/s10658-014-0480-z

Hossain AA, Ahmed A (1994) A monograph on mango varieties of Bangladesh. Bangladesh Agricultural Research Institute, Dhaka

IPCC (2007) Climate change 2007: impacts, adaptation and vulnerability. In: Working group II contribution to the fourth assessment report of the intergovernmental panel on climate change. Cambridge University Press, Cambridge, UK

IPCC (2014) Climate change 2014: synthesis report. In: Pachauri RK, Meyer LA (eds) Contribution of working groups I, II and III to the fifth assessment report of the intergovernmental panel on climate change. IPCC, Geneva, Switzerland, 151 p

Javaid A (2005) Major diseases of mango in Pakistan and their management—a review. Pak J Phytopathol 17:108–112

Javaid A, Shoaib A, Khan S (2012) Mango cultivation in Pakistan. In: Valavi SG, Mohan R, Govil JN, Peter KV, Thottappilly G (eds) Mango cultivation in different countries. Studium Press LLC, USA, pp 385–394

Jawanda JS (1963) Studies on mango malformation. P Hortic J 3:281

Kamle M, Kumar P (2016) *Colletotrichum gloeosporioides*: pathogen of anthracnose disease in mango (*Mangifera indica* L.). Curr Trends Plant Dis Diagn Manag Pract 207–219. https://doi.org/10.1007/978-3-319-27312-9_9

Kang Y, Khan S, Ma X (2009) Climate change impacts on crop yield, crop water productivity and food security—a review. Prog Nat Sci 19:1665–1674

Kankam F, Larbi-Koranteng S, Adomako J, Kwodaga JK, Akpatsu IB, Danso Y et al (2022) Anthracnose disease of mango: epidemiology, impact and management optionsn. In: Current and emerging challenges in the diseases of trees. IntechOpen, London

Karimi V, Karimi E, Keshavarz M (2018) Climate change and agriculture: Impacts and adaptive responses in Iran. J Integr Agric 17:1–15

Kostermans AJGH, Bompard JM (1993) The mangoes, their botany, nomenclature, horticulture and utilization. Academic Press, London

Kumar J, Beniwal SPS (1992) Mango malformation. In: Kumar J, Chaube HS, Singh US, Mukhopadhyay AN (eds) Plant diseases of international importance. Prenttice-Hall Inc., New Jersey, pp 357–393

Kumar J, Singh US, Beniwal SPS (1993) Mango malformation: one hundred years of research. Ann Rev Phytopathol 31:217–222

Kumari P, Singh R, Punia R (2017) Survival of *Colletotrichum gloeosporioides* causing anthracnose of mango at different depths and durations in soil. J Pharmacogn Phytochem 6:2194–2198

Leadbeater AJ (2014) Plant health management: fungicides and antibiotics, encyclopedia of agriculture and food systems. Biology. https://doi.org/10.1016/B978-0-444-52512-3.00179-0

Li Q, Bu J, Shu J, Yu Z, Tang L, Huang S et al (2019) *Colletotrichum* species associated with mango in southern China. Sci Rep 9:18891. https://doi.org/10.1038/s41598-019-54809-4

Luck J, Spackman M, Freeman A, Trebicki P, Griffiths W, Finlay K, Chakraborty S (2011) Climate change and diseases of food crops. Plant Pathol 60:113–121

Mahato A (2014) Climate change and its impact on agriculture. Int J Sci Res Publ 4:1–6

Malhi GS, Kaur M, Kaushik P (2021) Impact of climate change on agriculture and its mitigation strategies: a review. Sustainability 13:1318

Malo SE, Mc Millan RT Jr (1972) A disease of *Mangifera indica* L. in Florida similar to mango malformation. Fla Sta Hort Soc 85:254–268

Mendelsohn R (2009) The impact of climate change on agriculture in developing countries. J Nat Res Policy Res 1:5–19

Mirshekari A, Ding P, Kadir J, Ghazali HM (2012) Effect of hot water dip treatment on postharvest anthracnose of banana var Berangan. Afr J Agric Res 7:6–10. https://doi.org/10.5897/AJAR11.056

Molua EL, Lambi CM (2007) The economic impact of climate change on agriculture in Cameroon. Policy Research Working Paper, The World Bank, Washington, DC, USA, pp 1–31

Mukherjee SK (1951) Origin of mango. Indian J Genet Plant Breed 11:49–56

Mukherjee SK (1953) Origin, distribution and phylogenetic affinities of the species of *Mangifera* L. J Linn Soc Bot 55:65–83

NASA Earth Observatory. Goddard Space Flight Centre, United States. www.earthobservatory.nasa.gov. Accessed 15 May 2020

Nazir N, Bilal S, Bhat KA, Shah TA, Badri ZA, Bhat FA, Dorjey S (2018) Effect of climate change on plant diseases. Int J Curr Microbiol Appl Sci 7:250–256

Noriega-Cantú DH, Téliz D, Mora-Aguilera G, Rodríguez-Alcazar J, Zavaleta-Mejía E, Otero-Colinas G, Campbell CL (1999) Epidemiology of mango malformation in Guerrero, Mexico, with traditional and integrated management. Plant Dis 83(3):223–228. https://doi.org/10.1094/PDIS.1999.83.3.223

Padron Soroa J (1983) Factors responsible for low yield of mango. Boletin de Resenas Protection de Plantas 10:37

Patel B, Tandel YN, Patel AH, Patel BL (2016) Chilling injury in tropical and subtropical fruits: a cold storage problem and its remedies: a review. Int J Environ Sci Technol 5:1882–1887

Peterson RA, Schipke LG, Clarkson PC (1991) Significance of two mango flower diseases in the dry tropics. Proc. 3rd Int. Mango Symp. held at Darwin, 24–29th Sept, 1989:338–343

Popenoe W (1932) Manual of tropical and sub-tropical fruits. Macmillan Co., New York

Prakash Om, Raoof MA (1994) Studies on Powdery Mildew (Oidium mangiferae) disease of mango: distribution, perpetuation, losses and chemical control. Bio Memoirs 207:31–45

Prakash Om, Srivastava KC (1987) Mango diseases and their management. A world review Today and tomorrow's Printers and publisher, New Delhi, pp 180

Prakash Om, Misra AK, Panday BK (1996) Anthracnose disease of tropical and sub-tropical fruits. In: Agnihotri VP, Om parkash, Ram kishun, Misra AK (eds) Disease scenario in crop plants, vol I. Fruits and vegetables, intern, Books and Periodical services, Delhi, pp 1–27

Purseglove JW (1972) Mangoes west of India. Acta Hortic 24:107–174

Qin LP, Zhang Y, Su Q, Chen YL, Nong Q, Xie L et al (2019) First report of anthracnose of *Mangifera indica* caused by *Colletotrichum scovillei* in China. Plant Dis 103:1043. https://doi.org/10.1094/PDIS-11-18-1980-PDN

Ramírez F, Kallarackal J (2018) Phenological growth stages of Feijoa [Acca sellowiana (O. Berg) Burret] according to the BBCH scale under tropical Andean conditions. Scientia Horticulturae 232:184–190. https://doi.org/10.1016/j.scienta.2017.12.059

Sardrood BP, Goltapeh EM (2018) "Effect of agricultural chemicals and organic amendments on biological control Fungi". In: Lichtfouse E (ed) Sustainable agriculture reviews (Vol. 12), Springer, Berlin, pp 217–359

Sardsud U, Sardsud V, Singkaew S (2003) Postharvest loss assessment of mango cv. Nam Dok Mai. Warasan Witthayasat Kaset. Available at: https://agris.fao.org/agris-search/search.do?recordID=TH2005000687

Sathaye J, Shukla PR, Ravindranath NH (2006) Climate change, sustainable development and India: global and national concerns. Curr Sci 90:314–325

Schoeman MH, Manicom BQ, Wingfield MJ (1995) Epidemiology of powdery mildew on mango blossoms. Plant Dis 79:524–528

Schwartz A (1968) A new mango pest. Farming S Africa 9:7

Sharma A, Verma KS (2007) In vitro cross pathogenicity and management of Colletotrichum gloeosporioides causing anthracnose of mango. Ann Plant Prot Sci 15:186–188

Shi N, Ruan H, Jie Y, Chen F, Du Y (2020) Sensitivity and efficacy of fungicides against wet bubble disease of Agaricus bisporus caused by Mycogone perniciosa. Eur J Plant Pathol 157:873–885. https://doi.org/10.1007/s10658-020-02047-0

Sosnowski MR, Fletcher JD, Daly AM, Rodoni BC, Viljanen-Rollinson SLH (2009) Techniques for the treatment, removal and disposal of host material during programmes for plant pathogen eradication. Plant Pathol 58:621–635. https://doi.org/10.1111/j.1365-3059.2009.02042.x

Strange RN, Scott PR (2005) Plant disease: a threat to global food security. Annu Rev Phytopathol 43:83–116

Thind TS, Hollomon DW (2018) Thiocarbamate fungicides: reliable tools in resistance management and future outlook. Pest Manag Sci 74:1547–1551. https://doi.org/10.1002/ps.4844

Tovar-Pedraza JM, Mora-Aguilera JA, Nava-Diaz C, Lima NB, Michereff SJ, Sandoval-Islas JS et al (2020) Distribution and pathogenicity of Colletotrichum species associated with mango anthracnose in Mexico. Plant Dis 104:137–146. https://doi.org/10.1094/PDIS-01-19-0178-RE

Uddin MN, Shefat SHT, Afroz M, Moon NJ (2018) Management of anthracnose disease of mango caused by *Colletotrichum gloeosporioides*: a review. Acta Sci Agric 2:169–177

Usman M, Fatima B, Khan MM, Chaudhry MI (2003) Mango in Pakistan: a chronological review. Pak J Agric Sci 40:151–154

Watt G (1891) The mango tree. In: A dictionary of the economic products of India, Govt. Printing Press, Calcutta, India, vol. 5, pp. 149

World Meteorological Organization (1992) International meteorological vocabulary, 2nd edn. WMO, Geneva, Switzerland

Yáñez-López R, Torres-Pacheco I, Guevara-González RG, Hernández-Zul MI, Quijano-Carranza JA, Rico-García E (2012) The effect of climate change on plant diseases. Afr J Biotechnol 11:2417–2428

Zin NA, Badaluddin NA (2020) Biological functions of *Trichoderma* spp. for agriculture applications. Ann Agric Sci 65:168–178. https://doi.org/10.1016/j.aoas.2020.09.003

Chapter 24
Insect Pests of Mango: Identification, Biology, and Management Strategies

Tauseef Khan Babar

Abstract Mango (*Mangifera indica* L.), popularly known as "The King of Fruits", is a native of Southeast Asia that has been cultivated for more than 4000 years. Currently grown in over 120 countries, India and Pakistan dominate global production, contributing 76.5% of the total. In Pakistan, Punjab and Sindh provinces are key contributors to mango cultivation, with significant regional variations in yield. Although the potential yield of mangoes is 21.6 tons per hectare, the average farm yields are significantly lower at 9.96 tons per hectare, primarily due to stress induced by both biotic and abiotic factors. In terms of biotic factors, insect pests, numbering over 400 species, pose a significant threat to mango production. These pests infest various parts of the mango plant at different growth stages, causing significant damage. The mango industry in Pakistan also faces challenges in international markets, with pest infestations, particularly fruit flies, fruit borers, stone weevils, and white scales, affecting export quality and prices. This chapter sheds light on the ecological dynamics of mango pests, highlighting the need for sustainable, eco-friendly pest management strategies to protect mango orchards and improve yield and quality. By understanding pest biology and refining control methods, this chapter aims to enhance the resilience of mango farming systems in the face of evolving pest pressures.

Keywords Biology · Identification · Insect pests · King of fruits · Management strategies · Mango

24.1 Introduction

Mango, *Mangifera indica* L. (Sapindales: Anacardiaceae), generally referred as "The King of Fruits" has been under cultivation for more than 4,000 years on this earth. Historically, its origin can be found in the South East Asian or Indo-Burman territory

T. K. Babar (✉)
Department of Entomology, Faculty of Agricultural Sciences and Technology, Bahauddin Zakariya University, Multan, Pakistan
e-mail: tauseefkhan@bzu.edu.pk

S. A. H. Naqvi et al. (eds.), *Climate Change and Mango Production*,
https://doi.org/10.1007/978-3-031-87035-4_24

(Rajan 2021). Currently, mango is being cultivated in more than 120 countries in the world, but India and Pakistan where it is revered as the national fruit hold 76.5% share in the global production (FAO 2023). Pakistan is the world's 5th largest mango grower with a production of 1.8 MT; 75% of the produce originates from the province of Punjab with the 63% area under cultivation while the province of Sindh occupying 37% area provides 25% of the fruit (Karar et al. 2021; Anonymous 2023). In Punjab, the major mango producing districts include Bahawalpur, Multan, Muzaffargarh, and Rahim Yar Khan while Hyderabad, Mirpur Khas, and Thatta are from the Sindh province (Shah et al. 2022).

Pakistan has a potential yield of 21.6 tons/ha but unfortunately at farm level average yield is 9.96 tons/ha (Lodhi et al. 2006). Numerous abiotic (environmental) and biotic (pests) factors are associated with this vast yield gap (Laxman et al. 2016; Maiti and Kumari 2016). Among biotic factors, 400 kinds of insect pests have been identified to trouble mango farms (Peña et al. 1998). Of these diversified insect pests, various species are found infesting the mango plantations in Pakistan at various phenological stages or parts of the plant like at inflorescence, leafhopper (*Amritodus atkinosoni* Lethierry, *Idioscopus clypealis* Lethierry, *I. niveosparus* Lethierry); on foliage and buds, thrips (*Scirtothrips dorsalis* Hood), mealy bug (*Drosicha mangiferae* Green, *D. stebbingii* Green), midges (*Erosomyia mangiferae* Felt), scale insects (*Aulacaspis tubercularis* Newstead), bud mites (*Eriophyes mangiferae* Sayed); fruit fly (*Bactrocera zonata* Saunders and *B. rubus* Linnaeus), stone weevil (*Sternochetus mangiferae* Fabricius), and fruit borer (*Deanolis albizonalis* Hampson) on fruits; and stem borer (*Batocera rufomaculata* DeGeer and *B. rubus* Linnaeus) and bark beetles (*Hypocryphalus mangiferae* Stebbing) (Peña and Mohyuddin 1997; Chapman 1999) (Table 24.1). Significantly, Pakistan is also ranked at 7th position among the mango exporting nations and its mangoes are exported to more than 60 countries around the world (FAO 2018; Ayyaz et al. 2019). Unfortunately, the country often losses the due price of its exported mango fruit due to multiple factors including the infestation of obnoxious pests like fruit fly (Badar 2015).

To fetch high profit from the export of this delicious and marvellous fruit, it is essential to use environment friendly pest control technologies in addressing the problems caused by insect pests. This chapter aims to provide valuable insights into the complex relationships between mango pests and their environment, offering practical solutions to safeguard mango orchards from the ever-evolving pest pressures. By advancing our understanding of pest biology and refining management strategies, we can enhance the resilience of mango farming systems and ensure their continued success in a changing agricultural landscape.

Table 24.1 Commonly occurring insect pest species at various phenological stages of mango plant

Insect pests at various phenological stages of mango plant	Pest species		Taxonomic classification	
	Common name	Scientific name	Family	Order
Insect pests of inflorescence and tender shoots	Mango leaf hopper	*Amritodus atkinosoni* Lethierry *Idioscopus clypealis* Lethierry *I. niveosparus* Lethierry	Cicadellidae	Hemiptera
Insect pests of foliage and buds	Mago thrips	*Frankliniella schultzai* Trybom *Thrips subnudula* Karny *Rhipiphorothrips cruentatus* Hood *Scirtothrips dorsalis* Hood	Thripidae	Thysanoptera
	Mango mealybug	*Drosicha mangiferae* Green *D. stebbingi* Green *Rastrococcus iceryoides* Green	Pseudococcidae	Hemiptera
	Mango midges	*Dasineura amaramanjarae* Grover *Erosomyia mangiferae* Felt *Procontarinia matteiana* Kieffer and Cecconi	Cecidomyiidae	Diptera
	Mango white scale	*Aulacaspis tubercularis* Newstead	Diaspididae	Hemiptera
Insect pests of stem	Mango stem borer	*Batocera rufomaculata* DeGeer *B. rubus* Linnaeus	Cerambycidae	Coleoptera
	Mango bark beetle	*Hypocryphalus mangiferae* Stebbing	Scolytidae	Coleoptera
Insect pests of fruits	Mango fruit borer	*Deanolis albizonalis* Hampson *Autocharis albizonalis* Hampson *Noorda albizonalis* Hampson	Pyralidae	Lepidoptera
	Mango fruit fly	*Bactrocera zonata* Saunders *B. dorsalis* Hendel	Tephritidae	Diptera
	Mango stone weevil	*Sternochetus mangiferae* Fabricius *S. gravis* Fabricius *S. frigidus* Fabricius	Curculionidae	Coleoptera

24.2 Insect Pests of Inflorescence and Tender Shoots

24.2.1 Mango Leafhopper

24.2.1.1 Species and Distribution

Mango hopper (Hemiptera: Cicadellidae) is the most devastating pest of mango (Sohi and Sohi 1990) which can inflict a loss of 80–100% at flowering and fruiting stages (Rahman et al. 2016). Though it is a monophagous pest of mango, there are a few reports of its feeding on fig (*Ficus carica*) and cesiman (*Thrunstira deliciose*) (Pruthi and Batra 1960). About 15 species of this notorious pest have been reported to exist in the mango growing regions of Asia (Viraktamath 2007) and *Amritodus atkinsoni* Lethierry, *Idioscopus clypealis* Lethierry, and *I. niveosparus* Lethierry are the three species in question (Fig. 24.1a–c) are of economic importance and are known to inflict serious damage in Pakistan (Soomro 1988; Mohyuddin and Mahmood 1992).

24.2.1.2 Identification

Eggs: Female lays oval shaped egg singly into the midrib of young leaf, flower stocks or unopened flower buds. Eggs are dull white at first, but they eventually turn light brown (Kannan and Rao 2006).

Nymphs: Newly born nymphs have two tiny red eyes and are whitish in hue. With each moulting, a gradual change in their coloration occurs which alters to yellow, yellowish green, green, and ultimately greenish brown. The wingless nymphs are very active moving rapidly around the inflorescence and shoots. Hopper's presence can easily be noticed, if we place our hand at the opposite side of the trunk or inflorescence, they tend to move in front of us (Pezhman and Radjabi 2002).

Adults: They have a wide head and slender abdomen in the direction of anal end which looks like a wedge-shaped body. All the species (Fig. 24.1a–c) are distinguishable to the naked eye on account of certain body features. For example, dark grey colour *A. atkinsoni* is considerably greater in size and accommodates two noticeable spots on the abdomen and scutellum, while *I. clypealis* is the smallest with the two spots on the scutellum (Fig. 24.1a, b) (Jindal et al. 2016). In a review of the *Idiocerine* family of leafhoppers from Pakistan, Khatri and Webb (2014) provided a description of these species (Fig. 24.2a, b) along with the three new species.

Fig. 24.1 Different mango hopper species found in the Pakistan include *Amritodus atkinosoni* (**a**; https://idtools.dpi.nsw.gov.au/keys/cicadell/species/amritodus.htm), *Idioscopus clypealis* (**b**; https://inaturalist.ca/taxa/447090), and *I. niveosparus* (**c**; https://uk.inaturalist.org/photos/252480274) (Soomro 1988). All the figures were accessed on 11th Jan 2024

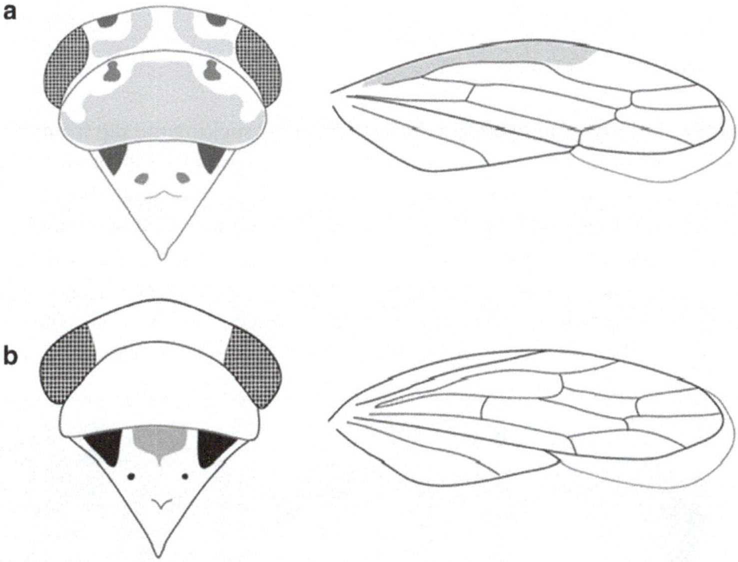

Fig. 24.2 Different morphological features differentiating between *A. atkinosoni* (**a**) and *I. clypealis* (**b**). The former species possesses a shagreen pronotum and forewing without outer subapical cells (**a**) whereas in the later species, fore margin of head is without spot in male (**b**) (Khatri and Webb 2014)

24.2.1.3 Biology

The *Idioscopus* species found in mango orchards generally share similar biological traits. However, *I. niveosparsus* typically prefers to breed on new leaf flushes and inflorescences, while *I. clypealis* is more commonly observed breeding on the inflorescences (Hiremath and Hugar 1989). Adults are present year-round in cracks and crevices, with their population rising as flowering begins (Talpur et al. 2002). The female deposits individual eggs on delicate branches, flower buds, and fresh leaves. An adult female can lay up to 200 eggs, which typically hatch within 4–7 days. The nymphs undergo three to four moults before reaching adulthood in approximately 8–13 days. The adult insects overwinter in cracks and crevices (Fig. 24.3a, b). In South Asia, usually a generation of hoppers develops within 15–19 days and annually 2–3 generations have been reported (Reddy et al. 2020; Awan et al. 2021). Abiotic conditions, such as temperature and relative humidity, have a significant effect in the population variations of this unpleasant bug (Anant et al. 2019). Mango hoppers are particularly sensitive to temperature increases, so their population may be influenced by the shade in plantations, with higher numbers likely in shaded areas compared to more exposed, sunny locations (Narasimha Murthy 1981). In the sub-continent region, the peak population of this pest are noticed in month of March and April (Tandon et al. 1983; Talpur et al. 2002).

24.2.1.4 Nature of Damage

Nymphs and adults of mango hopper cause damage by sucking the sap from different plant parts and fruits (Talpur et al. 2002). In general, hopper's infestation develops in two ways: First, primary damage occurs on the inflorescence and other vulnerable parts of the plant owing to the insertion of an ovipositor for egg laying, which causes physical harm and resulting in the drying and withering of such plant parts. Second, secondary harm occurs when both nymph and adult hoppers penetrate and drain the cell sap from sensitive leaves, shoots, inflorescences, flowers, and tender fruits (Fig. 24.3c, d). This high draining of plant sap causes flower buds to wither and

Fig. 24.3 *Amritodus atkinosoni* adult sitting on a bark of mango tree (**a**, **b**; Accessed on 11th Jan 2024, https://inaturalist.ala.org.au/photos/86038595). Typical symptoms of hopper's infestation on the tender shoots (**c**), inflorescence (**d**), and appearance of sooty mould due to deposition of honeydew on leaves (**e**), and fruits are shown (**f**) (Ganeswaran 2017; Ansari et al. 2019)

shed, as well as shoots and leaves to wilt and dry up (Reddy and Sreedevi 2016). The heavy feeding by mango hoppers leads to the secretion of honeydew on various parts of the plant, which fosters the growth of the black sooty mold fungus, *Capnodium mangiferae* Cooke. As the fungus spreads, it not only impedes the plant's growth and development but also disrupts the tree's normal photosynthetic processes. This fungal growth can reduce the plant's ability to produce energy, further weakening its health and productivity. Additionally, the accumulation of honeydew can attract other pests, exacerbating the overall damage to the mango tree (Hamid and Jalaluddin 2006). When the infestation is severe, the entire orchard seems sickly, and the damaged plants' shoots and leaves have hundreds of hoppers' cast skins on them. "Honey Dew Disease" or "Hopper Burn" are terms used to describe this kind of injury (Fig. 24.3e, f) (Backus et al. 2005). This pest thrives in old, neglected, tightly spaced orchards that are shaded and have high humidity levels (Namni et al. 2017).

24.2.1.5 Management

(1) Weeds like custard apple (*Annona squamosa* L.), and hibiscus (*Hibiscus trionum* L.) act as alternate host plants of hoppers therefore it is necessary to perform the regular cultural operations like ploughing and manual removal of weeds to avoid their proliferation in the field (Peña 2002).
(2) Timely pruning of mango orchards is useful to destroy the breeding places of hoppers. In this regard, priority should be given to prune the central branches of fully grown trees for facilitating the adequate light penetration (Rahman 2007).
(3) Host plant features play a critical role in imparting resistance against invasion of the different insect pests. Karar and Bakhsh (2018) found that "Fajri Khan" was the most susceptible and "Anwar Retaul" as the resistant variety in South Punjab, Pakistan. The authors ranked the several mango cultivars according to their susceptibility after evaluating the insect population density. Fajri Klan had the highest pest population density among the cultivars, followed by Chaunsa, White Chaunsa, Black Chaunsa, Langra, Dusehri, Retaul-12, and Sobey Di Ting. Among the assessed cultivars, Anwar Retaul had the lowest pest population density, suggesting that it might be the least vulnerable. Similar to this, in the past Talpur and Khuhro (2003) reported the relative susceptibility of "Langra" and "Sarolee" cultivars from the Sindh, Pakistan whereas the least population of the pest was recorded on Neelum, Zafran, and Dashehari.
(4) The weaver ant, *Oecophylla smaragdina* Fabricius, has been recognized as an effective biological control agent for managing the mango hopper. Research has shown that the abundance of these ants plays a crucial role in their success as a pest management tool (Peng and Christian 2005).
(5) Application of botanical pesticides like azadirachtin is recommended when the population of the pest is low (< 4/panicle) especially in the initial growth stages (Adnan et al. 2014). However, if it is > 4 hoppers/panicle, the use of systemic insecticides like imidacloprid or thiamethoxam or lambda-cyhalothrin at the panicle initiation stage can be undertaken. During the full bloom stage, the

use of agrochemicals is avoided to sustain the pollinators activity in mango orchards (Verghese and Devi Thangam 2011). Therefore, in this context, different modules have been developed to manage the infestation of mango hoppers in the sub-continent (Anithakumari et al. 2010; Ghaffar et al. 2022). For example, to manage the population of the *I. clypealis* Ghaffar et al. (2022) designed three modules. After the fruit harvest, thiamethoxam (Actara 25WG) was sprayed in module I. In module II, however, a coat of clothianidin (Trunk 20SC) was applied as a second spray exclusively on tree trunks in December. Following fruit harvest, module III involved a first spray of thiamethoxam (Actara 25WG), a second spray of clothianidin (Trunk 20SC) exclusively on tree trunks in December, and a third spray of spinetoram (Delegate 25WG) on mango trees prior to flowering season in January. Based on their findings, the authors concluded that the perusal of Module III resulted in a notable decline in the population density of *I. clypealis*.

(6) Entomopathogenic fungi including *Clonostachys rosea* Schroers and *Metarhizium anisopliae* Sorokin have demonstrated their potency against this pest under the laboratory conditions (Rahman et al. 2019; Tamta et al. 2022). Additionally, the infestation of *I. clypealis* in mango orchards has been reduced by the foliar application of other fungal isolates, *Lecanicillium lecanii* Zare and Gams and *Beauveria bassiana* (Balsamo) Vuillemin (Tumbada et al. 2018).

24.3 Insect Pests of Foliage and Buds

24.3.1 Mango Thrips

24.3.1.1 Distribution and Species

Thrips (Thysanoptera: Thripidae) infest leaves, blooms, and fruits of mango and are the escalating challenge to mango farmers (Grove et al. 2001). According to an estimate, 6000 species of thrips are reported to inflict damage to various fruits and vegetables around the world (Grimaldi et al. 2004; Mirabbalou 2013). Various species of the pest have been found feeding on different parts of the mango plants such as *Frankliniella schultzai* Trybom and *Thrips subnudula* Karny are limited to mango inflorescence, *Rhipiphorothrips cruentatus* Hood feeds on leaves, and *Scirtothrips dorsalis* Hood sucks the inflorescence and young fruits (Krishnamoorthy and Visalakshi 2012; Affandi et al. 2017) (Table 24.1). While the thrips species present on mango rootstocks include *Pantachaetothrips* spp., *Selenothrips rubrocinctus* Giard, *Caliothrips impurus* Priesner, and *S. dorsalis* (Patel et al. 1997).

Generally, they are polyphagous and cause damage to both horticultural (almond, apple, cashew nut, custard, guava, mango, pomegranate) and agricultural crops.

Furthermore, thrips also act as vector of different viral, bacterial, and fungal diseases of the crops (Aliakbarpour and Che 2011).

24.3.1.2 Identification

Eggs: Typically, white or yellowish in colour with bean-shape.

Larvae: Thrips have two larval instars. First instar larvae are transparent having a short, swollen, and seven segmented antennae and longer legs. Reticulated head and thorax without sclerotisation are the prominent identifying features. In contrast, the larvae in their second instar have larger body setae, a reticulated head and thorax, with sclerotization on the head, an enlarged mouth cone, three-segmented maxillary palpi, and elongated, cylindrical, seven-segmented antennae (Hoddle 2022).

Prepupa: Yellowish; each meso- and meta-thorax has two pairs of external wing buds; antennae are enlarged, short, and clearly segmented. Forward, in front of the head, are antennae (Hoddle 2022).

Pupa: Dark yellow, with red-pigmented eyes and ocelli; elongated wing buds; short, reflecting antennae; female pupae have a wider, pointed abdomen than males, who have a blunter, smaller abdomen. Curving back over the head are antennae (Hoddle 2022).

Adults: Adult thrips have dark bodies with a red band on the first abdominal segment. In their immature stages, the first two abdominal segments and the anal segments are bright red, while the rest of the body appears pale orange (Hoddle 2022).

24.3.1.3 Biology

Female *S. dorsalis* lay between 30 and 300 eggs in the tissues of young plants, with the eggs hatching within 4 to 26 days. The larval stage lasts from 2 to 16 days and includes two instars (Lowry et al. 1992; Murai et al. 2000). The second instar stops feeding on the plant as it reaches maturity and searches for a safe location to change into the pre-pupal and pupal stages (Lewis 1973). *S. dorsalis* needs 10–30 days to finish its life cycle, but this depends on the host plant and abiotic variables including humidity and temperature (Fig. 24.4) (Lewis 1997).

24.3.1.4 Nature of Damage

Flower thrips cause damage (Fig. 24.5a) by sucking sap and scraping the tissues of the plant's panicle including petals, anthers, pollen, and floral nectaries. Under severe infestation, leaf tips may turn brown and curl (Fig. 24.5b) (Aliakbarpour and Che 2010). Furthermore, the florets of the panicle are also affected due to oviposition

Fig. 24.4 Digrammatic illustration of the different life stages (egg, larva, prepupa, pupa, adult) of thrips. Accessed on 25th Jan 2024; https://biobee.co.za/pests/thrips/thrips-life-cycle-2

Fig. 24.5 An adult of mango thrips (**a**) with its characteristic infestation symptoms on leaves (**b**) and mango fruit (**c**) (Choi et al. 2013)

and feeding of the pest which results in a decrease in the fruit set and vigour of the panicle (Fig. 24.5c) (Sierra-Baquero et al. 2020).

24.3.1.5 Management

(1) Regular monitoring of the nursery plants is helpful in timely management of the pest. It can be accomplished by installation of the coloured sticky traps (Xia et al. 2019). While assessing the potential of different coloured (blue, green, orange, purple, white, yellow) sticky traps, Carrillo-Arámbula et al. (2022), found that white coloured traps attracted more population of the thrips.
(2) Application of plant based pesticides like neem can prevent the establishment of thrips (Patel et al. 2013).
(3) Use of new chemistry insecticides like chlorfenapyr, nitenpyram + chlorfenapyr, and imidacloprid on mango plants against *S. dorsalis* can lower the incidence of thrips infestation in nursery plantation (Karar et al. 2022).

24.3.2 *Mango Mealybug*

24.3.2.1 Species and Distribution

Following mango hoppers, the mango mealybug, *Drosicha mangiferae* Green (Hemiptera: Pseudococcidae), is the second most destructive pest affecting mangoes (Atwal 2008). Since 1907, it has been reported to have caused significant financial losses to mango growers of the Indo-Pakistani subcontinent (Green 1908) In some cases, yield losses of 50–80% have been observed (Karar et al. 2012). More than 20 species of this pest are known to infest mango plantations, with *Drosicha mangiferae* and *Drosicha stebbingi* identified as key pests in Bangladesh, Bhutan, China, India, Nepal, and Pakistan. Meanwhile, *Rastrococcus iceryoides* Green has been documented in Malaysia (Ben-Dov 1994; Muniappan et al. 2009) (Table 24.1). Among these, the giant mealybug, *D. mangiferae*, is a polyphagous pest which is known to feed on 71 different host plants (Subramanian et al. 2021).

24.3.2.2 Identification

Mealybugs are soft-bodied plant sap sucking insects that are commonly known as "scale insects." (Downie and Gullan 2004). The mealy wax secretion that typically covers their body is the source of the term "mealybug." (Kosztarab and Kozár 1988). Due to their appearance, mealybugs are often mixed with woolly aphids or cushion scale insects, but their lifelong leg retention sets them apart (Mani and Shivaraju 2016).

Eggs: Females lay the oval shaped pinkish colour eggs which later transform into yellow colour.

Nymphs: Chestnut brown in the colour, first instar nymphs are highly energetic and negatively geotropic. The nymphs in their second instar are similar to those in their first, except they have duller bodies with grey-brown proboscis and antennae. The female nymph in her third instar has more waxy material secreted by its dermal glands. The third-instar male nymph, known as a "prepupa," usually looks like the third-instar female nymph, but it is smaller in stature. For males, the pupal stage is located in the bark's nooks and crannies like a cocoon (Franco et al. 2009).

Adult: The mature adult has a clear sexual dimorphism, with males having one pair of wings (Fig. 24.6c) and females lacking wings (Fig. 24.6a, b). Typically large (0.4–0.8 mm in length), the neotenic wingless female can continue to feed and grow until mating (Gullan and Kosztarab 1997).

Fig. 24.6 Wingless females of mango mealybug observed around the canopy of mango tree (**a**, **b**) and winged male mango mealy bug (**c**) are shown (Mani and Shivaraju 2016)

24.3.2.3 Biology

The gravid females descend along the tree-trunk to the ground from March to May for laying 300–500 eggs in clusters at a depth of about 80–100 mm (Kumar et al. 2009). The eggs hatch as early as December of the same year or as late as March of the following year (Chandra et al. 1987). After hatching, nymphs begin to ascend tree trunks, and groups of them can be observed draining the cell sap off panicles and new shoots. During the bright and sunny days, mealybugs are most active and in the months of March and April, the highest population of pest is seen (Butani 1979). In contrast to female mango mealybugs, which have three nymphal instars, males have two (Gullan and Martin 2009). For females, the first instar lasts 45–71 days, the second 18–38 days, and the third 15–26 days. For females, the average duration is 77–135 days, whereas for males, it is 67–119 days (Rahman and Latif 1944). Figure 24.7 shows the lifecycle of this pest.

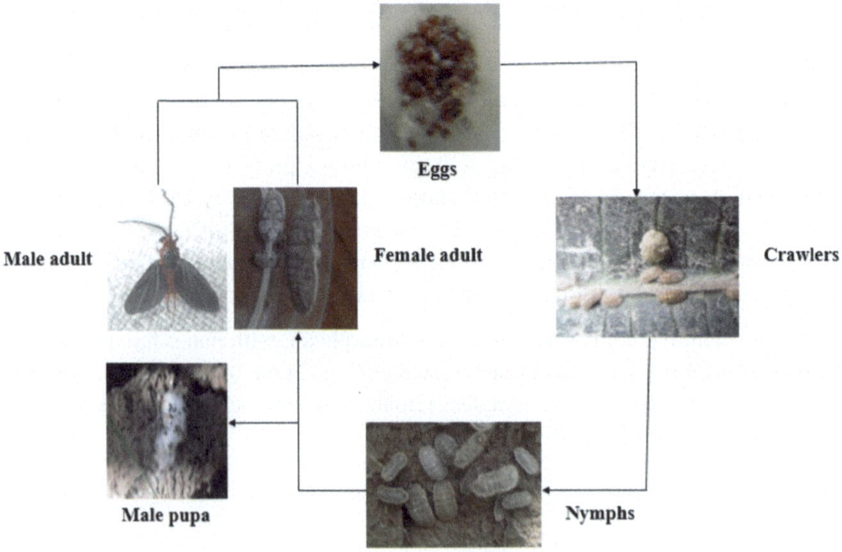

Fig. 24.7 Life cycle of mango mealybug (Mani and Shivaraju 2016)

24.3.2.4 Nature of Damage

Mealybugs feed on phloem. Usually, female adults and nymphs of both sexes harm the mango tree during its growth, and the infestation affects the leaves, inflorescence, trunk, and fruits, among other areas (Fig. 24.8a–f). However, after 2nd nymphal instar stage, males terminate their feeding (Mani and Shivaraju 2016). Mealybugs use their stylets to puncture plant tissue during feeding in order to extract sap from either mesophyll or phloem, or both. Following stylet penetration, a hardened saliva is secreted, encasing the stylets in a sheath (Franco et al. 2009). Excessive feeding causes the afflicted panicles to shrivel and dry out, which reduces the size of the fruit and causes it to fall too soon (Tobih et al. 2002). Exudation of honey dew, which causes a sooty mould fungus on various plant parts and impacts the photosynthetic activity of plants when this pest is present on leaves, is another noticeable sign of its infestations (Pitan et al. 2002; Ashfaq et al. 2011).

24.3.2.5 Management

(1) The mango mealybug lays eggs in the soil which passes their diapause stage there by remaining dormant for nearly six months. From June to December, the eggs may be destroyed if the orchards are deep ploughed or hoed to a depth of 15 cm (Mohyuddin and Mahmood 1992). While managing mango mealybugs using a variety of cultural operations, Karar et al. (2010) found that mounding the tree with debris on a plastic sheet once a year was the best way to reduce the pest's nymphal population (Fig. 24.9a).

(2) Removal of weed plants like *Clerodendrum infortunatum* Linn. that serves as an alternate host plant is useful in managing the incidence of giant mealybugs (Tandon 1995).

Fig. 24.8 Damage symptoms of mango mealybug shown on inflorescence (**a**, **b**), leaves (**c**), fruit (**d**), and tree trunk (**e**, **f**) are shown (Ahmed et al. 2015; Karar et al. 2020)

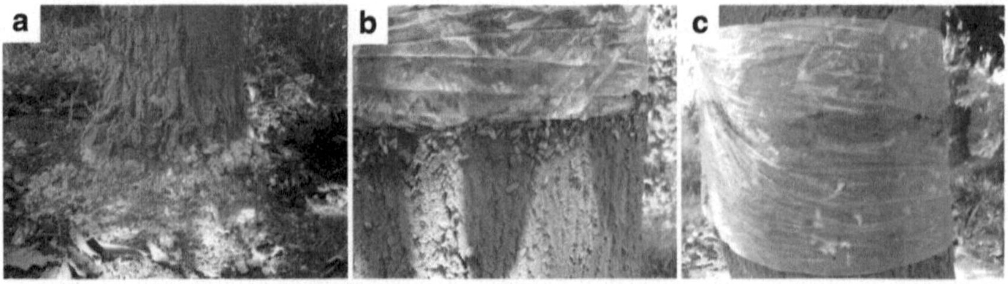

Fig. 24.9 An illustration of the cultural practices being used in the mango orchards for the management of mango mealybug. The common practices include collection and removal of its eggs (**a**), the use of sticky bands (**b**), and the development of a greasy layer in the middle of sticky band (**c**) around the tree trunks (Karar et al. 2010, 2020)

(3) In the past, various types of bands (mechanical control) have been placed around the tree trunk to prevent the upward movement of the overwintering and surviving mango mealybug population (Bindra et al. 1970; Srivastava 1980). While assessing the effectiveness of various bands (Fig. 24.9b, c) against this notorious pest in the South Punjab, Pakistan, Karar et al. (2010) noted that only 0.8% of nymphal instars could pass through their newly created band, demonstrating its tremendous effectiveness. The band was made by applying a 1:1 mud and manure mixture to the tree trunk, making it 25.4 cm wide and 60 cm above the ground. After covering the mixture with a 22.9 cm wide plastic sheet, it was wrapped around the trunk and fastened with nails. Lastly, to stop gravid females from moving higher, a 3.81 cm thick coating of grease was put on the plastic sheet (Fig. 24.9c).

(4) Heavy irrigation of mango orchard in October also helps in destruction of mango mealybug eggs (Kaul et al. 2009).

24.3.3 Mango Midges

24.3.3.1 Distribution and Species

Midges (Diptera: Cecidomyiidae) are one of the significant sucking pest of mango orchards across the world (Tokuda and Uechi 2021). Approximately 13 species of midges have been identified in association with 16 different types of gall makers (Mani 1944). Others claim that 16 different kinds of gall midges prey on mangoes throughout Asia (Peña et al. 2002; Jiao et al. 2018). *Dasineura amaramanjarae* Grover, the inflorescence midge, *Erosomyia mangiferae* Felt, the gall midge or blister midge, and *Procontarinia matteiana* Kieffer and Cecconi are the main species affecting mango (Reddy et al. 2020) (Table 24.1). Mango gall or blister midge, *E. mangiferae*, is one of these species that has drawn a lot of attention globally because it poses a major risk to mango growers (Ahmed et al. 2005; Shimizu et al. 2023). On mango farms, it is reported to destroy 70% of the fruit set and destroy the blossoms

(Verghese and Rao 1987). Due to *D. amaramanjarae* infestation, yield reductions of up to 100% have been documented in Pakistan (Rehman et al. 2014). Similarly, in countries like India, Indonesia, Kenya, Mauritius, Oman, South Africa, and the United Arab Emirates, *P. matteiana*, the leaf gall midge, is a significant mango pest (Augustyn et al. 2013; Henri et al. 2015).

24.3.3.2 Identification

Eggs: Mango midges effecting mango lay small translucent, elongated and cylindrical shaped eggs during the daytime.

Larvae: Mango midge passes through four larval instar stages. The first larval instar is transparent, 2nd is whitish while 3rd and 4 thare yellowish-orange in color.

Pupa: Prior to its pupation in the bud, 4th instar larva spins a cocoon around itself.

Adult: Adult females are larger in size than males and are orange in color (Fig. 24.10).

24.3.3.3 Biology

Up to 150 eggs can be laid by a female on delicate floral components, such as the inflorescence axis, recently set fruits, or delicate leaves that encircle the inflorescence (Nakahara and Branch 1993). Within two to three days, the eggs hatch, and the young larvae begin feeding on the delicate areas where the eggs were deposited (Abbas et al. 1985). Galls are created as a result of this feeding, and occasionally pupation occurs inside of them. In order to overwinter, mature larvae (fourth instar) fall to

Fig. 24.10 An adult of mango midge sitting on a leaf of mango plant (Hasnain et al. 2013)

the earth. When favourable conditions arise the following season and the flowering stage begins, this diapause phase and the pupation inside the galls come to an end (Verghese et al. 1988; Carrillo et al. 2017). In general, the pupal stage lasts for five to seven days, whereas the larval period ranges from seven to ten days. The pest's entire life cycle is completed in 14–25 days (Pezhman and Askari 2004).

24.3.3.4 Nature of Damage

Throughout the year, midges frequently attack the mango plant at several stages of growth, such as the burst stage of the floral buds, the juvenile fruiting stage, and on the foliage (Kaushik et al. 2012; Rehman et al. 2014). They inflict damage by feeding on inflorescence (Fig. 24.11a) and fruit tissues and making galls on leaves (Fig. 24.11b–d). A maggot is present in each of the 2–4 mm long galls (Fig. 24.11e). These maggots enter the bud, eat on the internal material, and then drill into the inflorescence's axis to totally destroy it (Pezhman and Askari 2004; Rehman et al. 2014). The appearance of a right-angled bend with an exit hole, from which last instar's (4th) maggots emerge, makes infested panicles easy to identify. The most noticeable sign of the infection is the appearance of black spots on the inflorescence and tiny blister-like galls (either alone or in clusters) on the infected buds, shoots, and early fruits (Fig. 24.11b–e) (Reddy et al. 2020). Abiotic factors like rainfall and temperature are the most significant to affect the population fluctuations of midges in the mango orchards. Harris and Schreiner (1992) associated the presence of greater number of galls during rainy periods to the relative humidity (RH) as it improves the larval and pupal survival.

Fig. 24.11 Infestation of mango midges on inflorescence (**a**), leaves (**b–d**), and the formed galls is shown (**e**) (CABI 2019)

24.3.3.5 Management

(1) While screening the different mango cultivars (Anwar Ratol, Dusehri, Kala Chaunsa, Sufaid Chaunsa) in different mango growing regions of the Punjab, Pakistan for their susceptibility to mango midge, *Procontarinia mangicola* Shi, Rehman et al. (2014) found "Sufaid Chaunsa" as the most susceptible variety. Sial et al. (2015) also evaluated the different mango cultivars (Anwar Ratol, Chaunsa, Dusehri, Langra, Sindhri) for their susceptibility to the *P. mangicola* and the results revealed the highest susceptibility of "Anwar Ratol" to the pest.
(2) Cultural practices like soil preparation, weeds eradication, manuring, and timely irrigation are the key components of pest management that can limit the population of mango midges (Prasad 1966; Grover 1985). During the month of November, it is recommended to perform deep ploughing in the orchard. In addition to providing protection from the pest, this will expose the pupae and overwintering larvae to the heat of the sun and their natural enemies (Nayanthara 2020).
(3) Insects have been endowed by nature to respond differently to colours when choosing their host plants (Plažanin et al. 2012). While monitoring the population of *P. mangicola*, Muhammad et al. (2013) found that among the eight coloured traps (blue, green, orange, pink, red, violet, white, yellow), mango midges were significantly attracted towards orange coloured trap. Rehman et al. (2014) in a similar study used three types coloured traps (blue, green, yellow) and noted the more effectiveness of yellow-coloured traps in attracting the population of midges. This coloured preference of mango gall midges can be useful in the development of traps for monitoring and management purposes in future.
(4) Use of systemic chemicals belonging to new chemistry insecticides with low toxicity to the natural enemies like bifenthrin, dimethoate, dinotefuran, imidacloprid, nitenpyram, and pyrifluquinazon is the most common strategy among the mango growers of Asia (Muhammad et al. 2017; Shimizu et al. 2023).
(5) Biopesticides derived from plant sources like neem also proved useful in managing the population of mango midges. Rehman et al. (2016) observed that using neem seed kernel extract (NSKE) in conjunction with soil racking was more successful in reducing the number of *Procontarini mangiferae* Felt in Pakistan.

24.3.4 Mango White Scale

24.3.4.1 Distribution and Species

Scale insects belong to a group of sap-sucking insect pests that are divided into three main families i.e., Coccidae (soft scales); Diaspididae (armoured scales); and Pseudococcidae (mealybugs) (Makale et al. 2020). Globally, more than 70 scale insects have been documented to infest mango but, *Aulacaspis tubercularis* Newstead

(Hemiptera: Diaspididae), is known to cause a significant damage to mango nursery and orchard plantations (Abate and Dechassa 2021; Dako 2023) (Table 24.1). Importantly, the earlier studies have shown that an increase of one individual of *A. tubercularis* reduced the fruit yield by 4.28 kg/tree per year (Daneel and Joubert 2009) and the presence of 4–5 scales per fruit is sufficient to deteriorate the quality of mango fruit (EFSA Panel on Plant Health et al. 2022). White mango scale can result in a 50–100% financial loss in cases of severe infestation. (Fita et al. 2020).

24.3.4.2 Identification

Eggs: Oval shape eggs with reddish-brown to purple colour (Hamdy 2020).

Larvae: The first nymphal instar, or crawler, is flattened, elongated-oval, reddish-brown, and has only partially formed legs, antennae, and eyes. First instars of both sexes usually have similar appearances, but by the second instar, sexual dimorphism is noticeable (Moharum 2012).

Pupae: The winged adult emerges after the male goes through two pupa-like phases. However, the established female nymph stays affixed to the host plant for the remainder of its life after moulting first into a pre-ovipositional immature and subsequently into an ovipositional adult (Van Halteren 1970).

Adult: A thin, almost round body with white armour and dark, oval terminal exuviae is a defining characteristic of the adult female, which is larger than the male (Fig. 24.12a). The little adult male has three elevated parallel dorsal ridges on its elongated body cover (Fig. 24.12b), whereas the elongated, dark brown adult female is typically seen beneath the scale cover (Fig. 24.12a) (Abo-Shanab 2012).

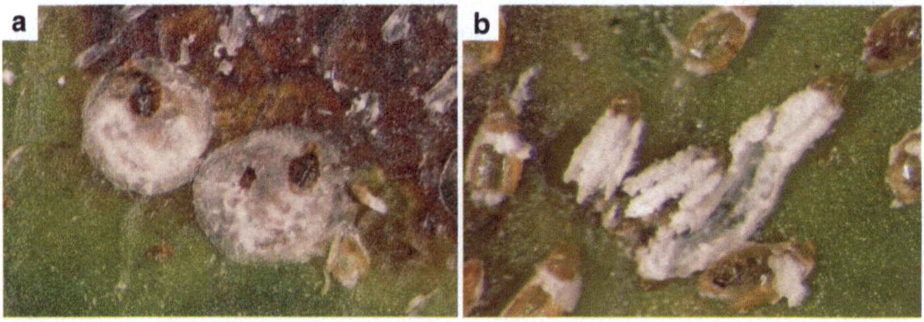

Fig. 24.12 Adult female (**a**) and male (**b**) of mango white scale (*Aulacaspis tubercularis*) (Carrillo et al. 2018)

24.3.4.3 Biology

Sex pheromones are released by adult female *A. tubercularis* to entice males to mate (Ortiz et al. 2017). When the gravid female lays between 80 and 200 eggs beneath its cover, the pest's life cycle starts. The eggs hatch in 7–8 days, and the ovipositional stage typically lasts 8–12 days. Within 24 h of hatching, the crawlers, or nymphs, travel several centimetres in search of a suitable location to settle. After settling in, the first nymphal instars begin sexual differentiation and insert their stylets into the plant for feeding (Juárez-Hernández et al. 2014; Hamdy 2020). These first-stage nymphal instars in males turn from orange to white and yellowish, then secrete fine white threads of wax all over their bodies before finally losing their legs and antennae (del Pino et al. 2020). Before transforming into the prepupa, when the waxy ridges complete their development, their second nymphal instars create three waxy ridges. The male scale cover does not change form from prepupa to pupa; it stays the same till the adult winged male departs. Females develop a greyish, round, waxy covering measuring 3–4 mm^2 during the second nymphal instar. The insect matures into an adult female after initiating a second waxy cover. Nymphs remain in their permanent home until they become adults. (Belachew and Jenber 2022). While the male crawlers live in colonies of 10–80 individuals, usually surrounding the female mother, the female crawlers are dispersed randomly on the upper and, to a lesser extent, lower sides of the leaves as well as the surface of the fruits (Van Halteren 1970). From egg laying to adulthood, a female's life cycle takes 35–40 days, while a male's takes 23–28 days, and there have been reports of three–four generations of the pest (Hamdy 2020). An illustration of the life cycle of the pest is shown in Fig. 24.13.

24.3.4.4 Mode of Damage

A. tubercularis is a polyphagous pest that attacks the 18 families with over 30 genera of crops (Malumphy 2014; García Morales et al. 2016). Numerous abiotic factors and the host plant features of the mango cultivars have been associated with the varying intensity of its infestation (Bautista-Rosales et al. 2013). Contrary to other sap feeding insect species that depend on phloem sap, females of white mango scale prefer the parenchymal cell contents for nutrition. For this purpose, they insert their stylets into tender parts of the plant to suck the cell contents (Fig. 24.14a) (Juárez-Hernández et al. 2014). It results in the appearance of blemishes on the skin of ripened fruits (Fig. 24.14b), yellow scars on leaves which drop off, and finally leads to the dieback of the mango plant and death of young seedlings (Ayalew et al. 2015). Appearance of blemishes on the fruit (Fig. 24.14b) has not only an aesthetic value but also economic dimensions as it causes market resistance, especially from the export market, which hurts the foreign producers (Raza et al. 2023).

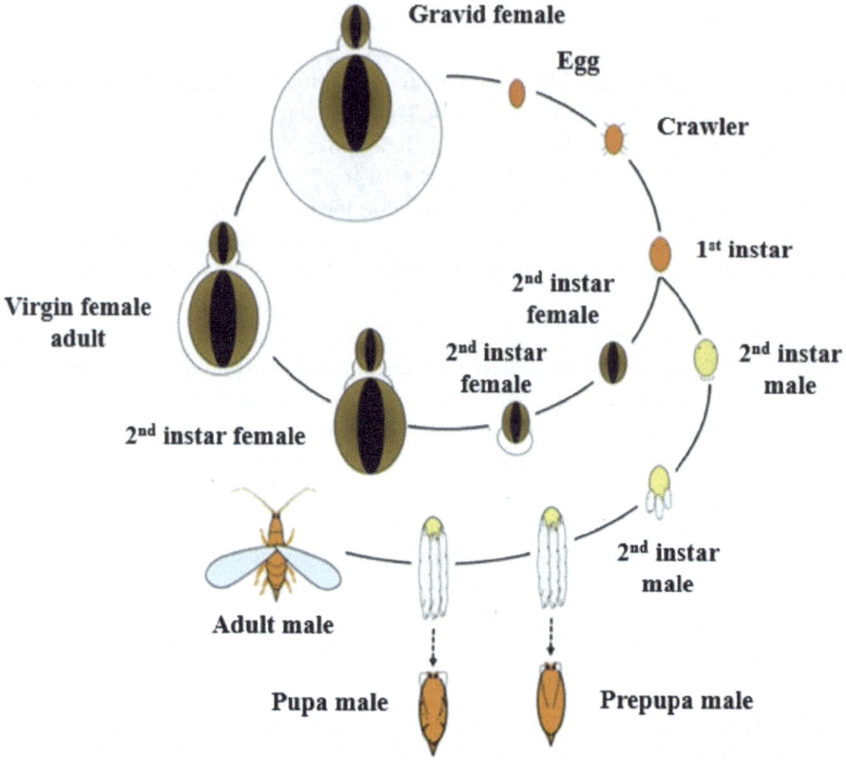

Fig. 24.13 Life cycle of mango white scale (del Pino et al. 2020)

Fig. 24.14 Images showing the infestation of mango white scale on leaves (**a**) and fruit (**b**) (Raza et al. 2023)

24.3.4.5 Management

(1) Use of planting material free from the infestation of white mango scales (Terefe et al. 2014).
(2) Mango plants develop resistance as a result of host plant characteristics. Cultivars with dense foliage and larger trees, for instance, offer a favourable habitat for *A. tubercularis* growth (Urías-López et al. 2010). Similarly, biochemical composition of fruits (sugar contents, acidity) is also critical in imparting resistance/susceptibility of mango cultivars against this pest (Urías et al. 2006).
(3) Cultural practices like pruning which involves the elimination of old dried branches, overlapped shoots, and infested parts can be useful in the management of this pest. Pruning improves the exposure of the tree to the entry of air which decreases humidity and discourages hiding and ovipositioning space to the *A. tubercularis* (Pérez Barraza et al. 2016).
(4) Mango fruit can be shielded from insect pests (fruit flies and scale insects), mechanical harm (scratches and scars from harvesting), sunlight exposure, and diseases (anthracnose, stem end rot, etc.) by using bagging material. Bagging is getting popular among the mango growers (Hofman et al. 1997; Sarker et al. 2009). Earlier studies have also shown that this tool significantly reduced the fruit damage as compared to the un-bagged fruits due to multiple factors, including scales (Chonhenchob et al. 2011).
(5) Application of systemic insecticides in the form of drenching have been found more effective in managing the population of *A. tubercularis* (Habtegebriel et al. 2020). In addition, the application of mew chemistry insecticides like movento and dimethoate is recommended to decrease the incidence of this pest (Atnafu 2020).

24.3.5 Mango Bud Mite

24.3.5.1 Distribution and Species

Mites belonging to the class Arachnida are one of the non-insect arthropod pest species that cause damage to the mango nursery and orchard plantations (Melo et al. 2023). Earlier, they were considered as occasional pest but due to indiscriminate and irrational application of chemical insecticides against other sucking pests became a serious threat to the farmers (Singh et al. 2016). Four families of mites including Eriophyidae, Tarsonemidae, Tenuipalpidae, and Tetranychidae are found to be associated with the phytophagous activity (Abou-Awad et al. 2010; Khan et al. 2022). Mites colonise leaves, buds, and fruits. The bud mite, *Eriophyes mangiferae* Sayed (Acari: Eriophyidae), is a severe problem in mango-producing regions around the world (Peña et al. 1998).

24.3.5.2 Identification

Adult mango bud mites are white in colour with cylindrical body having a length of about 0.20 mm (Fig. 24.15a, b). The prodorsum bears a pair of backwards-pointing set of setae on tubercles with three strong ridges in between them. The feather claw is six-rayed (Sternlicht and Goldenberg 1976).

24.3.5.3 Nature of Damage

Mango bud mite sucks the sap internally from terminal buds of young and old trees, causing bud malformation and stunting of inflorescences. In addition, it causes proliferation of stem at terminal shoot which is a typical symptom of "witches broom or mango malformation" (Fig. 24.15c, d) (Kumar and Misra 2016; Garcia-Lopez et al. 2020). Globally, mango malformation is a deadly disease that poses serious economic implications to the mango growers (Chakrabarti 2011). This condition has been attributed to a variety of causes, including nutritional inadequacies, hormonal imbalance, viruses, phytoplasmas, fungus, and mites (Garcia-Lopez et al. 2023). However, the presence of higher population of mango bud mites in malformed trees and appearance of symptoms of witches broom and galls formation (Fig. 24.15c, d) in other plant species due to activity of some members of the Eriophyidae mites (Westphal and Manson 1996) validates their prominent role. Gamliel-Atinsky et al.

Fig. 24.15 Mango bud mite (*Aceria mangiferae*) inhibiting the mango leaves (**a, b**) with its characteristic infestation symptoms of stunted inflorescence (**c**) and malformed buds (**d**) are shown (Carrillo et al. 2018)

(2010) in their preliminary studies also showed the potential of mango white bud mite, *E. mangiferae*, to act as a vector of the pathogenic fungi of the genus *Fusarium* in mango malformation.

24.3.5.4 Management

(1) Timely pruning of the affected twigs can limit the extent of their damage (Pena et al. 2005).
(2) Essential oils derived from plants due to their low toxicity to our ecosystem and non-target organisms could be a potential alternative for mites control. Their killing mechanism involves the release of fumes that suppress the population of mites (Isman and Machial 2006; Abo-Shnaf et al. 2022).
(3) Spray of acaricides, like dicofol proved effective against mango mites (Al-Azzazy 2005).

24.4 Insect Pests of Stem

24.4.1 Mango Stem Borer

24.4.1.1 Distribution and Species

Mango stem borers of the genus *Batocera* (Coleoptera: Cerambycidae) are one of the deadliest pests of the subcontinent (Vasanthi and Raviprasad 2015). *Batocera rufomaculata* DeGeer and *B. rubus* Linnaeus are the most destructive and prevalent borers in mango orchards. Aside from mango, they also target apple, fig, jackfruit, mulberry, and papaya (Urca and Ribak 2021) (Table 24.1). *B. rufomaculata* used to be a minor pest in South Asia but in the past poor farm management practices such as orchard sanitation and early removal of dry shoots from trees increased severity of its infestation (Venkata Rami Reddy et al. 2018).

24.4.1.2 Identification

Eggs: Female beetles lay shiny creamy white oval shaped eggs.

Larvae: Mature grubs (larvae) are cream in colour having a dark brown head provided with strong mandibles.

Pupae: Pupae are brownish yellow to dark and approximately 50–55 mm in length.

Adults: Adults are provided with a hard exoskeleton which is dark brown in colour and accommodates two pink dots and lateral spines (Fig. 24.16). Typically, female

Fig. 24.16 An adult of mango stem borer (Magar et al. 2022)

beetles are larger in size and measure between 55 to 60 mm while males are 50–55 mm (Orkweha et al. 2020).

24.4.1.3 Biology

With the commencement of monsoon season, mango stem borers emerge, and mating takes place. The mated female deposits about 200 eggs under the loose bark or in the crevices of the stem. After hatching, the young grubs emerge which are typically 85–95 mm long upon maturity and have well-defined mandibles for boring the tree stem (Magar et al. 2022). The grub has seven larval instars where 1st larval instar stage lasts for 2–3 days and all other instars (2nd–7th) have 10–18 days (Veeresh 1985). Grubs spend the winter in the tree tunnel, and the pupal stage lasts around 20–25 days. Adults emerge from pupae and live for 60 to 100 days (Reddy and Rajan 2021). An illustrated diagram of the life cycle of mango stem borer (*B. rufomaculata*) is presented below (Fig. 24.17).

24.4.1.4 Mode of Damage

After hatching, grubs use strong mandibles to make about 23 cm wide tunnel through the sapwood (Fig. 24.18a) which disturbs the flow of sap in plant and affects its foliage and production. These tunnels can be found deep within the tree's trunk or around its periphery (Sudhi-Aromna et al. 2007). Initially, the symptoms of damage are not obvious, but the oozing of sticky material at various points on the tree trunk and

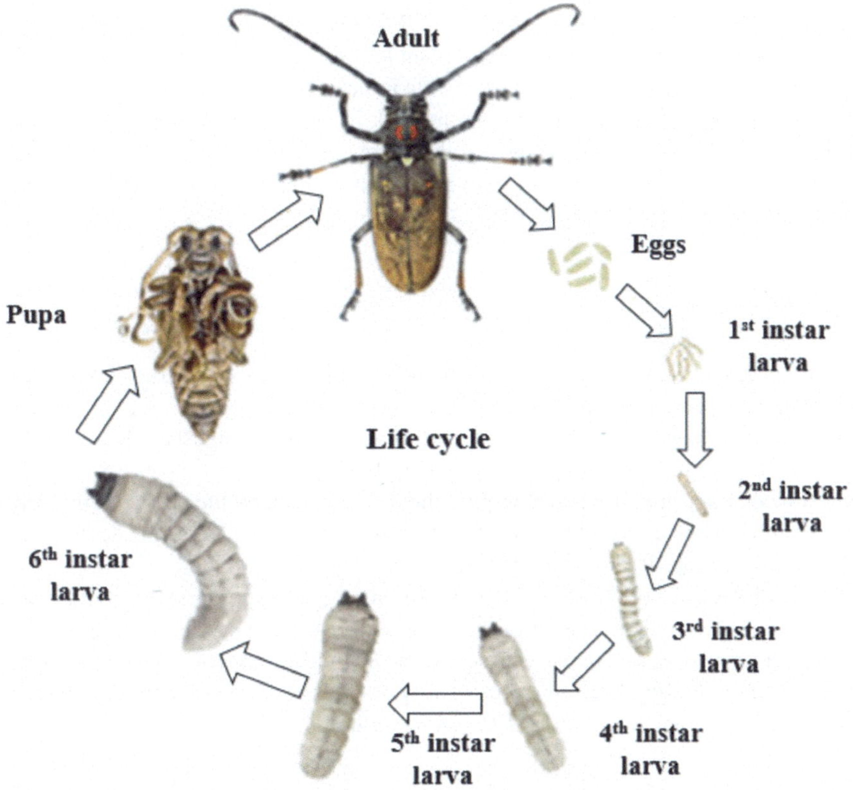

Fig. 24.17 Life cycle of mango stem borer (*Batocera rufomaculata*) (Magar et al. 2022)

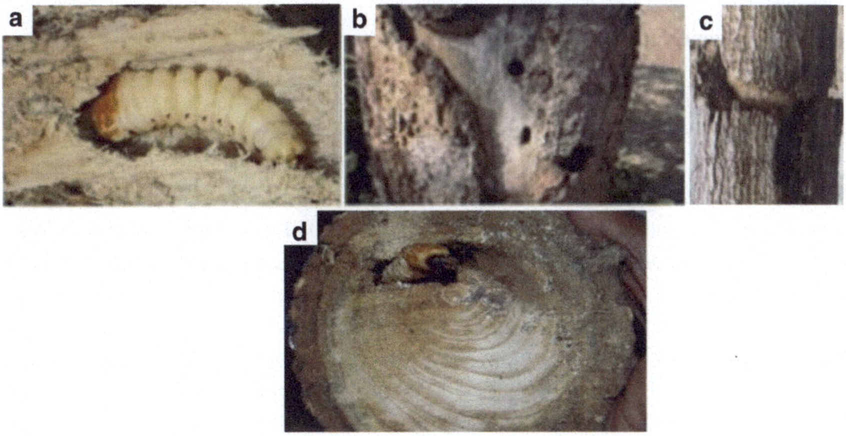

Fig. 24.18 Mango stem borer larva (grub) feeding inside the woody part of the stem (**a**) with characteristic symptoms of its infestation including appearance of holes and splitted bark are shown (**b**–**d**) (Magar et al. 2022)

branches can observed. The infestation of *B. rufomaculata* remains largely unnoticeable until the shedding and drying of plant leaves occurs (Shivananda et al. 2012; Upadhyay et al. 2013). Another, more visible indicator of its attack is the emergence of holes with exuding sap and frass on the bark in late stages of infestation, which changes the colour of the branches to yellow (Fig. 24.18b–d). This colour transition then causes drying and dieback of terminal shoots and branches, eventually leading to the tree's death (Reddy et al. 2014).

24.4.1.5 Management

(1) Immediately prune and eradicate the infested branches with grubs and pupae within (Upadhyay et al. 2013).
(2) Using a nylon mesh to wrap tree stems throughout May–August can help capture newly emerged adult beetles (Reddy et al. 2014).
(3) It is also a common practice to plug the holes created by the pest with the help of cotton soaked in dichlorovos or petrol to inactivate the larvae inside the stem (Venkata Rami Reddy et al. 2018).
(4) Upadhyay et al. (2013) found that the employment of appropriate cultural practices such as orchard sanitation and destruction of plant dry shoots followed by application of new chemistry insecticides like imidacloprid or thiamethoxam for five times starting from 2nd week of July at 15 days intervals is more impactful in managing the population of *B. rufomaculata.*

24.4.2 Mango Bark Beetle

24.4.2.1 Species and Distribution

Bark beetle (Scolytidae: Coleoptera) is the most devastating and economically significant pest of forests with the description of 247 genera and 5990 species (Hulcr et al. 2015; Gomez et al. 2023). Among these beetles, mango bark beetle, *Hypocryphalus mangiferae* Stebbing, is an exotic and secondary pest species that has seriously lowered mango fruit yield potential in Pakistan and Oman (Van Wyk et al. 2007). Masood et al. (2008) reported the presence of four species of bark beetles, including *H. mangiferae*, *Sinoxylon* sp., *Xyleborus* sp., and *Nitidulidae* sp. in the major mango-growing areas of Pakistan. However, *H. mangiferae* was found to be the most commonly encountered species affecting mango plantations. In addition, *H. mangiferae*, has been implicated as a vector of "Mango Sudden Death Disease (MSDD)" (Galdino et al. 2017; Pereira et al. 2019). It transmits the fungal pathogen, *Ceratocystis fimbriata* Ellis and Halst, to a healthy plant (Pereira et al. 2021) which ultimately leads to plant death in periods as short as two months after initial infection (Masood et al. 2010; Oliveira et al. 2015). Unfortunately, the poor management practices for the control of *H. mangiferae* often lead to massive proliferation of MSDD

Fig. 24.19 Images of the different life stages of mango bark beetle (*Hypocryphalus mangiferae*) including eggs (**a**), larval instars (**b**), pupae (**c**), and adult (**d**) are shown (Masood et al. 2009)

(Bibi et al. 2023) and a significant loss of 20–60% has been noted (Masood et al. 2012).

24.4.2.2 Identification

Eggs: Female beetles lay cylindrical, shining, and creamy white eggs in batches which are broadly rounded on both the ends (Fig. 24.19a) (Masood et al. 2009).

Larvae: After hatching, it passes through five larval instar stages and the 1st and last (5th) larval instars can be easily distinguished due to the length and width of head capsule (Fig. 24.19b) (Al Adawi et al. 2013).

Pupa: Pupae are soft and creamy white which change to pale tan colour near maturity (Fig. 24.19c) (Masood et al. 2009).

Adult: *H. mangiferae* adults are cylindrical in shape with emerginate type eyes. Initially, the freshly emerged adults have light brown colour which later transform into darker brown colour. Mango bark beetle can also be identified due to presence of fine inconspicuous hairs over its whole body (Fig. 24.19d) (Bright and Skidmore 2002).

24.4.2.3 Biology

After mating, each mated female lays between 41–139 eggs inside the egg galleries. The eggs hatch in 4–5 days, releasing tiny, cylindrical, wrinkled, legless larvae that are creamy white in color. These larvae begin to tunnel into the phloem, creating their own galleries. The larval stage consists of five instars, lasting 16–18 days. Pupation occurs near the exit hole, with newly formed pupae being soft and creamy white, eventually changing to a pale tan as they mature. The pupal stage lasts 4–5 days (Masood et al. 2009; Al Adawi 2012). Upon emerging from the pupae, the immature adults (white in color) begin maturation feeding to fully sclerotize. This feeding process is crucial in the life cycle of the mango bark beetle, as it facilitates the beetle's role in carrying the spores of the *C. fimbriata* fungus, which are responsible for the transmission of MSDD (Lieutier et al. 2004).

24.4.2.4 Nature of Damage

Female *H. mangiferae* begin their attack on mango trees by boring into the bark to create egg galleries. (Fig. 24.20a). Shortly after, males arrive, and mating occurs, after which the females lay their eggs inside the galleries (Al Adawi 2012). This pest prefers to attack weakened and stressed mango trees. Appearance of the tiny holes and powdery material at the boring point are the distinctive signs of its attack (Fig. 24.20b). In addition, irregular galleries of black colour in the cambium region of the affected tree can be visualised upon peeling off the bark of the woody stem (Fig. 24.20c) (Ploetz et al. 1996).

The role *H. mangiferae* as a vector of MSDD (Fig. 24.21a) has already been established (Abbasi et al. 2008; Masood and Saeed 2012). Host plant features like the emission of certain chemical compounds (geraniol and isobutyl acetate) from the mango plant attract mango bark beetles. Once beetles have established themselves in mango orchards, they produce species-specific aggregating pheromones (Byers 1995; Peña and Mohyuddin 1997). This increase in populations of bark beetles produces a biotic load on healthy trees, and the intensity of infestation grows (Peña et al. 2009). *Scolytus* beetles are notable for their capacity to fly across great distances (45 km/day) for several days (Hill 1994; Byers 2000). Therefore, *H. mangiferae* is considered an active carrier of pathogenic fungal inoculum from infected to healthy

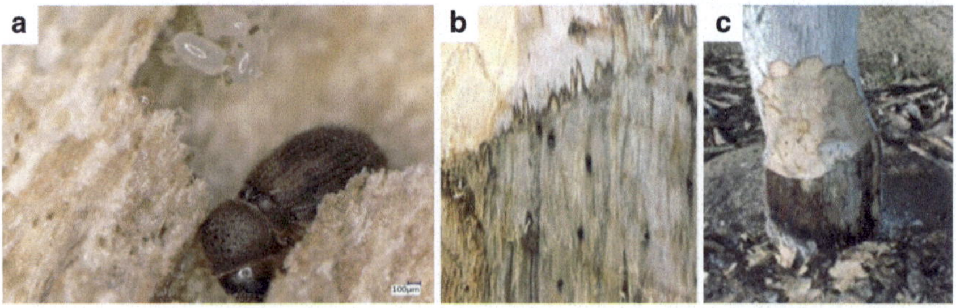

Fig. 24.20 Typical symptoms and the infesting mechanism of mango bark beetle (*Hypocryphalus mangiferae*) is shown (a–c). Infestation of the plant begins with the boring of *H. mangiferae* into the bark of a mango tree (**a**) with the characteristic symptoms of holes appearance (**b**) and formation of irregular galleries (**c**) in the trunk of infested mango plant (Carrillo et al. 2018)

Fig. 24.21 Mango sudden death disease (MSDD) affected tree (**a**) with characteristic symptoms of gummosis (**b**) and bark splitting (**c**, **d**) is shown (Al Adawi et al. 2006)

plants, which contributes to the spread of MSDD in a relatively short period of time (Fig. 24.21a–d) (Al Adawi et al. 2013; Galdino et al. 2016).

24.4.2.5 Management

(1) To monitor the population of *H. mangiferae* in mango orchards, it is recommended to install yellow colour sticky traps. This strategy is helpful to plan a rationalized pest management strategy (Saeed et al. 2010).
(2) Remove the dead or severely injured mango trees due to MSDD as these provide a safe breeding place for bark beetles and the fungal spores that they carry with them. Therefore, for the timely management of the pest, it is necessary to destroy such kind of protecting places. Masood et al. (2014) found that the injection of fungicides to the mango tree trunks lessened the intensity of the MSDD.
(3) Provision of healthy and balanced nutrients to the mango trees enhances its vigour to sustain the infestation of different pathogens especially *C. fimbriata* as the bark beetles prefer to attack on the weakened unhealthy plants and in this way our orchards are indirectly protected from this pest (Al-Subhi et al. 2006; Umar et al. 2022).

(4) Chemical control of bark beetle is often problematic due to the cryptic feeding habit of the pest. However, Saeed et al. (2011) demonstrated that the use of systemic insecticide, Lorsban 40EC (chloropyrifos), proved more effective in lowering the population of *H. mangiferae.*

24.5 Insect Pests of Fruit

24.5.1 Mango Fruit Borer

24.5.1.1 Distribution and Species

Mango, unlike other fruits, is largely resistant to fruit borers, with the exception of a few sporadic and localized borer pests. The red-banded caterpillar, *Deanolis albizonalis* Hampson (Lepidoptera: Pyralidae), is the most common borer pest of mango fruit (Tenakanai et al. 2006; Nayanathara and Narayana 2022). It is anonymous to the *Autocharis albizonalis* Hampson and *Noorda albizonalis* Hampson (Sahoo and Jha 2009) (Table 24.1). This pest was reported for the first time in the major mango growing regions of India as early as 1955 (Sengupta and Behura 1955) It is now extensively distributed across Australia, Bangladesh, Burma, Brunei, China, Indonesia, Pakistan, Papua New Guinea, the Philippines, and Thailand (Royer 2009; Rao et al. 2020). Previously, it used to be a rare and minor pest of mango (Butani 1979) but in the last few years it gained the status of a major pest and is known to cause an extensive damage of 10–55% (Krishnamoorthy et al. 2014). Mango is the main host of this fruit borer (Krishnarao et al. 2022) although it also infests other wild hosts, such as *Cyperus rotundus* L. (coco grass) (Baradevanal et al. 2021).

24.5.1.2 Identification

Eggs: Adult female lays oval shaped eggs in clusters, from one to four eggs near the apex of fruit and sometimes under the sepals or in small crevices in the fruit. Freshly laid eggs are milky white in colour and turn into pinkish within 2–3 days (Krull and Basedow 2006).

Larva: Mango fruit borer larva is distinguishable due to appearance of alternating red and white bands on its body and a blackish "collar" on the 1st segment (Fig. 24.22a, b). Head is brown or black in colour (Fenner 1997).

Pupa: The larva exits the fruit to pupate. It is blue-green in colour at the pre-pupal stage (Fig. 24.22c) while fully developed pupa turns to dark brownish in colour and attains an obtect shape (Krull 2004).

Fig. 24.22 Mango fruit borer larva is characterized with the presence alternate red and white bands on its body (**a**, **b**) and pupates in the crevices of bark (**c**), from where the greyish/fawn coloured moths (**d**) emerge (Royer 2009)

Adult: Generally, adults are nocturnal and during the day they can be found in resting position under mango leaves (Fig. 24.22d). Morphologically, both the sexes look alike, but the mid-leg of the male accommodates a dark brown tibial scale brush which is absent in females. This differentiating feature between the two sexes indicates the prevalence of "sexual dimorphism" in mango fruit borer. In addition, male adult moths can also be distinguished from the female by an abdomen extending beyond the hind wings (Leefmans and van der Vecht 1930).

24.5.1.3 Biology

Adult female moths lay their eggs in large groups near the fruit's distal end or apex, where they are shielded by a waxy coating. This egg stage lasts for 2 to 4 days. (Tenakanai et al. 2006). Typically, oviposition begins as early as 45 days after the fruit set and continues until the fruit reaches maturity (Sahoo and Das 2004). Golez (1991) reported that the pest's larvae undergo five stages. After hatching, the larvae enter the young fruit through the distal end, where they begin to consume the fruit's contents and create tunnels, eventually blocking the entrance hole with their excreta (Krull and Basedow 2006). The larvae travel between fruits using "silk" threads, lowering themselves onto a new fruit. The larval stage lasting for 11–16 days is followed by, a pre-pupal stage of 2–3 days. During this stage, the larvae stop feeding and exhibit minimal movement (Rao et al. 2020). Next, pupation occurs inside earthen cocoons or in debris under the tree. However, there are reports about the larval pupation in the fruit (Sengupta and Behura 1955). The pupal stage lasts 9–14 days, and the entire life cycle is completed within 30–41 days (Sahoo and Das 2004).

24.5.1.4 Mode of Damage

The fruit borer can affect all growth stages of mango fruit, with damage most commonly observed on the jointed fruits (Fig. 24.23a, b). The caterpillars feed both on the mango pulp and seeds (Fig. 24.23c). Typically, the infestation process begins after 45–55 days of flower formation and continues up to fruit maturation (Sujatha and Zaheruddeen 2002). The initial larval instars (1st and 2nd) remain in the fruit

Fig. 24.23 Salient symptoms of mango fruit borer infestation are shown (a–d). Typically, mango fruit borer infests the jointed fruits (**a**, **b**) and its larvae can be found upon dissection of the fruit feeding on the internal contents of fruit (**c**). Splitting of fruit skin (**d**) is the most common symptom associated with its damage (Royer 2009)

flesh while the later instars feed and complete their development inside the seed. Furthermore, the early instars tend to prefer small tender fruits and the large size fruits are favoured by the later instars (Bhattacharya 2014). To feed on the internal fruit contents, early larval instars have to make a hole in the fruit skin to obtain their nutrition (Fig. 24.23b). The appearance of a small hole at the distal end of the fruit is a typical symptom of fruit borers infestation (Krull 2004). As a result of the caterpillar's activity, sap begins to ooze from the entrance hole and accumulates at the fruit apex. Over time, this sap darkens and becomes more noticeable (Fig. 24.23b). Another clear indication of their infestation is the cracking of the affected fruits (Fig. 24.23d) (Royer 2009). The infested fruits become unfit for human consumption, and the situation worsens as secondary bacterial and fungal infections cause fruit rotting and necrosis. This exacerbates the problem, posing a significant threat to the commercial production of mangoes (Golez 1991).

24.5.1.5 Management

(1) Periodic collection and destruction of the fruit borer infested fruits to prevent further proliferation of the pest in the orchard and complete its life cycle (Peña et al. 1998).
(2) Eradicate alternate host plants and destroy the pupae present inside the dry branches of mango fruit (Joubert et al. 1999).
(3) Fruit bagging, an eco-friendly technique of covering the mango fruits with paper bags (Buthelezi et al. 2021). This physical barrier prevents the pests like fruit

borers from reaching the fruit and enhances the colour and weight of the fruits (Karar et al. 2019).

(4) Cultivation of resistant cultivars in mango orchards is a useful tool to counter the threat of fruit borers. While screening the different mango cultivars for their susceptibility to the fruit borer, Bhattacharyya (2018) noted that the phenolic compounds of the plants are crucial in developing the resistance against the invasion of this pest.

(5) Application of new chemistry insecticides (indoxacarb) at fruit set or pyrethroids (cypermethrin, deltamethrin, lambda cyhalothrin) at late stage is often undertaken by the mango growers to manage the incidence of this pest (Sahoo 2016).

24.5.2 Mango Fruit Fly

24.5.2.1 Species and Distribution

Globally, Tephritid fruit flies (Diptera: Tephritidae) constitute a large group of insects with more than 500 genera and 5000 species (Vargas et al. 2015; Scolari et al. 2021). Fruit fly species of six genera, *Anastrepha*, *Bactrocera*, *Ceratitis*, *Zeugodacus* and *Rhagoletis* are designated as harmful (Doorenweerd et al. 2018). Among the worldwide distributed species, 43 belong to the genus *Bactrocera* (Khalid et al. 2002) and in Pakistan 11 species including *B. papaya, B. zonata, B. cucurbitae, B. dorsalis, B. carambolae, B. minax, B. occipitalis, B. philippinensis, B. tau, B. latifrons*, and *B. umbrosa* have been identified (Stonehouse et al. 2002; Dhillon et al. 2005). However, two fruit fly species, *B. zonata* Saunders and *B. dorsalis* Hendel (Table 24.1; Fig. 24.24a, b), are recognized as major contributors to both qualitative and quantitative losses in various fruit plantations. (Khan et al. 2021). Numerous factors such as high fenudity power (Rwomushana and Tanga 2016), polyphagy (Khan et al. 2005), several generation (Choudhary et al. 2019), greater mobility, and climatic adoptability (Sarwar et al. 2013; Khan and Akram 2018) of the genus *Bactrocera* members make them economically important among the horticultural crops. Fruit fly is native to South and Southeast Asia (White and Elson-Harris 1992) and is now widespread in Pakistan where it has been estimated to cause 90–100% yield losses depending on species population, genotypes, hosts and locality (Stonehouse et al. 1998; Sharma et al. 2015). According to an estimate, every year Pakistan has to bear a loss of US$ 200 million due to infestation of fruit flies on different fruits and vegetables (Murtaza et al. 2021), whereas the seasonal yield losses fluctuate between 30–100% (Syed et al. 1970; Khan et al. 2005). Due to their detrimental effects on the horticultural crops, fruit flies are also called "quarantine pests" as their damage or infestation can cause severe trade restrictions and loss of market access to traders, producers, and governments (Follett and Neven 2006).

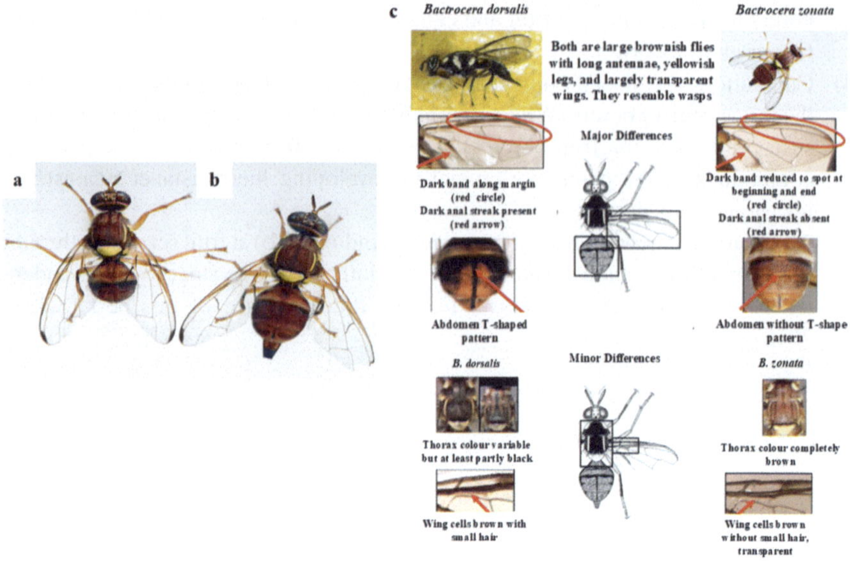

Fig. 24.24 Identifying features of two commonly occurring fruit fly species, *Bactrocera dorsalis* (**a**) and *B. zonata* (**b**) in mango orchards of Pakistan with their salient differentiating characters (**c**) are shown. Accessed on 04th Feb 2024; https://www.iaea.org

24.5.2.2 Identification

Fruit fly is a holometabolous insect that passes through four stages (egg, larva (maggots), pupa, adult) in its life to complete its development.

Eggs: Elongated, elliptical, and white coloured eggs are somewhat rounded at the posterior end and slightly pointed at the anteriorly margin with a distinct micropyle.

Larvae: Generally, the larvae have typical features of a maggot such as an involuted (rolled inward, spirally) head, three thoracic, and eight abdominal segments. There are three larval instars.

Pupae: They are barrel-shaped, yellowish to yellowish-brown.

Adults: Adults are about 6 mm long and reddish brown with yellowish thoracic markings. They have transparent wings with a small brown spot on the tip. Thorax colour is variable but at least partly black. Wing cells are brown with small hair. Abdomen has a characteristic T-shaped pattern (Fig. 24.24a). Salient morphological features of the two prominent fruit fly species found in mango orchards of Pakistan (Khan et al. 2021) are shown in Fig. 24.24a–c.

24.5.2.3 Biology

Female fly punctures the skin of mature fruits using the long and sharp ovipositor to lay eggs in batches of 1–20 per fruit (Fig. 24.25) (Salmah et al. 2017). The chemical released from ripe fruits also plays a decisive role in the selection of a suitable host for ovipositioning (Jayanthi et al. 2012). Under optimal conditions, larvae (maggots) hatch from the eggs within 1–2 days and find their way to the fleshy part of the fruit. Fruit fly larva passes through three stages within 2–3 weeks. Fully developed larva is cylindrical and dull deep reddish yellow (Vayssieres et al. 2008). During its development, 3rd larval instar moults twice before jumping or dropping for pupation in the soil (Fig. 24.25). However, mature larvae that do not leave the fruits in time can overwinter inside the host fruits (Huan et al. 2019). The larvae pupating in the soil form a yellow–brown to dark brown "barrel-like" puparia, and pupal development is completed within 10–14 days (Mutamiswa et al. 2021). After pupation, adults emerge from the soil and become sexually mature within a week (Fig. 24.25). For their survival and growth, they feed on nectar, honeydew, pollen, and rotting fruits. Adult fly is brown or dark brown in colour with hyaline wings and yellow legs (Fletcher 1987; Huan et al. 2019).

Fig. 24.25 Diagrammatic presentation of the life cycle of a mango fruit fly (COLEACP 2013)

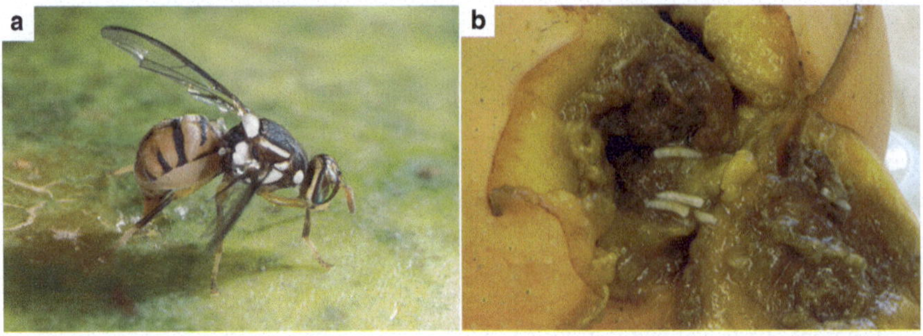

Fig. 24.26 Mode of infestation of mango fruit fly in mango is illustrated (a–b). Mango fruit fly inserts her sharp ovipositor into the fruit skin to lay eggs (**a**) which after hatching turn into the larvae (maggots) and starts devouring the internal contents (**b**) and spoils the fruit which ultimately causes it to fall down (Mahoo and Mtembei 2012)

24.5.2.4 Nature of Damage

Female fruit flies puncture the physiologically ripened mango fruit's mesocarp with their long and sharp ovipositor to lay eggs (Fig. 24.26a) (Vayssières et al. 2009). This is followed by the entry of bacteria through the ovipositional point and infected fruit starts to decay. Next, upon hatching, larvae (maggots) start boring inside the fruit by making tunnels and feeding on the decaying fruit pulp (Fig. 24.26b). Due to rotting of fruit and feeding by maggots, mango fruit losses its colour and drops prematurely. Unfortunately, the infested fruit becomes unfit for human consumption and market acceptance is significantly reduced (Clarke et al. 2005).

24.5.2.5 Management

Fruit flies management has always been critical and challenging as mentioned above that its 3rd instar larvae leave after infesting the fruits, fall down for pupation in the soil therefore it is likely that both larvae and pupae in fruits and soils are protected from surface-applied insecticides (Heve et al. 2017). In addition, many countries of the world have formerly taken out the effective broad-spectrum and systemic insecticides from their markets which has further deteriorated the situation (Böckmann et al. 2014). However, different pre-harvest (cultural, biological, mechanical) and post-harvest tools are being practised by the farmers to keep this pest away from the orchards (Riaz et al. 2000; Hossain et al. 2020).

(1) Cultural control of the pest involves the regular collection of the fallen infested fruits and dispose them properly to minimize the further spread of infestation (Verghese et al. 2022). Since, pest pupates in soil, deep ploughing during the months of November and December can expose the pupae to the solar radiation of the sun and natural enemies (predators) (Karar et al. 2010).

(2) Fruit bagging, one of the pre-harvest strategies that is being used around the world for improving the visual quality of fruit crops, reduce the incidence of insect pests, mechanical damage, sunburn of the skin, agrochemical residues on fruits, and birds damage (Buthelezi et al. 2021). Though, various mango growing countries are employing this ecofriendly tool for the management of fruit flies yet there is a need to use this method on a larger scale (Sarker et al. 2009; Karar et al. 2019).
(3) The selection and cultivation of resistant mango cultivars against the infestation of fruit flies can provide a sigh of relief to the mango growers especially at the time of establishment of new orchards (Rattanapun et al. 2009; Mani et al. 2022). While assessing the resistance/susceptibility of fourteen mango cultivars in the South Punjab, Pakistan, Karar et al. (2021) found that the early maturing mango cultivars (Dusehri, Malda, Langra) escaped from the fruit flies infestation whereas medium varieties (Chaunsa, Fajri Klan, Sensation) suffered less, and late varieties (Sanglakhi, Retaul-12, Mehmood Khan, Tukhmi, Kala Chaunsa, Chitta Chaunsa, Dai Wala, Sobey De Ting) had more infestation. Overall, Retaul-12, was found to be the most susceptible to fruit flies.
(4) Adoption of male annihilation technique (MAT) by establishing methyl eugenol based pheromone traps in the mango orchards (Shelly 2020). MAT exploits the attraction of males to parapheromone lures, locally eradicating males so that unmated females do not reproduce (Bagheri et al. 2017). Typically, in such kind of traps, fruit fly males are attracted to methyl eugenol and are controlled by the use of plastic traps which contain the wicks soaked in the liquid pheromones (methyl eugenol) and the males die inside due to overheating (Sim et al. 2022). Mango growers in the sub-continent usually erect six traps per acre throughout the fruiting season not only for monitoring purpose but also reducing the population density of male fruit flies in orchards (Mohyuddin and Mahmood 1992; Singh et al. 2014).
(5) Bait application technique (BAT) involving both bait placement and its spray is being used to suppress infestation of the pest (Zida et al. 2023). Through this technology, fruit flies are attracted to a mixture of protein and insecticide applied in spots which results in their killing (Maung et al. 2020). BAT technique successfully demonstrated its utility in the management of fruit flies under laboratory and field environment (Dominiak 2006; Hasnain et al. 2023). BAT can be initiated 45 days prior to fruit maturity (Reddy and Sreedevi 2016).
(6) Post-harvest management of fruit flies involves immersion of mango fruits in hot water (48 °C ± 1 °C for 60 min) within 24 h of harvesting. This process gives 100% mortality of eggs in the fruit without affecting its quality (Anwar and Malik 2007; Mwando et al. 2021). Mango growers who intend to export their produce, it is mandatory to follow the instructions of World Trade Organisation (WTO) and the destined country (Labuschagne et al. 1996). For example, the countries exporting to Japan and USA are required to subject their fruit to pass through Vapour Heat Treatment (VHT) and gamma irradiation (Hallman 1999; Javed et al. 2022).

24.5.3 Mango Stone Weevil

24.5.3.1 Distribution and Species

Mango stone weevil, *Sternochetus mangiferae* (Fabricius), belonging to one of the largest families of insects, Curculionidae, is a worldwide threat to mango fruit (Peña et al. 1998; Oberprieler et al. 2007) (Table 24.1). Three species of the pest, *S. mangiferae*, *S. gravis*, and *S. frigidus*, are known to infest the mango fruits. Among these, *S. mangiferae* is the most widely distributed, *S. gravis* prevails in India, Indonesia, Malaysia, Myanmar, and Pakistan, and *S. frigidus* is exclusively found in Malaysia, Pakistan, and Thailand (Bhattacharyya et al. 2016). The losses associated with the *S. mangiferae* vary from 5–80% in different mango varieties (Verghese 1999).

24.5.3.2 Identification

Eggs: The freshly laid eggs are elongate and creamy white which are surrounded by a protective sheath (brown) provided with two tiny tails at one end (Pinese and Holmes 2005).

Larva: The newly emerged white grub (larva) is elongate and without legs. Initial larval instars do not appear like a typical weevil larva. However, older instars (4th, 5th) exhibit a compact and C-shaped look similar to a common weevil larva (Pinese and Holmes 2005).

Pupa: The newly formed pupa is white which experiences a change in colouration to a very pale-red just before the adult eclosion stage. Notably, abdominal apex accommodates paired urogomphi (Pinese and Holmes 2005).

Adult: Mango stone weevil adult is provided with a typical curved snout and the greyish-brown body that accommodates paler markings (Fig. 24.27a). Due to presence of well-developed elytra, these weevils rarely fly but at night can fly short distances, and camouflage when touched or disturbed (Pinese and Holmes 2005).

24.5.3.3 Biology

Life cycle of the pest commences about two weeks after fruit setting when the female deposits an eggs on the immature fruit (de Roffignac et al. 2007). Immediately, the weevil (female) carves a slit in the fruit to exudate sap for covering and gluing the egg to the fruit (Schoeman 1987). Balock and Kozuma (1964) noted that the eggs are laid singly, and a female can lay up to 15 eggs per day on different fruits. After hatching, the neonate larvae burrow through the pulp into the developing seed, where they feed on leaves. Typically, the larva passes through 5–7 instar stages (Hansen

Fig. 24.27 An adult of mango stone weevil (**a**) with its characteristic damage symptoms (**b**) is shown (CABI 2014)

et al. 1989). Pupation takes place in the seed, rarely in the flesh, and lasts about 7 days. The adult emerges by cutting the seed coat up to 2 months after the fruit has fallen on the ground and the flesh has decayed. Overall, it takes 5–8 weeks for the pest to transform into an adult from the egg stage (Devi et al. 2011).

24.5.3.4 Nature of Damage

Mango fruits infested with the *S. mungiferae* do not exhibit any visible external symptoms as the incision developed by the female seed weevil and the larval penetration wound heals with the increase in fruit size (Hansen 1991). However, the presence of hard, amber-coloured secretion on the fruit skin can differentiate it from the uninfested fruit (Verghese et al. 2005). As mentioned above, larva is the most destructive stage of the pest, this is time when it burrows through the flesh and starts feeding on the seeds (Fig. 24.27b) which not only causes a decrease in the viability of seeds but also the availability of seed as a rootstock source for propagation (Follett and Gabbard 2000). The presence of weevils in fruit is always annoying therefore various countries have designated *S. mangiferae* as a "quarantine pest" and do not like to import the infested mangoes (EFSA Panel on Plant Health et al. 2018). Furthermore, severe infestation of mango stone weevil is also believed to play a critical role in the premature drop of fruit (Verghese et al. 2005).

24.5.3.5 Management

(1) Cultural control of the pest involves sanitation of field by collecting all the fallen fruit and seeds and destroying them to manage the pest infestation for the following year (Hansen and Armstrong 1990).
(2) Use of predatory insects of this pest like weaver ant, *O. smaragdina*, can lower the incidence of infestation (Abdulla et al. 2016; Sahayaraj and Hassan 2023).

(3) Insecticides with expeditious knock down effect are being used for the management of this harmful pest include thiamethoxam, carbaryl, deltamethrin, dimethoate, endosulfan, fenvalerate, and imidacloprid applied at least twice during the flowering and early fruiting stage have been found effective in lowering the populations of the stone weevil and infestation of fruits (Verghese et al. 2003).

24.6 Conclusion

Overall, the successful management of insect pests in mango cultivation hinges on a profound understanding of their biology, behaviour, and ecological interactions, forming the cornerstone of sustainable and effective control strategies. By fostering a deeper understanding of pest dynamics and advancing management techniques, we can ensure the long-term health of mango crops, enhancing productivity while minimizing environmental impact. As we have seen, the diverse range of pests ranging from fruit flies and scale insects to mealybugs and caterpillars presents significant challenges to both yield and fruit quality. Insect pests like mango thrips, white scales, fruit borers which were once regarded as minor pests have now assumed the status of being a "major pest" and the cultivation of decades old mango varieties along with the looming scenario of climate change has seriously affected the pest community structure inhabiting our mango orchards. Furthermore, the irrational application of toxic chemicals for the management of these insect pests caused a tremendous increase in the development of resistance in these species and loss of natural enemies. Therefore, to compete in the international market and enhance our export, there is a desire to grow pest free and pesticides residue-free produce. All this can be materialized only by adopting novel and ecofriendly insect pest management tools such as use of drones, coloured sticky traps, natural enemies, semiochemicals, microbial pesticides, and development of insect resistant cultivars. Hence there is a need to strengthen good agricultural practices (GAP) and research in this direction is essential to develop future integrated pest management (IPM) strategies for our mango orchards.

References

Abate B, Dechassa N (2021) White mango scale, *Aulacaspis tubercularis* Newstead (Hemiptera: Diaspididae): a challenging mango productions in Ethiopia: a review. Bioprocess Eng 5:17–22

Abbas S, Verghese A, Fasih M (1985) Studies on the mango inflorescence midge, *Erosomyia indica* Grover. In: Chadha KL, Pal RN (eds) II International symposium on Mango 231. Bangalore, India, pp 593–596

Abbasi QD, Jan ND, Mahar AN, Panhwar A, Khuhro RD (2008) Etiology of mango tree mortality in Pakistan. Int J Fruit Sci 8:237–250

Abdulla NR, Rwegasira GM, Jensen KM, Mwatawala MW, Offenberg J (2016) Control of mango seed weevils (*Sternochetus mangiferae*) using the African weaver ant (*Oecophylla longinoda* Latreille) (Hymenoptera: Formicidae). J Appl Entomol 140:500–506

Abo-Shanab A (2012) Suppression white mango scale, *Aulacaspis tubercularis* Newstead (Hemiptera: Diaspididae) on mango trees in El-Beheira Governorate. Egypt Acad J Biol Sci 5:43–50

Abo-Shnaf R, Allam SFM, El-Sobky ML, Abdul-Shafc AF, El-Tony AG (2022) Biodiversity of mites in mango orchards (*Mangifera indica* L.) and evaluation of some mineral and essential oils against *Cisaberoptus kenyae* Keifer (Acari: Eriophyidae) management. Acarologia 62:130–142

Abou-Awad BA, Metwally A-SM, Al-Azzazy MM (2010) Effect of different eriophyid and tetranychid mango mite species on development, longevity, fecundity and predation of *Typhlodromus mangiferus* Zaher and El-Brolossy (Acari: Phytoseiidae). Arch Phytopathol Plant Protect 43:390–403

Adnan SM, Uddin MM, Alam MJ, Islam MS, Kashem MA, Rafii MY, Latif MA (2014) Management of mango hopper, *Idioscopus clypeali*s, using chemical Insecticides and neem oil. Sci World J 2014:709614

Affandi A, Velasco LRI, Javier PA, Depositario DPT (2017) Development and survivorship of *Scirtothrips dorsalis* Hood (Thysanoptera: Thripidae) in different growth stages of mango and selected weeds. AGRIVITA, J Agric Sci 40:101–106

Ahmed F, K. Shazad, S. Naseer, M. U. Saleem, Hussain SI (2015) Mango mealybug. In: Fact Sheet for Farmers. CABI, UK

Ahmed W, Nawaz MA, Saleem BA, Asim M (2005) Incidence of mango midge and its control in different mango growing countries of the world. In: Afzal M, Nawaz M, Hussain Z (eds) International conference on mango and date palm: culture and export, Faisalabad, Pakistan, pp 98–101

Al-Azzazy M (2005) Integrated management of mites infesting mango trees. Al-Azhar University, Egypt

Al-Subhi AM, Al-Adawi AO, Van Wyk M, Deadman ML, Wingfield MJ (2006) *Ceratocystis omanensis*, a new species from diseased mango trees in Oman. Mycol Res 110:237–245

Al Adawi AO (2012) Studies on *Ceratocystis* spp. Associated with mango die-back in Oman. University of Pretoria

Al Adawi AO, Al Jabri RM, Deadman ML, Barnes I, Wingfield B, Wingfield MJ (2013) The mango sudden decline pathogen, *Ceratocystis manginecans*, is vectored by *Hypocryphalus mangiferae* (Coleoptera: Scolytinae) in Oman. Eur J Plant Pathol 135:243–251

Al Adawi AO, Deadman ML, Al Rawahi AK, Al Maqbali YM, Al Jahwari AA, Al Saadi BA, Al Amri IS, Wingfield MJ (2006) Etiology and causal agents of mango sudden decline disease in the Sultanate of Oman. Eur J Plant Pathol 116:247–254

Aliakbarpour H, Che SMR (2010) Diurnal activity of four species of thrips (Thysanoptera: Thripidae) and efficiencies of three nondestructive sampling techniques for thrips in mango inflorescences. J Econ Entomol 103:631–640

Aliakbarpour H, Che SMR (2011) Seasonal abundance and spatial distribution of larval and adult thrips (Thysanoptera) on weed host plants in mango orchards in Penang, Malaysia. Appl Entomol Zool 46:185–194

Anant AK, Pandey N, Ray A, Behera SK (2019) Influence of abiotic factors on the incidence of mango hopper (Hemiptera: Cicadellidae) in Chhattisgarh. J Entomol Zool Stud 7:1067–1070

Anithakumari D, Lakshmi B, Sathyanarayana RG, Lakshmi NRM (2010) Influence of abiotic factors on the incidence of hopper and chemical control strategies in mango. Karnataka J Agric Sci 22

Anonymous (2023) Chapter 2 Agriculture, Economic Survey of Pakistan 2022–2023. In: Finance Do (ed). Govt.of Pakistan, Islamabad, pp 19–30

Ansari MS, Basri R, Shekhawat SS (2019) Insect pests infestation during field and storage of fruits and vegetables. In: Malik A, Erginkaya Z, Erten H (eds) Health and safety aspects of food processing technologies. Springer International Publishing, Cham, pp 121–207

Anwar R, Malik AU (2007) Hot water treatment affects ripening quality and storage life of mango (*Mangifera indica* L.). Pak J Agric Sci 44:304–311

Ashfaq M, Ara J, Noor A, Hebert PD, Mansoor S (2011) Molecular phylogenetic analysis of a scale-like insect infesting mango orchards in Pakistan. Eur J Entomol 5759:553–559

Atnafu Y (2020) Management of white mango scale, *Aulacaspis tubercularis* (Homoptera: Diaspididae) using pruning, oils and pesticides in eastern Wellega, Ethiopia. Doctoral Dissertation, Jimma University, Ethiopia

Atwal AS (2008) Pests of sub-tropical and tropical fruits. In: Agricultural pests of South Asia and their management, 6th edn. Kalyani Publishers, Ludhiana, India, pp 373–379

Augustyn WA, Plooy Wd, Botha BM, Wilpe Ev (2013) Infestation of *Mangifera indica* by the mango gall gly, *Procontarinia matteiana*, (Kieffer & Cecconi) (Diptera: Cecidomyiidae). Afr Entomol 21:79-88, 10

Awan RRH, Lanjar AG, Bukero A, Nizamani IA, Nahiyoon SA, Kumar D (2021) Life table studies of three mango hopper species in field conditions. Pure Appl Biol 11:284–290

Ayalew G, Fekadu A, Sisay B (2015) Appearance and chemical control of white mango scale (*Aulacaspis tubercularis*) in Central Rift Valley. Sci Technol Arts Res Journal 4:59–63

Ayyaz S, Bonney L, Akmal N (2019) Competitiveness in mango trade: a comparative analysis between Pakistan and other mango exporting nations. Int J Food Agric Econom 7:341–349

Backus EA, Serrano MS, Ranger CM (2005) Mechanisms of hopperburn: an overview of insect taxonomy, behavior, and physiology. Annu Rev Entomol 50:125–151

Badar H (2015) Value chain performance improvement for sustainable mango industry development in Pakistan. Doctoral Dissertation, The University of Queensland, Australia

Bagheri A, Kolyaee R, Askari Seyahooei M, Modaress Najafabadi SS, Faraji G (2017) Efficacy of methyl eugenol bait traps for controlling the mango fruit fly *Bactrocera zonata* (Diptera: Tephritidae). J Crop Prot 6:181–189

Balock J, Kozuma T (1964) Notes on the biology and economic importance of the mango weevil, *Sternochetus mangiferae* (Fabricius), in Hawaii (Coleoptera: Curculionidae). In: Proceedings of the Hawaiian Entomological Society 18, Hawaii, pp 353–364

Baradevanal G, Chander S, Jayanthi PDK, Singh HS, Reddy DS (2021) Assessing the risk of mango quarantine pest *Deanolis sublimbalis* Snellen under different climate change scenarios. J Plant Dis Prot 128:853–863

Bautista-Rosales PU, Ragazzo-Sánchez JA, Calderón-Santoyo M, Cortéz-Mondaca E, Servín-Villegas R (2013) *Aulacaspis tubercularis* Newstead in mango orchards of Nayarit, Mexico, and relationship with environmental and agronomic factors. Southwest Entomol 38(221–230):210

Belachew ZG, Jenber AJ (2022) Status, importance, and management of the White Mango Scale (*Aulacaspis tubercularis* Newstead) In Ethiopia: a review. Entomol Appl 9:50–59

Ben-Dov Y (1994) A systematic catalogue of the mealybugs of the world (Insecta: Homoptera: Coccoidea: Pseudococcidae and Putoidae) with data on geographical distribution, host plants, biology and economic importance. Intercept Limited, Andover, UK, 686 p

Bhattacharya M (2014) A review on the biology and symptoms of attack of mango red banded catterpillar (*Autocharis albizonalis* Hampson). J Agric Vet Sci 7:01–05

Bhattacharyya B, Pujari DdM, Das M (2016) Mango weevils: identification, biology and integrated management. In: Insect Pests Management in Fruit Crops. Biotech Books, pp 35–49

Bhattacharyya M (2018) Varietal preference of mango fruit borer *Autocharis albizonalis* Hampson in regard to the phenol content of the fruit. Int J Biores Stress Manag 9:164–166

Bibi F, Hameed A, Muhammad N, Shahzad K, Ahmad I, Shah TA, Z. Gaafar A-R, Hodhod MS, Bourhia M, Nafidi H-A (2023) Potential of integrated nutrient management to rehabilitate the dieback-affected mango cultivar Sammer Bahisht Chaunsa. Sustain 15:11118

Bindra O, Varma G, Sandhu G (1970) Studies on the relative efficacy of banding materials for the control of mango mealybug, *Drosicha stebbingi* (Green). J Res (Punjab Agric Univ) 7:491–494

Böckmann E, Köppler K, Hummel E, Vogt H (2014) Bait spray for control of European cherry fruit fly: an appraisal based on semi-field and field studies. Pest Manage Sci 70:502–509

Bright DE, Skidmore RE (2002) A catalog of Scolytidae and Platypodidae (Coleoptera). Supplement 2 (1995–1999). NRC Research Press, Canada

Butani DK (1979) Insects and fruits. Periodical Expert Book Agency, New Dehli, India

Buthelezi NMD, Mafeo TP, Mathaba N (2021) Preharvest bagging as an alternative technique for enhancing fruit quality: a review. HortTechnology 31:4–13

Byers JA (1995) Host-tree chemistry affecting colonization in Bark beetles. In: Cardé RT, Bell WJ (eds) Chemical ecology of insects 2. Springer, Boston, pp 154–213

Byers JA (2000) Wind-aided dispersal of simulated bark beetles flying through forests. Ecol Model 125:231–243

CABI (2014) *Sternochetus mangiferae* (Fabricius, 1775), mango seed weevil (pest/pathogen). In. CABI PlantwisePlus, UK

CABI (2019) *Erosomyia mangiferae* (Mango gall midge). In: Technical Fact Sheet. CABI, UK

Carrillo-Arámbula L, Infante F, Cavalleri A, Gómez J, Ortiz JA, Fanson BG, González FJ (2022) Colored sticky traps for monitoring phytophagous thrips (Thysanoptera) in mango agroecosystems, and their impact on beneficial insects. PLOS One 17:e0276865

Carrillo D, Birke A, Guillen L, Peña J (2017) Pests of mango. In: Muhammad Siddiq, Jeffrey K. Brecht, Sidhu JS (eds) Handbook of mango fruit: Production, postharvest science, processing technology and nutrition. Wiley, New York, pp 61–90

Carrillo D, Duncan R, Pena J (2018) Mango pests and beneficial insects: Insects on flowers and fruits EDIS

Chakrabarti DK (2011) Economic importance. In: Chakrabarti DK (ed) Mango malformation. Springer The Netherlands, pp 15–16

Chandra A, Bhati D, Singh K (1987) Bionomics of mango mealy bug, *Drosicha mangiferae* Green (Margarodidae: Hemiptera). Bull Entomol 28:145–152

Chapman KR (1999) Mango-special challenges in Asia and Oceania. In: VI International Symposium on Mango 509, pp 95–106

Choi KS, Yang JY, Park YM, Kim S, Choi H, Lyu D, Kim D-S (2013) Pest lists and their damages on mango, dragon fruit and atemoya in Jeju, Korea. Korean J Appl Entomol 52:45–51

Chonhenchob V, Kamhangwong D, Kruenate J, Khongrat K, Tangchantra N, Wichai U, Singh SP (2011) Preharvest bagging with wavelength-selective materials enhances development and quality of mango (*Mangifera indica* L.) cv. Nam Dok Ma. J Sci Food Agric 91:664–671

Choudhary JS, Mali SS, Mukherjee D, Kumari A, Moanaro L, Rao MS, Das B, Singh A, Bhatt B (2019) Spatio-temporal temperature variations in MarkSim multimodel data and their impact on voltinism of fruit fly, *Bactrocera* species on Mango. Sci Rep 9:9708

Clarke AR, Armstrong KF, Carmichael AE, Milne JR, Raghu S, Roderick GK, Yeates DK (2005) Invasive phytophagous pests arising through a recent tropical evolutionary radiation: the *Bactrocera dorsalis* complex of fruit flies. Annu Rev Entomol 50:293–319

COLEACP (2013) How to control the mango fruit fly. In: CTA practical guide series. COLEACP; CTA, Brussels, Belgium, p 8

Dako OD (2023) White mango scale (*Aulacaspis tubercularis*) infestation on mango: distinct scenarios of the pest status in Ethiopia and Kenya. Heliyon 9

Daneel MS, Joubert PH (2009) Biological Controlof the mango scale *Aulacspis tubercularis* NEWSTEAD (Coccidae: Diaspididae) by a parasitoid *Aphytis chionaspis* REN (Hymenoptera: Aphelinidae), 820 edn. International Society for Horticultural Science (ISHS), Leuven, Belgium, pp 567–574

de Roffignac L, LeBellec F, Herzog D (2007) *Sternochetus mangiferae* (Fabricius) (Coleoptera: Curculionidae), charançon du noyau de la mangue, en vergers de manguiers à la Guadeloupe. Fruits 62:279–293

del Pino M, Bienvenido C, Boyero JR, Vela JM (2020) Biology, ecology and integrated pest management of the white mango scale, *Aulacaspis tubercularis* Newstead, a new pest in southern Spain-a review. Crop Protect 133:105160

Devi MB, Devi AK, Devi AB, Singh KM (2011) Biology of mango pulp weevil, *Sternochetus gravis* (Fabr.)(Curculionidae: Coleoptera) in the agro-ecosystem of Manipur. J Exp Sci 2

Dhillon MK, Singh R, Naresh JS, Sharma HC (2005) The melon fruit fly, *Bactrocera cucurbitae*: a review of its biology and management. J Insect Sci 5

Dominiak B (2006) Review of the use of protein food based lures in McPhail traps for monitoring Queensland fruit fly'*Bactrocera tryoni'* (Froggatt) (Diptera: Tephhtidae). General Appl Entomol J Entomol Soc New South Wales 35:7–12

Doorenweerd C, Leblanc L, Norrbom AL, San Jose M, Rubinoff D (2018) A global checklist of the 932 fruit fly species in the tribe Dacini (Diptera, Tephritidae). ZooKeys:19

Downie D, Gullan P (2004) Phylogenetic analysis of mealybugs (Hemiptera: Coccoidea: Pseudococcidae) based on DNA sequences from three nuclear genes, and a review of the higher classification. Syst Entomol 29:238–260

EFSA Panel on Plant Health, Bragard C, Baptista P, Chatzivassiliou E, Di Serio F, Gonthier P, Jaques Miret JA, Justesen AF, Magnusson CS, Milonas P (2022) Pest categorisation of *Aulacaspis tubercularis*. EFSA J 20:e07307

EFSA Panel on Plant Health, Bragard C, Dehnen-Schmutz K, Di Serio F, Gonthier P, Jacques MA, Jaques Miret JA, Justesen AF, Magnusson CS, Milonas P (2018) Pest categorisation of *Sternochetus mangiferae*. EFSA J 16:e05439

FAO (2018) FAO statistics. Food and agriculture organization of the United Nations. In. FAOSTAT Database, Rome, Italy

FAO (2023) Food and Agriculture Organization of the United Nations. In. FAOSTAT Database, Rome, Italy

Fenner T (1997) Red-banded mango caterpillar: biology and control prospects. Mango Care Newsletter, 2pp. Queensland

Fita T, Getu E, Wakgari M, Woldetsadike K (2020) Efficacy of *Azadirachta indica* (A. Juss) seed powder water extract against *Aulacuspis tubercularis* Newsteed (Homoptera: Diaspididae) on mango (*Mangifera indica* l.) In East Wollega, Ethiopia. SINET Ethiopian J Sci 43:11–20

Fletcher B (1987) The biology of dacine fruit flies. Annu Rev Entomol 32:115–144

Follett PA, Gabbard Z (2000) Effect of mango weevil (Coleoptera: Curculionidae) damage on mango seed viability in Hawaii. J Econ Entomol 93:1237–1240

Follett PA, Neven LG (2006) Current trends in quarantine entomology. Annu Rev Entomol 51:359–385

Franco JC, Zada A, Mendel Z (2009) Novel Approaches for the management of mealybug pests. In: Ishaaya I, Horowitz AR (eds) Biorational Control of arthropod pests: application and resistance management. Springer, Netherlands, pp 233–278

Galdino TVdS, Ferreira DdO, Santana Júnior PA, Arcanjo LdP, Queiroz EA, Sarmento RA, Picanço MC (2017) The role of the beetle *Hypocryphalus mangiferae* (Coleoptera: Curculionidae) in the spatiotemporal dynamics of mango wilt. J Econ Entomol 110:865–874

Galdino TVdS, Kumar S, Oliveira LSS, Alfenas AC, Neven LG, Al-Sadi AM, Picanço MC (2016) Mapping global potential risk of mango sudden decline disease caused by *Ceratocystis fimbriata*. PLOS One 11:e0159450

Gamliel-Atinsky E, Freeman S, Maymon M, Belausov E, Ochoa R, Skoracka A, Pena J, Palevsky E (2010) The role of eriophyoids in fungal pathogen epidemiology, mere association or true interaction? Exp Appl Acarol 51:190–204

Ganeswaran R (2017) *Idioscopus clypealis* (Mango leafhopper). In: CABI Digital Library. CABI, UK

Garcia-Lopez E, Batista-Marte C, Serra C, Sosa-Natta A, Villegas-Monter A, Hernandez-Castro E, Camacho-Tapia M, Mora-Aguilera J (2023) Mango malformation: etiology, symptoms, distribution and cultivar susceptibility in the Dominican Republic. Can J Plant Sci 103:300–311

Garcia-Lopez E, Teliz-Ortiz D, Villegas-Monter Á, Hernandez-Castro E, Valdovinos-Ponce G, Mora-Aguilera JA (2020) Tolerance of five mango cultivars to malformation disease in Mexico. J Plant Pathol 102:1145–1151

García Morales M, Denno BD, Miller DR, Miller GL, Ben-Dov Y, Hardy NB (2016) ScaleNet: a literature-based model of scale insect biology and systematics. Database 2016:bav118

Ghaffar A, Khan A, Hameed A, Iqbal J, Ahmad M, Raza S, Imran M, Muhammd N, SHAH S, RAZA M (2022) Establishment of strategy for mango hopper *Idioscopus clypealis* (Lethierry) management and impact of weather factors on its population in South Punjab, Pakistan. Biol Clin Sci Res J 2022

Golez HG (1991) Bionomics and control of the mango seed borer, *Noorda albizonalis* Hampson (Pyralidae, Lepidoptera). 291 edn. International Society for Horticultural Science (ISHS), Leuven, Belgium, pp 418–424

Gomez DF, Riggins JJ, Cognato AI (2023) Bark beetles. In: Forest entomology and pathology. Entomology, vol 1. Springer International Publishing Cham, pp 299–337

Green EE (1908) Remarks on Indian scale insects (Coccidae), Part III. With a catalogue of all species hitherto recorded from the Indian continent. Remarks on Indian scale insects (Coccidae), part III. With a catalogue of all species hitherto recorded from the Indian continent. 2:15–46

Grimaldi D, Shmakov A, Fraser N (2004) Mesozoic thrips and early evolution of the order Thysanoptera (Insecta). J Paleontol 78:941–952

Grove T, Giliomee J, Pringle K (2001) Thrips (Thysanoptera) species associated with mango trees in South Africa. Afr Entomol 9:153–162

Grover P (1985) Protection of mango crop against newly discovered midge pests. Cecidol Int 6:35–40

Gullan PJ, Kosztarab M (1997) Adaptations in scale insects. Annu Rev Entomol 42:23–50

Gullan PJ, Martin JH (2009) Sternorrhyncha:(jumping plant-lice, whiteflies, aphids, and scale insects). In: Resh VH, RT C (eds) Encyclopedia of insects. Elsevier, Amsterdam, pp 957–967

Habtegebriel B, Melisie D, Daba T, Hailu T, Azerefegn F (2020) Integrated control of the white mango scale through tree management and soil drenching with a systemic insecticide in Western Ethiopia. Ethiop J Agric Sci 30:25–32

Hallman GJ (1999) Ionizing radiation quarantine treatments against tephritid fruit flies. Postharvest Biol Technol 16:93–106

Hamdy NM (2020) Life table and morphometric studies of *Aulacaspis tubercularis* infesting mango trees in Egypt (Diaspididae: Hemiptera). J Plant Proct Pathol 11:613–620

Hamid M, Jalaluddin M (2006) Recurring incidence of sooty mould of mango in Karachi and its control. Int J Biol Biotechnol 3:561–565

Hansen J (1991) Mango weevil biology, field control, and postharvest technology: a review. Int J Trop Agric 9:234–244

Hansen J, Armstrong J (1990) The failure of field sanitation to reduce infestation by the mango weevil, *Cryptorhynchus mangiferae* (F.) (Coleoptera: Curculionidae). Int J Pest Manage 36:359–361

Hansen JD, Armstrong JW, Brown SA (1989) The distribution and biological observations of the mango weevil, *Cryptorhynchus mangiferae* (Coleoptera: Curculionidae), in Hawaii. In: Proceedings of Hawaiian Entomological Society, Hawaii, pp 31–39

Harris K, Schreiner I (1992) A new species of gall midge (Diptera: Cecidomyiidae) attacking mango foliage in Guam, with observations on its pest status and biology. Bull Entomol Res 82:41–48

Hasnain M, M. Naeem Aslam, M. Ashiq Sangi, Khan S (2013) Mango midge. In: Fact sheet For Farmers. CABI, UK

Hasnain M, Saeed S, Ullah UN, Ullah S, Zaka SM (2023) Synergist response of the Peach fruit fly, *Bactrocera zonata* (Saunders) to some ammonium based proteinaceous food bait attractants. BMC Zool 8:17

Henri DC, Jones O, Tsiattalos A, Thébault E, Seymour CL, van Veen FJF (2015) Natural vegetation benefits synergistic control of the three main insect and pathogen pests of a fruit crop in southern Africa. J Appl Ecol 52:1092–1101

Heve WK, El-Borai FE, Carrillo D, Duncan LW (2017) Biological control potential of entomopathogenic nematodes for management of Caribbean fruit fly, *Anastrepha suspensa* Loew (Tephritidae). Pest Manage Sci 73:1220–1228

Hill DS (1994) Agricultural entomology. Timber Press, USA

Hiremath S, Hugar P (1989) Effects of injecting systemic insecticides into mango during off season on the mango hopper population. Madras Agric J 76:638–640

Hoddle M (2022) *Scirtothrips dorsalis* (Chilli Thrips). CABI Compendium, Wallingford, CT, USA

Hofman PJ, Smith LG, Joyce DC, Johnson GI, Meiburg GF (1997) Bagging of mango (*Mangifera indica* cv.Keitt') fruit influences fruit quality and mineral composition. Postharvest Biol Technol 12:83–91

Hossain M, Sarkar B, Hossain M, Mian M, Rajotte E, Muniappan R, O'Rourke M (2020) Comparison of biorational management approaches against mango fruit fly (*Bactrocera dorsalis* Hendel) in Bangladesh. Crop Protect 135:104807

Huan L, Zhang D-j, Xu Y-j, Lei W, Cheng D-f, Qi Y-x, Ling Z, Yongyue L (2019) Invasion, expansion, and control of *Bactrocera dorsalis* (Hendel) in China. J Integr Agric 18:771–787

Hulcr J, Atkinson TH, Cognato AI, Jordal BH, McKenna DD (2015) Morphology, taxonomy, and phylogenetics of bark beetles. In: Vega FE, Hofstetter RW (eds) Bark beetles. Elsevier, San Diego, pp 41–84

Isman MB, Machial CM (2006) Pesticides based on plant essential oils: from traditional practice to commercialization. In: Rai M, Carpinella MC (eds) Advances in phytomedicine. Elsevier, pp 29–44

Javed S, Fu H, Ali A, Nadeem A, Amin M, Razzaq K, Ullah S, Rajwana IA, Nayab S, Ziogas V (2022) Comparative response of mango fruit towards pre-and post-storage quarantine heat treatments. Agronomy 12:1476

Jayanthi KPD, Woodcock CM, Caulfield J, Birkett MA, Bruce TJ (2012) Isolation and identification of host cues from mango, *Mangifera indica*, that attract gravid female oriental fruit fly, *Bactrocera dorsalis*. J Chem Ecol 38:361–369

Jiao K-L, Wang H, Wei D-W, Mo J-Y, Wang Y-H, Bu W-J, Kolesik P (2018) A new species of *Procontarinia* (Diptera: Cecidomyiidae) damaging fruit of mango, *Mangifera indica* (Anacardiaceae), in China. Zootaxa 4413:368–376

Jindal V, Banta G, Singh M (2016) Molecular identification of mango hoppers infesting mango trees in Punjab through DNA barcoding. Indian J Hortic 73:192–196

Joubert PH, Daneel MS, Grove T (1999) Progress towards integrated pest management (IPM) on mangoes in South Africa. VI International Symposium on Mango 509:811–818

Juárez-Hernández P, Valdez-Carrasco J, Valdovinos-Ponce G, Mora-Aguilera JA, Otero-Colina G, Téliz-Ortiz D, Hernández-Castro E, Ramírez-Ramírez I, González-Hernández VA (2014) Leaf penetration pattern of *Aulacaspis tubercularis* (Hemiptera: Diaspididae) stylet in mango. Fla Entomol 97:100–107

Kannan M, Rao NV (2006) Ecological studies on mango hopper, *Amritodus atkinosoni* Lethier in Andhra Pradesh as a basis for IPM. Crop Res 32:235–238

Karar H, Ahmad M, Ullah H, Wajid M, Zubair M, Raza H (2019) Effectiveness of fruit bagging for the control of insect-pests complex and its impact on quality of mango fruits. J Hortic Sci Technol 2:45–48

Karar H, Arif MJ, Ali A, Hameed A, Abbas G, Abbas Q (2012) Assessment of yield losses and impact of morphological markers of various mango (*Mangiferae indica*) genotypes on mango mealybug (*Drosicha mangiferae* Green)(Homoptera: Margarodidae). Pak J Zool 44

Karar H, Bakhsh MA (2018) Effect of host plant on abundance of mango hoppers, *Idioscopus clypealis* (Lethierry)(Hemiptera: Cicadellidae). Pak Entomol 40:57–61

Karar H, Bashir MA, Abbas G, Nisar MS, Khan AK, Batool M, Iqbal J, Khan KA, Ghramh HA, Al Kahtani S (2020) Safe, economical and effective technology for the management of mango mealy bug (*Drosicha mangiferae* Green) in infested mango orchards and other plantations. Fresenius Environ Bull 29:7034–7040

Karar H, Bashir MA, Basit A, Atta S, Anjum AA, Bakhsh A, Hussain A, Hameed A, Wang Y, Alajmi RA, Metwally DM, Imran M (2021) Effect of host plant on cornucopia of mango fruit flies (Diptera: Tephritidae) and their triumphant management in context of climate change. Saudi J Biol Sci 28:2366–2373

Karar H, Javed MU, Yaseen M, Bashir MA, Sajjad A, Essa M, Wajid M, Mubashir M, Mustafa G, Zubair M (2022) Comparative efficacy of conventional vs new chemistry insecticides against mango thrips, *Scirtothrips dorsalis* Hood (Thripidae: Thysanoptera). J King Saud Univ-Sci 34:102233

Karar H, Sayyed AH, Arif MJ, Ashfaq M, Aslam M (2010) Integration of cultural and mechanical practices for management of the mango mealybug *Drosicha mangiferae*. Phytoparasitica 38:223–229

Kaul V, Shankar U, Khushu M (2009) Bio-intensive integrated pest management in fruit crop ecosystem. In: Peshin R, Dhawan AK (eds) Integrated pest management: innovation-development process. Springer, Dordrecht. https://doi.org/10.1007/978-1-4020-8992-3_21

Kaushik DK, Baraiha U, Thakur B, Parganiha O (2012) Pest complex and their succession on mango (*Mangifera indica*) in Chhattisgarh, India. Plant Arch 12:303–306

Khalid A, Khan A, Mohammad A (2002) Non-traditional method of fruit fly control in guava orchards in Dera Ismail Khan. Pak J Agric Res 17:195–196

Khan BS, Arslan M, Ghaffar A, Farooq M, Ahmad S, Ullah S, Rasool A (2022) Soil inhabiting mite diversity (Acari) of mango orchards from Dera Ghazi Khan, Pakistan. Pak J Agric Res 35

Khan HAA, Akram W (2018) Trichlorfon and spinosad resistance survey and preliminary determination of the resistance mechanism in Pakistani field strains of *Bactrocera dorsalis*. Sci Rep 8:11223

Khan MA, Ashfaq M, Akram W, Lee J-J (2005) Management of fruit flies (Diptera: Tephritidae) of the most perishable fruits. Entomological Res 35:79–84

Khan MH, Khuhro NH, Awais M, Asif MU, Muhammad R (2021) Seasonal abundance of fruit fly, *Bactrocera* species (Diptera: Tephritidae) with respect to environmental factors in guava and mango orchards. Pak J Agric Res 34:266

Khatri I, Webb M (2014) Review of the Idiocerine leafhoppers of Pakistan (Hemiptera, Cicadellidae) with a description of a new species. Zootaxa 3860:280–290

Kosztarab M, Kozár F (1988) Scale insects of Central Europe. Junk, Dordrecht

Krishnamoorthy A, Reddy P, Verghese A (2014) Borer pests: status and challenges. Indian Institute of Horticultural Research, Bengaluru, pp 14–15

Krishnamoorthy A, Visalakshi PG (2012) Record of thrips on mango. J Hortic Sci 7:110–111

Krishnarao G, Sujatha A, Sunitha P, Vyas M, Jayanthi PDK (2022) A transcriptomic approach reveals the molecular basis of pre-pupal diapause of Red Banded Mango Caterpillar, *Deanolis sublimbalis*. Curr Sci 123:471

Krull S, Basedow T (2006) Studies on the biology of *Deanolis sublimbalis* Snellen (Lepidoptera, Pyralidae) and its natural enemies on mango in Papua New Guinea. MITT DTSCH GES ALLG ANGEW ENT 15:273–276

Krull SME (2004) Studies on the mango-ecosystem in Papua New Guinea with special reference to the ecology of *Deanolis sublimbalis* Snellen (Lepidoptera, Pyralidae) and to the biological control of *Ceroplastes rubens* Maskell (Homoptera, Coccidae). Universitätsbibliothek Giessen

Kumar A, Pandey SK, Kumar R (2009) Population dynamics of mango mealy bug, *Drosicha mangiferae* Green from Jhansi, Uttar Pradesh. Biol Forum Int J 1:77–79

Kumar P, Misra AK (2016) Current scenario of mango malformation and its management strategies: an overview. In: Kumar P, Gupta V, Tiwari A, Kamle M (eds) Current trends in plant disease diagnostics and management practices. Springer, Switzerland, pp 221–236

Labuschagne T, Brink T, Steyn W, De Beer M (1996) Fruit flies attacking mangoes-their importance and post harvest control. Yearbook-South Afr Mango Growers' Assoc 16:17–19

Laxman RH, Annapoornamma CJ, Biradar G (2016) Mango. In: Rao NKS, Shivashankara KS, Laxman RH (eds) Abiotic stress physiology of horticultural crops. Springer India, New Delhi, pp 169–181

Leefmans S, van der Vecht J (1930) The red ringed mango caterpillar, *Noorda albizonalis* Hamps. Korte Mededeelingen Van Het Instituut Voor Plantenziekten 14:1–8

Lewis T (ed) (1973) Thrips, their biology, ecology and economic importance. Academic, New York

Lewis T (1997) Pest thrips in perspective. In: Lewis T (ed) Thrips as crop pests. CABI, Wallingford, United Kingdom, pp 1–13

Lieutier F, Day KR, Battisti A, Grégoire J-C, Evans HF (2004) Bark and wood boring insects in living trees in Europe: a synthesis. Springer, The Netherlands

Lodhi TE, Abbas M, Asif M, Hussain DA (2006) To analyze the future perspectives of agricultural extension services in mango production and marketing with special reference to world trade organization (WTO) in district Multan, Pakistan. Pakistan J Life Soc Sci 4:68–75

Lowry VK, Smith JW Jr, Mitchell FL (1992) Life-fertility tables for *Frankliniella fusca* (Hinds) and *F. occidentals* (Pergande) (Thysanoptera: Thripidae) on Peanut. Ann Entomol Soc Am 85:744–754

Magar BR, Joshi M, Poudel S (2022) Mango stem borer: a serious pest and management strategies. Rev Food Agric 3:54–57

Mahoo mbogmdLK, Mtembei K (2012) Come together to mass trap Mango fruit fly: *Bactrocera* spp. In: Fact sheet for farmers. CABI, UK

Maiti R, Kumari A (2016) Biotic Stress: Insect Pests. In: Maiti R, Kumari A, Thakur AK, Sarkar NC (eds) Bioresource and stress management. Springer, Singapore, pp 101–129

Makale F, Watson G, Kibwage P, Kansiime M (2020) Biodiversity and agriculture: addressing scale insect threats in Kenya: scale insects photo guide. Biodiversity and agriculture: addressing scale insect threats in Kenya: scale insects photo guide

Malumphy C (2014) An annotated checklist of scale insects (Hemiptera: Coccoidea) of Saint Lucia, Lesser Antilles. Zootaxa 3846:69–86–69–86

Mani M (1944) Studies on Indian Itonididae (Cecidomyiidae: Diptera). VII. Indian J Entomol 5:151–164

Mani M, Natarajan N, Hegde RD, Tej MK (2022) Host plant resistance to insect pests in horticultural crops. In: Mani M (ed) Trends in Horticultural Entomology. Springer Nature Singapore, Singapore, pp. 335–386

Mani M, Shivaraju C (2016) Biology. In: Mani M, Shivaraju C (eds) Mealybugs and their management in agricultural and horticultural crops. Springer India, New Delhi, pp 87–106

Masood A, Saeed S (2012) Bark beetle, *Hypocryphalus mangiferae* Stebbing (Coleoptera: Curculionidae: Scolytinae) is a vector of mango sudden death disease in Pakistan. Pak J Bot 44:813–820

Masood A, Saeed S, Erbilgin N, Jung Kwon Y (2010) Role of stressed mango host conditions in attraction of and colonization by the mango bark beetle *Hypocryphalus mangiferae* Stebbing (Coleoptera: Curculionidae: Scolytinae) and in the symptom development of quick decline of mango trees in Pakistan. Entomological Res 40:316–327

Masood A, Saeed S, Mahmood A, Malik SA, Hussain N (2012) Role of nutrients in management of mango sudden death disease in Punjab, Pakistan. Pak J Zool 44:675–683

Masood A, Saeed S, Sajjad A (2008) Characterization and damage patterns of different bark beetle species associated with mango sudden death syndrome in Punjab, Pakistan. Pak Entomol 30:163–168

Masood A, Saeed S, Sajjad A, Ali M (2009) Life cycle and biology of mango bark beetle, *Hypocryphalus mangiferae* (Stebbing), a possible vector of mango sudden death disease in Pakistan. Pak J Zool 41

Masood A, Salman M, Saeed S (2014) Fungicide injection, an efficient management technique of mango sudden death disease in Punjab, Pakistan. Pak J Phytopathol 26:259–263

Maung KL, Mon YY, Khine MP, Chan KN, Phyoe A, Soe AT, Han TYY, Myo WW, San SSS, Khai AA (2020) Current knowledge of mango and fruit fly (Diptera: Tephritidae) control in myanmar: a review. Adv Entomol 9:49–58

Melo AS, Melo JWS, Paz-Neto AA, Gondimjunior MGC (2023) Mite diversity and intra-plant distribution in mango crop. Syst Appl Acarol 28(864–875):812

Mirabbalou M (2013) A checklist of Iranian thrips (Insecta: Thysanoptera). Far Eastern Entomol 1–27

Moharum FA-E (2012) Description of the first and second female and male instars of white mango scale *Aulacaspis tubercularis* Newstead (Coccoidea: Diaspididae). J Basic Appl Zool 65:29–36

Mohyuddin A, Mahmood R (1992) Integrated control of mango pests in Pakistan. IV International Mango Symposium 341:467–483

Muhammad W, Javed H, Saeed S, Kassi AK, Iqbal N, Ahmad I (2017) Evaluation of different insecticides against mango midges (Diptera: Cecidomyiidae). J Entomol Zool Stud 5:1888–1890

Muniappan R, Shepard B, Watson G, Carner G, Rauf A, Sartiami D, Hidayat P, Afun J, Goergen G, Rahman AZ (2009) New records of invasive insects (Hemiptera: Sternorrhyncha) in Southeast Asia and West Africa. J Agric Urban Entomol 26:167–174

Murai T, Kawai S, Chongratanameteekul W, Nakasuji F (2000) Damage to tomato by *Ceratothripoides claratris* (Shumsher)(Thysanoptera: Thripidae) in central Thailand and a note on its parasitoid, *Goetheana shakespearei* Girault (Hymenoptera: Eulophidae). Appl Entomol Zool 35:505–507

Murtaza G, Ramzan M, Bilal H, Ejaz A, Khan MAA, Riaz T, Waqas M (2021) Monitoring of fruit fly, *Bactrocera zonata* (Diptera: Tephritidae) population by installing traps in mango orchard Bahawalnagar, Pakistan. J Appl Res Plant Sci 2:148–151

Mutamiswa R, Nyamukondiwa C, Chikowore G, Chidawanyika F (2021) Overview of oriental fruit fly, *Bactrocera dorsalis* (Hendel) (Diptera: Tephritidae) in Africa: from invasion, bio-ecology to sustainable management. Crop Protect 141:105492

Mwando NL, Ndlela S, Meyhöfer R, Subramanian S, Mohamed SA (2021) Hot water treatment for post-harvest disinfestation of *Bactrocera dorsalis* (Diptera: Tephritidae) and its effect on cv. tommy atkins mango. Insects 12:1070

Nakahara LM, Branch PQ (1993) Mango blossom midge. In: Proceedings: conference on mango in Hawaii

Namni S, Amin M, Miah M, Rahman M, Suh S (2017) Role of weather parameters on seasonal abundance of insects in a mangobased agroforestry in Bangladesh, with particular reference to mango hopper. Bangladesh J Agric Res 42:197–205

Narasimha Murthy B (1981) Susceptibility of mango (*Mangifera indica* Linnaeus) varieties to infestation by (Idiocerinae: Jassidae: Homoptera) of hoppers. College of Horticulture, Vellanikkara

Nayanathara J, Narayana R (2022) Caterpillar complex: a rise in threat to mango panicle. Indian J Ecol 49:252–256

Nayanthara J (2020) Incidence and management of insect pests infesting inflorescence of mango. Department of Agricultural Entomology, College of Agriculture, Vellayani

Oberprieler R, Marvaldi A, Anderson RS (2007) Weevils, weevils, weevils everywhere. Zootaxa 491–520

Oliveira L, Guimarães L, Ferreira M, Nunes A, Pimenta L, Alfenas A (2015) Aggressiveness, cultural characteristics and genetic variation of *Ceratocystis fimbriata* on *Eucalyptus* spp. For Pathol 45:505–514

Orkweha K, Chankhamrian W, Thipsrirach S (2020) A study of methods for detecting *Batocera rufomaculata*. In: International Conference on Power, Energy and Innovations (ICPEI). IEEE, pp. 229–232

Ortiz A, Vela J, Bienvenido C, Campos B, Rodríguez C, Boyero J (2017) Datos preliminares para la identificación de la feromona sexual en la cochinilla blanca del mango (*Aulacaspis tubercularis*). In: X Congreso Nacional de Entomología Aplicada, Valencia, La Rioja

Patel J, Patel M, Radadia G, Shah A, Pandya H (1997) Insect pest management in mango nursery. J Appl Hort Navs 3:125–128

Patel K, Saxena S, Patel K, Gajre N (2013) Biorational pest management in mango. Bioinfolet 10:947–951

Peña J, Aluja M, Wysoki M (2009) Pests. In: The mango: botany, production and uses. CABI Wallingford, UK, pp. 317–366

Peña J, Mohyuddin A (1997) Insect pests. In: Litz RE (ed) The mango: botany, production and uses. CABI, New York, p 587

Peña JE (2002) Integrated pest management and monitoring techniques for mango pests. In: VII International Mango Symposium 645, pp 151–161
Peña JE, Mohyuddin AI, Wysoki M (1998) A review of the pest management situation in mango agroecosystems. Phytoparasitica 26:129–148
Pena JE, Palevsky E, Otero-Colina G, Ochoa R, Meister CW (2005) Mango bud mite, *Aceria mangiferae* bionomics and control under Florida conditions. In: Proceedings of the Florida State Horticultural Society, pp 228–234
Peña JE, Sharp JL, Wysoki M (2002) Tropical fruit pests and pollinators: biology, economic importance, natural enemies and control. CABI Publishing, Walingford
Peng RK, Christian K (2005) The control efficacy of the weaver ant, *Oecophylla smaragdina* (Hymenoptera: Formicidae), on the mango leafhopper, *Idioscopus nitidulus* (Hemiptera: Cicadellidea) in mango orchards in the Northern Territory. Int J Pest Manage 51:297–304
Pereira RM, da Silva Galdino TV, Rodrigues-Silva N, da Silva RS, de Souza TC, Picanço MC (2019) Spatial distribution of beetle attack and its association with mango sudden decline: an investigation using geostatistical tools. Pest Manage Sci 75:1346–1353
Pereira RM, Martins GF, da Silva RS, Galdino TVdS, Ferreira DdO, Flechtmann CAH, Picanço MC (2021) Evidence of fungal structures in the mandible and elytra of *Hypocryphalus mangiferae* collected from mango trees showing symptoms of mango sudden decline disease. Int J Pest Manage 67:315–319
Pérez Barraza MH, Urías López MA, Osuna García JA, Pérez Luna AI, Nolasco González Y, García Álvarez NC (2016) Effect of pruning on white mango scale and production of mango 'Ataulfo.' Rev Mexicana Cienc Agric 7:1841–1853
Pezhman H, Askari M (2004) A study on the biology of mango inflorescence midge in Hormozgan province. Appl Entomol Phytopathol 72
Pezhman H, Radjabi G (2002) Study on biology of mango hopper" *Idioscopus clypealis* Letheirry"(Hom: Cicadellidae) in Hormozgan Province. Appl Entomol Phytopathol 70
Pinese B, Holmes R (2005) Managing mango seed weevil. Horticulture and Forestry Science Newsletter. Queensland, Australia, p 23
Pitan OOR, Mwansat G, Akinyemi SOS, Adebayo OS, Akinlosotu TA (2002) Effect of mango mealybug and sooty mould attack on mango and the impact of the released *Gyranusoidea tebygi* Noyes on yield. Fruits 57:105–113
Ploetz R, Benscher D, Vázquez A, Colls A, Nagel J, Schaffer B (1996) Mango decline: research in florida on an apparently wide-spread disease complex. V Int Mango Sympos 455:547–557
Prasad S (1966) The relation of mango blossom midge (*Dasineura amaramanjarae*) to the yield of mango. Cecidol. Ind 3:171–184
Pruthi HS, Batra H (1960) Important fruit pests of North-West India. Bull. No. 80 (ICAR, New Dehli, India) 80
Rahman KA, Latif MA (1944) Description, bionomics and control of the giant mealybug, *Drosicha stebbingi*, Green (Homoptera: Coccidae). Bull Entomol Res 35:197–209
Rahman MS, Rahman MA, Mia MM, Jahan M (2019) Comparative study on microbial and botanical pesticides as biorational control of mango hopper. Asian J Med Biol Res 5:330–335
Rahman S, Kuldeep S, Ramesh K, Gajendra S (2016) Loss assessment by releasing hoppers on young shoots and flowering and fruited panicles of mango. Hort Res Spectr 5:78–80
Rahman SMA (2007) Mango hopper: bioecology and management—a review. Agric Rev 28:49–55
Rajan S (2021) Mango: The King of Fruits. In: Kole C (ed) The Mango Genome. Springer International Publishing, Cham, pp 1–11
Rao GK, Sujatha A, Suneetha P, Jayanthi P (2020) Behavioural insights of red banded mango caterpillar deanolis sublimbalis snellen. Indian J Entomol 82:756–765
Rattanapun W, Amornsak W, Clarke AR (2009) *Bactrocera dorsalis* preference for and performance on two mango varieties at three stages of ripeness. Entomol Exp Appl 131:243–253
Raza ST, Khan AH, Hameed A, Muhammad N, Grewal AG, Malik MT, Imran M, Mustafa G, Iqbal A (2023) A review on White Mango Scale biology, ecology, distribution and management. Agriculture 13:1770

Reddy P, Rajan VV (2021) Laboratory rearing of mango stem borer, *Batocera rufomaculata* De Geer using drumstick twigs. Pest Manage Hortic Ecosyst 27:88–91

Reddy PR, Chakravarthy A, Sudhagar S, Kurian R (2014) A simple technique to capture, contain and monitor the fresh-emerging beetles of tree borers. Current Biotica 8:191–194

Reddy PVR, Rashmi M, Sreedevi K, Singh S (2020) Sucking pests of mango. In: Sucking pests of crops, pp 411–424

Reddy PVR, Sreedevi K (2016) Arthropod communities associated with mango (Mangifera indica L.): diversity and interactions. In: Chakravarthy AK, Sridhara S (eds) Economic and ecological significance of arthropods in diversified ecosystems: sustaining regulatory mechanisms. Springer, Singapore, pp 271–298

Rehman HM, Mahmood R, Razaq M, Reddy GVP (2016) Damage patterns, monitoring, and management of *Procontarinia mangiferae* (Diptera: Cecidomyiidae) in Pakistan. J Econ Entomol 109:2446–2453

Rehman HMU, Mahmood R, Razaq M (2014) Occurrence, monitoring techniques and management of *Dasineura amaramanjarae* Grover (Diptera: Cecidomyiidae) in Punjab, Pakistan. Pak J Zool 46

Riaz M, Baloch K, Makhdum A, Chaudhary Z, Ghulam M, Stonehouse J (2000) Fruit flies in Pakistan: damage and control in farmers' fields. In: Proceedings of the Indian Ocean Commission, Regional Fruit Fly Symposium, Flic en Flac, Mauritius, 5–9th June 2000. Indian Ocean Commission, pp 83–85

Royer J (2009) Spread of red-banded mango caterpillar',*Deanolis sublimbalis*' Snellen (Lepidoptera: Pyralidae), in Cape York Peninsula, Australia. Aust Entomol 36:119–130

Rwomushana I, Tanga CM (2016) Fruit fly species composition, distribution and host plants with emphasis on mango-infesting species. Fruit fly research and development in Africa-towards a sustainable management strategy to improve horticulture, pp 71–106

Saeed S, Masood A, Sajjad A, Zahid DM (2010) Monitoring the dispersal potential of bark beetle, *Hypocryphalus mangiferae* Stebbing (Scolytidae: Coleoptera) in mango orchards. Pak J Zool 42:473–479

Saeed S, Masood A, Sayyed AH, Kwon YJ (2011) Comparative efficacy of different pesticides against mango bark beetle *Hypocryphalus mangiferae* Stebbing (Coleoptera: Scolytidae). Entomol Res 41:142–150

Sahayaraj K, Hassan E (2023) Distribution and diversity of predatory insects in agroecosystems. In: Worldwide predatory insects in agroecosystems. Springer, Berlin, pp 25–70

Sahoo A, Das B (2004) Incidence and biological observations of mango fruit borer, *Deanolis albizonalis* (Hampson)(Pyralidae: Lepidoptera) in West Bengal. Environ Ecol 22:180–183

Sahoo S (2016) Chemical control of red banded mango fruit borer, *Autocharis albizonalis* Hampson (Pyralidae: Lepidoptera). Indian J Entomol 78:10–14

Sahoo SK, Jha S (2009) Bio-ecology of mango fruit borer, *Autocharis* (=*Noorda*) *Albizonalis* Hampson (Pyralidae; Lepidoptera)—a recent threat to mango growers in West Bengal, India, 820 edn. International Society for Horticultural Science (ISHS), Leuven, Belgium, pp 601–610

Salmah M, Adam N, Muhamad R, Lau W, Ahmad H (2017) Infestation of fruit fly, *Bactrocera* (Diptera: Tephritidae) on mango (*Mangifera indica* L.) in Peninsular Malaysia. J Fund Appl Sci 9:799–812

Sarker D, Rahman M, Barman J (2009) Efficacy of different bagging materials for the control of mango fruit fly. Bangladesh J Agric Res 34:165–168

Sarwar M, Hamed M, Rasool B, Yousaf M, Hussain M (2013) Host preference and performance of fruit flies *Bactrocera zonata* (Saunders) and *Bactrocera cucurbitae* (Coquillett)(Diptera: Tephritidae) for various fruits and vegetables. Int J Sci Res Environ Sci 1:188–194

Schoeman A (1987) Observations on the biology of the mango weevil, *Sternochetus mangiferae* (F). Yearbook-South African Mango Growers' Association 7

Scolari F, Valerio F, Benelli G, Papadopoulos NT, Vaníčková L (2021) Tephritid fruit fly semiochemicals: current knowledge and future perspectives. Insects 12:408

Sengupta G, Behura B (1955) Some new records of crop pests from india. Indian J Entomol 17

Shah MH, Rafique R, Almas M, Usman M, Yasin S, Bibi S (2022) A review: fruit production industry of Pakistan trends, issues and way forward. Pak J Agric Res 35

Sharma D, Adhikari D, Tiwari D (2015) Fruit fly surveillance in Nepal. Agric Biol Sci J 1:121–125

Shelly TE (2020) Evaluation of a genetic sexing strain of the oriental fruit fly as a candidate for simultaneous application of male annihilation and sterile insect techniques (Diptera: Tephritidae). J Econ Entomol 113:1913–1921

Shimizu Y, Moriya N, Jahana O, Matsumura M, Takeuchi M (2023) Effects of sixteen insecticides on the mango gall midge *Procontarinia mangicola* (Shi) (Diptera: Cecidomyiidae) infesting mango leaves. Int J Trop Insect Sci 43:601–608

Shivananda T, Verghese A, Hegde R, Kumar N, Singh A (2012) Management of mango trunk borers using sealer cum healer. Techn Bull 72:12

Sial M, Saeed Q, Saeed S, Jaleel W, Naqqash MN, Majeed F (2015) Susceptibility of mango cultivars against larvae of mango midge *Procontarinina mangicola* Shi (Cecidomyiidae: Diptera). Appl Sci Business Econ 2:1–7

Sierra-Baquero PV, Varón-Devia EH, Gomes-Dias L, Monje-Andrade B (2020) Economic injury level for the flower thrips *Frankliniella* cf. *gardeniae* Moulton (Thysanoptera: Thripidae) in mango. Revista Facultad Nacional De Agronomía Medellín 73:9213–9220

Sim SB, Curbelo KM, Manoukis NC, Cha DH (2022) Evaluating *Bactrocera dorsalis* (Hendel)(Diptera: Tephritidae) response to methyl eugenol: Comparison of three common bioassay methods. J Econ Entomol 115:556–564

Singh P, Singh R, Srivastava C (2016) Phytophagous mites of Indian fruit plants. Insect Pests Manage Fruit Crops 30:621–649

Singh S, Sharma D, Kular J, SINGH P, Gill P, SINGH N, Bons M, SINGH B, KAUR A, SAINI MK (2014) Eco-friendly management of fruit flies, Bactrocera spp. in pear with methyl eugenol based traps at different locations in Punjab. J Insect Sci 27:57-62

Sohi A, Sohi A (1990) Mango leafhoppers (Homoptera: Cicadellidae)—a review. J Insect Sci 3:1–12

Soomro AH (1988) Integrated control of mango insect pest in Sindh. In. Sindh Agriculture University, Tandojam

Srivastava R (1980) Efficacy of alkathene bands to prevent ascent of mango mealy bug nymphs on mango trees. Ind J Entomol 42(1):22–129

Sternlicht M, Goldenberg S (1976) Mango eriophyid mites in relation to inflorescence. Phytoparasitica 4:45–50

Stonehouse J, Mahmood R, Poswal A, Mumford J, Baloch KN, Chaudhary ZM, Makhdum AH, Mustafa G, Huggett D (2002) Farm field assessments of fruit flies (Diptera: Tephritidae) in Pakistan: distribution, damage and control. Crop Protect 21:661–669

Stonehouse JM, Mumford JD, Mustafa G (1998) Economic losses to tephritid fruit flies (Diptera: Tephritidae) in Pakistan. Crop Protect 17:159–164

Subramanian S, Boopathi T, Nebapure SM, Yele Y, Shankarganesh K (2021) Mealybugs. In: Omkar (ed) Polyphagous pests of crops. Springer, Singapore, pp. 231–272

Sudhi-Aromna S, Jumroenma K, Chaowattanawong P, Plodkornburee W, Sangchote Y (2007) Studies on the biology and infestation of stem borer, *Batocera rufomaculata*, in Durian. Int Workshop Trop Subtrop Fruits 787:331–338

Sujatha A, Zaheruddeen S (2002) Biology of pyralid fruit borer, *Deanolis albizonalis* Hampson: a new pest of mango. J Appl Zool Res 13:1–5

Syed R, Ghani M, Murtaza M (1970) Studies on the Trypetids and their natural enemies in West Pakistan. III. Dacus (Strumeta) zonatus (Saunders). Technical Bulletin, Commonwealth Institute of Biological Control, pp 1–16

Talpur MA, Khuhro R, Nizamani I (2002) Phenological relationship between mango hoppers *Idioscopus* spp. and mango inflorescence/fruit. Pak J Appl Sci 2:533–536

Talpur MA, Khuhro RD (2003) Relative population of mango hopper species on different mango varieties. J Asia-Pacif Entomol 6:183–186

Tamta AK, Pandey R, Sharma JR, Rai R, Barman M, MG D, Mitra D, Mohapatra PKD, Sami R, Al-Mushhin AA (2022) First record of *Clonostachys rosea* (Ascomycota: Hypocreales) entomopathogenic fungus in the mango hopper *Amritodus atkinsoni* (Hemiptera: Cicadellidae). Pathogens 11:1447

Tandon P (1995) Integrated insect pest management in mango. In: Proceedings national symposium on integrated pest management and environment organized by Plant Protection Association of India and FAO at Madras, India, pp 2–4

Tandon P, Lal B, Rao G (1983) Prediction of the mango hopper, *Idioscopus clypealis* Leth. population in relation to physical environmental factors. Entomon 8:257–261

Tenakanai D, Dori F, Kurika K (2006) Red-banded mango caterpillar, *Deanolis sublimbalis* Snellen (Lepidoptera: Pyralidae: Odontinae), in Papua New Guinea. Pest and disease incursions: risks, threats and management in Papua New Guinea

Terefe T, Tsegaye S, Wakuma T (2014) White mango scale insect's infestations and its implications in guto gida and digadistrcts of East Wellega Zone. ABC Res Alert 2(2)

Tobih F, Omoloye A, Ivbijaro M, Enobakhare D (2002) Effects of field infestation by *Rastrococcus invadens* Williams (Hemiptera: Pseudococcidae) on the morphology and nutritional status of mango fruits, *Mangifera indica* L. Crop Protect 21:757–761

Tokuda M, Uechi N (2021) Invasive Species. In: Yukawa J, Tokuda M (eds) Biology of gall midges: evolution, ecology, and biological interactions. Springer, Singapore, pp 255–267

Tumbada RD, Pokhrkar D, Datkhile RV (2018) Efficacy of entomopathogenic fungi against mango hoppers. J Pharmacogn Phytochem 7:3198–3202

Umar Ud, Ahmed N, Zafar MZ, Rehman A, Naqvi SAH, Zulfiqar MA, Malik MT, Ali B, Saleem MH, Marc RA (2022) Micronutrients foliar and drench application mitigate mango sudden decline disorder and impact fruit yield. Agronomy 12:2449

Upadhyay SK, Chaudhary B, Sapkota B (2013) Integrated management of mango stem borer (*Batocera rufomaculata* Dejan) in Nepal. Global J Biol Agric Ealth Sciences 2:132–135

Urca T, Ribak G (2021) The relationship between body size and flight power output in the mango stem borer (*Batocera rufomaculata*). J Insect Physiol 133:104290

Urías-López M, Osuna-García J, Vázquez-Valdivia V, Pérez-Barraza M (2010) Population dynamics and distribution of the white mango scale (*Aulacaspis tubercularis* Newstead) in Nayarit, Mexico. Rev Chapingo Ser Hortic 16:77–82

Urías L, Vázquez V, Pérez B (2006) El cultivo del mango: Principios y tecnología de producción

Van Halteren P (1970) Notes on the biology of the scale insect *Aulacaspis mangiferae* Newst. (Diaspididae, Hemiptera) on mango. Ghana J Agric Sci 3:83–85

Van Wyk M, Al Adawi AO, Khan IA, Deadman ML, Al Jahwari AA, Wingfield BD, Ploetz R, Wingfield MJ (2007) *Ceratocystis manginecans* sp. nov., causal agent of a destructive mango wilt disease in Oman and Pakistan. Fungal Divers 27:213–230

Vargas RI, Piñero JC, Leblanc L (2015) An overview of pest species of *Bactrocera* fruit flies (Diptera: Tephritidae) and the integration of biopesticides with other biological approaches for their management with a focus on the Pacific region. Insects 6:297–318

Vasanthi P, Raviprasad T (2015) Incidence of *Batocera rufomaculata* De Geer (Coleoptera: Cerambycidae) on cashew. J Entomol Zool Stud 3:174–177

Vayssieres J-F, Korie S, Coulibaly O, Temple L, Boueyi SP (2008) The mango tree in central and northern Benin: cultivar inventory, yield assessment, infested stages and loss due to fruit flies (Diptera Tephritidae). Fruits 63:335–348

Vayssières J-F, Korie S, Coulibaly O, Van Melle C, Temple L, Arinloye D (2009) The mango tree in central and northern Benin: damage caused by fruit flies (Diptera Tephritidae) and computation of economic injury level. Fruits 64:207–220

Veeresh VGK (1985) Pest problems in mango—world situation. In: II International symposium on mango 231, pp 551–565

Venkata Rami Reddy P, Gundappa B, Chakravarthy AK (2018) Pests of Mango. In: Omkar (ed) Pests and their management. Springer, Singapore, pp 415–440

Verghese A (1999) Recent studies on the management of mango stone weevil *Sternochetus mangiferae* Fab.(Coleoptera: Curculionidae) in South India. In: VI International symposium on mango 509, pp. 819–822

Verghese A, Devi Thangam S (2011) Mango hoppers and their management. Extension Folder 31-11

Verghese A, Jayanthi Mala B, Rashmi M (2022) Managing major insect pests of mango in the southern coastal belts of India. In: Transforming coastal zone for sustainable food and income security: proceedings of the international symposium of ISCAR on Coastal Agriculture, 16–19 Mar 2021. Springer, Berlin, pp 165–180

Verghese A, Nagaraju D, Jayanthi PK, Madhura H (2005) Association of mango stone weevil, *Sternochetus mangiferae* (Fabricius) (Coleoptera: Curculionidae) with fruit drop in mango. Crop Protect 24:479–481

Verghese A, Nagaraju D, Kamala Jayanthi P, Anil Kumar H, Madhura H (2003) Management of the nut weevil, *Sternochetus mangiferae* (Fabricius) on Totapuri, a mango variety of processing industry. Pest Manage. Horticultural Ecosystems 9:93–96

Verghese A, Rao G (1987) Determination of relevant critical stages for the management of the mango hopper, *Idioscopus clypealis* Lethierry. Indian J Hortic 44:280–283

Verghese A, Tandon P, Rao GP (1988) Spatial distribution pattern and sampling plan for the blister midge, *Erosomyia indica* Grover (Cecidomyiidae: Diptera) in India. Int J Trop Insect Sci 9:515–518

Viraktamath C (2007) New genera and species of *Idiocerine* leafhoppers (Hemiptera: Cicadellidae) from India, Sri Lanka, and Myanmar. Biosyst 1:21–30

Westphal E, Manson D (1996) Feeding effects on host plants: Gall formation and other distortions. In: World Crop Pests. Elsevier, Amsterdam, pp 231–242

White IM, Elson-Harris MM (1992) Fruit flies of economic significance: their identification and bionomics. CAB international in association with ACIAR, Wallingford, Oxon

Xia H, Li J, Shi M, Wang Q, Zheng L, Fu J, You M (2019) Field trapping effect of different colors sticky cards to thrips in mango orchard. Fujian J Agric Sci 34:818–823

Zida I, Nébié K, Sawadogo A, Tassembédo B, Kiénou T, Dabiré RA, Nacro S (2023) Effectiveness of four integrated pest management approaches in the control of fruit flies (Diptera: Tephritidae) in mango agro-ecosystems in the South-Sudanian zone of Burkina Faso. Adv Entomol 11:124–142

Chapter 25
"King of Fruits" Under Siege: Unravelling the Implications of Climate Change on Insect Pest Dynamics

Tauseef Khan Babar

Abstract Since the era of industrialization, our planet Earth has been warming at an alarming rate, with a projected rise of 1.4 to 5.8 °C rise in temperature by the end of the twenty-first century. This warming, primarily due to increased concentrations of greenhouse gases (GHG), is disrupting our ecosystems and intensifying extreme weather events in the disguise of climate change, with far-reaching implications for agriculture. Poikilothermic organisms such as insects, mites, and nematodes, which rely on ambient temperature to regulate their bodily functions, are particularly sensitive to temperature fluctuations. Rising temperatures are altering the behaviour, distribution, and abundance of insect pests besides distorting the interactions of deadly crop pests with their hosts, natural enemies, and competitors, potentially jeopardizing food security worldwide. This chapter delves into the broader implications of climate change on insect pests, with a specific focus on mango, *Mangifera indica* L., a key tropical fruit crop, popularly known as "The King of Fruits". It examines how climate-induced changes in phenology and environmental conditions are deemed to alter the insect pest dynamics and challenge mango cultivation in the future. By providing a detailed overview of these implications, this chapter aims to enhance understanding of how climate change will influence pest management strategies and crop production, highlighting the urgent need for adaptive approaches in the face of evolving agricultural threats.

Keywords Climate change · Insect pests · King of fruits · Mango

T. K. Babar (✉)
Department of Entomology, Faculty of Agricultural Sciences and Technology, Bahauddin Zakariya University, Multan, Pakistan
e-mail: tauseefkhan@bzu.edu.pk

S. A. H. Naqvi et al. (eds.), *Climate Change and Mango Production*,
https://doi.org/10.1007/978-3-031-87035-4_25

25.1 Introduction

Since the era of industrialization, our planet Earth has been warming at an alarming rate and experienced a rise of about 1 °C. The Intergovernmental Panel on Climate Change (IPCC) reports that global temperature has been increasing progressively over the past three decades, with the 2000s standing out as the warmest decade on record. (IPCC 2021; World Meterological Organization 2024). Unfortunately, the projections for twenty-first-century global warming far exceed the natural climate variability of the past 100 decades and over the course of next century, further warming of 1.4 to 5.8 °C is anticipated (Szyniszewska et al. 2024). It is generally accepted that the primary deriver behind this unforeseen surge in temperature is increased concentrations of greenhouse gases (GHG) including carbon dioxide (CO_2), methane (CH_4), and nitrous oxide (N_2O) in the atmosphere. Numerous human activities like deforestation, fossil fuels burning, and urbanization directly or indirectly emit these GHGs into our environment (Yoro and Daramola 2020). Consequently, the released GHGs trap the Sun's energy in our atmosphere like a greenhouse and gradually raise the temperature of our planet (Fig. 25.1) (Lashof and Ahuja 1990; Shah et al. 2024). Of all the GHGs, CO_2 is the most significant and the most prevalent (Rosenzweig 1989), its concentration in the atmosphere has increased dramatically from 280 ppm (pre-industrial period) to 416 ppm today, and is expected to double in 2100 (Pachauri and Reisinger 2007). Climate change, apart from raising the mean temperatures is also disturbing the local weather patterns, leading to drier conditions in some areas and wetter ones in other. Furthermore, the intensity and frequency of extreme weather events such as floods, droughts, and heat waves are likely to increase (Trenberth et al. 2015; IPCC 2021).

Poikilothermic organisms including insect, mite, mollusc, and nematode plant pests are those whose body temperature varies with environment (Kerkut and Taylor 1958; Cloudsley-Thompson 1968). Temperature is a key factor that is known to influence the growth, distribution, and reproductive behaviour of insect pests (Kocmánková et al. 2009). In addition, changes in temperature can disrupt the dynamics between insect pests, their environment, and other insect pest species, including natural predators, competitors, vectors, and mutualists (Prakash et al. 2014; Harvey et al. 2023). Rising temperatures and CO_2 can profoundly affect interactions between agricultural crops and insect pests, potentially jeopardizing global food security (Caulfield and Bunce 1994; Roth and Lindroth 1995). Therefore, farmers have to face more severe and new pest management challenges in future due to climate change (Fand et al. 2012; Gvozdenac et al. 2022).

Agricultural crop production has remained in a state of stagnation, continually besieged by the mounting threat of insect pests, with an estimated loss of 18% in a year (Oerke 2006; Das et al. 2011). Farmers have to spend an enormous amount of money in managing these deadly pests (Waterfield and Zilberman 2012). Annually, US$ 4–5 billion are spent in worldwide efforts to manage the diamondback moth, *Plutella xylostella* L. (Zalucki et al. 2012). Unfortunately, in the context of climate change, the future is bleak as the challenges posed by phytophagous insect pests

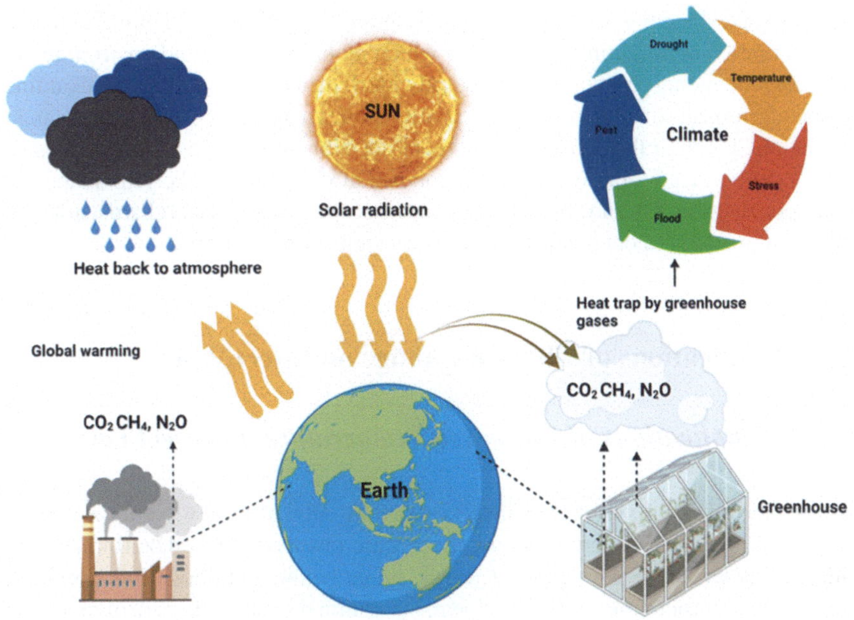

Fig. 25.1 Schematic presentation of the warming mechanism of Earth due to emission of the green house gases (CO_2, CH_4, N_2O) into the atmosphere and implications of altered climate on agricultural crops (Shah et al. 2024)

are expected to intensify in future (Das et al. 2011; IPCC 2021). Horticultural crops including fruits and vegetables that play a vital role in the economic development of a country are not an exception (Datta 2013; Bhattacharjee et al. 2022). Among the fruit crops, mango (*Mangifera indica* L.), popularly known as "The King of Fruits" is a highly valuable plant species cultivated in the tropical belt of the planet (Tharanathan et al. 2006; Singh 2016). According to FAO (2022), it is ranked at the 3rd position among the most cultivated fruit crop of the world on the basis of the harvested area and at the 5th position on yield basis. Out of the total out put of the world, the highest share of 76.5% comes from South Asia, where India and Pakistan are the major contributors (FAO 2023). Mango is an evergreen tree whose growth cycle can be divided into four main phenological phases: vegetative, quiescence, flowering, and fruiting (Delgado et al. 2011; Normand and Lauri 2018). Climate-related changes have profoundly disturbed the phenological cycles of different crops by shifting their habitats, while crop habitats depend on genotype and phenotype (Chmielewski et al. 2004; Makhmale et al. 2016). Research has shown the occurrence of widespread changes in flowering and fruiting patterns of mango (Anwar et al. 2011; Rajan 2012), leading to the emergence of new insect pests or resurgence of earlier pests with more intensity, which will have serious implications in future mango production (Talpur et al. 2002; Mayamba et al. 2014).

This chapter begins with an overview of the broader implications of climate change on insect pests, highlighting the ways in which global warming and environmental shifts are influencing pest populations worldwide. It then narrows its focus to examine the specific impacts of climate change on the dynamics of insect pests affecting mango cultivation. Through this exploration, the author aims to provide a comprehensive understanding of how these global environmental changes are likely to shape the behaviour, distribution, and abundance of key insect pests in mango orchards, offering insights into potential challenges for pest management in the future.

25.2 Implications of Climate Change on Insect Pests

25.2.1 Implications of Increased Temperature on Insect Pests

All the physiological systems of insects function optimally only within a limited range of temperatures where several enzymes are involved (González-Tokman and Villada-Bedoya 2024). However, at temperatures below and above this optimum value, enzymatic activities will slow down, but within the range, the rate of reactions will increase with increasing temperatures and vice versa (Ma et al. 2021). According to an estimate, an increase of 10 °C can double the rate of their metabolic reactions (Clarke and Fraser 2004). Given that, increased temperatures tend to accelerate insect consumption, growth, and movement, which can affect their population dynamics by influencing factors such as fecundity, survival, generation time, population size, and geographic distribution (Fig. 25.2) (Bale et al. 2002). Therefore, in this context, an increase in the insect herbivory is anticipated. Generally, species that cannot acclimatize to elevated temperatures have to struggle for their survival, while others that are temperature-tolerant can thrive and reproduce more rapidly (Shrestha 2019; Skendžić et al. 2021). Rising temperature has strong controlling effect upon the population density of insects that live above ground, compared to those that spend most of their life cycle underground. For these subterranean hexapod species, soil serves as a thermal insulating medium to neutralize the effect of temperature changes (Bale et al. 2002). Pathania et al. (2020) exhibited that the population of whitefly, *Bemisia tabaci* Gennadius, is mainly regulated by abiotic factors such as temperature, precipitation, and relative humidity. High temperature coupled with high humidity correlates positively with the population build-up of *B. tabaci*. Keeping in view these facts, it is clear that in future changes occurring in insect populations will depend on the extent of rise in global temperatures.

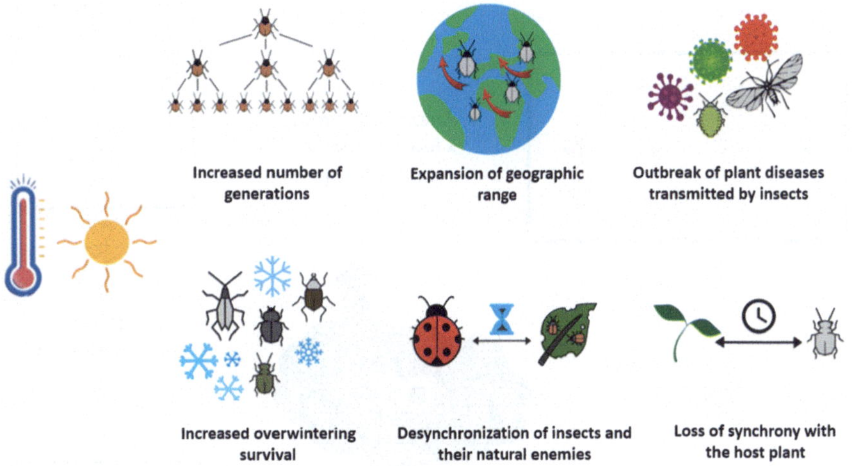

Fig. 25.2 Diagrammatic illustration of the effects of increasing temperature on various biological features of insects life (Skendžić et al. 2021)

25.2.2 Implications of Increased CO_2 Concentration on Insect Pests

Numerous effects including the abundance, distribution, and performance of herbivorous insects due to increased concentrations of atmospheric CO_2 have been documented. Such effects can significantly affect feeding efficiency, growth rate, fecundity potential, and population densities of insect pests (Fuhrer 2003). The effects of elevated CO_2 concentration on insect pests are largely dependent on the type of host plants they feed on. C3 crops (such as wheat, rice, and cotton) are more greatly affected by increased CO_2 concentration than C4 crops (such as corn and sorghum). This differential response means that herbivory may be more pronounced on C3 plants, while C4 plants, being less responsive to elevated CO_2, are less likely to experience changes in insect feeding patterns (Lincoln et al. 1984; Rezaei et al. 2023). As mentioned earlier, increased concentration of CO_2 can boost plant photosynthesis, leading to better plant growth and higher productivity. However, plants grown under elevated level of CO_2 experience a change in their leaves chemistry, which can affect their nutritional quality and palatability to leaf-feeding insects (Fig. 25.3) (Lincoln 1993). Half of all insects, including most Orthoptera, Hemiptera, and Lepidoptera, feed on plants during their larval stages. Therefore, the quantity of plant material ingested by insects is dependent upon the nutritional composition of the plant. Furthermore, crops grown under higher CO_2 levels often store more sugars and starches in their leaves, which decreases palatability by altering the carbon-to-nitrogen ratio (C:N) (Cotrufo et al. 1998). Insects, particularly herbivores, essentially need nitrogen for their development, and therefore elevated CO_2 levels can enhance herbivory in some crop pests (Bezemer et al. 1998). In this context, agricultural crops

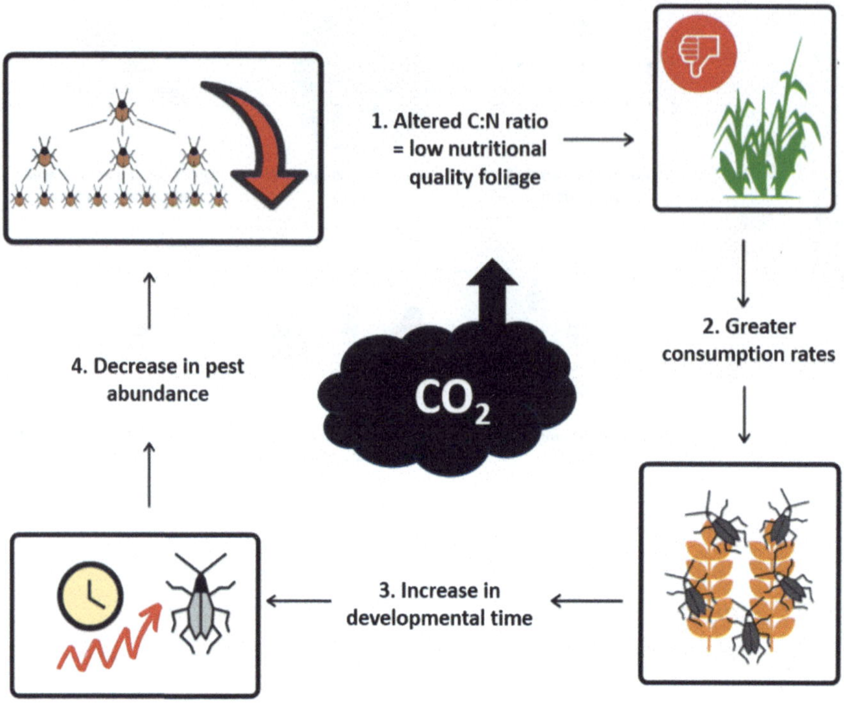

Fig. 25.3 Diagramatic illustration of the effect of increased concentration of CO_2 on various biological features of insects life (Skendžić et al. 2021)

are more likely to suffer due to higher consumption of plant tissues for obtaining an equivalent level of food (Hamilton et al. 2005; DeLucia et al. 2008). Overall, the changes in the quality and quantity of host plants due to elevated levels of CO_2 can affect the egg laying capacity, growth rate, reproductive potential, and longevity of adult insect herbivores (Fig. 25.3) (Yoro and Daramola 2020).

25.2.3 Implications of Changed Precipitation Pattern on Insect Pests

Climate change has led to significant shifts in global precipitation patterns, altering the amount, intensity, and frequency of rainfall (Subedi et al. 2023). Under this phenomenon, a decrease in the frequency of precipitation and an increase in the intensity of precipitation is commonly observed. Such type of rainfall pattern is believed to cause droughts and floods. Heavy rainfall not only washes away the insect eggs and larvae but also the small-sized pests like aphids, jassids, mites, whiteflies etc. (Fig. 25.4) (Kocmánková et al. 2009; Field and Barros 2014). Insect pest species overwintering in the soil in the changed precipitation scenario are found to be at high risks. The survival and reproduction of several insect pest species are highly vulnerable to this altered precipitation pattern and significantly changes their population dynamics (Skendžić et al. 2021). Due to varying rainfall pattern, the nutritional quality of plants is also affected which is critical for the performance of herbivorous insects (Holopainen et al. 2018). Typically, water stressed (drought) environment distorts various biological processes of plants, making them more vulnerable to diseases and pests (Fig. 25.4) (Sahar Abdou 2019). Such effects may include; provision of a congenial environment for the growth of herbivorous insects, attraction of some insect species towards disease-stressed plants, and taking advantage of the poor defence system of disease-stressed plants due to reduced production of secondary metabolites (Gely et al. 2020; Subedi et al. 2023). Ahmed et al. (2017) found that aphids raised under drought conditions have a lower rate of parasitism due to the reduced size or availability of their host. Importantly, sap-feeding insects tend to perform better when feeding on trees experiencing intermittent water stress, compared to those under constant water stress (Sconiers and Eubanks 2017). Overall, the changing precipitation patterns may have positive or negative effects on the biological parameters of insect pests, their host, and natural enemies as shown in Fig. 25.5.

25.2.4 Implications of Climate Change on Insect Distribution

Global implications of climate change on the life-table parameters (growth, reproduction, survival) of insects or their interactions with the host plants or natural enemies are well documented (Fig. 25.5) (Liliane and Mutengwa 2020; Subedi et al. 2023). However, among the climatic variables (humidity, precipitation, temperature), the temperature changes are found to be more critical than others (Frank 2020). Insects are projected to migrate polewards and higher elevations on account of warmer climate which will put global food security at high risk (Fig. 25.6). For instance, the changing precipitation pattern in East Africa is believed to have a major role in the outbreak of Desert Locust, *Schistocerca gregaria* Forsskal (Peng et al. 2020; Babar 2023). The changing climate is anticipated to expand the geographic distribution of insects and in this context the low temperatures are known to be more significant than high temperatures (Robbins et al. 2022).

Unfortunately, in the case of agricultural insect pests, shifts in dispersal patterns due to climate change are particularly significant for crop production (Chen and Ma 2010). All the essential requirements for the growth, development, reproduction, and survival of insect species are dependent on the climate and are primarily species-specific. Therefore, any change happening in temperature and precipitation patterns with the foreseeable changes in climate is likely to effect the distribution, survival and reproduction of species in the future (Fig. 25.6) (Manderscheid et al. 2016). Due to the spread of insect pests to new areas, along with the shift in the growing areas of their host plants, farmers will face new and more severe pest problems (Lehmann

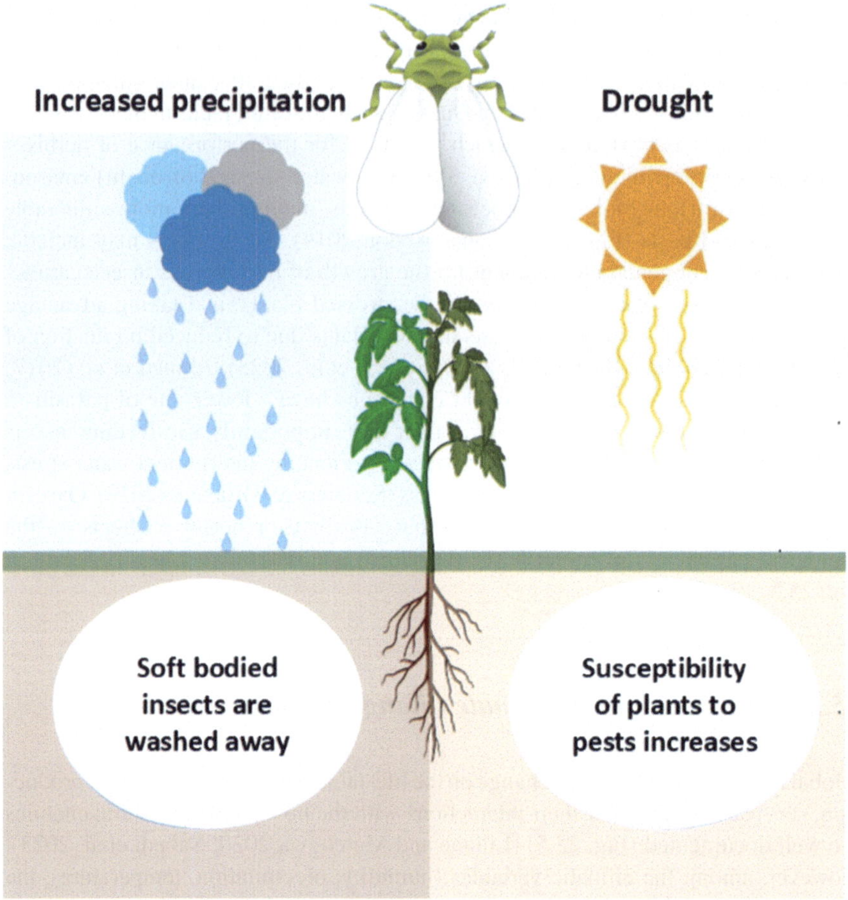

Fig. 25.4 Digrammatic illustration of the effect of changing precipitation pattern on the population dynamics of insect pests (Dave et al. 2024)

et al. 2020). For instance, it is projected that by 2055, the ranges of insect pests in Central Europe will move to higher altitudes along with an increase in the number of generations. A notable example of this trend is the > 1000 km northward spread of the European corn borer, *Ostrinia nubilalis* Hubner, observed by Evans et al. (2013). In USA, pink bollworm, *Pectinophora gossypiella* Saunders, a serious threat to cotton production, is presumed to be outspreading its present range from the frost-free zone in southern Arizona and California into the cotton growing areas of Central California (Fand et al. 2012).

The Mediterranean fruit fly (*Ceratitis capitata* Wiedemann), commonly known as Medfly, is a major agricultural pest that feeds on over 300 plant species, including apple, citrus, mango, peach, and persimmon (Shelly et al. 2014; Liquido et al. 2019). This pest is widely distributed across the world in various climatic zones (El Keroumi

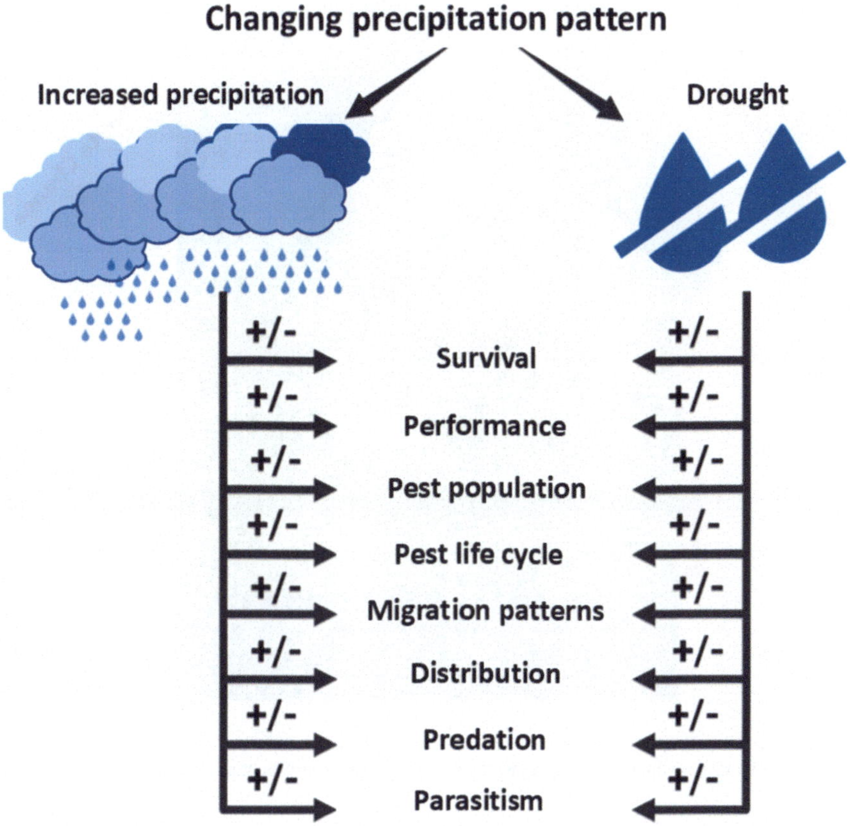

Fig. 25.5 Digrammatic illustration of the effect of changing precipitation pattern on various biological features of the insects life (Subedi et al. 2023)

et al. 2010). Over the time, it has outspread from its native Sub-Saharan Africa to temperate zones (Ricalde et al. 2012), and now well-established in other parts of the world such as Australia, Central America, Europe, Hawaii, North Africa, and South America (Deschepper et al. 2021). In China, where citrus is the most widely grown fruit and the country ranks second in global citrus production, the citrus industry faces significant threats from another fruit fly species, *Bactrocera tsuneonis* Miyake (Diptera: Tephritidae) (Mao et al. 2024). The symptom of its infestation begins to appear in the first week of October, and each maggot devours between two to ten carpels. By early November, the mature larvae (maggot) cause the infested fruit to drop to the ground, and they pupate in the upper two inches of soil (Weems and Fasulo 2012). Unfortunately, the expansion of *B. tsuneonis* has contributed to a steady decline in China's citrus production. This species is native to Japan, but in 1959, was first recorded in the Ningming and Pingxiang regions of Guangxi province (Cai et al. 2020). In 2016, *B. tsuneonis* was captured for the first time in Huaiji county,

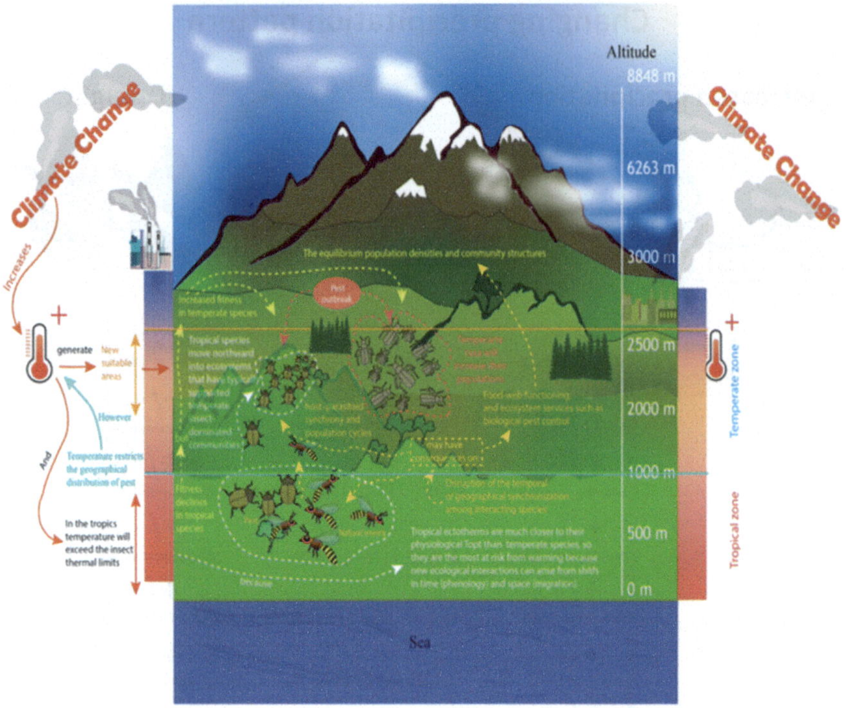

Fig. 25.6 Schematic presentation of the effects of increasing temperature on the distribution and species interaction of the temperate regions inhabiting pests (Aguila et al. 2023)

Guangdong province, signalling its potential to spread further (Hou et al. 2018). Using the MaxEnt (Maximum Entropy) model to predict the future distribution of this dipterous pest under the climate change scenario, an expansion in the suitability of the habitats is anticipated to take place in future which may shift its habitats toward higher latitudes relative to that under current climate conditions, thereby exacerbating the threat to citrus production in China and beyond (Mao et al. 2024).

25.2.5 Implications of Climate Change on Insect Overwintering

Since insects are poikilothermic or cold-blooded organisms, their ability to maintain homeostasis in the face of changes in ambient temperature is very limited. For their survival in the thermally stressful environments, insects have evolved different mechanisms (González-Tokman and Villada-Bedoya 2024). Among these stresses, cold temperature poses a serious threat by its direct effect on their physiological processes and finally culminates in their mortality (Overgaard and MacMillan 2017). To cope with the detrimental effects of low temperatures, insects have evolved a variety of adaptive strategies such as diapause, which reduces their metabolic activity to endure cold temperatures or freeze avoidance, where they eliminate potential nucleators to prevent freezing (Bale and Hayward 2010). Diapause is an important part of the insect life cycle and survival (Boyes et al. 2021), being obligate for insects that produce one brood per year and facultative for multivoltine species that produce multiple broods (Numata and Shintani 2023). For the induction of diapause, environmental signals such as photoperiod or day length play a vital role as short day lengths points the arrival of harsh winter conditions (Saunders et al. 2004). Climate change, however, is disrupting theses environmental cues, leading to shifts in the timing of diapause (Bale and Hayward 2010). Additionally, these altered photoperiods driven by rising temperatures and changing precipitation patterns are also affecting the ability of insects to synchronize their life cycles with seasonal transitions (Fig. 25.7) (Van Dyck et al. 2015; Kerr et al. 2020).

An early onset of diapause can be harmful, draining excessive energy reserves and causing broader ecological imbalances, with potential negative consequences for agricultural productivity (Tougeron et al. 2020). While an enhanced survival during overwintering period could lead to an increase in overwintering population and therefore to a greater number of insects on plants during the warmer period of a year (Fig. 25.7) (Bale and Hayward 2010). This is evident from the examples of corn earworm (*Helicoverpa zea* Boddie) and the cotton bollworm (*Helicoverpa*

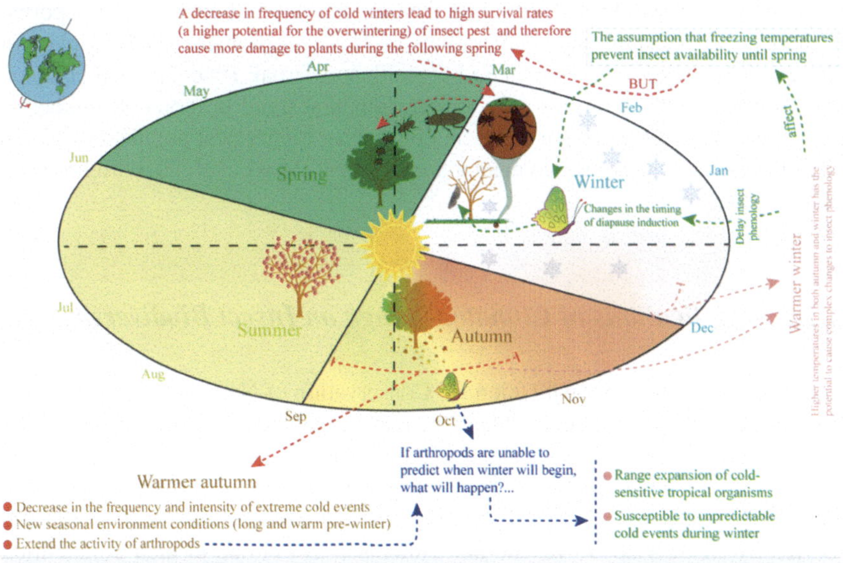

Fig. 25.7 Reduced frequency and intensity of extreme winter cold events (long and warmer pre-winter periods) that led to altered environmental conditions, extended activity of arthropods, expanded cold-sensitive tropical species, and increased overwintering potential of pests is shown (Aguila et al. 2023)

armigera Hubner) where warming not only caused an expansion in their range but also increased overwintering survival of these pests (Diffenbaugh et al. 2008; Fand et al. 2012). Insect pests of horticultural plants cultivated in and limited to greenhouse environments will find it easy to survive outdoors as mean values of winter temperatures increase, as shown by the increased overwintering success of *Liriomyza huidobrensis* Blanchard outside greenhouses in the United Kingdom (Baker et al. 1996).

25.2.6 Implications of Climate Change on the Number of Generations

Temperature is one of the most critical climatic variables that strongly influences the phenology of insects (Figs. 25.2 and 25.6) (Gutiérrez and Wilson 2021). The current unusual increase in global temperature may catalyse the development rates of insects within the shorter periods of time (Jha et al. 2024). Results of several models have shown that an increase of 2 °C can result one to five extra generations of insects in a year such as in aphids (Yamamura and Kiritani 1998; Menéndez 2007). Effects of increasing temperature on insect species are highly dependent on whether they are univoltine or multivoltine (Porter et al. 1991; Skendžić et al. 2021). An interesting example of multivoltine species is cabbage white butterfly, *Pieris brassicae* L. (Bale et al. 2002). In the past, various authors have also reported the increase in the number of generations of *H. zea*, *Oscinella frit* L., *Javasella pellucida* Fabricius, *Cydia pomonella* L. (Smith and Tirpak 1988; Hill and Dymock 1989; Kurppa 1990). Unfortunately, this incremental trend in terms of number of generations of insect pests can cause an eruption of higher pest populations which makes the effective pest management a gigantic task (Moss et al. 2010; Renner and Zohner 2018).

25.2.7 Implications of Climate Change on Insect Biodiversity

According to the International Union for Conservation of Nature (IUCN) "Invasive alien species (IAS) are animals, plants or other organisms that are introduced by humans, either intentionally or accidentally, into places outside of their natural range, negatively impacting native biodiversity, ecosystem services or human economy and well-being" (IUCN 2021). These IAS are one of the major cause of the loss of global biodiversity (IPBES 2019; Venette and Hutchison 2021) due to their competition with the endemic species for resources, predation or transmission of diseases to native species, and finally resulting in major ecosystem changes (Clavero and García-Berthou 2005; Bellard et al. 2016). Given that insect species acting as pests and disease vectors has assumed a particular importance as their spread has immensely increased now a days due to increased travel, trade, and agriculture. According to an estimate, US$ 76.9 billion are needed every year to tackle the threat of invasive pest species (Bradshaw et al. 2016) and up to 2020, US$ 712.4 billion have been utilized (Fig. 25.8) (Renault et al. 2022). A small proportion of the introduced IAS are thought to establish themselves, and even fewer of these species propagate and turn into profitable pests (Pérez et al. 2022). This is often referred to as the "rule of 10" which states that around one in ten introduced species escape into the ecosystem, one in ten of these imported species establish themselves in the environment, and one in ten of these established species turn into profitable pests (Vander Zanden 2005). However, now it is widely believed that the invasive insect pest species under the warming climate are expected to undergo a range expansion, increased population density, and increased generation time, which would have detrimental consequences for the future agricultural production (Cannon 1998; Ratnayake 2014) such as the massive spread of the invasive Fall Armyworm, *Spodoptera frugiperda* J.E. Smith, originating from the African maize fields (Wudil et al. 2022).

Climate change triggered extreme weather events provoking biological (insects) invasions have various negative threats on the biodiversity of an area (Diez et al. 2012). One such example is *Thrips parvispinus* Karny, a thrips species native to Thailand (Mound and Collins 2000), which has become a destructive pest of pepper (*Capsicum solanum* L.) and other solanaceous crops in Indonesia (Murai 2002; Sartiami and Mound 2013). However, it is believed that the rapid spread of *T. parvispinus* to other parts of the world including Africa, Europe, North America, Southeast Asia, and India occurred due to accelerated movement of plant materials through international trade (Waterhouse 1998; Zhang et al. 2011). Over the past two decades, a drastic expansion in the geographic distribution and host range of this pest has been noticed (Johari and Natalia 2018). For the first time, infestation of *T.*

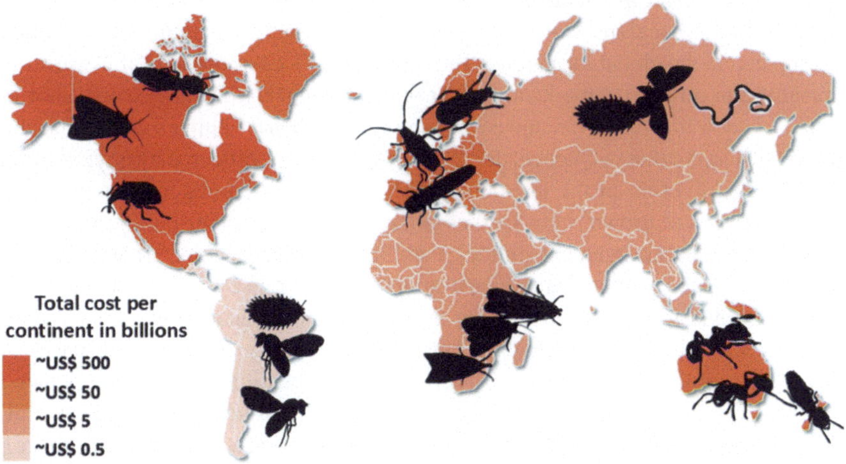

Fig. 25.8 Geographical distribution of the total cost of invasive pest species with three top species in each continent is shown (Renault et al. 2022)

parvispinus was reported from India on papaya, *Carica papaya* L. (Tyagi et al. 2015) and subsequently it outspread to other plants (Rachana et al. 2022). The changing climate in India may provide conditions favourable for the establishment and further spread of such exotic pests, particularly those native to tropical regions. As one of the world's top producers of chili, papaya, and vegetables, India is highly vulnerable to the economic impacts of *T. parvispinus* outbreaks (Hulagappa et al. 2022).

25.2.8 Implications of Climate Change on Insect Transmitted Plant Diseases

Insect vectors, particularly sap-feeders such as aphids, leafhopper, and whiteflies will experience an expansion in geographical ranges, changes in their life cycles, increased number of generations, and population densities (Roos et al. 2011; Bragard et al. 2013). All these effects are likely to affect the feeding and reproductive activity, disrupting their relationships with plants they feed on, or increasing overwintering survival (Canto et al. 2009; Blanc and Michalakis 2016). This expansion in their niches, populations, and performance is likely to favour the occurrence and spread of insect-transmitted plant diseases with a serious effect on their epidemiology (Skendžić et al. 2021). Leafhoppers (Cicadellidae), which include over 20,000 species and are the second-largest family of hemipteran insects (Stiller 2009), are major vectors of viral and bacterial diseases that affect many important crops (Weintraub and Beanland 2006; Brochu et al. 2023). Although leafhopper diversity is high globally, only about 5% of all leafhopper species and 35% of Nearctic leafhoppers (those found from Mexico to the Arctic) are found in Canada, primarily due to harsh winters and limited food resources (Omar 1949; Maw 2000). Climate change is expected to affect the distribution and reproductive potential of these insects in Canada's northern regions where the mean values of annual temperature have doubled as compared to the global mean rates (Firlej et al. 2019). Consequently, this warming is resulting in more extreme heat, milder winters, longer growing seasons, shorter periods of snow and ice cover, and earlier springs (Bush and Lemmen 2019). Brochu et al. (2023) found that in Quebec, the number of host plants affected by phytoplasma has increased fivefold in the last decade. This suggests that either the population of leafhoppers has grown, new species of leafhoppers are contributing to disease spread, or current management strategies are failing. Insect vectors of plant diseases are highly sensitive to changes in temperatures (Juroszek and Von Tiedemann 2013) as demonstrated in Uganda, where the increase in the population of insect vector, *B. tabaci*, caused by climate change, triggered an outbreak of "cassava disease" in cassava plantations (Kriticos et al. 2015). Similarly, as mentioned earlier that elevated levels of CO_2 tend to influence the quality and quantity of host plants (Agrell et al. 2006). These physiological distortions in the context of insect vectors transmitting plant diseases hold great significance as shown in Fig. 25.9 (Mauck et al. 2012; Yoro and Daramola 2020).

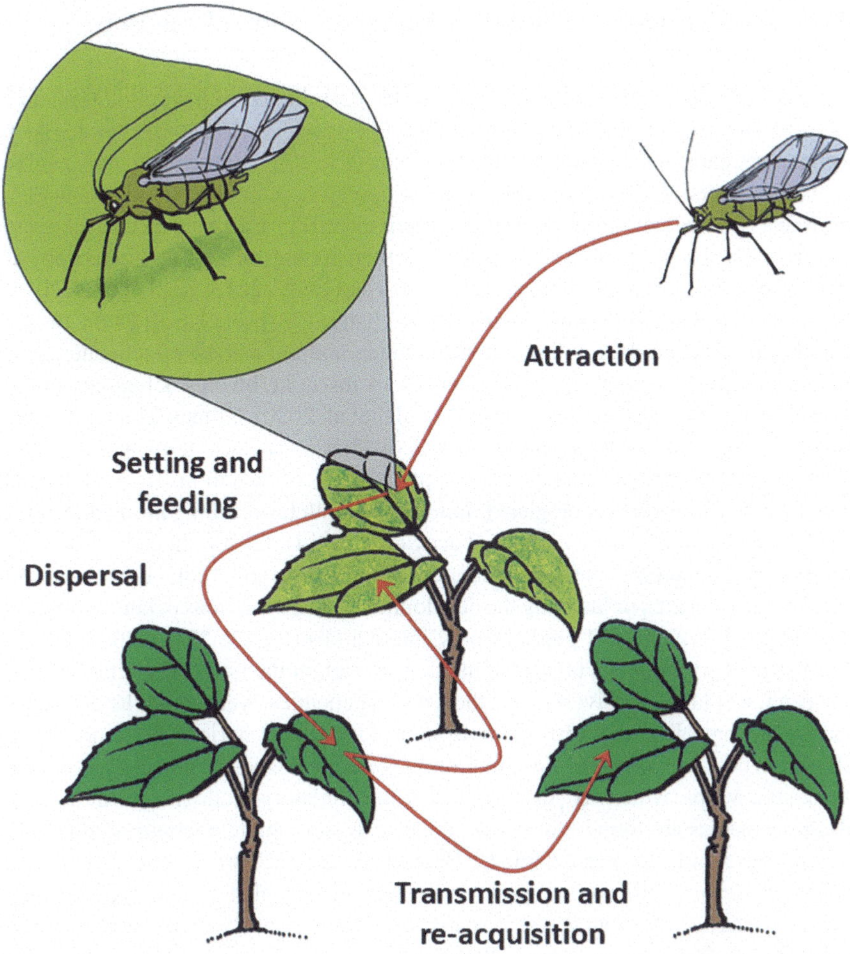

Fig. 25.9 Schematic presentation of insect vector, pathogen, and plant interaction (Mauck et al. 2012). For the pathogen transmission, the vector (aphid) must locate and orient toward the infected plant (attraction), which is mediated by chemical signals cues such as volatile molecules released by the plant detected by vectors using the visual signals cues. After locating the infected host, the vector must interact trophically with the host to acquire the pathogen. Duration and nature of trophic interaction depends on the infection-induced changes in host defences, host nutritional status, or morphological or chemical features. Once the pathogen is acquired, the vector must disperse to a susceptible host for transmission to occur. Pathogen-induced changes in the cues mediating host selection behaviour can influence dispersal in the short term, and effects of infected hosts on vector performance (development, reproduction, and survival) can influence dispersal and transmission in the long term. However, under the influence of climate change, this tritrophic interaction is likely to be altered

25.3 Implications of Climate Change on Natural Enemies

Numerous extrinsic and intrinsic factors of insect pests linked with the climate change are known to influence the interaction of insect pests with their natural enemies such as predators, parasites, and parasitoids (van Doan et al. 2021). Intrinsic factors involve the direct effects of abiotic variables such temperature, humidity, wind speed, etc., on insects (Robinet and Roques 2010; Cornelissen 2011). Extrinsic factors take into consideration the indirect effects such as shift in ecological niches, altered physiology of herbivores, and their host plants (Jeffs and Lewis 2013; Kaplan et al. 2016; Chidawanyika et al. 2019). Given that climate change is expected to affect the timing of both the pests and their natural enemies which undermines the effectiveness of a biocontrol initiative (Nechols 2021). Density of insect herbivores is regulated both by natural enemies and the host plants (Mandal et al. 2020). Climate change has the potential to reshape predator-prey dynamics by altering species' traits and by influencing both consumptive and non-consumptive effects of predation. These shifts can fundamentally transform ecological relationships, with far-reaching implications for biodiversity and ecosystem stability (Facey et al. 2014). Factors like elevated CO_2 concentration, changing rainfall pattern, and rising temperatures alter the timing and productivity of plants, influencing the development and populations of herbivores and impacting the dynamics of predator-prey interactions (Fig. 25.10) (Jamieson et al. 2012). This disruption can have a significant impact on the tri-trophic relationships between host plants, herbivores, and their natural enemies, weakening the effectiveness of biocontrol (Fig. 25.10) (Thomson et al. 2010). In agricultural landscapes, different types of natural enemies are found regulating the population of aphids such as parasitic wasps (which lay eggs in aphids), and predatory beetle (ladybird beetles). All these species are found sensitive to the effects of rising temperatures and could respond differently to these changes (Bale et al. 2002; Sharma et al. 2016). The effects of warming climate are more prominent on specialist natural enemies than generalists, as the latter can adjust to spatial mismatches with their host communities (Gilman et al. 2010; Selvaraj et al. 2013).

25.4 Implications of Climate Change on Insect Pests of Mango

25.4.1 Insect Pests of Inflorescence and Tender Shoots

Mango hoppers (Hemiptera: Cicadellidae) are recognized as one of the most significant insect pests of mango (*M. indica*), primarily infesting the inflorescence and tender shoots of the plant (Reddy et al. 2018). This pest can cause severe damage by feeding on the sap, resulting in reduced fruit yield and quality (Rahman 2007). In light of the ongoing changes in global climate patterns, understanding how climate change may influence the population dynamics, behaviour, and distribution of mango

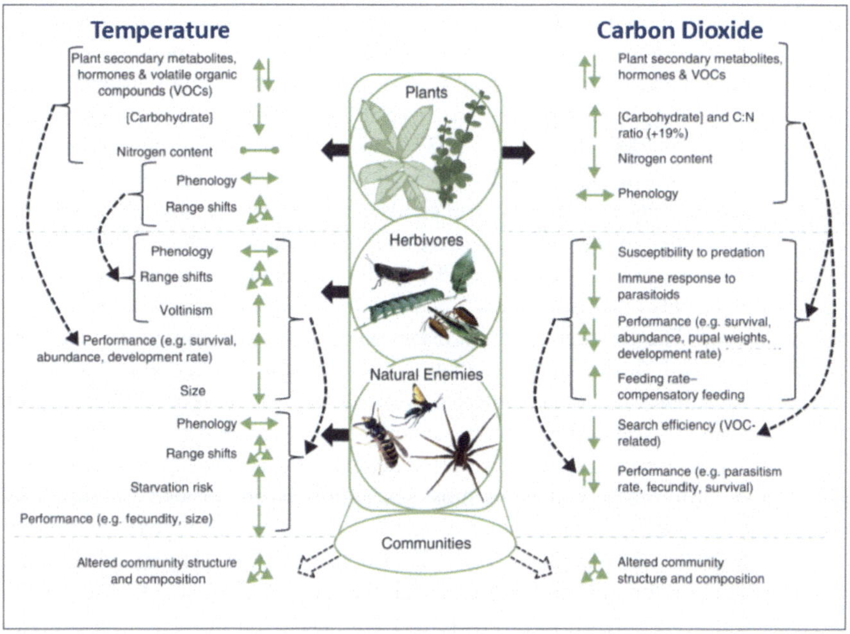

Fig. 25.10 Schematic presentation of the effects of increased CO_2concentration and average air temperature on tri-trophic interactions and communities based on literature (Facey et al. 2014). Green arrows denote the general direction in which the given variables respond. Broken black arrows joining findings indicate indirect effects, for example, direct effects of increased CO_2 concentration on plant quality (solid thick arrow) indirectly affect herbivore success, as well as the searching efficiency of natural enemies (broken thin arrows)

hoppers is crucial. The following section explores the potential impacts of climate change on the ecological interactions between this pest and mango crops, highlighting shifts in pest pressure and their broader implications for mango production.

25.4.1.1 Mango Hoppers

Leaf hoppers are the most serious and widespread monophagous pests of mango across the world's mango growing regions (Jayanthi and Verghese 2014). *Amritodus atkinsoni* Lethierry, *Idioscopus nitidulus* Walker, *I. nagpurensis* Pruthi, *I. clypealis* Lethierry are the most commonly found species in these areas (Joshi and Kumar 2012; Reddy et al. 2018). Their infestation during flowering season causes a loss of 25–60% in yield (Patil et al. 1988). Climate change has emerged as a formidable force, reshaping the natural world in profound ways. One of the most significant yet often overlooked consequences is its impact on the phenology of horticultural crops, including the mango (Rajan 2012; Makhmale et al. 2016). Phenology- the study of the timing of biological events such as flowering, fruiting, and leafing- plays an important role in the intricate interactions between plants, their environment, and the myriads of organisms that rely on them (Piao et al. 2019; Inouye 2022). For mango plants, changes in phenological patterns can disrupt not only growth cycles but also plant-pest dynamics, particularly in relation to invertebrate pests like mango hoppers (Parmesan 2006; Bajpai et al. 2023). Over the years, numerous studies have illuminated how climate change is altering the flowering mechanisms of mango trees, with far-reaching consequences for crop yields, pest populations, and ultimately, the sustainability of mango cultivation (Legave et al. 2013; da Silva et al. 2014; Ramírez et al. 2014; Yadav et al. 2023; Raju et al. 2024). Hoppers tend to congregate during the vegetative (on newly emerging leaves) and reproductive (on inflorescence) phases of the mango plant. The emergence of new leaves take place several times in a year from the terminal growth flushes and only mature terminal branches accommodate pyramidal flower panicles that bear numerous white flowers (Davenport 2007; Rahman 2007; Anwar et al. 2011). Mango trees generally flower during the cooler months of the year viz., December to March, as a period of cold stress is needed before flowering (Fig. 25.11). Unfortunately, it is the most susceptible stage because a higher level of leaf hoppers infestation takes place during this phenological stage (flowering) of the plant (Jayanthi and Verghese 2014). Enormous numbers of nymphs and adult hoppers cluster on the inflorescence and suck the sap during spring. Consequently, the infested flowers shrivel, turn brown, and ultimately fall off. The pest excretes honeydew that covers the inflorescence, leaves, and fruits promoting the growth of black sooty mold, *Meliola mangiferae* Earle, which hampers the photosynthesis and deteriorates the quality of fruits (Verghese and Jayanthi 2001). After flowering, the hoppers move to the new leaves and main trunk, but their peak activity remains during the flowering season (Patil et al. 1988), where swarms of adults are seen hovering in mango groves, sitting on all plant parts. Though the peak activity of mango hoppers is limited to mango flowering season, hoppers remain in cracks and crevices of trunk throughout the year and start infesting the emerging young shoots/inflorescence in the last week of the December (Babu et al. 2002).

Scientists are of the view that the changing climate scenario will cause more precipitation to occur with more intensity and frequency in certain areas (Field et al. 2014). Although mango trees are moderately flood-tolerant, excessive flooding can reduce transpiration and stomatal conductance, negatively impacting tree growth (Bhattacharjee et al. 2022). Additionally, in the light of altered precipitation patterns, increasing flood intensity is expected to affect the growth of mango seedlings and accelerate the incidence of pests and diseases (Asare-Nuamah et al. 2022). This high humid environment coupled with the shady conditions provides favourable conditions for the multiplication of mango hoppers (Fig. 25.12) (Chowdhury 2015). In the recent past, Horgan et al. (2021) illustrated that the feeding activity of leafhoppers increases with temperatures between 15–35 °C and declines when the temperature goes beyond 35 °C.

The IPCC (2021) suggests that climate change could also reduce winds speed which would have serious implications on the development of pests and diseases. Wind patterns can affect pest dynamics by altering resource distribution and insect

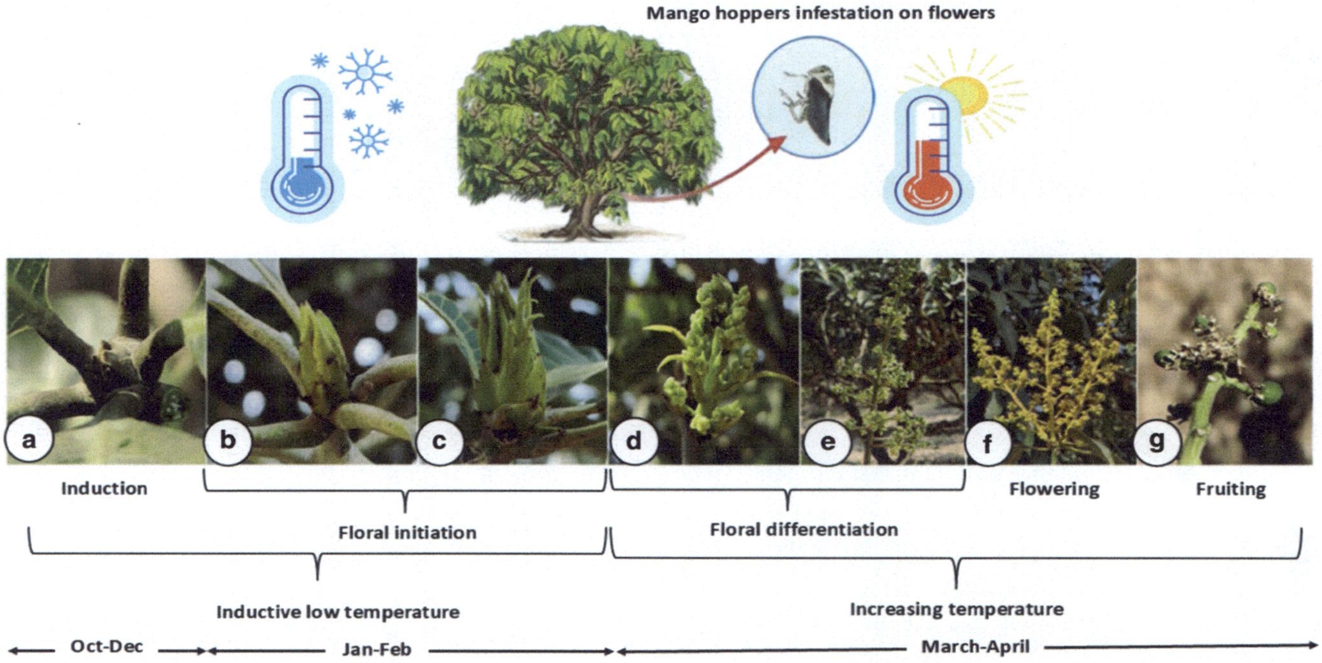

Fig. 25.11 Different phenological stages of mango fruit development starting from the dormancy stage (**a**) and including bud swelling (**b**), bud burst (**c**), panicle emergence (**d**), early anthesis (**e**), full bloom (**f**), and fruit setting (**g**) occurring at various times and seasons of the year shown. Climate change related parameters especially the rising temperature has significantly affected these phenological stages, posing serious implications on the infestation intensity of mango hoppers at the flowering stage. Modified from Bajpai et al. (2023) and UNIDO (2015)

Fig. 25.12 Schematic presentation of the effects of increased precipitation on the population dynamics of mango hoppers under the influence of climate change. The increased precipitation provides congenial environment for the proliferation of mango hoppers by raising the humidity in mango orchards. Modified from Subedi et al. (2023) and MRS (2013)

pests density or by drying out plant surfaces, thereby reducing insect pest susceptibility (Fig. 25.13) (Agrios 2005). Hence, in this context, this changing wind pattern is expected to affect insect pest management strategies at the farms. Mango being an evergreen tree can grow more profusely and forms a dense canopy which acts as a windbreak (Zhang et al. 2018). However, pruning is recommended to manage the canopy, promote new growth, and enhance fruit yield. This practice helps increased air flow and light penetration, reducing pest and disease pressure (Fig. 25.13) (Gopu et al. 2014). Raut et al. (2018) noted a decline in the population of hoppers with an increase in wind speed. The effect of airflow on mango pests and diseases after pruning is not well understood. However, farmers can improve the shapes of mango tree canopies for optimal yield and better pest and disease control by understanding airflow velocity (Leng et al. 2023).

Rising atmospheric CO_2 concentration is anticipated to decrease the foliar nitrogen contents, affecting the insects host preference (Sun et al. 2016). Most plant hoppers have also been negatively affected by the increased CO_2 concentrations due to decline in foliar nitrogen and increase in C:N ratio (Senthil-Nathan 2021). However, feeding patterns of different plant hopper species may vary (Lv et al. 2019). The increased CO_2 concentration not only raises the ambient temperature, but also affects insect health. Notably, elevated atmospheric CO_2 and associated increases in

Fig. 25.13 Schematic presentation of the effects of increased winds under the influence of climate change on the population dynamics of mango hoppers. In case of densely populated mango orchards, more infestation of insect pests especially hoppers occur due to less movement of air and difficulties in penetration of light among the plants (**a**), while pruned trees offer better fruit production due to free movement of air and better light penetration (**b**). Modified from MRS (2013)

global mean temperatures would have a negative impact on plant yield and quality, as extreme temperatures can cause damage to the plant's photosynthesis machinery (Senthil-Nathan 2021; Leng et al. 2023).

25.4.2 Insect Pests of Foliage and Buds

Mango cultivation is heavily impacted by a range of insect pests, including thrips, mealybugs, blossom midges, white scales, and bud mites, all of which target the foliage and buds of the plant (Reddy et al. 2018). However, the ongoing shifts in climate patterns have significantly influenced the population dynamics and behaviour of these pests (Leng et al. 2023). This section explores the effects of climate change on the prevalence and severity of some of the pests infestations during these phenological stages of the mango plant, highlighting how altered environmental conditions are reshaping pest interactions and threatening crop health.

25.4.2.1 Mango Thrips

Thrips (Thysanoptera: Thripidae) are one of the emerging problems in mango nurseries and orchards (Grove et al. 2001). Globally, 6000 species of thrips have been found to be associated in causing infestation to different kind of horticultural crops (Grimaldi et al. 2004; Reddy 2016). For example, in Malaysia, *Thrips hawaiiensis* Morgan, *Scirtothrips dorsalis* Capuz, *Frankliniella schultzei* Trybom, and *Megalurothrips usitatus* Bagnall have been reported (Aliakbarpour and Rawi 2012); in Israel, *Frankliniella occidentalis* Pergande (Wysoki et al. 1992); in India, *T. palmi*, *T. hawaiensis*, and *T. subnudula* Karny (Krishnamoorthy and Visalakshi 2012); *F. bispinosa* Morgan and *F. kelliae* Sakimura, in the United States (Pena et al. 1998) and, in Australia, *Selenothrips rubrocinctus* Giard (Peng and Christian 2004). In Brazil, *S. rubrocinctus*, *F. schultzei*, *F. ardenia* Moulton, and *F. brevicaulis* Hood species have been reported (Barbosa and Paranhos 2005). However, *Scirtothrips dorsalis* Hood, in the recent past has assumed the status of a key pest with serious economic implications to the mango fruit production (Lin et al. 2015; Affandi et al. 2017).

Thrips is one of the universal pests of mango fruit that attacks on leaves, inflorescences, and fruits (Aliakbarpour and Che 2010). Both immatures and adults are "scraper-suckers," feeding on pollen, sap from petals, and flower ovaries, which causes wilting of inflorescence and decreased fruit set (Matos et al. 2019). The scrapping of the plant tissues causes development of a silvery colouration which later evolves to give a ferruginous appearance and roughening of fruit surfaces (Peña 2002; Kumar et al. 2012).

Global climate change is likely to have both direct and indirect effects on key agricultural thrips species. Direct effects include changes in thrips development rates, reproductive potential, distribution pattern, migration/dispersal, and their ability to adapt to new environments, while indirect effects involve interactions with host plants, natural enemies, and other insect species (Porter et al. 1991). A key factor in the spread of thrips species is dispersal, whether natural or artificial. Natural dispersal can be active, where thrips use their wings with cilia to fly, or passive, where they are carried by wind currents without active flight (Fig. 25.14) (Garrick and Liburd 2017). Artificial dispersal occurs when thrips are unintentionally moved via plant materials, soil, or shipping containers during international trade. Both the forms are not mutually exclusive, as thrips will use them simultaneously to move around for short or long range movements (Vierbergen 1995).

Numerous environmental factors have been cited to play a decisive role in the dispersal of thrips species (Porter et al. 1991). Climate change is expected to increase the frequency of hurricanes, which could further aid in thrips distribution (Garrick and Liburd 2017). Strong winds or wind currents derived from hurricanes, or a storm can hasten dispersal by moving thrips infested plant materials far from the source of infestation. In addition, thrips can take advantage of these wind currents by using their cilia and body orientation to glide long distances. A notable example of this was observed in 1988 in Trinidad, when winds from a tropical depression moved *T. palmi* from neighbouring Caribbean islands, leading to its establishment as a pest in Trinidad (Cooper 1991). Additionally, at elevated temperatures, the flying capacity of

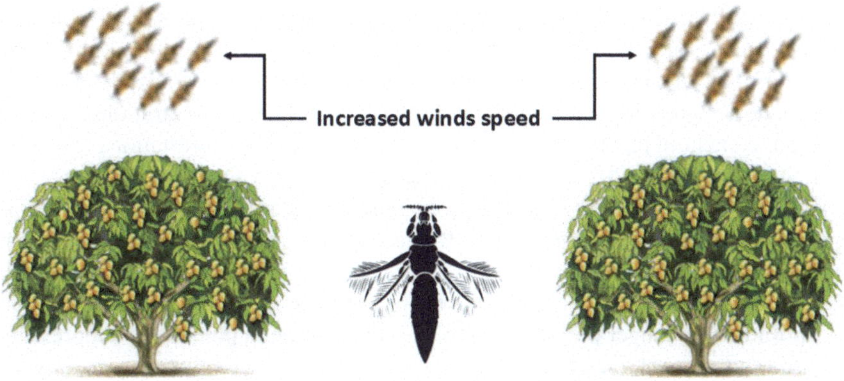

Fig. 25.14 Schematic presentation of the effects of increased winds speed in the context of climate change on the population dynamics of mango thrips. Increased wind speeds tend to disperse the population of thrips over the long distances which results in their further spread in the mango orchards. Modified from Lin et al. (2015) and MRS (2013)

insects is expected to increase as their flight muscles will be more efficient in causing them to move more rapidly (Machin et al. 1962). In the past, Carter (1939) noted the mass movement and dispersal of thrips with the aid of prevailing wind currents. The authors further explained that species, *Thrips tabaci* Lindeman, which is usually not a migratory pest species, but can migrate *en masse* due to non-provision of good quality food. There are concerns among the scientific community that this massive migration of thrips populations poses a serious threat and is likely to become a more common phenomena due to increased intensity of droughts under the changing climate (Garrick and Liburd 2017). Furthermore, an increased global transportation of goods and commodities has also facilitated the spread of thrips as evidenced by *T. tabaci* being one of the most frequently intercepted pest species at United States ports in the period between 1994–1999 (Nickle 2008, 2009).

Female thrips oviposit in the mango plant tissues including stems, foliage, flowers, and fruits (Mouratidis et al. 2024). Stuart et al. (2011) noted an increase in the egg laying capacity and development of adults between 25–30 °C. During their life cycle, some thrips species might undergo reproductive diapause triggered by critical photoperiod, but the effects of photoperiod on their reproduction remains poorly understood. Climate change is likely to exacerbate the risks posed by the crop pests, particularly with the extended summer seasons. One such example is thrips, whose population is reported to expand in the summer (Leng et al. 2023). High atmospheric concentrations of CO_2 can significantly influence the insect population dynamics (Lu et al. 2022). Research on thrips species, *F. occidentalis* and *F. intonsa*, has shown that elevated concentrations of CO_2 decreases developmental time, and increases the egg lying capacity, and oviposition capacity of a female per day, implying a higher infestation rate in future (Prakash et al. 2014; Liu et al. 2022).

As mentioned earlier, CO_2 is the most prominent impetus for the current global warming and a tremendous rise of its concentration in the atmosphere has been documented. This rising CO_2 concentration not only affects the plants photosynthesis rate, but it also affects the feeding behaviour and growth of herbivorous insects (Leng et al. 2023). Elevated CO_2 concentrations have been shown to significantly affect C3 plants like mango, with studies indicating that higher CO_2 concentrations increase the sugar content in plants, which in turn may intensify thrips infestations (Tubiello et al. 2002; Liu et al. 2022). However, exposure to high CO_2 concentrations such as 60% CO_2 at 30 °C for 12 h has been linked to increased mortality of thrips eggs (Seki and Murai 2012). Similarly, Bumroongsook and Kilaso (2018) found that extended exposure to CO_2 at 25 °C can lower the thrips population. In a nutshell, the effectiveness of CO_2 in controlling thrips populations depends on the duration of exposure, with longer periods of high CO_2 concentration potentially suppressing thrips development (Leng et al. 2023).

25.4.2.2 White Mango Scale

Climate change and the expanding international trade of crops and ornamental plants have facilitated the emergence, introduction, and establishment of some of the deadly pests around the world (del Pino et al. 2020; Raza et al. 2023). Among them, the white mango scale (WMS), *Aulacaspis tubercularis* Newstead (Hemiptera: Diaspididae) is presently taken as one of the most destructive pests of mango groves, whose control poses serious problems to the mango growers (López-Guillén et al. 2017). This polyphagous tropical species is found in Asia, Africa, Oceania, Central and South America, parts of Europe, and the Caribbean (Fig. 25.15) (García Morales et al. 2016).

A. tubercularis devours the mango plant by sucking the vital contents from leaves, twigs, and fruits. The most common signs of its infestation include chlorosis of leaves, pink blemishes on fruit, dieback, and mortality of young trees. Notably, the

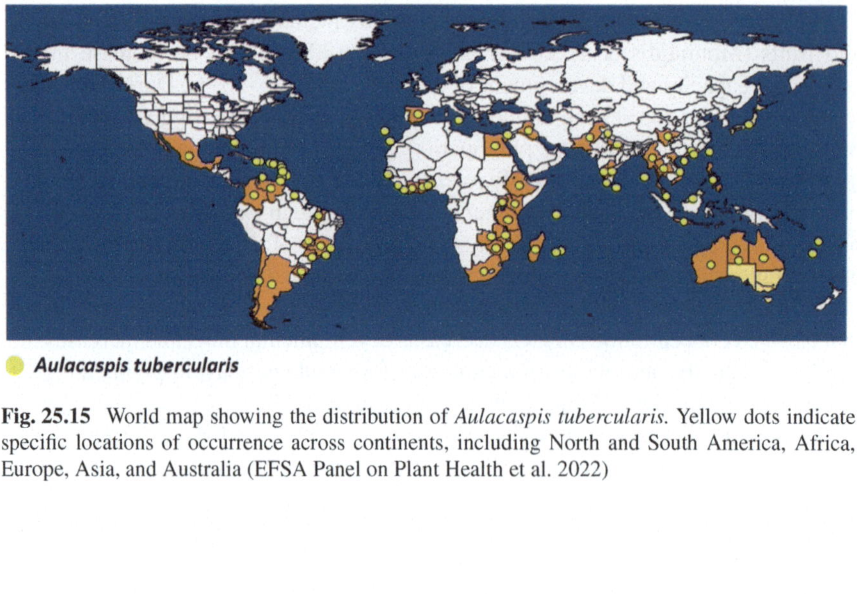

Fig. 25.15 World map showing the distribution of *Aulacaspis tubercularis*. Yellow dots indicate specific locations of occurrence across continents, including North and South America, Africa, Europe, Asia, and Australia (EFSA Panel on Plant Health et al. 2022)

sap-sucking activity of females and crawlers of WMS while feeding dwindles the aesthetic value of mango fruit which greatly affects their export potential in the global markets (Pablo et al. 2014; Azrag et al. 2022). Pakistan is the 7th largest exporter of mangoes in the world (Aitzaz et al. 2024), but presently all the grown mango varieties especially the late ones such as "Sufaid Chaunsa" (a prominent exportable variety in Pakistan) have been found to be vulnerable to the increasing infestation of *A. tubercularis* (Raza et al. 2023). All stages of mango growth, from seedlings to mature trees, are vulnerable to *A. tubercularis*, and severe infestations can cause losses of up to 50–100% (Fig. 25.16) (Roge 2021; Dako 2023).

In Europe mango fruit is exclusively produced in Spain where its cultivation is on rise due to the high European demand for this delicious and nutritive fruit (González et al. 2018). This pest was first recorded in 1997 on mango orchards of Canary Islands and by 2010, it had spread to the major mango-growing regions of Andalusia, including Malaga and Granada (Fig. 25.17) (Arteaga et al. 2003; Boyero et al. 2017). Importantly, the major proportion of mangoes production in Spain originates from these areas (del Pino et al. 2020). *A. tubercularis* is believed to be a non-native pest in Spain, likely introduced through infested plant material (Pellizzari and Porcelli 2014). However, Alemu et al. (2024) emphasized the potential role of climatic variables (temperature, humidity) and crop husbandry practices (canopy structure, planting pattern, plant spacing, pruning) in the establishment of *A. tubercularis* in a new locality.

Fig. 25.16 Diagrammatic illustration of the infestation of *Aulacaspis tubercularis* occuring at different growth phases of mango plant. Climate change triggered phenological changes of mango plant are likely to influence the dynamics of *A. tubercularis*. Modified from https://sqmnutrition.com (Accessed on 15 November 2024) and del Pino et al. (2020)

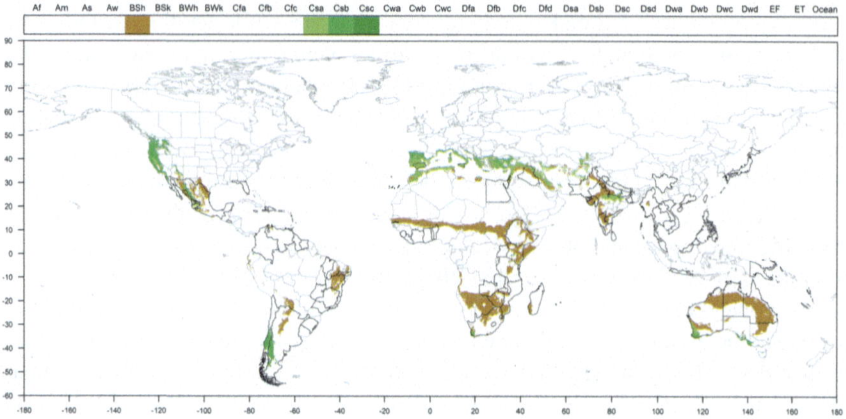

Fig. 25.17 Global distribution of Köppen-Geiger climate types prevailing in the European Union (EU) and in other countries from where *Aulacaspis tubercularis* has been reported. According to Köppen-Geiger climate, BSh represents (Hot semi-arid climate), Csa (Hot-summer mediterranean climate), Csb (Warm-summer mediterranean climate), and Csc (Cold-summer mediterranean climate) (EFSA Panel on Plant Health et al. 2022)

Among the African countries, infestation of *A. tubercularis* was first reported in Ethiopia in 2010 and since then it has outspread across the country (Ayalew et al. 2015; Abera et al. 2021). While analysing the role of various environmental factors in affecting the infestation of *A. tubercularis*, Abera et al. (2023) found a prominent role of temperature and low humidity on their population density. In Mexico, *A. tubercularis* was first detected in 1999 in one of the important mango producing states of Mexico, Nayarit, and hereafter it stretched out to other states (Colima, Jalisco, Sinaloa) and is now posing a serious threat to the Mexico's mango production (Aquino et al. 2011). Bautista-Rosales et al. (2013) in a study examined the effect of environmental changes on the population density of *A. tubercularis* and found that the warming temperature alone could not be held responsible for the proliferation of this pest unless accompanied with the varying intensity of rainfall. Furthermore, it was noted that both sexes required relatively different environmental conditions; females thrive at lower temperatures (18–22 °C) and higher humidity (73–78%), while males are most abundant at higher temperatures (25–28 °C) and humidity levels > 70%. Previously, in a similar work in the California red scale, *Aonidiella urantia* Mask, the authors noted the same observations and associated the high degree of males tolerance to the warm temperatures as compared to the females as it allows them to have a different time-space distribution on the hosts (Asplanato and García-Marí 2001).

Insects, including *A. tubercularis*, have evolved survival strategies in response to harsh environments, such as adjusting their sex ratios through mechanisms like heterogametic or haplodiploid reproduction, where environmental conditions, especially temperature, can influence the expression of these traits (Gempe and Beye 2011). In the case of scales that are considered the most specialized hemipterans where a marked sexual dimorphism occurs in which adult males are winged, small, non-feeding, and short-lived, while females are larger and longer-living (Gullan and Kosztarab 1997). Reproduction in certain winged scales is intricate, with 10 distinct identified forms (Gavrilov 2007). Life cycle stages and their duration have been documented for WMS (Urías-Lopez 2006) and unusually higher male-to-female ratio has been associated with the worst droughts occurring over the last seven decades in the northwestern region of Mexico (Bautista-Rosales et al. 2013). One possible explanation is that females in harsh environments are likely to produce more males than in favourable conditions as they possess wings and have relatively short life span (2–14 h) (Nur 1980). Environmental factors especially temperature and humidity also influence the most of the biological and ecological attributes of *A. tubercularis*, which are crucial for its population dynamics (Labuschagne et al. 1995). Various authors have reported the population growth rates of *A. tubercularis* in mango groves located in different parts of the world. For instance, 2–4 generations per year have been documented from Egypt (El-Metwally et al. 2011; Abo-Shanab 2012), while 3 generations per year from Mexico (Urías-López et al. 2010) and Ethiopia (Ofgaa et al. 2018), and 5–6 generations per year have been recorded in South Africa (Labuschagne et al. 1995). It is pertinent to mention here that this knowledge of phenological aspects such as the number of generations per year, the population fluctuations, and the critical developmental stages when the pest is more vulnerable to parasitism or insecticides are essential for the implementation of a successful integrated pest management (IPM) program (De Villiers and Pringle 2007).

25.4.2.3 Insect Pests of Stem

Mango stem borer and mango bark beetle are two of the most destructive pests that attack the stems of mango plants, posing significant threats to the health and productivity of mango orchards (Reddy et al. 2018). The dynamics of these pests have been increasingly influenced by shifts in climate, which have altered their behaviour, distribution, and severity (Masood and Saeed 2012; Baradevanal et al. 2023). This section provides the potential impacts of climate change on the prevalence and intensity of these pests, offering insights into how changing environmental conditions are reshaping the challenges faced by mango growers.

25.4.2.4 Mango Bark Beetle

Organisms known as "*vectors*" can introduce pathogens like bacteria or fungi or viruses into plants, causing infections (Purcell and Almeida 2005). Insects, mites, and nematodes act as vectors by facilitating the movement of plant pathogens among stationary plants. While many insects and arthropods may harbour plant pathogens, they are not considered vectors as they cannot transmit these pathogens to plants (Butter 2018). Over the past several decades, as mentioned earlier an increase in the frequency of severe windstorms and drought like conditions has led to the emergence and outbreaks of deadly insect pests including the insect vectors of plant diseases (Rosenzweig et al. 2000; Singh et al. 2024). For instance, in the Northern Hemisphere and Europe, forests have seen unprecedented mortality of tree plantations, at a double rate over the last 40 years (Allen et al. 2015; Senf et al. 2018). Bark beetle species in the genus *Cryphalus* (formerly *Hypocryphalus*) (Johnson et al. 2017) have become a focus of scientific research in these forests and other tree plantations around the world due to their detrimental effects (Masood et al. 2010; Faccoli et al. 2016; Zheng et al. 2019).

Bark beetles belonging to sub-family of weevils (Coleoptera: Curculionidae, Scolytinae) with a worldwide population of about 6,000 species (Johnson et al. 2017). Among these, most of the species have recently emerged as notable pests of various tree commodities, including mango, *M. indica* (Anacardiaceae) (Al Adawi et al. 2013), fig, *Ficus carica* L. (Moraceae) (Faccoli et al. 2016), pines, *Pinus* spp. L. (Pinaceae) (Dong et al. 2022) and mulberry, *Morus* spp. L. (Moraceae) (Johnson et al. 2017). These beetles breed only in dead trees and tree parts and contribute to ecological processes like nutrient cycling and as a food source for other animals (Raffa et al. 2015). However, a few species are found in the stressed and dying trees at low populations density but tend to attack *en masse* and kill the healthy trees with a rise in their population (Raffa et al. 2008; Kausrud et al. 2012). Typically, adult bark beetles locate and enter suitable host plants where mating takes place and then the females oviposit under the bark; the neonate larvae (grubs) feed and develop to maturity in the phloem and the brood adults emerge to seek out new trees (Hlásny et al. 2021). Consequently, a successful beetle colonization leads to the mortality of trees, because hundreds of simultaneously attacking beetles damage the inner bark and shatter nutrient transport to the roots. Additionally, the beetles acting as a vector transmit different fungal species that degrade tree defense mechanism and block water transport in the sapwood (Krokene 2015). A schematic presentation of the life cycle and mode of transmission of fungal pathogen in the tree plants by the *Xyleborus* beetles belonging to the subfamily Scolytinae within the family Curculionidae is shown in Fig. 25.18.

Generally, the bark beetle species that kill trees are specialized to breed in only one type of trees and can utilize a tree for just one or two generations before depleting the resources in the bark. Trees have evolved different defense strategies including anatomical, chemical, and physiological that allow them to resist bark beetle attacks most of the time. The formation of necrotic lesions around beetle attacks in the phloem, production of terpenes and other harmful chemicals, and the flow of resin are some of the typical signs of tree defenses. Adult beetles, their offspring, and the beetles' fungal partners can be killed by these defenses (Raffa et al. 2008; Celedon and Bohlmann 2019). Despite these defenses, there are two main ways that beetles can reproduce: by escaping the defenses or by exhausting them. In the former case, the beetles will only infest trees that are already stressed or dying, often due to drought or other factors (Huang et al. 2020). Species employing this strategy are non-aggressive or semi-aggressive and are resistant to outbreaks only in stressed stands (Raffa et al. 2008). Other species overwhelm tree defenses through mass attacks coordinated by powerful chemical signals (aggregation pheromones) that rapidly direct hundreds

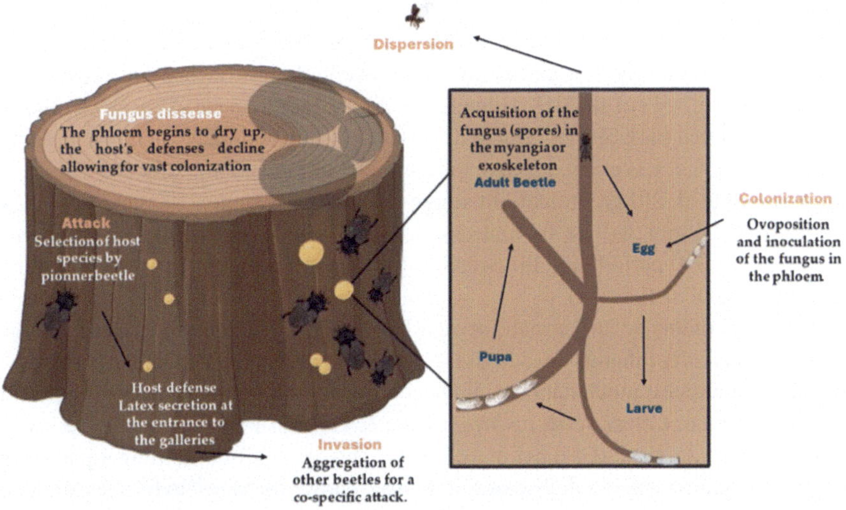

Fig. 25.18 Digrammatic illustration of the life cycle, reproduction, and fungal transmission mechanism of *Xyleborus* beetles (Rodríguez-Becerra et al. 2024)

of beetle attacks onto a single tree, leading to the death of the tree when it exceeds its ability to fend off the beetles (Krokene 2015). In addition, large-scale climatic events, like windstorms or droughts, can aggravate these outbreaks by weakening tree defenses or increasing beetle populations (Wallin and Raffa 2004; Marini et al. 2017).

In the Northern Hemisphere, the European spruce bark beetle beetle (ESBB), *Ips typographus* L., and the mountain pine beetle (MPB), *Dendroctonus ponderosae* Hopkins are prominent pests. Unfortunately, ESBB outburst in Europe caused by extreme weather events have reached a supranational scale (Seidl et al. 2016; Senf and Seidl 2018), and in western North America, MPB eruption affected extensive area of pine forests in a single decade (2000–2010) (Meddens et al. 2012; Cooke and Carroll 2017). Both the outbreaks have overwhelmingly highlighted the lack of preparedness by policymakers and managers in the light of these aggravating forest disturbances (Soderberg et al. 2021). For instance, MPB is native to pine forests of western North America, although the current distribution is not as extensive as its pine hosts which are found further north and east in Canada and further south into mainland Mexico than the current MPB distribution (Dowle et al. 2017). Due to the warming effect, a rapid northward expansion in northern British Columbia and western Alberta, Canada was witnessed (Janes et al. 2014), but new infestations have not been observed south of the current distribution in the southwest United States (USA) (Soderberg et al. 2021). Scientific community believes that the cotemporary climate change scenario has contributed to the outbreaks of bark beetles in Europe by reducing winter mortality (Raffa et al. 2008), allowing beetles to complete more generations per year (Baier et al. 2007; Griess et al. 2012), expanding suitable habitats at higher latitudes and altitudes (Jönsson et al. 2009; Jakoby et al. 2019), and increasing the likelihood of extreme weather events like droughts that reduce tree resistance (Matthews et al. 2018; Huang et al. 2020). Due to these likely mechanisms, it is expected that bark beetles related disturbances in Europe especially in Belgium, Denmark, France, Germany, Luxemburg, and Netherlands, are likely to increase in future (Seidl et al. 2011). Projections suggest that average annual damage from bark beetles in this region will be six times higher in the 2021–2030 decade as compared to the 1971–2010 period, and this trend will likely continue under future warming scenarios (Seidl et al. 2014).

In mango orchards, the minute mango bark beetle, *Hypocryphalus mangiferae* Stebbing, typically inhabits dead or dying twigs and branches of mango trees and by itself does not cause much damage. However, due to its potential role as a vector of deadly pathogen, *Ceratocystis fimbriata* Ellis & Halst., of mango attains the status of a pest (Fig. 25.19) (Al Adawi et al. 2013). This is a soil-borne fungal pathogen responsible for the lethal wilt diseases of mango, known as "*seca*" in Brazil (Abrahão and Wegmuller 1969; Galdino et al. 2016), and mango sudden decline (MSD) or mango sudden death disease (MSDD) in Oman and Pakistan (Van Wyk et al. 2007; Masood et al. 2008). *H. mangiferae* is found in all the tropical regions, with the exception of New Guinea (Wood 1992). With the help of mycangia in elytra and mouthparts, beetles carry the fungal bodies over long distances (Fig. 25.19). Infestations in new areas usually begin on the branches of the trees where the beetles normally initiate attack (Pereira 2013). Over time, the fungus may infect other parts of the plant, such as the trunk and roots, and afterwards, may remain in the soil. Once the soil is infested with the fungus, it can result in the loss of an entire orchard, rendering the area unsuitable for mango cultivation (Van Wyk et al. 2007; Al Adawi et al. 2013). Due to persistence and viability of fungal pathogen in the soil for long periods of time, its eradication is extremely difficult (van Wyk et al. 2005; Syazwan et al. 2024). The proliferation and introductions of this pathogen in new areas takes place through these contaminated soils, infected tissues or by the vector beetle, *H. mangiferae* (Masood and Saeed 2012; Al Adawi et al. 2013).

In the light of climate change, plants may lose moisture through the process of transpiration, water columns in the xylem break apart or cavitate, producing an ultrasonic acoustic emission that is detected by harmful bark beetles (Scolytidae) (Staley et al. 2007). Therefore, infestation of bark beetles is more commonly found in drought stressed plants (Fig. 25.20) (Netherer et al. 2019), while moderately stressed trees are generally more resistant to the infestation of bark beetles (Netherer et al. 2015). Furthermore, the population of mango bark beetle, *H. mangiferae*, like other insect pests is expected to undergo a shift in its niche (Rossetto and Ribeiro 1990) as witnessed by an increase in the number of invasions of bark beetles and associated microorganisms (Li et al. 2016; Marincowitz et al. 2020; Schebeck et al. 2023). Using various computational programs and models, there are projections that the dissemination of the fungal spores is anticipated to be altered in the warmer climate which can seriously affect the geographic distribution of MSDD in the coming days (Biber-Freudenberger et al. 2016; Galdino et al. 2016).

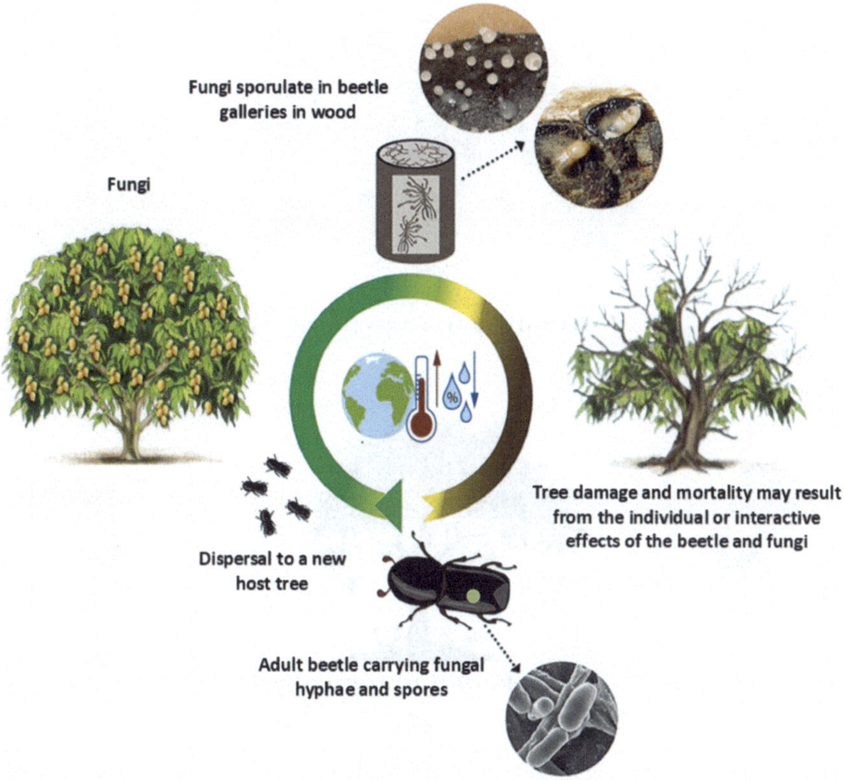

Fig. 25.19 Schematic presentation of the life cycle of mango bark beetle (*Hypocryphalus mangiferae*) and its role in fungal (*Ceratocystis fambriata*) transmission. Life cycle of *H. mangiferae* begins following hatching from an egg in a maternal gallery (tunnel) under the tree bark. The neonate larvae (grubs) remain in the tunnels, feeding on the phloem and burrowing further into the wood. At this stage, fungi begin to grow and produce spores, which can provide nutrition to the developing larvae (grubs). Larvae upon maturation, pupate under the tree bark. Once fully grown, the adults emerge and fly off in search of new host trees, carrying fungi (*C. fambriata*) with them. Fungal spores are carried on their exoskeleton, in specialized structures called mycangia or via other associated organisms, and then inoculated into a new host tree. Modified from Linnakoski and Forbes (2019)

25.4.3 Insect Pests of Fruits

Insect pests pose a significant threat to mango crops, with species such as the mango fruit fly, fruit borer, and stone weevil specifically infesting the fruit (Reddy et al. 2018). These pests not only reduce the quality and yield of mangoes but also complicate pest management strategies. However, like many other agricultural pests, these species are increasingly vulnerable to the impacts of climate change. Shifts in temperature, precipitation patterns, and extreme weather events can alter their life cycles, distribution, and behaviour, potentially exacerbating the challenges they present to

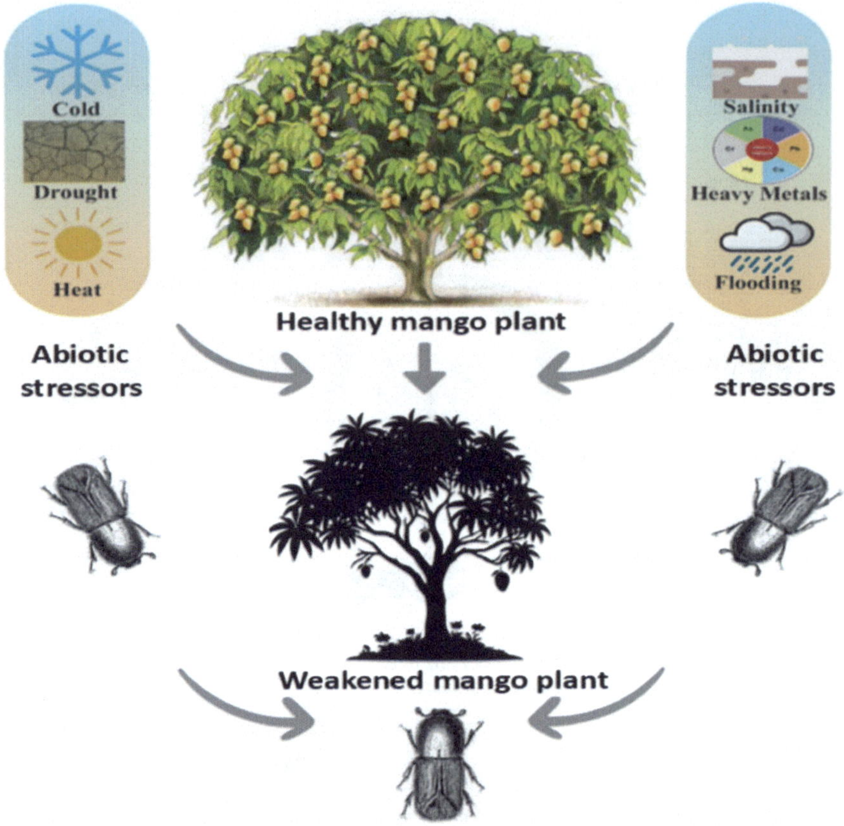

Fig. 25.20 Climate change associated effects such increasing temperatures, CO_2 concentrations, and changing precipitation patterns are likely to influence the growth of mango plants in orchards along with the increased incidence of numerous diseases. Consequently, the plants face numerous stresses and becomes weaker or susceptible. These weakened (stressed) plants provide an ideal environment for the proliferation of mango bark beetle which results in an increased incidence of mango sudden death disease (MSDD). Modified from Singh et al. (2024)

mango cultivation (Leng et al. 2023; Cao and Feng 2024). The preceding section explores the interplay between climate change and the dynamics of these key insect pests, highlighting the potential consequences for mango production and the need for adaptive pest management approaches.

25.4.3.1 Mango Fruit Fly

Insects being the poikilotherms are liable to be influenced with the changing global climate (Andrewartha and Birch 1986). As mentioned above, the potential impacts of elevated temperatures or warming on agriculturally important pests include an expansion in geographic ranges, overwintering ability in new areas, exposure to heat or cold stresses, and number of generations (Fig. 25.2) (Wagner et al. 2023). Fruit flies belonging to the family *Tephritidae* are one of the most damaging and economically important pests that attack a wide variety of fruits and vegetables (Sultana et al. 2020). Over 5000 species are found in this family and among these about 350 pose a serious threat to the global horticultural production and market access (Aluja and Norrbom 1999; Plant Health Australia 2018; Drew and Hancock 2022). Within this family, *Bactrocera* species are of particular concern due to the substantial damage they cause (Zhao et al. 2024), with 73 species declared as pests and documented to have established themselves as invasive species in other parts of the world (Vargas et al. 2015). Out of the 11 species of the genus *Bactrocera* found in Pakistan (Stonehouse et al. 2002), two fruit fly species, *Bactrocera zonata* Saunders and *B. dorsalis* Hendel are considered to cause both qualitative and quantitative losses in different horticultural crops (Khan et al. 2021). Female fruit flies lay eggs in ripening fruit, followed by larvae and other opportunist secondary microorganisms feeding on the fruit pulp leading to decomposition. This damage reduces fruit marketability, causes fruit drops, and ultimately lowers yields (Delrio and Cocco 2010).

B. zonata is also known as "peach fruit fly" feeds on more than 50 commercial and wild host plants including apricot, citrus, fig, guava, mango, peach, and prickly pear (Allwood et al. 1999). However, its host range swelled to include crops such as apple, eggplant, loquat, potato, and tomato (Abd El-Samea et al. 2006). Given that availability of host plants year-round in tropical and subtropical regions provides a favourable environment for its spread, resulting in significant economic losses (Rauf et al. 2013). *B. zonata* is native to South and South-East Asia but it has invaded and now well-established in many countries across the Arabian Peninsula, North Africa, and some of the Indian Ocean Islands like Mauritius and Réunion (Fig. 25.21) (De Meyer et al. 2010). Its spread has largely been driven by favourable climatic conditions and availability of abundant hosts (Delrio and Cocco 2010). Similarly, *B. zonata* of the same origin and *C. capitata*, native of Sub-Saharan Africa have now penetrated through introduction via overland routes, natural dispersal (through air or sea transport), and imported fruits into Europe (Stonehouse et al. 2002; Ricalde et al. 2012) (Fig. 25.22). Scientific community is more concerned about the prevalence of these invasive pest species in Europe due to their potential injurious effects on the horticultural crops including mango (Lopes et al. 2023). Studies by Duyck et al. (2004) found that the optimum temperature for adult *B. zonata* development varied from 25–30 °C whilst egg, larval, and pupal survival was highest at 25 °C. However, none of its life stages survived at temperatures below 12.6 °C, and the upper temperature tolerance was close to 35 °C. Based on this and similar findings, scientific community believes that the warming climate is likely to alter the distribution pattern of important horticultural pests including the *B. zonata* (Walther et al. 2009; Delrio and Cocco 2010).

In Australia, The National Fruit Fly Strategy, identified 46 species as "high priority pests" but these are yet to establish (Plant Health Australia 2018). Of the 11 species that are presently existing in Australia, seven (*B. aquilonis* May, *B. jarvisi* Tryon, *B. neohumeralis* Hardy, *B. musae* Tryon, *B. tryoni* Froggatt, *C. capitata*, *Zeugodacus*

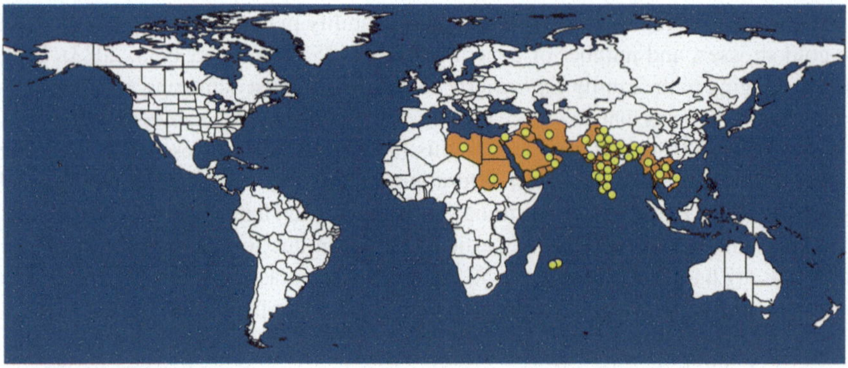

Fig. 25.21 World map showing the distribution of *Bactrocera zonata*. Regions highlighted in orange indicate areas where the species is present, covering parts of Africa, the Middle East, and South Asia. Yellow dots represent specific locations within these regions. The rest of world is shown in white, illustrating absence of the species (European Food Safety Authority (EFSA) et al. 2021)

ucumis French) are known to inflict an annual loss of ~ $ 4.8 billion to different valuable crops (Horticultural Policy Council 1991). Importantly, the predominant population of these species are non-native to Australia with their origins located in South-East Asia and the South Pacific (Plant Health Australia 2008, 2018). Abdalla et al. (2012) estimated that the annual likelihood of the incursion by an exotic fruit fly species is 21% with the associated cost of ~ $ 13 million to get rid of these incursions (Plant Health Australia 2009). However, the distribution of fruit fly species like other arthropod species in Australia is primarily regulated through the prevailing climatic variables and the availability of the hosts (Fig. 25.23). For example, the geographic range of the three species (*B. frauenfeldi* Schiner, *B. kraussi* Hardy, *B. musae*) is currently limited to north-east Queensland (Royer and Hancock 2012) whereas in case of *B. neohumeralis,* it is extended along eastern Australia, from Queensland to central New South Wales (Hancock et al. 2000). Other species are found across broader regions of the country, thanks to their wider climate tolerances (Sultana et al. 2020). Historically, among these mango fruit fly species, *B. frauenfeldi*, had its origin from the Papua New Guinea region and was first detected in Cape York Peninsula in 1974 and since then had spread to Townsville by 1997 (Drew 1976). Despite its invasive potential, availability of plenty of hosts, and lack of control measures to prevent its spread, it has not spread further south. Royer et al. (2016) in an exhaustive analysis of cue-lure trapping data of *B. frauenfeldi* species from 1997 to 2012 related it to the varying environmental variables. Similarly, the Food and Agriculture Organization (FAO) of the United Nations stated that the fruit fly species (*B. cucurbitae, B. dorsalis, B. latifrons*) in Japan may shift their geographical ranges due to warming temperatures (FAO 2008).

Fruit flies always attack mango fruits during fruit ripening stage (Bana et al. 2017), with infestations being more common in areas with heavy rainfall (Patel

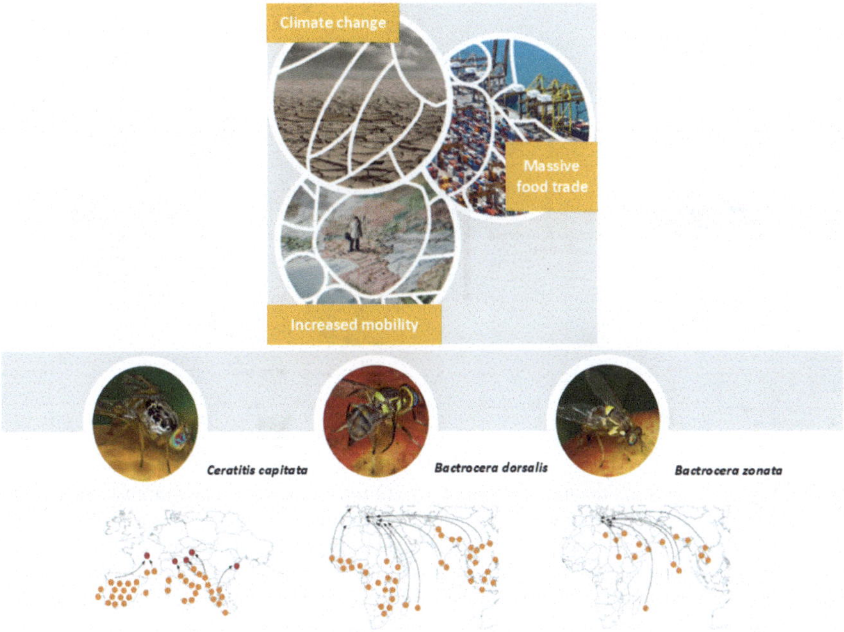

Fig. 25.22 Schematic presentation of the penetration and potential expansion of fruit fly species in Europe under the influence of climate change. The Mediterranean fruit fly (*Ceratitis capitata*) has long been established in the coastal and warmer temperate regions of the Mediterranean. However, under the threat of warming climate, its population has gradually expanded into northern regions, previously considered too cold for survival. Similarly, the oriental fruit fly (*Bactrocera dorsalis*) and the peach fruit fly (*Bactrocera zonata*), both native to Asia, have not yet established themselves in mainland Europe but are rapidly advancing towards it (Accessed on 15 November 2024; https://fruitflies-ipm.eu/challenge)

et al. 2013). Typically, fruit harvest in Pakistan begins from the province of Sindh in the month of May while the produce from the Punjab province sets to arrive in the market in the month of June, continuing up to September–October (Malik and Hasan 2019). Given that all the cultivated mango varieties have been categorised into early maturing, mid-season, and late season maturing varieties (Fig. 25.24) While assessing the preference of mango fruit fly against the various cultivated varieties in Pakistan, Karar et al. (2021) observed that the late maturing mango varieties were found to be more susceptible. Due to changing climate, phenological cycle of mango plants have been seriously affected (Anwar et al. 2011; Leng et al. 2023), it is anticipated that dynamics of fruit fly species and their interactions with the host plants may undergo some changes (Awarikabey et al. 2023). Warmer temperatures have been shown to improve the sexual performance of male fruit flies but also reduce their sperm storage at temperatures exceeding 28 °C (Weldon et al. 2022). Temperature changes can exhaust the conserved nutrients, affecting mating behaviour and survival. Kumar et al. (2010) observed that the development rate of fruit flies in mango variety "Chaunsa" increased with the increase in temperature from 20 to 35

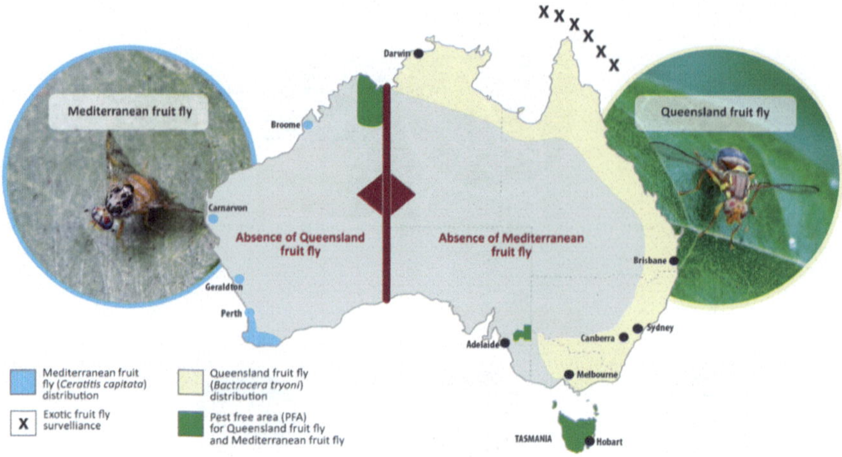

Fig. 25.23 Distribution of Mediterranean fruit fly (*Ceratitis capitata*) and Queensland fruit fly (*Bactrocera tryoni*) in Australia. *B. tryoni* is found in the Northern Territory, Queensland, New South Wales and Victoria and is endemic to Australia, whereas *C. capitata* is an introduced pest and is only present in Western Australia, mostly in the south-west of the state. Climatic variables (temperature, humidity) have played a critical role in defining these geographical boundaries between theses two important species (Accessed on 15 November 2024; https://www.agriculture.gov.au/biosecurity-trade/pests-diseases-weeds/fruit-flies-australia)

°C. The authors also noted a rise in the larval survival (%), adult emergence (%), and growth index with a rise in temperature from 20 to 27 °C and thereafter a decrease occurred up to 35 °C. Thus, a temperature of 27 °C was found to be ideal for survival and development of the immature stage and reproduction of *B. dorsalis*. On the other hand, elevated temperatures can reduce the mango fruit size, but this may increase the rate of parasitism by wasps used for biological control of fruit flies (Haran et al. 2019).

All over the world scientists are investigating the impact of elevated CO_2 concentrations on mango fruit fly during the post-harvest phase, as fruit fly infestation is closely tied to changes in fruit peel texture and volatile chemicals that facilitate oviposition (Grechi et al. 2021). High CO_2 concentrations (50–82%) for 4–5 days at low temperatures during storage have been found to deter fruit flies from infesting mangoes (Kumar and Vishwakarma 2018). However, there are also reports that low (< 0.5%) and high (> 50%) concentrations of CO_2 can effectively control insects in fresh fruit crops (Lacroix and Follett 2015; Tucker 2016).

Light intensity plays a key role in accumulating phenolic compounds in plants, which are important for plant defense against pests. Research has shown that ultraviolet (UV) radiation can induce production of phenolics, which may protect mango fruit from fruit fly infestation (Yildirim 2020; Martínez-Silvestre et al. 2022). However, there are conflicting views on the effectiveness of phenolic compounds in preventing fruit fly attacks. Verghese et al. (2012) found that high phenolic content in mangoes could reduce fruit fly infestations, while Rashmi et al. (2020) noted a

Fig. 25.24 Life cycle of mango fruit fly is shown (**a**). Temperature requirement of early maturing variety, Sindhri (**b**), mid-season maturing variety, Samar Bahisht Chaunsa (**c**), and late maturing variety, Sufaid Chaunsa (**c**) in Pakistan are shown. Climate change is anticipated to alter the phenological cycle of the mango plant which will have serious implications on the dynamics of fruit fly species. Modified from MRS (2013) and UNIDO (2015)

negative correlation between fruit fly infestation and phenolic content, suggesting that the protective effect may depend on mango variety and the stage of fruit ripening.

25.4.3.2 Mango Stone Weevil

In the contemporary world, invasive species are a serious threat to both the natural ecosystems and agricultural systems, potentially causing severe damage to global economies and biodiversity if not properly managed (Paini et al. 2016; Venette and Hutchison 2021). Unfortunately, the harm inflicted by these species is often irreversible, with both direct and indirect consequences that can endanger human health and food security (Huang McBeath et al. 2010; Mollot et al. 2017). Rising global trade and transportation has created new pathways for the introduction of invasive species so a shift in their biogeographic patterns was obvious. Therefore, it is believed that humans are now the primary agents in transporting thousands of species beyond their native habitats and facilitating the breakdown of natural barriers, allowing pests and potentially harmful organisms to spread (Venette et al. 2010). Furthermore, in the context of climate change, the detrimental effects of invasive species have gained a significant status (da Silva et al. 2023). Among the invasive pest species, mango stone weevil (MSW), *Sternochetus mangiferae* Fabricius (Curculionidae: Coleoptera), native to India, holds great value (Abdulla et al. 2016; EFSA Panel on Plant Health et al. 2018). *S. mangiferae*, is a highly specialized, monophagous pest of mangoes, and is found across major mango-growing regions worldwide (Reddy et al. 2018). In Asia, *S. mangiferae* has been reported in countries including Bangladesh, Bhutan, Hong Kong, India, Indonesia, Malaysia, Myanmar, Nepal, Oman, Pakistan, the Philippines, Sri Lanka, Thailand, the United Arab Emirates, and Vietnam (Fig. 25.25) (EFSA Panel on Plant Health et al. 2018). Symptoms of MSW damage are subtle, with the only visible indication of a brown, hardened secretion at the oviposition sites (Follett 2002). The neonate larvae (grubs) primarily feed within the stone of the fruit, drastically affecting the germination of mango seeds, but rarely feeds on the pulp (Pena et al. 1998; Follett and Gabbard 2000). In addition, the pest has been intercepted in mangoes and seeds during international trade. Given that, this weevil has been designated as the "quarantine pest" (Yahia 2006; Bhattacharyya et al. 2016). Long-distance spread of *S. mangiferae* primarily occurs through the movement of infested mangoes and seeds, which can contain larvae, pupae, or adult weevils (Braimah and van Emden 2010). However, there have been limited attempts to identify the role of climatic variables in the context of climate change on the biological attributes of this pest (Baradevanal et al. 2023; da Silva et al. 2023; Aidoo et al. 2024).

In India, Bhattacharyya et al. (2016) observed that life cycle of the pest takes 40–45 days during June-July, while adult weevils overwinter from July–August until

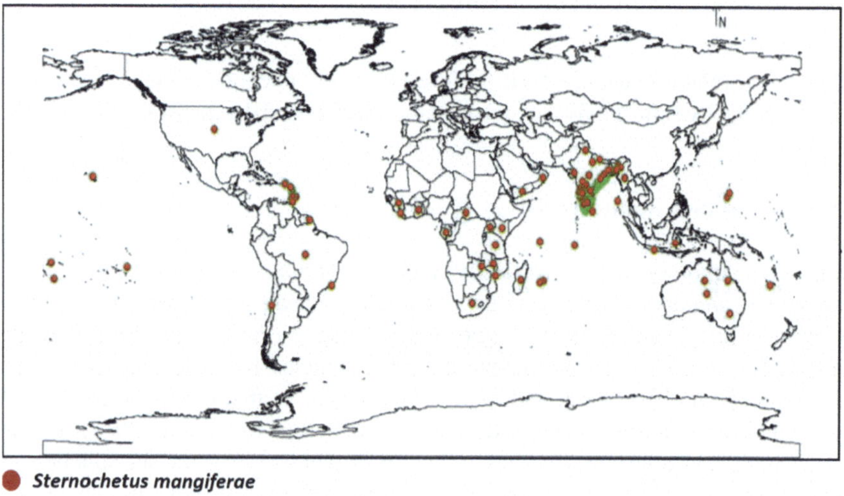

Fig. 25.25 World map showing the distribution of *Sternochetus mangiferae,* marked by dark red dots (Baradevanal et al. 2023)

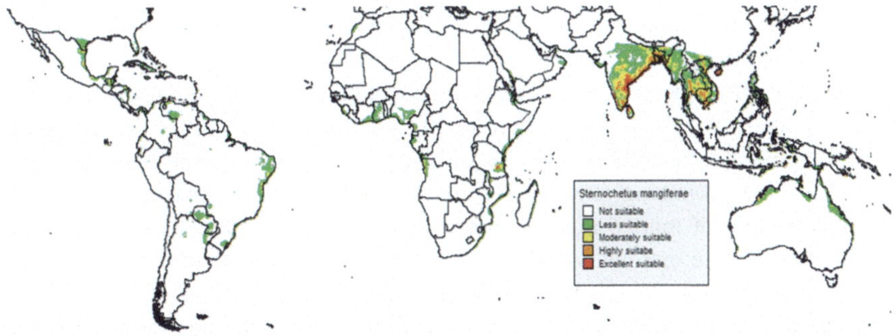

Fig. 25.26 World map showing the distribution suitability of *Sternochetus mangiferae* using the program MaxEnt (Baradevanal et al. 2023)

the next season. Thus, there is one generation yearly. Using the MaxEnt model, Baradevanal et al. (2023) predicted the high suitability areas for the distribution of MSW in South Asia (Bangladesh, India, Myanmar, Thailand, Vietnam), Australia, and Africa (Tanzania). In India, projections showed that regions with higher pest suitability include Andhra Pradesh, the coastal areas of Orissa, the southern parts of West Bengal, and portions of Kerala and Karnataka. Furthermore, the authors also noted the high-suitability regions around the world, where MSW may potentially establish itself (Fig. 25.26). Similarly, among the top ten mango producing nations of the world, Brazil is ranked at the seventh position, where mango cultivation is limited to a few municipalities of São Francisco (FAO 2022). In February, 2014, for the first time MSW was recorded in Brazil from some areas (Silva et al. 2022) but recently, da Silva et al. (2023) found more suitable areas for the development of this invasive pest.

25.5 Future Outlook

Climate change induced changes such as increasing temperatures and changing precipitation patterns have seriously disturbed the pest management strategies that were once effective. Consequently, there is an urgent need to explore and implement adaptive mitigation strategies to protect mango orchards from the growing pest pressures under climate change. Mitigation strategies, including strict quarantine laws, IPM, climate-smart agricultural practices, and the use of pest-resistant and climate-resilient varieties are crucial for sustaining mango production. In the preceding lines a brief description of these mitigation strategies is given.

1. Establish physical and commercial barriers to prevent invasive pests from spreading into new environments. Invasive pest species overwhelmingly occupy the new areas due to absence of biological constraints such as natural enemies which were present in their native habitats. Therefore, removal of undesirable or

invasive pests should be a top priority along with the strict quarantine measures (Nasution 2024).

2. Research efforts need to be geared in the development of climate-resilient crops. The simplest way is to screen high-quality seedlings and varieties that are more resistant to abiotic and biotic (pests and diseases) stresses (Asare-Nuamah et al. 2022). One promising strategy for adapting to climate change is the genetic improvement of mango plants. This involves selecting cultivars or rootstocks that can tolerate extreme weather conditions and possess a robust root system with efficient water absorption (Atlin et al. 2017; Bura et al. 2023).
3. Improvement in management practices in mango orchards such as use of intercropping can be helpful to mitigate the climate change induced stresses (Asare-Nuamah et al. 2022). By planting diverse crop varieties, farmers can reduce the risk of pest outbreaks. For decades, mango farmers have relied on intensive monocropping, which, while efficient in terms of land use, creates an environment conducive to pest, disease, and weed infestations (Singh et al. 2011; Deguine et al. 2018). Unfortunately, to address these problems our farmers extensively apply pesticides. This overuse of pesticides presents significant health risks to both farmers and consumers, while also disrupting local ecosystems. A promising alternative is intercropping mango trees with aromatic plants (Zhang et al. 2021). This novel approach has shown potential not only in improving soil quality but also reducing the density of certain mango pests. The aromatic compounds released by the fragrant leaves of these plants can help repel pests and mitigate biotic stress (Swain et al. 2012; Tripathi et al. 2019).
4. Ecofriendly pest management strategies such as IPM should be followed in the mango orchards (Skendžić et al. 2021). This approach focuses on preventive and indirect measures for plant protection, aiming to reduce environmental harm while optimizing crop yields. Although IPM methods are effective for managing certain pests, many mango farmers still prefer insecticides for pest control (Peña 2002). Adopting low-pesticide pest management strategies requires a thorough understanding of pest biology and their interactions with both the crop and the surrounding environment. Recognizing the factors that influence crop infestation, including the dynamics of host-pest relationships, is essential for effective pest management (Mwungu et al. 2020).
5. Use of models to predict the effects of climate change on mango production, primarily focusing on aspects such as photosynthesis, stomatal conductance, and fruit growth, and quality have been in practice now a days (Léchaudel et al. 2013; Boudon et al. 2020). Given that there is a dire need to develop models that can project the effects of climate change on mango pests and diseases. However, creating accurate models based on phenological data requires extensive long-term monitoring of pestiferous species, their natural predators, and their interactions with both current and altered environmental conditions. Furthermore, the development of reliable, well-fitting models necessitates the active involvement of decision-makers and other stakeholders (Gaydos et al. 2019; Jones et al. 2021). Long-term surveillance of pest dynamics, particularly in regions highly sensitive to climate variability, is essential for enhancing the effectiveness of pest

management strategies and informed decision-making processes (Leng et al. 2023).

25.6 Conclusion

Overall, this chapter explored the potential impacts of escalating climate change on the insect pests with significant challenges to agricultural systems, particularly in the context of pest management. The warming climate is not only altering temperature regimes but also reshaping the dynamics between crops like mango and their associated insect pests. These changes are likely to result in increased pest populations, new pest invasions, and altered pest behaviour, further complicating efforts to protect "The King of Fruits" and ensure food security. As the frequency and intensity of climate-driven disruptions grow, farmers will face mounting difficulties in managing these pests. The need for innovative, adaptive pest management strategies becomes more urgent than ever, requiring a combination of climate-resilient agricultural practices, integrated pest management, and ongoing research to predict and mitigate the effects of climate change on crop-pest interactions. Unfortunately, there is a scarcity of scientific research on the biological attributes and dynamics of insect pests of mango plant across the world in the light of global warming. Addressing these challenges proactively will be crucial to sustaining global food production in the face of a rapidly changing climate.

References

Abd El-Samea SA, Fetoh, Badr El-Sabah A (2006) Scienific note new record of *Bactrocera zonata* (Saunders)(Diptera: Tephritidae) on potatoes in Egypt. Egypt J Agric Res 84:61–63

Abdalla A, Millist N, Buetre B, Bowen B (2012) Benefit-cost analysis of the national fruit fly strategy action plan. Australian Bureau of Agricultural and Resource Economics and Sciences, Canberra

Abdulla NR, Rwegasira GM, Jensen KM, Mwatawala MW, Offenberg J (2016) Control of mango seed weevils (*Sternochetus mangiferae*) using the African weaver ant (*Oecophylla longinoda* Latreille) (Hymenoptera: Formicidae). J Appl Entomol 140:500–506

Abera M, Jemaneh B, Tesfaye A (2021) Survey of white mango scale (*Aulacaspis tubercularis*) distribution on mango (*Mangifera indica*) production at Assosa and Bambasi districts, in Benishangul Gumuz region, western Ethiopia. International Journal of Entomology Research. ISSN:2455-4758

Abera M, Jemaneh B, Tesfaye A (2023) Factors influencing white mango scale (*Aulacaspis tubercularis*) infestation status at Assosa and Bambasi districts, in Benishangul Gumuz region, Western Ethiopia. Agric Sci Dig 43:681–687

Abo-Shanab AS (2012) Suppression of white mango scale, *Aulacaspis tubercularis* (Hemiptera: Diaspididae) on mango trees in El-Beheira Governorate, Egypt. Egypt Acad J Biol Sci A Entomol 5:43–50

Abrahão J, Wegmuller O (1969) Criação do inseto *Hypocryphalus mangiferae* em culturas puras de *Ceratocystis fimbriata*. Anais da II reunião anual da Sociedade Brasileira de Fitopatologia

Affandi A, Velasco LRI, Javier PA, Depositario DPT (2017) Development and survivorship of *Scirtothrips dorsalis* Hood (Thysanoptera: Thripidae) in different growth stages of mango and selected weeds. AGRIVITA J Agric Sci 40:101–106

Agrell J, Anderson P, Oleszek W, Stochmal A, Agrell C (2006) Elevated CO_2 levels and herbivore damage alter host plant preferences. Oikos 112:63–72

Agrios GN (2005) Plant pathology, 5th edn. San Diego, Elsevier

Aguila R, Carlos L, Li X, Akutse KS, Bamisile BS, Sánchez Moreano JP, Lie Z, Liu J (2023) Host–parasitoid phenology, distribution, and biological control under climate change. Life 13:2290

Ahmed SS, Liu D, Simon J-C (2017) Impact of water-deficit stress on tritrophic interactions in a wheat-aphid-parasitoid system. PLOS One 12:e0186599

Aidoo OF, Amaro GC, Souza PGC, Picanço MC, Awuah-Mensah KA, Silva RSd (2024) Climate change impacts on worldwide ecological niche and invasive potential of *Sternochetus mangiferae*. Pest Manage Sci 81(2):667–677

Aitzaz M, Aatizaz M, Aatzaz G (2024) Climate change: threats to agricultural sustainability in Pakistan. Pak Soc Sci Rev 8:354–362

Al Adawi AO, Al Jabri R, Deadman M, Barnes I, Wingfield B, Wingfield MJ (2013) The mango sudden decline pathogen, *Ceratocystis manginecans*, is vectored by *Hypocryphalus mangiferae* (Coleoptera: Scolytinae) in Oman. Eur J Plant Pathol 135:243–251

Alemu K, Tsedaley B, Ejeta D (2024) Current status of white mango scale (*Aulacaspis tubercularis* Newstead) and associated factors in Assosa district western Ethiopia. Discov Appl Sci 6:485

Aliakbarpour H, Che SMR (2010) Diurnal activity of four species of thrips (Thysanoptera: Thripidae) and efficiencies of three nondestructive sampling techniques for thrips in mango inflorescences. J Econ Entomol 103:631–640

Aliakbarpour H, Rawi CSM (2012) The species composition of thrips (Insecta: Thysanoptera) inhabiting mango orchards in Pulau Pinang, Malaysia. Trop Life Sci Res 23:45

Allen CD, Breshears DD, McDowell NG (2015) On underestimation of global vulnerability to tree mortality and forest die-off from hotter drought in the Anthropocene. Ecosphere 6:1–55

Allwood A, Chinajariyawong A, Kritsaneepaiboon S, Drew R, Hamacek E, Hancock D, Hengsawad C, Jipanin J, Jirasurat M, Krong CK (1999) Host plant records for fruit flies (Diptera: Tephritidae) in Southeast Asia

Aluja M, Norrbom A (1999) Fruit flies (Tephritidae): phylogeny and evolution of behavior. CRC Press

Andrewartha HG, Birch LC (1986) The ecological web: more on the distribution and abundance of animals. University of Chicago Press

Anwar R, Ahmad S, Rajwana IA, Khan AS, Memon N, Nafees M (2011) Phenological growth patterns and floral malformation of mango (*Mangifera indica* L.) tree under subtropical climate. Pak J Agric Sci 48:107–113

Aquino NI, Martínez OG, Canales RJF, Díaz M, Heredia CR, Álvarez RE (2011) El cultivo de mango en Nayarit, acciones e impactos en materia fitosanitaria 1993–2010. Revista Fuente Año 2

Arteaga A, Lorenzo C, Prendes C (2003) *Aulacaspis tubercularis* Newstead (Coccoidea, Diaspididae); nueva plaga en los cultivos de mango (*Mangifera indica* L.) en la isla de La Gomera, Canarias. Phytoma España 151:40–47

Asare-Nuamah P, Antwi-Agyei P, Dick-Sagoe C (2022) Mitigating the risks of climate variability and change on mango seedlings in Ghana: evidence from mango seedlings producers in the Yilo Krobo Municipality. Environ Chall 8:100594

Asplanato G, García-Marí F (2001) *Aonidiella aurantii* (Homoptera: Diaspididae): Piojo rojo de California, cochinilla roja californiana, "poll roig" de California. Universidad Politécnica de Valencia, Valencia, España. In. Universidad Politécnica de Valencia Valencia, España

Atlin GN, Cairns JE, Das B (2017) Rapid breeding and varietal replacement are critical to adaptation of cropping systems in the developing world to climate change. Glob Food Sec 12:31–37

Awarikabey EN, Afun JVK, Osekre EA, Billah MK (2023) Mango phenology and fruit fly population dynamics in the transition zone of Ghana. Bull Entomol Res 113:169–179

Ayalew G, Fekadu A, Sisay B (2015) Appearance and chemical control of white mango scale (*Aulacaspis tubercularis*) in Central Rift Valley. Sci Technol Arts Res J 4:59–63

Azrag AG, Mohamed SA, Ndlela S, Ekesi S (2022) Predicting the habitat suitability of the invasive white mango scale, *Aulacaspis tubercularis*; Newstead, 1906 (Hemiptera: Diaspididae) using bioclimatic variables. Pest Manage Sci 78:4114–4126

Babar TK (2023) Risks of deserts locust and its mitigation. In: Mukhtar A, Shakeel A (eds) Disaster risk reduction in agriculture, 1st Edition. Springer, Singapore, pp 361–392

Babu L, Maheswari T, Rao N (2002) Seasonal incidence and biology of the mango hopper *Amritodus atkinsoni* Lethier (Homoptera: Cicadellidae). Entomon 27(1):35–42

Baier P, Pennerstorfer J, Schopf A (2007) PHENIPS—a comprehensive phenology model of *Ips typographus* (L.)(Coleoptera, Scolytinae) as a tool for hazard rating of bark beetle infestation. For Ecol Manage 249:171–186

Bajpai Y, Soni SK, Bajpai A, Tiwari S, Rastogi L, Trivedi M (2023) Patterns of shifting phenostages and flowering time in mango (*Mangifera indica* L.) due to climate change. Preprints. https://doi.org/10.20944/preprints202306.1600.v1

Baker R, Cannon R, Walters K (1996) An assessment of the risks posed by selected non-indigenous pests to UK crops under climate change. Asp Appl Biol (United Kingdom)

Bale JS, Hayward SAL (2010) Insect overwintering in a changing climate. J Exp Biol 213:980–994

Bale JS, Masters GJ, Hodkinson ID, Awmack C, Bezemer TM, Brown VK, Butterfield J, Buse A, Coulson JC, Farrar J (2002) Herbivory in global climate change research: direct effects of rising temperature on insect herbivores. Global Change Biol 8:1–16

Bana J, Sharma H, Kumar S, Singh P (2017) Impact of weather parameters on population dynamics of oriental fruit fly, *Bactrocera dorsalis* (Hendel) (Diptera: Tephritidae) under south Gujarat mango ecosystem. J Agrometeorol 19:78–80

Baradevanal G, Chander S, Singh HS, Reddy DS, Rajan S (2023) Mapping the risk of quarantine pest *Sternochetus mangiferae* under different climate change scenarios through species distribution modelling. Int J Trop Insect Sci 43:919–932

Barbosa F, Paranhos B (2005) Artrópodes-pragas associados à cultura da mangueira no Brasil e seu controle

Bautista-Rosales P, Ragazzo-Sánchez J, Calderón-Santoyo M, Cortéz-Mondaca E, Servín-Villegas R (2013) *Aulacaspis tubercularis* Newstead in mango orchards of Nayarit, Mexico, and relationship with environmental and agronomic factors. Southwest Entomol 38:221–230

Bellard C, Leroy B, Thuiller W, Rysman JF, Courchamp F (2016) Major drivers of invasion risks throughout the world. Ecosphere 7:e01241

Bezemer TM, Jones TH, Knight KJ (1998) Long-term effects of elevated CO_2 and temperature on populations of the peach potato aphid *Myzus persicae* and its parasitoid *Aphidius matricariae*. Oecologia 116:128–135

Bhattacharjee P, Warang O, Das S, Das S (2022) Impact of climate change on fruit crops- a review. Curr World Environ 17:319

Bhattacharyya B, Pujari D, Das M (2016) Mango weevils: identification, biology and integrated management. In: Insect pests management in fruit crops biotech books. Academic Press, Cambridge, MA, USA, pp 35–49

Biber-Freudenberger L, Ziemacki J, Tonnang HE, Borgemeister C (2016) Future risks of pest species under changing climatic conditions. PLOS One 11:e0153237

Blanc S, Michalakis Y (2016) Manipulation of hosts and vectors by plant viruses and impact of the environment. Curr Opin Insect Sci 16:36–43

Boudon F, Persello S, Jestin A, Briand A-S, Grechi I, Fernique P, Guédon Y, Léchaudel M, Lauri P-É, Normand F (2020) V-Mango: a functional–structural model of mango tree growth, development and fruit production. Ann Bot 126:745–763

Boyero JR, González J, Vela JM (2017) Plagas del mango en España. Phytoma España 287:23–28

Boyes DH, Evans DM, Fox R, Parsons MS, Pocock MJO (2021) Is light pollution driving moth population declines? A review of causal mechanisms across the life cycle. Insect Conserv Divers 14:167–187

Bradshaw CJ, Leroy B, Bellard C, Roiz D, Albert C, Fournier A, Barbet-Massin M, Salles J-M, Simard F, Courchamp F (2016) Massive yet grossly underestimated global costs of invasive insects. Nat Commun 7:12986

Bragard C, Caciagli P, Lemaire O, Lopez-Moya J, MacFarlane S, Peters D, Susi P, Torrance L (2013) Status and prospects of plant virus control through interference with vector transmission. Annu Rev Phytopathol 51:177–201

Braimah H, van Emden HF (2010) Prospects and challenges for sustainable management of the mango stone weevil, *Sternochetus mangiferae* (F.) (Coleoptera: Curculionidae) in West Africa: a review. Int J Pest Manage 56:91–101

Brochu A-S, Dionne A, Fall ML, Pérez-López E (2023) A decade of hidden phytoplasmas unveiled through citizen science. Plant Dis 107:3389–3393

Bumroongsook S, Kilaso M (2018) Modified atmosphere for thrip disinsection on cut lotus flowers. Appl Ecol Environ Res 16(4):5237–5247

Bura S, Jasrotia A, Sharma S, Sharma A, Tutlani A (2023) Recent advances in breeding of mango (*Mangifera indica*): a review. Int J Environ Clim Chang 13:521–538

Bush E, Lemmen DS (2019) Canada's changing climate report. Government of Canada= Gouvernement du Canada

Butter NS (2018) Insect vectors and plant pathogens, 1st edn. CRC Press, Boca Raton

Cai P, Song Y, Huo D, Lin J, Zhang H, Zhang Z, Xiao C, Huang F, Ji Q (2020) Chemical cues induced from fly-oviposition mediate the host-seeking behaviour of *Fopius arisanus* (Hymenoptera: Braconidae), an effective egg parasitoid of *Bactrocera dorsalis* (Diptera: Tephritidae), within a tritrophic context. Insects 11:231

Cannon RJC (1998) The implications of predicted climate change for insect pests in the UK, with emphasis on non-indigenous species. Global Change Biol 4:785–796

Canto T, Aranda MA, Fereeres A (2009) Climate change effects on physiology and population processes of hosts and vectors that influence the spread of hemipteran-borne plant viruses. Global Change Biol 15:1884–1894

Cao R, Feng J (2024) Future Climate change and anthropogenic disturbance promote the invasions of the world's worst invasive insect pests. Insects 15:280

Carter W (1939) Populations of *Thrips tabaci*, with special reference to virus transmission. J Anim Ecol 8(2):261–276

Caulfield F, Bunce JA (1994) Elevated atmospheric carbon dioxide concentration affects interactions between *Spodoptera exigua* (Lepidoptera: Noctuidae) larvae and two host plant species outdoors. Environ Entomol 23:999–1005

Celedon JM, Bohlmann J (2019) Oleoresin defenses in conifers: chemical diversity, terpene synthases and limitations of oleoresin defense under climate change. New Phytol 224:1444–1463

Chen Y, Ma C (2010) Effect of global warming on insect: a literature review. Acta Ecol Sin 30:2159–2172

Chidawanyika F, Mudavanhu P, Nyamukondiwa C (2019) Global climate change as a driver of bottom-up and top-down factors in agricultural landscapes and the fate of host-parasitoid interactions. Front Ecol Evol 7

Chmielewski F-M, Müller A, Bruns E (2004) Climate changes and trends in phenology of fruit trees and field crops in Germany, 1961–2000. Agric for Meteorol 121:69–78

Chowdhury SK (2015) Diversity and nature of damage of mango insect pests at Kaliachak-II Block of Malda, West Bengal, India. J Entomol Zool Stud 3:307–311

Clarke A, Fraser K (2004) Why does metabolism scale with temperature? Funct Ecol 18:243–251

Clavero M, García-Berthou E (2005) Invasive species are a leading cause of animal extinctions. Trends Ecol Evol 20(3):1–110

Cloudsley-Thompson J (1968) Hot blood or cold? Thermoregulation in terrestrial poikilotherms. Sci Prog (1933–), 499–509

Cooke BJ, Carroll AL (2017) Predicting the risk of mountain pine beetle spread to eastern pine forests: considering uncertainty in uncertain times. For Ecol Manage 396:11–25

Cooper B (1991) Status of *Thrips palmi* (Karny) in Trinidad FAO Plant Prot Bull 39 (1): 45–46

Cornelissen T (2011) Climate change and its effects on terrestrial insects and herbivory patterns. Neotrop Entomol 40

Cotrufo MF, Ineson P, Scott A (1998) Elevated CO_2 reduces the nitrogen concentration of plant tissues. Global Change Biol 4:43–54

da Silva AC, de Souza AP, Leonel S, de Souza ME, Ramos DP, Tanaka AA (2014) Growth and flowering of five mango cultivar under subtropics conditions of Brazil. Am J Plant Sci 5(3):393–402

da Silva JMM, Ramos RS, Souza PGC, da Silva PJ, Picanço MC, Silva GA, da Silva RS (2023) Mapping Brazilian expansion risk levels of mango weevil (*Sternochetus mangiferae* Fabricius) based on MaxEnt. Neotrop Entomol 52:760–771

Dako OD (2023) White mango scale (*Aulacaspis tubercularis*) infestation on mango: distinct scenarios of the pest status in Ethiopia and Kenya. Heliyon 9

Das D, Singh J, Vennila S (2011) Emerging crop pest scenario under the impact of climate change–a brief review. J Agric Phys 11:13–20

Datta S (2013) Impact of climate change in Indian horticulture-a review. Int J Sci Environ Technol 2:661–671

Dave K, Kumar A, Dave N, Jain M, Dhanda PS, Yadav A, Kaushik P (2024) Climate change impacts on legume physiology and ecosystem dynamics: a multifaceted perspective. Sustainability 16:6026

Davenport TL (2007) Reproductive physiology of mango. Braz J Plant Physiol 19:363–376

De Meyer M, Robertson MP, Mansell MW, Ekesi S, Tsuruta K, Mwaiko W, Vayssières J-F, Peterson AT (2010) Ecological niche and potential geographic distribution of the invasive fruit fly *Bactrocera invadens* (Diptera, Tephritidae). Bull Entomol Res 100:35–48

De Villiers M, Pringle KL (2007) Seasonal occurrence of vine pests in commercially treated vineyards in the Hex River Valley in the Western Cape Province, South Africa. Afr Entomol 15(241–260):220

Deguine J-P, Jacquot M, Allibert A, Chiroleu F, Graindorge R, Laurent P, Lambert G, Albon B, Marquier M, Gloanec C (2018) Agroecological protection of mango orchards in La Réunion. Sustain Agric Rev 28 Ecol Agric, 249–307

del Pino M, Bienvenido C, Boyero JR, Vela JM (2020) Biology, ecology and integrated pest management of the white mango scale, *Aulacaspis tubercularis* Newstead, a new pest in southern Spain—a review. Crop Protect 133:105160

Delgado PH, Aranguren M, Reig C, Galvan DF, Mesejo C, Fuentes AM, Sauco VG, Agustí M (2011) Phenological growth stages of mango (*Mangifera indica* L.) according to the BBCH scale. Sci Hortic 130:536–540

Delrio G, Cocco A (2010) The peach fruit fly, *Bactrocera zonata*: a major threat for Mediterranean fruit crops? In: XXVIII International Horticultural Congress on Science and Horticulture for people (IHC 2010): International Symposium on the 940, pp 557–566

DeLucia EH, Casteel CL, Nabity PD, O'Neill BF (2008) Insects take a bigger bite out of plants in a warmer, higher carbon dioxide world. Proc Natl Acad Sci 105:1781–1782

Deschepper P, Todd TN, Virgilio M, De Meyer M, Barr NB, Ruiz-Arce R (2021) Looking at the big picture: worldwide population structure and range expansion of the cosmopolitan pest *Ceratitis capitata* (Diptera, Tephritidae). Biol Invasions 23:3529–3543

Diez JM, D'Antonio CM, Dukes JS, Grosholz ED, Olden JD, Sorte CJ, Blumenthal DM, Bradley BA, Early R, Ibáñez I (2012) Will extreme climatic events facilitate biological invasions? Front Ecol Environ 10:249–257

Diffenbaugh NS, Krupke CH, White MA, Alexander CE (2008) Global warming presents new challenges for maize pest management. Environ Res Lett 3:044007

Dong Y, Lucky A, Hulcr J (2022) *Cryphalus lipingensis* Tsai & Li, 1963: EENY-788/IN1362, 06/2022. EDIS

Dowle EJ, Bracewell RR, Pfrender ME, Mock KE, Bentz BJ, Ragland GJ (2017) Reproductive isolation and environmental adaptation shape the phylogeography of mountain pine beetle (Dendroctonus ponderosae). Mol Ecol 26:6071–6084

Drew R (1976) New exotic fruit fly introductions to Australia. Qld J Agric 102
Drew R, Hancock D (2022) Biogeography, speciation and taxonomy within the genus *Bactrocera* Macquart with application to the *Bactrocera dorsalis* (Hendel) complex of fruit flies (Diptera: Tephritidae: Dacinae). Zootaxa 5190:333–360
Duyck PF, Sterlin JF, Quilici S (2004) Survival and development of different life stages of *Bactrocera zonata* (Diptera: Tephritidae) reared at five constant temperatures compared to other fruit fly species. Bull Entomol Res 94:89–93
EFSA Panel on Plant Health, Bragard C, Baptista P, Chatzivassiliou E, Di Serio F, Gonthier P, Jaques Miret JA, Justesen AF, Magnusson CS, Milonas P, Navas-Cortes JA, Parnell S, Potting R, Reignault PL, Stefani E, Thulke H-H, Van der Werf W, Civera AV, Yuen J, Zappalà L, Grégoire J-C, Malumphy C, Antonatos S, Kertesz V, Maiorano A, Papachristos D, MacLeod A (2022) Pest categorisation of *Aulacaspis tubercularis*. EFSA J 20:e07307
EFSA Panel on Plant Health, Bragard C, Dehnen-Schmutz K, Di Serio F, Gonthier P, Jacques MA, Jaques Miret JA, Justesen AF, Magnusson CS, Milonas P (2018) Pest categorisation of *Sternochetus mangiferae*. EFSA J 16:e05439
El-Metwally MM, Moussa SFM, Ghanim NM (2011) Studies on the population fluctuations and distribution of the white mango scale insect, *Aulacaspis tubercularis* Newstead within the canopy of the mango trees in eastern of Delta region at the north of Egypt. Egypt Acad J Biol Sci A Entomol 4:123–130
El Keroumi A, Naamani K, Dahbi A, Luque I, Carvajal A, Cerda X, Boulay R (2010) Effect of ant predation and abiotic factors on the mortality of medfly larvae, *Ceratitis capitata,* in the Argan forest of Western Morocco. Biocontrol Sci Technol 20:751–762
European Food Safety Authority (EFSA), Mertens J, Schenk M, Delbianco A, Graziosi I, Vos S (2021) Pest survey card on *Bactrocera zonata*. EFSA Support Publ 18:1999E
Evans EW, Carlile NR, Innes MB, Pitigala N (2013) Warm springs reduce parasitism of the cereal leaf beetle through phenological mismatch. J Appl Entomol 137:383–391
Faccoli M, Campo G, Perrotta G, Rassati D (2016) Two newly introduced tropical bark and ambrosia beetles (Coleoptera: Curculionidae, Scolytinae) damaging figs (*Ficus carica*) in southern Italy. Zootaxa 4138:189–194
Facey SL, Ellsworth DS, Staley JT, Wright DJ, Johnson SN (2014) Upsetting the order: how climate and atmospheric change affects herbivore–enemy interactions. Curr Opin Insect Sci 5:66–74
Fand BB, Kamble AL, Kumar M (2012) Will climate change pose serious threat to crop pest management: a critical review. Int J Sci Res Publ 2:1–14
FAO (2008) Climate-related transboundary pests and diseases. Technical background document from the expert consultation held on 25 to 27 February 2008. Rome, Italy
FAO (2022) Food and Agriculture Organization of the United Nations. In FAOSTAT database. Rome, Italy
FAO (2023) Food and Agriculture Organization of the United Nations. In: FAOSTAT database. Rome, Italy
Field CB, Barros VR (2014) Contribution of working group II to the fifth assessment report of the intergovernmental panel on climate change (IPCC). In: Climate change 2014: impacts, adaptation, and vulnerability, Part A: Global and sectoral aspects. Cambridge University Press, Cambridge, UK and New York, NY, USA
Field CB, Barros VR, Dokken DJ, Mach KJ, Mastrandrea MD, Bilir TE, Chatterjee M, Ebi KL, Estrada YO, Genova RCea (2014) Part A global and sectoral aspects; contribution of working group II to the fifth assessment report of the intergovernmental panel on climate change (IPCC). In: Field CB, Barros VR (eds) IPCC summary for policymakers. In climate change 2014: impacts, adaptation, and vulnerability. Cambridge University Press, New York, pp 1–32
Firlej A, Ricard M-P, Dieni A, Ménard É, Bourgeois G, Grenier P (2019) Adaptation des mesures phytosanitaires pour les ravageurs et maladies des cultures fruitières à l'égard des impacts des changements climatiques. In IRDA
Follett PA (2002) Mango seed weevil (Coleoptera: Curculionidae) and premature fruit drop in mangoes. J Econ Entomol 95:336–339

Follett PA, Gabbard Z (2000) Effect of mango weevil (Coleoptera: Curculionidae) damage on mango seed viability in Hawaii. J Econ Entomol 93:1237–1240

Frank SD (2020) Review of the direct and indirect effects of warming and drought on scale insect pests of forest systems. For Int J For Res 94:167–180

Fuhrer J (2003) Agroecosystem responses to combinations of elevated CO_2, ozone, and global climate change. Agric Ecosyst Environ 97:1–20

Galdino TVdS, Kumar S, Oliveira LS, Alfenas AC, Neven LG, Al-Sadi AM, Picanco MC (2016) Mapping global potential risk of mango sudden decline disease caused by *Ceratocystis fimbriata*. PLOS One 11:e0159450

García Morales M, Denno BD, Miller DR, Miller GL, Ben-Dov Y, Hardy NB (2016) ScaleNet: a literature-based model of scale insect biology and systematics. Database

Garrick TA, Liburd OE (2017) Impact of climate change on a key agricultural pest: thrips. In: Environmental sustainability and climate change adaptation strategies. IGI Global, pp 232–254

Gavrilov IA (2007) A catalog of chromosome numbers and genetic systems of scale insects (Homoptera: Coccinea) of the world. Isr J Entomol 37:1–45

Gaydos DA, Petrasova A, Cobb RC, Meentemeyer RK (2019) Forecasting and control of emerging infectious forest disease through participatory modelling. Philos Trans R Soc B Biol Sci 374:20180283

Gely C, Laurance SG, Stork NE (2020) How do herbivorous insects respond to drought stress in trees? Biol Rev 95:434–448

Gempe T, Beye M (2011) Function and evolution of sex determination mechanisms, genes and pathways in insects. BioEssays 33:52–60

Gilman SE, Urban MC, Tewksbury J, Gilchrist GW, Holt RD (2010) A framework for community interactions under climate change. Trends Ecol Evol 25:325–331

González-Tokman D, Villada-Bedoya S (2024) Physiological mechanisms of heat tolerance in insects. In: González-Tokman D, Dáttilo W (eds) Effects of climate change on insects: physiological, evolutionary, and ecological responses. Oxford University Press

González JMH, Sánchez EG, Massip JMF (2018) Introducción al cultivo del mango en el sur peninsular. Caja Rural de Granada

Gopu B, Balamohan T, Soman P, Jeyakumar P (2014) Canopy management in mango (*Mangifera indica* L.) cv. Alphonso with reference to flowering, yield and quality characters under ultra high density planting. J Appl Hortic 16

Grechi I, Preterre A-L, Caillat A, Chiroleu F, Ratnadass A (2021) Linking mango infestation by fruit flies to fruit maturity and fly pressure: a prerequisite to improve fruit fly damage management via harvest timing optimization. Crop Protect 146:105663

Griess VC, Acevedo R, Härtl F, Staupendahl K, Knoke T (2012) Does mixing tree species enhance stand resistance against natural hazards? A case study for spruce. For Ecol Manage 267:284–296

Grimaldi D, Shmakov A, Fraser N (2004) Mesozoic thrips and early evolution of the order Thysanoptera (Insecta). J Paleontol 78:941–952

Grove T, Giliomee J, Pringle K (2001) Thrips (Thysanoptera) species associated with mango trees in South Africa. Afr Entomol 9:153–162

Gullan PJ, Kosztarab M (1997) Adaptations in scale insects. Annu Rev Entomol 42:23–50

Gutiérrez D, Wilson RJ (2021) Intra- and interspecific variation in the responses of insect phenology to climate. J Anim Ecol 90:248–259

Gvozdenac S, Dedić B, Mikić S, Ovuka J, Miladinović D (2022) Impact of climate change on integrated pest management strategies. In: Benkeblia N (ed) Climate change and agriculture. Wiley, pp 311–372

Hamilton JG, Dermody O, Aldea M, Zangerl AR, Rogers A, Berenbaum MR, Delucia EH (2005) Anthropogenic changes in tropospheric composition increase susceptibility of soybean to insect herbivory. Environ Entomol 34:479–485

Hancock DL, Hamacek EL, Lloyd AC, Elson-Harris MM (2000) The distribution and host plants of fruit flies (Diptera: Tephritidae) in Australia. Queensland Department of Primary Industries

Haran J, Grove T, Van Noort S, Benoit L, Addison P (2019) Natural biocontrol of fruit flies in indigenous hosts: a perspective for population control in the agroecosystem. Biol Control 137:104022

Harvey JA, Tougeron K, Gols R, Heinen R, Abarca M, Abram PK, Basset Y, Berg M, Boggs C, Brodeur J (2023) Scientists' warning on climate change and insects. Ecol Monogr 93:e1553

Hill M, Dymock J (1989) Impact of climate change: agricultural/horticultural systems. DSIR Entomology Division Submission to the New Zealand Climate Change Program. Department of Scientific and Industrial Research, Auckland, New Zealand, p 16

Hlásny T, König L, Krokene P, Lindner M, Montagné-Huck C, Müller J, Qin H, Raffa KF, Schelhaas M-J, Svoboda M (2021) Bark beetle outbreaks in Europe: state of knowledge and ways forward for management. Curr For Rep 7:138–165

Holopainen JK, Virjamo V, Ghimire RP, Blande JD, Julkunen-Tiitto R, Kivimäenpää M (2018) Climate change effects on secondary compounds of forest trees in the Northern Hemisphere. Front Plant Sci 9

Horgan FG, Arida A, Ardestani G, Almazan MLP (2021) Elevated temperatures diminish the effects of a highly resistant rice variety on the brown planthopper. Sci Rep 11:262

Horticultural Policy Council (1991) The impact of fruit flies on Australian horticulture: report to the minister for Primary Industries and Energy. Australia, Canberra

Hou B-H, Ouyang G-C, Lu H-L, Ma J, Lu Y-Y, Xia Y (2018) First detection of *Bactrocera tsuneonis* (Diptera: Tephritidae) in Guangdong province of China. Fla Entomol 101:533–535

Huang J, Kautz M, Trowbridge AM, Hammerbacher A, Raffa KF, Adams HD, Goodsman DW, Xu C, Meddens AJ, Kandasamy D (2020) Tree defence and bark beetles in a drying world: carbon partitioning, functioning and modelling. New Phytol 225:26–36

Huang McBeath J, McBeath J, McBeath JH, McBeath J (2010) Invasive species and food security. Environ Chang Food Secur China, 157–176

Hulagappa T, Baradevanal G, Surpur S, Raghavendra D, Doddachowdappa S, Shashank PR, Mallaiah KK, Bedar J (2022) Diagnosis and potential invasion risk of *Thrips parvispinus* under current and future climate change scenarios. PeerJ 10:e13868

Inouye DW (2022) Climate change and phenology. Wiley Interdiscip Rev Clim Chang 13:e764

IPBES W (2019) Intergovernmental science-policy platform on biodiversity and ecosystem services. Summary for policy makers of the global assessment report on biodiversity and ecosystem services of the intergovernmental science-policy platform on biodiversity and ecosystem services. IPBES Secretariat, Bonn, Germany

IPCC (2021) Climate change 2021—the physical science basis: working group I contribution to the sixth assessment report of the intergovernmental panel on climate change (IPCC). Cambridge University Press, Cambridge, UK and New York, NY, USA

IUCN (2021) Invasive alien species and climate change. Accessed on 02 January 2025: https://iucn.org/resources/issues-brief/invasive-alien-species-and-climate-change

Jakoby O, Lischke H, Wermelinger B (2019) Climate change alters elevational phenology patterns of the European spruce bark beetle (*Ips typographus*). Global Change Biol 25:4048–4063

Jamieson MA, Trowbridge AM, Raffa KF, Lindroth RL (2012) Consequences of climate warming and altered precipitation patterns for plant-insect and multitrophic interactions. Plant Physiol 160:1719–1727

Janes JK, Li Y, Keeling CI, Yuen MMS, Boone CK, Cooke JEK, Bohlmann J, Huber DPW, Murray BW, Coltman DW, Sperling FAH (2014) How the mountain pine beetle (*Dendroctonus ponderosae*) breached the Canadian rocky mountains. Mol Biol Evol 31:1803–1815

Jayanthi PK, Verghese A (2014) Migratory behaviour of mango hoppers, *Idioscopus* spp. In relation to host plant flowering phenology: a synchronous shift. Bioscan 9:639–642

Jeffs CT, Lewis OT (2013) Effects of climate warming on host–parasitoid interactions. Ecol Entomol 38:209–218

Jha PK, Zhang N, Rijal JP, Parker LE, Ostoja S, Pathak TB (2024) Climate change impacts on insect pests for high value specialty crops in California. Sci Total Environ 906:167605

Johari A, Natalia D (2018) The abundance of *Thrips parvispinus* Karny (Thysanoptera: Thripidae) on various crops in Jambi region, Sumatera, Indonesia. J Entomol Res 42:237–244

Johnson AJ, Knížek M, Atkinson TH, Jordal BH, Ploetz RC, Hulcr J (2017) Resolution of a global mango and fig pest identity crisis. Insect Syst Divers 1:ixx010

Jones CM, Jones S, Petrasova A, Petras V, Gaydos D, Skrip MM, Takeuchi Y, Bigsby K, Meentemeyer RK (2021) Iteratively forecasting biological invasions with PoPS and a little help from our friends. Front Ecol Environ 19:411–418

Jönsson AM, Appelberg G, Harding S, Bärring L (2009) Spatio-temporal impact of climate change on the activity and voltinism of the spruce bark beetle, *Ips typographus*. Global Change Biol 15:486–499

Joshi P, Kumar S (2012) Effect of some meteorological factors on seasonal abundance of Idioscopus nitidulus (Walker)(Hemiptera: Cicadellidae) in mango orchards of Haridwar (India). New York Science Journal 5:101–103

Juroszek P, Von Tiedemann A (2013) Plant pathogens, insect pests and weeds in a changing global climate: a review of approaches, challenges, research gaps, key studies and concepts. J Agric Sci 151:163–188

Kaplan I, Carrillo J, Garvey M, Ode PJ (2016) Indirect plant–parasitoid interactions mediated by changes in herbivore physiology. Curr Opin Insect Sci 14:112–119

Karar H, Bashir MA, Basit A, Atta S, Anjum AA, Bakhsh A, Hussain A, Hameed A, Wang Y, Alajmi RA, Metwally DM, Imran M (2021) Effect of host plant on cornucopia of mango fruit flies (Diptera: Tephritidae) and their triumphant management in context of climate change. Saudi J Biol Sci 28:2366–2373

Kausrud K, Økland B, Skarpaas O, Grégoire JC, Erbilgin N, Stenseth NC (2012) Population dynamics in changing environments: the case of an eruptive forest pest species. Biol Rev 87:34–51

Kerkut G, Taylor B (1958) The effect of temperature changes on the activity of poikilotherms. Behaviour, 259–279

Kerr NZ, Wepprich T, Grevstad FS, Dopman EB, Chew FS, Crone EE (2020) Developmental trap or demographic bonanza? Opposing consequences of earlier phenology in a changing climate for a multivoltine butterfly. Global Change Biol 26:2014–2027

Khan MH, Khuhro NH, Awais M, Asif MU, Muhammad R (2021) Seasonal abundance of fruit fly, *Bactrocera* species (Diptera: Tephritidae) with respect to environmental factors in guava and mango orchards. Pak J Agric Res 34:266

Kocmánková E, Trnka M, Juroch J, Dubrovský M, Semerádová D, Možný M, Žalud Z (2009) Impact of climate change on the occurrence and activity of harmful organisms. Plant Prot Sci 45:S48–S52

Krishnamoorthy A, Visalakshi PG (2012) Record of thrips on mango. J Hortic Sci 7:110–111

Kriticos DJ, Maywald GF, Yonow T, Zurcher EJ, Herrmann NI, Sutherst R (2015) Exploring the effects of climate on plants, animals and diseases. Climex Version 4:184

Krokene P (2015) Conifer defense and resistance to bark beetles. In: Bark beetles. Elsevier, pp 177–207

Kumar R, Omkar, Shukla R (2010) Effect of temperature on the growth, development and reproduction of fruit fly, *Bactrocera dorsalis* Hendel (Diptera: Tephritidae) in mango JEFA 5(2): 50–153

Kumar R, Vishwakarma R (2018) Preharvest approaches to control insect infestation in fruit. In: Siddique MW (ed) Preharvest modulation of postharvest fruit and vegetable quality. Elsevier, pp 301–357

Kumar V, Seal DR, Kakkar G, McKenzie CL, Osborne LS (2012) New tropical fruit hosts of *Scirtothrips dorsalis* (Thysanoptera: Thripidae) and its relative abundance on them in south Florida. Fla Entomol 95:205–207

Kurppa S (1990) Insect pest damage, predicting and control in Finnish cereal cultivation during the 1980s. Agriculture Research Centre, Institute of Plant Protection, Jokioinen, Finland, p 53

Labuschagne T, Van Hamburg H, Froneman I (1995) Population dynamics of the mango scale, *Aulacaspis tubercularis* (Newstead)(Coccoidea: Diaspididae), in South Africa. Isr J Entomol, 29

Lacroix M, Follett P (2015) Combination irradiation treatments for food safety and phytosanitary uses. Stewart Postharvest Rev 11

Lashof DA, Ahuja DR (1990) Relative contributions of greenhouse gas emissions to global warming. Nature 344:529–531

Léchaudel M, Lopez-Lauri F, Vidal V, Sallanon H, Joas J (2013) Response of the physiological parameters of mango fruit (transpiration, water relations and antioxidant system) to its light and temperature environment. J Plant Physiol 170:567–576

Legave J-M, Normand F, Lauri P-E (2013) Climate change and its probable effects on mango production and cultivation. X Int Mango Symp 1075:21–31

Lehmann P, Ammunét T, Barton M, Battisti A, Eigenbrode SD, Jepsen JU, Kalinkat G, Neuvonen S, Niemelä P, Terblanche JS, Økland B, Björkman C (2020) Complex responses of global insect pests to climate warming. Front Ecol Environ 18:141–150

Leng L, Norawanis M, Musah A, Shahlehi S (2023) Brief review: Climate change and its impact on mango pests and diseases. J Agric Crop, 391–399

Li G, Du S, Wen Z (2016) Mapping the climatic suitable habitat of oriental arborvitae (*Platycladus orientalis*) for introduction and cultivation at a global scale. Sci Rep 6:30009

Liliane TN, Mutengwa SC (2020) Factors affecting yield of crops. In: Amanullah (ed) Agronomy. IntechOpen, Rijeka, p Ch. 2

Lin C-N, Wei M-Y, Chang N-T, Chuang Y-Y (2015) The occurrence of *Scirtothrips dorsalis* Hood in mango orchards and factors influencing its population dynamics in Taiwan. J Asia-Pacif Entomol 18:361–367

Lincoln D (1993) The influence of plant carbon dioxide and nutrient supply on susceptibility to insect herbivores. Vegetatio 104:273–280

Lincoln D, Sionit N, Strain B (1984) Growth and feeding response of *Pseudoplusia includens* (Lepidoptera: Noctuidae) to host plants grown in controlled carbon dioxide atmospheres. Environ Entomol 13:1527–1530

Linnakoski R, Forbes KM (2019) Pathogens- the hidden face of forest invasions by wood-boring insect pests. Front Plant Sci, 10

Liquido NJ, McQuate GT, Suiter KA, Norrbom AL, Yee WL, Chang CL (2019) Compendium of fruit fly host plant information: the USDA primary reference in establishing fruit fly regulated host plants. In: Area-wide management of fruit fly pests. CRC Press, pp 363–368

Liu X, Liu H, Wang Y, Qian L, Chen F (2022) Elevated CO_2 aggravates invasive thrip damage by altering its host plant nutrient and secondary metabolism. Environ Pollut 295:118736

Lopes D, de Andrade E, Egartner A, Beitia F, Rot M, Chireceanu C, Balmés V, Loomans A, Konefal T, Radonjić S (2023) FRUITFLYRISKMANAGE: a Euphresco project for *Ceratitis capitata* Wiedemann (Diptera: Tephritidae) risk management applied in some European countries. EPPO Bulletin 53:354–371

López-Guillén G, Noriega Cantú D, Urías López MA (2017) Geographic distribution and damage estimate of white scale of mango in Chiapas, Mexico. Revista Mexicana de Ciencias Agrícolas 8:1851–1866

Lu Z, Sun Z, Li Y, Hao R, Chen Y, Chen B, Qin X, Tao X, Gui F (2022) Effects of elevated CO(2) concentration on host adaptability and chlorantraniliprole susceptibility in *Spodoptera frugiperda*. Insects 13

Lv C, Wang H, Wang J, Ma X, Xia C (2019) The influence of elevated CO_2 on *Vallisneria spiralis*, R*adix auricularia*, and their plant–herbivore interaction. Aquat Ecol 53:137–150

Ma C-S, Ma G, Pincebourde S (2021) Survive a warming climate: insect responses to extreme high temperatures. Annu Rev Entomol 66:163–184

Machin K, Pringle JWS, Tamasige M (1962) The physiology of insect fibrillar muscle-IV: the effect of temperature on a beetle flight muscle. Proc R Soc Lond Ser B Biol Sci 155:493–499

Makhmale S, Bhutada P, Yadav L, Yadav B (2016) Impact of climate change on phenology of mango-the case study. Ecol Environ Conserv 22:S127–S132

Malik A, Hasan M (2019) Road testing the mango quality manual in Pakistan- a case study of enhancing mango fruit quality in Asian mango chains Technical report 2017–19 ACIAR Mango Agribusiness Research Program in Pakistan

Mandal P, Mondal F, Hossain M (2020) Factors influences selection and adaptation of aphid to their host plant. J Plant Sci Crop Prot

Manderscheid R, Erbs M, Burkart S, Wittich K-P, Löpmeier F-J, Weigel H-J (2016) Effects of free-air carbon dioxide enrichment on sap flow and canopy microclimate of maize grown under different water supply. J Agron Crop Sci 202:255–268

Mao J, Meng F, Song Y, Li D, Ji Q, Hong Y, Lin J, Cai P (2024) Forecasting the expansion of *Bactrocera tsuneonis* (Miyake)(Diptera: Tephritidae) in China using the MaxEnt model. Insects 15:417

Marincowitz S, Duong TA, Taerum SJ, De Beer ZW, Wingfield MJ (2020) Fungal associates of an invasive pine-infesting bark beetle, *Dendroctonus valens*, including seven new *Ophiostomatalean* fungi. PersIa Mol Phylogeny Evol Fungi 45:177–195

Marini L, Økland B, Jönsson AM, Bentz B, Carroll A, Forster B, Grégoire JC, Hurling R, Nageleisen LM, Netherer S (2017) Climate drivers of bark beetle outbreak dynamics in Norway spruce forests. Ecography 40:1426–1435

Martínez-Silvestre KE, Santiz-Gómez JA, Luján-Hidalgo MC, Ruiz-Lau N, Sánchez-Roque Y, Gutiérrez-Miceli FA (2022) Effect of UV-B radiation on flavonoids and phenols accumulation in tempisque (*Sideroxylon capiri* Pittier) callus. Plants 11:473

Masood A, Saeed S (2012) Bark beetle, *Hypocryphalus mangiferae* Stebbing (Coleoptera: Curculionidae: Scolytinae) is a vector of mango sudden death disease in Pakistan. Pak J Bot 44:813–820

Masood A, Saeed S, Erbilgin N, Jung Kwon Y (2010) Role of stressed mango host conditions in attraction of and colonization by the mango bark beetle *Hypocryphalus mangiferae* Stebbing (Coleoptera: Curculionidae: Scolytinae) and in the symptom development of quick decline of mango trees in Pakistan. Entomol Res 40:316–327

Masood A, Saeed S, Sajjad A (2008) Characterization and damage patterns of different bark beetle species associated with mango sudden death syndrome in Punjab, Pakistan. Pak Entomol 30:163–168

Matos STSd, Andrade DJd, Zanata RMP, Azevedo EBd, Lima ÉFB (2019) New records of thrips species on mango and natural enemies associated. Revista Brasileira de Fruticultura 41:e-123

Matthews B, Netherer S, Katzensteiner K, Pennerstorfer J, Blackwell E, Henschke P, Hietz P, Rosner S, Jansson P-E, Schume H, Schopf A (2018) Transpiration deficits increase host susceptibility to bark beetle attack: Experimental observations and practical outcomes for *Ips typographus* hazard assessment. Agric for Meteorol 263:69–89

Mauck K, Bosque-Pérez NA, Eigenbrode SD, De Moraes CM, Mescher MC (2012) Transmission mechanisms shape pathogen effects on host-vector interactions: evidence from plant viruses. Funct Ecol 26:1162–1175

Maw HEL (2000) Checklist of the Hemiptera of Canada and Alaska. NRC Research Press

Mayamba A, Nankinga CK, Isabirye B, Akol AM (2014) Seasonal population fluctuations of *Bactrocera* invadens (Diptera: Tephritidae) in relation to mango phenology in the Lake Victoria Crescent, Uganda. Fruits 69:473–480

Meddens AJH, Hicke JA, Ferguson CA (2012) Spatiotemporal patterns of observed bark beetle-caused tree mortality in British Columbia and the western United States. Ecol Appl 22:1876–1891

Menéndez R (2007) How are insects responding to global warming? Tijdschr Entomol 150:355

Mollot G, Pantel J, Romanuk T (2017) The effects of invasive species on the decline in species richness: a global meta-analysis. In: Advances in ecological research. Elsevier, pp 61–83

Moss RH, Edmonds JA, Hibbard KA, Manning MR, Rose SK, van Vuuren DP, Carter TR, Emori S, Kainuma M, Kram T, Meehl GA, Mitchell JFB, Nakicenovic N, Riahi K, Smith SJ, Stouffer

RJ, Thomson AM, Weyant JP, Wilbanks TJ (2010) The next generation of scenarios for climate change research and assessment. Nature 463:747–756
Mound AL, Collins WD (2000) A south east Asian pest species newly recorded from Europe: *Thrips parvispinus* (Thysanoptera: Thripidae), its confused identity and potential quarantine significance. EJE 97:197–200
Mouratidis A, Bastin S, Pomposo M, Marrero E, Goldarazena A, Hernández-Suárez E (2024) First report of *Scirtothrips dorsalis* Hood in the Canary Islands. EPPO Bull 54:61–63
MRS (2013) Codes of prcatice for mango farm management: Manual for the training of master trainers. In: Mango Research Station (MRS) S, Multan, Govt. of Punjab Agriculture Department (ed), Lahore, p 127
Murai T (2002) The pest and vector from the East: *Thrips palmi*. Marullo R and Mound LA (eds), pp 19–32
Mwungu CM, Muriithi B, Ngeno V, Affognon H, Githiomo C, Diiro G, Ekesi S (2020) Health and environmental effects of adopting an integrated fruit fly management strategy among mango farmers in Kenya. Afr J Agric Resour Econ 15:14–26
Nasution S (2024) The explosion of pests and diseases due to climate change. In: IOP conference series: earth and environmental science. IOP Publishing, p 012072
Nechols JR (2021) The potential impact of climate change on non-target risks from imported generalist natural enemies and on biological control. Biocontrol 66:37–44
Netherer S, Matthews B, Katzensteiner K, Blackwell E, Henschke P, Hietz P, Pennerstorfer J, Rosner S, Kikuta S, Schume H, Schopf A (2015) Do water-limiting conditions predispose Norway spruce to bark beetle attack? New Phytol 205:1128–1141
Netherer S, Panassiti B, Pennerstorfer J, Matthews B (2019) Acute drought Is an important driver of bark beetle infestation in Austrian Norway spruce tands. Front For Glob Chang 2
Nickle DA (2008) Commonly intercepted thrips at US ports-of-entry from Africa, Europe, and the Mediterranean. III. The genus Thrips Linnaeus, 1758 (Thysanoptera: Thripidae). Proc Entomol Soc Wash 110:165–185
Nickle DA (2009) Commonly intercepted thrips at US ports-of-entry from Africa, Europe, and the Mediterranean. IV. Miscellaneous thripine genera excluding *Frankliniella*, *Iridothrips*, and *Thrips* (Thysanoptera: Thripidae). Proc Entomol Soc Wash 111:215–238
Normand F, Lauri P (2018) É (2018) Advances in understanding mango tree growth and canopy development. Achieving sustainable cultivation of mangoes. Burleigh Dodds series in agricultural science. Burleigh Dodds Science Publishing, Cambridge, pp 87–120
Numata H, Shintani Y (2023) Diapause in univoltine and semivoltine life cycles. Annu Rev Entomol 68:257–276
Nur U (1980) Evolution of unusual chromosome systems in scale insects (Coccoidea: Homoptera). Insect Cytogenet, 97–117
Oerke EC (2006) Crop losses to pests. J Agric Sci 144:31–43
Ofgaa D, Emana G, Kahuthia-Gathu R (2018) Population dynamics of white mango scale, *Aulacaspis tubercularis* Newstead (Hemiptera: Diaspididae) in Western Ethiopia. Afr J Agric Res 13:1598–1605
Omar PW (1949) The nearctic leafhoppers (Homoptera: Cicadellidae) a generic classification and check list. Entomol Soc Wash 3:253
Overgaard J, MacMillan HA (2017) The integrative physiology of insect chill tolerance. Annu Rev Physiol 79:187–208
Pablo JH, Jorge VC, Guadalupe VP, Mora-Aguilera JA, Gabriel OC, Daniel TO, Elías HC, Iván RR, Víctor AGH (2014) Leaf penetration pattern of *Aulacaspis tubercularis* (Hemiptera: Diaspididae) stylet in Mango. Fla Entomol 97:100–107
Pachauri RK, Reisinger A (2007) Climate change 2007: synthesis report. Contribution of working groups I, II and III to the fourth assessment report of the intergovernmental panel on climate change (IPCC). IPCC, Geneva, Switzerland
Paini DR, Sheppard AW, Cook DC, De Barro PJ, Worner SP, Thomas MB (2016) Global threat to agriculture from invasive species. Proc Natl Acad Sci 113:7575–7579

Parmesan C (2006) Ecological and evolutionary responses to recent climate change. Annu Rev Ecol Evol Syst 37:637–669

Patel K, Saxena S, Patel K (2013) Fluctuation of fruit fly oriented damage in mango in re lation to major abiotic factors. HRS 2:197–201

Pathania M, Verma A, Singh M, Arora PK, Kaur N (2020) Influence of abiotic factors on the infestation dynamics of whitefly, *Bemisia tabaci* (Gennadius 1889) in cotton and its management strategies in North-Western India. Int J Trop Insect Sci 40:969–981

Patil P, Giraddi R, Goudreddy B (1988) Tree hopper infesting mango seedlings and its control. CABI 30(2):2–14

Pellizzari G, Porcelli F (2014) Alien scale insects (Hemiptera Coccoidea) in European and Mediterranean countries: the fate of new and old introductions. Phytoparasitica 42:713–721

Pena J, Mohyuddin A, Wysoki M (1998) A review of the pest management situation in mango agroecosystems. Phytoparasitica 26:129–148

Peña JE (2002) Integrated pest management and monitoring techniques for mango pests. VII Int Mango Symp 645:151–161

Peng R, Christian K (2004) The weaver ant, *Oecophylla smaragdina* (Hymenoptera: Formicidae), an effective biological control agent of the red-banded thrips, *Selenothrips rubrocinctus* (Thysanoptera: Thripidae) in mango crops in the Northern Territory of Australia. Int J Pest Manage 50:107–114

Peng W, Ma NL, Zhang D, Zhou Q, Yue X, Khoo SC, Yang H, Guan R, Chen H, Zhang X, Wang Y, Wei Z, Suo C, Peng Y, Yang Y, Lam SS, Sonne C (2020) A review of historical and recent locust outbreaks: links to global warming, food security and mitigation strategies. Environ Res 191:110046

Pereira RM (2013) Identificação e distribuição de coleobrocas em mangueiras com seca causada por *Ceratocystis fimbriata*

Pérez G, Vilà M, Gallardo B (2022) Potential impact of four invasive alien plants on the provision of ecosystem services in Europe under present and future climatic scenarios. Ecosyst Serv 56:101459

Plant Health Australia (2009) Economic assessment of the implementation of the proposed national fruit fly strategy: Part 1 (Plant Health, August). Australia, Canberra

Piao S, Liu Q, Chen A, Janssens IA, Fu Y, Dai J, Liu L, Lian X, Shen M, Zhu X (2019) Plant phenology and global climate change: current progresses and challenges. Global Change Biol 25:1922–1940

Plant Health Australia (2008) Draft national fruit fly strategy plant health Australia. In: DEAKIN, ACT, 2600

Plant Health Australia (2018) The Australian handbook for the identification of fruit flies. Australia, Canberra

Porter J, Parry M, Carter T (1991) The potential effects of climatic change on agricultural insect pests. Agric for Meteorol 57:221–240

Prakash A, Rao J, Mukherjee AK, Berliner J, Pokhare SS, Adak T, Munda S, Shashank P (2014) Climate change: impact on crop pests. Applied Zoologists Research Association (AZRA), Central rice research ..., Odisha, India

Purcell AH, Almeida RP (2005) Insects as vectors of disease agents. Encycl Plant Crop Sci 10:1–5

Rachana R, Roselin P, Amutha M, Sireesha K, Reddy GN (2022) Invasive pest, *Thrips parvispinus* (Karny)(Thysanoptera: Thripidae) a looming threat to Indian agriculture. Curr Sci 122:211–213

Raffa KF, Aukema BH, Bentz BJ, Carroll AL, Hicke JA, Turner MG, Romme WH (2008) Cross-scale drivers of natural disturbances prone to anthropogenic amplification: the dynamics of bark beetle eruptions. Bioscience 58:501–517

Raffa KF, Gregoire J-C, Lindgren BS (2015) Natural history and ecology of bark beetles. In: Bark beetles. Elsevier, pp 1–40

Rahman SMA (2007) Mango hopper: bioecology and management- a review. Agric Rev 28:49–55

Rajan S (2012) Phenological responses to temperature and rainfall: a case study of mango. Trop Fruit Tree Species Clim Chang, 71

Raju C, Pazhanivelan S, Perianadar IV, Kaliaperumal R, Sathyamoorthy N, Sendhilvel V (2024) Climate change as an existential threat to tropical fruit crop production—a review. Agriculture 14:2018

Ramírez F, Davenport TL, Fischer G, Pinzón JCA, Ulrichs C (2014) Mango trees have no distinct phenology: the case of mangoes in the tropics. Sci Hortic 168:258–266

Rashmi M, Verghese A, Shivashankar S, Reddy PR, Kandakoor S (2020) Chemical defense and herbivory: a case study of phenolics versus *Bactrocera dorsalis* (Hendel) infestation in mango. Pest Manag Hortic Ecosyst 26

Ratnayake R (2014) Why plant species become invasive? Characters related to successful biological invasion. In: Proceedings of national symposium on invasive alien species, pp 22–41

Rauf I, Ahmad N, Rashdi S, Ismail M, Khan MH (2013) Laboratory studies on ovipositional preference of the peach fruit fly *Bactrocera zonata* (Saunders) (Diptera: Tephiritidae) for different host fruits. Afr J Agric Res 8:1300–1303

Raut P, Desai V, Narangalkar A, Naik KV, Mehendale S, Karmarkar M (2018) Effect of weather parameters on mango hoppers population. J Entomol Zool Stud 6:112–114

Raza ST, Khan AH, Hameed A, Muhammad N, Grewal AG, Malik MT, Imran M, Mustafa G, Iqbal A (2023) A review on White Mango Scale biology, ecology, distribution and management. Agriculture 13:1770

Reddy PP (2016) Thrips and their management. In: Reddy PP (ed) Sustainable crop protection under protected cultivation. Springer Singapore, Singapore, pp 213–220

Reddy VR, Poluru, Gundappa B, Chakravarthy AK (2018) Pests of Mango. In: Omkar (ed) Pests and their management. Springer Singapore, Singapore, pp 415–440

Renault D, Angulo E, Cuthbert RN, Haubrock PJ, Capinha C, Bang A, Kramer AM, Courchamp F (2022) The magnitude, diversity, and distribution of the economic costs of invasive terrestrial invertebrates worldwide. Sci Total Environ 835:155391

Renner SS, Zohner CM (2018) Climate change and phenological mismatch in trophic interactions among plants, insects, and vertebrates. Annu Rev Ecol Evol Syst 49:165–182

Rezaei EE, Webber H, Asseng S, Boote K, Durand JL, Ewert F, Martre P, MacCarthy DS (2023) Climate change impacts on crop yields. Nat Rev Earth Environ 4:831–846

Ricalde MP, Nava DE, Loeck AE, Donatti MG (2012) Temperature-dependent development and survival of Brazilian populations of the Mediterranean fruit fly, *Ceratitis capitata*, from tropical, subtropical and temperate regions. J Insect Sci 12:33

Robbins ZJ, Xu C, Aukema BH, Buotte PC, Chitra-Tarak R, Fettig CJ, Goulden ML, Goodsman DW, Hall AD, Koven CD, Kueppers LM, Madakumbura GD, Mortenson LA, Powell JA, Scheller RM (2022) Warming increased bark beetle-induced tree mortality by 30% during an extreme drought in California. Glob Change Biol 28:509–523

Robinet C, Roques A (2010) Direct impacts of recent climate warming on insect populations. Integr Zool 5:132–142

Rodríguez-Becerra SH, Vázquez-Rivera R, Ventura-Hernández KI, Pawar TJ, Olivares-Romero JL (2024) The biology, impact, and management of *Xyleborus* beetles: a comprehensive review. Insects 15:706

Roge GN (2021) White mango scale: a threat to mango production in Ethiopia. Am J Entomol 5:47–50

Roos J, Hopkins R, Kvarnheden A, Dixelius C (2011) The impact of global warming on plant diseases and insect vectors in Sweden. Eur J Plant Pathol 129:9–19

Rosenzweig C (1989) Global climate change: predictions and observations. Am J Agric Econ 71:1265–1271

Rosenzweig C, Iglesias A, Yang X, Epstein PR, Chivian E (2000) Climate change and US agriculture: the impacts of warming and extreme weather events on productivity, plant diseases, and pests. Center for Health and the Global Environment, Harvard Medical School, Boston, MA, USA, pp 13–14

Rossetto C, Ribeiro I (1990) Seca da mangueira. 12: recomendacoes de controle. Revista de Agricultura 65

Roth SK, Lindroth RL (1995) Elevated atmospheric CO_2: effects on phytochemistry, insect performance and insect-parasitoid interactions. Global Change Biol 1:173–182

Royer JE, Hancock DL (2012) New distribution and lure records of Dacinae (Diptera: Tephritidae) from Queensland, Australia, and description of a new species of *Dacus* Fabricius. Aust J Entomol 51:239–247

Royer JE, Wright CL, Hancock DL (2016) *Bactrocera frauenfeldi* (Diptera: Tephritidae), an invasive fruit fly in Australia that may have reached the extent of its spread due to environmental variables. Austral Entomology 55:100–111

Sahar Abdou Z (2019) Impact of climate change on plant diseases and IPM strategies. In: Snježana T-P (ed) Plant diseases. IntechOpen, Rijeka, p Ch. 6

Sartiami D, Mound LA (2013) Identification of the *Terebrantian thrips* (Insecta, Thysanoptera) associated with cultivated plants in Java, Indonesia. ZooKeys, 1

Saunders D, Lewis R, Warman G (2004) Photoperiodic induction of diapause: opening the black box. Physiol Entomol 29:1–15

Schebeck M, Schopf A, Ragland GJ, Stauffer C, Biedermann PHW (2023) Evolutionary ecology of the bark beetles *Ips typographus* and *Pityogenes chalcographus*. Bull Entomol Res 113:1–10

Sconiers WB, Eubanks MD (2017) Not all droughts are created equal? The effects of stress severity on insect herbivore abundance. Arthropod Plant Interact 11:45–60

Seidl R, Müller J, Hothorn T, Bässler C, Heurich M, Kautz M (2016) Small beetle, large-scale drivers: how regional and landscape factors affect outbreaks of the European spruce bark beetle. J Appl Ecol 53:530–540

Seidl R, Rammer W, Lexer MJ (2011) Climate change vulnerability of sustainable forest management in the Eastern Alps. Clim Change 106:225–254

Seidl R, Schelhaas MJ, Rammer W, Verkerk PJ (2014) Increasing forest disturbances in Europe and their impact on carbon storage. Nat Clim Chang 4:806–810

Seki M, Murai T (2012) Insecticidal effect of high carbon dioxide atmospheres on thrips eggs oviposited in plant tissue. Appl Entomol Zool 47:433–436

Selvaraj S, Ganeshamoorthi P, Pandiaraj T (2013) Potential impacts of recent climate change on biological control agents in agro-ecosystem: a review. Int J Biodivers Conserv 5:845–852

Senf C, Pflugmacher D, Zhiqiang Y, Sebald J, Knorn J, Neumann M, Hostert P, Seidl R (2018) Canopy mortality has doubled in Europe's temperate forests over the last three decades. Nat Commun 9:4978

Senf C, Seidl R (2018) Natural disturbances are spatially diverse but temporally synchronized across temperate forest landscapes in Europe. Global Change Biol 24:1201–1211

Senthil-Nathan S (2021) Effects of elevated CO2 on resistant and susceptible rice cultivar and its primary host, brown planthopper (BPH), *Nilaparvata lugens* (Stål). Sci Rep 11:8905

Shah IH, Manzoor MA, Jinhui W, Li X, Hameed MK, Rehaman A, Li P, Zhang Y, Niu Q, Chang L (2024) Comprehensive review: Effects of climate change and greenhouse gases emission relevance to environmental stress on horticultural crops and management. J Environ Manage 351:119978

Sharma HC, War AR, Pathania M, Sharma SP, Akbar SM, Munghate RS (2016) Elevated CO_2 influences host plant defense response in chickpea against Helicoverpa armigera. Arthropod Plant Interact 10:171–181

Shelly T, Epsky N, Jang EB, Reyes-Flores J, Vargas R (2014) Trapping and the detection, control, and regulation of tephritid fruit flies. Springer Dordrecht Heidelberg. New York 10:978–994

Shrestha S (2019) Effects of climate change in agricultural insect pest. Acta Sci Agric 3:74–80

Silva AC, Ricalde MP, Scalzer RRC, Zilli JE, Lopes RB (2022) Natural occurrence of Beauveria bassiana on adults of the invasive mango seed weevil Sternochetus mangiferae (Coleoptera: Curculionidae) in Brazil. J Plant Dis Prot 129:79–84

Singh NK (2016) Origin, diversity and genome sequence of mango (Mangifera indica L). Indian J Hist Sci 51:355

Singh V, Kumar Soni M, Singh K (2011) Increasing profitability of mango (*Mangifera indica* L.) orchard through intercropping. Glob Conf Augment Prod Util Mango Biot Abiotic Stress 1066:151–157

Singh VV, Naseer A, Mogilicherla K, Trubin A, Zabihi K, Roy A, Jakuš R, Erbilgin N (2024) Understanding bark beetle outbreaks: exploring the impact of changing temperature regimes, droughts, forest structure, and prospects for future forest pest management. Rev Environ Sci Bio/Technol, 1–34

Skendžić S, Zovko M, Živković IP, Lešić V, Lemić D (2021) The impact of climate change on agricultural insect pests. Insects 12:440

Smith J, Tirpak D (1988) The potential effects of global climate change on the United States. In: Technical report. Environmental Protection Agency (Report to Congress)

Soderberg DN, Mock KE, Hofstetter RW, Bentz BJ (2021) Translocation experiment reveals capacity for mountain pine beetle persistence under climate warming. Ecol Monogr 91:e01437

Staley JT, Hodgson CJ, Mortimer SR, Morecroft MD, Masters GJ, Brown VK, Taylor ME (2007) Effects of summer rainfall manipulations on the abundance and vertical distribution of herbivorous soil macro-invertebrates. Eur J Soil Biol 43:189–198

Stiller M (2009) Leafhoppers associated with grasslands of South Africa-grassland biome endemics. Grassroots 9:13–15

Stonehouse J, Mahmood R, Poswal A, Mumford J, Baloch KN, Chaudhary ZM, Makhdum AH, Mustafa G, Huggett D (2002) Farm field assessments of fruit flies (Diptera: Tephritidae) in Pakistan: distribution, damage and control. Crop Protect 21:661–669

Stuart RR, Gao Y-l, Lei Z-r (2011) Thrips: pests of concern to China and the United States. Agric Sci China 10:867–892

Subedi B, Poudel A, Aryal S (2023) The impact of climate change on insect pest biology and ecology: implications for pest management strategies, crop production, and food security. J Agric Food Res 14:100733

Sultana S, Baumgartner JB, Dominiak BC, Royer JE, Beaumont LJ (2020) Impacts of climate change on high priority fruit fly species in Australia. PLOS One 15:e0213820

Sun Y, Guo H, Ge F (2016) Plant–aphid interactions under elevated CO_2: some cues from aphid feeding behavior. Front Plant Sci 7:502

Swain S, Dora D, Sahoo S, Padhi S, Sanyal D (2012) Influence of mango-based intercropping systems on improvement of soil health under rainfed situation. Commun Soil Sci Plant Anal 43:2018–2026

Syazwan SA, Mohd-Farid A, Lee SY, Mohamed R (2024) Comparative analysis of mitochondrial genomes in *Ceratocystis fimbriata* complex across diverse hosts. Gene 921:148539

Szyniszewska AM, Akrivou A, Björklund N, Boberg J, Bradshaw C, Damus M, Gardi C, Hanea A, Kriticos J, Maggini R, Musolin DL, MacLeod A (2024) Beyond the present: how climate change is relevant to pest risk analysis. EPPO Bull 54:20–37

Talpur MA, Khuhro R, Nizamani I (2002) Phenological relationship between mango hoppers *Idioscopus* spp. and mango inflorescence/fruit. Pak J Appl Sci 2:533–536

Tharanathan RN, Yashoda HM, Prabha TN (2006) Mango (*Mangifera indica* L.), "The king of fruits"- an overview. Food Rev Int 22:95–123

Thomson LJ, Macfadyen S, Hoffmann AA (2010) Predicting the effects of climate change on natural enemies of agricultural pests. Biol Control 52:296–306

Tougeron K, Brodeur J, Le Lann C, van Baaren J (2020) How climate change affects the seasonal ecology of insect parasitoids? Ecol Entomol 45:167–181

Trenberth KE, Fasullo JT, Shepherd TG (2015) Attribution of climate extreme events. Nat Clim Chang 5:725–730

Tripathi P, Shah S, Kashyap S, Tripathi A (2019) Fruit yield and quality characteristics of high density *Prunus persica* (L.) Batsch plantation intercropped with medicinal and aromatic plants in the Indian Western Himalayas. Agrofor Syst 93:1717–1728

Tubiello F, Rosenzweig C, Goldberg R, Jagtap S, Jones J (2002) Effects of climate change on US crop production: simulation results using two different GCM scenarios. Part I: wheat, potato, maize, and citrus. Clim Res 20:259–270

Tucker GS (2016) Food preservation and biodeterioration. Wiley, India

Tyagi K, Kumar V, Singha D, Chakraborty R (2015) Morphological and DNA barcoding evidence for invasive pest thrips, *Thrips parvispinus* (Thripidae: Thysanoptera), newly recorded from India. J Insect Sci 15:105

UNIDO (2015) Codes of practice for mango farming and processing. Mango Research Station Shujabad, Multan, Govt. of Punjab, Lahore

Urías-López M, Osuna-García J, Vázquez-Valdivia V, Pérez-Barraza M (2010) Fluctuación poblacional y distribución de la escama blanca del mango (*Aulacaspis tubercularis* Newstead) en Nayarit, Mexico. Revista Chapingo. Serie Horticultura 16:77–82

Urías-Lopez MA (2006) Principales plagas del mango en Nayarit. El cultivo del mango: Principios y tecnología de producción. V. Vázquez V. y MH Pérez B.(Eds.). Instituto Nacional de Investigaciones Forestales, Agrícolas y Pecuarias. Santiago Ixcuintla, Nayarit, 211–234

van Doan C, Pfander M, Guyer AS, Zhang X, Maurer C, Robert CAM (2021) Natural enemies of herbivores maintain their biological control potential under short-term exposure to future CO(2), temperature, and precipitation patterns. Ecol Evol 11:4182–4192

Van Dyck H, Bonte D, Puls R, Gotthard K, Maes D (2015) The lost generation hypothesis: could climate change drive ectotherms into a developmental trap? Oikos 124:54–61

van Wyk M, Al-Adawi AO, Wingfield BD, Al-Subhi AM, Deadman ML, Wingfield MJ (2005) DNA based characterization of *Ceratocystis fimbriata* isolates associated with mango decline in Oman. Australas Plant Pathol 34:587–590

Van Wyk M, Al Adawi AO, Khan IA, Deadman ML, Al Jahwari AA, Wingfield BD, Ploetz R, Wingfield MJ (2007) *Ceratocystis manginecans* sp. nov., causal agent of a destructive mango wilt disease in Oman and Pakistan. Fungal Divers 27:213–230

Vander Zanden MJ (2005) The success of animal invaders. Proc Natl Acad Sci 102:7055–7056

Vargas RI, Piñero JC, Leblanc L (2015) An overview of pest species of *Bactrocera* fruit flies (Diptera: Tephritidae) and the integration of biopesticides with other biological approaches for their management with a focus on the Pacific region. Insects 6:297–318

Venette RC, Hutchison WD (2021) Invasive insect species: Global challenges, strategies & opportunities. Frontiers Media SA, p 650520

Venette RC, Kriticos DJ, Magarey RD, Koch FH, Baker RH, Worner SP, Gómez Raboteaux NN, McKenney DW, Dobesberger EJ, Yemshanov D (2010) Pest risk maps for invasive alien species: a roadmap for improvement. Bioscience 60:349–362

Verghese A, Jayanthi PK (2001) Integrated pest management in fruits. Pest management in horticultural ecosystems, 1st edn. Capital Publishing Company, New Delhi, pp 1–23

Verghese A, Soumya C, Shivashankar S, Manivannan S, Krishnamurthy S (2012) Phenolics as chemical barriers to female fruit fly, *Bactrocera dorsalis* (Hendel) in mango. Curr Sci, 563–566

Vierbergen G (1995) International movement, detection and quarantine of *Thysanoptera* pests. In: Thrips biology and management. Springer, pp 119–132

Wagner T, Schliep EM, North JS, Kundel H, Custer CA, Ruzich JK, Hansen GJ (2023) Predicting climate change impacts on poikilotherms using physiologically guided species abundance models. Proc Natl Acad Sci 120:e2214199120

Wallin KF, Raffa KF (2004) Feedback between individual host selection behavior and population dynamics in an eruptive herbivore. Ecol Monogr 74:101–116

Walther G-R, Roques A, Hulme PE, Sykes MT, Pyšek P, Kühn I, Zobel M, Bacher S, Botta-Dukát Z, Bugmann H, Czúcz B, Dauber J, Hickler T, Jarošík V, Kenis M, Klotz S, Minchin D, Moora M, Nentwig W, Ott J, Panov VE, Reineking B, Robinet C, Semenchenko V, Solarz W, Thuiller W, Vilà M, Vohland K, Settele J (2009) Alien species in a warmer world: risks and opportunities. Trends Ecol Evol 24:686–693

Waterfield G, Zilberman D (2012) Pest management in food systems: an economic perspective. Annu Rev Environ Resour 37:223–245

Waterhouse DF (1998) Biological control of insect pests: Southeast Asian prospects
Weems HV, Fasulo TR (2012) Japanese orange fly, *Bactrocera tsuneonis* (Miyake) (Insecta: Diptera: Tephritidae). EDIS
Weintraub PG, Beanland L (2006) Insect vectors of phytoplasmas. Annu Rev Entomol 51:91–111
Weldon CW, Terblanche JS, Bosua H, Malod K, Chown SL (2022) Male Mediterranean fruit flies prefer warmer temperatures that improve sexual performance. J Therm Biol 108:103298
Wood SL (1992) A catalog of Scolytidae and Platypodidae (Coleoptera). part 2: taxonomic index. Gt Basin Nat Mem 13:1–1553
World Meterological Organization (2024) State of the climate in Asia 2023. WMO, 38
Wudil AH, Usman M, Rosak-Szyrocka J, Pilař L, Boye M (2022) Reversing years for global food security: a review of the food security situation in Sub-Saharan Africa (SSA). Int J Env Res Public Health 19:14836
Wysoki M, Ben-Dov Y, Swirski E, Izhar Y (1992) The arthropod pests of mango in Israel. IV Int Mango Symp 341:452–466
Yadav S, Korat J, Yadav S, Mondal K, Kumar A, Kumar S (2023) Impacts of climate change on fruit crops: a comprehensive review of physiological, phenological, and pest-related responses. Int J Environ Clim Chang 13:363–371
Yahia EM (2006) Effects of insect quarantine treatments on the quality of horticultural crops. Stewart Post Harvest Rev 2
Yamamura K, Kiritani K (1998) A simple method to estimate the potential increase in the number of generations under global warming in temperate zones. Appl Entomol Zool 33:289–298
Yildirim AB (2020) Ultraviolet-B-induced changes on phenolic compounds, antioxidant capacity and HPLC profile of in vitro-grown plant materials in Echium orientale L. Ind Crops Prod 153:112584
Yoro KO, Daramola MO (2020) CO_2 emission sources, greenhouse gases, and the global warming effect. Advances in carbon capture. Elsevier, Sawston, UK, pp 3–28
Zalucki MP, Shabbir A, Silva R, Adamson D, Shu-Sheng L, Furlong MJ (2012) Estimating the economic cost of one of the world's major insect pests, *Plutella xylostella* (Lepidoptera: Plutellidae): just how long is a piece of string? J Econ Entomol 105:1115–1129
Zhang H, Xie Y, Li Z (2011) Identification key to species of *Thrips* genus from China (Thysanoptera, Thripidae), with seven new records. Zootaxa 2810:37–46
Zhang J, Bo L, Mingyue W (2018) Study on windbreak performance of tree canopy by numerical simulation method. J Comput Multiph Flows 10:259–265
Zhang Y, Han M, Song M, Tian J, Song B, Hu Y, Zhang J, Yao Y (2021) Intercropping with aromatic plants increased the soil organic matter content and changed the microbial community in a pear orchard. Front Microbiol 12:616932
Zhao Z, Carey JR, Li Z (2024) The global epidemic of *Bactrocera* pests: mixed-species invasions and risk assessment. Annu Rev Entomol 69:219–237
Zheng S, Johnson AJ, Li Y, Chu C, Hulcr J (2019) *Cryphalus eriobotryae* sp. nov. (Coleoptera: Curculionidae: Scolytinae), a new insect pest of loquat *Eriobotrya japonica* in China. Insects 10:180

Chapter 26
Management of Mango Diseases Under Global Climate Shifts

Muhammad Atiq, Nasir Ahmed Rajput, Azeem Akram, Salman Ahmad, Irfan Ahmad, Shakeel Ahmad, and Yasir Ali

Abstract Mango fruit among other horticultural produce invariably suffers postharvest diseases that are of concern given international trade, storage, marketing and cause significant economic losses to the producers and exporters, especially in the disadvantaged countries. This chapter attempts to consider the most crucial and fresh unaddressed research questions related to the control of postharvest diseases of mangoes with particular regard to anthracnose, stem-end rot and bacterial black spot, which deteriorate mango in its quality and make storage a challenge. Pre-harvest conditions, environmental factors and storage methods commonly used in mango-producing countries, including Pakistan are identified as key contributors to disease development. The existing management tools, including chemical treatments, cold storage and modified atmospheric packaging are reviewed for their efficacy and limitations. Also, some of the alternatives to traditional postharvest management strategies, such as biological control, phytosanitary measures and novel approaches like nanotechnology and sensor based disease detection are examined for their applicability in controlling postharvest diseases of mango. This chapter stresses the necessity of ongoing global collaboration and adequate funding of research for the improvement of mango postharvest quality by introducing new dimensions, such as early disease diagnosis and improved long-term storage.

M. Atiq (✉) · N. A. Rajput · A. Akram
Department of Plant Pathology, Faculty of Agriculture, University of Agriculture, Faisalabad, Pakistan
e-mail: muhammad.atiq@uaf.edu.pk

S. Ahmad
Department of Plant Pathology, College of Agriculture, University of Sargodha, Sargodha, Pakistan

I. Ahmad
Department of Forestry, Range Management and Wildlife, University of Agriculture, Faisalabad, Pakistan

S. Ahmad
Institute of Agronomy, Faculty of Agricultural Sciences and Technology, Bahauddin Zakariya University, Multan, Pakistan

Y. Ali
Department of Plant Pathology, College of Agriculture, University of Layyah, Layyah, Pakistan

S. A. H. Naqvi et al. (eds.), *Climate Change and Mango Production*,
https://doi.org/10.1007/978-3-031-87035-4_26

26.1 Introduction

Mango is a highly imperative horticultural crop in Pakistan and greatly contributes to the country's income. Pakistan is the world's 4th largest mango producer and the 10th largest exporter. The mango farming in the country yields 9–10 tons/ha (Khan et al. 2023). About 2.3 million tons of mangos are produced annually in Pakistan. India produces approximately 56.6% of the total mango in the entire world (Nayak et al. 2023). India tops the globe regarding mango producers, and it produces approximately 24.7 million tons every year. Indonesia follows as the second largest producer with productions of about 3.6 million tons produced annually. China for instance cultivates 2.4 million tons per year they consume the fruit locally and export to other countries. Mexico has a warm climate, and consequently mango grows well in this environment, likewise for Brazil. Approximately 2.1 million tons are produced in Brazil each year, of which the majority is intended for exportation (World Population Review 2024). Mango is one of the fruits that are produced on large areas that exported more than 32% out of the world exports go to the United States. Many infectious diseases resulting from different phytopathogens are adversely affecting both the quality and yield of mango fruit.

In present world, diseases of mango trees and fruits are 83 while 27 diseases have been found important in Pakistan (Nazir et al. 2024). Some of the major problems still faced by farmers desiring to achieve upright yields include disease control, especially arising from pest resistance to chemicals. Even though the production of mangoes is relatively significant in Pakistan then the sector has several problems particularly concerning the biotic disease affecting post-harvest mango fruits suitable for export market (Wickrama et al. 2024). Some of the dominant diseases affecting the mango fruit are anthracnose, stem-end rot as well as fungal infections due to *Colletotrichum gloeosporioides*, and *Botryodiplodia theobroma* according to Karunanayake and Adikaram (2020). Anthracnose is a common disease and is manifested on mango trees in all development zones of the countries where this fruit is grown. This disease affects formatively different parts of the plant like flowers, leaves, fruits, and etc., resulting in low yield and quality. High humidity is suitable conducive which is found commonly; it can reach up to 100 percent in affected crop at such areas (Haggag 2010). On fruits, the disease shows symptoms of black spot, shriveled, and fruit drop, further proceeding to the subsequent stages during storage and transportation resulting to high postharvest losses (Yenjit et al. 2004). Stem-end rot is also another manner of Postharvest diseases in mango fruits that affects up to 40% of the fruits in Pakistan. This disease is caused by a complex of fungal pathogens such as *Lasiodiplodia theobromae*, *Dothiorella* species and *Colletotrichum gloeosporioides* (Alam et al. 2017). The initial infection of these pathogens occurred at the stem-end, causing an internal rot that only shows on the ripened fruit surface. The disease is most challenging to manage since it may start during storage or transit of produce after having been harvested. Fungal diseases are common on mango and in addition to these, mango bacterial black spot (MBBS) caused by *Xanthomonas campestris* pv. *Mangiferae indicae* is increasing being reported to be serious where temperatures and rainfall are high. MBBS infects all aerial portions of mango plant and results to formation of black water sores on the leaves and fruits (Evans et al. 2017; Fang et al. 2012). However, severe infection of fruits leads to losses in both quantity and quality of fruits as indicated by Kaur et al. (2024). The impact of climate change on mango diseases intensity and mango tree health mentioned in Table 26.1.

Mango fruit losses are also experienced during and after harvesting, due to poor management practices including; harvesting at incorrect ripeness, physical damage while in transit, and poor storage conditions. Mango fruit bears a very short shelf-life and is highly sensitive to chilling injury when stored at a temperature band of 7 °C to 13 °C. CI leads to such unfavorable effects like peel pitting, uneven ripening, susceptibility to decay making the fruit less marketable. Postharvest handling, particularly the temperature management is important in order to have longer postharvest period of mango fruits. Optimizing storage temperature at 8–13°C and relative humidity at 90–95% were also recommended to minimize spoilage impact and retain the fruit quality during export as stated by Zul Helmey et al. (2021), Kiran et al. (2022). However, achieving the right balance in temperature control is essential However, achieving the right balance in temperature control is essential: excessively low temperatures can result in chilling injuries, while higher temperatures accelerate ripening and decay (Ghaffar et al. 2022). Improving postharvest handling and disease management is

Table 26.1 Impact of global climate shifts on mango diseases

Climate factor	Impact on mango diseases	Disease severity and spread
Rising temperatures	Heat stress weakens mango trees, making them more susceptible to diseases like anthracnose, fusarium wilt, and bacterial diseases	Increased fungal and bacterial growth leading to higher infection rates
Increased humidity	Higher moisture levels favor the development of fungal diseases such as powdery mildew and anthracnose	Enhanced disease proliferation, especially in humid regions, leading to crop loss
Altered rainfall patterns	Excessive rainfall creates favorable conditions for root rot and stem canker diseases	Wet conditions favor the spread of waterborne pathogens such as *Phytophthora* spp.
Extreme weather events	Storms and heavy winds cause physical damage to trees, increasing exposure to pathogens	Post-storm conditions lead to an increase in opportunistic infections
Extended growing seasons	Longer growing seasons increase the period of exposure to pathogens	Increased chances for disease buildup and repeated infection cycles
Increased CO_2 concentration	Elevated CO_2 can alter mango tree metabolism, impacting their immune response to pathogens	May either enhance or reduce disease susceptibility depending on disease type and environmental conditions

Fig. 26.1 Proper tree cutting to avoid fungal diseases

essential to enhancing Pakistan's mango export potential. By addressing these challenges, the country can increase the proportion of mango fruit suitable for export, thereby boosting its revenues and reinforcing its position as a leading mango exporter on the global stage (Figs. 26.1 and 26.2).

Fig. 26.2 Infection of powdery mildew causing improper fruit setting and leaves showing the salt injuries

26.2 Post-harvest Disease Development and Investigating Pre-harvest Conditions that Effect Susceptibility

Mangoes are highly perishable due to their climacteric nature, which leads to the production of large amounts of ethylene, commonly referred to as the "ripening hormone." However, the quality and marketability of mangoes are significantly reduced by anthracnose, a fungal disease caused by *Colletotrichum gloeosporioides*. This disease is responsible for post-harvest losses of approximately 35 to 40%, making it one of the most serious post-harvest issues for mangoes. Various pre-harvest and post-harvest strategies, including chemical treatments, have been employed to manage mango anthracnose (Jeevanantham et al. 2024). The mineral composition of mango flesh and exposure to heat after harvest (such as sunlight, phytosanitary heat treatments, and elevated storage temperatures) are strongly linked to the expression of internal disorders (IDs). High levels of nitrogen (N), high N/Ca, Mg/Ca, and K/Ca ratios, along with low levels of calcium (Ca) and boron (B), have been associated with most of these disorders (Ullah et al. 2024). Anthracnose (*Colletotrichum gloeosporioides*) and stem-end rot caused by *Lasiodiplodia theobromae* are postharvest diseases of economic importance, causing a loss in quality and quantity after harvesting (Anusha et al. 2023). Mango research has generally focused on maturity at harvest and postharvest management, which determines the final quality at the consumer level. However, preharvest cultural techniques have a significant impact on postharvest performance and overall quality, just like they do for other stone fruits. These practices affect the environmental circumstances during fruit growth. Mango quality encompasses various attributes, including intrinsic factors such as texture, sweetness, acidity, aroma, shelf life, and nutritional value, as well as extrinsic factors like fruit size (Léchaudel and Joas 2007). Pakistan is well known for its mangoes' exceptional qualities and ranks as the 6th largest mango producer globally. Mango exports from Pakistan are still restricted, though, because of post-harvest illnesses such anthracnose and stem-end rot (SER), which are mostly brought on *by Colletotrichum gloeosporioides* and *Lasiodiplodia theobromae*. These diseases are becoming more prevalent due to changing environmental conditions in Pakistan's mango-growing regions. Foreign markets prefer quarantine methods like Hot Water Treatment (HWT) because of its high effectiveness in controlling post-harvest diseases (Malik et al. 2018).

Fruits harvested during the humid season had higher levels of infestation, although fewer fungal agents were involved, with *Colletotrichum gloeosporioides* and, to a lesser extent, *Phoma mangiferae*, being the primary culprits. Cultural practices significantly influenced mango infection rates. Orchard sanitation, particularly cleaning and pruning, helps to reduce infection. However, poorly maintained orchards produced mangoes with the highest levels of infestation. Additionally, the practice of placing harvested fruits upside down on the soil, to eliminate sap, increased the risk of contamination by pathogenic fungi (Diedhiou et al. 2007).

26.3 Importance of Pre-harvest Management in Ensuring Mango Quality

High postharvest losses pose a significant challenge for commercial mango supply chains. Once harvested, the quality of the fruit cannot be improved, only preserved. Enhancements in quality must occur during the production stage. Therefore, it is essential to understand the factors at the production level and the corresponding management strategies to ensure top-quality mangoes at harvest and throughout the supply chain (Rehman et al. 2015). Postharvest disease management is one of the key challenges in commercial mango supply chains (Fiaz et al. 2014). Fungal disease stem end rot (SER) is complex; it significantly impacts the quality of mangoes exported from Pakistan. The dynamics of the disease are not fully understood, and further clarification is needed to enhance the effectiveness of management strategies. This study sought to characterize the SER disease cycle and infection process in the context of Pakistan's expanding population (Alam et al. 2020).

26.4 Existing Management Strategies for Postharvest Diseases

26.4.1 Hot Water Quarantine Treatment (HWQT)

Postharvest diseases and disorders degrade mango quality, often leading to significant losses, with some fruit becoming entirely unmarketable. Additionally, the risk of fruit fly infestation has made hot water quarantine treatment (HWQT) a mandatory requirement for market access in countries such as China and Iran (Rajwana et al. 2011). Extended hot water quarantine treatment (HWQT) is a widely used commercial phytosanitary method for disinfesting mangoes of fruit flies. It involves immersing the fruit in hot water, either in static or flow-through systems, which allows heat to transfer to the skin and flesh of the fruit. A short-term hot water dip, lasting 3–5 min, is used to control fungal infections. Through hot water dips or sprays, postharvest applications of fungicides are also common for disease control. To eliminate fruit fly larvae and eggs, HWQT has gained significant global recognition for its effectiveness. It is applied for around an hour. Due to its economic benefits and competitive performance compared to other advanced techniques, this method is crucial for mango-producing countries like Pakistan (Amin et al. 2020).

26.4.2 Effectiveness of Chemical Treatments in Mango Disease Control

Pyraclostrobin and mefentrifluconazole (160 mg active component per kg) were applied three times to effectively control the disease. A QuEChERS-based technique utilizing high-performance liquid chromatography-tandem mass spectrometry (HPLC–MS/MS) was created to find mefentrifluconazole and pyraclostrobin residues on mangoes. Mefentrifluconazole and pyraclostrobin initial concentrations varied from 0.18 to 0.34 mg/kg. The half-lives of these substances were 5.6–10.8 days and 5.5–9.0 days, respectively. After 21 days of foliage application, final residue levels were 0.02–0.04 mg/kg for mefentrifluconazole and 0.01–0.04 mg/kg for pyraclostrobin, both below the maximum residue limit (MRL) for pyraclostrobin (Wang et al. 2023).

26.4.3 Cold Storage: PCM-Based Systems for Energy-Efficient Preservation

The effects on the physical and chemical characteristics of mango fruits of a PCM-based cold storage system operating under periodic (P12) and continuous (P24) circumstances were investigated. The fruits underwent weight loss, peel color changes, and soluble solid concentration under both P12 and P24 storage settings. However, they retained high levels of firmness, titratable acidity, and ascorbic acid, which are comparable to those kept in a traditional cold storage system (G24). The PCM-based storage systems (P12 and P24) maintained the temperature within the allowed range and consumed less energy than the conventional system, according to the temperature response and energy consumption analysis, which revealed that energy consumption decreased by 39% in P12 and 18% in P24 storage circumstances. These findings suggest that the periodic PCM-based cold storage (P12) is a viable energy-saving alternative to conventional cold storage, effectively preserving mango fruit quality during power outages and off-peak load periods (Karthikeyan et al. 2022).

26.4.4 Modified Atmosphere Packaging (MAP) for Mango Export

In order to better protect the quality of fully mature mangoes during the lengthy transit times necessary for export from South America to the USA, the viability of implementing modified atmosphere packaging (MAP) technology in the current mango export systems was investigated. The mangoes "Ataulfo," "Tommy Atkins," and "Kent" were packaged at a commercial packing house in Nayarit, Mexico, in either Apio Breatheway MAP bags or NOW Plastics perforated bags (control), each of which held around 4 kg of fruit. For "Ataulfo," "Tommy Atkins," and "Kent," the average initial maturity stages of the mangoes were 3.0, 2.5, and 3.1 (on a scale of 1 to 5). The corresponding soluble solids content (SSC) readings were 11.4%, 9.6%, and 9.3%, respectively. Commercial trucks then delivered the fruit to an importer in Hidalgo, Texas, and then to Gainesville, Florida (8–13 °C), where it was stored for a further three weeks at the University of Florida (12.5 °C), mimicking the length of time it would take to ship from South America to the United States. After 3 weeks, the MAP conditions in the bags were: 'Ataulfo' at 4.7% O_2 + 13.1% CO_2, 'Tommy Atkins' at 3.6% O_2 + 11.4% CO_2, and 'Kent' at 8.9% O2 + 9.0% CO_2. Across all three cultivars, the mangoes in the MAP bags remained less ripe and retained better quality in terms of flesh color, firmness, SSC, and titratable acidity (TA) following three weeks of shipping simulation and three more days at 20 °C compared to those in the perforated control bags (Brecht et al. 2021).

26.4.5 Traditional Management of Anthracnose and Stem End Rot in Mangoes

The occurrence of anthracnose during both pre- and post-harvest stages of mango production is a frequent issue, leading to significant economic losses each year (Jenny et al. 2019). Controlling anthracnose can be effectively managed with pre-harvest, post-harvest, or a combination of both approaches. The main strategies involve fungicide application, proper cultural practices, and selecting suitable cultivars (Dofuor et al. 2023). Carbendazim was found to be the most effective fungicide against *Colletotrichum gloeosporioides*, completely inhibiting mycelial growth at a 100 ppm dose. Propiconazole also achieved 100% inhibition at a 500 ppm dose, while mancozeb was highly effective at 1000 ppm. Copper oxychloride, however, was the least effective fungicide, showing only 55% inhibition even at the highest dose (Iqbal et al. 2022). The widespread use of synthetic fungicides to control mango anthracnose disease (MAD) can result in environmental pollution, water contamination, soil compaction, and ecological issues, ultimately impacting the sustainability of mango production (Dofuor et al. 2023).

26.4.6 Limitations and Sustainable Approaches to Stem End Rot Management

The individual and combined use of biocontrol agents such as *Trichoderma harzianum*, *Bacillus subtilis*, and *Pichia anomala* through foliar sprays or fruit dips effectively controlled mango anthracnose. These bio-formulations could serve as alternatives to chemical fungicides, contributing to environmental protection and integrated disease management (Sharma et al. 2021). Several pathogens associated with stem end rot have been identified globally, including *Lasiodiplodia theobromae*, *Neofusicoccum parvum*, *Fusicoccum aesculi*, *Nattrassia mangiferae*, and *Fusicoccum* species in Australia and Brazil. In Egypt, the main pathogens are *Lasiodiplodia theobromae*, *Colletotrichum gloeosporioides*, *Alternaria alternata*, and *Botrytis cinerea* (Alam et al. 2017). Carbendazim was highly effective in reducing fungal infections in mango plants, preventing symptoms like gum exudation, dieback, and wilting, while also enhancing vegetative growth (Khanzada et al. 2005). Researchers are developing sustainable alternatives to chemical pesticides, with plant extracts like NECO, ASTOUN, FERCA, and *Moringa oleifera* effectively controlling mango stem-end rot. These eco-friendly options show promise for managing postharvest diseases, but further testing on naturally infected fruits is needed (Yeo et al. 2023).

26.4.6.1 Effectiveness of Microbial Antagonists as Biocontrol Agents

As biocontrol agents, microbial antagonists are sprayed on soil, seeds, leaves, or fruits. As found by scientists, these microbes can decrease plant illnesses brought on by phytopathogens. These microbes release biological compounds that are safe for the environment and stop harmful bacteria from growing (Lugtenberg and Kamilova, 2009). They can also shield plants from diseases as they induce systemic resistance in the plant (Schmidt et al. 2019) or can do so by depriving the plant pathogens of their necessary nutrients and helping the plant absorb nutrients and water more efficiently (Ezrari et al. 2021). These microorganisms produce biodegradable biological compounds that inhibit the growth of pathogens (Lugtenberg and Kamilova 2009). Because they cause the plant to develop systemic resistance, they can also lessen plant infections (Schmidt et al. 2019) or by competing for essential nutrients, aiding in the plant's nutrient uptake (Ezrari et al. 2021). Enhancing germination and growth of plants is another aspect associated with biocontrol agents. Phosphorus (P), potassium (K), iron (Fe), zinc (Zn), and ambient nitrogen can all be dissolved by microorganisms with biocontrol properties (Olanrewaju et al. 2017). They can also produce cytokinins, siderophores, gibberellic acid, indole acetic acid, and many other compounds that promote plant growth (Glick, 2014).

Furthermore, they can also alleviate the impact of various abiotic stress factors, for example, salinity, drought, high and low temperatures, and high concentrations of heavy metals (Khan et al. 2023). Nonetheless, their ability to biocontrol can be influenced by the characteristics of the soil and fertility, the regality of the habitat of rhizobacteria, and abiotic stresses. These factors make their biocontrol specificity high (Tewari et al. 2019). Thus, it is crucial to locate new, non-exploited microbial biocontrol agents with the potential for fertility enhancement and abiotic stress alleviation while promoting citrus. Both pre-harvest and post-harvest microbial infections can be decreased with the help of biocontrol products. Therefore, when certified as generally regarded as safe, the use of biological control agents in crops enhances plant health and is safe for consumers (Zhao et al. 2020), enhances the plant's resistance to phytopathogens, and can even stimulate plant growth.

26.5 Impact of Phytosanitary Regulations on Mango Exporting Countries

Sanitary and phytosanitary (SPS) issues have become increasingly important in global trade, driven by rising demand for fresh and processed foods in developed countries since the 1990s. Growing public awareness and concerns about food safety have led these countries to impose stricter SPS regulations. However, these stringent standards often create significant barriers for exports from developing countries, which face challenges due to weaker SPS controls and fragmented systems. Developing nations often struggle to comply with these regulations due to limited resources and inadequate infrastructure, making it harder to access developed markets (Das 2013).

26.5.1 Traditional Phytosanitary Measures

Conventional phytosanitary methods include chemical dip treatments (usually with dimethoate solutions), heat treatments (using hot air or water at 43–48 °C), cold disinfestation (0–3 °C), and fumigation (using ethylene dibromide and methyl bromide). These treatments must be applied consistently and effectively to achieve the required efficacy for each pest species being targeted (Akter et al. 2023). Pesticide use for disinfestation is being limited due to concerns over residues and food safety. Fumigation is also restricted, as ethylene dibromide has been banned and methyl bromide is a known greenhouse gas contributor (Haynes and Dominiak 2018). Anthracnose, caused by *Colletotrichum gloeosporioides*, is a significant pre- and post-harvest disease affecting mangoes worldwide, including in Bangladesh (Uddin et al. 2018). Effective control of anthracnose can be achieved through both pre- and post-harvest fungicide applications. Organic sulfur fungicides like dithio-carbamates, along with heterocyclic nitrogen compounds such as mancozeb, zineb, and captan, are sufficient to manage mango anthracnose. Among benzimidazoles, benomyl and carbendazim have been particularly successful in controlling anthracnose under in vivo conditions (Iqbal et al. 2022). Additionally, stem end rot (SER) is a fungal disease complex that significantly threatens the export of high-quality mangoes from Pakistan to international markets (Alam et al. 2020). When it came to managing fungal infections in mango plants, carbendazim outperformed thiophanate-methyl and Alliete. This resulted in a significant improvement in the plants' vegetative development by lowering gum exudation, dieback, and wilting (Khanzada et al. 2005).

26.5.2 International Phytosanitary Standards and Certification in Mango Trade

Horticultural trade significantly contributes to the economies of both developing and developed countries, accounting for 9 to 29% of gross income. The USDA's Animal and Plant Health Inspection Service (APHIS) and the adoption of ISPM No. 18 have advanced the use of irradiation as an effective strategy to address quarantine restrictions. These policies have opened opportunities for both developed and developing nations to access profitable U.S. markets (Bustos-Griffin et al. 2012). The new rule enables countries in Asia, Africa, and the Americas to access fresh agricultural markets with minimal investment and risk, making it feasible to use PI for low-volume or short-season crops without the need for large-scale facilities (Bustos-Griffin et al. 2015).

26.5.3 Quarantine Requirements in Major Importing Countries

Irradiation at doses below 1 kGy is an effective phytosanitary treatment that maintains the quality of most fresh produce. It is the only method with an internationally recognized generic dose to sterilize all fruit flies and is being considered for other pests. This non-chemical, cost-effective alternative reduces reliance on methyl bromide, which harms the environment and human health. Notable laws pertaining to phytosanitary treatments exist in two nations. To sterilize or stop the adult emergence of all insects on all produce, with the exception of pupae and adult Lepidoptera, a dose of 400 Gy is approved in the USA (USDA-APHIS 2006). Australia's ICA National Protocol allows 150 Gy for Tephritid fruit flies and 400 Gy for all insects, except adult Lepidoptera that pupate internally, provided the treatment is approved by Food Standards Australia and New Zealand (Roberts 2016). Effective management of mango diseases such as anthracnose and stem end rot is crucial for ensuring the export quality of mangoes, with fungicides and phytosanitary treatments playing vital roles. However, the increasing restrictions on pesticide use and fumigation due to concerns over food safety and environmental impact highlight the need for alternative solutions. Phytosanitary regulations, especially in developed countries, pose challenges for mango exporters from developing nations, requiring them to meet stringent standards with limited resources. International standards like irradiation offer a promising non-chemical treatment option, allowing for safer and more sustainable exports. As global demand for safe, high-quality produce rises, addressing these challenges will be key to expanding market access and maintaining competitiveness in the mango trade.

26.5.4 The Inadequate Nature of Conventional Management Approaches

There are still significant gaps in postharvest disease management, particularly regarding the areas that follow, despite the use of several strategies:

26.5.5 Dependency on Agrochemicals

Agrochemicals are widely used to control pests and diseases, but this has a detrimental effect on fruit quality, the environment, and the safety of consumers. The main drawbacks of microbial pesticides, which are living organisms, are that they have a high specificity against the diseases and pathogens they are meant to combat, which may require the use of multiple microbial pesticides, and that their effectiveness is frequently inconsistent because of the influence of various biotic and abiotic factors (Fenta et al. 2023).

26.5.6 Limitations of Biological Control

Biological Control Agents may eliminate infections and insect pests that harm horticultural crops (Heydari and Pessarakli 2010). Biological agents for control (BCAs) are one of the best ways to stop sickness from spreading (Fu et al. 2017). In competition with others for nutrition, plants generate secondary metabolites, including volatile organic compounds (VOCs) and lytic enzymes, and their defense mechanisms are triggered (Lastochkina et al. 2019). By suppressing harmful aerobic microbial species while encouraging helpful microbes, microbial antagonists drastically alter the fruit microbiome, thereby improving fruit health (Berg et al. 2020). Introducing gene-transfer-produced resistant species and cultivars (Dale et al. 2017) is an inexpensive and environmentally beneficial method to reduce pathogen-related losses (Panth et al. 2020; Xu et al. 2022). Additionally, the selection of somaclonal variation and the generation of mutations are crucial methods for disease control (Xu et al. 2017). Research has demonstrated the significance of genes like smut and peroxidase that include antifungal chemicals in promoting disease resistance (Brogden 2005).

26.6 Suggested Novel Field of Research

26.6.1 Early Detection of Disease

Environmental variables, host tissue properties, and physiological stages are all highly associated with the development of disease. Biosensors convert biological reactions into electrical signals that may be observed (Qiao et al. 2020). Microbes, enzymes, cells, organelles, antimicrobial peptides, antibodies, and other biosensors can all be used to track the microbial flora of fruits and vegetables. The good properties of antibodies and antimicrobial peptide-based biosensors, such as stability, many sites for bacterial capture, and high specificity, are being considered for the detection of bacteria (Islam et al. 2022). To efficiently prevent spoiling, sensors based on maturing genes or phytochemicals have been devised. Carbon nanotubes and a metal catalyst are currently being utilized to create a sensor that tracks ethylene levels to determine when fruits and vegetables start to decay (Fong et al. 2020).

26.6.2 Genetic Improvement and Nano Tools

Introducing gene-transfer-produced resistant species and cultivars (Dale et al. 2017) is an inexpensive and environmentally beneficial method to reduce pathogen-related losses (Panth et al. 2020; Xu et al. 2022). Additionally, an essential technique to control diseases is the selection of soma clonal variation and mutation induction (Xu et al. 2017). True genetic resistance to anthracnose, caused by *Colletotrichum* species, has yet to be fully established in mangoes. However, some cultivars show varying degrees of partial resistance or tolerance. Research indicates that certain *Mangifera indica* cultivars experience a notable delay in lesion formation during fruit ripening, highlighting their potential for breeding programs aimed at addressing postharvest challenges (Grice et al. 2023). Nanotechnology has increasingly demonstrated its potential in managing post-harvest diseases, particularly through its antimicrobial properties, which are highly effective in food processing, storage, and transportation. Innovations like nano-sensors, nano-coatings, nano-bubbles, and nano-packaging are key applications of nanomaterials in the food industry (Mahmoud and Gaber 2024).

26.6.3 Long-Term Storage Strategies

26.6.3.1 Modified Atmosphere Packaging (MAP)

Modified atmosphere packaging (MAP) technology is being used more and more in the healthy and perishable-cut food industry globally as a value addition because it provides the potential to slow down the respiration rate and increase the shelf life of fresh produce. However, the public's interest in how MAP technology affects the survival and growth of harmful organisms has increased due to foodborne disease outbreaks and the advent of resistant foodborne pathogens in MAP (Caleb et al. 2013).

26.6.3.2 Coatings Based on Chitosan

Adding cinnamon essential oil to the bioactive film made of chitosan and Arabic gum resulted in an entwined structure, enhanced fluid barrier qualities, and an increase in antioxidant potential (Xu et al. 2019). It has been studied how well edible coatings work as a biological control technique (Salgado-Cruz et al. 2021), and the antifungal effect of using probiotics and edible coatings together was evaluated (Barragán-Menéndez et al. 2020).

26.6.3.3 Control of Temperature and Humidity

Improving storage conditions by carefully controlling the temperatures and humidity can greatly lower postharvest losses. To improve storage practices, the efficacy of hot water and other postharvest treatments can be enhanced by optimizing the combination of temperature, relative humidity, and airflow to manage pathogen development. Research on temperature and relative humidity's impact on fruit quality and decay caused by various pathogens is critical to preserving mango quality during postharvest.

26.6.3.4 Innovative Postharvest Handling and Distribution Techniques:

Adopting cutting-edge postharvest handling and distribution methods is critical to ensuring that the quality and safety of fruits are preserved. This includes developing, implementing, and evaluating creative and advanced postharvest handling, packaging, and distribution techniques, as well as a deep understanding of the relationships among storage conditions, biological control agents, and pathogens. A comprehensive approach that integrates these technologies with biological controls will pave the way for sustainable mango production and exportation. Table 26.2 summarizers the strategies to cope with mango diseases under climate change. By embracing innovation and collaboration, mango exporting countries can effectively navigate the challenges posed by phytosanitary regulations and position themselves for success in the global market.

Table 26.2 Management strategies for mango diseases under climate change

Management strategy	Description	Climate-adaptation benefits
Resistant mango varieties	Developing and planting varieties resistant to diseases like anthracnose, bacterial wilt, and fusarium wilt	Reduces the need for chemical treatments, providing long-term disease control
Early disease detection	Use of advanced diagnostic tools such as molecular techniques and remote sensing for early disease detection	Early intervention minimizes the spread and reduces the cost of treatment
Timely pruning and sanitation	Removal of infected plant parts and maintenance of proper orchard sanitation	Reduces inoculum sources and prevents the spread of diseases in the orchard
Biological control agents	Use of beneficial microorganisms like *Trichoderma* spp., *Bacillus subtilis*, and *Pseudomonas* spp. to control pathogens	Eco-friendly solution that minimizes the use of chemical fungicides and pesticides
Integrated disease management (IDM)	A combination of chemical, biological, and cultural practices tailored to the specific disease and climatic conditions	Enhances flexibility and effectiveness in disease control while reducing environmental impact
Water management	Efficient irrigation systems and soil drainage techniques to reduce waterlogging and prevent root rot diseases	Mitigates the risk of waterborne diseases, especially in regions with erratic rainfall patterns
Use of fungicides and biocides	Application of targeted fungicides and biocides based on weather forecasts and disease pressure	Prevents overuse of chemicals, focusing on when the disease risk is highest
Postharvest disease control	Implementing postharvest treatments such as hot water treatment and controlled atmosphere storage	Reduces the impact of diseases like anthracnose during fruit transport and storage

26.7 Conclusion

The present study is made in the conclusion that, managing postharvest diseases of mango remains a great challenge and so far traditional approaches appear to be their limited effective ways for preventing economic as well as quality losses. Furthermore, factors like pre-harvest conditions and poor storage and handling further worsen postharvest disease setup. However, the fact is that although chemical treatments and cold storage initially successfully controlled postharvest diseases of fruit to an extent physiologically safe levels during transportation and marketing throughout long distances until handing over them at final destination, these methods very soon reached their limitations indicating a global movement towards safer posts-harvesting practices using biological control agents (BCAs) or through eco-friendly statuses with different novel approaches as well demonstrated by use nanotechnology.

Further, there has been a direct impact on mango trade due to the enforcement of strict phytosanitary regulations necessitating new management solutions. For the next steps, future research should target early detection of disease (using drones), non-chemical treatments and storage conditions. The use of hi-tech methods to be combined with eco-friendly techniques might bring relief in this matter and make our food last a little longer than the days mentioned.

References

Akter H, Cunningham N, Rempoulakis P, Bluml M (2023) An overview of phytosanitary irradiation requirements for Australian pests of quarantine concern. Agric 13(4):771

Alam MW, Rehman A, Sahi ST, Malik AU (2017) Exploitation of natural products as an alternative strategy to control stem end rot disease of mango fruit in Pakistan. Pak J Agric Sci 54(3)

Alam MW, Rehman A, Malik AU, Ahmad S, Haider MS, Amin M et al (2020) Dynamics of stem end rot disease of mango fruit and its management. Pak J Agric Sci 57(1)

Amin M, Malik AU, Rajwana IA, Razzaq K, Faried HN, Ullah S, Zafar MS et al (2020) Postharvest respiration rate, physiological weight loss and physico-chemical quality of Mango fruit as influenced by different hot water quarantine treatments under simulated shipment conditions. J Hortic Sci Technol 3(1):12–18

Anusha S, Indra N, Kannan R, Geetha P, Rahale CS (2023) Morphological and molecular characterization of major postharvest diseases of mango. Int J Plant Soil Sci 35(19):1291–1299

Barragán-Menéndez C, Gálvez-López D, Rosas-Quijano R, Salvador-Figueroa M, Ovando-Medina I, Vázquez-Ovando A (2020) Films of chitosan and Aloe vera for maintaining the viability and antifungal activity of Lactobacillus paracasei TEP6. Coatings 10(3):259

Berg G, Rybakova D, Fischer D, Cernava T, Vergès MCC, Charles T, Schloter M et al (2020) Microbiome definition re-visited: old concepts and new challenges. Microbiome 8:103

Brecht JK, Sargent SA, Ristow N, Osuna García JA, Singh S (2021) Demonstrating modified atmosphere packaging (MAP) feasibility for mango export. In: XIII International controlled and modified atmosphere research conference-CAMA2021 1386, pp 161–166

Brogden KA (2005) Antimicrobial peptides: pore formers or metabolic inhibitors in bacteria? Nat Rev Microbiol 3(3):238–250

Bustos-Griffin E, Hallman GJ, Griffin RL (2012) Current and potential trade in horticultural products irradiated for phytosanitary purposes. Radiat Phys Chem 81(8):1203–1207

Bustos-Griffin E, Hallman GJ, Griffin RL (2015) Phytosanitary irradiation in ports of entry: a practical solution for developing countries. Int J Food Sci Technol 50(1):249–255

Caleb OJ, Mahajan PV, Al-Said FAJ, Opara UL (2013) Modified atmosphere packaging technology of fresh and fresh-cut produce and the microbial consequences—a review. Food Bioproc Technol 6:303–329

Dale J, James A, Paul JY, Khanna H, Smith M, Peraza-Echeverria S, Harding R et al (2017) Transgenic Cavendish bananas with resistance to Fusarium wilt tropical race. Nat Commun 8(1):1496

Das K (2013) The impact of sanitary and phytosanitary measures on India's exports and the challenges/opportunities of the SPS Agreement. In: Research handbook on environment, health and the WTO. Edward Elgar Publishing, pp 761–824

Diedhiou PM, Mbaye N, Drame A, Samb PI (2007) Alteration of post-harvest diseases of mango Mangifera indica through production practices and climatic factors. Afr J Biotechnol 6(9):1033–1041

Dofuor AK, Quartey NK-A, Osabutey AF, Antwi-Agyakwa AK, Asante K, Boateng BO et al (2023) Mango anthracnose disease: the current situation and direction for future research. Front Microbiol 14:1168203

Evans EA, Ballen FH, Siddiq M (2017) Mango production, global trade, consumption trends, and postharvest processing and nutrition. In: Handbook of Mango Fruit: Production, Postharvest Science, Processing Technology and Nutrition, 1–16

Ezrari S, Mhidra O, Radouane N, Tahiri A, Polizzi G, Lazraq A, Lahlali R (2021) Potential role of rhizobacteria isolated from citrus rhizosphere for biological control of citrus dry root rot. Plants 10(5):872

Fang NH, Ren YH, Shyang CR, Fen LR, Hsuan HT (2012) New Botryosphaeriaceae fruit rot of mango in Taiwan: identification and pathogenicity

Fenta L, Mekonnen H, Kabtimer N (2023) The exploitation of microbial antagonists against postharvest plant pathogens. Microorganisms 11(4):1044

Fiaz M, Malik AU, Amin M, Khan AS, Rehman A, Alam MW, Johnson P et al (2014) Production locality influences postharvest disease development and quality in mangoes. In: XXIX International horticultural congress on horticulture: sustaining lives, livelihoods and landscapes (IHC2014): IV 1111, pp 369–376

Fong D, Luo SX-L, Andre RS, Swager TM (2020) Trace ethylene sensing via wacker oxidation. ACS Cent Sci 6(4):507–512

Fu L, Penton CR, Ruan Y, Shen Z, Xue C, Li R, Shen Q (2017) Inducing the rhizosphere microbiome by biofertilizer application to suppress banana Fusarium wilt disease. Soil Biol Biochem 104:39–48

Ghaffar A, Khan A, Hameed A, Iqbal J, Ahmad M, Raza S, Imran M, Muhammad N, Shah S, Raza M (2022) Establishment of strategy for mango hopper Idioscopus clypealis (Lethierry) management and impact of weather factors on its population in south Punjab Pakistan. Biol Clin Sci Res J 2022

Glick BR (2014) Bacteria with ACC deaminase can promote plant growth and help to feed the world. Microbiol Res 169(1):30–39

Grice KRE, Bally ISE, Wright CL, Maddox C, Ali A, Dillon NL (2023) Mango germplasm screening for the identification of sources of tolerance to anthracnose. Australas Plant Pathol 52:27–41

Haggag WM (2010) Mango diseases in Egypt. Agric Biol J North Am 1:285–289

Haynes FE, Dominiak BC (2018) Irradiation for phytosanitary treatment of the Queensland fruit fly Bactrocera tryoni Froggatt benefits international trade. Crop Prot 112:125–132

Heydari A, Pessarakli M (2010) A review on biological control of fungal plant pathogens using microbial antagonists. J Biol Sci 10:273–290

Iqbal S, Khan MA, Atiq M, Rajput NA, Usman M, Nawaz A, Ahmad H (2022) Mango anthracnose: global status and the way forward for disease management. J Innov Sci 8:222–235

Islam MA, Karim A, Ethiraj B, Raihan T, Kadier A (2022) Antimicrobial peptides: promising alternatives over conventional capture ligands for biosensor-based detection of pathogenic bacteria. Biotechnol Adv 55:107901

Jeevanantham S, Praveen A, Livitha R, Balamurugan K (2024) Post-harvest anthracnose of mango caused by Colletotrichum gloeosporioides: a review. Arch Curr Res Int 24:106–115

Jenny F, Sultana N, Islam M, Khandaker MM, Bhuiyan MAB (2019) A review on anthracnose of mango caused by Colletotrichum gloeosporioides. Bangladesh J Plant Phytopathol 35:65–74

Karthikeyan K, Mariappan V, Sarafoji P, Uma Bharathi K, Loganathan M, Jaya Bharata Reddy M, Anish R (2022) Effect of PCM-based cold storage system under periodic and continuous operations on physico-chemical characteristics of mango (Mangifera indica L.) fruit and performance evaluation of mango cold storage systems. Int J Ambient Energy 43:7678–7693

Karunanayake KOLC, Adikaram NK (2020) Stem-end rot in major tropical and sub-tropical fruit species. Ceylon J Sci 49:327–336

Kaur G, Sharma N, Malhotra S, Devliyal S, Gupta R (2024) Mango leaf disease detection using VGG16 convolutional neural network model. In: 2024 3rd international conference for innovation in technology (INOCON). IEEE, pp 1–6

Khan MW, Hussain Z, Ghafoor I (2023) Mango fruit orchards infestation: a case study on mango fruits in Pakistan

Khanzada MA, Lodhi AM, Shahzad S (2005) Chemical control of Lasiodiplodia theobromae, the causal agent of mango decline in Sindh. Pak J Bot 37:1023

Kiran S, Iqbal J, Iqbal A, Bukhari SIUS, Iqbal M, Mustafa G, Majeed T (2022) Development of sustainable mango pot media nutrition for healthy and clean mango nursery plants. J Glob Innov Agric Sci 10:267–271

Lastochkina O, Seifikalhor M, Aliniaeifard S, Baymiev A, Pusenkova L, Garipova S, Maksimov I (2019) Bacillus spp.: efficient biotic strategy to control postharvest diseases of fruits and vegetables. Plants 8:97

Léchaudel M, Joas J (2007) An overview of preharvest factors influencing mango fruit growth, quality and postharvest behaviour. Braz J Plant Physiol 19:287–298

Lugtenberg B, Kamilova F (2009) Plant-growth-promoting rhizobacteria. Annu Rev Microbiol 63:541–556

Mahmoud GAE, Gaber SH (2024) Nanotechnology applications in postharvest disease management. In: Recent advances in postharvest technologies, Vol 2, Postharvest applications. Springer Nature Switzerland, pp 225–245

Malik MT, Tariq T, Khan AH, Ullah H, Imran M, Iqbal J, Zainab A (2018) Outbreak of Anthracnose and Stem End Rot diseases of mango in changing climate and their management through hot water treatment. Pak J Phytopathol 30:91–98

Nayak D, Yadav A, Kumar S, Lalhruaitluangi N, Mandal D (2023) Mango. In: Tropical and subtropical fruit crops. Apple Academic Press, pp 559–616

Nazir W, Naqvi SAH, Ahmed N, Alrefaei AF, Zulfiqar MA (2024) Optimizing micronutrient supplementation in mango orchards for the suppression of mango anthracnose (Colletotrichum gloeosporioides). Appl Fruit Sci 1–14

Olanrewaju OS, Glick BR, Babalola OO (2017) Mechanisms of action of plant growth promoting bacteria. World J Microbiol Biotechnol 33:1–16

Panth M, Hassler SC, Baysal-Gurel F (2020) Methods for management of soilborne diseases in crop production. Agric 10:16

Qiao Z, Fu Y, Lei C, Li Y (2020) Advances in antimicrobial peptides-based biosensing methods for detection of foodborne pathogens: a review. Food Control 112:107116

Rajwana IA, Amin M, Khan A, Saeed M (2011) Effect of combined application of fungicides and hot water quarantine treatment on postharvest diseases and quality of mango fruit. Pak J Bot 43(1):65–73

Rehman A, Malik AU, Ali H, Alam MW, Sarfraz B (2015) Preharvest factors influencing the postharvest disease development and fruit quality of mango. J Environ Agric Sci 3:42–47

Roberts PB (2016) Food irradiation: standards, regulations and world-wide trade. Radiat Phys Chem 129:30–34

Salgado-Cruz MDLP, Salgado-Cruz J, García-Hernández AB, Calderón-Domínguez G, Gómez-Viquez H, Oliver-Espinoza R, Yáñez-Fernández J et al (2021) Chitosan as a coating for biocontrol in postharvest products: a bibliometric review. Membranes 11(6):421

Schmidt TM, Thomé AHE, Sperotto RA, Granada CE (2019) Effect of rhizobia inoculation on the development of soil-borne pathogens infecting common bean plants. Eur J Plant Pathol 153(3):687–694

Sharma A, Sharma IM, Sharma M, Sharma K, Sharma A (2021) Effectiveness of fungal, bacterial and yeast antagonists for management of mango anthracnose (Colletotrichum gloeosporioides). Egypt J Biol Pest Control 31:1–11

Tewari S, Shrivas VL, Hariprasad P, Sharma S (2019) Harnessing endophytes as biocontrol agents. In: Plant health under biotic stress: Volume 2: Microbial interactions, pp 189–218

Uddin M, Shefat S, Afroz M, Moon N (2018) Management of anthracnose disease of mango caused by Colletotrichum gloeosporioides: a review. Acta Sci Agric 2(10):169–177

Ullah MA, Kiloes AM, Aziz AA, Joyce DC (2024) Impact of factors contributing to internal disorders of mango (Mangifera indica L.) fruit—a systematic literature review. Scientia Hortic 331:113150

Wang S, Wang X, He Q, Lin H, Chang H, Liu Y, Song X et al (2023) Analysis of the fungicidal efficacy, environmental fate, and safety of the application of a mefentrifluconazole and pyraclostrobin mixture to control mango anthracnose. J Sci Food Agric 103(1):400–410

Wickrama SP, Kandangama NB, Wickramaarachchi T, Weerahewa J (2024) Assessing the impact of non-tariff measures on Sri Lankan mango exports: Insights, challenges, and recommendations. Front Sustain Food Syst 8:1293263

World Population by Country 2024 (Live) (2024–10–19) World Population Review. https://worldpopulationreview.com/

Xu ST, Bai TT, Zhang L, Fan HC, Yang PW, Yin KS, Zheng SJ et al (2017) Evaluation of different banana varieties on Fusarium wilt TR4 resistance by phenotypic symptom and real-time quantitative PCR. Southwest China J Agric Sci 30:1997–2002

Xu T, Gao C, Feng X, Yang Y, Shen X, Tang X (2019) Structure, physical and antioxidant properties of chitosan-gum Arabic edible films incorporated with cinnamon essential oil. Int J Biol Macromol 134:230–236

Xu J, Xian Q, Wang K, Dong J, Zhang C, Du S, Chen X et al (2022) Screening and identification of candidate Fusarium wilt-resistance genes from pumpkin. Horticult Plant J 8(5):583–592

Yenjit P, Intanoo W, Chamswarng C, Siripanich J, Intana W (2004) Use of promising bacterial strains for controlling anthracnose on leaf and fruit of mango caused by Colletotrichum gloeosporioides. Walailak J Sci Technol 1(2):56–69

Yeo YS, Dembele DD, Camara B, Ouattara S, Rey JY, Fernandez D, Kone D (2023) Effect of some biopesticides based on essential oil and plant extracts on postharvest mango stem-end rot disease caused by Lasiodiplodia theobromae. J Agric Food Res 14:100798

Zhao F, Chen L, Yang L, Sun L, Li S, Li M, Feng Q (2020) Effects of land use and rainfall on sequestration of veterinary antibiotics in soils at the hillslope scale. Environ Pollut 260:114112

Zul Helmey MS, Muhamad Hafiz MH, Wan Mahfuzah WI, Shaidatul Azdawiyah AT, Mohd Aziz R, Mohd Farid AF, Siti Afiqah J et al (2021) Effect of paclobutrazol (PBZ) application induction on vegetative, reproductive growth and fruit quality of harumanis mango under greenhouse conditions. Acta Hortic 1317:209–216

Chapter 27
Climate Change Nexus and Its Impact on Mango Cultivation: Novel Strategies to Combat Pests and Diseases, Unraveling Phytosanitary Issues and Insights into Global Trade Trends

Muhammad Atiq, Nasir Ahmed Rajput, Azeem Akram, Salman Ahmad, Irfan Ahmad, Mukhtar Ahmed, Shakeel Ahmad, and Amjad Malik

Abstract Mango (*Mangifera indica*) holds cultural, nutritional, medicinal, industrial and economic value across the globe. This fruit is a main contributor in international trade but its cultivation experiences evolving constraints due to climate change. Key mango-producers have witnessed changes in flowering, fruiting and yield because of these environmental shifts. The emergence of new pests and diseases impends mango production. Traditional farming tools are no longer ample in handling these new risks. It is the most pressing need for the mango industry to prefer sustainable practices. Use of precision agriculture, smarter irrigation techniques and bio-based pest intervention can serve as a strategic approach to minimize the impact of environmental shifts on mango productivity. To address the challenges

M. Atiq (✉) · N. A. Rajput · A. Akram
Department of Plant Pathology, Faculty of Agriculture, University of Agriculture, Faisalabad, Pakistan
e-mail: muhammad.atiq@uaf.edu.pk

S. Ahmad
Department of Plant Pathology, College of Agriculture, University of Sargodha, Sargodha, Pakistan

I. Ahmad
Department of Forestry, Range Management and Wildlife, University of Agriculture, Faisalabad, Pakistan

M. Ahmed
Department of Agronomy, Pir Mehr Ali Shah Arid Agriculture University, Rawalpindi, Pakistan

S. Ahmad (✉)
Institute of Agronomy, Faculty of Agricultural Sciences and Technology, Bahauddin Zakariya University, Multan, Pakistan
e-mail: shakeelahmad@bzu.edu.pk

A. Malik
Faculty of Education, University College of the North (UCN), 304-Ed 7, Thompson, MB, Canada

S. A. H. Naqvi et al. (eds.), *Climate Change and Mango Production*,
https://doi.org/10.1007/978-3-031-87035-4_27

generated by climate change, incorporating breeding programs, biotechnology and nanotechnology based solutions into pest management can represent a transformative approach. Innovations in logistics like improved cold chain systems can prove fruitful for ensuring a stable global mango supply, especially as climate change negatively impacts production patterns. Globally, strict phytosanitary measures including upgraded inspections and certifications are needed to uphold the high standards required for mango export. This chapter focuses on novel strategies to combat rising pest threats, address phytosanitary challenges in global trade, examine climate change impacts on mango production and supply chains and stresses the need for sustainable practices to heighten mango production.

27.1 Introduction

Mango (*Mangifera indica* L.), the main fruit crop of the family Anacardiaceae, is mainly grown in tropical and subtropical regions of the world because of its taste, aroma, and nutritional value. The genus name composed of two breads: 'mango' derived from the Spanish language and 'fera' the Latin word to bear. The species name is 'Indica', which indicates India (Gautam et al. 2023). Mangoes have always been revered in the tropical parts of regions as the "king of fruits" and is grown in various regions in the world (Murtaza et al. 2021). With respect to mango production on the world scale, Pakistan is at 4th rank and in terms of exports it stands at 10th (Khan et al. 2023). Mangoes are produced most in India with about 24.7 million tons every year. The second producer is Indonesia, with an estimated production of 3.6 million tons of mangoes in a year's time. Daily, China is responsible for the production of some two and a half million diversified treated mangoes on the local and external market. In Pakistan, the annual production of mangoes is about 2.3 million tons. Mexico's almost tropical climate is quite conducive to the farming of mangoes. For similar reasons, the warm climate of these countries is conducive for mango farming, Brazil too has ideal features that enhance the production of damn good mangoes (World Population Review 2024). It is an impressively important fruit in warm temperate and tropical climates (Amin et al. 2019; Bhattacharya et al. 2018; Cotrim et al. 2011; Da Cunha et al. 2018; DeRosa et al. 2010). This fruit is very much appreciated for its unique taste. This fruit is thought to be an excellent source of nutrients and phytochemical substances. This exotic crop is intertwined within the social structure of the rural area in Maltan (Ekesi et al. 2016; Evans et al. 2017; Kah et al. 2019; Khan and Naveed 2017; Khodakovskaya et al. 2018). It is generating many folk stories which depict the happiness, grief and reminiscence associated with the growing of it (Nasir et al. 2014; Nayak et al. 2023; Normand et al. 2015; Raza et al. 2020; Sarwar et al. 2019; Tharanathan et al. 2006; Wang et al. 2019). Its influence permeates around production, commerce and financial aspects. It has always occupied a place of high prestige in all calendar events of the citizens. Such a fruit also possesses spiritual, cultural factors in the day-to-day activities of the people apart from its corporate significance (Saadat et al. 2022).

Mango production is seriously affected and reduced by several pathological problems, severe diseases, and the use of pesticides in agriculture. Such diseases also attack the mango leaves thereby reducing the quantity and the quality of mango production to a considerable extent. (Shaik and Swamykan 2023). Numerous diseases brought on by certain phytopathogens lower both the quality and productivity of the mangoes. Approximately 83 diseases worldwide have been documented that target mango tree and fruit. And of these, 27 were considered as important in Pakistan. Farmers pursuing maximum production continue to face fighting disease as a hurdle, more so since this is attributable to pesticide resistance. There is a current restriction that impedes further control of plant diseases and this is the ability of pathogens to become resistant to borrowed-among other strategies-fungi. Nonetheless, efforts are being made in the research work to increase the movable and push back the immovable backgrounds of infections. The acquisition of the above resistance in pathogens makes the control of plant diseases even complex (Khaskheli 2020). Mango anthracnose disease (MAD) irrefutably is a very severe disease. In unmanaged mango plantations, the disease is causing up to 100% loss of their yield. There are different strains in the *Colletotrichum* genus which are responsible for global MAD. They are all able to least resistant to this disease. These days there is no cultivator in the market who is resistant to all the strains (Dofuor et al. 2023). Governmental policy, climate change and globally increasing temperature rise are among the main aggravating factors for agriculture today. Climate change, such as increased atmospheric CO_2 levels, has an impact on agricultural production as well as insect pest populations. Climate change can affect pests in a variety of ways, including expanding their geographical bounds, increasing pest survivability during overwintering, and producing more generations. Helps to promote the dissemination of plant diseases that are transmitted by driving insects while lessening the efficacy of biological control, especially natural agents of predation (Skendžić et al. 2021). Increasing level of CO_2 in the air is the leading contributor towards climate change. It has a bearing on agriculture. CO_2 is very important in the main activities of plants, photosynthesis included. Intermittent cropping is one of the major problems facing mango farming. Unpredictable weather conditions such as too cool nights and uncharacteristically warm winters adversely affect mango flowering. Extreme climate is contributing to changing mango production. Thus even though all these climates are changing very rapidly, it is not good news for mango buyers, producers and scientists. Nevertheless, this suggests the great potential of using the existing diverse mango genetic resources for breeding programs and selection for resistance (Halder and Hasan 2020). Climate change, therefore, carries risk disease outbreaks and modulates pathogen evolution. It is changing host–pathogen interactions. It promotes the occurrence of vaginal templates with new virulence factors. Attending such situations can also alter the geographic range of the disease. This may cause new plant disease outbreaks in non-endemic countries (Singh et al. 2023).

27.2 Climate Change and Its Impact on Cultivation of Mango

27.2.1 Impact of Temperature Fluctuations

Temperature fluctuations affect mango flowering, fruiting and overall productivity, significantly influencing the mango growth cycle, including flowering, fruit development, taste, and appearance in most production areas. Mango trees need higher temperatures to grow, while flowering starts after the coldest winter period and increases as temperatures rise. Vegetative growth is induced if the average monthly temperature rises above 15 °C, although the temperature requirements for vegetative growth and flowering are clearly distinguishable. Regions with summer temperatures below 30 °C or winter temperatures below 6 °C cannot support commercial production of mango fruits. Winter temperatures in mango-growing areas range between 6 and 22 °C. Scientists found that climate scenarios predict a 3–6 °C increase in temperature, with the rise being greater in Northern India than in southern regions. Climate change could harm hot regions of the troposphere where temperatures are already high. Though mango trees are well-established in hot, semiarid regions, as well as monsoon tracts and tropical regions, they are sensitive to heat, harshness, and evaporation. Mango trees can survive in pH ranges of 0–48, while prolonged exposure to temperatures below 6 °C is lethal. Most mango varieties can withstand short temperature dips of 1–2 °C, as long as the cold is not prolonged and plant growth remains active for several hours.

27.2.2 Water Scarcity and Irrigation Challenges in Mango Farming

It is well-known that mango is a valuable commodity in addressing the challenges of water scarcity in agriculture. As water sources become increasingly scarce worldwide, it is crucial to ensure efficient irrigation practices and proper water management in crop production. While mango trees can withstand certain drought conditions, water is nonetheless essential for fruit development and quality. Fruits and plant organs require water for proper cell enlargement and division, and a shortage of water can hinder growth. Since mangoes typically grow in dry seasons, farmers need to irrigate the trees to ensure high yields and fruit quality. Managing soil water effectively is critical to encourage optimal mango production. To enhance regulated water supply, it is important to assess the quantitative relationship between water consumption, mango production, and fruit quality.

Varying levels of soil moisture affected the growth and quality of the Guifei mango variety, using soil moisture sensors. Fruit development is initially controlled, increasing during the growth phase and then decreasing during the expansion stage, with a single growth spurt. The first 42 days of fruit development are particularly vulnerable, as water shortages during this period can lead to fruit abscission, reduced fruit mass, and fewer cells. There is an inverse link between the number of fruits and their average weight, but irrigation has no meaningful influence. Dry years negatively impact mango development and production yield, as reduced water content in the leaves affects the plant's drought tolerance and turgor. Drought causes stomatal closure, which inhibits cell enlargement and growth processes (Jaleel et al. 2008). Prolonged drought also reduces the rate of photosynthesis and disrupts normal plant metabolism, potentially becoming lethal in extreme cases. Under water-deficit (WD) conditions in mangoes, vegetative growth slows significantly. Although short-term drying of rivers may induce flowering, it has a strongly negative effect on vegetative activity, characterized by a decrease in the number of leaves, their size, and hydration potential, depending on the specifics of the situation (Schaffer et al. 1994). Regulated deficit irrigation (RDI) is a novel strategy that improves water-use efficiency. Mangoes are relatively tolerant to water stress, and their growth and productivity are not as severely impacted by water scarcity (Santos et al. 2014).

27.2.3 Climate Change Effects on Soil Structure and Nutrients Cycling

Climate change directly and indirectly impact on soil functions. Direct effects include alterations in soil processes such as the transformation of organic carbon and nutrient cycling, driven by changes in moisture, temperature, and increased soil erosion due to more frequent and intense rainfall events. Blum scientist was a pioneer in creating the idea of relating soil processes and functions to environmental and societal benefits. Soil management strategies in response to climate change can also have an impact on the conditions required to support these activities.

Many studies have used data to analyses the impacts of climate change on soil functioning (Coyle et al. 2016). For example, in soils rich in organic matter in the UK, higher temperatures and lower soil moisture conditions, which arise from increased temperatures or reduced humidity, have been shown to reduce soil carbon (C) sequestration potential (Ostle et al. 2009). Climate change can have a direct and indirect impact on soil, impacting related functions that support adaptation efforts. Agricultural management strategies may offset some of these climate-induced consequences by accelerating soil organic carbon (SOC) sequestration (Haddaway et al. 2015). Farmers may employ adaptive tactics for a range of closely related reasons, including swings in market prices, technology advancements, greater understanding, and the effects of climate change (Reidsma et al. 2015). In modern European agriculture, numerous practical strategies have been proposed to cope with climate change. These include developing irrigation systems in areas of concern, adjusting crop calendars, improving fertilizer application practices on agricultural land, modifying ploughing systems, and reclaiming regions affected by thawing permafrost (Mandryk et al. 2017).

27.3 Climate Change and Adaptation in Mango Production

A review of information indicates that an important adaptation technique for mango cultivation involves shifting production from areas experiencing high levels of climatic stress to regions with more suitable climatic conditions. This strategy helps mango trees thrive in environments that better meet their needs for growth and fruit production. Additionally, there is a more urgent need for research on various aspects of stress physiology, particularly in relation to high temperatures, water deficits, light, temperature stress, and salinity stress. These areas of research are critical for creating solutions to reduce the effects of climate change on mango production.

27.4 Adaptation Strategies for Mango Cultivation Under Climate Stress

One adaptation strategy involves relocating mango cultivation to areas with more favorable climates. Another urgent need is research on various levels, particularly focusing on stress physiology related to high temperatures, water scarcity, light deprivation, and salt stress. Additionally, the genetic enhancement of mango cultivars is a promising approach, taking into account factors such as drought tolerance, high-temperature resilience, and the ability to induce flowering in hot conditions. Rootstock selection should prioritize drought and salinity tolerance while identifying low-vigor rootstocks with robust root systems capable of accessing water from a large soil volume and withstanding strong winds. Moreover, tailored orchard management practices should be established, including improving irrigation methods to enhance water efficiency, regulating foliage canopy and blooming, and planning orchards to create an ideal microenvironment for the plants. Mango (*Mangifera indica* Linn.), a popular evergreen fruit crop in tropical and subtropical regions, faces threats from around 400 insect and mite pests (Venkata Rami Reddy et al. 2018). Despite the numerous pests affecting mango crops, only a few are of significant economic concern, including leafhoppers, stem borers, fruit flies, stone weevils, mealybugs, and gall midges. Fruit flies and stone weevils are particularly important for quarantine regulations, hindering international mango trade. Additionally, increasing atmospheric CO_2 levels and climate change may indirectly affect crops by influencing pest dynamics (Skendžić et al. 2021). Evidence indicates that pests like aphids and weevil larvae thrive in conditions of elevated CO_2, and changing climate parameters heighten the risk of new pest invasions (Rajatiya et al. 2018).

In arid and high-temperature regions, the likelihood of fungal infections such as anthracnose and powdery mildew in mangoes is minimized because spores are quickly deactivated by sunlight, low moisture levels, and temperature fluctuations (under 18 °C or beyond 35 °C). Nevertheless, as the global emphasis on premium-quality fruit for domestic use and export increases, pests that harm mangoes by feeding, puncturing, or depositing eggs in the flesh or seed can result in substantial crop damage. This highlights the need for effective pest management strategies to protect fruit quality (Ansari et al. 2019). Only fruit flies, seed weevils, and lepidopterous larvae are able to penetrate the fruit flesh and seed. Among mango pests, hoppers are the most major and ubiquitous, harming mango (*Mangifera indica* L.) plants nationwide. They cause tremendous damage during the blossoming stage, resulting in significant losses (Rai et al. 2019). Hoppers are active from October to January and from May to February.

Mango anthracnose (MAD) is a globally damaging disease that significantly impacts mangoes both during the growing season and after harvest. Infected mango trees suffer damage, resulting in lower yield and fruit quality. In poorly managed orchards with favorable conditions for the disease, MAD can lead to total yield loss. This disease has spread in almost all mango-growing areas, and the considerable financial losses it generates have led to extensive studies on postharvest wastage. Of all the diseases affecting mangoes, powdery mildew is the most prevalent. Its onset typically depends on the atmospheric factors of the region. New shoots appear and develop rapidly right before blooming. At this stage, tender leaves and flower clusters are covered with a pale, powdery layer, preventing the flowers on the affected cluster from being pollinated and leading to their early drop.

27.5 Traditional Management Practices of Anthracnose and Stem End Rot and Their Limitations

The occurrence of anthracnose during both pre- and post-harvest stages of mango production is a frequent issue, leading to significant economic losses each year. Controlling anthracnose can be effectively managed with pre-harvest, post-harvest, or a combination of both approaches. The main strategies involve fungicide application, proper cultural practices, and selecting suitable cultivars (Dofuor et al. 2023). Carbendazim was found to be the most effective fungicide against *Colletotrichum gloeosporioides*, completely inhibiting mycelial growth at a 100 ppm dose. Propiconazole also achieved 100% inhibition at a 500 ppm dose, while mancozeb was highly effective at 1000 ppm. Copper oxychloride, however, was the least effective fungicide, showing only 55% inhibition even at the highest dose (Iqbal et al. 2022).

The indiscriminate use of synthetic fungicides to control mango anthracnose can result in environmental pollution, water contamination, soil compaction, and ecological issues, ultimately impacting the sustainability of mango production (Dofuor et al. 2023). The individual and combined use of biocontrol agents such as *Trichoderma harzianum*, *Bacillus subtilis*, and *Pichia anomala* through foliar sprays or fruit dips effectively controlled mango anthracnose. These bio-formulations could serve as alternatives to chemical fungicides, contributing to environmental protection and integrated disease management. Several diseases linked to stem-end rot have been identified worldwide, including *Lasiodiplodia theobromae, Neofusicoccum parvum, Fusicoccum aesculi, Nattrassia mangiferae,* and *Fusicoccum* species in Australia and Brazil. In Egypt, the main pathogens are *Colletotrichum gloeosporioides, Alternaria alternata, Botrytis cinerea and Lasiodiplodia theobromae* (Alam et al. 2020). Carbendazim was highly effective in reducing fungal infections in mango plants, preventing symptoms like gum exudation, dieback, and wilting, while also enhancing vegetative growth (Khanzada et al. 2005). Researchers are developing sustainable alternatives to chemical pesticides, with plant extracts like NECO, ASTOUN, FERCA, and *Moringa oleifera* effectively controlling mango stem-end rot. These eco-friendly options show promise for managing postharvest diseases, but further testing on naturally infected fruits is needed.

27.6 Challenges Faced in Mango Trade

Some of the main challenges in mango trading include high international freight costs, issues with customs clearance, and difficulties meeting sanitary and phytosanitary (SPS) standards. There's also a shortage of good-quality mangoes, problems with certification, high local transport costs, and a lack of standard post-harvest handling. Other challenges are poor infrastructure, competition from other exporters, and limited institutional support for credit (Padaliya and Pundir 2022). Small farmers in developing countries face difficulties in meeting strict food safety standards due to their lack of technical and financial resources, which affects their ability to export (Hou et al. 2015).

27.6.1 Quarantine Pests as Barriers to Mango Trade

Mango is affected by around 400 insect and mite pests, though only a few cause serious damage. Major pests include leafhoppers, stem borers, fruit flies, stone weevils, and others. Among them, fruit flies and stone weevils are especially important because they affect the international trade of mangoes due to quarantine concerns (Venkata Rami Reddy et al. 2018). Fruit flies (Tephritidae: Diptera) are among the notorious pests of horticultural crops worldwide, causing both yield losses and marketability issues. They also pose barriers to international trade of fresh produce. The biggest risk for fruit fly spread comes from unauthorized transport of untreated fruits and vegetables by sea or airline passengers (Verghese nd Rashmi 2016). Moreover, the mango stone weevil (*Sternochetus mangiferae* Fabricius 1775) is an insect that exclusively infests mangoes and is widespread in major mango-cultivating regions around the world. In Asia, it is present in countries such as India, Pakistan, Bangladesh, Thailand, and others. The infestation rate of mango stone weevil differs significantly, ranging from 5 to 90%, influenced by the mango variety, management practices used, and location (Baradevanal et al. 2023). The USA and Australia have important regulations regarding phytosanitary treatments. The US allows a conventional dose of 400 Gy to sterilise or control adult insects on any crops, with the exception of pupae and mature Lepidoptera. Australia's Interstate Certificate Assurance (ICA) procedure mandates 150 Gy for fruit flies and 400 Gy for other insects, excluding internally pupating mature Lepidoptera. Food Standards Australia and New Zealand have approved the use of irradiation on produce (Roberts 2016). Furthermore, the USDA's Animal and Plant Health Inspection Service recently changed its rules to allow irradiation treatment at U.S. ports of entry. This change benefits exporting countries by giving them access to U.S. markets without needing to invest heavily in building irradiation facilities in their own countries. It opens up opportunities for countries in Asia, Africa, and the Americas to explore market potential with lower costs and risks, especially for products with low volumes or short seasons that wouldn't justify building a facility (Bustos-Griffin et al. 2015).

27.6.2 Post-harvest Diseases of Mango Impacting Mango Trade and Their Management

Mango Anthracnose is caused by *Colletotrichum gloeosporioides*. This disease is a significant disease affecting mangoes on pre harvest and post-harvest stages worldwide, including in Bangladesh (Uddin et al. 2018). The disease harms mango trees, resulting in reduced fruit yield and quality. In poorly maintained orchards with favorable conditions for the disease, it can lead to total yield loss (Dofuor et al. 2023). Effective control of anthracnose can be achieved through the use of fungicides both before and after harvest. Applying organic sulfur fungicides, such as dithiocarbamates, along with heterocyclic nitrogen compounds like mancozeb, zineb, and captan can provide adequate protection against mango anthracnose (Iqbal et al. 2022). Additionally, stem end rot (SER) is a complex fungal disease that significantly threatens the quality of mango exports from Pakistan (Alam et al. 2020). *Lasiodiplodia theobromae, Neofusicoccum parvum, Fusicoccum aesculi, Nattrassia mangiferae*, and *Fusicoccum* spp. Have been found to linked with SER in Australia and Brazil. In Egypt, the *Lasiodiplodia theobromae, Colletotrichum gloeosporioides, Alternaria alternata*, and *Botrytis cinerea* has been reported as the etiology of disease (Alam et al. 2020). SER pathogens access the mango stem by natural holes such as stomata, lenticels, and wounds. Once inside, they can persist in the stem tissue until the fruit ripens, at which point they detect physiological changes and convert from an endophytic stage to a damaging necrotrophic stage that produces watery rot at button of mango (Diskin et al. 2017). Carbendazim has proven to be effective in reducing fungal infections in mango plants, preventing symptoms such as gum exudation, dieback, and wilting, while significantly boosting vegetative growth (Khanzada et al. 2005).

27.7 Limitations Caused by Varying Production

Climate change is posing a greater challenge for the production of mangoes (*Mangifera indica*), as it affects the supply chain and destabilizes yield stability (Rajvardhan Kiran et al. 2023). Production fluctuations can lead to excesses or shortages of goods, exacerbating the shifting dynamics of the market (Hadachek et al. 2024). Extreme conditions, especially droughts or excessive precipitation, can drastically reduce yields, resulting in cost spikes and potential losses for exporters (Asare-Nuamah et al. 2022). Conversely, overproduction may flood the market, lowering prices and negatively impacting the profitability of growers (van der Maden et al. 2021).

27.8 Solution to Overcoming Such Obstacles is to Get into Emerging Export Markets

Countries such as India, Mexico, and Pakistan are actively seeking new markets in North America and Europe, where the demand for fresh mangoes is rising (Siddiqui and Paul 2018). To sell their fruit in these markets, growers must adhere to stringent phytosanitary laws to ensure that the fruit is free of diseases and pests (Weerahewa and Wickramaarchchi 2023). To promote safe exports, the International Plant Protection Convention (IPPC) is creating standardized guidelines for the mango trade (Stubbings 2005).

27.9 The Use of Technology in the Sustainable Trade of Mangoes

Technology plays an essential role in sustaining a viable mango trade, particularly through advances in cold chain logistics that protect mango quality and extend shelf life. Harvesting, packaging, preserving, and transporting mangoes require proper temperature control to prevent post-harvest losses, which can amount to 25–30% in the absence of adequate management (Le et al. 2022). To address these challenges, technological advancements are vital, with breakthroughs in cold chain logistics playing a key role (Han et al. 2021). One such innovation is modified atmosphere packaging (MAP), which surrounds mangoes with a high-CO_2 and low-O_2 environment to sustain their quality and extend their shelf life (Ntsoane et al. 2020;

Fig. 27.1 A well-maintained mango orchard following sanitary and phytosanitary measures to avoid biotic and abiotic stresses to tress

Wei et al. 2021). Additionally, mangoes are preserved at optimal temperatures even when external conditions vary due to the use of phase change materials (PCMs) in packaging, which absorb or release thermal energy to maintain constant temperatures during transit (Amwoka et al. 2021). Enhanced packing methods and insect-proof packaging also help reduce pest threats during international trade (Thakur et al. 2021). By utilizing these technologies, exporters can comply with the strict phytosanitary requirements set by importing nations (Wickrama et al. 2024). Digital technologies are transforming supply chain management by providing real-time data on market trends and consumer preferences, enabling cultivators to make informed decisions about production and marketing strategies, ultimately enhancing their competitiveness in the global market (Attaran 2020) (Figs. 27.1 and 27.2).

27.10 Challenges in Mango Cultivation Due to Climate Change

The cultivation of mangoes faces intricate challenges from climate change, which directly affects crop yields, the prevalence of diseases and pests, and global trade patterns. Changes in temperature, water availability, and soil health significantly

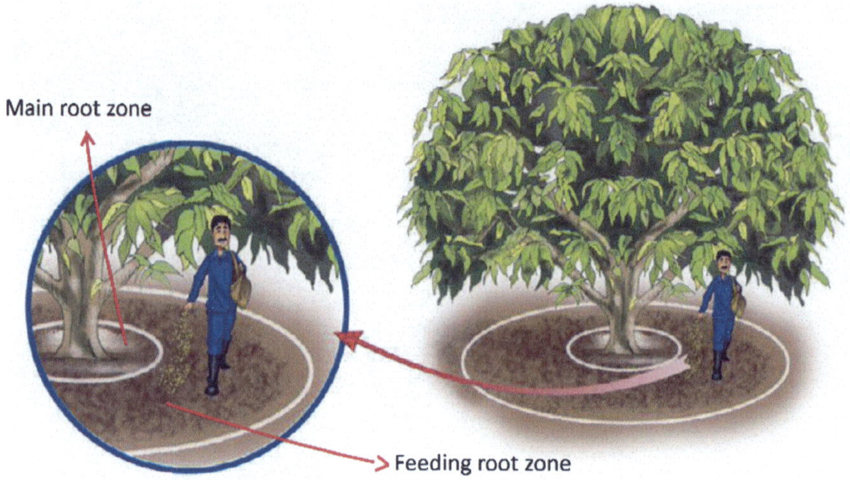

Fig. 27.2 Proper sanitation help farmers to apply proper fertilizer at proper place

impact mango flowering, fruiting, and overall production, particularly in tropical regions like Pakistan. Sustainable mango cultivation is further complicated by the emergence of new pests and diseases, which are exacerbated by climate fluctuations (Table 27.1). To address these challenges, innovative approaches such as Integrated Pest Management (IPM) combined with biotechnology-based solutions are needed. Mango-exporting economies also confront additional hurdles, including phytosanitary regulations and stringent limitations on global trade due to threats posed by diseases like anthracnose and pests like fruit flies, which can compromise the quality of their exports (Table 27.2).

Table 27.1 Impact of climate change on mango pests, diseases, and phytosanitary issues

Climate change factor	Impact on pests and diseases	Phytosanitary issues
Rising temperatures	Accelerated pest life cycles, increased disease incidences (e.g., anthracnose, bacterial wilt)	Spread of quarantine pests and pathogens to new regions, complicating trade regulations
Erratic rainfall patterns	Increased humidity promoting fungal diseases (e.g., powdery mildew)	Poor postharvest handling due to excess moisture, leading to the spread of pathogens
Increased CO_2 concentration	Enhanced plant growth, but increased susceptibility to pests like mealybugs and fruit flies	Higher levels of pest infestation during critical harvest periods, challenging phytosanitary measures
Drought and water scarcity	Stress on mango trees reducing resistance to pests and diseases	Difficulty in maintaining adequate water quality for pest control in processing plants
More frequent storm events	Physical damage to mango trees, making them more vulnerable to pest attacks	Increased risk of contamination from external sources due to storm-related damages
Shifts in pest distribution	New pests or pathogens emerging in regions previously unaffected by mango cultivation	Changing pest profiles require adaptation of quarantine standards and surveillance systems

Table 27.2 Novel strategies to combat pests, diseases, and phytosanitary issues in mango cultivation

Strategy	Description	Impact on mango cultivation and trade
Climate-smart integrated pest management (IPM)	Combines biological, cultural, and minimal chemical control	Reduces pesticide resistance, promotes sustainability, and aligns with global trade standards
Biocontrol agents	Use of natural predators, parasitoids, or microbial agents	Minimizes reliance on chemical pesticides, improving the phytosanitary status of exports
Resistant mango varieties	Development of varieties with enhanced resistance to pests and diseases	Reduces pest and disease impact, ensuring high-quality mangoes that meet international standards
Climate-adaptive disease forecasting	Use of weather data and predictive models for pest and disease outbreaks	Early intervention allows for timely management, reducing damage and enhancing trade compliance
Optimized irrigation and water management	Implementing efficient irrigation systems and drought-resistant crops	Ensures consistent tree health, reducing vulnerability to pests and diseases in stress conditions
Phytosanitary surveillance	Enhanced monitoring and quarantine measures to detect and control pests	Facilitates smooth international trade by maintaining the export quality and safety of mangoes
Postharvest disease management	Application of biocides, hot water treatments, and controlled atmosphere storage	Reduces disease spread during transport, preserving mango quality and extending shelf life
Global trade compliance	Adapting to evolving phytosanitary standards and export protocols	Ensures mango exports meet international market regulations, increasing market access

27.11 Conclusion

Climate change has caused the emergence of new pests and diseases which threatens mango industry. Conventional farming tools are no longer plentiful in managing these new risks. It is pivotal to prefer sustainable practices because use of precision agriculture, smarter irrigation techniques and bio-based pest intervention can serve as a strategic approach to minimize the impact of environmental shifts on mango productivity. To address the constraints posed by climate change, incorporating breeding programs, biotechnology and nanotechnology based solutions into pest management can embody a transformative approach. Novelties in logistics such as amended cold chain systems can prove productive for ensuring a stable global mango supply, particularly as climate change adversely impacts production patterns. Worldwide, strict phytosanitary measures including upgraded inspections and certifications are obligatory to maintain the high standards required for mango export.

References

Alam MW, Rehman A, Malik AU, Ahmad S, Haider MS, Amin M (2020) Dynamics of stem end rot disease of mango fruit and its management. Pak J Agric Sci 57:1–8

Amin M, Ali Q, Ashraf M, Abbas T, Ali M (2019) Soil management for sustainable mango production: a review. Int J Agric Biol 22:1–8

Amwoka EM, Ambuko JL, Jesang' HM, Owino WO, (2021) Effectiveness of selected cold chain management practices to extend the shelf life of mango fruit. Adv Agric 2021:8859144

Ansari MS, Basri R, Shekhawat SS (2019) Insect pest infestation during field and storage of fruits and vegetables. In: Malik A, Erginkaya Z, Erten H (eds) Health and safety aspects of food processing technologies. Springer, Cham, Switzerland, pp 121–207

Asare-Nuamah P, Antwi-Agyei P, Dick-Sagoe C (2022) Mitigating the risks of climate variability and change on mango seedlings in Ghana: evidence from mango seedlings producers in the Yilo Krobo Municipality. Environ Challenges 8:100594

Attaran M (2020) Digital technology enablers and their implications for supply chain management. Supply Chain Forum Int J 21:158–172

Baradevanal G, Chander S, Singh HS, Reddy DS, Rajan S (2023) Mapping the risk of quarantine pest Sternochetus mangiferae under different climate change scenarios through species distribution modelling. Int J Trop Insect Sci 43:919–932

Bhattacharya S, Hazra S, Jana BB (2018) Management of old and senile mango orchards. The mango: botany, production and uses, pp 411–424

Bustos-Griffin E, Hallman GJ, Griffin RL (2015) Phytosanitary irradiation in ports of entry: a practical solution for developing countries. Int J Food Sci Technol 50:249–255. https://doi.org/10.1111/ijfs.12619

Cotrim CE, Coelho-Filho MA, Coelho EF, Ramos MM, Cecon PR (2011) Regulated deficit irrigation and Tommy Atkins mango orchard productivity under micro-sprinkling in Brazilian semi-arid. Eng Agric 31:1052–1063

Coyle C, Creamer RE, Schulte RPO, O'sullivan L, Jordan P (2016) A functional land management conceptual framework under soil drainage and land use scenarios. Environ Sci Policy 56:39–48

Da Cunha T, Ferraz LP, Wehr PP, Kupper KC (2018) Antifungal activity and action mechanisms of yeast isolates from citrus against *Penicillium italicum*. Int J Food Microbiol 276:20–27

DeRosa MC, Monreal C, Schnitzer M, Walsh R, Sultan Y (2010) Nanotechnology in fertilizers. Nat Nanotechnol 5:91–91

Diskin S, Feygenberg O, Maurer D, Droby S, Prusky D, Alkan N (2017) Microbiome alterations are correlated with the occurrence of postharvest stem-end rot in mango fruit. Phytobiomes 1:117–127

Dofuor AK, Quartey NKA, Osabutey AF, Antwi-Agyakwa AK, Asante K, Boateng BO, Ninsin KD (2023) Mango anthracnose disease: the current situation and direction for future research. Front Microbiol 14:1168203

Ekesi S, De Meyer M, Mohamed SA, Virgilio M, Borgemeister C (2016) Taxonomy, ecology, and management of native and exotic fruit fly species in Africa. Annu Rev Entomol 61:219–238

Evans EA, Ballen FH, Siddiq M (2017) Mango production, global trade, consumption trends, and postharvest processing and nutrition. In: Siddiq M, Brecht JK, Sidhu JS (eds) Handbook of mango fruit: production, postharvest science, processing technology, and nutrition. John Wiley and Sons, United Kingdom, pp 1–16

Gautam P, Kumar A, Prakash O, Singh J, Ali I (2023) Advances in production technology of mango. In: Gautam P, Kumar A, Prakash O, Singh J, Ali I (eds) A textbook on advances in production technology of tropical and subtropical fruits. New Vishal Publication, New Delhi, pp 9–59

Hadachek J, Ma M, Sexton RJ (2024) Market structure and resilience of food supply chains under extreme events. Am J Agric Econ 106:21–44

Haddaway NR, Hedlund K, Jackson LE, Kätterer T, Lugato E, Thomsen IK, Söderström B (2015) What are the effects of agricultural management on soil organic carbon in boreo-temperate systems? Environ Evid 4:23

Halder S, Hasan MA (2020) Climate change and mango production. Chem Sci Rev Lett 9:1–7

Han JW, Zuo M, Zhu WY, Zuo JH, Lü EL, Yang XT (2021) A comprehensive review of cold chain logistics for fresh agricultural products: current status, challenges, and future trends. Trends Food Sci Technol 109:536–551

Hou MA, Grazia C, Malorgio G (2015) Food safety standards and international supply chain organization: a case study of the Moroccan fruit and vegetable exports. Food Control 55:190–199

Iqbal S, Khan M, Atiq M, Rajput N, Usman M, Nawaz A, Kachelo GA, Akram A, Ahmad H (2022) Mango anthracnose: global status and the way forward for disease management. J Innov Sci 8:222–235

Jaleel CA, Manivannan P, Wahid A, Farooq M (2008) Drought stress in plants: a review on morphological characteristics and pigment composition. Int J Agric Biol 11:100–105

Kah M, Kookana RS, Gogos A, Bucheli TD (2019) A critical evaluation of nano pesticides and nano fertilizers against their conventional analogs. Nat Nanotechnol 13:677–684

Khan RA, Naveed M (2017) Occurrence and seasonal abundance of fruit fly, *Bactrocera zonata* Saunders (Diptera: Tephritidae) in relation to meteorological factors. Pak J Zool 49:3

Khan MW, Hussain Z, Ghafoor I (2023) Mango fruit orchards infestation: a case study on mango fruits in Pakistan. J xi'an Shiyou Univ, Natural Science Edition 19:1222–1230

Khanzada MA, Lodhi AM, Shahzad S (2005) Chemical control of *Lasiodiplodia theobromae*, the causal agent of mango decline in Sindh. Pak J Bot 37:1023

Khaskheli MI (2020) Mango diseases: impact of fungicides. In: Baimey HK, Hamamouch N Yao Kolombia A (eds) Horticultural crops. IntechOpen, London, pp 143

Khodakovskaya MV, de Silva K, Biris AS, Dervishi E, Villagarcia H (2018) Carbon nanotubes induce growth enhancement of tobacco cells. ACS Nano 6:2128–2135

Le TD, Viet Nguyen T, Muoi NV, Toan HT, Lan NM, Pham TN (2022) Supply chain management of mango (*Mangifera indica* L.) fruit: a review with a focus on product quality during postharvest. Front Sustain Food Syst 5:799431

Mandryk M, Reidsma P, van Ittersum MK (2017) Crop and farm-level adaptation under future climate challenges: an exploratory study considering multiple objectives for Flevoland, the Netherlands. Agric Syst 152:154–164

Murtaza G, Ramzan M, Bilal H, Ejaz A, Khan MAA, Riaz T, Waqas M (2021) Monitoring of fruit fly, Bactrocera zonata (Diptera: Tephritidae) population by installing traps in mango orchard Bahawalnagar, Pakistan. J Appl Res Plant Sci 2:148–151

Nasir M, Mughal SM, Mukhtar T, Awan MZ (2014) Powdery mildew of mango: a review of ecology, biology, epidemiology, and management. Crop Prot 64:19–26

Nayak D, Yadav A, Kumar S, Lalhruaitluangi N, Mandal D (2023) Mango. In: Debashis MD, Wermund U, Phavaphutanon L, Cronje R (eds) Tropical and subtropical fruit crops. Apple Academic Press, New York, pp 559–616

Normand F, Lauri PE, Legave JM (2015) Climate change and its probable effects on mango production and cultivation. Acta Hortic 1075:21–31

Ntsoane ML, Sivakumar D, Mahajan PV (2020) Optimisation of O_2 and CO_2 concentrations to retain quality and prolong shelf life of 'shelly' mango fruit using a simplex lattice mixture design. Biosyst Eng 192:14–23

Ostle NJ, Levy PE, Evans CD, Smith P (2009) UK land use and soil carbon sequestration. Land Use Policy 26:274–283

Padaliya S, Pundir R (2022) Challenges faced by mango exporters. Guj J Ext Educ 34:15–20

Rai D, Rajkumar MB, Punithavalli M (2019) Insect pests of mango and their management. In: Pandey AK, Mall P (eds) Insect pest's management of fruit crops. Biotech Books, New Delhi, pp 19–33

Rajatiya J, Varu DK, Gohil P, Solanki M, Halepotara F, Gohil M, Solanki R (2018) Climate change: impact, mitigation, and adaptation in fruit crops. Int J Pure Appl Biosci 6:1161–1169

Rajvardhan Kiran P, Mani I, Parrey RA, Srivastav M (2023) Mapping of supply chain and assessment of pre- and postharvest losses of Alphonso mango in India. Environ Conserv J 24:64–74

Raza MF, Yao Z, Bai S, Cai Z, Zhang H (2020) Insights into pest-host interactions in mango orchards and their management. Bull Entomol Res

Reidsma P, Wolf J, Kanellopoulos A, Schaap BF, Mandryk M, Verhagen J, van Ittersum MK (2015) Climate change impact and adaptation research requires integrated assessment and farming systems analysis: a case study in The Netherlands. Environ Res Lett 10:045004

Roberts PB (2016) Food irradiation: standards, regulations, and worldwide trade. Radiat Phys Chem 129:30–34

Saadat MN, Jahangir S, Gupta K (2022) Mango culture as depicted in Gambhira folklore. GeoJournal 87:1499–1509

Santos MR, Martinez MA, Donato SLR, Coelho EF (2014) Tommy Atkins mango yield and photosynthesis under hydric deficit in the semiarid region of Bahia. Rev Bras Eng Agric Ambient 18:899–907

Sarwar M, Hamed M, Yousaf M, Hussain M (2019) Monitoring of population density and fruit infestation intensity of Tephritid Fruit Flies (Diptera: Tephritidae) in *Citrus reticulata* blanco orchard. J Vet Res 2:1–5

Schaffer B, Whiley AW, Crane JH (1994) Mango. In: Schaffer B, Andersen PC (eds) Handbook of environmental physiology of fruit crops, vol 2. subtropical and tropical crops. CRC Press, Boca Raton, FL, pp 165–197

Shaik T, Swamykan SI (2023) Identification of diseases affecting mango leaves using deep learning models. In: International conference on artificial intelligence: towards sustainable intelligence. Springer Nature, Cham, pp 132–144

Siddiqui AA, Paul T (2018) Present status and future opportunities in processed food exports including SPS-related issues in international markets. Int J Manag IT Eng 8:294–314

Singh BK, Delgado-Baquerizo M, Egidi E, Guirado E, Leach JE, Liu H, Trivedi P (2023) Climate change impacts on plant pathogens, food security and paths forward. Nat Rev Microbiol 21(10):640–656

Skendžić S, Zovko M, Živković IP, Lešić V, Lemić D (2021) The impact of climate change on agricultural insect pests. Insects 12:440

Stubbings G (2005) The International Plant Protection Convention (IPPC)

Thakur K, Sharma A, Sharma K (2021) Management of agricultural insect pests with physical control methods. Pharma Innov J 10:306–314

Tharanathan RN, Yashoda HM, Prabha TN (2006) Mango (*Mangifera indica* L.), "The king of fruits"—an overview. Food Rev Int 22:95–123

Uddin M, Shefat S, Afroz M, Moon N (2018) Management of anthracnose disease of mango caused by *Colletotrichum gloeosporioides*: a review. Acta Sci Agric 2:169–177

van der Maden E, Ringo E, Likoko E (2021) Scoping study on fruits and vegetables: results from Tanzania (No. 2021-111). Wageningen Economic Research

Venkata Rami Reddy P, Gundappa B, Chakravarthy A (2018) Pests of mango. In: Omkar (eds) Pests and their management. Springer Nature, Singapore. pp. 415–440.

Verghese A, Rashmi M (2016) Management of pre-harvest insect infestation for quality assurance of fruit crops to quarantine and export. National Bureau of Agriculturally Important Insects, pp 428–450

Wang H, Ramnani P, Pham T, Villarreal CC, Yu X, Liu G, Mulchandani A (2019) Gas biosensor arrays based on single-stranded DNA-functionalized single-walled carbon nanotubes for detecting volatile organic compound biomarkers released by *huanglongbing* disease-infected citrus trees. Sensors (Basel) 19:4795

Weerahewa J, Wickramaarchchi T (2023) Food losses in international trade of agricultural commodities: a case study in Sri Lanka

Wei S, Mei J, Xie J (2021) Effects of different carbon dioxide-modified atmosphere packaging and low-temperature storage at 13 °C on the quality and metabolism in mango (*Mangifera indica* L.). Agriculture 11:636

Wickrama SP, Kandangama NB, Wickramaarachchi T, Weerahewa J (2024) Assessing the impact of non-tariff measures on Sri Lankan mango exports: insights, challenges, and recommendations. Front Sustain Food Syst 8:1293263

World Population Review (2024) Mango production by country. https://worldpopulationreview.com/country-rankings/mango-production-by-country. Accessed 14 Aug 2014

Chapter 28
Mitigation Strategies and Options Forward for the Management of Mango Malformation Disease Under Changing Climate Scenarios

Abdurrehman, Talha Mansoor, Muhammad Hamza, Muhammad Tariq Malik, Fahad Hakim, Aimal Waheed, Muhammad Shabbir, Ahsan Amin, and Muhammad Nasir Subhani

Abstract Mango malformation disease (MMD), caused primarily by *Fusarium* species, remains one of the most economically devastating diseases of mango (*Mangifera indica*), significantly reducing yield and fruit quality. Climate change exacerbates the challenges associated with MMD by influencing the pathogen's life cycle, host susceptibility, and disease dynamics through altered temperature, humidity, and rainfall patterns. This chapter explores mitigation strategies and

Abdurrehman (✉) · T. Mansoor
Department of Plant Protection, College of Food and Agriculture Sciences, King Saud University, Riyadh, Kingdom of Saudi Arabia
e-mail: arafiquf@ksu.edu.sa

M. Hamza
Department of Forestry, Faculty of Agriculture Sciences and Technology, Bahauddin Zakariya University, Multan, Pakistan

M. T. Malik
Mango Research Institute, Multan, Pakistan

F. Hakim
Lithuanian Research Centre for Agriculture and Forestry, Institute of Horticulture, Babtai, Kaunas Reg., Lithuania

A. Waheed
Biochemistry Research Group, Faculty of Natural Sciences, Vytautas Magnus University, Kaunas, Lithuania

M. Shabbir · A. Amin
Department of Plant Pathology, Faculty of Agriculture Sciences and Technology, Bahauddin Zakariya University, Multan, Pakistan

M. N. Subhani
Department of Plant Pathology, Faculty of Agriculture Sciences, University of Punjab, Lahore, Pakistan

S. A. H. Naqvi et al. (eds.), *Climate Change and Mango Production*,
https://doi.org/10.1007/978-3-031-87035-4_28

forward-looking options to manage MMD under changing climate scenarios. Integrated management approaches are discussed, including resistant cultivar development, optimized cultural practices, and judicious use of fungicides. The role of biological control agents and microbiome engineering is highlighted in suppressing pathogens and enhancing plant resilience. Additionally, the chapter emphasizes predictive disease modeling using advanced climatic data and artificial intelligence to enable proactive management. Strengthening farmer awareness, capacity building, and policy support are critical for implementing these strategies effectively. By leveraging traditional knowledge and scientific innovations, this chapter aims to provide a roadmap for sustainable management of MMD in a rapidly changing climate, ensuring the long-term health and productivity of mango orchards.

Keywords Mango malformation disease · Climate resilience · Microbiome engineering · Sustainable management

28.1 Introduction

A fungus is responsible for the malformation disease impacting mangos. This phenomenon occurs in multiple regions globally known for mango production. The abnormal growth of flowers and leaves due to mango malformation disease leads to reduced plant growth and fruit yield. The only identified hosts of mango malformation infections are the mango (*Mangifera indica*). Several species within the genus Fusarium are implicated in the malformation disease affecting mangoes.

Atypical, dense growth patterns in flowers and shoots often signify the presence of mango malformation infection. In a plant, growth can simultaneously exhibit both normal characteristics and be impacted by mango deformity.

Stem buds and Leave, serving as growth points, result in distorted shoots characterized by short internodes and delicate leaves. The leaves take on a compact, clustered appearance due to their significantly smaller size compared to those of healthy plants, curling inward toward the stem. Established trees can endure shoot damage, whereas younger plants exhibit more pronounced symptoms, often resulting in significant stunting (Darvas, 1987; Sternlicht and Goldenberg, 1976).

Affected flowering stems or panicles can yield up to three times the usual number of blooms due to their increased thickness and branching. The sterile, enlarged blossoms fail to yield any fruit. In comparison to perfect blooms, there are often a greater number of male flowers present. Furthermore, panicles may yield leaves that are distorted and reduced in size.

Mango malformation sickness does not pose any adverse effects on the health of people, and consuming mango fruit is considered safe.

Infected plant material serves as the main vector for disease transmission. In compromised orchards, the condition referred to as mango malformation progresses at a gradual pace. Mango malformation disease is associated with the *Aceria mangiferae* mango bud mite, as the feeding activity of these mites results in lesions that can promote fungal infection.

Establish new trees utilizing pest-free nursery stock to mitigate the transmission of mango malformation disease throughout the orchards. It is crucial to avoid removing marcot or grafting material from an orchard that shows signs of being affected. Establishing nurseries in orchards impacted by mango malformation disease is inadvisable. It is essential to clean pruning tools thoroughly between trees to prevent the spread of diseases and pests. Ensure that you and your farm laborers are well-informed about the signs and symptoms of mango malformation disease, along with other pests. Verify that the staff, visitors, vehicles, and tools are:-

- Climate change has a profound effect on both social and natural environments. To enhance the accuracy of predictions regarding the potential impacts of warming temperatures on plant diseases, specifically mango malformation, it is essential for decision-makers and resource managers to obtain information about anticipated changes in climate averages and variability. The quantity and quality of water resources, as well as resources obtained from land and marine environments and agricultural productivity, are all notably influenced by fluctuations in seasonal to interannual weather patterns.

This chapter succinctly examines the roles associated with climatic scenario in pathogen adaptation strategies for mango malformation, including suggestions for reducing the chance of mango malformation disease across different climate scenarios. Impacts of climate change on mango tree growth and development.

Precipitation and temperature have the most effects on the mango's phenological and physical phases. Whiley et al. (1989) assert that a mango tree thrives optimally at temperatures ranging from twenty to 39 °C and requires temperatures below 15 degrees Celsius to facilitate blooming. Dambreville et al. (2012) indicate that temperatures between 20 and 28 °C significantly influence leaf development. This is likely due to the tendency of dry materials to move to the roots at elevated levels of CO2 (Whiley et al. 1989). Mangoes have a distinct response to increasing temperatures compared to annual crops. This enables it to endure circumstances that would lead to desiccation, which is beneficial for productivity in subsequent growing seasons. Conversely, stress may hinder plant development and result in a significant decrease in yield (Morison et al. 2006). Stress induced by floods and drought results in reduced growth of mango plants. For mango production to be profitable in both established and developing countries, a choice of crops suitable to the ecosystem is essential. Both seasons, dry and wet, are essential for mango blooming, with the wet season promoting vegetative growth output (Ramírez and Davenport 2010; Sukhvibul et al. 1999). Precipitation often occurs during the mango blooming season, which may be detrimental and lead to complete crop failure (Makhmale et al. 2016). Mango trees persistently thrive despite elevated temperatures, frequent precipitation, and humidity, which inhibits their blossoming. Nevertheless, the fast closing of a mango plant's pore openings due to climatic variations results in severe aridity and vapor pressure deficit, which are detrimental to photosynthesis. Typically, flower buds emerge near the terminal branch of development. Research conducted in grow chambers on mango trees confirms that nighttime temperatures between 8 and 15 °C and daytime temperatures below 20 °C induce flowering immediately after shoot initiation (Shu and Sheen 1987).

28.2 Effect of Global Warming on Mango Tree Yield and Quality

Climatic fluctuations induce stress that impacts the architecture, metabolic processes, structure, and physiology of mango trees, hence influencing their growth, production, and quality. The yield and quality of mangoes are adversely affected by unseasonal rainfall and sudden temperature fluctuations. The number and quality of mangoes diminish as the plant's physical features (vigor and canopy), traits for reproduction (fruiting capacity and fruit size), and quality features (juice content, duration of storage, and susceptibility to pests and diseases) fluctuate. High temperatures and low humidity exposure in mango plants reduces their capacity for transpiration, photosynthesis, and water storage in leaves. (Faria et al. 2016; Khanum et al. 2020). To mitigate the adverse impacts of these stresses, the mango plant decreases its canopy size, so hindering stomatal conductance and perhaps affecting yield (Shanthala et al. 2021). Pre-flowering and blooming rainfall diminish mango fruit output. It also obstructs fruit development and pollination. Likewise, in the concluding phases of fruit development, storms significantly diminish fruit yield. The dimensions, yield, and quality metrics of mangoes, including sugar, vitamin C content, starch, and solubility of solids, are influenced by soil water amount (Wei et al. 2017). Numerous studies have proved the crucial function of irrigation in enhancing the productivity of rainfed agriculture and improving the quality of mango harvests amid fast climatic changes (Hahn and García 2022; Wei et al. 2017).

Climatic variations affect the maturation and quality of mango fruits. The prevalence of illnesses and pests, such as anthracnose, in areas characterized by elevated humidity and rainfall partially elucidates this phenomenon (Makhmale et al. 2016; Ploetz 2003). During this period, mango fruits may diminish in visual appeal owing to many factors, including the discoloration of the peel to black (Makhmale et al. 2016). Conversely, low humidity and rainfall promote fruit development, harvesting, and the alleviation of biotic stressors (Zuazo et al. 2021).

The variety of mango genetic materials facilitates the adaptability of breeding and collecting methods to changing circumstances. Mango production is dependent upon genotype in both humid and arid climates. Employing these features will subsequently mitigate the effects of climate change on mango production and quality. The quantity and quality of mangoes are enhanced by adaptation measures, which are essential in the fight against climate change. In both traditional and contemporary agricultural locations, climate-compatible crop types provide better mango yields.

28.3 Mango Malformation Disease

In 1891, MMD was founded in Bihar, India. MMD cases have been reported in the African nations of Sudan, the Kingdom of eSwatini, the Republic of South Africa, the Arab Republic of Egypt, and the Republic of Uganda. Later reports came from Central America, Oman, Israel, Europe, particularly Southern Spain, and Asia, including the Federation of Malaysia, the Islamic Republic of Pakistan, and Southern Zhonguo; and from continental America, encompassing Venezuela, Nicaragua, South Florida, El Salvador, the USA, Mexico, and Brazil. Reports ultimately emerged from several countries worldwide, including the United States, Australia, China, Israel, Brazil, El Salvador, the Arab Republic of El Salvador, the Federation of Mexico, Pakistan, Malaya, Myanmar, Sri Lanka, South Africa, Oman, Senegal, Sudan, Switzerland, and Uganda (Darvas, 1987; Sternlicht and Goldenberg, 1976).

Mango malformation disease is recognized as one of the most detrimental pathosystems for mangoes in almost all mango cultivation areas, especially in the tropical and subtropical parts of Asia. Pathogenicity experiments have shown that the principal fungal agent associated with mango malformation disease is *Fusarium mangiferae*. Comprehensive studies have been undertaken into the genesis, treatment, and manifestations of mango malformation disease. Nevertheless, research focusing on epidemiological aspects are the least prevalent. Thus far, doctors have devoted little effort to conducting a comprehensive epidemiological study of Mango malformation disease, which would facilitate appropriate management of the illness. The most effective but brief management approach for distorted terminal shoots seems to be pruning.

Recently climate change has become the serious concern among the scientist as it altered the cropping patterns as well as the dynamics of biotic and abiotic stresses faced by the plant the impact of climate change on the mango malformation disease development is given in Table 28.1.

28.3.1 Symptomology

The infected seedlings exhibit the bunchy top stage due to the minute shootlets that bear scaly leaves, creating a clustered appearance. Witches' broom manifests on the vegetative growth which are always found on internodes, causing seedlings to display the plant's characteristics. Severe damage from vegetative malformation in early development results in either the death or stunting of seedlings; conversely, if malformation develops later, the afflicted seedlings resume normal growth in the impacted regions. On the damaged planting, superficial root and a limited quantity of secondary roots were produced. Vegetative buds, seen on tree shoots, grow into tufts or clusters that rapidly desiccate yet reemerge the next season.

Branches of trees afflicted by vegetative malformation (VM) either fail to bud or exhibit reduced flowering frequency. The malformed inflorescences are generated

Table 28.1 Impact of changing climate on mango malformation disease

Climate factor	Impact on mango malformation disease	Effect on disease dynamics and severity
Increased temperatures	Higher temperatures may stress mango trees, enhancing susceptibility to fungal infections causing malformation	Fungal pathogens like *Fusarium* spp. and *Colletotrichum* spp. thrive in warmer conditions, leading to increased infection rates
Erratic rainfall	Unpredictable rainfall patterns create favorable conditions for fungal spore germination and growth	Excessive moisture during flowering periods can increase the incidence of malformation
Higher humidity	Increased humidity promotes fungal proliferation, which can cause significant deformities in mango inflorescences	Higher humidity levels lead to more severe and widespread malformation symptoms
Drought conditions	Water stress weakens mango trees, making them more vulnerable to pathogen attack, including those causing malformation	Drought-induced stress can exacerbate malformation by affecting tree health and immune response
Extended growing seasons	Longer growing seasons can prolong exposure to malformation-inducing pathogens, leading to an extended disease cycle	A longer period of infection may result in more severely deformed fruit and lower yields

by the VM-infected branch. Indications of abnormality in blooms or florals emerge subsequent to the appearance of the flowers. The markedly enlarged flowers of the infected the inflorescence constrict the hypertrophied the panicle axes, leading to premature fruit abortion or complete fruit absence. The quantity of affected panicles has risen, and their lush, green branches has made them more substantial. Severely distorted inflorescences provide a greater number of blooms, however most remain unopened.

Infected inflorescences often have a diminished sex ratio, however this may sometimes vary. Malformed hermaphroditic flowers possess nonfunctional pistils, and their pollen suggests limited viability. The degree of infection inside a tree differs among branches and individual plants. At the pinnacle of the same stem, both malformed and healthy flowers were produced.

Moreover, a substantial, robust panicle yielded few defective branches or a limited quantity of deformed flowers. There is an increase in fertility and aborted flowers, which are usually hermaphroditic, in malformed trees. The affected flower spikes bear flowers and stay greenish after the fruit develops in a typical panicle. Internal indicators comprise: the emergence of hyperplastic and hypertrophied the cells in the aberrant segments of the flower; ovules in infected hermaphroditic flowers exhibit poor differentiation or lack differentiation; degenerated or underdeveloped embryos

Fig. 28.1 Floral malformation in mango tree (**a–d**), and vegetative malformation on mango tree (**e–h**), Colony of *Fusarium solani* (**i**, **j**) Spores of *F. solani* (**k**, **l**) (Abdurrehman et al. 2024)

in affected fruits; and inflated structures resembling flowers are generated. Malformations are sometimes misidentified as deficiencies in nutrition, such as zinc shortages in trees and seedlings, which are difficult to identify in their first stages (Fig. 28.1).

28.3.2 Etiology

According to (Summanwar et al. 1966) It was first determined that the blossoming phase of MMD was caused by a fungus species belonging to the Gibberella fujikuroi (Sawada) species complex, which was formerly known as Fusarium moniliforme Shield. *F. moniliforme* also promoted the development of the disease's the vegetative symptoms, according to a later investigation (Varma et al. 1974).

Despite having many synonyms in the literature, such as Nelson, F. subglutinans, and Toussoun, the pathogen was reclassified as a new species in 2002, Wingfield Marsas & F. mangiferae Britz, based on samples from South Africa, Egypt, Israel, Florida (USA), and Malaysia (Wollenweb. & Reinking), and Marsas (Britz et al. 2002; Marasas et al. 2006; Steenkamp et al. 2000). F. mangiferae was later found in India (O'Donnell et al. 1998; Zheng and Ploetz 2002), Oman (Kvas et al. 2008),

Spain (Crespo et al. 2012), Sri Lanka (Sinniah et al. 2013), and China (Zhan et al. 2012), and is recognized as the predominant causative agent of MMD globally.

F. sterilihyphosum The secondary MMD causative agent Britz, Wing-field, and Marasas was first discovered in isolates from a small region of the African continent (Britz et al. 2002). In Brazil, deformity induction was seen after artificial inoculation (Lima et al. 2009). Lima and associates (2012). Another MMD causative agent, F. mexicanum sp. nov., was shown to originate in México (Otero-Colina et al. 2010). Multilocus sequencing has shown that F. mexicanum differs significantly compared to other MMD taxa within the G. fujikuroi species complex. (Fig. 4; Rodríguez-Alvarado et al. 2008; Otero-Colina et al. 2010).

The newly identified fourth species, *F. tupiense* sp. *nov.*, which is phylogenetically different from both *F. mangiferae* and *F. sterilihyphosum* but shares morphological similarities with *F. sterilihyphosum*, creates a separate teleomorph in the *G. fujikuroi* complex. *F. tupiense* causes malformation in Senegal (Senghor et al. 2012) and Brazil (Lima et al. 2012) (Fig. 4). MMD has also been linked to two other species: *F. pseudocircinatum* (Freeman et al. 2014) and F. proliferatum (Matsushima) Nirenberg (Lv et al. 2010; Zhan et al. 2010).

Having only been found to be the cause of MMD in the Western Hemisphere, specifically in Florida, F. mangiferae is the most common MMD pathogen in the Eastern Hemisphere (Britz et al. 2002; Zheng and Ploetz 2002). However, the Western Hemisphere's MMD agent descriptions have only been accessible since Otero-Colina et al. (2010). Florida, Brazil (Lima et al. 2012), and Mexico (Otero-Colina et al. 2010).

If *F. mangiferae* is found farther in the West, further research is needed to find out this. Several of the taxa mentioned above were found by employing PCR primer pairs. For F. mangiferae, Zheng and Ploetz (2002) created a pair of enhanced 608-bp fragments known as 1-3F/R, which have been applied to diagnostic procedures (Youssef et al. 2007).

When the amplification techniques were adjusted, an isolated pair, 61-2F/R, intended to determine an isolate of F. verticillioides (previously identified as *F. moniliforme* in 46), was able to successfully amp up a 445-bp segment for isolates of *F. mexicanum* and *F. sterilihyphosum*. (Zheng and Ploetz 2002; Rodríguez-Alvarado et al. 2008). For F. mangiferae, additional pairs of primers have been created (Newman et al. 2012).

28.3.3 Is the Mango Bud Mite Implicated in the Growth of MMD?

Large populations of the mango bud mites, Aceriamangiferae Sayed (Eriophyidae), are sometimes seen on damaged plants. Certain individuals hypothesized that the mite was responsible for MMD because to this association and its recorded ability to induce hypertrophied buds on mangoes. (Narasimhan 1954; Nariani and Seth 1962). Nonetheless, this idea lacked support from the subsequent evidence.

The mite was identified in Australia, although rare occurrences of MMD have only been newly recorded there; substantial mite populations can be found in trees that show no signs or symptoms in other mango-producing regions; and acaricides effectively diminished mite populations but did not alleviate MMD. However, the mite could have a part in MMD epidemiology. The pathogen was previously found in the mite on media for culture (Crookes and Rijkenberg 1985), but it has now been shown to adhere to its body (Gamliel-Atinsky et al. 2010).

Gamliel-Atinsky et al. (2010) asserted that the mite's little mouth precluded it from ingesting the infection. Nonetheless, it successfully disseminated F. mangiferae conidia to disease sites inside mango buds using experimental methods, presumably via body attachment. Buds may become more vulnerable to pathogen infection because to injuries inflicted by mite feeding. (Crookes and Rijkenberg 1985; Gamliel-Atinsky et al. 2010).

In Israel, *A. mangiferae* didn't appear to play a major role in the infection's spread among trees. Although several *F. mangiferae* conidia were caught, mites were not found in traps intended to track their migration in an MMD-affected orchard (Gamliel-Atinsky et al. 2007). It is essential to determine if and under what circumstances the mite contributes to the dissemination of MMD across plants and orchards in other mango-producing regions, taking into account how these factors can affect MMD management tactics. Other arthropods that are often drawn to infected panicles may serve as dispersers, even while windborne mango bud mites that originated from contaminated panicles didn't display conidia (Gamliel-Atinsky et al. 2009a, b, 2010). Examining the airborne dissemination of this virus by arthropods and other mechanisms is crucial.

Fusarium strilihyphosum has been found as an additional mycopathogen related to MMD in Brazil and South Africa due to artificial inoculation. A multilocus sequencing investigation identified Fusarium mexicanum as a new mycopathogen linked to mixed mugen disease (MMD) in Mexico, apart from other species within the *Gibberella-Fujikuroi* complex. *Fusarium tupiense*, morphologically similar to *Fusarium strilihyphosum* in Brazil and Senegal, exhibits phylogenetic divergence from *Fusarium mangiferae* and *Fusarium strilihyphosum*, serving as a teleomorph within the *Gibberella fujikuroi* complex. *Fusarium pseudocircanatum* and *Fusarium proliferatum* have been identified as supplementary MMD mycopathogens.

Zheng and Ploetz (2002) established the previously listed taxa and identified them via PCR primers. Fusarium mangiferae may be identified by amplifying a 608 base pair region using a single set of forward and reverse primers 1–3, used for molecular diagnostics. An new set of primers was developed to target *Fusarium verticillioides*, which cannot amplify *Fusarium mangiferae* DNA. Subsequently, modifications were implemented to facilitate the amplification of all associated mycoflora types to MMD, and further efforts were undertaken to develop primers for the identification of *Fusarium mangiferae*. Fragments of 445 base pairs from *Fusarium mexicanum* and *Fusarium strilihyphosum* were successfully amplified using a newly designed primer pair.

Published research indicates that mango mites of the species *Aceria mangiferae* are often present in significant quantities near locations of malformation, leading to hypertrophied terminals in mangoes, indicating that these mites may contribute to the etiology of malformation. Nonetheless, it was also contradicted by supplementary data from Australia, a nation with a substantial mite population and seldom documented outbreaks. Acaricides significantly reduced the population of *Aceria mangiferae*, although did not alleviate the symptoms of mango deformation. *Fusarium mangiferae* has been recognized as the etiological agent of MMD in South African orchards. Afterwards, the same the microbes was identified in Egypt and Israel. Several Fusarium taxa have been associated with MMD. *Fusarium proliferatum* from the Federation of Malaya and *Fusarium strilihyphosum* from South Africa were also identified.

Fusarium mangiferae generates polyphialides instead of chlamydospores or microconidia. *F mangiferae* exhibited distinct morphological characteristics compared to the other species. *Fusarium mangiferae* is the final etiological factor of mango malformation disease, and its role as the causative agent of mango deformation was confirmed using self-programmed inoculation. Alongside fungus, the aetiology include mite species, viral strains, and hormonal abnormalities. The toxin produced by Fusarium is a significant contributor to the malformation symptoms seen in mangos. *Fusarium mangiferae* has been mostly isolated from malformed tissues. *Fusarium mangiferae* is extensively found in all deformed tissues obtained from diverse sources, with an infection prevalence of up to 97%. The pathogen's conidia were seen to be reduced in the soil. Mango Malformation Disease (MMD) poses a substantial obstacle to the establishment of economically viable orchards. Researchers exhibit considerable interest in the disease.

28.4 Disease Cycle and Epidemiology

Aceria mangiferae and *Fusarium mangiferae* were involved into a disease cycle. A compilation of early inoculums, consisting of microconidia capable of infiltrating the colonization and infection site via three distinct mechanisms, comprises deformed vegetative and floral tissues. It will disseminate in two manners: firstly, by wind, and secondly, via bud mites associated with mycoflora. Thirdly, via the opening of the basal bud, contact is made with sick plant components. Conidial transfer may occur due to watering water or rain splashes. Pathogen reproduction and establishment may transpire within a temperature range of 5–41 °C, provided there is water retention in the leaves for a minimum duration of two hours. When moisture exceeds three hours, this illness will significantly escalate within the 15–30 °C range. Apical buds infested with mites will exacerbate and increase the frequency of the illness. subsequent to penetration and colonization at the bud site. The extent of colonization will dictate the progression of symptoms. Vegetation exhibiting the lowest infected threshold and flowering, as indicated by colonizing mycelium tissues, will not manifest malformations or symptoms.

Environmental variables significantly affect conidia viability, with the maximum mortality seen after two hours or more of strong light exposure, suggesting that

MMD propagates slowly in orchards. The dissemination of *Fusarium mangiferae* in compromised vegetation suggests that the tree's apical buds are the primary infection source, rather than a widespread distribution across the orchard. *Fusarium mangiferae* mostly inhabits malformed vegetation and floral tissues, exhibiting greater prevalence in symptomatic tissues compared to asymptomatic ones. Increased residual infection of *Fusarium mangiferae* in dormant lateral buds. Primary buds serve as the principal source of infection, as shown by Freeman's artificial inoculation of altered isolates of *Fusarium mangiferae.* Youssef et al. (2007) proposed that while conidia exhibited greater survival in summer, enhanced survival was expected in the soil owing to plant disease detritus. Inoculation of roots with *Fusarium mangiferae* does not lead to systemic infection, indicating that MMD cannot be propagated by seeds. A new management strategy may be developed to address the detrimental aspects of mango malformation illness by considering all previously provided information.

28.5 Economic Impact

The Mango Malformation Disease poses a threat to Pakistan's economy. The direct correlation between yield loss and disease severity complicates the precise quantification of yield loss in MMD. Mango trees with deformed terminals that have remained unproductive and resulted in substantial production loss should be swiftly pruned to prevent more losses. Reports indicate an 86–90% yield decrease over the last three to four years. Indian orchards experienced significant damage because to a 50% infestation of deformities. Mango malformation diseases is resulting in major declines in South Africa and Egypt. South Africa and India saw the most significant yield reductions, at 73% and 86%, accordingly.

28.6 Management Strategies

28.6.1 EarlierTactics for Managing MMDisease

Many strategies for managing cases of multiple myelitis (MM) have been tested and studied, including using fungicides in and other pesticides as well as upholding hygiene. However, none of these measures have been able to lessen the illness's severity enough.

Sanitation. The most widely used approach to treating MMD once it has spread around an orchard is cleanliness, which entails cutting symptomatic tissues off trees and discarding them. (Manicom 1989; Narasimhan 1959; Noriega-Cantú et al. 1999; Singh et al. 1974). The three extending from nodes and the symptomatic panicles should be removed as soon as feasible, according to advice (Manicom 1989). No matter how stringent the laws are, MMD is reduced by cleanliness. Most likely by reducing the orchard's inoculum. Even if it is complex to administer this management to big trees with difficult-to-reach panicles, growers may not be willing to put in the effort required to ensure the success of this technique. However, we think that an all-encompassing strategy for managing this illness must include proper cleanliness.

Chemicals: Singh et al. (1974) discovered a moderate control in India using cadmium (Cd), and nickel (Ni) cobalt (Co), sulphates; however, it is uncertain if these materials could be safely applied to a food crop. By injecting fosetyl-Al, When a medication that is mainly used to treat oomycete-related diseases was injected into the trunks of Keitt plants, the percentage of malformed inflorescence decreased from 96 to 48 percent, and the fruit yield increased to 46 to 95 kg tree-1. However, the decrease was not statistically important ($P < 0.05$). The efficiency of fungicides in controlling MMD has been found to be typically inconsistent. For example, despite repeated testing, the efficacy of benzimidazoles against MMD has been variable, even in cases when positive results have been recorded. (Muhammad et al. 1999) stated that, when Topsin-M was sprayed once during the bud differentiating stage, deformation was reduced to 19.8% as opposed to 35.4% in the control, untreated group; however, their results were not statistically evaluated. However, benomyl inhibited the infection in vitro, as per (Ibrahim et al. 1975) however when spray on mango trees, had no impact on the development of MMD. (Dieckman et al. 1982) and (Chadha et al. 1979) further discovered that benomyl has no field effect on MMD.

Additionally, tests on combinations of measures have been carried out. (Chadha et al. 1979; Chakrabarti 2011; Covarrubias 1980; Iqbal et al. 2011). Despite the absence of statistical significance, trimming the terminal 45 cm of damaged shoots along with a benomyl management lowered the probability of deformities by 70.4% as compared to the untreated control group. (Otero-Colina et al. 2010) found that a combined approach to management that included spraying many fungicides, trimming the terminal 80 cm of hard shoots, and spraying a sulphur acaricide five times significantly increased mango yields.

There are conflicting opinions regarding fungicides' efficacy in treating MMD, but they should still be considered as potential management strategies. Therefore, it would be necessary to identify efficient fungicides and optimal application intervals in order to maximise their use. Most of difficulties that cured MMD with fungicides treated the fruit either before or during bloom, up to the time the fruit set, but not after (Iqbal et al. 2011; Noriega-Cantú et al. 1999). There are two options, even if the basic assumption that guided the window's usage and the reason for the spray treatment's time were not made clear in these studies. Fungicides presumably protect buds from inoculum dispersed by malformed panicles while the buds are still on the tree, assuming they are not infected. The alternate belief that blooms have been infected before and that fungicides would eradicate the virus inside appears incorrect given the dearth of systemic and highly active fungicides. In general, specialist fungicide application has not been routine for managing MMD globally. For instance, fungicide treatment of mango MMD was prohibited until 2013, despite the fact that Israel discovered the disease as early as 1975. Since then, MMD needs evacuated.

28.6.2 New Insights into MMD Epidemiologyand Disease Management

28.6.2.1 Inoculum Availability and Dissemination, and Host Organs that Are Infected by the Pathogen

As was previously stated, a lot of research has been done to find out the cause of MMD. However, little research has been done on the epidemiology of this illness until lately (Gamliel-Atinsky et al. 2009a, b; Kumar et al. 1993; Ploetz 2001; Ploetz and Freeman 2009). The data presented lower pertain only to MMD induced by *F. mangiferae*. In an orchard, healthy and malformed inflorescences often appear together. From bud break to bud maturity, the quantity of conidia visible on the surface of deformed panicles grew dramatically. Early in the formation of the panicle (Gamliel-Atinsky et al. 2009a, b) a limited quantity of conidia was captured in the affected orchards; however, as the panicles reached the cauliflower stage (Gamliel-Atinsky et al. 2009a, b), these quantities significantly escalated. Microconidia are the most probable infectious propagules of *F. mangiferae*. This species does not produce chlamydospores or ascospores, and the fungus's main spores are found on both living and necrotic tissues that show deformation (macroconidia are created less often) (Leslie and Summerell 2006). The lengthy spread of MMD in fruit orchards might be explained by the conidia's quick death when exposed to direct sunshine (Manicom 1989). According to a recent research, 100% of the *F. mangiferae* conidia died after being exposed to sunlight for two to four hours.

There is no systemic colonisation of branch tissues, and the presence of *F. mangiferae* in the afflicted trees indicates that the apical buds are the main sites of infection (Gamliel-Atinsky et al. 2009a, b, 2009a; Ploetz 1994). In Florida, *F. mangiferae* was only found in malformed vegetative and floral tissues (Ploetz 1994). Disease was rare on such branches, even when they supported malformed panicles or shoots. It was primarily observed in deformed vegetative and floral Youssef et al. (2007). To verify that the fruit's apical buds were the main infection sites, Freeman et al. (1999) created a synthetic inoculation technique employing isolates of F. mangiferae that had been altered with the GUS sensor genes (β-glucuronidase). While wounds may serve as entrance routes for the virus, the investigation indicated that damage was not a necessity for disease (Freeman et al. 1999). According to Youssef et al. (2007), conidium survival in soil significantly decreased over the summer; after ten days, just fifty percent of the initial population persisted. The pathogen may persist in the soil for an extended duration if plant tissues are infected. For example, the survival rate after six months was less than 20% when infection-prone clusters were buried 30 cm below the soil surface (Gamliel-Atinsky et al. 2009a, b; Youssef et al. 2007). Claims that Fusarium oxysporum (29) or *F. mangiferae* may induce malformations in seedlings lack empirical validity. (Abdel-Sattar 1973) infecting the roots of mango trees. According to Youssef et al. (2007) *F. mangiferae* infections in mango roots were asymptomatic and not systematic. Since there have been no reports of seed infections, it is unlikely that the illness may spread via seeds. *F. mangiferae* may infect plants and budwood to spread over great distances (Prakash and Srivastava 1987). Materials that are potentially contaminated may covertly traverse inside and across nations, evading detection by production managers or quarantine officials. The prevalence of symptomatic plants, which ought to be readily recognised, in international trade is astounding (R. Campbell's, private communication). In summary, the available data suggests that *F. mangiferae* is mostly found in the mango host's apical, outside meristematic cell layer, and that diseases of these buds are localized as opposed to systemic. Outside of these sensitive areas, the virus is either absent or unable to proliferate effectively. This restricted geographic range has considerable consequences for MMD management.

28.6.2.2 Development of a New Integrated Management Strategy

A theoretical framework has been developed to illustrate the progression of MMD across spatial and temporal dimensions, grounded on the previously referenced sickness cycle. According to the model, tissues impacted by MMD act as reservoirs for the infection of new locations devoid of pathogens. These tissues may arise from disorders that are either spontaneous or induced (such as via tainted graft material) (Ploetz 2001). Aerial microconidia are generated on infected panicles and disseminated over orchards. They are considered the primary infectious agent of the illness. The formation of aberrant tissues in the orchard correlates with the pathogen's disease and the spread of the inoculum. In Israel, the first deformed panicles and vegetative development often manifest in April and last until the conclusion of August. The precise dates vary annually based on temperature and the flowering seasons of various cultivars. Conidia accumulate on dormant buds of an individual tree or on other plants in similar or different orchards. Recently established plantations are often situated next to heavily infected orchards, which provide as a source of inoculum for the newly planted trees. The model's practical significance lies in the fact that the infection window, which coincides with the protection window, is the period during which the disease should be addressed with fungicides or other protective agents. The need to save trees will persist in the future, but the duration of protection will fluctuate based on certain weather and industrial conditions, as well as the periods when infection occurs. Minimizing the main inoculum is another facet of treatment. Consequently, defects are promptly excised upon detection and concealed until the completion of flowering—a process referred to as "strict sanitary." Infected buds may ultimately develop into vegetative development or exhibit deformed or normal inflorescences after a period of dormancy lasting many months. In mature areas of the plant, such as the branches and trunks, some unconsciously infected buds may develop into defective tissues if severe pruning occurs. In more mature regions of the plant, some bud may also enter prolonged dormancy.

28.6.2.3 Evaluation of a New Integrated Management Strategy

A novel integrated management approach required the development of fungicides that could protect buds against *F. mangiferae* conidia. Accordingly, in vitro tests showed that prochloraz was the most effective fungicide at reducing *F. mangiferae* (Freeman et al. 2014). Prochloraz-Zn had an enzyme-disease-half (ED50) of 0.01 μg a.i./ml, carbendazim 0.7 μg a.i./ml, pyraclostrobin and boscalid 1.5 μg a.i./ml, and famoxadone and mancozeb 10 μg a.i./ml. In contrast to the aforementioned fungicides, supplemental fungicides such as triadimenol, bupirimate, tebuconazole, thiophanate methyl, triforine, mancozeb, and flutolanil did not inhibit the pathogen's in vitro development (Freeman et al. 2014).

Prochloraz's impact on bud infection was assessed. Even at a concentration of 0.1 mg/ml, the fungicide reduced infection in greenhouse studies. In a different series of studies, it was shown that the fungicide, administered over a period of four weeks, decreased disease incidence by 90% both up to 14 days before to inoculation (protective effect) and post-inoculation (curative effect). In 2005, six investigations were commenced in Israel's economically infested orchards based on prior results. The research indicated that prochloraz should be administered promptly to mitigate MMD.

Prochloraz administration, whether post-flowering, when inocula were already affecting exposed buds, or pre-flowering, when inocula were absent in the orchard, did not diminish MMD incidence (Freeman et al. 2014). Conversely, the use of prochloraz sprays with inoculum on infected panicles resulted in a substantial reduction in MMD incidence in treated plots relative to control plots. Prompt elimination of main inoculum by thorough cleaning is an essential element of the new integrated MMD management strategy. A supplementary study shown that effective sanitation coupled with prochloraz treatment resulted in a substantial yield enhancement (39.3%) over three years, compared to the unsprayed control procedure that just eliminated symptomatic tissue.

Neither strict sanitary standards nor fungicide treatment resulted in substantial production enhancements, yielding 17.8% and 14.4%, respectively. Consequently, when coupled with rigorous cleanliness, prompt applications decreased the occurrence of MMD and enhanced yield. These findings reinforce the existing Israeli protocol for prochloraz applications and stringent sanitation practices during the infection window, commonly referred to as the protection window. In 2013, prochloraz was approved for emergency treatment of MMD in Israel. In treated orchards, it is anticipated that disease would progressively diminish to minimal levels when this novel integrative management strategy is consistently used over time. Prochloraz treatment further reduces mango flower powdery mildew (Lonsdale and Kotze 1993).

As a result, because both conditions arise at the same stage of panicle growth, they may be treated concurrently. Alternatives must be investigated since *F. mangiferae* may develop resistance to prochloraz. (Guarnaccia et al. 2014; Mavroeidi and Shaw 2005; Navarro and Tello 2005). Numerous fungal diseases have been shown to exhibit decreased sensitivity and field resistance to prochloraz (Guarnaccia et al.

2014; Mavroeidi and Shaw 2005; Navarro and Tello 2005). Therefore, if Prochloraz were to be widely employed in the treatment of MMD, substitutes would be required.

28.6.3 Comparison of the New Integrated Strategy with MMD Management Practices Worldwide

Various nations cultivating mangoes have different MMD management strategies. The South African Agricultural Research Council (ARC) advises that impacted terminals must have the terminal's three nodes excised and destroyed of in plastic bags for incineration (De Villiers and Joubert 2008). The ARC article claims that if cleaning is conducted annually, the infection will eventually diminish to insignificant proportions. The ARC does not reference any chemical therapies. Recent surges in sickness rates to epidemic proportions in that nation (S. Freeman, personal communication) suggest that integrating various measures with sanitation may be essential for successful management strategies.

Standard protocols encompass: (i) incinerating any aberrant inflorescences or lateral twigs extending 50 cm beyond infected panicles; (ii) inspecting trees post-harvest to eliminate any symptomatic material beyond the production cycle; (iii) disinfecting pruning instruments by immersing them in a hypochlorite solution; (iv) applying copper paste to pruned sections of the plant; and (v) treating trees with fungicides and insecticides during the dormant period (Moacir Brito Oliveira, unpublished communication). The scope of the planned sanitary changes in South Africa, Brazil, and Mexico remains undetermined. No evidence support the efficacy of these treatments, except for the report of Noriega-Cantú et al. (1999).

Producers may find it easier to implement this advice if there is statistically substantial backing, which would mitigate the impact of this debilitating illness. The prolonged execution of the new management plan is anticipated to provide cumulative decreases in diseases. The elements of the new complete control plan—fungicide spraying and sanitation—were likely included in the previously outlined management plans from other nations. The new integrated management strategy employs distinct thinking concepts and is implemented differently than its predecessor. In the new comprehensive management approach, sanitation aims to eliminate the diseased plant tissues that produce conidia, or inoculum, which subsequently infect healthy buds. Cleansing must begin immediately upon the observation of distorted tissues and must keep going for as long as Assuming these branches get colonized and generate inoculum (mycelia) that internally contaminate healthy buds, the primary objective of sanitation in most other nations is to eliminate the systemically sick branches containing infected tissues. Sanitation should be performed whenever feasible, preferably after all malformed tissues have been detected. The objective of the new holistic control strategy's fungicide application is to inhibit airborne conidia from damaging vulnerable host tissues (buds). Spraying must commence immediately upon the identification of no infected buds and continue until the production of symptomatic tissues ceases, in order to safeguard them against Inoculum originating from deformed panicles. Spraying was recommended at different intervals in many other countries; however the rationale for those timings was not elucidated.

28.7 Climatic Scenarios

According to IPCC (2007b), there are a number of significant effects of climate change on several sectors, including ecosystems, water supplies and their management, food, fibre, and forest products; low-lying areas and coastal systems; industry, settlement, and culture; and health. To better predict the possible effects of climate change, managers and resource managers need information on future changes in the average and variability of the climate. Future climatic trends are hard to forecast, though. In particular, it is challenging to estimate the future warming caused from greenhouse gases because these emissions are dependent on a wide range of speculative and uncertain factors, including growing populations, the use of carbon-based fuel as a source of energy, technological advancement, economic development, environmental policy, and attitudes. Climatic scenarios have been created in order to study the possible effects of global climate change. As such, a climate scenario is not the same as a forecast or prediction of the climate (IPCC 2007a). In this context, climate research has made a substantial contribution to our understanding of both the short-term fluctuations in climate that are caused by natural processes and the long-term changes that are projected to occur due to human and natural impacts (IPCC 2007b).

28.7.1 Climatic Impact on Occurrence of Mango Malformation Disease

In areas with mild temperatures (ranging from 15 to 20 °C), the occurrence of mango malformation is relatively low. The disease severity remains minimal, and mango trees are less susceptible to malformation. Flowering responses may vary, but overall, the disease impact is not significant in these conditions. Sporadic climates, characterized by temperatures between 20 and 25 °C, occasional occurrences of malformation can be seen. Mango trees may experience sporadic symptoms, but the disease does not reach epidemic levels. Proper management practices are still essential to prevent further spread. Beyond 25 °C, the occurrence of mango malformation is practically nil. High temperatures inhibit the disease development, and mango trees remain largely unaffected. However, other factors (such as genotype and plant age) also play a role in disease susceptibility. Arid and semiarid regions, characterized by low humidity and limited rainfall, favor mango malformation. These climatic conditions create an environment conducive to the disease, leading to higher incidence rates. The favorable temperature for the appearance of mango malformation disease is approximately 26 °C + 2 °C (28–30 °C) along with a relative humidity (RH) exceeding 65%. These weather conditions create an environment conducive to the growth of the Fusarium pathogen responsible for mango malformation. When both temperature and humidity align within these ranges, the expression of malformation symptoms becomes more likely to happen. The disease is widespread in tropical and subtropical mango-growing countries. Early diagnosis of pathogens and proper management strategies are crucial to protect mango orchards and fruits from successive and secondary infections. Changing climate can alter the pattern of disease occurrence into various localities depending upon fulfillment of disease triangle conditions. Under altering climatic scenarios mango malformation disease causing fungi can adapt to various conditions to whom it was not fit before which can result in outbreak of disease in a locality.

28.8 Future Prospects

Mango malformation disease, a complex issue with debated causes, is expected to be further challenged by an uncertain climate. Here's a breakdown of potential management options for the future:

A multi-pronged approach is likely to be most effective. This could involve combining physical techniques like pruning to reduce disease spread, planting resistant cultivars (research into developing new ones is crucial), using biocontrol agents to suppress the fungus, and targeted fungicide applications (keeping in mind environmental impact and resistance).

Maintaining proper plant health through balanced fertilization can strengthen the mango tree's defenses. Efficient irrigation practices become even more critical under changing rainfall patterns. Additionally, developing monitoring systems to predict disease outbreaks based on weather conditions will allow for pre-emptive action.

Further research is needed to definitively identify the cause(s) of mango malformation, which can lead to more targeted control measures. Intensifying breeding programs to develop commercially viable mango varieties with strong resistance to malformation is another approach. Exploring the potential of techniques like gene editing to create mango trees with enhanced disease resistance is on the horizon.

Finding the right balance between effective disease control and eco-friendly practices will be crucial. Disseminating knowledge of these management options among mango growers is essential for widespread adoption. As climate change progresses, continuous adaptation of management strategies may be necessary.

By employing a combination of these approaches and ongoing research, mango growers can hope to mitigate the impact of mango malformation disease even under a changing climate (Table 28.2).

Table 28.2 Mitigation strategies for mango malformation disease under changing climate scenarios

Mitigation strategy	Description	Adaptation to changing climate and effectiveness
Pruning and removal of infected tissue	Regular removal of malformed inflorescences and infected branches	Prevents the spread of the pathogen and reduces the inoculum, maintaining tree health
Use of fungicides	Application of fungicides like *Copper-based fungicides* or *Carbendazim*	Provides effective control during high-risk periods, especially during humid weather
Resistant mango varieties	Development of mango cultivars with resistance to fungal infections causing malformation	Reduced susceptibility to malformation, especially in areas with increasing temperatures
Optimized irrigation practices	Implementation of efficient irrigation methods, such as drip irrigation, to reduce water stress	Mitigates the impact of drought and helps maintain tree vigor, reducing malformation risk
Integrated disease management (idm)	Combining cultural, biological, and chemical methods for comprehensive disease control	Flexibility to adapt to different climate conditions and mitigate disease under varied climates
Fungal biocontrol agents	Application of beneficial microorganisms like *Trichoderma* spp. to suppress fungal pathogens	Eco-friendly alternative that can reduce fungal infections causing malformation
Timely and targeted fertilization	Providing balanced nutrients to strengthen tree health, focusing on reducing stress factors	Enhances tree resistance to environmental stress, reducing vulnerability to malformation
Climate-resilient orchard design	Creating orchard designs that account for local climate patterns, such as spacing trees to reduce humidity	Increases airflow, reduces humidity around trees, and minimizes fungal growth that leads to malformation

28.9 Future Mitigation Strategies for Sustainable Mango Malformation Disease Management

Mango malformation disease, a persistent threat to global mango production, necessitates innovative and sustainable mitigation strategies. This chapter explores future-oriented approaches that can empower mango growers to navigate the challenges posed by a changing climate and evolving pathogens.

28.9.1 Leveraging Resistant Cultivars

Planting resistant mango cultivars is a cornerstone of sustainable disease management. Conventional breeding programs have yielded some success, but the emergence of new pathogens or strains necessitates continuous improvement. The future lies in:

- **Marker-assisted breeding**: Utilizing molecular markers to identify and select for disease-resistant genes during breeding programs, leading to faster development of resistant varieties.
- **Gene editing techniques like CRISPR**: Precisely modifying genes to enhance a mango tree's natural resistance to malformation disease. This approach holds immense promise but requires careful regulation and public acceptance.

28.9.2 Harnessing the Power of Biocontrol Agents

Biocontrol agents, organisms that naturally suppress plant pathogens, offer an eco-friendly approach to disease management. Future directions include:

- **Exploration of new biocontrol agents**: Researching the potential of novel bacteria, fungi, and viruses to combat mango malformation disease.
- **Mass production and formulation**: Developing cost-effective methods for mass producing and formulating biocontrol agents for ease of application by farmers.
- **Integration with other strategies**: Investigating how biocontrol agents can be effectively integrated with other disease management practices for a holistic approach.

28.9.3 Precision Agriculture for Disease Control

Precision agriculture utilizes technology to collect data and tailor management practices to specific needs. In the context of mango malformation disease, this could involve:

- **Advanced monitoring systems**: Deploying sensors and weather stations to monitor environmental conditions that favor disease outbreaks, allowing for preventive action.
- **Drone-based disease detection**: Utilizing drones equipped with multispectral imaging to identify early signs of malformation in orchards, enabling targeted interventions.
- **Variable-rate fungicide application**: Using precision technology to apply fungicides only where and when needed, minimizing environmental impact and fungicide resistance.

28.9.4 Enhancing Host Plant Resistance Through Management Practices

Several management practices can strengthen mango trees and improve their natural resistance to disease:

- **Optimizing nutrition**: Implementing balanced fertilization programs that ensure trees have the nutrients they need to fight off pathogens.
- **Water management**: Employing irrigation practices that avoid water stress, as stressed trees are more susceptible to disease.
- **Orchard sanitation**: Removing diseased plant material and maintaining good orchard hygiene to reduce the inoculum source for the pathogen.

28.9.5 The Role of Research and Collaboration

Sustainable management of mango malformation disease hinges on continuous research and collaboration:

- **Understanding the disease complex**: Further research is needed to definitively identify the cause(s) of mango malformation disease, which can inform the development of targeted mitigation strategies.
- **International collaboration**: Encouraging collaboration between researchers, growers, and policymakers across the globe to share knowledge and develop solutions applicable to diverse mango-growing regions.
- **Farmer education and training**: Providing extension services and training programs to equip mango growers with the knowledge and skills to implement sustainable disease management practices.

By embracing these future-oriented mitigation strategies, mango growers can navigate the complexities of a changing environment and pathogens. A combination of resistant cultivars, biocontrol agents, precision agriculture, and optimized management practices holds the key to sustainable mango malformation disease management and a secure future for mango production.

28.10 Conclusion

Mango malformation disease poses a significant threat to global mango production, with its severity expected to increase under changing climate conditions. This chapter underscores the importance of adopting a holistic and integrated approach to mitigate the impacts of MMD. Resistant cultivar development, optimized orchard management practices, and the strategic use of fungicides form the foundation of effective disease control. Moreover, emerging strategies such as biological control agents and microbiome engineering hold promise for enhancing host resistance and reducing pathogen load in an eco-friendly manner.

The application of predictive models, incorporating real-time climatic data and artificial intelligence, offers a proactive approach to disease management, enabling farmers to make informed decisions. However, the successful implementation of these strategies relies on empowering farmers through awareness campaigns, extension services, and capacity-building programs. Policy interventions that support research and provide financial incentives for adopting sustainable practices are equally vital.

By integrating traditional and modern approaches, the mango industry can build resilience against MMD, safeguarding the livelihoods of farmers and ensuring the sustainable production of this globally important fruit under the evolving climate scenarios.

References

Abdel-Sattar MM (1973) Histopathology of mango malformation. MSc thesis, Al-Azhar University, Cairo

Abdurrehman, Mansoor T, Subhani MN, Malik MT, Hasnain A, Abbas A, Alrefaei AF, Ali H, Hassan MZ, Naqvi SAH (2024) Unveiling intricacies of mango malformation dynamics caused by Fusarium solani through extensive field surveys and laboratory, epidemiology, and model-based insights. Applied Fruit Science 66(2):677–688.

Britz H, Steenkamp ET, Coutinho TA, Wingfield BD, Marasas WFO, Wingfield MJ (2002) Two new species of *Fusarium* section *Liseola* associated with mango malformation. Mycologia 94:722–730

Chadha KL, Pal RN, Prakash O, Tandon PL, Singh H, Singh NP, Rao MRK, Lal B (1979) Studies on mango malformation—its causes and control. Indian J Hortic 36:359–368

Chakrabarti DK (2011) Mango malformation. Springer

Covarrubias RA (1980) Control de la “deformación” o “escoba de bruja” del mango en México. Memorias Del Simposium “La Investigación El Desarrollo Experimental y La Docencia En CONAFRUT Durante 3:795–806

Crespo M, Cazorla FM, Hermoso JM, Guirado E, Maymon M, Torés JA, Freeman S, de Vicente A (2012) First report of mango malformation disease caused by *Fusarium mangiferae* in Spain. Plant Dis 96:286

Crookes CA, Rijkenberg FHJ (1985) A literature review of the distribution, symptomology, cause and control of mango blossom malformation. S Afr Mango Grow Assoc Res Rep 5:15–24

Dambreville A, Normand F, Lauri PÉ (2012) Plant growth co-ordination in natura: a unique temperature-controlled law among vegetative and reproductive organs in mango. Funct Plant Biol 40:280–291

Darvas JM (1987) Control of mango blossom malformation with trunk injection. S Afr Mango Grower’s Assoc Yearb 7:21–24

De Villiers EA, Joubert PH (2008) The cultivation of mango. ARC-Institute for Tropical and Subtropical Crops

Dieckman F, Manicom BQ, Coetzee K (1982) An attempt to control blossom malformation of mangoes with chemical sprays

Faria LN, Donato SLR, Santos MR, Castro LG (2016) Manejo da irrigação na indução floral da mangueira ‘Tommy Atkins’ no semiárido baiano. Eng Agrícola 36:387–398

Freeman S, Maimon M, Pinkas Y (1999) Use of GUS transformants of *Fusarium subglutinans* for determining etiology of mango malformation disease. Phytopathology 89:456–461

Freeman S, Maymon M, Biton A, Levin AG, Shtienberg D (2014) Management of mango malformation disease based on a novel strategy of timing of fungicide applications combined with sanitation. Crop Prot 61:84–91

Gamliel-Atinsky E, Freeman S, Maymon M, Belausov E, Ochoa R, Bauchan G, Skoracka A, Pena J, Palevsky E (2010) The role of eriophyoids in fungal pathogen epidemiology, mere association or true interaction? Eriophyoid Mites: Prog Progn 191–204

Gamliel-Atinsky E, Freeman S, Sztejnberg A, Maymon M, Belausov E, Palevsky E (2007) Interactions of the mango bud mite, *Aceria mangiferae*, with *Fusarium mangiferae*, the causal agent of mango malformation disease. IOBC WPRS Bull 30:23

Gamliel-Atinsky E, Sztejnberg A, Maymon M, Shtienberg D, Freeman S (2009a) Inoculum availability and conidial dispersal patterns of *Fusarium mangiferae*, the causal agent of mango malformation disease. Phytopathology 99:160–166

Gamliel-Atinsky E, Sztejnberg A, Maymon M, Vintal H, Shtienberg D, Freeman S (2009b) Infection dynamics of *Fusarium mangiferae*, causal agent of mango malformation disease. Phytopathology 99:775–781

Guarnaccia V, Aiello D, Polizzi G, Perrone G, Stea G, Vitale A (2014) Emergence of prochloraz-resistant populations of *Calonectria pauciramosa* and *Calonectria polizzii* in ornamental nurseries of Southern Italy. Plant Dis 98:344–350

Hahn F, García JA (2022) Mango stem response under different irrigation regimes. Int J Fruit Sci 22:35–56

Ibrahim AN, Satour MM, El-Tobshy ZM, Sattar MAA (1975) Pathological and histological note on mango malformation in Egypt

Iqbal Z, Akhtar N, Ghazanfar MU, Shehzad SM, Ahmad S, Asif M, Yasin M, Pervez MA, Dasti AA, Saleem A (2011) Management of mango malformation through physical alteration and chemical spray. Afr J Agric Res 6:1897–1901

Khanum Z, Tiznado-Hernández ME, Ali A, Musharraf SG, Shakeel M, Khan IA (2020) Adaptation mechanism of mango fruit (*Mangifera indica* L. cv. Chaunsa White) to heat suggest modulation in several metabolic pathways. RSC Adv 10:35531–35544

Kumar J, Singh US, Beniwal SPS (1993) Mango malformation: one hundred years of research. Annu Rev Phytopathol 31:217–232

Kvas M, Steenkamp ET, Al Adawi AO, Deadman ML, Al Jahwari AA, Marasas WFO, Wingfield BD, Ploetz RC, Wingfield MJ (2008) *Fusarium mangiferae* associated with mango malformation in the Sultanate of Oman. Eur J Plant Pathol 121:195–199

Leslie JF, Summerell BA (2006) Fusarium laboratory workshops—a recent history. Mycotoxin Res 22:73–74

Lima CS, Monteiro JHA, Crespo NC, Costa SS, Leslie JF, Pfenning LH (2009) VCG and AFLP analyses identify the same groups in the causal agents of mango malformation in Brazil. Eur J Plant Pathol 123:17–26

Lima CS, Pfenning LH, Costa SS, Abreu LM, Leslie JF (2012) *Fusarium tupiense* sp. nov., a member of the *Gibberella fujikuroi* complex that causes mango malformation in Brazil. Mycologia 104:1408–1419

Lonsdale JH, Kotze JM (1993) Chemical control of mango blossom diseases and the effect on fruit set and yield

Lv YC, Pu JJ, Qia YX, Xie YX, Lu Y, Zhang X, Zhang H, Zhang HQ (2010) *Fusarium proliferatum* caused mango malformation disease in Panzhihua and Huaping Provinces of China. IX Int Mango Symp 992:423–428

Makhmale S, Bhutada P, Yadav L, Yadav BK (2016) Impact of climate change on phenology of mango-the case study. Ecol Environ Conserv 22–S132

Manicom BQ (1989) Blossom malformation of mango. S Afr Mango Grow Assoc Yearb 10:11–12

Marasas WFO, Ploetz RC, Wingfield MJ, Wingfield BD, Steenkamp ET (2006) Mango malformation disease and the associated *Fusarium* species. Phytopathology 96:667–672

Mavroeidi VI, Shaw MW (2005) Sensitivity distributions and cross-resistance patterns of *Mycosphaerella graminicola* to fluquinconazole, prochloraz and azoxystrobin over a period of 9 years. Crop Prot 24:259–266

Morison IL, Growth MMP, Change C (2006) Blackwell Publishing Ltd. UK

Muhammad F, Ibrahim M, Pervez MA (1999) Effect of fungicides on mango malformation. Pak J Biol Sci 2:3

Narasimhan MJ (1954) Malformation of panicles in mango incited by a species of *Eriophyes*

Narasimhan MJ (1959) Control of mango [inflorescence] malformation disease

Nariani TK, Seth ML (1962) Role of eriophyid mites in causing malformation disease in Mango

Navarro MJ, Tello JC (2005) Reduced sensitivity of the mushroom pathogen *Verticillium fungicola* to prochloraz-manganese in vitro. Mycol Res 109:741–745

Newman Z, Freeman S, Biton I, Sa'ada D, Paz T, Maymon M, Lavi U (2012) Molecular diagnosis of mango malformation disease and phylogeny of *Fusarium mangiferae*. Phytoparasitica 40:287–297

Noriega-Cantú DH, Téliz D, Mora-Aguilera G, Rodríguez-Alcazar J, Zavaleta-Mejía E, Otero-Colinas G, Campbell CL (1999) Epidemiology of mango malformation in Guerrero, Mexico, with traditional and integrated management. Plant Dis 83:223–228

O'Donnell K, Cigelnik E, Nirenberg HI (1998) Molecular systematics and phylogeography of the *Gibberella fujikuroi* species complex. Mycol 90:465–493

Otero-Colina G, Rodríguez-Alvarado G, Fernández-Pavía S, Maymon M, Ploetz RC, Aoki T, O'Donnell K, Freeman S (2010) Identification and characterization of a novel etiological agent of mango malformation disease in Mexico *Fusarium mexicanum* Sp. Nov. Phytopathology 100:1176–1184

Ploetz RC (1994) Distribution and prevalence of *Fusarium subglutinans* in mango trees affected by malformation. Can J Bot 72:7–9

Ploetz RC (2001) Malformation: a unique and important disease of mango, *Mangifera indica* L

Ploetz RC (2003) Diseases of mango. In: Diseases of tropical fruit crops, pp 327–363

Ploetz RC, Freeman S (2009) Foliar, floral and soilborne diseases. In: The mango: botany, production and uses. CABI Wallingford UK, pp 231–302

Prakash O, Srivastava KC (1987) Mango diseases and their management. A world review

Ramírez F, Davenport TL (2010) Mango (*Mangifera indica* L.) flowering physiology. Scientia Hortic 126:65–72

Rodríguez-Alvarado G, Fernández-Pavía SP, Ploetz RC, Valenzuela-Vázquez M (2008) A *Fusarium* sp. different from *Fusarium oxysporum* and *F. mangiferae* is associated with mango malformation in Michoacán. Mexico. Plant Pathol 57:374–380

Senghor AL, Sharma K, Kumar PL, Bandyopadhyay R (2012) First report of mango malformation disease caused by *Fusarium tupiense* in Senegal. Plant Dis 96:1582

Shanthala L, Ramachandra RK, Mohan Kumar S, Swamy GSK, Honnabyraiah MK, Pallavi HM (2021) Evaluation of mango (*Mangifera indica* L.) rootstocks under induced moisture stress condition for various morpho-physiological traits. Int J Curr Microbiol Appl Sci 10:143–148

Shu ZH, Sheen TF (1987) Floral induction in axillary buds of mango (*Mangifera indica* L.) as affected by temperature. Scientia Hortic 31:81–87

Singh RN, Majumder PK, Sharma DK, Sinha GC, Bose PC (1974) Effect of de-blossoming on the productivity of mango. Scientia Hortic 2:399–403

Sinniah GD, Adikaram NKB, Vithanage ISK, Abayasekara CL, Maymon M, Freeman S (2013) First report of mango malformation disease caused by *Fusarium mangiferae* in Sri Lanka. Plant Dis 97:427

Steenkamp E, Britz H, Coutinho T, Wingfield B, Marasas WF, Wingfield MJ (2000) Molecular characterization of *Fusarium subglutinans* associated with mango malformation. Mol Plant Pathol 1:187–193

Sternlicht M, Goldenberg S (1976) Mango eriophyid mites in relation to inflorescence. Phytoparasitica 4:45–50

Sukhvibul N, Whiley AW, Smith MK, Hetherington SE, Vithanage V (1999) Effect of temperature on inflorescence and floral development in four mango (*Mangifera indica* L.) cultivars. Scientia Hortic 82:67–84

Summanwar AS, Raychaudhuri SP, Phatak SC (1966) Association of the fungus *Fusarium moniliforme* Sheld. with the malformation in Mango (*Mangifera indica* L.)

Varma A, Lele VC, Raychaudhuri SP, Ram A, Sang A (1974) Mango malformation: a fungal disease. Phytopathol Z 79:219–228

Wei J, Liu G, Liu D, Chen Y (2017) Influence of irrigation during the growth stage on yield and quality in mango (*Mangifera indica* L). PLoS One 12

Whiley AW, Rasmussen TS, Saranah JB, Wolstenholme BN (1989) Effect of temperature on growth, dry matter production and starch accumulation in ten mango (*Mangifera indica* L.) cultivars. J Hortic Sci 64:753–765

Youssef SA, Maymon M, Zveibil A, Klein-Gueta D, Sztejnberg A, Shalaby AA, Freeman S (2007) Epidemiological aspects of mango malformation disease caused by *Fusarium mangiferae* and source of infection in seedlings cultivated in orchards in Egypt. Plant Pathol 56:257–263

Zhan RL, Yang SJ, Liu F, Zhao YL, Chang JM, He YB (2012) First report of *Fusarium mangiferae* causing mango malformation in China. Plant Dis 96:762

Zhan R, Yang S, Ho H, Liu F, Zhao Y, Chang J, He Y (2010) Mango malformation disease in south China caused by *Fusarium proliferatum*. J Phytopathol 158:721–725

Zheng Q, Ploetz R (2002) Genetic diversity in the mango malformation pathogen and development of a PCR assay. Plant Pathol 51:208–216

Zuazo VHD, García-Tejero IF, Rodríguez BC, Tarifa DF, Ruiz BG, Sacristán PC (2021) Deficit irrigation strategies for subtropical mango farming: a review. Agronomy Sustain Dev 41:13

Chapter 29
Advanced Diagnostics Methods of Fungal Pathogens Associated with Mango in a Scenario of Climate Change

Riaz Ahmad and Aqleem Abbas

Abstract Mango, the "King of Fruits, is an economically significant fruit in the world, but the king of fruits is also facing numerous challenges due to a perennially shifting climate and developing biotic stresses under different agroecologies. Future climate predictions indicated a considerable increase in climatic patterns for climate extreme predictions. The world's minimum temperature is expected to climb by 2.8 degrees Celsius by mid-century (2040–2070). To address the negative effects of climate change scenarios, climate-smart technology for diagnosing and detecting plant diseases is required. For quantifying detrimental climate impacts on mango plant diseases, early detection of plant infections is an acceptable preventive technique for mango yield and quality control. As a result, effective diagnostic approaches and instruments that are simple, specific, fast, and economical have to be produced. This chapter provides a complete overview of the current plant pathogen diagnosis methodologies. Furthermore, it focuses on current advancements in the subject, beginning with an appreciation of fundamental research. Through a synthesis of recent findings, it illustrates the current fungal phylogenetic landscape and recommends important areas for future research. Fingerprinting techniques and next-generation sequencing approaches are thoroughly examined, showing their practicality and limitations in determining genetic diversity among fungal isolates. Furthermore, multi-omics methods, including genomics, transcriptomics, and proteomics, are investigated to provide insight into fungal pathogenesis and host interactions. The use of machine learning and artificial intelligence for data analysis is being investigated as a viable way to streamline research processes and improve the accessibility of results. Finally, the need for collaborating as researchers, and bioinformaticians to advance fungal phylogenetics and improve disease management approaches. This

R. Ahmad
Pest and Disease Research Laboratory, Department of Agriculture and Agribusiness Management, University of Karachi, Karachi, Pakistan

A. Abbas (✉)
Department of Agriculture and Food Technology, Karakoram International University (KIU), Gilgit, Pakistan
e-mail: aqlpath@gmail.com

S. A. H. Naqvi et al. (eds.), *Climate Change and Mango Production*,
https://doi.org/10.1007/978-3-031-87035-4_29

chapter aims to add to the most recent diagnostic techniques for plant fungal disease, such as fungal phylogenetics, and to motivate future research efforts in the field.

Keywords Plant diseases · Mango · Fungi · Genetic diversity · Fungal identification · DNA fingerprinting techniques · Sequencing technologies

29.1 Introduction

Mango (*Magnifera indica* L.) is an evergreen fruit tree that belongs to the *Anacardiaceae* family. It is grown widely in tropical and subtropical locations. It is also known as the "King of Fruits" because of its unique flavor and historical significance (Hassan et al. 2022; Leng et al. 2023). Mango cultivars vary greatly in size, shape, color, flavor, and texture, providing nutritional, medicinal, and industrial benefits all around the world (Muchiri et al. 2012). Mango-growing countries include India, Pakistan, China, Brazil, Mexico, the United Arab Emirates, Egypt, and Nigeria, where the crop is widely planted due to its nutritional and economic benefits. Furthermore, mango leaves play a key part in traditional treatments to treat various viral and noninfectious illnesses (Litz 2005; Barreto et al. 2008). Climate change in tropical and subtropical environments has dramatically increased mango pests and diseases (Abdurrehman et al. 2024). Changing climates are predicting more severe scorching heat, and Precipitation is an important factor in disease and pest outbreaks, as disease life cycles are dependent or influenced by humidity, heat, and wind, but heavy precipitation promotes the dissemination and growth of pests and diseases, causing agriculture crop production losses (Leng et al. 2019; Courson et al. 2022). According to the IPCC (IPCC 2019) assessment report, climate change effects on the agriculture sector due to cropping methods and farming diversification make agriculture production very susceptible to flooding and drought. It also harmed economic growth and development (Ahmed et al. 2019).

Mango cultivars are susceptible to significant losses caused by a variety of diseases pests and environmental factors, mango trees are vulnerable to over 83 types of infections comprising 52 fungal, three bacterial, and three plant parasitic nematodes all of which inflict considerable damage with change climatic conditions major diseases including dieback, anthracnose, stem-end rot, foliage blight, leaf spot, and red rust (Abdurrehman et al. 2024; Nazir et al. 2024). There's a requirement to create novel diagnostic techniques that can be simple, quick, inexpensive, dependable, sensitive, and, most significantly, portable for in-field identification of plant fungal infections (Luchi et al. 2020; Mansotra and Vakhlu 2021). The main reason we require early identification rather than chemical treatment of fungal diseases is that fungi are constantly evolving and expanding their host range, which is exacerbated by climate change and research demonstrating that climate change influences the appearance of infections in locations where the possibilities of pathogen survival were previously low, but climate change has increased (Chaloner et al. 2020; Mansotra and Vakhlu 2021).

Through this mysterious attraction, pathogenic microbe research has captivated scientists and nature lovers alike, enticing them to unravel the mysteries of their unique characteristics and diverse existences. These species have perfected the skill of nutrition acquisition through heterotrophy, which reverberates throughout ecosystems and influences human spheres Embark on a voyage to reconsider the fungus realm, where we additionally celebrate its known marvels but also venture into undiscovered areas of biological relevance (Manoharachary et al. 2010). They are extremely adaptable, taking on shapes such as elongated, polarized cells known as hyphae or yeast-like budding cells. Fungi reproduce by the generation of meiotic and/or mitotic spores, which allows them to thrive in a variety of environmental conditions (Hawksworth and Lücking 2017; Komur et al. 2021; Adnan et al. 2022; Zubek et al. 2022). Many ecosystems rely on fungi to degrade and recycle nutrients. Such organisms can act as parasites, population controllers, or form mutualistic partnerships with autotrophy (Ekowati et al. 2017; Shamim et al. 2017; Selvarajan et al. 2019). Fungi have a significant economic influence on agriculture and human health as pathogens, but they also have great applications in crop pest management, food production, and medications. The popularity of edible mushrooms emphasizes the importance of fungus. Furthermore, fungi act as indicators of environmental health, providing critical insights into ecosystem well-being (Borriello et al. 2014; Zubek et al. 2022; Chavhan et al. 2023; Rostas and Skelding 2023). However, accurately diagnosing plant fungal diseases remains difficult due to their basic body shapes and limited diagnostic markers, which frequently necessitate microscopic investigation. Furthermore, certain fungi are recognized primarily for their vegetative structures, rendering morphology-based classification impractical. (Sharma and Sharma 2016). Historically, mycologists depended on morphological identification to identify mango diseases caused by fungi according to phenotypic features, mainly spore-producing structures produced by asexual or sexual reproduction. Although this technique has produced useful insights into fungal variety, it has several drawbacks, particularly when dealing with ambiguous speciation, when separate species appear morphologically identical (Bhat and Browne 2010; Ling et al. 2010; Faria et al. 2011; Meghvansi et al. 2013; Raj et al. 2013). Molecular approaches have played a major role in transforming our understanding of fungal taxonomy, which has traditionally created considerable hurdles in categorizing fungi at higher levels and creating different species limitations for regulating mango fungal diseases. Within this realm, traditional approaches have frequently stumbled, especially concerning the characterization of significant fungal isolates used for extracting secondary metabolites—such as endophytes and endolichenic fungi—due to their inability to produce spores in cultivated conditions, complicating morphology-based identification endeavors. Likewise, even when sporulation occurs, the physical characteristics of spores and conidia can be highly variable, making exact classification difficult. Compounding these complexities is the fact that sexual features such as ascomata (teleomorphs) rarely develop under culture conditions, further limiting dependence on morphology as a solitary identifying technique. Considering these difficulties, the use of DNA sequence-centered techniques has emerged as a critical tool for determining the complex identification of many fungus species that cause diseases in mangoes and other fruit plants (Depotter et al. 2017; Faganello et al. 2017; Yang et al. 2017; Abdullah et al. 2018; Bilska et al. 2018; Pépin et al. 2018; Carnielli-Queiroz et al. 2019; Doyle et al. 2019; Li et al. 2019; Zhang et al. 2019; Cheng et al. 2020; Liang et al. 2020; Ndayihanzamaso et al. 2020; Prasannath et al. 2020). DNA sequencing methods, such as PCR (polymerase chain reaction) and sequencing of specific target genes, enable more accurate and reliable identification of fungi. As DNA sequencing methods advanced, particularly with parallel Sanger sequencing at the beginning of the 2000s, it became obvious that the diversity of fungus visible through fruiting bodies and culture was considerably surpassed by the diversity of unseen fungi, which may be difficult to investigate directly. Scientists can use sequenced DNA to detect genetic associations, address taxonomic problems, and identify previously unknown cryptic species. These molecular techniques complement and improve existing morphological techniques, enabling more detailed and accurate detection of fungi causing disease in mangoes (Damon et al. 2010; Liu et al. 2017; Shamim et al. 2017; Rajeshkumar et al. 2019; Wei et al. 2022). DNA barcoding has been marked as a breakthrough in fungal identification, although it isn't without drawbacks (Vilgalys and Hester 1990; Schoch et al. 2012; Ceballos-Escalera et al. 2022). DNA fingerprinting methods, which include ISSR (Inter-Simple Sequence Repeats) or VNTR (Variable Number Tandem Repeats), are currently used to distinguish closely related fungi (Rosenblad et al. 2022). Fungal communities is based on the second and subsequent generations of high-throughput sequencing (HTS) techniques (Van Dijk et al. 2018). The introduction of next-generation sequencing (NGS) technology brought in a new age of microbial taxonomy. This breakthrough strategy overturned the constraints of culture-based technologies and standard Sanger sequencing, ushering in a novel era of detecting and counting microbial taxa inside samples on a massive scale. However, amid this technical leap appears a maze of concerns. Biases in library preparation, the complex selection of an NGS platform, and the intricacies of PCR amplification have cast a pall over the otherwise dazzling canvas of NGS-based quantification of microbial communities. Nonetheless, the fungal research field has evolved like never before, with NGS technology, exemplified by Illumina sequencing, emerging as a symbol of speed, accuracy, and comprehensive genomic knowledge. Illumina's ability to overcome previous limits by organizing the simultaneous sequencing of many DNA fragments has revealed high-throughput and cost-effective paths for unraveling the secrets of fungal identification. With massive datasets at its disposal, NGS enables researchers to grasp entire genomic panoramas while providing simultaneous access to various markers. This orchestration creates a powerful foundation for the robust and exact identification of fungus, eliminating taxonomic concerns and uncovering the secretive realm of cryptic species. The rise of third-generation sequencing (TGS) technology heralds an era of change, with the ability to investigate single molecules and read lengths that dwarf those of its predecessors. Pacific Biosciences (PacBio) and Oxford Nanopore sequencing have risen to prominence, enriching fungal identification. With the ability to decode long sequences, these technologies enable the sequencing of entire genomes or large genomic regions. Such in-depth insights provide researchers with detailed genetic blueprints, allowing them to comprehend complex genomic landscapes, gene reshuffling, and structural changes. This heightened resolution serves as a beacon, guiding the way through problems associated with marker selection and the intricate tapestry of intraspecific interpretation (Cassia Garcia Simao and Lopes Gomes 2001; Adeniyi et al. 2018; Adnan et al. 2022; Ceballos-Escalera et al. 2022; Sun et al. 2022). Molecular markers and DNA barcodes for fungal identification are being thoroughly studied in several reviews (Dentinger et al. 2011; Schoch et al. 2012; Badotti et al. 2017; Robbertse et al. 2019; Adnan et al. 2022). However, fingerprinting approaches and NGS have yet to be thoroughly discussed. Several publications have emphasized the limits of DNA barcodes for fungi (Fišer Pečnikar and Buzan 2014; Choudhary et al. 2021). As a result, this chapter will not go into detail about molecular markers and DNA barcodes; instead, we will look at fingerprinting methods and sequencing technology. Considering the importance of DNA fingerprinting and sequencing technology in fungal disease identification under climate change, the objectives are: (1) To comprehensively discuss commonly employed DNA fingerprinting methods in fungal disease identification, (2) To explore the expanding applications of sequencing technologies as a robust tool for fungal diseases detection and identification, (3) To explore machine learning and artificial intelligence in fungal phylogenetics.

29.2 Traditional Fungal Phylogenetic Techniques

The significance of DNA sequences from various sections of DNA as powerful tools for quick fungal identification was highlighted, overcoming old criterion boundaries. DNA markers such as ITS, LSU, SSU, COX1, RPB1, RPB2, β-tubulin, MCM7, TEF1-α, γ-actin, ATP6, and CaM genes were extensively studied for fungal identification across different phyla. Strengths and shortcomings were discussed within Ascomycota, Basidiomycota, and other fungal taxa associated with plant diseases (Badotti et al. 2017). While various research was conducted to uncover a universal DNA barcode, prospective DNA barcodes have been identified to differ between species, with the ITS region appearing as a feasible option for most fungi due to its high interspecific diversity, conserved primer sites, and multiple-copy nature within the genome. Protein-coding areas such as *TEF1*-α, *RPB1*, *RPB2*, β-*tubulin*, and *CaM* genes can serve as secondary barcodes for species identification within Ascomycota genera. Traditional fungal phylogenetic approaches have been heavily criticized for their inability to reliably resolve evolutionary relationships amongst fungi. The use of single gene markers for phylogenetic research frequently fails to capture the entire complexity of fungal evolutionary history due to difficulties like as gene duplication, horizontal gene transfer, and insufficient lineage sorting, resulting in mistakes in phylogenetic reconstructions (Choudhary et al. 2021). Morphology-based classifications, another conventional approach, can be subjective and prone to misunderstanding, especially in cases of cryptic species or convergent evolution, emphasizing the importance of molecular data in fungal systematics. Furthermore, restricted taxon sampling in traditional approaches can lead to incomplete phylogenies and biased evolutionary interpretations, misplacing taxa and underestimating fungal diversity. To overcome these criticisms, recent fungal phylogenetic studies use a multi-locus strategy, combining data from different genes to increase resolution and accuracy.

This approach detects potential mistakes in phylogenetic reconstructions by identifying inconsistencies between gene trees. DNA sequencing technologies, such as high-throughput sequencing (HTS) and next-generation sequencing (NGS) platforms like Illumina, have transformed microbial taxonomy by providing comprehensive genetic insights, which have been augmented by third-generation sequencing (TGS) platforms like Pacific Biosciences and Oxford Nanopore. These modern technologies allow for longer read lengths and the elucidation of complicated genomic landscapes, resulting in the much-improved identification of cryptic species. By integrating molecular data, boosting taxon sampling, and adopting multilocus and advanced sequencing technologies, researchers might acquire a more thorough understanding of fungal evolution and diversity, overcoming the limits of old techniques.

29.2.1 Sequencing Technologies

Numerous sequencing technologies are used to identify and detect fungus. Sanger sequencing, one of the first technologies, is currently used for precise sequencing of particular genes or regions. High-throughput sequencing (HTS) technologies, such as Illumina, Ion Torrent, and 454 sequencing, allow for parallel sequencing of millions of DNA fragments, making them attractive for fungal identification and detection. Metabarcoding, an HTS-based method, amplifies certain DNA sections, such as the ITS region, in environmental samples to identify fungal taxa. Third-generation sequencing technologies, including, PacBio SMRT and nanopore sequencing, offer long-read capabilities that enable more complete genome assemblies and the investigation of complicated genomic regions of fungi. RNA sequencing (RNA-Seq) is used to analyze fungal transcriptomes and gene expression. These sequencing technologies, along with bioinformatics analysis, allow researchers to investigate fungal diversity, genetic makeup, and functional properties. The choice of sequencing technique is determined by the study objectives, available resources, and the complexity of the fungal samples. Next-generation sequencing (NGS) and third-generation sequencing (TGS) are particularly popular due to their efficiency and accuracy in researching fungal strains and people. Rapid advances in these technologies have greatly improved fungal identification and characterization, providing accurate ways for researching fungus. Traditional methods of fungal identification were time-consuming and unreliable, but the introduction of PCR and DNA sequencing techniques has resulted in considerable breakthroughs in fungal detection (Tsang et al. 2021).

In the 1990s, researchers used molecular identification methods to discover a wide range of difficult-to-cultivate fungus species. Their nutrition and ecological interactions were unknown since these species lacked fruiting bodies or other apparent traits. The advancement of DNA sequencing technology, particularly parallel Sanger sequencing in the early 2000s, revealed that the diversity of mycobiota that cannot be seen through fruiting bodies or culture is significantly more than previously assumed. False negatives, qualitative data without quantification, and an inability to distinguish between live and non-viable fungal cells are among the issues that these approaches face. The DNA–DNA hybridization test enables the analysis of genetic variation among fungal isolates, but it is labor-intensive and thus unsuitable for high-throughput studies. Although more time-consuming and expensive, nested PCR provides for more accurate detection of target sequences. Real-time PCR, while promising for multiplexing and measuring specific DNA targets, has limitations in sensitivity, repeatability, and specificity. The SCAR approach has excellent specificity for targeted fungal species, whereas fingerprinting techniques such as RFLP, RAPD, AFLP, and SNP analysis have strong discriminatory power but drawbacks such as labor-intensive procedures, limited reproducibility, and technical obstacles.

Transcriptomic technologies such as RNA-Seq and microarray analysis provide detailed gene expression patterns and insights into regulatory networks, but they necessitate high-quality RNA, specialized procedures, and extensive data analysis. Numerous comprehensive reviews have comprehensively explored the PCR, transcriptomic, and fingerprinting diagnostic approaches for fungus (McCartney et al. 2003; Sanzani et al. 2014; Terrero-Salcedo and Powers-Fletcher 2020). These include extensive evaluations of fungal identification techniques, such as primer design, amplification tactics, and detection methods. They also explain the vast applications of these technologies, such as diagnosing fungal infections, detecting fungal pathogens, researching antifungal resistance, and detecting diseases under climate change conditions. Given the extensive current literature, there is no imminent need to amend or give further information on these strategies in this context. For a more complete understanding, interested academics and readers should consult the existing studies. As a result, the parts that follow will concentrate on high-throughput sequencing (HTS) technologies used in fungal diagnostics, such as NGS and TGS. HTS, such as NGS, have evolved as methods for identifying and quantifying many species in a complicated biological sample. It entails targeting specific areas of DNA known as "barcodes" that contain genetic information unique to each species. These barcodes can be amplified with polymerase chain reaction (PCR) and sequenced with NGS technology. The resultant DNA sequences are compared to reference databases to identify the species' identities and relative abundances in the sample. Porter and Hajibabaei developed the concept of DNA metabarcoding, which is a highly sensitive technique for detecting numerous species in a single sample by amplifying a particular DNA region. Compared to traditional identification approaches, DNA metabarcoding enables the detection of a larger number of fungal taxa, particularly those with low relative abundance.

Furthermore, this method allows for the identification of uncultured microorganisms, resulting in a full understanding of the microbial community makeup. As a result, DNA metabarcoding has recently found a wide range of applications in numerous sectors (Porter and Hajibabaei 2018). DNA metabarcoding is especially beneficial for investigating microbial communities, including fungi, in environmental materials such soil, water, and air. It enables a complete assessment of biodiversity and ecosystem dynamics, revealing insights into species composition, community organization, and ecological relationships (Porter and Hajibabaei 2018). DNA metabarcoding, which uses next-generation sequencing (NGS), has transformed the study of fungal ecology, conservation biology, forensics, and microbiological investigation. Unlike traditional culture-based approaches and Sanger sequencing, NGS allows for large-scale identification and quantification of microbial species. However, biases in library preparation, NGS platform selection, and PCR amplification provide obstacles in correctly measuring microbial communities. For fungal identification, NGS frequently employs amplicon sequencing, which targets variable sections of the 28S rDNA or internal transcribed spacers (ITS) of rDNA. Despite its benefits, such as discovering uncommon species and giving quantitative data, NGS is challenged by the wide variation in rDNA copy number among species, which might impact abundance estimations. Third-generation sequencing (TGS) technologies such as PacBio SMRT and Oxford Nanopore provide long-read capabilities, allowing for thorough genome assemblies and deep phylogenetic analysis. These technologies increase fungal identification by enabling the sequencing of the whole ITS region and neighboring rRNA genes, resulting in higher taxonomic resolution. Challenges still exist, such as the requirement for high-fidelity PCR polymerases to reduce mistakes and the significance of precise primer selection to avoid biases. Furthermore, long-read sequencing methods have advanced, providing high-quality data for complex fungal genomes, facilitating de novo genome assembly, structural variation discovery, and fungal genomics research. NGS and TGS have also improved the identification of plant pathogenic fungus, which helps us understand their virulence, host specificity, and evolutionary history. Shotgun metagenomics and metatranscriptomics, for example, provide detailed insights into the genetic potential and activity of microbial communities without the need for pathogen separation. To summarize, NGS and TGS are potent techniques for fungal identification and characterization, overcoming numerous limitations of previous methods and providing precise, high-resolution data for research into fungal diversity, ecology, and genetics (Table 29.1, 29.2 and 29.3).

Over the last decade, next-generation sequencing (NGS) has become a useful method for gene sequencing and detecting microbial activity (Van Dijk et al. 2018). Microbiome research have become more precise thanks to innovations including higher template concentrations, fewer PCR cycles, and the use of high-fidelity polymerase enzymes. Computational methods such as SequenceErr further reduce errors, allowing for more exact identification of fungal species than previous procedures. Despite these advances, NGS technology continues to face obstacles, especially in fungi-specific applications. Biases in Internal Transcribed Spacer (ITS) primers, as well as the requirement for consistent results and publically available data, continue to exist. Targeted NGS processes, while effective, cannot completely replace classic culture approaches. High-throughput sequencing (HTS) investigations require careful selection of appropriate markers, primers, and amplification conditions. The ITS region of the nuclear ribosomal RNA (rRNA) operon is widely utilized as a DNA

Table 29.1 Advantages and Disadvantages of PCR based fungal detection methods

Diagnostic methods	Assays/platforms	Advantages	Disadvantages
Polymerase chain reaction (PCR)-based approaches	Conventional PCR	Rapid detection of fungal pathogens	False negatives due to PCR inhibitors
		Specific amplification of target fungal DNA	Qualitative data without quantification
		High sensitivity for detecting low levels of fungi	Inability to distinguish between viable and non-viable fungal cells
		Easy interpretation of results	Limitations in detecting low levels of inoculum
		Differentiation of closely related fungal species	Cost considerations for reagents and equipment
	BIO-PCR	Highly sensitive detection of fungal DNA	Higher cost compared to conventional PCR
		Detects fungal presence at very low levels	Requires additional steps and specialized equipment
		Eliminates PCR inhibitors	It may require more time and effort for optimization and validation
		Allows detection of viable fungal cells	Dependent on the appropriate selection of target sequences
		Reduces the risk of false positives	Primers and probes design may require sequence data
	DNA-DNA hybridization assay	Provides information on DNA relatedness	Time-consuming pairwise comparison
		Helpful for examining genetic diversity among fungal isolates	Requires the entire genome of the species
			Not suitable for high-throughput analysis
			Limited specificity for certain fungal species
			Selective media is used
	Nested PCR	Detection of target DNA at lower levels	More labor-intensive compared to conventional PCR

(continued)

Table 29.1 (continued)

Diagnostic methods	Assays/platforms	Advantages	Disadvantages
		Increased sensitivity	Higher cost compared to conventional PCR
		Amplification of specific DNA regions	Increased risk of contamination
		Improved specificity	Prone to false-positive results if not properly optimized
		Potential for multiplexing	Longer time is required due to nested amplification
	Real-time PCR	Quantification of specific DNA targets	Issues with sensitivity, repeatability, and specificity
		Reduced risk of false positives	Need appropriate target DNA fragments for primer and probe design
		Less time-consuming	Reliance on fluorescent probes or intercalating dyes for detection
		High sensitivity	Issues with sensitivity, repeatability, and specificity
		Potential for multiplexing	Cost considerations for probes and specialized equipment
			An additional optimization is required for multiplex assays
	Digital PCR droplet	Accurate and sensitive	Necessity to dilute DNA samples
		Absolute quantification	Expensive and time-consuming
		Without the need for standard curves or reference genes	Longer analysis time
		Detect even small amounts of a target gene	
		Low susceptibility to PCR inhibitors	

(continued)

Table 29.1 (continued)

Diagnostic methods	Assays/platforms	Advantages	Disadvantages
	Magnetic-capture hybridization PCR	Deal with PCR inhibitors in plant extracts	Cost
		Simultaneous detection of multiple target pathogens or DNA sequences in a single assay	Designing a capture probe specific to the target DNA sequence
		The magnetic beads used are coated with a biotinylated oligonucleotide that is specific to a DNA region of the fungal pathogen	Limited application
		Reduce the overall detection time	False-negative results if the target DNA sequence is present in low quantities
		Reduction of PCR inhibitors	
	SCAR approach	Amplification of specific fungal DNA region	Need for prior sequence data for primer design
		High specificity for targeted fungal species	Effort and expense in primer design and optimization
		Quick and easy implementation	Limited to known target organisms or groups
		Locus-specific amplification	Potential for cross-contamination during PCR
		Reproducible results	Dependent on the availability of sequence information
Isothermal amplification	Loop-mediated isothermal amplification (LAMP)	A simple and rapid method for specific detection of fungal genomic DNA	A relatively new technique with limited optimization and standardized protocols
		Highly sensitive and specific, capable of detecting fungal DNA at low levels	Heavy reliance on indirect detection methods, like turbidity and non-specific dyes, can lead to false-positive results

(continued)

Table 29.1 (continued)

Diagnostic methods	Assays/platforms	Advantages	Disadvantages
		All reactions can be performed under isothermal conditions, eliminating the need for complex temperature cycling	Optimization may be required for different fungal species
		Cost-effective due to the simplified reaction setup and reduced equipment requirements	May require optimization of primer design and selection for different fungal targets
		Suitable for field applications and point-of-care diagnostics	Limited availability of commercial kits and reagents
		Provides rapid and on-site detection of fungal pathogens	Limited information on genetic variations and detailed genotyping compared to sequencing-based approaches
		The amplification efficiency of LAMP is exceptionally high	Nonspecific amplification
	Rolling circle amplification	Exploits DNA or RNA polymerase to generate ssDNA or RNA molecules	Sensitivity to contamination
		Simplicity, efficiency, and no need of temperature cycling devices	Circular template limits the length of the amplified product
		Analyze gene expression, single nucleotide polymorphism, mRNA splicing and post-translational modification of protein molecules	Sensitivity to enzymatic inhibitors
DNA or RNA probe based assays	In situ hybridization	To detect mRNAs present in the fixed sample	Cost and hazardous nature of the radioactive probe

(continued)

Table 29.1 (continued)

Diagnostic methods	Assays/platforms	Advantages	Disadvantages
		Designing of an antisense ssRNA probe aimed to bind target mRNA	Cannot detect low concentrations of DNA and RNA
		Synthetic oligonucleotide probes and cDNA probes labeled with the radioactive isotopes	
		Very sensitive and easily quantified	
	Fluorescent in-situ hybridization	Detection of DNA or RNA sequences in cells or tissues using DNA or RNA probes, which are labeled directly or indirectly with fluorochromes	Limited sequence information
		Subcellular resolution	Requirement for specific sample preparation
		FISH can be combined with multiple probe sets labeled with different fluorescent dyes	False positive results with autofluorescence materials
		Compatibility with diverse sample types	Not well-suited for precise quantification
		Reproducibility, sensitivity, specificity, accuracy, and speed	Technical complexity and expertise requirements

White (1990), Begerow et al. (2004), Ling et al. (2010), Faria et al. (2011), Raj et al. (2013), Sanzani et al. (2014), Zhao et al. (2014), Yang et al. (2017), Abdullah et al. (2018), Bilska et al. (2018), Li et al. (2019), Zhang et al. (2019), Ndayihanzamaso et al. (2020)

barcode for species identification in both Sanger sequencing and HTS-based metabarcoding. However, ITS may not always give sufficient precision for species-level identification, necessitating the use of other markers for certain fungal groupings. The small subunit (SSU) and large subunit (LSU) nuclear rRNA genes are frequently utilized for aquatic and mycorrhizal fungi, however they may not resolve species-level differences in ascomycetes and basidiomycetes. Long-read third-generation sequencing (TGS) technologies, such as PacBio and Oxford Nanopore, improve taxonomic precision by sequencing the whole ITS region and surrounding rRNA genes. These developments are especially valuable for low-quality DNA samples,

Table 29.2 Advantages and disadvantages of sequencing methods in fungal diagnosis

Diagnostic methods	Assays/Platforms	Advantages	Disadvantages
Genomic approaches	Whole genome sequencing	Provides complete genomic information, including coding and non-coding regions	Requires high-quality DNA and specialized laboratory techniques
		Enables the identification of genes, gene clusters, and regulatory elements	Data analysis can be computationally intensive and require bioinformatics expertise
		Facilitates comparative genomics and evolutionary studies	Costlier compared to targeted sequencing approaches
		Allows the discovery of novel genes, pathways, and genetic variations	Requires genome assembly, which can be challenging for complex fungal genomes
		Provides insights into pathogenicity, virulence factors, and drug resistance mechanisms	Potential errors and gaps in genome assembly
		Using dideoxy chain termination, it is possible to study fungal populations and learn about their variations	Limited ability to capture dynamic changes in gene expression
First-generation sequencing	Traditional Sanger sequencing	Shotgun sequencing is more precise than second and TGS methods	Only one template can be detected at a time
			Slow and time-consuming
Next-generation sequencing	454 pyrosequencing (Roche)	Long read lengths allow for better assembly of complex fungal genomes and identification of structural variations	High error rates
		Makes use of parallel sequencing on a massive scale	
		Sequencing methods such as sequence by synthesis (SBS) and sequence by ligation (SBL)	

(continued)

Table 29.2 (continued)

Diagnostic methods	Assays/Platforms	Advantages	Disadvantages
		makes use of parallel sequencing on a massive scale Sequencing methods such as sequence by synthesis (SBS) and sequence by ligation (SBL)	Homopolymer errors, where consecutive identical bases are misread or skipped, leading to incorrect base calls
		Cost-effective, especially for smaller-scale fungal sequencing projects	Platform discontinuation; instruments and reagents are no longer commercially available
		Reduced GC bias compared to some other sequencing platforms	Limited read length for studying large fungal genomes or complex genomic regions
		Detection of rare variants within a fungal population	Have longer run times which can limit the speed at which fungal sequencing projects can be completed
		Capable of generating 80–120 Mb of sequence in 200–300 bp readings in a 4-h run	Homopolymer-rich DNA regions where insertions and deletions occurred
		1.2 million runs, with 800–1000 bases per read, are possible	
	AB/SOLiD technology	Sequencing by oligonucleotide ligation and detection, high accuracy in base calling, resulting in reliable sequence data,	Produces relatively short read lengths compared to some other sequencing technologies. This can pose challenges in assembling complex fungal genomes and analyzing regions with repetitive sequences

(continued)

Table 29.2 (continued)

Diagnostic methods	Assays/Platforms	Advantages	Disadvantages
		Offer scalability, allowing researchers to process a large number of samples simultaneously	Discontinued SOLiD sequencing technology in 2013. As a result, acquiring instruments, reagents, and support for AB/SOLiD sequencing can be challenging, restricting its accessibility for researchers
		Reduce systematic errors, such as the color code representation of DNA bases	Requires specific library preparation protocols, which can be more complex and time-consuming, errors from A-T bias
		It exhibits high coverage uniformity, meaning that the sequencing depth is relatively consistent across the genome	Relatively expensive compared to some other sequencing technologies
		Allows for sample multiplexing, where multiple samples can be barcoded and sequenced together in a single run	Produces relatively short read lengths compared to some other sequencing technologies
		Detection accuracy is higher	Only 40 bp can be read, which is only suitable for SNP mutation detection
	Illumina MiSeq, HiSeq and NovaSeq (Illumina)	Offer high-throughput capabilities, generating large sequencing data in a single run	Produces relatively short read lengths, which can pose challenges in assembling complex fungal genomes and analyzing regions with repetitive sequences
		Relatively cost-effective compared to some other sequencing technologies	Exhibits GC bias, where regions with high or low GC content may have biased sequencing coverage

(continued)

Table 29.2 (continued)

Diagnostic methods	Assays/Platforms	Advantages	Disadvantages
		Offer faster turnaround times compared to some other sequencing technologies	Can encounter challenges in accurately resolving repetitive regions in fungal genomes. These regions may lead to alignment errors,
		Compatible with a wide range of applications, including whole-genome sequencing, transcriptome analysis, and metagenomics	It can exhibit sequence-specific errors, such as misidentifying certain bases or sequencing artifacts
		It is widely adopted in the scientific community	Provides short-read data, which lacks long-range information. This can make resolving large-scale structural variations, chromosomal rearrangements, and haplotypes within fungal genomes challenging
		The typical run size is 40–50 Mb, with shorter single-end read lengths of up to 300 bases (read duration: 50–300 bp)	Errors in single nucleotide substitutions
		High throughput, good read quality, and low latency (20, 300, and 6000 GB/s, respectively)	
		For fungal metabarcoding research, Illumina's MiSeq sequencing platform is the gold standard	
		~ 550 bases in length (MiSeq 2 × 300), HiSeq: Up to 300 gigabases, NovaSeq: Up to 6000 gigabases	

(continued)

Table 29.2 (continued)

Diagnostic methods	Assays/Platforms	Advantages	Disadvantages
	Ion Torrent (PGM, Proton)	Semiconductor-based sequencing detects hydrogen ions released during DNA synthesis	Prone to errors in homopolymer regions
		Relatively inexpensive	Inaccuracies in base calling
		Straightforward library preparation	Read length limitations
		Suitable for targeted sequencing, exome sequencing, and amplicon sequencing	Limited use in mycology and fungal studies
		Read length "PGM (up to 400–600 bases), Proton (up to 200 bases)	Higher error rates compared to other platforms
		Throughput: (PGM: up to 5–12 million reads, Proton: up to 80 million reads)	Errors mainly come from short deletions
Third-generation sequencing technology (de novo sequencing technology)	PacBio RSII and Sequel (Pacific Biosciences)	With single-molecule real-time (SMRT) sequencing technology	Higher error rates compared to second-generation sequencing platforms
		Longer read lengths, produce long reads averaging 20–25 kilobases (up to 100 kilobases	Relatively higher costs
		Real-time sequencing	Longer turnaround time due to lower sequencing throughput
		Fewer biases and limitations related to GC content or sequence context	
		Potential for full-length transcript sequencing and isoform characterization	Ideal for sequencing short-to-medium length amplicons such as the full ITS region
		Long read lengths, ~ Sequel ~ 400,000 reads per SMRT cell	Errors from CG deletions

(continued)

Table 29.2 (continued)

Diagnostic methods	Assays/Platforms	Advantages	Disadvantages
	Oxford Nanopore MinION, GridION and PrometION (Oxford Nanopore Technologies)	Using long-read single-molecule electrical impulses, etc. Reasonably priced	High average error rate (6–12%)
		Read lengths of up to 2.4 megabases are possible while still being portable enough to operate on battery-powered computers	
		Using long-read single-molecule electrical impulses, etc. Reasonably priced	
		No GC bias or PCR amplification preference	
Transcriptomic approaches	RNA Sequencing (RNA-Seq)	Provides comprehensive and unbiased gene expression profiles	Requires high-quality RNA and specialized laboratory techniques
		Enables the detection of novel transcripts and isoforms	Data analysis can be computationally intensive and require bioinformatics expertise
		Quantifies gene expression levels accurately and sensitively	Costlier compared to other gene expression analysis methods
		Facilitates the identification of differentially expressed genes	Requires appropriate reference genome or transcriptome for alignment and analysis
		Allows the study of alternative splicing and post-transcriptional regulation	Potential challenges in data normalization and integration across different experiments
		Provides insights into regulatory networks and functional annotations	May miss low-abundance or rare transcripts due to library preparation and sequencing limitations
		Can be used to study fungal responses to environmental cues and host interactions	

Table 29.3 Comparison among traditional, NGS, and AI in the diagnosis of fungi

Description	Traditional methods	Next-generation sequencing (NGS)	Artificial intelligence (AI)
Time Efficiency	Time-consuming	Relatively fast but involves complex processing	Very fast, real-time potential
Expertise Required	High, requires skilled personnel	High, requires expertise in sequencing and bioinformatics	High, requires expertise in AI and data science
Sensitivity	Moderate, may miss low-abundance or slow-growing fungi	High, can detect a wide range of fungi	High, improves with more data
Specificity	Moderate to high, dependent on fungi viability in culture	High, provides detailed genetic profiles	High, capable of analyzing complex data patterns
Cost	Relatively low	High, due to equipment and processing	Varies, high initial setup cost but potential long-term savings
Data Handling	Low, manageable manually	High, large datasets requiring bioinformatics tools	High, large datasets requiring advanced AI tools
Detection Range	Limited to culturable fungi	Broad, includes unculturable fungi	Broad, can integrate various data types
Real-time Diagnostics	No	No	Potentially yes
Reliability	High but may miss certain fungi	Very high	High, improves with training data
Integration in Practice	Well-established	Growing, but requires infrastructure	Emerging requires significant development and validation

such as those obtained from old herbarium specimens. Bioinformatics tools such as QIIME, Mothur, FunGuild, SWARM, MED, and DADA2 are critical for data integration, visualization, and interpretation in HTS research. Quality control measures, such as screening out contaminants and low-quality sequences, are essential for accurate data processing. Reference databases such as UNITE, SILVA, and RDP help to classify fungal sequences, providing insights into fungi's ecological activities and relationships. To summarize, NGS and TGS have considerably increased fungal identification and characterization, providing high-resolution data that improves our understanding of fungal diversity, ecology, and genomes. However, continuing technological advancements and rigorous methodological considerations are crucial for realizing the full promise of these sequencing technologies. Metagenomic Next-Generation Sequencing (mNGS) is a high-throughput sequencing approach that enables the simultaneous sequencing of DNA or RNA from several organisms present in a complicated sample, such as environmental or clinical specimens (Gökdemir et al. 2022). mNGS technologies have developed as powerful tools for evaluating the phylogeny and functionality of non-cultivable microorganisms, particularly fungal plant diseases. Recently, research have been done to investigate the use of mNGS in detecting fungal diseases in agricultural settings, emphasizing its potential benefits as a tool for identifying fungal pathogens. It is important to emphasize that, while mNGS can help discover plant infections that cannot be cultured, it cannot establish pathogenicity on its own. Additional steps, such as detecting nucleic acids in host tissues and altering virulence-associated genes, are required to demonstrate pathogenicity. Despite the widespread use of NGS in many domains, metagenomic NGS analysis in agricultural sciences, notably for fungal disease diagnosis, is still in its early stages. As a result, this study emphasizes the importance of mNGS technologies in agronomic sciences and their potential as a widespread trend in the future. Several real-world examples demonstrate the adaptability of mNGS technologies across multiple fields although it has diverse application domains, mNGS shares similarities in barcode usage and scientific techniques. A significant use is to detect culturable and non-culturable substrates and organisms in a variety of conditions. Barcodes particular to target species make organism scanning easier, with ITS barcodes being used for fungal infections. Furthermore, mNGS is important in evolutionary and ecological research since it allows for a thorough investigation of microbial communities and ambient DNA. These findings help to understand metabolic pathways, genetic microheterogeneity, and microbial diversity in varied habitats (Fig. 29.1).

29.2.2 Usage of NGS and TGS in Fungal Detection

Recent breakthroughs in molecular techniques have transformed the study of fungal communities and genetic analysis. Metabarcoding, metagenomics, and metatranscriptomics that use the ITS region as a marker have greatly improved our understanding of fungal diversity and ecological relationships. NGS technologies, particularly Illumina MiSeq, have played critical roles in describing fungal communities in a variety of settings, providing high-resolution and extensive taxonomic coverage. Metabarcoding techniques that take advantage of the ITS region have enabled the characterization of fungal populations. This has enabled identifying the presence of fungal populations in a variety of environments, including soil, air, water, and plant surfaces. NGS technologies have helped to understand the biodiversity of fungi and their ecological functions. The ITS region's numerous copy numbers and broad taxonomic coverage have proven beneficial for the complete characterisation of fungal populations. NGS has transformed the understanding complex plant diseases, such as Fusarium head blight (FHB), by offering insight on the epidemiology and interactions between phytopathogens and beneficial microorganisms. Illumina MiSeq sequencing has revealed information about fungal populations' dynamics and ecological roles in wheat ears. Similarly, PacBio CCS long-read sequencing has allowed

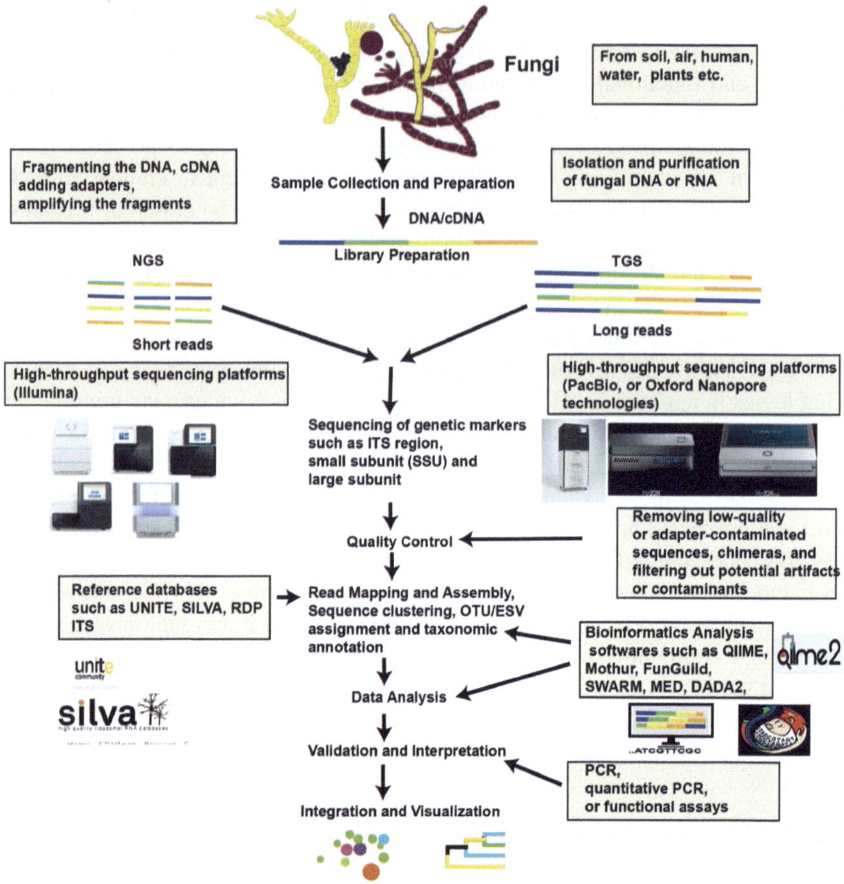

Fig. 29.1 Workflow of High-Throughput Sequencing (HTS) for Fungal Data Analysis. The figure illustrates the key steps involved in the integration, validation, interpretation, and data analysis of fungal high-throughput sequencing (HTS) data. The process begins with sample collection and preparation, involving the isolation and purification of fungal DNA or RNA from various sources such as soil, air, humans, water, and plants. Library preparation follows, where the genetic markers, such as the ITS region, small subunit (SSU), or large subunit, are sequenced using high-throughput sequencing platforms like Illumina or PacBio/Oxford Nanopore technologies

researchers to monitor changes in Fusarium spp. communities in crop residuals, providing a better understanding of how agricultural techniques affect fungal habitats. In the context of grapevine and apple diseases, NGS techniques have been successfully employed to identify microbial species connected with trunk diseases like Eutypa, Esca, Botryosphaeria, Phomopsis dieback, and Black foot. Illumina short-read sequencing targeting the ITS1 and ITS2 rDNA regions has been critical in identifying observed and previously unknown species associated with these disorders. NGS has also been used to examine phylloplane diseases in grapevines, as well as to investigate leaf epiphytic populations' responses to elicitors and biocontrol agents. Furthermore, NGS has shed light on the complexity and diversity of pathogenic and beneficial microbes linked with Apple Replant Disease (ARD) in apple plants. To overcome limits in taxonomic precision, investigations have looked into combining DNA metabarcoding with traditional isolation procedures or using long-read sequencing technologies like as PacBio and ONT. These methods have showed promise for enhancing taxonomic resolution, particularly for fungus groups like Phytophthora. The combination of metagenomics and transcriptomics has improved our understanding of fungal communities by allowing for detailed investigation of functional diversity and gene expression patterns. RNA-seq datasets were utilized to identify the taxonomic and functional diversity of the root microbiota. PacBio Single-Molecule Real-Time (SMRT) sequencing has enabled high-quality de novo assembly and annotation of genome complex, as demonstrated by Trichoderma reesei QM6a. The procedure for producing high-quality genomic DNA and the de novo assembly method has been described, stressing the advantages of PacBio SMRT sequencing in genomic analysis. SMRT sequencing's lengthy reads and excellent consensus reliability led to enhanced prediction of genes and functional annotations (Li and Wang 2021). The Illumina HiSeq platform's adaptability has made it the industry standard for RNA-Seq. The MiSeq desktop sequencer, released a few years ago, substantially expanded the potential applications of Illumina technology. RNA-Seq-based NGS allows for the rapid identification of novel fungal plant infections. Calonectria pseudonaviculata, a new fungal pathogen, was discovered in ornamental plants using a whole-genome sequencing technology called Illumina MiSeq. Two host isolates' 51.4 Mb genomes were studied, and a single nucleotide polymorphism (SNP) was discovered that was unique to each isolate, showing that they were *C. pseudonaviculata* (Kukurba and Montgomery 2015; Malapi-Wight et al. 2016). Another study used RNA-Seq-based NGS to analyze the pathogenomics of *Puccinia striiformis* f. sp. tritici-infected wheat leaves. The purpose was to understand population dynamics and new infectious pathogens. The investigation sought to understand any changes in the pathogen population and to determine potential explanations of these changes. The data revealed a huge and unexpected shift in the UK pathogen population. This transformation suggests the emergence of a new set of alien and emerging disease lineages. Capturing the expression patterns of fungal transcripts enabled RNA-Seq to provide a complete picture of the pathogen population. This method shed light on the pathogen population's genetic diversity and dynamics in its natural environment (Li and Wang 2021). The downy mildew pathogen Another study focused on *Pseudoperonospora cubensis*, which was diagnosed molecularly using RNA and DNA-based NGS approaches. The goal is to identify actionable genetic indicators for the quick and reliable identification of this infection. This was performed using comparative genomics on a closely related species, *P. humuli*, using RNA-Seq research. The researchers analyzed the transcriptomes of *P. humuli* and *P. cubensis*, discovering seven regions in the latter's genome that differed considerably from those of *P. humuli*. These locations demonstrated significant potential as future sites for diagnostic markers (Withers et al. 2016). *Monilinia fructicola* is the fungus that causes brown rot disease in stone and pome fruits, causing significant post- and pre-harvest damage. To further comprehend this pathogen's genetic characteristics, a hybrid and hierarchical de novo association technique was applied. The genome of the *M. fructicola* Mfrc123 strain was sequenced using this method, which employed both the Illumina short-read NGS and the Pacific Biosciences (PacBio) long-read third-generation sequencing platforms. *Hemileia vastatrix*, the fungus that causes coffee rust, has had its genome sequenced using a similar hybrid approach. The whole genome of *H. vastatrix* race XXXIII measured 547 Mb after being sequenced using the PacBio RS II and Illumina HiSeq technologies. Using this hybrid sequencing approach, the coffee rust fungus' genome was better defined, opening the door to research into its biology and potential pathogenic pathways. (De Miccolis Angelini et al. 2019; Porto et al. 2019). The hybrid strategy took advantage of the strengths of both sequencing technologies. Illumina NGS generated high-throughput, cost-effective short reads, whereas PacBio sequencing produced long reads with higher consensus accuracy. Merging the data from these two platforms resulted in a more comprehensive and accurate description of the *M. fructicola* genome.

The merging of Illumina and PacBio sequencing technologies resulted in a hybrid and hierarchical assembly technique, allowing for a more full and detailed understanding of the M. fructicola genome. This method overcome the restrictions of each platform individually and increased the quality and continuity of the assembled genome (Porto et al. 2019). Long-read sequencing technologies, such as PacBio and Oxford Nanopore Technologies (ONT), offer a solution to the issues posed by short-read high-throughput sequencing (HTS) when dealing with fungus (Furzer et al. 2021). Long-read approaches improve the precision of markers and barcodes, allowing for more accurate taxonomic classification. For example, ONT's tiny MinION portable device was used to test whether it is possible to correctly identify *Phytophthora* spp. by sequencing the entire ITS region of their ribosomal DNA. PacBio sequencing has also been used in studies on environmental fungal populations, such as Oomycetes. However, no long-read sequencing studies have been directed at the Phytophthora genus to yet. There are numerous types of aquatic mushrooms, many of which play vital roles in aquatic habitats ranging from freshwater to marine. The development of high-throughput sequencing (HTS) technologies has greatly expanded our understanding of the diversity, distribution, and ecological roles of aquatic fungi. HTS approaches including as metabarcoding and metagenomics have been used to study fungal communities in a variety of aquatic environments, including lakes, rivers, estuaries, and seas. HTS-based studies have shown the presence of numerous fungal species in aquatic habitats, including yeasts, molds, and filamentous fungi, as well as previously discovered fungal lineages. Researchers, for example, used metabarcoding methodologies targeting the ITS region to identify various fungal communities in freshwater settings, including chytrids, ascomycetes, and basidiomycetes. These research aided in the discovery of functional genes and metabolic pathways in aquatic fungus. Genes involved in the breakdown of complex organic materials, such as lignocellulose, have been discovered in aquatic fungi, emphasizing their importance in carbon cycling and ecosystem function. Furthermore, HTS has helped researchers understand the dynamics and succession of fungal communities in response to environmental changes. For example, studies of fungal populations in response to nutrient availability, temperature variations, and pollution have shed light on the resilience and adaptive capacities of aquatic fungi. These studies have also underlined the relevance of fungal-phytoplankton interactions, as well as fungi's role in nutrient cycling and organic matter breakdown in aquatic ecosystems. Recent research has also concentrated on the mycobiome of marine animals such as corals, sponges, and macroalgae, revealing the richness and functional potential of the associated fungal communities. HTS-based techniques have shown the presence of certain fungal taxa, as well as their possible roles in marine organism health and function. Overall, HTS approaches have greatly expanded our understanding of aquatic fungus, exposing their diversity, ecological activities, and reactions to environmental changes. These research have shed light on the roles of aquatic fungi in nutrient cycling, organic matter decomposition, and the overall operation of aquatic ecosystems. Despite the increased use of HTS approaches in the study of aquatic fungus, there is still a scarcity of multimarker studies that use numerous genetic markers. Most studies in this field use small subunit (SSU) ribosomal RNA gene sequences to identify taxa and analyze communities. However, the restricted usage of other markers like as LSU and ITS regions has hampered the categorization success in aquatic fungi (Wahl et al. 2018; Hodgson 2022). One of the most significant issues is the increased divergence rates reported in LSU and ITS markers compared to the SSU marker. This increased divergence makes it difficult to reliably categorize ITS sequences, even at the phylum level. Indeed, it has been observed that more than 60% of ITS sequences in aquatic fungi remain unclassifiable at the phylum level (Hodgson 2022). HTS has tremendously contributed to our understanding of the human mycobiome by shedding light on the fungal communities that live in diverse bodily locations and how they interact with the host. Invasive fungal diseases (IFDs) caused by a variety of fungus species, including Candida and Aspergillus, are associated with high morbidity and mortality rates. IFD investigations have used PCR testing and NGS to identify fungal species, examine genetic diversity, locate antibiotic resistance, and study outbreaks. NGS research has also considerably helped deadly IFDs like mucormycosis, with new insights acquired in areas such as fungus-host interactions, diagnostic advances, and genomic architecture. Furthermore, HTS has been used to evaluate the efficacy of antifungal therapies and to investigate genes important for virulence and pathogenesis in diseases such as Paracoccidioidomycosis. Metagenomic investigations have found fungal species associated with normal microbiota as well as those linked to specific health issues, such as fungal infections. HTS is also used to study antifungal resistance and the dissemination and mechanisms underlying resistance genes. HTS approaches have transformed our understanding of food-borne fungus and their role in brewing and fermentation processes, revealing their intricate structure and multi-organism interactions. Aside from their impact on food, fungus perform a variety of crucial roles in interactions with nonhuman mammals and animals. They can behave as parasites, live in the typical microbiota of the skin, nails, and claws, and function as symbionts in fermenting gut settings. HTS has contributed greatly to our understanding of saprotrophic and mycorrhizal fungi, revealing their various functions in nitrogen cycle and interactions between plants and microbes (Liu et al. 2022). Saprotrophic fungi help decompose organic debris, whereas mycorrhizal fungi create symbiotic relationships with plant roots, increasing nutrient uptake and plant growth (Purahong et al. 2019). HTS approaches have enabled a more thorough examination of these fungal communities, revealing information on their taxonomy, community structure, and functional potential. HTS-based studies have highlighted the extraordinary diversity of saprotrophic fungi and their contributions to ecosystem function. Researchers used metagenomic techniques to discover the genetic potential of saprotrophic fungus to digest complex organic molecules like lignocellulose and engage in nutrient cycling activities. These investigations have shown the succession patterns of fungal lineages during decomposition, as well as the impact of environmental conditions on community dynamics. Mycorrhizal fungi, which develop mutualistic relationships with plant roots, have also been widely examined using HTS. Sequencing the fungal ITS region provided researchers with information into mycorrhizal fungus diversity and its link to plant ecosystems. HTS methods have revealed relationships between mycorrhizal fungal diversity and variables such as tree richness, stand age, productivity, and soil parameters. These findings emphasize the complex connections between mycorrhizal fungi and plants, which influence nutrient exchange and ecosystem dynamics. Using HTS in saprotrophic and mycorrhizal fungus research has helped us understand their ecological functions, community structure, and functional potential. Metagenomic and metabarcoding techniques have improved our understanding of their variety, abundance, and interactions with their environments. Using the benefit of TGS, precise long-read sequencing were acquired using PacBio Sequel Systems to construct draft genomes of fungal endophytes isolated from Ginkgo biloba. Likewise, the whole genome of an endophytic strain of *Amphirosellinia nigrospora* (JS-1675) discovered in *Pteris cretica* was recently sequenced using PacBio RS II technology. Using the advantages of TGS, accurate long-read sequences were acquired utilizing PacBio Sequel Systems to assemble draft genomes of fungal endophytes isolated from Ginkgo biloba. Similarly, the genome of an endophytic strain of *Amphirosellinia nigrospora* (JS-1675) discovered in *Pteris cretica* was recently sequenced using PacBio RS II technology (Bielecka et al. 2022).

29.3 Machine Learning and Artificial Intelligence in Fungal Phylogenetic

Artificial intelligence (AI) has had a considerable impact on fungi/mycology, notably in terms of identifying and classifying different fungus species. AI technologies, including Machine Learning (ML) and Deep Learning (DL), have been used to automate and improve traditional methods of fungal identification, which are sometimes time-consuming and require specialized knowledge. For example, Smith et al. used an ML-based technique to automate slide imaging and Gram stain interpretation, obtaining 92.5% classification accuracy for bloodstream infections, including fungi. This highlights Artificial Intelligence's (AI's) ability to improve throughput while reducing subjectivity in microbiological diagnoses. Furthermore, deep neural networks have been used to classify fungal infections based on microscopic pictures, eliminating the need for biochemical tests and accelerating the diagnosis, which is critical for commencing prompt and targeted treatment. AI applications have resulted in significant advances for specific fungus species such as Candida and Aspergillus. Genome-wide gene essentiality predictions for *Candida albicans* have discovered antifungal drug targets using ML, and ML models modified to MALDI-TOF MS data have qualitatively detected fluconazole resistance in azole-tolerant *C. albicans* with 85% accuracy. Similarly, deep convolutional neural networks (CNNs) were used to classify distinct Aspergillus species. The Xception model, for example, has a detection accuracy of 98.2% on fresh raw pictures, demonstrating AI's capacity to distinguish fungal species based on colony form and other cultural features. These developments highlight AI's importance in improving the accuracy and efficiency of fungal diagnosis, resulting in better patient outcomes. AI has also played an important role in developing fungal infections. The COVID-19 pandemic and mucormycosis outbreak, for example, motivated the development of novel AI applications for the quick detection and investigation of Mucor infections. Hybrid learning-based neural network classifiers (HLNNC) utilizing CNN and support vector machines (SVM) have been designed to detect mucormycosis from ocular pictures of infected patients with good accuracy (99.5%) (Wainaina and Taherzadeh 2023). Furthermore, AI-powered studies have investigated public perceptions of the virus using sentiment analysis on social media platforms, yielding insights into public attitudes and improving public health policies. These applications demonstrate AI's ability to handle both clinical and epidemiological issues in fungal diseases. Fungal diagnostics have progressed dramatically over time, with classical methods, Next-Generation Sequencing (NGS), and Artificial Intelligence (AI) providing distinct approaches, as illustrated in Table 29.3. Microscopy and culture techniques have long been used to identify fungal organisms. These procedures, while dependable, are time-consuming and frequently limited by the knowledge required and the survivability of the fungi in culture. They may not detect all the fungal species, especially those that grow slowly or are in low numbers. Traditional fungal diagnostic methods, including microscopy and culture, have long been used to diagnose fungal diseases. While they are well-established and cost-effective, their reliance on culturing can be time-consuming and may exclude certain fungi, particularly those that are slow-growing or unculturable. Next-generation sequencing (NGS) has transformed fungal diagnosis by allowing for comprehensive, high-throughput study of fungal communities. NGS may detect a wide variety of fungi, including ones that cannot be cultured, by evaluating their genetic material straight from clinical samples. This method has more sensitivity and specificity than standard procedures, resulting in a more thorough and precise picture of fungal presence. However, NGS is costly, necessitates sophisticated equipment, and creates massive volumes of data that require specialist bioinformatics tools to analyze. Artificial intelligence (AI) is a new tool in fungal diagnosis that uses machine learning algorithms to improve the accuracy and speed of identification. AI can use complicated data patterns from diverse diagnostic approaches, such as pictures and genetic sequences, to predict fungal species with high accuracy. AI-powered devices can digest data faster than humans, potentially enabling real-time diagnosis. They also improve with additional data, becoming more precise with time. However, the introduction of AI into clinical practice remains in its earliest phases, necessitating significant computer resources and training data to create effective models.

29.4 Future Directions and Perspectives

The future of fungal pathogen diagnoses in mango production looks promising, especially with the inclusion of next-generation sequencing (NGS) technology. These high-throughput and precise identification methods have unprecedented promise for detecting both known and novel fungal diseases. As NGS technology becomes more affordable and accessible, its use in routine agricultural diagnostics is projected to skyrocket. Researchers are increasingly working on metagenomic and metatranscriptomic techniques to gain a full understanding of fungal community dynamics, including pathogenic and non-pathogenic species. Such in-depth understanding is critical in the context of climate change, which modifies interaction networks and influences disease prevalence. Another significant move ahead is the development of portable and quick diagnostic equipment. Portable sequencers, such as the Oxford Nanopore MinION, provide real-time sequencing and analysis, which is critical for early illness identification and management in the field. Future study must focus on optimizing these techniques for mango-associated fungal infections, assuring their accuracy and dependability across a variety of environmental situations. Furthermore, point-of-care diagnostic kits that employ technologies such as loop-mediated isothermal amplification (LAMP) and CRISPR-based detection techniques are poised to transform fungal pathogen diagnoses. These approaches are not only fast and inexpensive, but they also require little laboratory equipment, making them ideal for use in distant and resource-constrained places (Aragona et al. 2022; Raihan 2024). Future research should focus on cross-sector collaborations between plant pathologists, climate scientists, agronomists, and technology developers to combat fungal infections in the face of climate change. These interdisciplinary initiatives will make it easier to implement integrated treatments and manage diseases holistically. Capacity building and training for researchers, extension workers, and farmers is critical for the successful implementation of sophisticated diagnostic tools and climate-resilient practices (Piombo et al. 2021). Supportive policies and regulatory frameworks are required to promote these approaches. Engaging policymakers will boost innovation and uptake, while standardized data collecting and sharing protocols will improve diagnostic tool effectiveness (Clark et al. 2018; Lockie et al. 2020) Finally, advances in NGS technology, portable diagnostics, and collaborative initiatives point to a promising future for managing fungal infections in mangoes.

29.5 Conclusion

In conclusion, fingerprinting techniques and next-generation sequencing (NGS) approaches have made major contributions to fungal pathogen identification. Despite limitations such as low resolution, NGS, particularly Whole Genome Sequencing and third-generation systems, offer complete genetic insights. Recent advances include improved fingerprinting techniques, the incorporation of multi-omics methodologies, and the use of artificial intelligence (AI) and machine learning (ML) for better fungal diseases of mango identification and analysis concerning climate changes. Future directions emphasize the importance of coordinated efforts among researchers, Pathologists/mycologists, and bioinformaticians in harnessing technology improvements, ultimately strengthening our ability to successfully combat fungal diseases of plants. These breakthroughs have the potential to change fungal pathogen research and management, allowing for more accurate and effective therapies.

References

Abdullah AS, Turo C, Moffat CS, Lopez-Ruiz FJ, Gibberd MR, Hamblin J, Zerihun A (2018) Real-time PCR for diagnosing and quantifying co-infection by two globally distributed fungal pathogens of wheat. Front Plant Sci 9:1086

Abdurrehman MT, Subhani MN, Malik MT, Hasnain A, Abbas A, Alrefaei AF, Ali H, Hassan MZ, Naqvi SAH (2024) Unveiling intricacies of mango malformation dynamics caused by *Fusarium solani* through extensive field surveys and laboratory, epidemiology, and model-based insights. Appl Fruit Sci 66:677–688

Adeniyi M, Titilawo Y, Oluduro A, Odeyemi O, Nakin M, Okoh AI (2018) Molecular identification of some wild Nigerian mushrooms using internal transcribed spacer: polymerase chain reaction. AMB Express 8:1–9

Adnan M, Islam W, Gang L, Chen HYH (2022) Advanced research tools for fungal diversity and its impact on forest ecosystem. Environ Sci Pollut Res 29:45044–45062

Ahmed AU, Appadurai AN, Neelormi S (2019) Status of climate change adaptation in South Asia region. In: Alam M, Lee J, Sawhney P (eds) Status of climate change adaptation in Asia and the Pacific. Springer Nature, Switzerland, pp 125–152

Aragona M, Haegi A, Valente MT, Riccioni L, Orzali L, Vitale S, Luongo L, Infantino A (2022) New-generation sequencing technology in diagnosis of fungal plant pathogens: a dream comes true? J Fungi 8:737

Badotti F, de Oliveira FS, Garcia CF, Vaz ABM, Fonseca PLC, Nahum LA, Oliveira G, Góes-Neto A (2017) Effectiveness of ITS and sub-regions as DNA barcode markers for the identification of Basidiomycota (Fungi). BMC Microbiol 17:1–12

Barreto JC, Trevisan MT, Hull WE, Erben G, De Brito ES, Pfundstein B, Würtele G, Spiegelhalder B, Owen RW (2008) Characterization and quantitation of polyphenolic compounds in bark, kernel, leaves, and peel of mango (*Mangifera indica* L.). J Agric Food Chem 56:5599–5610

Begerow D, John B, Oberwinkler F (2004) Evolutionary relationships among beta-tubulin gene sequences of basidiomycetous fungi. Mycol Res 108:1257–1263. https://doi.org/10.1017/s0953756204001066

Bhat RG, Browne GT (2010) Specific detection of *Phytophthora cactorum* in diseased strawberry plants using nested polymerase chain reaction. Plant Pathol 59:121–129

Bielecka M, Pencakowski B, Nicoletti R (2022) Using next-generation sequencing technology to explore genetic pathways in endophytic fungi in the syntheses of plant bioactive metabolites. Agriculture 12:187

Bilska K, Kulik T, Ostrowska-Kołodziejczak A, Buśko M, Pasquali M, Beyer M, Baturo-Cieśniewska A, Juda M, Załuski D, Treder K (2018) Development of a highly sensitive FcMito qPCR assay for the quantification of the toxigenic fungal plant pathogen *Fusarium culmorum*. Toxins 10:211

Borriello R, Bianciotto V, Orgiazzi A, Lumini E, Bergero R (2014) Sequencing and comparison of the mitochondrial COI gene from isolates of Arbuscular Mycorrhizal Fungi belonging to *Gigasporaceae* and *Glomeraceae* families. Mol Phylogenet Evol 75:1–10. https://doi.org/10.1016/j.ympev.2014.02.012

Carnielli-Queiroz L, Fernandes PMB, Fernandes AAR, Ventura JA (2019) A rapid and reliable method for molecular detection of *Fusarium guttiforme*, the etiological agent of pineapple fusariosis. Braz Arch Biol Technol 62:e19180591

Cassia Garcia Simao Rd, Lopes Gomes S (2001) Structure, expression, and functional analysis of the gene coding for calmodulin in the chytridiomycete *Blastocladiella emersonii*. J Bacteriol 183:2280–2288

Ceballos-Escalera A, Richards J, Arias MB, Inward DJG, Vogler AP (2022) Metabarcoding of insect-associated fungal communities: a comparison of internal transcribed spacer (ITS) and large-subunit (LSU) rRNA markers. MycoKeys 88:1

Chaloner TM, Gurr SJ, Bebber DP (2020) The global burden of plant disease tracks crop yields under climate change. Nat Clim Chang 11:710–715

Chavhan RL, Sable S, Narwade AV, Hinge VR, Kalbande BB, Mukherjee AK, Chakrabarty PK, Kadam US (2023) Multiplex molecular marker-assisted analysis of significant pathogens of cotton (*Gossypium* sp.). Biocatal Agric Biotechnol 47:102557

Cheng Y, Tang X, Gao C, Li Z, Chen J, Guo L, Wang T, Xu J (2020) Molecular diagnostics and pathogenesis of fungal pathogens on bast fiber crops. Pathogens 9:223

Choudhary P, Singh BN, Chakdar H, Saxena AK (2021) DNA barcoding of phytopathogens for disease diagnostics and bio-surveillance. World J Microbiol Biotechnol 37:1–20

Clark B, Jones G, Kendall H, Taylor J, Cao Y, Li W, Zhao C, Chen J, Yang G, Chen L (2018) A proposed framework for accelerating technology trajectories in agriculture: a case study in China. Front Agr Sci Eng 5:485–498

Courson E, Petit S, Poggi S, Ricci B (2022) Weather and landscape drivers of the regional level of pest occurrence in arable agriculture: A multi-pest analysis at the French national scale. Agric, Ecosyst Environ 338:108105

Damon C, Barroso G, Férandon C, Ranger J, Fraissinet-Tachet L, Marmeisse R (2010) Performance of the COX1 gene as a marker for the study of metabolically active Pezizomycotina and Agaricomycetes fungal communities from the analysis of soil RNA. FEMS Microbiol Ecol 74:693–705

De Miccolis Angelini RM, Romanazzi G, Pollastro S, Rotolo C, Faretra F, Landi L (2019) New high-quality draft genome of the brown rot fungal pathogen *Monilinia fructicola*. Genome Biol Evol 11:2850–2855

Dentinger BTM, Didukh MY, Moncalvo JM (2011) Comparing COI and ITS as DNA barcode markers for mushrooms and Allies (Agaricomycotina). PLoS ONE 6:e25081

Depotter JRL, Rodriguez-Moreno L, Thomma BPHJ, Wood TA (2017) The emerging British *Verticillium longisporum* population consists of aggressive *Brassica* pathogens. Phytopathol 107:1399–1405

Doyle VP, Tonry HT, Amsden B, Beale J, Dixon E, Li H, Szarka D, Ward Gauthier N (2019) First report of *Cercospora* cf. *flagellaris* on industrial hemp (*Cannabis sativa*) in Kentucky. Plant Dis 103:1784–1784

Ekowati Y, van Diepeningen AD, Ferrero G, Kennedy MD, de Roda Husman AM, Schets FM (2017) Clinically relevant fungi in water and on surfaces in an indoor swimming pool facility. Int J Hyg Environ Health 220:1152–1160

Faganello FDS, Carrer R, Dias VD, Morello RMSC, Cunha MGD (2017) Molecular diagnosis of *Guignardia citricarpa* in asymptomatic sweet orange tissue. Rev Bras Frutic 39:e518

Faria CB, Abe CAL, Silva CNd, Tessmann DJ, Barbosa-Tessmann IP (2011) New PCR assays for the identification of *Fusarium verticillioides*, *Fusarium subglutinans*, and other species of the *Gibberella fujikuroi* complex. Int J Mol Sci 13:115–132

Fišer Pečnikar Ž, Buzan EV (2014) 20 years since the introduction of DNA barcoding: from theory to application. J Appl Genet 55:43–52

Furzer OJ, Cevik V, Fairhead S, Bailey K, Redkar A, Schudoma C, MacLean D, Holub EB, Jones JDG (2021) An improved assembly of the *Albugo candida* Ac2V genome reveals the expansion of the "CCG" class of effectors. Mol Plant-Microbe Interact 35:39–48. https://doi.org/10.1094/MPMI-04-21-0075-R

Gökdemir FŞ, İşeri ÖD, Sharma A, Achar PN, Eyidoğan F (2022) Metagenomics next generation sequencing (mNGS): an exciting tool for early and accurate diagnostic of fungal pathogens in plants. J Fungi 8:1195

Hassan AH, Bebawy AS, Saad MT, Mosaad GS, Saad BT, Eltayeb WN, Aboshanab KM (2022) Metagenomic nanopore sequencing versus conventional diagnosis for identification of the dieback pathogens of mango trees. Biotechniques 73:261–272

Hawksworth DL, Lücking R (2017) Fungal diversity revisited: 2.2 to 3.8 million species. Microbiol Spectr 5:5–4

Hodgson R (2022) Next-generation sequencing shows increasing temperatures affect stream fungal communities. Coastal Carolina University

Komur P, Chachula P, Kapusta J, Wierzbowska IA, Rola K, Olejniczak P, Mleczko P (2021) What determines species composition and diversity of hypogeous fungi in the diet of small mammals? A comparison across mammal species, habitat types and seasons in Central European mountains. Fungal Ecol 50:101021

Kukurba KR, Montgomery SB (2015) RNA sequencing and analysis. Cold Spring Harb Protoc, 951–969

Leng LY, Ahmed OH, Jalloh MB (2019) Brief review on climate change and tropical peatlands. Geosci Front 10:373–380

Leng LY, Ahmed OH, Jalloh MB, Awang A, Razak NA, Musah AA, Shahlehi S (2023) Brief review: climate change and its impact on mango pests and diseases. J Agric Crops 9:391–399

Li L, Zhang SY, Zhang C-Q (2019) Establishment of a rapid detection method for rice blast fungus based on one-step loop-mediated isothermal amplification (LAMP). Plant Dis 103:1967–1973

Li W-C, Wang T-F (2021) PacBio Long-Read sequencing, assembly, and funannotate reannotation of the complete genome of *Trichoderma reesei* QM6a. In: Mach-Aigner AR, Martzy R (eds) *Trichoderma reesei:* methods and protocols. Springer, US, New York, NY, pp 311–329

Liang K, Wang H-b, Wang S-s, Xu P-p, Zhang R-f, Suomeng D, Zheng X-b (2020) Rapid detection of potato late blight using a loop-mediated isothermal amplification assay. J Integr Agric 19:1274–1282

Ling KS, Wechter WP, Somai BM, Walcott RR, Keinath AP (2010) An improved real-time PCR system for broad-spectrum detection of *Didymella bryoniae*, the causal agent of gummy stem blight of cucurbits. Seed Sci Technol 38:692–703

Litz RE (2005) Biotechnology of fruit and nut crops. CAB International

Liu J, Li J, Tao Y, Sellamuthu B, Walsh R (2017) Analysis of bacterial, fungal and archaeal populations from a municipal wastewater treatment plant developing an innovative aerobic granular sludge process. World J Microbiol Biotechnol 33:1–8

Liu Y, Hu H, Cai M, Liang X, Wu X, Wang A, Chen X, Li X, Xiao C, Huang L (2022) Whole genome sequencing of an edible and medicinal mushroom, *Russula griseocarnosa*, and its association with mycorrhizal characteristics. Gene 808:145996

Lockie S, Fairley-Grenot K, Ankeny R, Botterill L, Howlett B, Mcbratney A, Probyn E, Sorrell T, Sukkarieh S, Woodhead I (2020) The future of agricultural technologies. Australian Council of Learned Academies (ACOLA)

Luchi N, Ioos R, Santini A (2020) Fast and reliable molecular methods to detect fungal pathogens in woody plants. Appl Microbiol Biotechnol 104:2453–2468

Malapi-Wight M, Salgado-Salazar C, Demers JE, Clement DL, Rane KK, Crouch JA (2016) Sarcococca blight: use of whole-genome sequencing for fungal plant disease diagnosis. Plant Dis 100:1093–1100

Manoharachary C, Kunwar I, Reddy SV (2010) Biodiversity, phylogeny and evolution of fungi. In: Sharma VP (ed) Nature at work: ongoing saga of evolution. Springer, India, pp 141–158

Mansotra R, Vakhlu J (2021) Comprehensive account of present techniques for in-field plant disease diagnosis. Arch Microbiol 203:5309–5320

McCartney HA, Foster SJ, Fraaije BA, Ward E (2003) Molecular diagnostics for fungal plant pathogens. Pest Manage Sci 59:129–142

Meghvansi MK, Khan MH, Gupta R, Veer V (2013) Identification of a new species of *Cercospora* causing leaf spot disease in *Capsicum assamicum* in northeastern India. Res Microbiol 164:894–902

Muchiri DR, Mahungu SM, Gituanja SN (2012) Studies on mango (*Mangifera indica,* L.) kernel fat of some Kenyan varieties in Meru. J Am Oil Chem Soc 89:1567–1575

Nazir W, Naqvi SAH, Ahmed N, Alrefaei AF, Zulfiqar MA (2024) Optimizing micronutrient supplementation in mango orchards for the suppression of mango anthracnose (*Colletotrichum gloeosporioides*). Appl Fruit Sci, 1–14

Ndayihanzamaso P, Karangwa P, Mostert D, Mahuku G, Blomme G, Beed F, Swennen R, Viljoen A (2020) The development of a multiplex PCR assay for the detection of *Fusarium oxysporum* f. sp. *cubense* lineage VI strains in East and Central Africa. Eur J Plant Pathol 158:495–509

Pépin N, Punja ZK, Joly DL (2018) Occurrence of powdery mildew caused by *Golovinomyces cichoracearum* sensu lato on *Cannabis sativa* in Canada. Plant Dis 102:2644–2644

Piombo E, Abdelfattah A, Droby S, Wisniewski M, Spadaro D, Schena L (2021) Metagenomics approaches for the detection and surveillance of emerging and recurrent plant pathogens. Microorganisms 9:188

Porter TM, Hajibabaei M (2018) Scaling up: a guide to high-throughput genomic approaches for biodiversity analysis. Mol Ecol 27:313–338

Porto BN, Caixeta ET, Mathioni SM, Vidigal PMP, Zambolim L, Zambolim EM, Donofrio N, Polson SW, Maia TA, Chen C (2019) Genome sequencing and transcript analysis of *Hemileia vastatrix* reveal expression dynamics of candidate effectors dependent on host compatibility. PLoS ONE 14:e0215598

Prasannath K, Galea VJ, Akinsanmi OA (2020) Characterisation of leaf spots caused by *Neopestalotiopsis clavispora* and *Colletotrichum siamense* in macadamia in Australia. Eur J Plant Pathol 156:1219–1225

Purahong W, Pietsch KA, Bruelheide H, Wirth C, Buscot F, Wubet T (2019) Potential links between wood-inhabiting and soil fungal communities: evidence from high-throughput sequencing. MicrobiologyOpen 8:e00856

Raihan A (2024) A review of recent advances, challenges, and potential future directions of climate-smart agriculture. In: Proceedings of the international conference on climate-smart agriculture

Raj M, Jeeva ML, Nath VS, Sankar S, Vidhyadharan P, Archana PV, Hegde V (2013) A highly sensitive nested-PCR method using a single closed tube for the detection of *Colletotrichum gloeosporioides* causing greater yam anthracnose. J Root Crops 39:163–167

Rajeshkumar KC, Yilmaz N, Marathe SD, Seifert KA (2019) Morphology and multigene phylogeny of *Talaromycesa myrossmaniae*, a new synnematous species belonging to the section Trachyspermi from India. MycoKeys 45:41–56

Robbertse B, Strope PK, Schoch C, Brover V (2019) Evaluating ITS length statistics and an interactive distance tree as tools to guide NCBI RefSeq ITS curation. In: 30th fungal genetics conference, Pacific Grove, CA

Rosenblad MA, Larsson E, Walker A, Thongklang N, Wurzbacher C, Nilsson RH (2022) Evidence for further non-coding RNA genes in the fungal rDNA region. MycoKeys 90:203–213. https://doi.org/10.3897/mycokeys.90.84866

Rostas JAP, Skelding KA (2023) Calcium/Calmodulin-stimulated protein Kinase II (CaMKII): different functional outcomes from activation, depending on the cellular microenvironment. Cells 12:401

Sanzani SM, Li Destri Nicosia MG, Faedda R, Cacciola SO, Schena L (2014) Use of quantitative PCR detection methods to study biocontrol agents and phytopathogenic fungi and oomycetes in environmental samples. J Phytopathol 162:1–13

Schoch CL, Seifert KA, Huhndorf S, Robert V, Spouge JL, Levesque CA, Chen W, Fungal Barcoding C, Fungal Barcoding Consortium Author L, Bolchacova E (2012) Nuclear ribosomal internal transcribed spacer (ITS) region as a universal DNA barcode marker for fungi. Proc Natl Acad Sci 109:6241–6246

Selvarajan R, Sibanda T, Sekar S, Nel WAJ (2019) Industrial effluents harbor a unique diversity of fungal community structures as revealed by high-throughput sequencing analysis. Pol J Environ 28:2353–2362

Shamim M, Kumar P, Kumar RR, Kumar M, Kumar RR, Singh KN (2017) Molecular markers in mycology. In: Singh BP, Gupta VK (eds) Molecular markers in mycology: diagnostics and marker developments. Springer, Switzerland

Sharma P, Sharma S (2016) Paradigm shift in plant disease diagnostics: a journey from conventional diagnostics to nano-diagnostics. In: Kumar P, Gupta K, Tiwari K, Kamle M (eds) Current trends in plant disease diagnostics and management practices. Springer International Publishing, Switzerland, pp 237–264

Sun X-R, Xu M-Y, Kong W-L, Wu F, Zhang Y, Xie X-L, Li D-W, Wu X-Q (2022) Fine identification and classification of a novel beneficial Talaromyces fungal species from Masson pine rhizosphere soil. J Fungi 8:155

Terrero-Salcedo D, Powers-Fletcher MV (2020) Updates in laboratory diagnostics for invasive fungal infections. J Clin Microbiol 58:10–1128

Tsang C-C, Teng JL, Lau SK, Woo PC (2021) Rapid genomic diagnosis of fungal infections in the age of next-generation sequencing. J Fungi 7:636

Van Dijk EL, Jaszczyszyn Y, Naquin D, Thermes C (2018) The third revolution in sequencing technology. Trends Genet 34:666–681

Vilgalys R, Hester M (1990) Rapid genetic identification and mapping of enzymatically amplified ribosomal DNA from several *Cryptococcus* species. J Bacteriol 172:4238–4246

Wahl HE, Raudabaugh DB, Bach EM, Bone TS, Luttenton MR, Cichewicz RH, Miller AN (2018) What lies beneath? Fungal diversity at the bottom of Lake Michigan and Lake Superior. J Great Lakes Res 44:263–270

Wainaina S, Taherzadeh MJ (2023) Automation and artificial intelligence in filamentous fungi-based bioprocesses: a review. Bioresour Technol 369:128421

Wei Y, Li L, Liu Y, Xiang S, Zhang H, Yi L, Shang Y, Xu W (2022) Identification techniques and detection methods of edible fungi species. Food Chem 374:131803

White T (1990) PCR protocols: a guide to methods and applications. Academic Press, London

Withers S, Gongora-Castillo E, Gent D, Thomas A, Ojiambo PS, Quesada-Ocampo LM (2016) Using next-generation sequencing to develop molecular diagnostics for *Pseudoperonospora cubensis*, the cucurbit downy mildew pathogen. Phytopathol 106:1105–1116

Yang X, Hameed U, Zhang A-F, Zang H-Y, Gu C-Y, Chen Y, Xu Y-L (2017) Development of a nested-PCR assay for the rapid detection of *Pilidiella granati* in pomegranate fruit. SciRep 7:40954

Zhang SY, Dai DJ, Wang HD, Zhang CQ (2019) One-step loop-mediated isothermal amplification (LAMP) for the rapid and sensitive detection of *Fusarium fujikuroi* in bakanae disease through NRPS31, an important gene in the gibberellic acid bio-synthesis. Sci Rep 9:3726

Zhao YQ, Wu YH, Zhao XX, An MN, Chen JG (2014) Study on the TaqMan Real-time PCR to the detection and quantification of *Rhizoctonia solani* AG-3 of tobacco target spot. Adv Mater Res 1010:80–83

Zubek S, Kapusta P, Roek K, Baszkowski J, Gielas I, Nobis M, Wierszcz S, Nowak A (2022) Fungal root colonization and arbuscular mycorrhizal fungi diversity in soils of grasslands with different mowing intensities. Appl Soil Ecol 172:104358

Chapter 30
Climate-Induced Changes in Mango Production: Phenological Impacts, Pest Threats, and Mitigation Approaches

Muhammad Ahmad, Muhammad Shabbir, Haider Ali, Muhammad Behroz Fazal Hashmi, Mukhtar Ahmed, Shakeel Ahmad, and Amjad Malik

Abstract Global agricultural systems have been greatly influenced by climate change, and tropical crops like mango (*Mangifera indica*) is increasingly facing the repercussions. All phases of the mango's growth, including fruit development, are being impacted by climate change. The behavior, range, and severity of mango pest are changing due to changes in the worldwide climate, which makes things more difficult for mango growers. The effect of typical weather change on mango phenology is examined in this chapter, with particular attention paid to how important developmental phases such as blooming, fruit set, and maturity are impacted by increasing temperatures and changed precipitation patterns. We look at how these changes impact biochemical characteristics, such sugar concentration, acidity, and nutritional profiles, which in turn impact consumer acceptability and fruit quality. The chapter also discusses how shifting temperatures and humidity may make infestations and outbreaks worse, as well as how climate change is making mango crops more vulnerable to illnesses and pests. Cultivar selection and integrated pest control are two sustainable management techniques that are emphasized in order to improve the resilience of mango production systems and address these issues. This chapter

M. Ahmad (✉) · M. Shabbir
Department of Plant Pathology, Faculty of Agricultural Sciences and Technology, Bahauddin Zakariya University, Multan, Pakistan
e-mail: muhmmadahmadam@gmail.com

H. Ali · M. B. F. Hashmi
Department of Entomology, Faculty of Agricultural Sciences and Technology, Bahauddin Zakariya University, Multan, Pakistan

M. Ahmed
Department of Agronomy, Pir Mehr Ali Shah Arid Agriculture University, Rawalpindi, Pakistan

S. Ahmad
Institute of Agronomy, Faculty of Agricultural Sciences and Technology, Bahauddin Zakariya University, Multan, Pakistan

A. Malik
Faculty of Education, University College of the North (UCN), Thompson, MB, Canada

S. A. H. Naqvi et al. (eds.), *Climate Change and Mango Production*,
https://doi.org/10.1007/978-3-031-87035-4_30

offers a thorough overview of the relationships between mango phonology, sustainable farming methods and climate change by combining the most recent research with useful tactics. It emphasizes the pressing need for adaptable solutions to guarantee the continued survival of this significant fruit crop.

Keywords Climate change · Mango pest · Integrated pest management · Mango phenology

30.1 Introduction

Mango (*Mangifera indica* L.) is a prominent species within the genus *Mangifera*, belonging to the Anacardiaceae family. It is an evergreen fruit tree known for its gorgeous, sweet fruit, and it is one of the most popular tropical fruits in the world (Bally 2006). Mango is believed to have originated in regions such as northwestern Myanmar, Bangladesh, and northeastern India. Historical and linguistic evidence suggests that mango cultivation originated over 4000 years earlier in that region (Yadav and Singh 2017). Domestication is believed to have first occurred in Southeast Asia before spreading into India. The geographic range of mangoes spans predominantly tropical and subtropical regions, supported by trade activities. In the era of the Mughal Empire, which governed a significant portion of what is now India and Pakistan, mangoes enjoyed immense popularity and were extensively cultivated (Warschefsky and von Wettberg 2019). Subsequently, they expanded into Africa, the Middle East, and ultimately during the colonial era, reached the Americas, Sri Lanka, India, Thailand, Bangladesh, Pakistan, Burma, Venezuela, South Africa, Mauritius, and Mexico are just a few of the many places where mango is now found. North America (Florida, Hawaii, Puerto Rico), Argentina, Brazil, the West Indies Islands, Vietnam, Thailand, Malaysia, the Indonesia, Somalia, Philippines, the Fiji Islands, Tropical Australia, Niger, Nigeria, Zaire, Madagascar, Kenya, Uganda, Egypt, Israel, Sudan, and Tanzania (Vasanthaiah et al. 2007).

The Food and Agriculture Organization (FAO) reports that mango is cultivated globally across an area of 6 million hectares, resulting in a total production of 59.15 million tons. India stands as the foremost producer of mangoes, yielding 26.29 million tons, which constitutes around 44% of the global total. Following India are China, Indonesia, and Pakistan in terms of production. Mango production is predominantly centered in Asia, accounting for almost 75% of the global total. India, China, the Philippines, Indonesia, Thailand, and Pakistan significantly contribute to this substantial production (FAO 2023). The global mango trade is concentrated among a limited number of countries. The leading exporters of mangoes include Mexico, Thailand, India, Peru, Pakistan, Ecuador, Brazil, the Netherlands, Yemen. This was and the Philippines, among others. Approximately 85% of total worldwide mango exports are attributed to these countries (Kiloes et al. 2024). The United States holds the position of the largest importer, representing 35.70% of its overall trade. Following this, Saudi Arabia accounts for 4.87%. The most frequently exchanged mango varieties consist of Tommy Atkins, Haden, Francis, Kent, Keitt, Ataulfo, Alphonso, Kesar, Edward, and Manila. Among the other well-known varieties in the market are Glenn, Valencia Nam Dok Mai Sindhri, Badami, and Pride (Evans et al. 2017).

Climate change affects ecosystems, human health, the economy, and society, making it one of the most significant global concerns of the twenty-first century (Bandh et al. 2021). Climate change is the phrase used to describe long-term, mostly human-caused changes in temperature, weather patterns, and atmospheric conditions. These changes are largely caused by an increase in greenhouse gases (GHGs), such as carbon dioxide (CO_2), nitrous oxide (N_2O), and methane (CH_4), in the atmosphere (Pachauri et al. 2014). The most prevalent of the greenhouse gases, carbon dioxide, increased from 22.15 billion metric tonnes in 1990 to 36.14 billion metric tonnes in 2014 (Sathaye et al. 2006; Abeydeera et al. 2019). Global warming and major environmental changes are the result of the natural greenhouse effect being exacerbated by the combustion of fossil energy sources (coal, oil, and gas), forest loss, and industrial activities (Yoro and Daramola 2020). Main drivers of climate change have been shown in Fig. 30.1.

Climate change will have different effects on agriculture across the world. The effects of climate change on agriculture are unpredictable, although a variety of outcomes are anticipated (Mendelsohn and Dinar 2009). Temperature variations, shifting patterns of rainfall, and the predicted increase in CO levels that will accompany climate change, especially in tropical regions, will all have a substantial influence on global agriculture. These changes in the climate, weather patterns, and

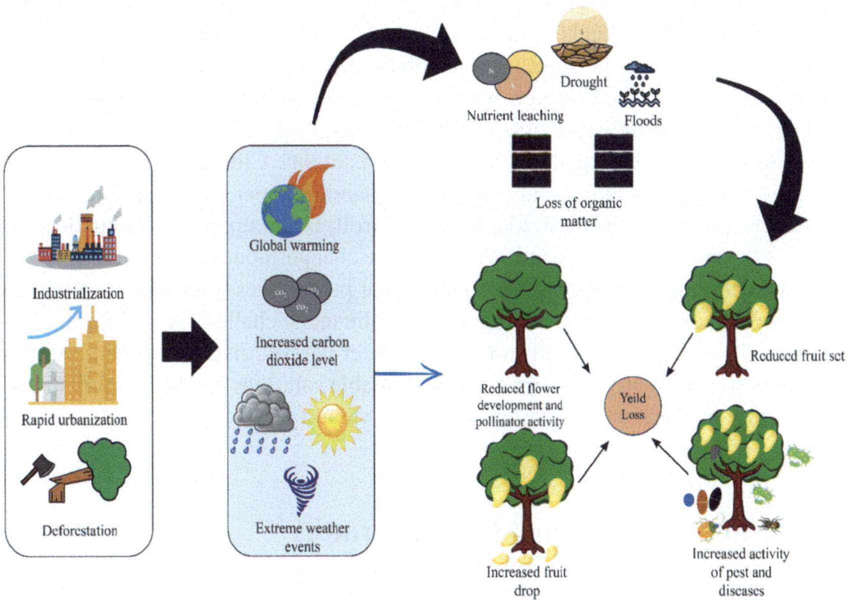

Fig. 30.1 Drivers of climate change and its possible impact on mango phenology

disease and pest patterns are expected to affect crop productivity (Skendžić et al. 2021). The areas that are suitable for producing significant staple crops may shift geographically as a result of climate change.

Climate change directly jeopardizes global food security and indirectly impacts plant-associated microbiota, including plant diseases, which influence crop physiology and production (Lahlali et al. 2024). Plant disease outbreaks reduce primary output and biodiversity globally, posing a serious danger to environmental sustainability and global food safety. By affecting pathogen proliferation and changing host–pathogen interactions, climate change exacerbates these risks and makes it easier for new pathogenic strains to emerge (Singh et al. 2023). The impact of rising temperatures on plant diseases has been extensively studied during the last ten years (Yáñez-López et al. 2012). Projected climatic changes will undoubtedly affect pathogen growth and survival rates, while also modifying host vulnerability, leading to fluctuations in disease effect on crops. The intricate impacts of climate change will differ according to the particular pathosystem and geographic location, affecting optimal infection circumstances, host specificity, then and infection mechanisms. Climate change is profoundly affecting the post-harvest pathology of agricultural produce, altering their nutritional value, storage, and vulnerability to pathogens. The interaction of elevated temperatures, unpredictable precipitation, heightened humidity, and severe weather phenomena is transforming the management of fruit crops post-harvest. Elevated temperatures may broaden the distribution of some pests that impact fruit crops, resulting in heightened damage both before to and during harvest (Yuan et al. 2024). Alterations in precipitation patterns, especially significant rainfall, result in heightened humidity levels in the atmosphere. Elevated humidity levels facilitate the development of mold and fungus on harvested crops. Contaminants like aflatoxins generated by molds present significant health hazards and may lead to product rejection in both domestic and foreign markets. Irregular and severe precipitation may result in heightened moisture absorption in inadequately stored crops, leading to deterioration, diminished quality, and insect infestations. Climate change is greatly impacting mango agriculture and post-harvest handling operations, particularly for post-harvest pathology. Mango, like a tropical fruit, is susceptible to fluctuations in humidity, temperature, and precipitation. Climate change-induced alterations in these parameters affect the emergence and intensity of post-harvest illnesses, resulting in heightened losses, diminished fruit quality, and financial challenges for farmers and exporters. This chapter seeks to examine the after harvest disease of mango in the context of climate change and to provide sustainable solutions to alleviate this stress, so enhancing its duration of storage and export viability.

30.2 Climate Change and Its Accelerators

Climate change is the outcome of intricate interactions between human activity and natural occurrences. The main human activities that worsen deforestation, particularly in tropical areas, include the burning of fossil fuels, intensive agricultural practices, and deforestation. Deforestation for farming reduces the number of trees that can absorb CO_2, which causes a significant amount of carbon stored in trees to be released into the atmosphere. Furthermore, greenhouse gas emissions are a result of the manufacture and use of fertilizers and pesticides (Nda et al. 2018). Both have the potential to worsen global warming and the ensuing climate change by affecting environment health and their ability to store carbon and emit greenhouse gases into the atmosphere (Trenberth 2018). By storing sunlight in the earth's atmosphere, greenhouse gas emissions contribute directly to climate change and have a variety of negative effects on the ecosystem (Kweku et al. 2018). The Intergovernmental Panel on Climate Change has verified that human activities, especially greenhouse gas emissions, have a major impact on Earth's climate (IPCC). Emissions from human activity, particularly carbon dioxide (CO_2) and methane (CH_4) have caused the average global temperature to rise, disrupting natural cycles, raising sea levels, and increasing the frequency of extreme weather events (Kabir et al. 2023). Both locally and globally, the effects of deforestation are apparent. Deforestation is thought to be responsible for 10–20% of the world's greenhouse gas emissions, mostly via the release of CO_2 into the atmosphere (Robalino and Herrera 2010). When combined with a notable net rise in temperature brought on by both biological and CO_2 In particular, after forest degradation during the industrial period, it affects precipitation levels and increases the average annual local temperature by around 1 °C (Lawrence et al. 2022). Concerns have been raised worldwide about the ozone layer's thinning, which is mostly attributed to human activities. As global temperatures rise and CO_2 concentrations rise, the ozone layer is being destroyed. The main causes of the ozone layer's thinning, in contrast to the majority of other types of pollution that originate from many sources, are (Demers et al. 1996). It is believed that rising greenhouse gas emissions are the primary cause of the issue. Perfluorocarbons (PFC), sulphur hexafluoride (SF_6), hydrofluorocarbons (HFC), methane (CH_4), nitrous oxide (N_2O) and carbon dioxide (CO_2), are among the greenhouse gases that are known to have a significant influence on the environment (Schaefer et al. 2006). Most people agree that CO_2 is the main cause of climate change worldwide. Over the last 125 years, the quantity of this gas, which is generated by both natural and human activity, has grown by 25% (Kabir et al. 2023). Because of human industrial activity, particularly since the Industrial Revolution, the amount of CO_2 in the atmosphere has increased dramatically. From 280 ppm in 1750 to 400 ppm in 2013, the amount of CO_2 in the stratosphere has grown by 30–31% over the last three decades (Diallo et al. 2017). The mostly constant CO_2 levels that have been maintained for more than a thousand years have been upset by this considerable rise (Kabir et al. 2023).

30.3 Impact of Climate Change on Mango Fruit and Flower Development

Weather and climate have a significant impact on agricultural output and growth. While crop habitat depends on genetics and phenotype, climate determines crop habitat. Two of the most important factors determining whether a location is suitable for mango growth are air temperature and rainfall, which have an influence on the vegetative and phenological phases of mangos. Mango flowering and fruiting cycles have historically undergone significant shifts due to climate change. In certain regions, this is having a major impact on fruit production (Makhmale et al. 2016). Codes and detailed crop-specific descriptions are presented for the growth stages of the mango tree, contributing to the standardization of international testing systems in fruit growing by Hernández Delgado et al., (2011) as shown in Fig. 30.2. Based on the general BBCH-scale, the one for mango uses 7 of the 10 principal stages (0–9), thus growth stages for bud, leaf and shoot development, inflorescence emergence, flowering, fruit development and fruit maturity are described. Secondary stages (also from 0 to 9) are numbered related to ordinal or percentage values of growth. The scale also uses mesostages, between the principal and secondary stages, to distinguish the different vegetative flushes and the principal terminal inflorescence emergence from the later axillary one, subsequent to terminal flowering failure. Mango has a reproductive propensity and as the temperatures rise it bears a higher number of leaves thereby altering the phenology of blooming. It may be noticed that the hermaphrodite flower (containing both stamen and carpel) percent in late emerging panicles was greater which also correlated with excessive temperatures. In producing or non-bearing trees of mango, burning of leaves and the death of twigs are the major indications of heat stroke. In mango, the principal impacts detected during climatic change were duplication of reproduction flushes, early flowering or delayed blooming, poor fruit set, reproduction buds transformation into vegetative ones, alterations in the fruit ripening etc. (Rajan et al. 2011). Rainfall during flowering removes pollen from the flower's stigma, which leads to inadequate or nonexistent fruit setting. Gujarat had an 80–90% reduction in mango production as a result of unseasonable rains followed by a severe dew assault, which reduced fruit setting, increased fruit drop, and increased the frequency of sooty rust and powdery mildewing in mangos (Rajatiya et al. 2018). Unpredictable rainfall occurring before to the blooming time led to a drop in pollination activities (Chandio et al. 2024). In tropical places, over cast heavy and unseasonal rains, have been reported to limit the flying activity of pollinators. This comprises bees (*Apis mellifera* L.), wild honey bees and flies (Diptera), which often seek nectar and pollen near flowers. In mango farms in Colombia (Ramírez and Kallarackal 2018).

Pollination is a crucial ecological function that facilitates plant reproduction and food production for both humans and animals (Samanta et al. 2024). Approximately 90% of all angiosperms depend partially or entirely on pollination for reproduction (Abrol 2013; Aizen et al. 2019). Moreover, around 80% of all flowering plant species have adapted to be pollinated by animals, mostly insects (Abrol 2013). The majority

Fig. 30.2 Primary and secondary phenological growth stages of mango tree according to BBCH-scale. *Source with permission* Hernández Delgado et al. (2011)

of pollinators, including 25,000–30,000 bee species, together with moths, beetles, wasps, and butterflies, are considered efficient pollinators (Dale et al. 2001; Hooper et al. 2012). Insects exhibit reactions to climate change that include shifts toward early seasonal activity; species may extend their activity periods or, conversely, experience delays (Forrest 2016).

Moreover, atypically high temperatures may either diminish or enhance synchronization in plant–insect interactions (Christmas et al. 2016). Reduced nocturnal temperatures and atypical precipitation in subtropical regions may lead to diminished fruit set, while decreased diurnal temperatures hinder pollination activities. Elevated temperatures during panicle growth in mango accelerate growth, resulting in fewer days available for successful pollination, perhaps leading to reduced yield. Unseasonal precipitation combined with fluctuating temperatures may trigger off-season flowering of mangoes in some instances; nevertheless, fruit set will be suboptimal. Elevated temperatures induce pollen desiccation. The elevation in carbon dioxide content alone enhances mango output; but, when coupled with elevated temperatures and variations in solar radiation, it results in a decline in production (Subedi 2019). The time and volume of precipitation and irrigation significantly influence fruit development and size (Spreer et al. 2009). The time and volume of precipitation and irrigation significantly influence fruit development and size. (Schulze et al. 2013).

Water shortage stress from blooming to the midpoint of the mango fruit's fast cell division phase significantly impacts growth rate and ultimate size of the fruit (Simmons et al. 1995). A 34% drop in fruit size was found by Simmons et al. (1998) for water-stressed 'Kensington Pride' mango trees, attributed to a reduced cell number, in comparison to non-stressed fruit. Consequently, a reduced cell count and restricted water availability may diminish calcium concentrations in growing fruit, thereby impacting postharvest quality characteristics. Conversely, excessive water input before to fruit maturity during and after the cell growth phase results in enhanced fruit size, yield, and greater vulnerability to physiological diseases (dos Santos and Martinez 2013; Schulze et al. 2013). Various irrigation methods may affect productivity and quality, whereas sufficient water can enhance average size of fruits and boost protein soluble and titratable acid levels. Soil water content substantially affects fruit productivity, terms of size, and value, including solids that dissolve, glucose, and vitamin C concentrations. Maintaining 65–70% of field water storage yields the best production, optimum fruit quality, and highest irrigation efficiency, thereby establishing it as the preferred watering strategy for mango farming (Wei et al. 2017) (Figure 2) [Note: Paste figure 2 under this citation.

30.4 Impact of Climate on Physiological and Biochemical Parameters of Fruits

30.4.1 Ripening

Elevated temperatures resulting from global warming may accelerate the process of ripening of mangoes, resulting in faster maturity. This is often linked to an increase in the production of ethylene, a hormone that initiates ripening. Early ripening facilitates earlier harvesting but may diminish the time allocated for transportation and marketing, so impacting logistics of supply chains and market accessibility. Optimal temperatures over 24 °C may expedite ripening; but, they may also diminish the fruit's value and duration of storage by accelerating softening and elevating respiration rates. Elevated temperatures and heightened breathing rates in warmer areas might diminish the efficacy of controlled environment storage methods, which seek to prolong the life cycle of mangoes by regulating carbon dioxide and oxygen concentrations. This may result in difficulties in long-distance transportation and export, when preserving the quality of mangoes for an extended duration is essential for entering international markets. Temperature variations and humidity alterations generated by climate may affect the functioning of enzymes like as a enzyme and pectinase, which degrade cell walls during the ripening process. This may accelerate the softening of mango fruit, resulting in less firmness and increased susceptibility to bruising during careing. Such alterations might diminish the fruit's period of shelf life, raising the probability of post-harvest losses, particularly in areas without sufficient cold storage capabilities.

30.4.2 Total Soluble Solids and Micronutrients

Variations in temperature within mango plantations may substantially influence the micro and macronutrient profile of mango fruit. The ambient temperature during the growth season affects nutrient absorption by the mango tree, the production and storage of vitamins and minerals in the fruit, and the overall quality of the fruit. The physical weight loss was greater for fruits held under cold storage conditions for an extended duration. Mangoes maintained at low temperatures failed to attain complete color development and optimal total soluble solids (TSS) content. Chilling damage was noted and it progressed with the time of storage (Ayele and Bayleyegn 2017).

Various irrigation methods may affect productivity and quality, whereas sufficient water can enhance mean fruit size and boost protein soluble and titratable acid levels. Soil water content substantially affects fruit productivity, size, and value, including solids that are soluble, glucose, and vitamin C concentrations. An irrigation strategy of 65–70% of field water storage yields the best output, optimum quality of fruits, and highest irrigation efficiency, therefore making it the preferred approach for mango agriculture (Wei et al. 2017).

30.5 Impact of Climate on the Post-harvest Pest and Diseases

30.5.1 Stem End Rot (SER)

Stem end Rot (SER) is a serious post-harvest problem for mango fruit in semi-arid areas. Certain harmful fungi undergo a shift from endophytic colonization to a necrotrophic phase as fruit ages, which leads to SER (Galsurker et al. 2020). Numerous fungal species are responsible for this condition. The most common fungal species linked to mango SER disease were *Lasiodiplodia* species (74%) and included *Lasiodiplodia theobromae, L. euphorbicola,* and *L. caatinguensis, Diaportheendophytica, Curvularia Colletotrichum gloeosporioides, Fusarium mangiferae and pseudobrachyspora,*. They were identified by Yeo et al. (2024). For example, SER is the most common postharvest pathogen in dry places and is considered the second most harmful disease in the world for mangoes, behind anthracnose, which is caused by *Colletotrichum gloeosporioides* (Galsurker et al. 2018). *L. theobromae* has a significant adaptability to several settings, including woody plants, wet clay soils, and people, demonstrating the ability to develop and infect hosts throughout a broad temperature range (9–39 °C) (Félix et al. 2016). Dianda et al. (2023) observed that increased prevalence correlated with higher annual mean temperatures ranging from 26 to 28.5 °C and less rainfall (900–1100 mm). Heat stress and drought are recognized to influence plant relationships with the microbiota including *Botryosphaeraceae* fungus (Slippers and Wingfield 2007). Transcriptomic investigations indicated that *Lasiodiplodia* does not exhibit increased aggressiveness with rising temperatures; instead, it appears like the plant host's defense diminishes, rendering it more vulnerable to the fungal endoscopic (Félix et al. 2019).Virulence assays on avocados and mango demonstrated significant disparities across individuals with genome-wide sequencing of two cultures (Avo62 and Man7) underscoring genetic variability in enzymes, metabolic pathways, and stress responses. A more virulent strain exhibited accelerated growth under stress conditions, including osmotic stress, an alkaline pH, and elevated temperature (Gunamalai et al. 2023). The study by Asman et al. (2024) indicates that drought circumstances increase the vulnerability of cocoa plants to an infectious agent particularly when inoculation occurs simultaneously with the start of water stress. The heightened susceptibility under water-restricted situations is alarming in the present context of climate change (Table 30.1).

30.5.2 Anthracnose

Variations in temperature influence the multiplication and development of mango floral disease pathogens, complicating predictions for their attacks and management (Varakumar et al. 2011). *C. gloeosporioides* induces mango anthracnose. It destroys mango panicles, diminishing mango fruit yield and fruit output decreased to 80% as

Table 30.1 Impact of climate-driven shifts in weather parameters on mango fruit quality

Weather parameter	Physical impact	Physiological impact	Biochemical impact
Increased temperature	Alters flowering patterns, leading to irregular fruiting seasons Sunburn on fruit surface Reduces fruit size and weight due to heat stress	Accelerated fruit ripening Reduced fruit set	Decrease in acidity Altered sugar accumulation Reduced vitamin C content
Erratic rainfall	Increased fruit cracking Waterlogging causing fruit drop	Disruption in flowering Inconsistent fruit development	Reduced dry matter Dilution of sugars and other phytochemicals
Drought	Smaller fruit size Increased fruit drop	Impaired physiological processes like photosynthesis and respiration	Reduced starch to sugar conversion Decreased antioxidant levels
High humidity	Increased fungal infections (e.g., anthracnose, powdery mildew)	Increased susceptibility to diseases	Decreased secondary metabolites Reduced phytochemical content
Heatwaves	Scorching and necrosis of fruit tissues	Impaired fruit set Disruption in hormonal balance	Imbalance in sugar-acid ratio Decrease in carotenoids
Cold Spells	Reduced fruit size	Delayed ripening Reduced photosynthetic efficiency	Reduced accumulation of sugars and organic acids
Increased CO_2 levels	Potential increase in fruit size (under optimal conditions)	Enhanced photosynthesis (depending on water and nutrient availability)	Increased carbohydrate content Potential changes in secondary metabolites May alter sugar and acid content, affecting taste
Extreme wind events	Increased physical damage (abrasions, drop)	Stress-induced fruit drop	Reduced phytochemical accumulation Potential oxidative stress
Unpredictable weather	Irregular fruit development and uneven ripening	Hormonal imbalances—Disrupted flowering and pollination	Fluctuating sugar, acid, and antioxidant levels
Increased ozone	Leads to leaf chlorosis and necrosis	Reducing photosynthetic efficiency Results in premature fruit senescence and yield reduction	

a result of anthracnose infestation. Environmental changes, including rising temperatures, light and water temperatures, soil fertility and water associability and altered precipitation, are directly associated with variations in plant-pathogen incidence and severity. Consequently, climate change is likely to modify the geographical and temporally distribution of plant diseases, leading to the emergence of new challenges. Meteorological factors that forecast crop disease incidence including temperature of the air, leaf moisture, precipitating, and relative humidity. *Colletotrichum* development is promoted with temperatures above 20 °C and relative humidity over 95% (Velásquez et al. 2018). *C. gloeosporioides* spores may endure without leaf wetness for a maximum of 48 h, and their infectious potential is restored upon exposure to high temperatures of up to 40 °C, contingent upon the strain (Guyader 2017). Misra (2012) posited that diminishing rainfall, leading to reduced humidity, would adversely affect the anthracnose disorder of mango. The past few decades have seen a significant rise in atmospheric CO_2 levels, posing a primary issue for agricultural output under warming temperatures (Zhang et al. 2021). Certain studies indicate that elevated CO_2 levels may enhance agricultural productivity; nevertheless, they might also alter host–pathogen interactions and associated disease outbreaks in varying ways (Raza et al. 2019). The fungal infections pathogen *C. gloeosporioides* exhibits altered growth and disease severity in response to increasing CO_2 concentrations, which is significantly influenced by the plant host type and the level of CO_2. In order to lessen the severity of anthracnose, for example, 700 ppm of CO_2 hindered and decreased the germination process, germ tube growth, and Colletotrichum appressoria production (Chakraborty et al. 2000). Changes in CO_2 concentrations from 700 parts per million to ambient levels had an impact on the host's (*Stylosanthes scabra*) susceptibility in the field. Additionally, the resistant version has been the target of increased antagonism, whereas the susceptible variant has not (Chakraborty et al. 2000). Later studies revealed that the karyotype and genetic fingerprints of *C. gloeosporioides* strains differed for certain CO_2 cultivar combinations, but these changes were not linked to increased aggression (Cogliati et al. 2013). Furthermore, Cogliati et al. (2013) demonstrated that temperatures between 22 and 26 °C combined with a mixture of 800 parts per million CO_2 produced as shown by the percentage of damaged plant leaves, among other environmental factors (Cogliati et al. 2013). Koo et al. (2016) We out a three-year study showing that, after 100 inoculation cycles, the destructive capacity of the anthrax infection pathogen in hot pepper was significantly decreased in an environment with 700 parts per million CO_2 and a temperature rise of 5 °C. According to research, higher CO_2 levels may increase agricultural output, but they may also change host–pathogen interactions and related disease outbreaks in different ways (Velásquez et al. 2018). Tea's susceptibility to *C. gloeosporioides* was greatly increased by higher concentrations of CO_2-reduced caffeine, an alkaloid found in tea known to play a role in plant defense, and jasmonic acid, a crucial signaling element in plant defense. Furthermore, it is important to acknowledge that an expanded plant-canopy structure under increased CO_2 conditions can capture a greater number of spores, resulting in exacerbated anthracnose during conducive weather conditions (Velásquez et al. 2018) (Table 30.2).

Table 30.2 Incidence and severity of mango stem end rot and anthracnose under climate change

Country/region	Disease	Climate change impact	References
India	Anthracnose	Increased rainfall and humidity exacerbate disease severity during flowering and fruiting stages	Singh et al. (2020)
	Stem end rot	Higher temperatures promote fungal colonization in postharvest conditions	Sharma et al. (2021)
Thailand	Anthracnose	Prolonged wet seasons enhance spore dispersal and infection rates on fruit surfaces	Rattanakreetakul et al. (2023)
	Stem end rot	Warm, humid conditions support the growth of latent infections in harvested fruit	Lim et al. (2014)
Australia	Anthracnose	Unpredictable rainfall patterns result in variable disease outbreaks, complicating control measures	Abasi et al. (2025)
	Stem end rot	High temperatures during harvest promote fungal growth in storage, reducing shelf life	Gelaye et al. (2024)
Brazil	Anthracnose	Rising temperatures and irregular rainfall lead to earlier and more severe infections	Costa et al. (2017)
	Stem end rot	Warm, humid storage conditions increase postharvest losses	Alves et al. (2021)
South Africa	Anthracnose	Heavy rainfall during the growing season promotes disease outbreaks	Honger et al. (2014)
	Stem end rot	Extended wet seasons increase disease prevalence in export mangoes	Honger et al. (2014)
Kenya	Anthracnose	Higher humidity levels extend the disease window, making disease management more challenging	Kimaru et al. (2020)
	Stem end rot	Postharvest losses increase due to inadequate cooling facilities and fungal growth	Kathimba et al. (2022)
Philippines	Anthracnose	Frequent typhoons and rainstorms increase anthracnose incidence on unprotected fruits	Dodd et al. (1991)

(continued)

Table 30.2 (continued)

Country/region	Disease	Climate change impact	References
Mexico	Stem end rot	Hot, dry spells during the growing season followed by humid conditions during storage lead to disease outbreaks	Pineda-Vaca et al. (2025)
Pakistan	Anthracnose	Monsoon season causes widespread infections, reducing marketable yields	Nazir et al. (2024)
Bangladesh	Anthracnose	Excessive rain and humidity during flowering increase disease severity	Khan et al. (2015)
	Stem end rot	Poor postharvest management under humid conditions elevates fungal decay	Islam et al. (2018)
Indonesia	Anthracnose	Rising temperatures and humidity support rapid fungal spore germination	Kailaku et al. (2023)
	Stem end rot	Increased frequency of storage diseases due to poor infrastructure	Kailaku et al. (2023)
Ecuador	Anthracnose	Climate variability disrupts flowering, increasing infection windows	Riera et al. (2019)
	Stem end rot	Export losses due to improper postharvest cooling under warm storage conditions	Riera et al. (2019)

30.5.3 Bacterial Cankers

Mango farming areas in Asia, Africa, Oceania, and the Americas are being affected by the growing illness known as bacterial black spot (BBS), which is caused by the pathogen *Xanthomonas citri* pv. *mangiferaeindicae*. The disease disseminates mostly by wind-driven precipitation and agricultural tools, with a restricted host range, impacting mango, cashew, and pepper plants. BBS is distinguished by black, angular lesions with raised margins on affected fruit and leaf surfaces in both the lower and higher sections of the tree canopy. Infected plants exhibit several small, water-soaked black lesions on fruit, leaves, and stem. Initial symptoms on leaves manifest as small, veined, water-soaked areas with rectangular black dots and, at times, chlorosis spots (Sossah et al. 2024). Temperatures between about 23 and 30 °C, a high relative humidity above 80%, and sporadic rainfall may substantially influence the spread of the virus (Manicom 1986; Pruvost and Luisetti 1991).

30.6 The Insect Major Pests of Mango Fruit

Mangoes are susceptible to several pests that harm the fruit, leaves, and stems. The following pests are included:

- Fruit Flies (*Bactrocera dorsalis* and *Bactrocera zonata*): These pests oviposit inside the fruit, resulting in deterioration and diminished marketability of the produce. Fruit flies are among the most detrimental pests of mangoes, particularly in areas where climatic circumstances promote their multiplication (Baloch et al. 2023).
- Mealybugs (*Drosicha mangiferae*): Mealybugs extract sap from the plant, so diminishing the tree's vitality and increasing its vulnerability to further pests and illnesses. This bug is especially troublesome during arid seasons.
- Thrips (*Scirtothrips dorsalis*): Thrips infest the tender leaves and blossoms of the mango tree, resulting in less fruit set and worse fruit quality. Thrips infections often correlate with arid conditions or irregular precipitation.

Mango Stem Borers (*Batocera rufomaculata*) penetrate the branches and stems of mango trees, compromising their structural integrity and often leading to early mortality (Venkata Rami Reddy et al. 2018).

Each among these pests has distinct climatic requirements for optimal development, and changes in the climate is anticipated to exacerbate the circumstances that promote their proliferation and dispersal (Lalitha 2020).

30.6.1 Effects of Climate Change on Mango Insect Pests

30.6.1.1 Temperature Changes and Pest Development

Temperature is a fundamental component affecting pest biology. Studies indicate that several mango pests, including fruit fly and mealybugs, thrive at elevated temperatures. Elevated temperatures accelerate the life cycles of these pests, allowing an increased number of generations during the mango cultivation period (Dhutraj et al. 2022). Jena et al. (2022) indicate that fruit flies exhibit a significant increase in reproductive rates when temperatures exceed 28 °C, a situation that is more prevalent due to climate change. In Pakistan, where summer temperatures may surpass 40 °C, fruit fly infestation have increasingly plagued mango-producing areas like Rahim Yar Khan and Multan (Rajwana et al. 2010).

Similarly, milder winters allow pests such as mealy bugs to survive and proliferate earlier in the season. As winters get milder, the natural decline of insect populations diminishes, enabling them to inflict more harm in the subsequent growing season. Prolonged growth seasons further increase the duration mango plants are susceptible to insect infestations (Subramanian et al. 2021).

30.6.1.2 Changes in Rainfall Patterns

Precipitation patterns significantly influence pest infestations. In areas with erratic or heavy precipitation, the incidence of the mango pests has risen. Irregular precipitation may provide humid circumstances that facilitate the proliferation of fungal infections, thereby impairing the fruit of the mango and increasing its vulnerability to insect infestations (Chakrabarti 2011).

In Pakistan, altered monsoon cycles have been associated with a rise in insect infestations. Research indicated that districts in Sindh experiencing late or irregular monsoon rains had an increase in thrips infections, since the humid circumstances provided an ideal hatching habitat for these pests. Thrips were shown to proliferate fast after substantial rainfall, resulting in significant damage to mango blooms and immature fruit (Vuppalapati 2023).

30.6.1.3 Elevated CO_2 Levels

Increased atmospheric CO_2 concentrations have direct as well as indirect impacts on mango pests. Elevated CO_2 levels influence the nutritional value of mango trees, often diminishing the nitrogen contents in both leaves and fruit. Consequently, pests such as aphids and leaf hoppers, which get sustenance from plant sap, must ingest greater quantities of plant matter to satisfy their nutritional requirements, resulting in increased damage (Rajatiya et al. 2018).

Recent research indicate that increased CO_2 concentrations may affect the effectiveness of plant defenses. Plants synthesize certain chemical compounds as a defense mechanism against pests, and variations in atmospheric CO_2 may influence the synthesis of these molecules. Research by Ahammed and Li (2022) indicates that under increased CO_2 circumstances, mango plants may synthesize less defense compounds, rendering them more susceptible to insect infestations.

30.6.2 Case Studies of Mango Pest Dynamics in Changing Climates

30.6.2.1 Pakistan

Pakistan is among the foremost global growers of mangoes, and the nation's mango sector has been significantly impacted by climate change. Growers in Punjab and Sindh have recently reported an escalation in insect outbreaks attributed to elevated temperatures and altered monsoon patterns.

Research by Khan et al. (2021) revealed the effects of climate change on a fruit fly infestation in the mango cultivation areas of Sindh. The researchers discovered that elevated temperatures in early spring enabled fruit fly colonies to develop sooner in the season, resulting in increased crop losses. The delayed arrival of monsoon rains in some regions led to prolonged dry and hot conditions, which facilitated the proliferation of pests such as mealybugs and thrips.

A report by the University of Agriculture, Faisalabad, indicated that more than 70% of mango producers in Punjab, particularly in places such as Multan, had seen a rise in pest pressure over the last decade. Numerous farmers ascribed this rise to changes in climatic patterns, notably milder winters and more unpredictable precipitation (Usman 2024).

30.6.2.2 India

India, the foremost worldwide producer of mangoes, has also seen alterations in pest dynamics as a result of climate change. Yadav et al. (2023) observed an increasing prevalence of fruit flies in north Indian mango plantations, coinciding with a steady increase in temperatures over the previous two decades. The proliferation of fruit fly populations into formerly colder areas, including the Himalayan foothills, illustrates the influence of climate changes on pest dispersion.

30.7 Impact on Mango Production and Economy

The economic ramifications of global warming on mango production are significant. Mangoes constitute a major export item in Pakistan, and losses from pests may result in considerable economic difficulties for growers. The Pakistan Horticulture Development and Export Company (PHDEC) reported that pest-related losses in mango production reached 20–30% of total output in some locations in 2022. This adversely impacts farmers' lives and diminishes Pakistan's competitive in the global mango market. In India, where mango export substantially impact the national economy, increasing insect infestations have resulted in elevated production costs owing to the need for frequent pesticide treatments. Moreover, rigorous international rules about pesticide residues in exported mangoes have complicated growers' ability to control pests successfully without compromising their export potential (Bhattacharjee et al. 2022). The synergistic influence of climate change and insect infestations is adversely affecting food security in mango-producing areas. Smallholder farmers, sometimes without the means to adopt modern pest control techniques, are especially susceptible to the economic repercussions of climate-induced insect infestations (Mustafa 2017).

30.8 Adaptation and Mitigation

30.8.1 Breeding and Selection of Climate-Resilient Mango Varieties

Choosing weather-resilient mango varieties is crucial for ensuring sustained production under increasingly uncertain climatic circumstances. This procedure include finding and cultivating types capable of enduring conditions like as severe temperatures, drought, salt, and heightened disease pressure. Currently, seed selection, cross-breeding, and mutant breeding are the primary methods of breeding. The prolonged juvenile period of perennial fruit trees renders the selection of mango reproductive progeny via the aforementioned approaches labor-intensive and necessitates extensive screening, with hybrid offspring often exhibiting considerable phenotypic variation (Song et al. 2023). New developments in genetic manipulation and molecular technology have transformed plant breeding, facilitating the accurate and economical selection of cultivars with specific features. Mango breeding and selection efforts in recent years have focused on cultivars that have higher yields, vivid colors, resistance to disease, and longer shelf lives. The established needs of the growers and the international mango business support the specifications. Dwarf and semi-dwarf scion cultivars from different mango-producing regions that are renowned for their exceptional disease resistance, higher fruit quality, or specific ethnic preference have been the focus of the selection effort. Cogshall, Fairchild, Graham, Ice Cream, Mallika, Manila, Neelum, and Rosegold are some of the varieties. These varieties' outstanding horticultural qualities allow them to be successfully marketed as patio and estate trees in South Florida, USA, supporting this multimillion-dollar industry.

Plant DNA may now be precisely altered thanks to genome editing and CRISPR-Cas9 (clustered regularly interspaced short palindromic repeats), which have revolutionized genetic manipulation. The effectiveness and accuracy of genomic modifications have increased because to developments in CRISPR techniques, such as base editing and prime editing. Without introducing foreign genes, this makes it easier to include certain traits (Wada et al. 2020; Vilela 2021). The study of other CRISPR-Cas systems, such as Cas12, Cas13, and Cas14, for their potential in genome editing and gene regulation is included under CRISPR-Cas Technologies Beyond Cas9. These technologies provide improved capabilities, including the ability to target RNA (Cas13) for gene regulation and enable different approaches for precise genetic modifications (Hillary and Ceasar 2023). CRISPR-Cas9 technology enables breeders to improve mango resilience to environmental challenges such as salt, drought, and pests by targeting genes associated with stress response. It may also alter genes that affect fruit size, taste, ripening, and enhance disease resistance, including resistance to mango malformation disease. This method seeks to cultivate mango varieties with improved robustness, superior taste, and extended shelf life (Antwi-Boasiako et al. 2024).

Marker-assisted selection (MAS) use molecular markers to detect and select plants exhibiting desirable features, circumventing the need for labor-intensive and expensive phenotypic assessments. MAS has expedited breeding initiatives, facilitating the creation of enhanced mango cultivars with preferred characteristics. The research reveals an SNP (21,881,933) in the β-1,3-glucanase 2 (β-1,3-GLU2) gene associated with anthracnose tolerance in mango, whereby the I196V mutation affects the enzyme's structure and functionality. Mango genotypes with the 'A' allele exhibit resistance to *Colletotrichum gloeosporioides*, whilst those with the 'G' allele show increased susceptibility. An economical allele-specific PCR (AS-PCR) marker was created to distinguish these SNP alleles. This marker enhances marker-assisted selection (MAS) in breeding programs, allowing for early genotyping and increasing breeding efficiency for disease-resistant mango cultivars (Felipe et al. 2022).

30.8.2 Improved Cold Chain

Maintaining low but secure temperatures throughout the handling of perishable goods, such as vegetables and fruits, from harvest to end-user (cold chain) is essential for keeping quality. The cold chain for perishable items involves the uninterrupted management of produce at low temperatures throughout postharvest activities, including harvesting, collecting, transportation, storage, processing, and marketing, until they reach the end consumers (Kitinoja 2013). The elements of an effective cold chain are categorized into seven components: on-farm chilling, initial cooling down, storage, shipping, transportation, retail, and consumer, all of which may include temperature control. Effective cold chain management diminishes respiration, thereby decreasing perish ability, reducing transpiration and consequently minimizing the loss of water and shriveling, lowering ethylene production and enhancing resistance to ethylene effects, which slows maturing and natural senescence, curtailing microbial activity, and mitigating browning along with the loss of flavor, texture, and nutrients. In research comparing standard harvesting practices Adhering to proper cold chain techniques resulted in reduced fruit pulp temperature (11 °C), decelerated ripening, minimized weight loss, and enhanced firmness preservation compared to conventional approaches. Consequently, effective cold chain management prolonged the shelf life of mangoes by 18 days. This method enhances the marketability and processing potential of mangoes (Amwoka et al. 2021).

30.8.3 Bagging

The research indicated that on-tree fruit bagging significantly enhances the quality and durability of mangoes (cv. 'Samar Bahisht Chaunsa') during freezing at 12 °C and subsequent ambient storage. Bagged fruits demonstrated reduced weight loss, enhanced firmness, and elevated concentrations of antioxidants, vitamin C, and total soluble solids relative to non-bagged fruits. Following 10 days of chilling, both bag and non-bagged fruits maintained their freshness for 5 days under atmospheric conditions. Nonetheless, fruits held for 20 days, irrespective of bagging, deteriorated during ambient storage. Bagging is advised as an efficacious method to improve postharvest quality of mangoes (Nadeem et al. 2022).

30.8.4 Strategies to Reduce the Infestation of Major Post-harvest Pest

30.8.4.1 Inhibition of Fruit Ripening

Ethylene is the primary phytohormone regulating the majority of processes linked to the climacteric fruit-ripening process. Auxin and abscisic acid, among other phytohormones, are significantly linked to the process of fruit ripening (Kumar et al. 2014). Numerous studies have looked at the potential use of the phytohormones to reduce postharvest decay because of the close correlation between fruit maturation and decay after harvest. In citrus and mango, the administration of the auxin derivative 2,4-dichlorophenoxyacetic acid (2,4-D) reduced SER by inhibiting fruit ripening and preventing stem-end abscission (Kobiler et al. 2001). The effects of postharvest treatment with 1-MCP, an ethylene receptors inhibitor that delays fruit senescence and prolongs preservation (Watkins and Nock 2008), on the reduction of postharvest deterioration were assessed in other studies. However, since ethylene plays two roles in the maturation of fruit and the plant's defence response (Alkan and Fortes 2015), The outcomes of the research were incongruous. 1-MCP therapy reduced fungal pathogen deterioration and SER in avocado and Indian jujube fruit (Diskin et al. 2015). On the other hand, 1-MCP increases fruit susceptibility to SER infections, according to earlier studies on citrus (Porat et al. 1999), mango, and avocado (Hofman et al. 2001; Woolf et al. 2005). 1. MCP seems to have a concentration- and time-dependent effect on fruit susceptibility (Ku et al. 1999; Dou et al. 2005; Blanco-Ulate et al. 2013). Consequently, minimal quantities of ethylene are likely essential for sustaining fruit resistance to pathogens (Alkan and Fortes 2015), whereas elevated levels of 1-MCP likely impede the ripening process and obstruct the fruit's inherent defenses.

30.8.4.2 Harvesting with Stem

A compelling method to diminish SER involves a straightforward harvesting technique: using secateurs to collect fruit with short pedicels, rather than the conventional approach of removing the fruit whole, which results in the absence of any stem. This little modification in harvesting technique substantially reduced the prevalence of SER and anthracnose in mango and avocado fruits (Diskin et al. 2017). Harvesting

Fig. 30.3 **a, b** Infestation of Stem end rot in transit and storage (C&D) effect of harvesting mangoes with peduncle on the infestation of SER and anthracnose

using elongated stems prevents the production of anacardic acid, hence decreasing the likelihood of fruit damage and subsequent infections. Likewise, the harvesting of mangoes with elongated pedicels reduced SER relative to harvesting with abbreviated pedicels (Sangchote 1989). Notably, the sap contains chemicals with antibacterial capabilities, which may account for the variation in SER occurrence. Mango sap, for example, includes alk(en)ylresorcinols (5-n-heptadecenylresorcinol and 5-n-pentadecylresorcinol) that have antifungal and antibacterial effects, especially against Alternaria alternata (Negi et al. 2002). Additionally, 'Kensington Pride' mangoes that were kept with 2- to 3-cm long stems had a significantly higher concentration of resorcinol in their peel and fewer anthracnose lesions than de-sapped fruit (Hassan et al. 2007). Similarly, mango cultivars with increased sap flow showed a lower incidence of anthracnose (Krishnapillai and Wilson Wijeratnam 2017). These findings suggest that picking fruit with stems retains a higher percentage of sap containing antifungal chemicals in the fruit stems and peel, which leads to less postharvest side deterioration SER and anthracnose (Fig. 30.3).

30.8.4.3 Heat Treatment

By promoting the spreading of the fruit's waxy covering, removing unattached microorganisms, and strengthening the fruit's natural resistance, heat treatment helps reduce postharvest illnesses, extension of shelf life and reduction of chilling wound (Fallik 2004). Hot-water dipping and rinsing are two examples of heat-treatment techniques, while hot vapour and dry-air treatments have been suggested as ways to reduce postharvest infections by triggering a defence response. Furthermore, heat treatment reduces fruit fly larval infection on impacted fruits (Mwando et al. 2021; Patel et al. 2023). Although it was less successful than its effect on anthracnose, hot-water immersion also reduced the SER of papaya and mango (Spalding and Reeder 1986; Alvindia and Acda 2015; Rashid et al. 2015).

30.8.4.4 Integrated Pest Management (IPM)

Integrated Pest Management (IPM) stands out as a highly effective approach for addressing the challenges posed by climate change on mango pests. IPM combines chemical treatments, cultural practices and to effectively reduce pest damage and decrease dependence on chemical pesticides. Integrated pest management strategies have been effectively applied in various mango cultivation areas. The introduction of parasitoid wasps as biological control agents has effectively contributed to the reduction of fruit fly populations in orchards (Haggag et al. 2014). The quantity of chemical needed to prevent postharvest decay was significantly decreased when heat treatment and certain chemical compounds were combined to achieve a synergistic improvement in control efficacy. Mangoes in various countries underwent treatment by being immersed for 5 min at 52 °C along with benomyl. It has been reported that a benomyl dip in hot water effectively controls SER caused by both *Dothiorelladominicana* and *Lasiodiplodia theobromae* (Johnson et al. 1990; Coates et al. 1993). For the management of SER and anthracnose, carbendazim may also be used with hot water (52 °C) (Johnson and Hofman 2009). Postharvest disease development was significantly reduced by the use of Thiabendazole (TBZ) both alone and in combination with Sportak (0.5 mL/L) a.i. Prochloraz (tank mix) (Amin et al. 2011). Additionally, techniques like pheromone trapping, trimming, and proper watering have been used to manage the number of insects in an environmentally friendly manner (Karuppuchamy and Venugopal 2016).

30.8.4.5 Climate-Smart Agricultural Practices

Implementing climate-smart agriculture techniques to adapt agricultural practices to changing climatic conditions is essential for minimizing the vulnerability of mango crops to pests. The focus is on developing resistant to pests mango types, maximizing irrigation practices, and enhancing soil health.

Currently, efforts are being made in Pakistan to create mango varieties that can better withstand changing weather and pest challenges. The Punjab Agriculture Research Board (PARB) has initiated a project focused on developing new mango varieties that exhibit greater resistance to fruit flies and other prevalent pests (Paliwal et al. 2021). Furthermore, initiatives aimed at enhancing water management via drip irrigation methods have demonstrated potential in mitigating the factors that promote pest proliferation, including waterlogging and high humidity levels (Skendžić et al. 2021).

30.9 Conclusion

Climate change has emerged as a formidable challenge to global mango production, significantly altering its phenology, pest dynamics, and fruit quality. Shifts in temperature, erratic rainfall patterns, increased humidity, and extreme weather events have disrupted traditional flowering and fruiting cycles, while simultaneously intensifying pest and disease pressure. These changes pose substantial threats to the yield, marketability, and economic viability of mango cultivation, especially in tropical and subtropical regions where it serves as a critical agricultural commodity.

Adapting to these challenges requires a multifaceted approach. Technological interventions such as precision agriculture, microclimate management, and the use of pre-harvest practices like fruit bagging are essential to mitigate environmental stressors and improve fruit quality. Integrated Pest Management (IPM) strategies, resistant cultivars, and improved postharvest technologies are equally vital to counteract pest and disease threats. Moreover, leveraging advancements in genomics can aid in identifying climate-resilient mango varieties, while exploring sustainable practices like agroforestry and organic cultivation can bolster long-term resilience.

References

Abasi F, Hussain S, Raja NI (2025) Orchards management under changing climate. In Challenges and Solutions of Climate Impact on Agriculture. Academic Press pp. 145–162

Abeydeera L, Mesthrige JW, Samarasinghalage TI (2019) Global research on carbon emissions: a scientometric review. Sustain 11:3972

Abrol DP (2013) Asiatic honeybee *Apis cerana:* biodiversity conservation and agricultural production. Springer

Ahammed GJ, Li X (2022) Elevated carbon dioxide-induced regulation of ethylene in plants. Environ Exp Bot 202:105025

Aizen MA, Aguiar S, Biesmeijer JC, Garibaldi LA, Inouye DW, Jung C, Martins DJ, Medel R, Morales CL, Ngo H (2019) Global agricultural productivity is threatened by increasing pollinator dependence without a parallel increase in crop diversification. Glob Chang Biol 25:3516–3527

Alkan N, Fortes AM (2015) Insights into molecular and metabolic events associated with fruit response to post-harvest fungal pathogens. Front Plant Sci 6:889

Alves NE, Gomes MJ, Vasconcelos CM, Lima AC, de Lima SL, Brito ES, Bassinello PZ, Martino HS (2021) Six months under uncontrolled relative humidity and room temperature changes technological characteristics and maintains the physicochemical and functional properties of carioca beans (Phaseolus vulgaris L.). Food Chem 342:128390.

Alvindia DG, Acda MA (2015) Revisiting the efficacy of hot water treatment in managing anthracnose and stem-end rot diseases of mango cv. 'Carabao.' Crop Prot 67:96–101

Amin M, Malik AU, Khan AS, Javed N (2011) Potential of fungicides and plant activator for postharvest disease management in mangoes. Int J Agric Biol 13:671–676

Amwoka EM, Ambuko JL, Jesang HM, Owino WO (2021) Effectiveness of selected cold chain management practices to extend shelf life of mango fruit. Adv Agric 2021:8859144

Antwi-Boasiako A, Amponsah P, Opoku JA, Coulibaly D, Mintah P (2024) Increasing mango production efficiency under the fast-changing climate. In: Hasanuzzaman M, Nahar K (eds) Abiotic stress in crop plants—ecophysiological responses and molecular approaches. IntechOpen, London, pp 121–143

Asman A, Iwanami T, Rosmana A (2024) Effect of drought stress on dieback disease development under *Lasiodiplodia theobromae* infection in cocoa clone" MCC 02". Beverage Plant Res 4:1–12
Ayele L, Bayleyegn A (2017) Effect of low temperature storage on ripening and quality of Mango (*Mangifera indica* L.). J Agric Sci Res 5:570–576
Bally ISE (2006) *Mangifera indica* (mango). In: Species profiles for pacific island agroforestry. Permanent Agriculture Resources, Hawaii, pp 1–25
Baloch N, Memon RM, Khuhro NH, Shaikh I (2023) Population fluctuations of *Bactrocera* species (Diptera: *Tephritidae*) in guava and mango orchards at different climatic conditions of Sindh. Pak J Zool 55:37
Bandh SA, Shafi S, Peerzada M, Rehman T, Bashir S, Wani SA, Dar R (2021) Multidimensional analysis of global climate change: a review. Environ Sci Pollut Res 28:24872–24888
Bhattacharjee P, Warang O, Das S, Das S (2022) Impact of climate change on fruit crops-a review. Curr World Environ 17:319
Blanco-Ulate B, Vincenti E, Powell ALT, Cantu D (2013) Tomato transcriptome and mutant analyses suggest a role for plant stress hormones in the interaction between fruit and *Botrytis cinerea*. Front Plant Sci 4:142
Chakrabarti DK (2011) Mango malformation. Springer, Berlin
Chakraborty S, Pangga IB, Lupton J, Hart L, Room PM, Yates D (2000) Production and dispersal of *Colletotrichum gloeosporioides* spores on *Stylosanthes scabra* under elevated CO_2. Environ Pollut 108:381–387
Chandio AA, Gokmenoglu KK, Joyo MA, Jiang Y (2024) Modeling the climate change impacts on major fruits production: recent evidence from Pakistan. Sci Hortic 324:112618
Christmas MJ, Breed MF, Lowe AJ (2016) Constraints to and conservation implications for climate change adaptation in plants. Conserv Genet 17:305–320
Coates LM, Johnson GI, Cooke AW (1993) Postharvest disease control in mangoes using high humidity hot air and fungicide treatments. Ann Appl Biol 123:441–448
Cogliati E, Pugliese M, Gilardi G, Gullino ML, Garibaldi A (2013) Assessment of the possible effects of climate change on anthracnose of basil. Protezione Delle Colture 2:61–62
Costa JFO, Ramos-Sobrinho R, Chaves TP, Silva JRA, Pinho DB, Assunção IP, Lima GSA (2017) First report of colletotrichum fructicola causing anthracnose on Annona leaves in Brazil. Plant Disease 101(2):386
Dale VH, Joyce LA, McNulty S, Neilson RP, Ayres MP, Flannigan MD, Hanson PJ, Irland LC, Lugo AE, Peterson CJ (2001) Climate change and forest disturbances: climate change can affect forests by altering the frequency, intensity, duration, and timing of fire, drought, introduced species, insect and pathogen outbreaks, hurricanes, windstorms, ice storms, or landslides. BioSci 51:723–734
Demers S, Roy S, De Mora S (1996) The impact of ozone layer depletion on the marine environment. Ecodecision
Dhutraj S, Waghmare G, Gawade RN (2022) Chapter-8 The impact of climate change on fruit crops. Chief Editor Dr. RK Naresh, p 110
Diallo M, Legras B, Ray E, Engel A, Añel JA (2017) Global distribution of CO_2 in the upper troposphere and stratosphere. Atmos Chem Phys 17:3861–3878
Dianda OZ, Wonni I, Ouédraogo L, Sankara P, Tollenaere C, Del Ponte EM, Fernandez D (2023) *Lasiodiplodia* species associated with mango (*Mangifera indica* L.) decline in Burkina Faso and influence of climatic factors on the disease distribution. Physiol Mol Plant Pathol 126:102041
Diskin S, Feygenberg O, Maurer D, Prusky D, Droby S, Alkan N (2015) Postharvest treatments for stem-end rot in fruits. Alon Hanotea 70:36–38
Diskin S, Lurie S, Feygenberg O, Maurer D, Droby S, Prusky D, Alkan N (2017) Postharvest microbiota dynamics of mango fruit stem-end. In: Korsten L, Chidamba L, Bill M (eds) IV International symposium on postharvest pathology: next generation innovation and commercial solutions for postharvest, Skukuza, Kruger National Park, South Africa, pp 195–200

Dodd JC, Estrada AB, Matcham J, Jeffries P, Jeger MJ (1991) The effect of climatic factors on Colletotrichum gloeosporioides, causal agent of mango anthracnose, in the Philippines. Plant Pathol 40(4):568–575

Dou H, Jones S, Ritenour M (2005) Influence of 1-MCP application and concentration on post-harvest peel disorders and incidence of decay in citrus fruit. J Hortic Sci Biotechnol 80:786–792

Evans EA, Ballen FH, Siddiq M (2017) Mango production, global trade, consumption trends, and postharvest processing and nutrition. In: Siddiq M, Brecht JK, Sidhu JS (eds) Handbook of mango fruit: production, postharvest science, processing technology and nutrition. Wiley, Chichester, pp 1–16

FAO (2023) Crop production statistics

Fallik E (2004) Prestorage hot water treatments (immersion, rinsing, and brushing). Postharvest Biol Technol 32:125–134

Felipe JE, Lachica JA, Dela Cueva FM, Laurel NR, Alcasid CE, Sison MLJ, Valencia LDC, Ocampo ETM (2022) Validation and molecular analysis of β-1,3-GLU2 SNP marker associated with resistance to *Colletotrichum gloeosporioides* in mango (*Mangifera indica* L.). Physiol Mol Plant Pathol 118:101804

Forrest JRK (2016) Complex responses of insect phenology to climate change. Curr Opin Insect Sci 17:49–54

Félix C, Duarte AS, Vitorino R, Guerreiro ACL, Domingues P, Correia ACM, Alves A, Esteves AC (2016) Temperature modulates the secretome of the phytopathogenic fungus *Lasiodiplodia theobromae*. Front Plant Sci 7:1096

Félix C, Meneses R, Gonçalves MFM, Tilleman L, Duarte AS, Jorrín-Novo JV, Van de Peer Y, Deforce D, Van Nieuwerburgh F, Esteves AC (2019) A multi-omics analysis of the grapevine pathogen *Lasiodiplodia theobromae* reveals that temperature affects the expression of virulence- and pathogenicity-related genes. Sci Rep 9:13144

Galsurker O, Diskin S, Duanis-Assaf D, Doron-Faigenboim A, Maurer D, Feygenberg O, Alkan N (2020) Harvesting mango fruit with a short stem-end altered endophytic microbiome and reduced stem-end rot. Microorganisms 8:558

Galsurker O, Diskin S, Maurer D, Feygenberg O, Alkan N (2018) Fruit stem-end rot. Horticulturae 4:50

Gelaye Y (2024) A review of storage temperature and relative humidity effects on shelf life and quality of mango (Mangifera indica L.) fruit and implications for nutrition insecurity in Ethiopia. Open Agricul 9(1):20220296

Gunamalai L, Duanis-Assaf D, Sharir T, Maurer D, Feygenberg O, Sela N, Alkan N (2023) Comparative characterization of virulent and less-virulent *Lasiodiplodia theobromae* isolates. Mol Plant-Microbe Interact 36:502–515

Guyader S (2017) Resilience of *Colletotrichum gloeosporioides* strains infecting yams to dry period and elevated temperature stresses. In: Proceedings of the 53rd annual meeting of the Caribbean food crops society (CFCS), San Juan, Puerto Rico, pp 16–22

Haggag WM, Shabaan A, Nasr A, Abd El-Salam A (2014) Integrated pest management for sustainable mango production. Int J Pharm Sci Res 29:276–282

Hassan MK, Dann EK, Irving DE, Coates LM (2007) Concentrations of constitutive alk(en)ylresorcinols in peel of commercial mango varieties and resistance to postharvest anthracnose. Physiol Mol Plant Pathol 71:158–165

Hernández Delgado PM, Aranguren M, Reig C, Fernández Galván D, Mesejo C, Martínez Fuentes A, Galán Saúco V, Agustí M (2011) Phenological growth stages of mango (Mangifera indica L.) according to the BBCH scale. Sci Hortic 130:536–540

Hillary VE, Ceasar SA (2023) A review on the mechanism and applications of CRISPR/Cas9/Cas12/Cas13/Cas14 proteins utilized for genome engineering. Mol Biotechnol 65:311–325

Hofman PJ, Jobin-Decor M, Meiburg GF, Macnish AJ, Joyce DC (2001) Ripening and quality responses of avocado, custard apple, mango, and papaya fruit to 1-methylcyclopropene. Aust J Exp Agric 41:567–572

Honger JO, Offei SK, Oduro KA, Odamtten GT, Nyaku ST (2014) Identification and species status of the mango biotype of Colletotrichum gloeosporioides in Ghana. Eur J Plant Pathol 140:455–467

Hooper DU, Adair EC, Cardinale BJ, Byrnes JEK, Hungate BA, Matulich KL, Gonzalez A, Duffy JE, Gamfeldt L, O'Connor MI (2012) A global synthesis reveals biodiversity loss as a major driver of ecosystem change. Nature 486:105–108

Islam F, Shamsi S, Bashar MA (2018) Fungi associated with anthracnose of mango (Mangifera indica L.) fruits and their pathogenic potentiality. Dhaka Univ J Biol Sci 27(1):93–100

Jena M, Patel S, Patel Z (2022) Population dynamics of fruit fly, *B. dorsalis* (Hendle) on mango and correlation with weather parameters. Int J Environ Climate Change 12:1630–1638

Johnson GI, Hofman PJ (2009) Postharvest technology and quarantine treatments. In: Litz RE (ed) The mango: botany, production, and uses. CAB International, Wallingford, pp 529–605

Johnson G, Sangchote S, Cooke A (1990) Control of stem end rot (Dothiorella dominicana) and other postharvest diseases of mangoes (cv. Kensington pride) during short-and long-term storage. Trop Agric Trinidad Tobago 67:183–187

Kabir M, Habiba UE, Khan W, Shah A, Rahim S, Patricio R, Ali L, Shafiq M (2023) Climate change due to increasing concentration of carbon dioxide and its impacts on environment in 21st century; a mini review. J King Saud Univ Sci 35:102693

Kailaku SI, Arkeman Y, Purwanto YA, Udin F (2023) Appropriate harvest age of mango (Mangifera indica cv. Arumanis) for quality assurance in long distance transportation planning in Indonesia. J Agricult Food Res 14:100763

Karuppuchamy P, Venugopal S (2016) Integrated pest management. In: Omkar O (ed) Ecofriendly pest management for food security. Elsevier Science, pp 651–684

Kathimba FK, Kimani PM, Narla RD, Kiirika LM (2022) Effect of storage temperature on fruit firmness and weight loss of nine tomato lines. Af J Plant Sci 16(10):276–284

Khan MH, Khuhro NH, Awais M, Asif MU, Muhammad R (2021) Seasonal abundance of fruit fly, *Bactrocera* species (Diptera: Tephritidae) with respect to environmental factors in guava and mango orchards. Pak J Agric Res 34:266

Khan MAH, Hossain I, Chowdhury MSM, Ahmad MU (2015) Weather impact on nursery diseases of mango saplings in Bangladesh. International. J Appl Sci Biotechnol 3(1):42–49

Kiloes AM, Joyce DC, Abdul Aziz A (2024) Exploring the challenges and opportunities of mango export from Indonesia: insights from stakeholder interviews. Qual Rep 29:811–830

Kimaru KS, Muchemi KP, Mwangi JW (2020) Effects of anthracnose disease on avocado production in Kenya. Cogent Food Agricult 6(1):1799531

Kitinoja L (2013) Use of cold chains for reducing food losses in developing countries. La Pine, Oregon

Kobiler I, Shalom Y, Roth I, Akerman M, Vinokur Y, Fuchs Y, Prusky D (2001) Effect of 2,4-dichlorophenoxyacetic acid on the incidence of side and stem end rots in mango fruits. Postharvest Biol Technol 23:23–32

Koo TH, Hong SJ, Yun SC (2016) Changes in the aggressiveness and fecundity of hot pepper anthracnose pathogen (*Colletotricum acutatum*) under elevated CO_2 and temperature over 100 infection cycles. Plant Pathol J 32:260

Krishnapillai N, Wilson Wijeratnam RS (2017) Sap volatile components in relation to susceptibility of anthracnose and *Aspergillus* rot of mangoes (*Mangifera indica* L.). J Hortic Sci Biotechnol 92:206–213

Ku VVV, Wills RBH, Ben-Yehoshua S (1999) 1-Methylcyclopropene can differentially affect the postharvest life of strawberries exposed to ethylene. HortScience 34:119–120

Kumar R, Khurana A, Sharma AK (2014) Role of plant hormones and their interplay in development and ripening of fleshy fruits. J Exp Bot 65:4561–4575

Kweku DW, Bismark O, Maxwell A, Desmond KA, Danso KB, Oti-Mensah EA, Quachie AT, Adormaa BB (2018) Greenhouse effect: greenhouse gases and their impact on global warming. J Sci Res Rep 17:1–9

Lahlali R, Taoussi M, Laasli SE, Gachara G, Ezzouggari R, Belabess Z, Aberkani K, Assouguem A, Meddich A, El Jarroudi M, Barka EA (2024) Effects of climate change on plant pathogens and host-pathogen interactions. Crop Environ 3:159–170

Lalitha N (2020) Climate change and adaptation strategies in the Gir Kesar mango region of Gujarat. In: Socio-economic and eco-biological dimensions in resource use and conservation: strategies for sustainability, pp 379–398

Lawrence D, Coe M, Walker W, Verchot L, Vandecar K (2022) The unseen effects of deforestation: biophysical effects on climate. Front for Glob Change 5:756115

Lim KH, Gu M, Song JH, Cho YS, Kim WS, Kim B, Jung SK, Choi, HS (2014) Growth, fruit production, and disease occurrence of rain-sheltered Asian pear trees. Scientia Horticulturae 177:37–42

Makhmale S, Bhutada P, Yadav L, Yadav BK (2016) Impact of climate change on phenology of mango—the case study. Ecol Environ Conserv 22:132

Manicom BQ (1986) Factors affecting bacterial black spot of mangoes caused by *Xanthomonas campestris* pv. *mangiferaeindicae*. Ann Appl Biol 109:129–135

Mendelsohn RO, Dinar A (2009) Climate change and agriculture: an economic analysis of global impacts, adaptation and distributional effects. Edward Elgar Publishing

Misra AK (2012) Mango diseases in relation to climate change. In: Proceedings of the national conference on threatening diseases of horticultural, medicinal, aromatic, and field crops in relation to changing climatic situation & zonal meeting, New Delhi, India, 3–5 November, pp 3–5

Mustafa G (2017) Adaptation strategies to climate change and their impact on agricultural production in Punjab, Pakistan

Mwando NL, Ndlela S, Meyhöfer R, Subramanian S, Mohamed SA (2021) Hot water treatment for post-harvest disinfestation of Bactrocera dorsalis (Diptera: Tephritidae) and its effect on cv. Tommy Atkins Mango. Insects 12:1070

Nadeem A, Ahmed ZF, Hussain SB, Omar AEDK, Amin M, Javed S, Ali A, Ullah S, Razzaq K, Rajwana IA, Nayab S, Ziogas V, Alam-Eldein SM, Mira AM (2022) On-tree fruit bagging and cold storage maintain the postharvest quality of mango fruit. Horticulturae 8:814

Nazir W, Naqvi SAH, Ahmed N, Rehman A, Alrefaei AF, Zulfiqar MA Din Umar U (2024) Optimizing micronutrient supplementation in Mango orchards for the suppression of mango anthracnose (Colletotrichum gloeosporioides). Appl Fruit Sci 66(4):1417–1430

Nda M, Adnan MS, Ahmad KA, Usman N, Razi MAM, Daud Z (2018) A review on the causes, effects and mitigation of climate changes on the environmental aspects. Int J Integr Eng 10:4

Negi P, John SK, Rao PU (2002) Antimicrobial activity of mango sap. Eur Food Res Technol 214:327–330

Pachauri RK, Allen MR, Barros VR, Broome J, Cramer W, Christ R, Church JA, Clarke L, Dahe Q, Dasgupta P (2014) Climate change 2014: synthesis report. Contribution of working groups I, II and III to the fifth assessment report of the Intergovernmental Panel on Climate Change. IPCC

Paliwal R, Singh G, Mir RR, Gueye B (2021) Genomic-assisted breeding for abiotic stress tolerance in horticultural crops. In: Rai AC, Rai A, Rai KK, Rai VP, Kumar A (eds) Stress tolerance in horticultural crops. Elsevier, Amsterdam, pp 91–118

Patel PR, Prajapati VP, Sharma H, Pandya HV, Saxena SP, Patel SM, Mayani J, Tandel YN (2023) Management of post-harvest diseases and fruit fly of mango using hot water treatment. Pharma Innov J 12:1271–1275

Pineda-Vaca D, Flores-Méndez G, Fernández-Pavía SP, Santillán-Mendoza R, Montero-Castro JC, Garay-Serrano E, Ortega-Arreola R, Rodríguez-Alvarado G (2025) Genetic diversity of Colletotrichum siamense causing mango anthracnose in the state of Jalisco, Mexico, based on ISSR markers. Can J Plant Pathol 1–17

Porat R, Weiss B, Cohen L, Daus A, Goren R, Droby S (1999) Effects of ethylene and 1-methylcyclopropene on the postharvest qualities of 'Shamouti' oranges. Postharvest Biol Technol 15:155–163

Pruvost O, Luisetti J (1991) Effect of time of inoculation with *Xanthomonas campestris* pv. *mangiferaeindicae* on mango fruit susceptibility and epiphytic survival of *X. campestris* pv. *mangiferaeindicae* on mango fruits in relation to disease development. J Phytopathol 133:139–151

Rajan S, Tiwari D, Singh VK, Saxena P, Singh S, Reddy YTN, Upreti KK, Burondkar MM, Bhagwan A, Kennedy R (2011) Application of extended BBCH scale for phenological studies in mango (*Mangifera indica* L.). J Appl Hortic 13:108–114

Rajatiya J, Varu D, Gohil P, Solanki M, Halepotara F, Gohil M, Mishra P, Solanki R (2018) Climate change: impact, mitigation, and adaptation in fruit crops. Int J Pure Appl Biosci 6:1161–1169

Rajwana EA, Mahmood K, Rajwana IA, Malik AU, Bally ISE, Kazmi MR, Khan MI (2010) Trends and challenges in mango nursery production in Pakistan. In: Lu P (ed) IX International mango symposium 992, Hainan Island, China, pp 63–68

Ramírez F, Kallarackal J (2018) Climate change, tree pollination and conservation in the tropics: a research agenda beyond IPBES. Wiley Interdiscip Rev Clim Change 9:502

Rashid MHA, Grout BWW, Continella A, Mahmud TMM (2015) Low-dose gamma irradiation following hot water immersion of papaya (*Carica papaya* Linn.) fruits provides additional control of postharvest fungal infection to extend shelf life. Radiat Phys Chem 110:77–81

Rattanakreetakul C, Keawmanee P, Bincader S, Mongkolporn O, Phuntumart V, Chiba S, Pongpisutta R (2023) Two newly identified Colletotrichum species associated with mango anthracnose in central Thailand. Plants 12(5):1130

Raza A, Razzaq A, Mehmood SS, Zou X, Zhang X, Lv Y, Xu J (2019) Impact of climate change on crops adaptation and strategies to tackle its outcome: a review. Plants 8:34

Riera N, Ramirez-Villacis D, Barriga-Medina N, Alvarez-Santana J, Herrera K, Ruales C, Leon-Reyes A (2019) First report of banana anthracnose caused by Colletotrichum gloeosporioides in Ecuador. Plant Disease 103(4):763–763

Robalino L, Herrera LD (2010) Trade and deforestation: a literature review. World Trade Organization

Samanta S, Senapati SK, Ganaie MH, Maji A (2024) Impact of climate change and global warming on crop pollinators and their mitigation strategies: a review. Agric Rev 1:1–9

Sangchote S (1989) Botryodiplodia stem end rot of mango and its control. In: Chacko E, Lim TK, Kulkarni V (eds) III International mango symposium 291, Darwin, Australia, pp 296–303

dos Santos MR, Martinez MA (2013) Soil water distribution and extraction by 'Tommy Atkins' mango (*Mangifera indica* L.) trees under different irrigation regimes. IDESIA (Chile) 31:7–16

Sathaye J, Shukla PR, Ravindranath NH (2006) Climate change, sustainable development and India: global and national concerns. Curr Sci 90:314–325

Schaefer DO, Godwin D, Harnisch J (2006) Estimating future emissions and potential reductions of HFCs, PFCs, and SF6. Energy J 27:63–88

Schulze K, Spreer W, Keil A, Ongprasert S, Müller J (2013) Mango (Mangifera indica L. cv. Nam Dokmai) production in Northern Thailand—costs and returns under extreme weather conditions and different irrigation treatments. Agric Water Manag 126:46–55

Sharma A, Sharma IM, Sharma M, Sharma K, Sharma A (2021) Effectiveness of fungal, bacterial and yeast antagonists for management of mango anthracnose (Colletotrichum gloeosporioides). Egypt J Biol Pest 31:1–11

Simmons S, Hofman P, Hetherington S (1995) The effects of water stress on mango fruit quality. In: Proceedings of the mango 2000 marketing seminar and production workshop, Brisbane, Australia, 30 July–3 August, pp 191–197

Simmons S, Hofman P, Whiley A, Hetherington S (1998) Effects of preharvest calcium sprays and fertilizers, leaf: fruit ratios, and water stress on mango fruit quality. In: Coates LM, Hofman PJ, Johnson GI (eds) ACIAR proceedings. Australian Centre for International Agricultural Research, Bruce, Australia, pp 19–26

Singh BK, Delgado-Baquerizo M, Egidi E, Guirado E, Leach JE, Liu H, Trivedi P (2023) Climate change impacts on plant pathogens, food security and paths forward. Nat Rev Microbiol 21:640–656

Singh DP, Prakash S, Malik V, Singh KK, Singh N, Gupta S, Sharma P (2020) A study on socioeconomic status of mango growers in Bijnor district of Western Uttar Pradesh. Int J Environ Climate Change 10(12):13–19
Skendžić S, Zovko M, Živković IP, Lešić V, Lemić D (2021) The impact of climate change on agricultural insect pests. Insects 12:440
Slippers B, Wingfield MJ (2007) Botryosphaeriaceae as endophytes and latent pathogens of woody plants: diversity, ecology and impact. Fungal Biol Rev 21:90–106
Song M, Wang H, Fan Z, Huang H, Ma H (2023) Advances in sequencing and key character analysis of mango (*Mangifera indica* L.). Horticult Res 10:uhac259
Sossah FL, Aidoo OF, Dofuor AK, Osabutey AF, Obeng J, Abormeti FK, Duker RQ, Antwi-Agyakwa AK, Osei-Owusu J, Loh SK, Honger JO, Borgemeister C (2024) A critical review on bacterial black spot of mango caused by *Xanthomonas citri* pv. *mangiferaeindicae*: current status and direction for future research. For Pathol 54:12860
Spalding DH, Reeder WF (1986) Decay and acceptability of mangos treated with combinations of hot water, imazalil, and gamma radiation. Plant Dis 70:1149–1150
Spreer W, Ongprasert S, Hegele M, Wünsche JN, Müller J (2009) Yield and fruit development in mango (*Mangifera indica* L. cv. Chok Anan) under different irrigation regimes. Agric Water Manag 96:574–584
Subedi S (2019) Climate change effects of Nepalese fruit production. Adv Plants Agric Res 9:141–145
Subramanian S, Boopathi T, Nebapure SM, Yele Y, Shankarganesh K (2021) Mealybugs. In: Omkar (eds) Polyphagous pests of crops. Springer, Singapore. https://doi.org/10.1007/978-981-15-8075-8_5
Trenberth KE (2018) Climate change caused by human activities is happening and it already has major consequences. J Energy Nat Resour Law 36:463–481
Usman M (2024) Impact of global warming on crop production: example from farmers' insight and adaptation measure. J Soc Sci Econ 3:73–85
Varakumar S, Kumar YS, Reddy OVS (2011) Carotenoid composition of mango (Mangifera indica L.) wine and its antioxidant activity. J Food Biochem 35(5):1538–1547
Vasanthaiah HKN, Ravishankar KV, Mukunda GK (2007) Mango. In: Kole C (ed) Fruits and nuts. Genome mapping and molecular breeding in plants, vol 4. Springer, Berlin, pp 303–323
Velásquez AC, Castroverde CDM, He SY (2018) Plant–pathogen warfare under changing climate conditions. Curr Biol 28:619-R634
Venkata Rami Reddy P, Gundappa B, Chakravarthy AK (2018) Pests of mango. In: Omkar (ed) Pests and their management. Springer, Singapore, pp 415–440
Vilela A (2021) An overview of CRISPR-based technologies in wine yeasts to improve wine flavor and safety. Fermentation 7:5
Vuppalapati C (2023) Specialty crop mangoes. In: Specialty crops for climate change adaptation. steam-h: science, technology, engineering, agriculture, mathematics & health. Springer, Cham, pp 419–514
Wada N, Ueta R, Osakabe Y, Osakabe K (2020) Precision genome editing in plants: state-of-the-art in CRISPR/Cas9-based genome engineering. BMC Plant Biol 20:1–12
Warschefsky EJ, von Wettberg EJB (2019) Population genomic analysis of mango (*Mangifera indica*) suggests a complex history of domestication. New Phytol 222:2023–2037
Watkins CB, Nock JF (2008) Effects of delayed controlled atmosphere storage of apples after rapid 1-MCP treatment. HortScience 43:1087
Wei J, Liu G, Liu D, Chen Y (2017) Influence of irrigation during the growth stage on yield and quality in mango (*Mangifera indica* L). PLoS ONE 12:e0174498
Woolf AB, Requejo-Tapia C, Cox KA, Jackman RC, Gunson A, Arpaia ML, White A (2005) 1-MCP reduces physiological storage disorders of 'Hass' avocados. Postharvest Biol Technol 35:43–60
Yadav S, Korat J, Yadav S, Mondal K, Kumar A, Kumar S (2023) Impacts of climate change on fruit crops: a comprehensive review of physiological, phenological, and pest-related responses. Int J Environ Climate Change 13:363–371

Yadav D, Singh SP (2017) Mango: history, origin and distribution. J Pharmacogn Phytochem 6:1257–1262

Yeo YS, Kone Y, Dembele DD, Amari EL-NODG, Rey J-Y, Del Ponte EM, Fernandez D, Kone D (2024) Prevalence of mango stem-end rot disease in Côte d'Ivoire and identification of associated fungal pathogens. Trop Plant Pathol 49:367–383

Yoro KO, Daramola MO (2020) CO_2 emission sources, greenhouse gases, and the global warming effect. In: Rahimpour MR, Farsi M, Makarem MA (eds) Advances in carbon capture. Woodhead Publishing, pp 3–28

Yuan X, Li S, Chen J, Yu H, Yang T, Wang C, Huang S, Chen H, Ao X (2024) Impacts of global climate change on agricultural production: a comprehensive review. Agronomy 14:1360

Yáñez-López R, Torres-Pacheco I, Guevara-González RG, Hernández-Zul MI, Quijano-Carranza JA, Rico-García E (2012) The effect of climate change on plant diseases. Af J Biotechnol 11(10):2417–2428

Zhang Y, Niu H, Yu Q (2021) Impacts of climate change and increasing carbon dioxide levels on yield changes of major crops in suitable planting areas in China by the 2050s. Ecol Indic 125:107588

Chapter 31
Dynamics of Disease Infestation Under Changing Climate—Impact and Future Perspectives

Hafiz Muhammad Imran, Muhammad Tariq Malik, Ghulam Mustafa, Fahad Hakim, and Aimal Waheed

Abstract Climate change exacerbates the prevalence and severity of mango diseases due to rising temperatures, erratic rainfall, and humidity fluctuations, creating conducive environments for pathogenic outbreaks. Traditional disease management strategies, such as chemical fungicides, face limitations due to environmental concerns and resistance development. Nanotechnology offers a promising solution for managing mango diseases effectively and sustainably under these changing conditions. Nanoparticles, such as silver, copper, and zinc-based formulations, exhibit antimicrobial properties that target a wide range of pathogens while minimizing the ecological footprint. Their small size allows for better penetration and controlled release, enhancing their efficacy and reducing the quantity required. Recent studies have demonstrated the potential of nanoparticles in suppressing fungal pathogens like *Colletotrichum gloeosporioides* and *Lasiodiplodia theobromae*, which cause devastating diseases like anthracnose and stem-end rot. Additionally, nanoparticle applications can be integrated with existing disease management practices, such as resistant cultivars and biocontrol agents, to provide a holistic approach. This chapter explores the role of nanoparticles in mitigating mango diseases in the context of climate change, highlighting their mechanisms, efficacy, and environmental safety. By embracing nanotechnology, mango cultivation can achieve enhanced disease

H. M. Imran · M. T. Malik
Mango Research Institute, Multan, Pakistan

G. Mustafa (✉)
Department of Horticulture, MNS-University of Agriculture, Multan, Pakistan
e-mail: gmustafabzu@gmail.com

Department of Agriculture Extension, Government of the Punjab, Punjab, Pakistan

F. Hakim
Institute of Horticulture, Lithuanian Research Centre for Agriculture and Forestry, Babtai, Kaunas, Lithuania

A. Waheed
Biochemistry Research Group, Faculty of Natural Sciences, Vytautas Magnus University, Kaunas, Lithuania

S. A. H. Naqvi et al. (eds.), *Climate Change and Mango Production*,
https://doi.org/10.1007/978-3-031-87035-4_31

resistance and sustainable productivity amidst the challenges posed by global climate shifts.

Keywords Mango diseases · Nanoparticles · Climate change · Sustainable management

31.1 Introduction

The cultivation of crops is essential to the survival of animals and human life on Earth. The tremendous rise in global population has made crop productivity, crop protection, and food security increasingly crucial (Hlatshwayo et al. 2023). People's daily food needs are satisfied by agricultural practices, which are completely reliant on the climate. A shift in population, a change in consumer preferences, and the relative cost of production all contribute to a dynamic adjustment in crop requirements. Climate change, influenced by both natural and human influences, is altering ecosystems and disrupting the delicate balance between diseases and plants. As temperatures rise, precipitation patterns change, and extreme weather events become more common, plant diseases spread and emerge more frequently. These changing Environmental conditions can have an impact on a pathogen's ability to survive, reproduce, sporulate, and lengths of distance.

A big concern is about how climate change may affect agriculture, pests, and illnesses, as well as about its potential effects, hidden details, and mitigating actions (Singh et al. 2023). Mango (*Mangifera indica* L.) is one of the most widely cultivated and consumed fruits globally, particularly in tropical and subtropical regions. However, mango production is threatened by various diseases, which are exacerbated by climate change. Mango diseases cause significant losses in fruit production during pre-harvest and post-harvest handling. The disease affects plants part, damage flowers, fruit resulting in poor fruit set and ultimately lower yield, poor quality fruit in size, appearance and taste. The diseases in mango are due to non-living factors (abiotic) and pathogen (biotic). Abiotic disease's spread from plant to plant is not possible and are not contagious. Abiotic diseases in plants arise from non-living factors, such as prolonged exposure to air pollutants like nitrogen dioxide from vehicle emissions and sulfur dioxide from industrial activities, as well as nutritional imbalances, herbicide toxicity, or extreme light, moisture, or temperature conditions. In contrast, biotic diseases are caused by pathogens and are often termed infectious diseases due to their ability to spread within and between plants. Pathogens can induce diverse diseases, including powdery mildew, dieback, anthracnose, bacterial canker, mango malformation, and vascular wilt. While plants generally exhibit resistance to most pathogens, susceptibility to specific pathogens varies and can affect mango trees throughout their lifecycle. Various plant parts, including the trunk, branches, twigs, leaves, petioles, flowers, and fruits, are vulnerable to attack by bacteria, fungi, and nematodes. These pathogens result in multiple issues such as rotting, dieback, anthracnose, scabs, necrosis, blotches, spots, yellowing, wilting, and mildew. Certain diseases are particularly significant economically, causing substantial losses in mango fruit production, quality degradation, and reduced export potential (Normand et al. 2015).

31.2 Temperature and Disease Development

Plant productivity and health are the focus of plant pathology. Variations in crop types, changes in climate, and changes in plant growth regions present health and disease challenges. Throughout the year, all plants and their pathogens are present in the environment. Temperature, humidity, and rainfall are examples of climate variables that have a big impact on plant development and disease occurrence. Plants require conducive abiotic conditions, such as soil moisture, temperature, and cultural practices, for optimal growth (Singh et al. 2023). Abiotic stress in plants is caused by excess or deficiency of certain abiotic elements. These stresses do not spread from diseased plants to healthy ones, yet they have an impact on plants at every stage of development. The impact of these environmental variables is determined by their kind, severity, and degree of deviation from the norm (He et al. 2018). The development and spread of plant diseases are influenced by various factors. An epidemic takes place when a disease rapidly spreads and impacts a large population within a brief timeframe. The field of epidemiology focuses on understanding epidemics and the factors that drive their occurrence and progression. The study of epidemiology considers both biotic and abiotic variables. The kind of plant and stage of growth influence how differently it can tolerate these variations. Buds, blooms, and young fruit are examples of young, growing tissues that are highly sensitive.

Plant growth typically occurs in the range of 1 to 40 °C, with 15 to 30 °C being the ideal range for maximum growth. Temperature plays an important role in illness progression, impacting pathogen proliferation, survival, and infection. Temperature also influences the ratè, length, latency period, sporulation, inoculum distribution, and host resistance dynamics of the pathogen infection process (Zhang et al. 2023) Many pathogens have optimal temperature ranges for growth and reproduction, often between 20 and 30 °C. High temperatures usually promote the establishment and spread of plant diseases, especially bacteria and fungi, while also weakening host defence systems (Chen et al. 2024) Warmer temperatures promote spore formation in fungi like rusts and powdery mildews. For example, the fungal disease Fusarium graminearum, which causes Fusarium head blight in wheat, is distributed through wind-borne spores, with ideal temperatures for dissemination between 15 and 25 °C (Ghimire et al. 2020). Additionally, rising temperatures can promote the survival and overwintering of diseases and their vectors, including nematodes and insects, leading to prolonged activity periods and expanded geographic distributions (El-Sayed and Kamel 2020). Human activities such as topping pruning in extreme weather, irrigation and control measures adopted also play a significant role in disease development and management. Temperature increases facilitate bacterial growth, leading to increased populations of pathogens like *Xylella fastidiosa* (Farigoule et al. 2022).

Fungal pathogen *Magnaporthe oryzae*, responsible for rice blast disease, exhibits optimal growth at 25 °C (Paul et al. 2022). Similarly, the bacterial pathogen *Pseudomonas syringae*, causing bacterial leaf spot in tomatoes, thrives at temperatures between 20 and 28 °C. Temperature also influences dispersal of pathogen through various mechanisms. Warmer temperatures increase convection, facilitating wind-borne dispersal of pathogens like fungal spores. Temperature-driven changes in precipitation patterns and water cycles enhance pathogen dispersal through water. Temperature affects insect vector behavior, increasing the spread of pathogens like aphid-borne viruses (Lahlali et al. 2024).

31.3 Changing Precipitation Patterns and Disease Epidemiology

Rainfall plays a crucial role in shaping plant diseases by affecting water and nutrient accessibility, soil moisture levels, pH balance, pesticide leaching, runoff, and the development and spread of inoculum (Chen et al. 2023). Availability of moisture in the form of high humidity causes the development of disease caused by fungi, bacteria and nematodes. Climate change is affecting global precipitation patterns, with far-reaching effects for plant disease epidemiology. Higher levels of rainfall frequently diminish the effectiveness of chemical and biological control strategies while enhancing the establishment and spread of plant diseases, particularly those caused by fungi and oomycetes (Elad and Pertot 2014). Changes in rainfall intensity, frequency, and duration are disrupting the delicate balance between plants, pathogens, and the environment. There are complex interactions between changing precipitation patterns and epidemiology through various mechanisms. Plant diseases are impacted differently by the way that precipitation patterns have changed as a result of climate change (Garrett et al. 2021). Many pathogens, such as fungal and bacterial pathogens, require moisture to infect and spread. Higher rainfall may make soil-borne and foliar diseases like blight, leaf spot, damping-off, and root rot more common and severe (Lamichhane et al. 2024). Precipitation facilitates pathogen dispersal through water, increasing disease spread. Drought-stressed plants are more vulnerable to infection, while excessive rainfall can worsen disease severity. Less rainfall may minimise the prevalence and spread of some diseases, such as powdery mildew (PM) and downy mildew, but it may also render host plants more sensitive to other illnesses including Fusarium spp., Verticillium, and drought stress. *Phytophthora infestans*, fungal pathogen causing potato late blight, requires moisture to infect and spread. Climate-driven changes in precipitation patterns are altering disease epidemiology. Warmer and wetter conditions facilitate disease development and spread. Changing precipitation patterns alter disease dynamics, requiring adaptive management strategies. Emergence of new diseases: Changes in precipitation patterns create opportunities for new pathogens to emerge and spread. For instance, the bacterial pathogen

Xylella fastidiosa, causing Pierce's disease in grapes, is spreading due to changing precipitation patterns in California (Elad and Pertot 2014).

31.4 Shifts in Disease Pressure Due to Climate-Driven Changes in Humidity

Humidity is a critical environmental factor that significantly affects the dynamics of plant diseases. It influences essential physiological processes in host plants, including evapotranspiration, transpiration, and stomatal conductance, which in turn affect plant health and susceptibility. High humidity creates favorable conditions for the germination, infection, and survival of many plant pathogens, especially fungi and bacteria. Additionally, it contributes to the formation and retention of dew, providing free water essential for the development and proliferation of numerous diseases (Nath et al. 2022). Elevated levels of humidity can compromise the effectiveness of a plant's natural resistance mechanisms and reduce the efficacy of disease control measures. This creates an environment conducive to rapid pathogen growth and disease spread. Prolonged periods of high humidity, especially in enclosed or poorly ventilated areas like greenhouses, exacerbate disease outbreaks by creating microclimates that are ideal for pathogenic activity. The impact is particularly pronounced in diseases such as powdery mildew, downy mildew, leaf spots, and bacterial blights, where free water is a crucial factor for pathogen life cycles (Li et al. 2021). Furthermore, increasing global humidity trends due to climate change are expected to intensify the severity and frequency of plant disease epidemics, posing significant challenges for sustainable agricultural practices.

31.5 Shifts in Disease Pressure Due to Climate-Driven Changes in Plant Phenology

Climate change is profoundly disrupting the intricate interactions between plants, pathogens, and the environment, leading to significant shifts in disease dynamics and pressures. A key factor driving these changes is the alteration of plant phenology, which refers to the timing of recurring plant life cycle events such as flowering, fruiting, and dormancy. Rising global temperatures are causing earlier springs and extended autumns, thereby modifying growing seasons and altering the developmental stages of plants. These shifts impact the synchronization between plants and their pathogens, enabling some pathogens to exploit host plants more effectively.

Furthermore, erratic rainfall patterns and prolonged droughts are affecting soil moisture availability, nutrient uptake, and overall plant resilience, exacerbating disease susceptibility. Changes in light intensity and photoperiodism due to shifting climate conditions are also influencing plant development, potentially altering the expression of resistance genes and defense mechanisms. These disruptions not only affect existing disease cycles but also facilitate the emergence of new pathogens or the migration of existing ones to previously unaffected regions. The consequences extend beyond individual plant-pathogen interactions, impacting entire ecosystems and agricultural landscapes. For instance, a longer growing season may favor pathogen reproduction, leading to higher disease intensity, while extreme weather events such as floods and heatwaves can create conditions ideal for certain diseases. Such challenges underscore the urgency of integrating climate-resilient crop management strategies, advanced pathogen monitoring systems, and sustainable agricultural practices to mitigate the growing risks posed by climate change (Visser and Both 2005).

31.6 Disease Pressure and Plant Phenology

Plant phenology influences disease pressure through:

1. Host plant susceptibility: Changes in plant growth stages and physiology affect susceptibility to infection.
2. Pathogen life cycle: Climate-driven changes in plant phenology impact pathogen reproduction, dispersal, and survival.
3. Disease development and spread: Altered plant phenology influences disease development and spread rates.

For example, earlier springs can lead to increased susceptibility of wheat to powdery mildew.

Climate-Driven Changes in Plant Phenology and Disease Pressure

Climate-driven changes in plant phenology are altering disease pressure:

1. Extended growing seasons: Warmer temperatures and altered precipitation patterns lead to longer growing seasons, increasing disease pressure.
2. Shifts in disease dynamics: Changes in plant phenology alter disease dynamics, requiring adaptive management strategies.
3. Emergence of new diseases: Climate-driven changes in plant phenology create opportunities for new pathogens to emerge and spread.

31.7 Mango Diseases

Mango infections significantly reduce fruit production during pre-harvest and post-harvest handling. The disease damages plant parts, causing damage to flowers and fruit, resulting in poor fruit set and, as a result, poorer yield and poor-quality fruit in size, look, and flavor. Mango illnesses are caused by non-living causes (abiotic) as well as pathogens. Abiotic diseases cannot be transmitted between plants and are not infectious. Abiotic diseases are caused by non-living environmental factors that lead to plant damage and reduced vigor. These include prolonged exposure to air pollutants like nitrogen dioxide from vehicle emissions and sulfur dioxide released by industrial processes. Other contributors include imbalances in soil nutrients resulting in deficiencies or toxicities, herbicide overuse leading to chemical injury, and extreme environmental conditions such as excessive light, moisture, or temperature fluctuations. These factors can disrupt plant metabolic processes, weaken structural integrity, and make plants more vulnerable to secondary infections or stressors.

In contrast, biotic diseases are caused by living organisms, such as bacteria, fungi, viruses, nematodes, and other pathogens, and are often termed infectious diseases due to their ability to spread within a plant and between neighboring plants. These pathogens exploit host tissues, leading to symptoms like wilting, necrosis, leaf spots, and rotting, which can compromise plant health and productivity. Unlike abiotic diseases, which are not transmissible, biotic diseases can lead to epidemics, particularly under favorable environmental conditions.

Furthermore, the interaction between abiotic and biotic stress can exacerbate plant health challenges. For instance, air pollution may weaken plant defenses, making them more susceptible to infections, while high moisture levels caused by climate anomalies can create ideal conditions for pathogen proliferation. The management of both abiotic and biotic diseases requires an integrated approach, combining preventive measures such as improved cultural practices and soil management with targeted interventions like biocontrol agents or disease-resistant crop varieties to mitigate their impact effectively.

Pathogens can cause a wide range of diseases including powdery mildew, dieback, anthracnose, bacterial canker, mango deformation, and vascular wilt. Most plants are resistant to most diseases; nevertheless, the sensitivity of most plants to one or more pathogens differs. Mango plants are susceptible to a variety of illnesses during their life cycle. Pathogens such as bacteria, fungi, and nematodes affect all parts of the plant, including the trunk, branches, twigs, leaves, petioles, flowers, and fruit. Pathogens are responsible for a wide array of plant diseases in mango trees, including rotting, dieback, anthracnose, scab, necrosis, blotches, leaf spots, yellowing, wilting, mildew, and several others. These diseases can significantly impair both the yield and quality of mango fruits. Among these, some diseases, such as anthracnose and powdery mildew, are of particularly high economic importance as they lead to substantial reductions in mango production and post-harvest quality. This deterioration not only affects local markets but also results in a significant decline in export potential, thereby impacting the profitability of mango cultivation globally.

Recent studies highlight that climate change is intensifying the spread and severity of these diseases due to shifting weather patterns, including increased humidity and erratic rainfall, which create favorable conditions for many pathogens. Additionally, improper orchard management practices, the overuse of chemical inputs, and a lack of effective disease-resistant mango cultivars exacerbate the problem. In some regions, post-harvest losses due to pathogen infections, such as fruit rot, have been reported to reach up to 30–50% of total production.

Integrated disease management strategies, including improved orchard hygiene, the use of bio-control agents, resistant cultivars, and advanced diagnostic tools, are increasingly being emphasized to mitigate these losses. Moreover, innovative approaches such as microbiome engineering and precision agriculture are emerging as promising solutions to enhance mango tree resilience and ensure sustainable production in the face of growing disease challenges (González-Fernández and Hormaza 2020).

31.7.1 Causes of Mango Plant Diseases

Most mango diseases are caused by unfavorable farm conditions, soil salinity, sodicity and water stress from over irrigation or under irrigation. Use of poor-quality water, sandy and hard soil, lack of drainage, excessive ploughing and hoeing are predisposing factors which make plants sick and susceptible to diseases. Pathogen infection is caused by its entry into a plant through wounds caused by insect infestation, mechanical injury or due to environmental stress.

31.7.2 Spread of Plant Diseases

Most mango diseases are caused by fungi while remaining by bacteria and other pathogens. Fungal and bacterial diseases are mostly spread through air, water, wind and contact from humans and animals.

31.7.3 Sign and Symptoms of Mango Plant Diseases

Symptoms are the plant's response to the presence of pathogens, serving as indicators of underlying issues. Common symptoms include wilting, yellowing, or necrosis of leaves, as well as abnormalities in leaf and fruit morphology. Other visible signs may involve spots or lesions on the leaves and fruits, bark damage, fruit splitting, and gummosis (the exudation of sap). In addition to these, mango plants may exhibit stunted growth, distorted or deformed fruits, and premature leaf drop, which are often associated with fungal, bacterial, or viral infections. Further, the presence of pests such as scale insects or mealybugs can also exacerbate these symptoms, leading to reduced plant vigor and susceptibility to secondary infections. Proper diagnosis and management of these symptoms are essential for preventing widespread damage and ensuring the health of mango crops.

31.8 Common Mango Plant Diseases and Their Mode of Infection

Mango plants suffer from various diseases at all phenological stages of their growth and development. It is usually less important than other diseases in dry areas of production. Moist conditions in nurseries and green house can result in serious damage to young leaves. Conidia of this fungus are the most important inoculum, and their optimum production occurs 25 and 30 °C in free moisture and relative humidity 95–97%. Diseases like anthracnose, powdery mildew, mango malformation, powdery mildew, die back, stem end rot, body rot are of great economic importance as these cause heavy losses in mango production leading to tree decline.

31.9 Climate Change Impacts on Mango Diseases

Climate change alters the dynamics of mango diseases through:

1. Temperature increase: Warmer temperatures facilitate fungal growth and infection
2. Changing precipitation patterns: Altered rainfall and drought patterns enhance disease spread and severity
3. Increased CO_2 levels: Elevated CO_2 levels stimulate plant growth, increasing susceptibility to disease.

The impact of climate change on the dynamics of mango diseases is given in Table 31.1.

31.9.1 Anthracnose (Colletotrichum gloeosporioides, Alternaria sp.*)*

Anthracnose is a deadly mango disease that wreaks havoc on fruit grown in humid and high-rainfall conditions, inflicting significant losses. It is reported from every mango-producing country throughout the world. Losses from fruit infection have been estimated to range between 2 and 39%. Anthracnose symptoms occur on the leaf surface as oval or irregular dark to deep brown dots of varying sizes. Infected leaves show the appearance of shot holes. Younger leaves are more vulnerable to attack than older ones. During inflorescence emergence, its attack has an impact on fruit set and, eventually, yield. Blackisth brown patches emerge on panicles and flowers, which enlarge and coalesce, causing floral death—the disease's most damaging phase.

The rain during flowering is very helpful to cause infection and disease dissemination. It also affects fruit after harvesting and during storage. The pathogens remain on the plants through the year and attack on shoots, foliage and branches. Depending

Table 31.1 Impact of changing climate on disease infestation dynamics

Climate change factor	Impact on disease infestation	Future perspectives and adaptations
Rising temperatures	Increased temperature accelerates the life cycle of pathogens and pests	Shift in disease prevalence and emergence of new pathogens, requiring quicker response systems
Altered rainfall patterns	Irregular rainfall patterns can cause periods of both drought and excessive moisture	Increased frequency of fungal and bacterial diseases due to wet conditions and water stress in dry conditions
Higher CO_2 levels	Higher CO_2 enhances plant growth but also increases susceptibility to certain pathogens	Potential for higher disease incidences if pathogens exploit increased plant vigor
Increased humidity	Enhanced humidity supports fungal growth and spread of airborne diseases like rust and mildew	Greater challenges in managing disease outbreaks in humid environments, requiring advanced fungicide strategies
Storms and extreme weather	Physical damage to plants from storms creates open wounds, facilitating pathogen entry	Strengthened disease control strategies post-storm and development of storm-resistant plant varieties
Shifts in pest distribution	Warmer temperatures expand the range of many pests and pathogens to new regions	Expansion of monitoring and control efforts in previously unaffected areas

on prevailing environmental conditions and insect infestation, its severity varies on flowers. Black spots appear on panicles and fruits. In case of sever attack, entire inflorescence is destroyed resulting no fruit setting. Young, infected fruits develop black spots and drop off. The fruits become infected before maturity on tree carrying fungus and after harvest cause considerable loss during transit, marketing and storage. Rain during blossoming helps to spread illness and sickness. It also has an impact on fruits after they have been harvested and stored. The infections linger on the plants all year and affect the shoots, leaves, and branches. The severity of insect infestation on flowers varies depending on the prevailing environmental conditions. Black stains emerge on the panicles and fruits. In the event of a severe attack, the entire inflorescence is damaged, resulting in no fruit setting. Young, infected fruits acquire black patches and eventually drop off. The fruits become infected with fungus before maturity on the tree, and after harvest, they cause significant loss during transit, marketing, and storage. The pathogen lives on fallen leaves, twigs, and peduncles attached to trees, and under favorable conditions, spore generation occurs, serving as a source of infection for subsequent blooms. Humidity, rain, heavy dews, and hazy circumstances during the blossoming season are all causes of significant blossom blight. For pathogen infection and development on mango fruit, a relative humidity greater than 95% for 12 h is required. Continued rainfall during fruiting development disrupts

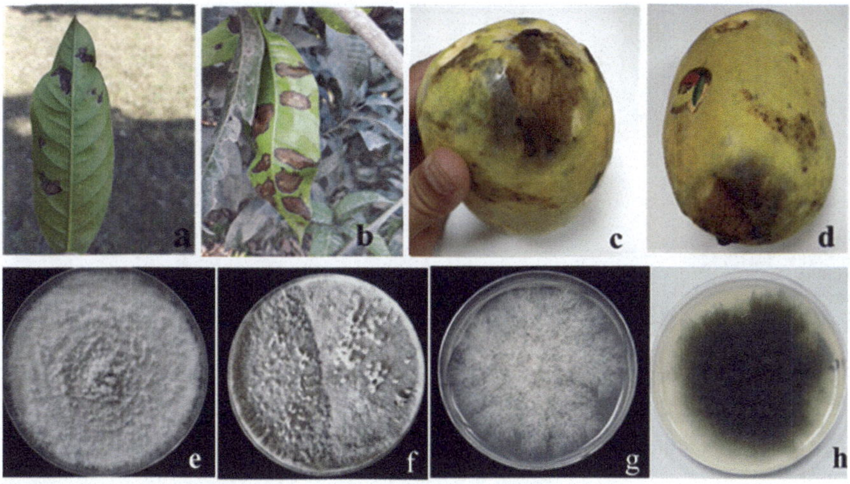

Fig. 31.1 Anthracnose of mango on leaves and fruits

control strategies and is regarded the most critical factor for disease development on mango fruits (Dofuor et al. 2023) (Fig. 31.1).

31.9.2 Powdery Mildew (Oidium mangiferae) and Blossom Blight

Powdery Mildew is a devastating disease that affects nearly all mango types. It accounts for around 20% of crop loss in the mango region, but can reach 70–80% loss on individual plants when environmental circumstances are favourable for the disease. The distinctive symptoms develop on leaves, pancile stalks, flowers, and early fruit in the form of a surface powdery fungal growth. The infected flowers and fruits drop in the earliest stage, preventing fruit setting and resulting in a significant reduction in crop production. Rain and dryness during flowering resulted in disease proliferation (Nasir et al. 2014) (Figs. 31.2 and 31.3).

31.9.3 Apical necrosis

Apical necrosis is a bacterial illness that was initially identified in Spain in 1991 and is also seen in Israel, Italy, Portugal, and Egypt. Apical buds, leaves, and panicles exhibit necrotic, water-soaked lesions. Milky bacterial exudates frequently form on infected apical buds. The bacterium (*Pseudomonas syringae* pv. *syringae* van Hall)

Fig. 31.2 Powdery mildew of mango on leaves

Fig. 31.3 Blossom Blight of mango

is an epiphyte, not an aggressive pathogen. Cold and damp temperatures promote the development of this illness (Cazorla et al. 1998) (Fig. 31.4).

31.9.4 Bacterial Black Spot

Bacterial black spot is a prevalent leaf, stem, and fruit disease in many mango-producing regions. Water-soaked patches develop on black, elevated, angular-shaped leaves. Lesions coalesce to produce vast necrotic areas, and germs may leak from lesions under damp conditions. Old lesions break and turn white or grey when they dry. The bacterium (*Xanthomonas campestris* pv. *mangiferae indica*) lives in diseased plant sections of the tree and is spread by wind-driven rain. Infected propagation material causes long-distance spread. Disease development is favoured by high relative humidity (over 90%) and moderate temperatures ranging from 25 to 30 °C. Diseases affecting leaves and fruit are intimately connected (Sossah et al. 2024).

Fig. 31.4 Apical necrosis of mango

31.9.5 Mango Malformation (Fusarium mangiferae)

Mango malformation is a fungal infection caused by *Fusarium mangiferae*. The disease targets the leaves and inflorescence, causing vegetative and floral malformations, respectively. Vegetative malformation attack is commonly found in immature nursery plants, although it can also afflict mature plants. In the event of a panicle infection, panicles stay tight with shorter internodes. The panicles remain unproductive, producing a dense clump of greenish flowers with irregular development. Diseased panicles induce crop loss owing to excessive water and nutrient absorption, and they also harm blooms and fruits on neighbouring branches (Freeman et al. 2014) (Fig. 31.5).

Fig. 31.5 Mango malformation disease

31.10 Mitigation Strategies

To address climate-driven mango disease

1. **Climate-resilient cultivars**: Develop and deploy cultivars with enhanced disease resistance and climate tolerance.
2. **Integrated pest management**: Implement IPM practices combining physical, cultural, biological, and chemical controls.
3. **Disease forecasting**: Utilize climate-driven disease forecasting models to predict and prepare for disease outbreaks.
4. **Agroecological practices**: Promote agroecological practices like conservation agriculture and agroforestry to enhance ecosystem resilience.

Future strategies to manage mango diseases is given in Table 31.2.

Table 31.2 Future strategies for managing disease infestation under changing climate

Strategy	Description	Future adaptations and effectiveness
Climate-resilient crops	Development of disease-resistant and climate-adapted crop varieties	Ensures crops can withstand new and emerging diseases under shifting climate conditions
Advanced disease forecasting	Use of climate modeling, satellite data, and early warning systems for predicting disease outbreaks	Enhances proactive management and early intervention, reducing crop losses
Biological control and biocontrol agents	Use of natural predators, parasites, and beneficial microbes to control diseases	Reduces reliance on chemicals, adapting to increased pest pressure due to climate change
Precision agriculture techniques	Use of technologies like drones, sensors, and AI to monitor plant health and disease progression	Real-time data allows for precise application of treatments, reducing waste and improving efficiency
Integrated pest and disease management (IPDM)	Combination of cultural, biological, and chemical controls tailored to climate variations	Provides adaptive management strategies for dealing with disease pressures across diverse climates
Soil health management	Enhancing soil microbiota and health to improve plant resistance to pathogens	Healthier soils support plants' immune systems, reducing the likelihood of disease outbreaks
Enhanced quarantine and surveillance systems	Strengthened national and international disease monitoring and quarantine measures	Helps prevent the spread of emerging diseases, especially in global trade and during climate-induced migration of pathogens
Resilient infrastructure and storage	Upgrading infrastructure to withstand extreme weather events and ensuring proper postharvest disease management	Reduces disease spread in the postharvest phase, ensuring quality and minimizing crop loss

References

Cazorla FM, Torés JA, Olalla L, Pérez-García A, Farré JM, de Vicente A (1998) Bacterial apical necrosis of mango in Southern Spain: a disease caused by *Pseudomonas syringae* pv. *syringae*. Phytopathology 88:614–620

Chen W, Modi D, Picot A (2023) Soil and phytomicrobiome for plant disease suppression and management under climate change: a review. Plants 12:2736

Chen DV, Slowinski SP, Kido AK, Bruns EL (2024) High temperatures reduce growth, infection, and transmission of a naturally occurring fungal plant pathogen. Ecology 105:e4373

Dofuor AK, Quartey NK-A, Osabutey AF, Antwi-Agyakwa AK, Asante K, Boateng BO, Ablormeti FK, Lutuf H, Osei-Owusu J, Osei JHN (2023) Mango anthracnose disease: the current situation and direction for future research. Front Microbiol 14:1168203

El-Sayed A, Kamel M (2020) Climatic changes and their role in emergence and re-emergence of diseases. Environ Sci Pollut Res Int 27:22336–22352

Elad Y, Pertot I (2014) Climate change impact on plant pathogens and plant diseases. J Crop Impr 28

Farigoule P, Chartois M, Mesmin X, Lambert M, Rossi JP, Rasplus JY, Cruaud A (2022) Vectors as sentinels: rising temperatures increase the risk of xylella fastidiosa outbreaks. Biology (Basel) 11

Freeman S, Shtienberg D, Maymon M, Levin AG, Ploetz RC (2014) New insights into mango malformation disease epidemiology lead to a new integrated management strategy for subtropical environments. Plant Dis 98:1456–1466

Garrett KA, Nita M, De Wolf ED, Esker PD, Gomez-Montano L, Sparks AH (2021) Chapter 24—Plant pathogens as indicators of climate change. In: Letcher TM (ed) Climate change, 3rd edn. Elsevier, United Kingdom, pp 499–513

Ghimire B, Sapkota S, Bahri BA, Martinez-Espinoza AD, Buck JW, Mergoum M (2020) Fusarium head blight and rust diseases in soft red winter wheat in the Southeast United States: state of the art, challenges and future perspective for breeding. Front Plant Sci 11

González-Fernández J, Hormaza JI (2020) Pest and diseases in mango (*Mangifera Indica* L.).

He M, He CQ, Ding NZ (2018) Abiotic stresses: general defenses of land plants and chances for engineering multistress tolerance. Front Plant Sci 9:1771

Hlatshwayo SI, Ngidi MSC, Ojo TO, Modi AT, Mabhaudhi T, Slotow R (2023) The determinants of crop productivity and its effect on food and nutrition security in rural communities of South Africa. Front Sustain Food Syst 7

Lahlali R, Taoussi M, Laasli S-E, Gachara G, Ezzouggari R, Belabess Z, Aberkani K, Assouguem A, Meddich A, El Jarroudi M, Barka EA (2024) Effects of climate change on plant pathogens and host-pathogen interactions. Crop Environ 3:159–170

Lamichhane JR, Barbetti MJ, Chilvers MI, Pandey AK, Steinberg C (2024) Exploiting root exudates to manage soil-borne disease complexes in a changing climate. Trends Microbiol 32:27–37

Li L, Mohd MH, Mohamed Nor NMI, Subramaniam S, Latiffah Z (2021) Identification of Botryosphaeriaceae associated with stem-end rot of mango (*Mangifera indica* L.) in Malaysia. J Appl Microbiol 130(4):1273–1284

Nasir M, Mughal SM, Mukhtar T, Awan MZ (2014) Powdery mildew of mango: a review of ecology, biology, epidemiology and management. Crop Protect 64:19–26

Nath R, Nath PD, Borah M (2022) Role of micronutrients in conferring resistance against different plant viruses. Plant Dis Res 37(2):225–226

Normand F, Lauri P, Legave JM (2015) Climate change and its probable effects on mango production and cultivation. Acta Hortic 1075:21–31

Paul SK, Mahmud NU, Gupta DR, Rani K, Kang H, Wang G-L, Jankuloski L, Islam T (2022) Oryzae pathotype of Magnaporthe oryzae can cause typical blast disease symptoms on both leaves and spikes of wheat under a growth room condition. Phytopathol Res 4:9

Singh BK, Delgado-Baquerizo M, Egidi E, Guirado E, Leach JE, Liu H, Trivedi P (2023) Climate change impacts on plant pathogens, food security and paths forward. Nat Rev Microbiol 21:640–656

Sossah FL, Aidoo OF, Dofuor AK, Osabutey AF, Obeng J, Abormeti FK, Duker RQ, Antwi-Agyakwa AK, Osei-Owusu J, Loh SK, Honger JO, Borgemeister C (2024) A critical review on bacterial black spot of mango caused by *Xanthomonas citri* pv. *mangiferaeindicae*: current status and direction for future research. For Pathol 54: e12860

Visser ME, Both C (2005) Shifts in phenology due to global climate change: the need for a yardstick. Proc R Soc Lond B Biol Sci 272:2561–2569

Zhang Y, Xu J, Li R, Ge Y, Li Y, Li R (2023) Plants' response to abiotic stress: mechanisms and strategies. Int J Mol Sci 24

Chapter 32
Innovative Technologies for Surveillance and Management of Mango Tree Diseases and Disorders in a Changing Climate

Syed Atif Hasan Naqvi, Aqleem Abbas, and Muhammad Ahmad

Abstract Mango (*Mangifera indica*) cultivation is a cornerstone of tropical agriculture, offering economic and nutritional benefits worldwide. However, the increasing frequency and intensity of diseases driven by climate change threaten its production. Advances in technology, from drone-based imaging to artificial intelligence (AI), have provided novel solutions for monitoring and managing these challenges. This chapter explores the latest innovations, including remote sensing, Internet of Things (IoT) devices, metagenomics, and robotics, and their role in enhancing resilience against climate-induced disease dynamics. These technologies offer scalable, sustainable strategies for disease management, bridging gaps between traditional practices and modern solutions.

Keywords Mango cultivation · Climate resilience · Disease management · Artificial intelligence · Remote sensing

32.1 Introduction

Mango, often called the "king of fruits," is cultivated in over 100 countries, with India, China, and Thailand as leading producers. The crop's global economic significance is matched by its susceptibility to a wide range of diseases, including anthracnose, powdery mildew, and bacterial canker. The vulnerability of mango orchards has

S. A. H. Naqvi (✉) · M. Ahmad
Department of Plant Pathology, Bahauddin Zakariya University, Multan, Pakistan
e-mail: atifnaqvi@bzu.edu.pk

S. A. H. Naqvi
Department of Biological Sciences, Brock University, St. Catharines, ON L2S 3A1, Canada

London Research and Development Center, Agriculture and Agri-Food Canada, Vineland Station, ON L0R 2E0, Canada

A. Abbas
Department of Agriculture and Food Technology, Karakoram International University (KIU), Gilgit, Pakistan

S. A. H. Naqvi et al. (eds.), *Climate Change and Mango Production*,
https://doi.org/10.1007/978-3-031-87035-4_32

been exacerbated by climate change, which alters pathogen lifecycles, increases pest pressures, and impacts tree physiology (Tharanathan et al. 2006).

To ensure sustainable mango production, there is an urgent need for innovative, precise, and eco-friendly solutions. This chapter examines cutting-edge technologies that facilitate early detection, continuous monitoring, and targeted management of mango diseases, addressing the challenges posed by a warming planet.

32.2 Impact of Climate Change on Mango Tree Health

32.2.1 Climate-Driven Disease Dynamics

Climate change significantly influences the health of mango trees, primarily through its impact on disease dynamics. Rising global temperatures, increased humidity, and erratic rainfall patterns create favorable conditions for many pathogens and pests that affect mango crops. For example, diseases like anthracnose (*Colletotrichum gloeosporioides*), bacterial canker, and powdery mildew are becoming more widespread and severe due to climate-induced stress on mango trees and prolonged wet periods that facilitate pathogen growth. Higher temperatures can also accelerate the lifecycle of pests, increasing their population density and the frequency of outbreaks. Furthermore, changing climate conditions disrupt the phenological synchronization of mango trees, such as flowering and fruiting, with the activity of pathogens, resulting in greater susceptibility and reduced yields (Nelson 2008).

Efforts to mitigate these impacts include developing climate-resilient mango cultivars, employing integrated pest management (IPM), and using predictive models for early disease detection. The application of advanced technologies, such as remote sensing and machine learning, also aids in monitoring and managing disease spread under changing climatic conditions. Moreover, adaptive agricultural practices like canopy management, optimized irrigation, and timely application of biopesticides are essential in combating these challenges (Ploetz 2003).

Climate change has reshaped the epidemiology of many mango diseases:

- **Temperature Increases**: Elevated temperatures accelerate pathogen reproduction and virulence, particularly in diseases such as bacterial canker and mango malformation (Misra 2012).
- **Altered Rainfall Patterns**: Excessive moisture promotes fungal diseases like anthracnose, while drought conditions weaken tree immunity (Delgado-Ospina et al. 2021).
- **Prolonged Wet Seasons**: Favorable for the growth of fungal pathogens and the spread of waterborne diseases (Hellberg and Chu 2016).
- **Emergence of Novel Pathogens**: Diseases previously restricted to specific regions are now appearing in new areas due to changing climatic conditions (Patz et al. 2005).

32.2.2 Implications for Mango Production

- **Yield Losses**: Widespread disease outbreaks can reduce fruit production by up to 40% (Anyamba et al. 2014).
- **Quality Degradation**: Diseases not only affect quantity but also compromise fruit quality, leading to market rejection (Ramírez-Gil et al. 2019).
- **Increased Costs**: The reliance on fungicides and bactericides drives up production costs, posing challenges for smallholder farmers (Khursheed et al. 2022).

32.3 Innovations in Disease Surveillance and Management

32.3.1 Remote Sensing and Drone Technologies

Innovations in disease surveillance and management are transforming forestry through the integration of remote sensing and drone technologies, providing powerful tools to monitor, predict, and mitigate the impacts of forest diseases. Remote sensing, using satellite and aerial imagery, enables large-scale monitoring of forest health by detecting changes in canopy color, density, and structure. Advanced multispectral and hyperspectral sensors can identify early signs of disease stress, such as water stress or pigment changes in leaves, long before symptoms are visible to the naked eye. This capability is critical for managing widespread pathogens, such as those causing sudden oak death or pine wilt disease, which can devastate forest ecosystems (Cotrozzi 2022).

Drones equipped with high-resolution cameras and sensors bring precision to forest health monitoring by offering localized and detailed data. They are particularly effective in hard-to-reach areas, providing insights into tree-specific conditions, mapping disease spread, and guiding intervention strategies. For instance, drones can be used to deploy biological control agents or monitor the effectiveness of treatments. The real-time data collected by drones can be integrated with GIS platforms and machine learning algorithms to model disease progression and optimize management plans (Quamar et al. 2023).

Together, these technologies enhance the efficiency and accuracy of forest disease surveillance and management, minimizing economic losses and preserving ecosystem services. Their use supports proactive decision-making, enabling forest managers to mitigate risks, protect biodiversity, and ensure the sustainability of forest resources in the face of climate change and emerging pathogens (Millar et al. 2007).

Remote sensing has revolutionized agricultural surveillance. Equipped with advanced cameras, drones can monitor mango orchards efficiently.

32.3.2 Applications

- **Multispectral Imaging**: Detects chlorophyll levels and plant stress, signaling early disease symptoms (Wang et al. 2018).
- **Thermal Cameras**: Identify heat signatures associated with infected areas (Gade and Moeslund 2014).
- **Hyperspectral Imaging**: Offers detailed spectral analysis to differentiate between healthy and diseased foliage (Bock et al. 2010).
- **Disease Mapping**: Creates spatial maps of disease incidence to guide targeted interventions (Lawson 2013).

32.3.3 Advantages

- **Scalability**: Covers large areas quickly, reducing the need for manual scouting (Gao et al. 2020).
- **Precision Agriculture**: Enables site-specific management practices, conserving resources (Oliver et al. 2013).
- **Cost-Effectiveness**: Though initially expensive, drones reduce long-term monitoring costs (Howell et al. 2021).

32.4 Internet of Things (IoT) and Sensor Networks

The Internet of Things (IoT) is revolutionizing forestry by enabling real-time monitoring and management of forest ecosystems. IoT involves a network of interconnected devices embedded with sensors that collect and transmit data through the internet. In forestry, IoT applications include tracking tree growth, monitoring environmental parameters, and detecting signs of disease or pest outbreaks. Sensors placed in forests measure temperature, humidity, soil moisture, and light intensity, providing valuable data for forest managers. These sensors communicate with centralized systems, allowing remote monitoring of forest health over vast areas (Singh et al. 2022). IoT also enhances wildfire management by deploying smart devices that detect smoke, temperature spikes, and wind patterns to provide early warnings. These systems reduce response times and improve firefighting strategies. Additionally, IoT-based systems are used in timber tracking and inventory management, ensuring sustainable logging practices. GPS-enabled devices attached to harvested timber allow real-time tracking, helping prevent illegal logging and ensuring compliance with regulatory standards (Zakari et al. 2024). By integrating IoT with advanced analytics and artificial intelligence, forestry operations can predict trends, such as disease spread or fire risks, and develop proactive solutions. This technology improves the efficiency of resource allocation, reduces environmental damage, and supports sustainable forest management practices. IoT is a transformative tool that aligns forestry with modern precision management approaches, ensuring ecosystem resilience in the face of growing environmental challenges (Mahule et al. 2024).

Sensor networks in forestry involve the deployment of interconnected sensors across forests to collect and transmit data about environmental and ecological conditions. These networks play a crucial role in monitoring forest health, biodiversity, and climate impacts. By measuring variables such as temperature, humidity, soil composition, and carbon dioxide levels, sensor networks provide detailed insights into forest dynamics (Bayne et al. 2017). In disease and pest management, sensor networks detect early signs of infestations by monitoring tree stress indicators and environmental factors that favor pathogen development. This allows forest managers to implement targeted interventions, minimizing damage and costs. For instance, pheromone-based sensors can monitor pest populations and predict outbreaks (Uma et al. 2024). Sensor networks are also essential for wildfire management. Distributed sensors detect changes in temperature, humidity, and gas concentrations, providing early warnings of potential fires. When integrated with weather stations, these networks improve predictions of fire behavior, aiding in strategic firefighting efforts (Molina-Pico et al. 2016).

In addition to environmental monitoring, sensor networks contribute to sustainable forestry practices by assessing soil health and water availability, ensuring that forest management activities do not deplete resources. The data collected from sensor networks can be analyzed using GIS tools and machine learning algorithms to model ecological processes and assess long-term changes in forest ecosystems (Kumar et al. 2024). By enabling continuous, real-time data collection over large and remote areas, sensor networks enhance decision-making, promote sustainable management, and protect forest ecosystems from emerging threats.

IoT devices provide real-time monitoring of orchard conditions, offering valuable insights into disease risks.

32.4.1 Applications

- **Environmental Monitoring**: Sensors track parameters such as humidity, temperature, and soil moisture to predict fungal disease outbreaks (Trilles Oliver et al. 2019).
- **Volatile Organic Compounds (VOCs)**: Sensors detect VOCs released by infected trees, enabling early intervention (Khelili 2022).
- **Integration with Smart Systems**: IoT devices trigger automated irrigation and pest control systems when specific thresholds are met (Chamara et al. 2022).

32.4.2 Advantages

- **Continuous Monitoring**: Reduces the risk of overlooking critical disease signs (Nong and Tarek 2023).
- **Data-Driven Decisions**: Enhances the accuracy of disease prediction models (Ijeh et al. 2024).

32.5 Artificial Intelligence (AI) and Machine Learning (ML)

Artificial Intelligence (AI) is revolutionizing forestry by enabling smarter, data-driven decision-making to address challenges in forest management, conservation, and resource utilization. AI leverages vast datasets collected from satellites, drones, IoT devices, and sensor networks to analyze patterns, predict outcomes, and optimize management strategies (Holzinger et al. 2022). One key application of AI in forestry is in disease and pest management. AI algorithms analyze data from remote sensing and ground-based sensors to detect early signs of stress in trees caused by pathogens or pests. This allows forest managers to take proactive measures to prevent large-scale damage. Additionally, AI-powered tools can identify species-specific characteristics in forests, aiding in biodiversity monitoring and conservation planning (Abdullah et al. 2023). AI is also critical in wildfire management. Machine learning models process weather data, vegetation conditions, and historical fire patterns to predict wildfire risks and model fire spread. These predictions enhance preparedness and resource allocation for firefighting efforts. Another transformative application of AI in forestry is in timber management. AI-driven systems can automate timber inventory assessments, optimize logging operations, and ensure sustainable harvesting practices. By using data from GPS and satellite imagery, AI helps track logging activities and detect illegal logging in real time (Mohan et al. 2021).

Machine Learning (ML), a subset of AI, is transforming forestry by providing advanced tools for analyzing complex datasets and uncovering hidden patterns in forest dynamics. ML algorithms learn from historical data and adapt to new information, making them ideal for modeling and predicting forestry-related phenomena. In disease and pest management, ML models analyze sensor data, drone imagery, and climatic variables to identify early signs of infestations and predict their spread. These insights enable targeted interventions, reducing costs and environmental impact. For example, ML can classify tree species and detect anomalies, such as diseased or stressed trees, from satellite or drone images with high accuracy (Jain et al. 2020).

ML is also widely used in wildfire prediction and management. By processing meteorological data, vegetation indices, and historical fire records, ML models predict fire risks and simulate fire behavior. These predictions improve preparedness and guide firefighting strategies, minimizing the loss of life and property. Another application of ML is in forest inventory and mapping. ML techniques analyze LiDAR and hyperspectral data to estimate tree biomass, canopy structure, and species diversity, which are critical for sustainable forest management and carbon accounting (Reddy et al. 2024).

ML also aids in monitoring deforestation and land-use changes by analyzing time-series satellite imagery. Algorithms detect unauthorized logging or encroachment activities, enabling timely enforcement actions. In ecological research, ML helps model complex relationships between climate factors, soil conditions, and forest growth, contributing to better understanding and management of forest ecosystems under changing environmental conditions. With its ability to process vast and diverse datasets, ML is a valuable tool for advancing precision forestry, enhancing sustainability, and conserving biodiversity (Haq et al. 2024).

32.5.1 Applications

- **Image Recognition**: AI models trained on vast datasets can identify diseases based on leaf patterns, spots, or discolorations (Tyagi and Chahal 2020).
- **Predictive Analytics**: ML algorithms analyze historical climate and disease data to forecast outbreaks (Murthy 2023).
- **Automated Advisory Systems**: AI-driven platforms provide farmers with real-time recommendations (Raman et al. 2024).

32.5.2 Advantages

- **Improved Efficiency**: Reduces reliance on human expertise for disease detection (Elshafei 2024).
- **Scalable Solutions**: Cloud-based AI systems can serve both small and large-scale farmers (Rane et al. 2024).

32.6 Metagenomics and Microbiome Profiling

Metagenomics, the study of genetic material recovered directly from environmental samples, is revolutionizing forestry by providing insights into the microbial communities that influence forest health and ecosystem dynamics. By sequencing DNA from soil, tree surfaces, or water samples, metagenomics allows researchers to analyze the diversity and function of microorganisms without the need to culture them in the lab (Rani et al. 2024). One key application of metagenomics in forestry is in understanding soil health. Forest soils host diverse microbial communities that drive nutrient cycling, organic matter decomposition, and carbon sequestration. Metagenomic analyses help identify microbial taxa involved in these processes, providing insights into how forest management practices influence soil health and fertility (Vieira et al. 2021). Metagenomics is also vital in studying plant–microbe interactions. Symbiotic relationships, such as those between trees and mycorrhizal fungi, are crucial for tree growth and resilience. Metagenomics enables the characterization of these symbiotic networks, helping foresters develop strategies to enhance forest productivity and stress tolerance (Verma et al. 2021).

In disease management, metagenomics can identify pathogens and track their spread in forest ecosystems. By analyzing environmental DNA, researchers can detect early signs of emerging diseases, enabling proactive interventions. Additionally, metagenomics helps uncover microbial responses to environmental changes, such as deforestation or climate change, providing valuable data for conservation strategies (Akoijam and Joshi 2022). The ability of metagenomics to generate comprehensive microbial profiles supports forest restoration efforts by guiding the re-establishment of beneficial microbial communities. As sequencing technologies advance, metagenomics is becoming an indispensable tool for sustainable forest management and the conservation of biodiversity (Wang et al. 2025).

Microbiome profiling, the study of microbial communities associated with specific environments or hosts, is an emerging field in forestry with transformative potential. Forest trees harbor diverse microbiomes in their roots, leaves, bark, and surrounding soil, which play critical roles in their health and productivity. Profiling these microbiomes provides insights into their structure, function, and interactions with trees and the broader ecosystem (Baldrian et al. 2023). In forestry, microbiome profiling is essential for understanding the role of beneficial microbes in promoting tree health. For example, endophytic bacteria and fungi within tree tissues enhance nutrient acquisition, pathogen resistance, and stress tolerance. By profiling these communities, researchers can identify microbial strains that could be harnessed for bioinoculants to improve forest resilience (Vacher et al. 2021).

Microbiome profiling is also crucial in disease management. Pathogenic microbes can disrupt forest ecosystems, but profiling helps detect and monitor these harmful organisms. It also reveals shifts in microbial community composition due to disease outbreaks, guiding targeted interventions. In soil health, microbiome profiling provides a detailed picture of the microbial consortia responsible for processes such as nitrogen fixation, organic matter breakdown, and carbon sequestration. This knowledge supports sustainable forestry practices by informing soil management strategies that preserve these vital functions (Baldrian et al. 2023). Another application is in forest restoration. Microbiome profiling aids in selecting microbial communities that can promote the growth of saplings and improve degraded soil conditions. This is particularly relevant for reforestation projects in areas impacted by deforestation or climate change. By integrating microbiome profiling with modern sequencing and bioinformatics tools, forestry professionals can unlock the potential of microbial communities to enhance forest productivity, resilience, and sustainability, ensuring the health of forest ecosystems for future generations (Robinson et al. 2023).

Advances in molecular biology have facilitated the study of mango tree-associated microbiomes.

32.6.1 Applications

- **Environmental DNA (eDNA) Analysis**: Detects pathogenic DNA in soil, water, or plant samples (Cheney 2022).
- **Microbiome Studies**: Identifies beneficial microbes that can suppress pathogens or enhance tree immunity (Vacher et al. 2021).
- **Pathogen Identification**: NGS technologies provide detailed pathogen profiles, enabling precise interventions (Jongman et al. 2020).

32.6.2 Advantages

- **Non-Destructive Testing**: Sampling does not harm trees or disrupt orchard ecosystems (Mazzola et al. 2019).
- **Pathogen Insights**: Helps design biocontrol strategies using natural antagonists (Berg et al. 2015).

32.7 Satellite Imaging and GIS Integration

Satellite imaging is a transformative technology in forestry, enabling large-scale, real-time monitoring of forests across diverse terrains. With the ability to capture high-resolution images and data on various spectral bands, satellites provide critical insights into forest health, composition, and dynamics. These capabilities are instrumental for applications ranging from biodiversity conservation to commercial forest management. One key application of satellite imaging in forestry is in mapping forest cover and detecting changes over time. Using multispectral and hyperspectral imaging, satellites can identify vegetation types, assess canopy density, and detect deforestation or illegal logging activities. Satellite imagery also plays a crucial role in monitoring forest degradation caused by pests, diseases, or wildfires by detecting subtle changes in vegetation health, such as stress-induced chlorophyll loss (Smigaj et al. 2024). By providing consistent, wide-scale coverage, satellite imaging is particularly useful for monitoring forests in remote or inaccessible regions. Advances in satellite technology, such as synthetic aperture radar (SAR), enhance data acquisition in all weather conditions and through dense canopies. Combined with modern analytics, satellite imaging empowers forest managers and policymakers to make informed decisions, ensuring sustainable forest management and conservation (Panda et al. 2016).

Geographic Information Systems (GIS) are indispensable tools in forestry, allowing the integration, visualization, and analysis of spatial data to support informed decision-making. GIS enables forest managers to create detailed maps of forest landscapes, analyze spatial patterns, and monitor changes over time, providing critical insights for sustainable management. One of the primary uses of GIS in forestry is mapping forest cover and land use. By combining spatial data from satellites, drones, and field surveys, GIS generates detailed maps showing tree species distribution, biomass, and soil types. These maps support timber inventory management, biodiversity conservation, and restoration planning (Marano et al. 2019). GIS integration facilitates landscape-level planning by assessing the impacts of human activities, such as logging or infrastructure development, on forest ecosystems. The ability to combine spatial data with socio-economic information ensures a balanced approach to forest management that aligns conservation with economic goals. By providing a platform for spatial analysis and data-driven insights, GIS enhances the efficiency, precision, and sustainability of forestry operations, contributing to the preservation of forest ecosystems in the face of growing environmental challenges (Fardusi et al. 2017).

Satellite imagery, combined with GIS tools, offers a macro perspective on mango orchard health.

32.7.1 Applications

- **Large-Scale Monitoring**: Tracks orchard conditions across regions (Mani and Varghese 2018).
- **Risk Assessment**: GIS maps overlay disease data with climate patterns to identify high-risk zones (Palaniyandi 2012).
- **Epidemiological Studies**: Supports long-term research on disease trends and climate interactions (Ozdenerol 2015).

32.7.2 Advantages

- **Accessibility**: Satellite data is increasingly affordable for large-scale agriculture (Mani and Varghese 2018).
- **Enhanced Planning**: Guides regional and national-level disease management programs (LoPresti et al. 2016).

32.8 Robotics and Automation

Robotics and automation are transforming forestry by enhancing efficiency, precision, and sustainability in forest management and operations. These technologies address challenges such as labor shortages, operational inefficiencies, and the need for sustainable practices in managing forest resources. Robotics is revolutionizing tree harvesting and logging operations. Autonomous and semi-autonomous machines, such as robotic harvesters and forwarders, can perform tasks like felling, delimbing, and transporting logs with remarkable precision. Equipped with advanced sensors and GPS, these robots navigate challenging terrains while minimizing damage to surrounding vegetation and soil. Automation reduces the physical strain on workers and enhances safety in hazardous environments (Picchio et al. 2019). Drones, a form of robotic technology, play a significant role in forest monitoring. They are equipped with cameras, multispectral sensors, and LiDAR to capture detailed images and data of forest canopies, tree health, and biodiversity. Drones can assess areas affected by pests, diseases, or fires and guide reforestation efforts by planting seeds in hard-to-reach areas, significantly accelerating restoration projects (Wich and Koh 2018).

Automation is also streamlining forest nursery operations. Automated systems control seed planting, watering, and monitoring, ensuring optimal growth conditions. These systems improve resource efficiency while reducing human labor. In pest and disease management, robotics enables precision application of pesticides or biological agents. Autonomous sprayers target infected areas, minimizing chemical use and environmental impact. Robots can also monitor forests for early signs of disease or pest outbreaks, using artificial intelligence to analyze data in real-time (Shamshiri et al. 2018) (Tables 32.1, 32.2 and 32.3).

32.8.1 Applications

- **Ground Robots**: Equipped with cameras and sensors, these robots inspect tree trunks and leaves for signs of disease (Shalal et al. 2015).
- **Autonomous Sprayers**: Target infected areas with precision, minimizing chemical use (Salamí et al. 2014).

32.8.2 Advantages

- **Labor Savings**: Reduces dependency on manual labor, especially in large orchards (He and Schupp 2018).
- **Eco-Friendly Practices**: Limits pesticide runoff and overapplication (Madiga Bala et al. 2020).

Table 32.1 Sensors mounted on drones for plant disease detection and monitoring

Remote sensors	Advantages	Disadvantages	Diseases
Digital camera (RGB)	Vegetation characteristics may capture grayscale or color pictures, and the visible spectrum allows for improved disease identification at the leaves' level. Lightweight, inexpensive, extremely easy to use, simple data processing, and minimal work environments	Reduced number of spectral bands and visibility of less light Vulnerable to environmental factors	Cotton bacterial angular, Ascochyta blight, grapefruit citrus canker, sugar beet Cercospora leaf spot, rust, blights, smuts, spots
Multispectral camera	Low cost, fast frame imaging and high, more robust than RGB cameras, work efficiency; electromagnetic spectrum ranging from the visible to the Near-Infrared (NIR), allowing the calculation of different robust vegetation indices; sensing and recording radiations from the visible and invisible portions of the electromagnetic spectrum	Few bands, discontinuous spectrum, and low spectral resolution	Blights, Blasts, viruses
Hyperspectral sensing	Capable of sensing and recording a wide variety of narrow bands and continuous spectra, giving researchers and farmers more insight into the spectral properties of illnesses and crops Hyperspectral cameras have a greater number of channels than multispectral ones, are more robust than RGB cameras, and can discern between various stress types in crops by using data collected from the visible to the Near-Infrared (NIR)	More expensive	Blasts, Blights, and nematodes, viruses, rots, scabs, rusts

(continued)

Table 32.1 (continued)

Remote sensors	Advantages	Disadvantages	Diseases
Thermal infrared cameras (InfraRed (IR) region consists of several spectral. Bands, including Near InfraRed (NIR), Short-Wave InfraRed (SWIR), Mid-Wave InfraRed (MWIR), Long-Wave InfraRed (LWIR), and Far InfraRed (FIR))	Sensitive to infrared spectrum, therefore it may be used day or night and is able to provide more data on plant health than other sensors	Problems with the images' temporal and geographic resolutions; issues with the weather and lighting; problems with the variety of crop species and their development stages; problems with the height at which the photographs were taken	Recently used to monitor diseases such as Cercospora leaf spot, scab and mildews
Fluorescence imaging	Can determine how plant responses to various stresses affect photosynthesis	Has been used to detect a few diseases, difficult to use in field and greenhouse conditions	Mildews, rust and cankers

References (Christiansen et al. 2017; Mahlein 2016; Chen et al. 2018; Torres-Sánchez et al. 2013; Bleecker and Kende 2000; Barbedo 2018: Bendig et al. 2014; Geipel et al. 2014; Yang et al. 2014; Rango et al. 2006; Laliberte et al. 2011)

32.9 Advanced Diagnostic Tools

32.9.1 CRISPR-Based Diagnostics

CRISPR-based diagnostics are transforming forestry by enabling fast, accurate detection of pathogens and pests that threaten forest health. Using CRISPR's precision to target specific DNA or RNA sequences, these diagnostics can identify harmful organisms such as fungi, bacteria, or invasive species at an early stage. Unlike traditional methods that are time-consuming and require lab facilities, CRISPR-based tools are portable, cost-effective, and provide results within minutes, making them ideal for on-site forest management (Pérez et al. 2024). This technology is particularly useful for monitoring emerging diseases, such as those caused by tree pathogens or beetle infestations, which can devastate forest ecosystems. Early detection through CRISPR allows forest managers to implement targeted interventions, reducing the spread of infections and minimizing economic and ecological losses. By enhancing disease surveillance, CRISPR-based diagnostics contribute to sustainable forestry practices, ensuring the health and resilience of forests in the face of increasing environmental challenges (Nahata 2020).

Table 32.2 Advantages and disadvantages of using drones for agricultural applications

Drones	Advantages	Disadvantages
Rotary-wing (Multirotor)	Take off and land vertically Return home capability High-detailed plant measurements Automatic recovery capability The lift generated by batteries Hovering and flying at low altitudes Depending on multiple rotors e.g., three rotors (tri-copters),four rotors (quad-copters), six rotors (hexacopters), and eight rotors (octocopters)	Small payload sensor capacity Most of the capacity taken by batteries Low endurance Low speed The area covered is limited
Fixed-wing	The lift created by wings Longer flight High speed High endurance High payload sensor capacity Some have automatic recovery capability Fly a longer time and are more suitable for covering larger areas High altitude	Fly at an airspeed above their stall speed; sometimes problems occur in generating the desired data about the crops Low flexibility makes small crop monitoring very difficult Require runways and space to land and take off
Hybrid vertical take-off and landing (VTOL)	Resolve issues with multirotor and fixed-wing drones Combine fixed-wing drone cruise flight with multirotor drone VTOL capabilities Longer flight times, extensive coverage, quick vertical takeoffs, and minimal energy requirements	Expensive and hovering issues

32.9.2 Biosensors

Biosensors are emerging as valuable tools in forestry, offering precise and real-time monitoring of forest health and environmental conditions. These devices combine biological recognition elements, such as enzymes or antibodies, with electronic components to detect specific biological or chemical markers. In forestry, biosensors are used to identify pathogens, pests, and pollutants that can harm trees and ecosystems (Wilson 2013). For example, biosensors can detect stress signals in trees caused by fungal infections, nutrient deficiencies, or drought. By monitoring volatile organic compounds (VOCs) emitted by stressed trees, biosensors enable early detection of health issues, allowing timely interventions. Additionally, they can assess soil

Table 32.3 Usage of drones and other devices for plant disease detection

S. No.	Drones	Sensors/cameras/ other devices	Disease	References
1.	DT-18 UAV platform	MicaSense RedEdge® sensor	*Flavescence dorée* (FD) and Grapevine Trunk Diseases (GTD)	Albetis et al. (2018)
2.	Hexa-rotor DJI S800 EVO	Headwall Nano-hyperspectral	Myrtle rust	Sandino et al. (2018)
3.	Phantom 2 Vision +	Micasense RedEdgeTM	Rice sheath blight	Martinelli et al. (2015)
4.	Mikrokopter OktoXL	Multispectral camera	Grapevine leaf stripe disease (GLSD)	Di Gennaro et al. (2016)
5.	Long-range DT-18	MicaSense RedEdgeTM sensor	Grapevine Disease	Albetis et al. (2017)
6.	HiSystems GmbH Mikrokopter	Sony NEX-5N	Potato late blight	Sugiura et al. (2016)
7.	MX-SIGHT	Multi-spectral sensors	Downy mildew of opium	Calderón et al. (2014)
8.	Coaxial quad-copter	MicaSense RedEdge 3 camera	Physiological stress in pine trees	de Souza et al. (2017)
9.	*QuestUAV Qpod*	Thermal, RGB (red, green and blue) and NIR (near-infrared) sensors	Red Band Needle Blight disease of pine trees	Smigaj et al. (2015)
10.	S800 EVO Hexacopter	Canon 5DsR camera, multispectral MicaSense RedEdge camera, Headwall Nano-Hyperspace	Grape Phylloxera	Vanegas et al. (2018)
11.	DJI Phantom 3	Sony EXMOR sensor	Soybean Foliar Diseases	Tetila et al. (2017)
12.	eBee	IXUS 127 HS Canon camera	Ceratocystis Wilt in Eucalyptus Crops	de Souza et al. (2017)
13.	HiSystems GmbH	MiniMCA6	Huanglongbing-infected citrus trees	Garcia-Ruiz et al. (2013)
14.	Modified Sig Rascal 110 RC airframe	General camera	Potato late blight	Techy et al. (2010)
15.	Quantalab, IAS-CSIC,	Multispectral and thermal	*Verticillium* wilt of olives	Calderón et al. (2014)
16.	DJI Phantom 3 quadcopter	MicaSense RedEdge multispectral camera	Tomato Spot Wilt Disease in Peanuts	Patrick et a. (2017)
17.	Phantom 4	RGB camera	*Fusarium* wilt of radish	Dang et al. (2020)
18.	Model TFX-11	Sporangia-sampling devices	Potato late blight	Aylor et al. (2011)
19.	Multi-rotor MikrokopterOktoXL	Multispectral and thermal	Water status within a vineyard	Santesteban et al. (2017)

(continued)

Table 32.3 (continued)

S. No.	Drones	Sensors/cameras/ other devices	Disease	References
20.	DJI Phantom 4	RGB camera	Potato Virus Y	Sugiura et al. (2018)
21.	3D Robotics	Gamaya OXI VNIR 40 camera	Beet Cyst Nematode in Sugar Beet	Joalland et al. (2018)
22.	DJI Matrice 600 pro	Pika L 2.4 (Resonon Inc., Bozeman MT, USA)	Target spot and bacterial spot diseases	Abdulridha et al. (2020)
23.	UAV system (S1000)	Hyperspectral imaging sensor (UHD 185)	Yellow rust	Guo et al. (2021)
24.	FeimaD200 quadrotor	Model RedEdge-MX	Pine Wood Nematode Disease	Qin et al. (2021)
25.	Multi-rotor DJI Mavic Pro	Mltispectral sensor (Parrot Sequoia)	Olive quick decline syndrome	Castrignanò et al. (2020)

quality by measuring levels of nutrients, toxins, or microbial activity, aiding sustainable forest management. Portable and highly sensitive, biosensors enhance forest monitoring by providing rapid, on-site analysis. Their application supports proactive decision-making, reducing environmental damage and promoting the long-term health and productivity of forest ecosystems (Wasilewski et al. 2024).

32.10 Blockchain and Digital Platforms

Blockchain and digital platforms are transforming forestry by enhancing transparency, traceability, and efficiency in forest management. Blockchain enables secure, tamper-proof recording of data related to timber harvesting, supply chains, and carbon credits, ensuring sustainable practices and combating illegal logging. Digital platforms integrate blockchain with tools like GIS, IoT, and remote sensing, providing real-time data on forest resources, health, and land use. These platforms facilitate decision-making, stakeholder collaboration, and monitoring of conservation efforts. By promoting accountability and sustainable practices, blockchain and digital platforms contribute to the effective management and preservation of forest ecosystems (Stopfer et al. 2024).

32.10.1 Applications

- **Blockchain**: Tracks interventions, ensuring transparency and accountability in disease management (Kete 2022).
- **Digital Platforms**: Connect farmers, researchers, and policymakers for collaborative problem-solving (van Ewijk et al. 2024).

32.10.2 Advantages

- **Community Engagement**: Encourages knowledge sharing and collective action (Diaz-Valdivia 2024).
- **Traceability**: Builds consumer trust in sustainably managed produce (Wu and Pullman 2020).

32.11 Case Studies and Success Stories

32.11.1 India: Drone-Based Surveillance of Anthracnose

Drone-based surveillance offers a promising tool for monitoring and managing anthracnose in India, particularly in crops and forestry. Using drones equipped with multispectral and hyperspectral sensors enables early disease detection by identifying stress patterns in plants that are invisible to the naked eye. This technology helps in identifying hotspots of infection and supports targeted interventions, reducing the unnecessary use of fungicides and minimizing environmental impact (Kouadio et al. 2023). In India, where anthracnose severely impacts crops like mangoes and cashews, drone technology has demonstrated significant potential in precision agriculture. Drones facilitate rapid assessment and mapping of disease outbreaks by capturing high-resolution images over vast agricultural fields. The integration of artificial intelligence further enhances this capability, as machine learning algorithms can process collected data to differentiate anthracnose symptoms from other plant stresses, such as nutrient deficiencies or pest infestations. Drone imaging has reduced the time required to detect anthracnose, leading to a 25% decline in disease-related losses. The adoption of drones in disease management aligns with India's push towards smart agricultural practices. Challenges, however, remain in terms of affordability for small-scale farmers and the regulatory framework for drone usage. Addressing these barriers through subsidies, cooperative models, and streamlined regulations can help expand the adoption of this technology across diverse agricultural landscapes (Guebsi et al. 2024).

32.11.2 Brazil: IoT and Precision Agriculture

In Brazil, the integration of Internet of Things (IoT) and precision agriculture technologies is transforming the agricultural sector by enabling more efficient, sustainable practices. Precision agriculture in Brazil has grown significantly through innovations such as GPS-based systems, remote sensors, and cloud computing, which allow farmers to monitor soil health, crop conditions, and environmental factors in real time. These technologies enhance decision-making, reduce resource waste, and improve productivity (Bolfe et al. 2020). A study surveying Brazilian farmers revealed that 84% of respondents have adopted at least one digital technology, with a significant portion utilizing mobile apps, digital platforms, and sensors to track crop conditions. Precision farming tools, such as IoT-enabled sensors, drones, and satellite imaging, provide valuable data that can optimize irrigation, fertilization, and pest control, leading to reduced resource consumption and higher yields. Additionally, IoT facilitates supply chain traceability, helping farmers manage their operations more effectively and sustainably (Cherubin et al. 2022) (Figs. 32.1, 32.2, 32.3, 32.4 and 32.5).

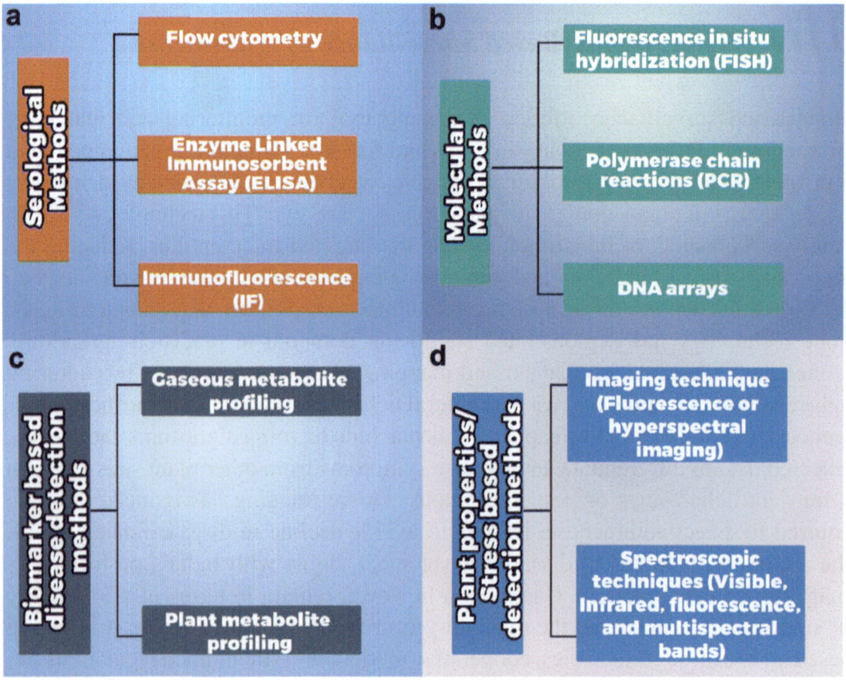

Fig. 32.1 Available methods for plant disease detection, viz., Direct (**a** and **b**) and indirect (**c, d**) [5]. Direct methods include traditional, serological, and molecular methods; indirect methods include biomarker-based methods, such as metabolite profiling from plant pathogens and plant interactions; and stress-based methods, such as remote drone sensing (Abbas et al. 2023)

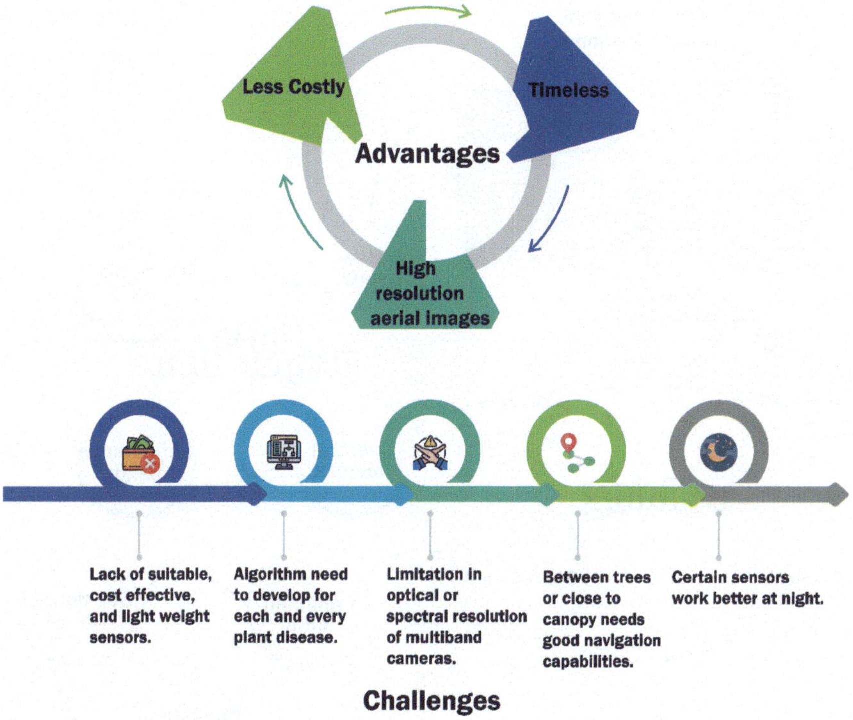

Fig. 32.2 Advantages and challenges/disadvantages of using drones for plant disease detection (Abbas et al. 2023)

32.12 Challenges and Future Directions

32.12.1 Technological and Financial Barriers

- High costs limit access for small-scale farmers.
- Need for capacity-building to train farmers in using these tools.

32.12.2 Policy and Governance Issues

- Lack of regulations on the integration of advanced technologies in agriculture.
- Need for subsidies and incentives to promote adoption.

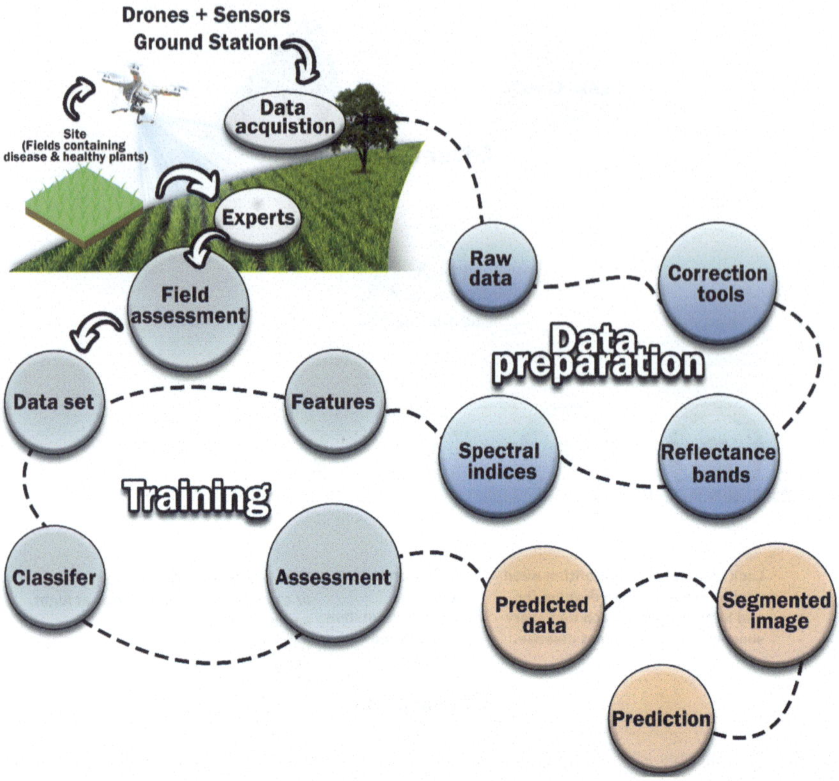

Fig. 32.3 Drone system to detect plant diseases comprises four sections, namely Data acquisition, Data Preparation, Training, and Prediction. The system interacts with the target area (fields, farms or forests) directly or indirectly to obtain information. The information is pre-processed and arranged into features, then sent to a supervised machine learning classifier that processes data and provides prediction reports via segmented images. The data acquisition process included drones, which indirectly acquire airborne data, and the expert acquired direct ground data through visual assessments of the fields. Drones consist of high-performance brushless rotors, specific load capacity, and dimensions. They are controlled automatically using aground station software, which controls the route, speed, and altitude at a distance (Abbas et al. 2023)

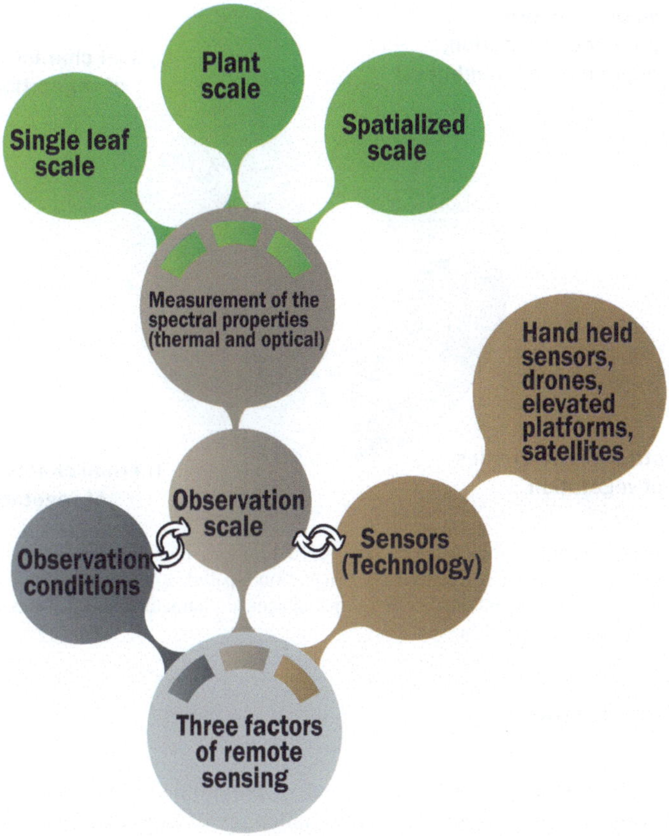

Fig. 32.4 Measurement of the spectral properties (thermal and optical) (Abbas et al. 2023)

32.12.3 Research Gaps

- Limited field validation of certain technologies.
- Insufficient studies on long-term impacts of climate change on mango disease dynamics.

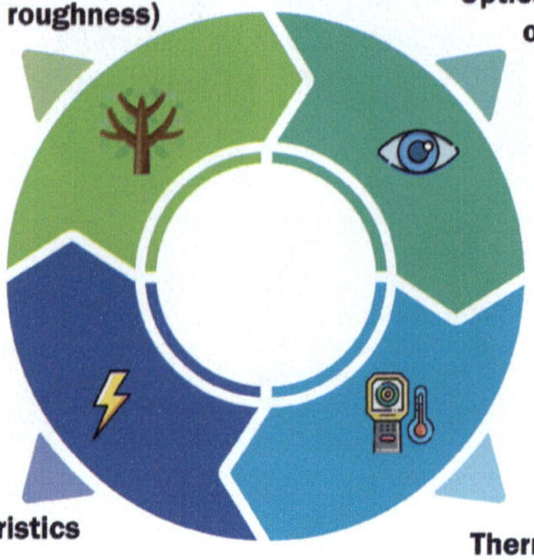

Fig. 32.5 Remote sensing using drone technology determines the plant's health status, such as canopy structure, including leaf areas and orientation, spatial arrangements, and roughness affected by diseases. Further optical, thermal, and dielectric characteristics of vegetation were also determined (Abbas et al. 2023)

32.13 Conclusion

Innovative technologies provide transformative potential for the management of mango tree diseases, addressing the dual challenges of climate change and food security. Integrating these tools into existing agricultural systems will require collaboration among scientists, policymakers, and growers. By harnessing these advancements, the mango industry can achieve sustainable growth while safeguarding against emerging threats.

References

Abdullah HM, Mohana NT, Khan BM, Ahmed SM, Hossain M, Islam KS, Redoy MH, Ferdush J, Bhuiyan M, Hossain MM (2023) Present and future scopes and challenges of plant pest and disease (P&D) monitoring: Remote sensing, image processing, and artificial intelligence perspectives. Remote Sens Appl: Soc Environ 32:100996

Abbas A, Zhang Z, Zheng H, Alami MM, Alrefaei AF, Abbas Q, Naqvi SAH, Rao MJ, Mosa WF, Abbas Q, Hussain A (2023) Drones in plant disease assessment, efficient monitoring, and detection: a way forward to smart agriculture. Agronomy 13(6):1524

Abdulridha J, Ampatzidis Y, Kakarla SC, Roberts P (2020) Detection of target spot and bacterial spot diseases in tomato using UAV-based and benchtop-based hyperspectral imaging techniques. Precis. Agric 21:955–978

Akoijam N, Joshi S (2022) Conservation metagenomics: understanding microbiomes for biodiversity sustenance and conservation. In: Molecular genetics and genomics tools in biodiversity conservation. Springer, pp 31–61

Albetis J, Duthoit S, Guttler F, Jacquin A, Goulard M, Poilvé H, Féret JB, Dedieu G (2017) Detection of *Flavescence dorée* grapevine disease using unmanned aerial vehicle (UAV) multispectral imagery. Remote Sens 9(4):308

Albetis J, Jacquin A, Goulard M, Poilvé H, Rousseau J, Clenet H, Dedieu G, Duthoit S (2018) On the potentiality of UAV multispectral imagery to Detect *Flavescence dorée* and Grapevine Trunk Diseases. Remote Sens 11(1):23

Anyamba A, Small JL, Britch SC, Tucker CJ, Pak EW, Reynolds CA, Crutchfield J, Linthicum KJ (2014) Recent weather extremes and impacts on agricultural production and vector-borne disease outbreak patterns. PLoS ONE 9:e92538

Aylor DE, Schmale III DG, Shields EJ, Newcomb M, Nappo CJ (2011) Tracking the potato late blight pathogen in the atmosphere using unmanned aerial vehicles and Lagrangian modeling. Agric For Meteorol 151(2):251–260

Baldrian P, López-Mondéjar R, Kohout P (2023) Forest microbiome and global change. Nat Rev Microbiol 21:487–501

Barbedo JG (2018) Factors influencing the use of deep learning for plant disease recognition. Biosyst Eng 172:84–91

Bayne K, Damesin S, Evans M (2017) The internet of things—wireless sensor networks and their application to forestry. N Z J for 61:37–41

Bendig J, Bolten A, Bennertz S, Broscheit J, Eichfuss S, Bareth G (2014) Estimating biomass of barley using crop surface models (CSMs) derived from UAV-based RGB imaging. Remote Sens 6(11):10395–10412

Berg G, Krause R, Mendes R (2015) Cross-kingdom similarities in microbiome ecology and biocontrol of pathogens. In. Frontiers Media SA, p 1311

Bleecker AB, Kende H (2000) Ethylene: a gaseous signal molecule in plants. Annu Rev Cell Dev Biol 16(1):1-8

Bock C, Poole G, Parker P, Gottwald T (2010) Plant disease severity estimated visually, by digital photography and image analysis, and by hyperspectral imaging. Crit Rev Plant Sci 29:59–107

Bolfe ÉL, Jorge LAdC, Sanches IDA, Luchiari Júnior A, da Costa CC, Victoria DdC, Inamasu RY, Grego CR, Ferreira VR, Ramirez AR (2020) Precision and digital agriculture: adoption of technologies and perception of Brazilian farmers. Agriculture 10:653

Calderón R, Montes-Borrego M, Landa BB, Navas-Cortés JA, Zarco-Tejada PJ (2014) Detection of downy mildew of opium poppy using high-resolution multi-spectral and thermal imagery acquired with an unmanned aerial vehicle. Precis Agric 15:639–631

Castrignanò A, Belmonte A, Antelmi I, Quarto R, Quarto F, Shaddad S, Sion V, Muolo MR, Ranieri NA, Gadaleta G, Bartoccetti E (2020) Semi-automatic method for early detection of *Xylella fastidiosa* in olive trees using UAV multispectral imagery and geostatistical-discriminant analysis. Remote Sens 13(1):14

Chamara N, Islam MD, Bai GF, Shi Y, Ge Y (2022) Ag-IoT for crop and environment monitoring: past, present, and future. Agric Syst 203:103497

Cheney CM (2022) Genomic characterization of extended-spectrum beta-lactamase producing escherichia coli in oregonian wastewater

Chen J-W, Lau YY, Krishnan T, Chan K-G and Chang C-Y (2018) Recent advances in molecular diagnosis of *Pseudomonas aeruginosa* infection by State-of-the-art genotyping techniques. Front Microbiol 9:1104

Cherubin MR, Damian JM, Tavares TR, Trevisan RG, Colaço AF, Eitelwein MT, Martello M, Inamasu RY, Pias OHdC, Molin JP (2022) Precision agriculture in Brazil: the trajectory of 25 years of scientific research. Agriculture 12:1882

Christiansen MP, Laursen MS, Jørgensen RN, Skovsen S, Gislum R (2017) Designing and testing a UAV mapping system for agricultural field surveying. Sensors 17(12):2703

Cotrozzi L (2022) Spectroscopic detection of forest diseases: a review (1970–2020). J For Res 33:21–38

Dang LM, Hassan SI, Suhyeon I, kumar Sangaiah A, Mehmood I, Rho S, Seo S, Moon H (2020) UAV based wilt detection system via convolutional neural networks. Sustain Comput: Inform Syst 28:100250

Delgado-Ospina J, Molina-Hernández JB, Chaves-López C, Romanazzi G, Paparella A (2021) The role of fungi in the cocoa production chain and the challenge of climate change. Journal of Fungi 7:202

Diaz-Valdivia A (2024) Blockchains and environmental governance: theoretical foundations towards collective action. Available at SSRN 4824458

Di Gennaro SF, Battiston E, Di Marco S, Facini O, Matese A, Nocentini M, Palliotti A, Mugnai L (2016) Unmanned Aerial Vehicle (UAV)-based remote sensing to monitor grapevine leaf stripe disease within a vineyard affected by esca complex. Phytopathol Mediterr 262–275

Elshafei AM (2024) Exploring the role of artificial intelligence in revolutionizing microbial diagnostics. Asian J Res Infect Dis 15:79–90

Fardusi MJ, Chianucci F, Barbati A (2017) Concept to practice of geospatial-information tools to assist forest management and planning under precision forestry framework: a review. Ann Silvic Res 41:3–14

Gade R, Moeslund TB (2014) Thermal cameras and applications: a survey. Mach vis Appl 25:245–262

Garcia-Ruiz F, Sankaran S, Maja JM, Lee WS, Rasmussen J, Ehsani R (2013) Comparison of two aerial imaging platforms for identification of Huanglongbing-infected citrus trees. Comput Electron Agric 91:106–115

Gao J, Yaseen N, MacDavid R, Frujeri FV, Liu V, Bianchini R, Aditya R, Wang X, Lee H, Maltz D (2020) Scouts: improving the diagnosis process through domain-customized incident routing. In: Proceedings of the annual conference of the ACM special interest group on data communication on the applications, technologies, architectures, and protocols for computer communication, pp 253–269

Geipel J, Link J, Claupein W (2014) Combined spectral and spatial modeling of corn yield based on aerial images and crop surface models acquired with an unmanned aircraft system. Remote Sens 6(11):10335–1055

Guebsi R, Mami S, Chokmani K (2024) Drones in precision agriculture: a comprehensive review of applications, technologies, and challenges. Drones 8:686

Guo A, Huang W, Dong Y, Ye H, Ma H, Liu B, Wu W, Ren Y, Ruan C, Geng Y (2021) Wheat yellow rust detection using UAV-based hyperspectral technology. Remote Sens 13(1):123

Haq B, Jamshed MA, Ali K, Kasi B, Arshad S, Kasi MK, Ali I, Shabbir A, Abbasi QH, Ur-Rehman M (2024) Tech-driven forest conservation: combating deforestation with Internet of Things, artificial intelligence, and remote sensing. IEEE Internet Things J

He L, Schupp J (2018) Sensing and automation in pruning of apple trees: a review. Agronomy 8:211

Hellberg RS, Chu E (2016) Effects of climate change on the persistence and dispersal of foodborne bacterial pathogens in the outdoor environment: a review. Crit Rev Microbiol 42:548–572

Holzinger A, Saranti A, Angerschmid A, Retzlaff CO, Gronauer A, Pejakovic V, Medel-Jimenez F, Krexner T, Gollob C, Stampfer K (2022) Digital transformation in smart farm and forest operations needs human-centered AI: challenges and future directions. Sensors 22:3043

Howell LG, Clulow J, Jordan NR, Beranek CT, Ryan SA, Roff A, Witt RR (2021) Drone thermal imaging technology provides a cost-effective tool for landscape-scale monitoring of a cryptic forest-dwelling species across all population densities. Wildl Res 49:66–78

Ijeh S, Okolo CA, Arowoogun JO, Adeniyi AO, Omotayo O (2024) Predictive modeling for disease outbreaks: a review of data sources and accuracy. Int Med Sci Res J 4:406–419

Jain P, Coogan SC, Subramanian SG, Crowley M, Taylor S, Flannigan MD (2020) A review of machine learning applications in wildfire science and management. Environ Rev 28:478–505

Joalland S, Screpanti C, Varella HV, Reuther M, Schwind M, Lang C, Walter A, Liebisch F (2018) Aerial and ground based sensing of tolerance to beet cyst nematode in sugar beet. Remote Sens 10(5):787

Jongman M, Carmichael PC, Bill M (2020) Technological advances in phytopathogen detection and metagenome profiling techniques. Curr Microbiol 77:675–681

Kete S (2022) Blockchain technology applications to the forestry sector–a scoping review

Khelili MA (2022) Deep learning and parallelization of meta-heuristic methods for IoT cloud. Université de mohamed kheider biskra

Khursheed A, Rather MA, Jain V, Rasool S, Nazir R, Malik NA, Majid SA (2022) Plant based natural products as potential ecofriendly and safer biopesticides: a comprehensive overview of their advantages over conventional pesticides, limitations and regulatory aspects. Microb Pathog 173:105854

Kouadio L, El Jarroudi M, Belabess Z, Laasli S-E, Roni MZK, Amine IDI, Mokhtari N, Mokrini F, Junk J, Lahlali R (2023) A review on UAV-based applications for plant disease detection and monitoring. Remote Sensing 15:4273

Kumar D, Pandey V, Dixit S (2024) Agronomic strategies for enhancing forest resilience to climate change. In: Forests and climate change: biological perspectives on impact, adaptation, and mitigation strategies. Springer, pp 385–420

Laliberte AS, Goforth MA, Steele CM, Rango A (2011) Multispectral remote sensing from unmanned aircraft: image processing workflows and applications for rangeland environments. Remote Sens 3(11):2529–2551

Lawson AB (2013) Statistical methods in spatial epidemiology. Wiley

LoPresti A, Heinis A, Wells D, Sweeney J, Plass L, Wiler L, Tilton L, Lubitz R, Kirby TL (2016) ASSESMENT RUBRIC FOR ADAPTATION: SUSTAINABLE CITIES GOAL 11

Madiga Bala D, Padigapati Venkata NS, Suresh R, Mude RN (2020) Agriculture, economics, ecology and trade: a way forward for better world. World Food Policy 6:96–104

Mahule A, Roy K, Sawarkar AD, Lachure S (2024) Enhancing environmental resilience: precision in air quality monitoring through AI-driven real-time systems. Artif Intell Air Qual Monit Predict: 48–74

Mahlein AK (2016) Plant disease detection by imaging sensors–parallels and specific demands for precision agriculture and plant phenotyping. Plant Dis 100(2):241–251

Martinelli F, Scalenghe R, Davino S, Panno S, Scuderi G, Ruisi P, Villa P, Stroppiana D, Boschetti M, Goulart LR, Davis CE (2015) Advanced methods of plant disease detection. A review. Agron Sustain Dev 35:1–25

Mani JK, Varghese A (2018) Remote sensing and GIS in agriculture and forest resource monitoring. Geospatial Technol Land Resour Mapp Monit Manag: 377–400

Marano G, Langella G, Basile A, Cona F, De Michele C, Manna P, Teobaldelli M, Saracino A, Terribile F (2019) A geospatial decision support system tool for supporting integrated forest knowledge at the landscape scale. Forests 10:690

Mazzola M, Hewavitharana SS, Fazio G, Robinson T, DeJong T, Esumi T, Tao R, Gasic K, Saski C, Musacchi S (2019) Achieving sustainable cultivation of temperate zone tree fruits and berries volume 1: physiology, genetics and cultivation. Burleigh Dodds Science Publishing

Millar CI, Stephenson NL, Stephens SL (2007) Climate change and forests of the future: managing in the face of uncertainty. Ecol Appl 17:2145–2151

Misra A (2012) Mango diseases in relation to climate change. In: Proceedings of the national conference on threatening diseases of horticultural, medicinal, aeromatic and field crops in relation to changing climatic situation & zonal meeting. New Delhi, India, pp 3–5

Mohan M, Richardson G, Gopan G, Aghai MM, Bajaj S, Galgamuwa GP, Vastaranta M, Arachchige PSP, Amorós L, Corte APD (2021) UAV-supported forest regeneration: current trends, challenges and implications. Remote Sens 13:2596

Molina-Pico A, Cuesta-Frau D, Araujo A, Alejandre J, Rozas A (2016) Forest monitoring and wildland early fire detection by a hierarchical wireless sensor network. J Sens 2016:8325845

Murthy RK (2023) Early detection and prediction of zoonotic disease events using event-based surveillance and machine learning. Washington State University

Nahata SS (2020) Sequence analysis and vegetative growth in transgenic eucalyptus with CRISPR-Cas9-induced mutations in the eucalyptus homologs of the floral genes EMBRYO DEVELOPMENT ARREST 33 (EDA33) and TAPETAL DEVELOPMENT AND FUNCTION1 (TDF1)

Nelson SC (2008) Mango anthracnose (Colletotrichum gloeosporiodes)

Nong NB, Tarek M (2023) Surveillance approaches to intrusion detection, crop health, and disease prevention in agriculture. Q J Emerg Technol Innov 8:1–17

Oliver MA, Bishop TF, Marchant BP (2013) Precision agriculture for sustainability and environmental protection. Routledge Abingdon, UK

Ozdenerol E (2015) GIS and remote sensing use in the exploration of Lyme disease epidemiology. Int J Environ Res Public Health 12:15182–15203

Palaniyandi M (2012) The role of remote sensing and GIS for spatial prediction of vector-borne diseases transmission: a systematic review. J Vector Borne Dis 49:197–204

Panda SS, Rao MN, Thenkabail PS, Misra D, Fitzgerald JP (2016) Remote sensing systems—platforms and sensors: aerial, satellite, UAV, optical, radar, and LiDAR. In: Remote sensing handbook, vol 1. CRC Press, pp 3–86

Patrick A, Pelham S, Culbreath A, Holbrook CC, De Godoy IJ, Li C (2017) High throughput phenotyping of tomato spot wilt disease in peanuts using unmanned aerial systems and multispectral imaging. IEEE Instrum Meas Mag 20(3):4–12

Patz JA, Campbell-Lendrum D, Holloway T, Foley JA (2005) Impact of regional climate change on human health. Nature 438:310–317

Pérez AA, Vazquez-Meves G, Hunter ME (2024) Early detection of wildlife disease pathogens using CRISPR-Cas system methods. CRISPR J

Picchio R, Proto AR, Civitarese V, Di Marzio N, Latterini F (2019) Recent contributions of some fields of the electronics in development of forest operations technologies. Electronics 8:1465

Ploetz R (2003) Diseases of mango. Dis Trop Fruit Crop: 327–363

Qin J, Wang B, Wu Y, Lu Q, Zhu H (2021) Identifying pine wood nematode disease using UAV images and deep learning algorithms. Remote Sens 13(2):162

Quamar MM, Al-Ramadan B, Khan K, Shafiullah M, El Ferik S (2023) Advancements and applications of drone-integrated geographic information system technology—a review. Remote Sensing 15:5039

R Shamshiri R, Weltzien C, Hameed IA, J Yule I, E Grift T, Balasundram SK, Pitonakova L, Ahmad D, Chowdhary G (2018) Research and development in agricultural robotics: a perspective of digital farming

Raman RK, Kumar A, Sarkar S, Yadav AK, Mukherjee A, Meena RS, Kumar U, Singh D, Das S, Kumar R (2024) Reconnoitering precision agriculture and resource management: a comprehensive review from an extension standpoint on artificial intelligence and machine learning. Indian Res J Ext Educ 24:108–123

Ramírez-Gil JG, López JH, Henao-Rojas JC (2019) Causes of Hass avocado fruit rejection in preharvest, harvest, and packinghouse: economic losses and associated variables. Agronomy 10:8

Rane J, Kaya O, Mallick S, Rane N (2024) Smart farming using artificial intelligence, machine learning, deep learning, and ChatGPT: applications, opportunities, challenges, and future directions. Gener Artif Intell Agric Educ Bus: 218–272

Rango A, Laliberte A, Steele C, Herrick JE, Bestelmeyer B, Schmugge T, Roanhorse A, Jenkins V (2006) Using unmanned aerial vehicles for rangelands: current applications and future potentials. Environ Pract 8(3):159–168

Rani J, Kumar D, Sharma V, Handique M (2024) Metagenomics and AI: a synergistic approach to data-driven insights. In: Genomic intelligence. CRC Press, pp 21–39

Reddy PV, Manikanta C, Singh H (2024) Forest inventory techniques and remote sensing. Exploring 55

Robinson JM, Hodgson R, Krauss SL, Liddicoat C, Malik AA, Martin BC, Mohr JJ, Moreno-Mateos D, Muñoz-Rojas M, Peddle SD (2023) Opportunities and challenges for microbiomics in ecosystem restoration. Trends Ecol Evol 38:1189–1202

Salamí E, Barrado C, Pastor E (2014) UAV flight experiments applied to the remote sensing of vegetated areas. Remote Sensing 6:11051–11081

Sandino J, Pegg G, Gonzalez F, Smith G (2018) Aerial mapping of forests affected by pathogens using UAVs, hyperspectral sensors, and artificial intelligence. Sensors 18(4):944

Santesteban LG, Di Gennaro SF, Herrero-Langreo A, Miranda C, Royo JB, Matese A (2017) High-resolution UAV based thermal imaging to estimate the instantaneous and seasonal variability of plant water status within a vineyard. Agric Water Manag 183:49–59

Shalal N, Low T, McCarthy C, Hancock N (2015) Orchard mapping and mobile robot localisation using on-board camera and laser scanner data fusion–Part A: tree detection. Comput Electron Agric 119:254–266

Singh R, Gehlot A, Akram SV, Thakur AK, Buddhi D, Das PK (2022) Forest 4.0: digitalization of forest using the Internet of Things (IoT). J King Saud Univ-Comput Inf Sci 34:5587–5601

Smigaj M, Gaulton R, Barr SL, Suárez JC (2015) UAV-borne thermal imaging for forest health monitoring: detection of disease-induced canopy temperature increase. Int Arch Photogramm, Remote Sens Spat Inf Sci 40:349–354

Smigaj M, Agarwal A, Bartholomeus H, Decuyper M, Elsherif A, de Jonge A, Kooistra L (2024) Thermal infrared remote sensing of stress responses in forest environments: a review of developments, challenges, and opportunities. Curr For Rep 10:56–76

de Souza CH, Lamparelli RA, Rocha JV, Magalhães PS (2017) Mapping skips in sugarcane fields using object-based analysis of unmanned aerial vehicle (UAV) images. Comput Electron Agric 143:49–56

Stopfer L, Kaulen A, Purfürst T (2024) Potential of blockchain technology in wood supply chains. Comput Electron Agric 216:108496

Sugiura R, Tsuda S, Tamiya S, Itoh A, Nishiwaki K, Murakami N, Shibuya Y, Hirafuji M, Nuske S (2016) Field phenotyping system for the assessment of potato late blight resistance using RGB imagery from an unmanned aerial vehicle. Biosyst Eng 148:1–0

Sugiura R, Tsuda S, Tsuji H, Murakami N (2018) Virus-infected plant detection in potato seed production field by UAV imagery. Am Soc Agric Biol Eng 1

Techy L, Schmale III DG, Woolsey CA (2010) Coordinated aerobiological sampling of a plant pathogen in the lower atmosphere using two autonomous unmanned aerial vehicles. J Field Robot 27(3):335–343

Tetila EC, Machado BB, Belete NA, Guimarães DA, Pistori H (2017) Identification of soybean foliar diseases using unmanned aerial vehicle images. IEEE Geosci Remote Sens Lett 14(12):2190–2194

Tharanathan R, Yashoda H, Prabha T (2006) Mango (Mangifera indica L.), "The king of fruits"—an overview. Food Rev Intl 22:95–123

Torres-Sánchez J, López Granados F, De Castro AI, Peña-Barragán JM (2013) Configuration and specifications of an unmanned aerial vehicle (UAV) for early site specific weed management. PloS one 8(3):e58210

Trilles Oliver S, González-Pérez A, Huerta Guijarro J (2019) Adapting models to warn fungal diseases in vineyards using in-field internet of things (IoT) nodes. Sustainability 11:416

Tyagi AK, Chahal P (2020) Artificial intelligence and machine learning algorithms. In: Challenges and applications for implementing machine learning in computer vision. IGI Global, pp 188–219

Uma GS, Mahanta DK, Sharma L (2024) Advancing forest insect pest management: a focus on biotechnological approaches. In: Biotechnological approaches for sustaining forest trees and their products. Springer, pp 347–383

Vacher C, Castagneyrol B, Jousselin E, Schimann H (2021) Trees and insects have microbiomes: consequences for forest health and management. Curr For Rep 7:81–96

Vanegas F, Bratanov D, Weiss J, Powell K, Gonzalez F (2018) Multi and hyperspectral UAV remote sensing: grapevine phylloxera detection in vineyards. In: 2018 IEEE aerospace conference. IEEE, pp 1–9

van Ewijk E, Ataa-Asantewaa M, Asubonteng KO, Van Leynseele YP, Derkyi M, Laven A, Ros-Tonen MA (2024) Farmer-centred multi-stakeholder platforms: From iterative approach to conceptual embedding. J Knowl Econ: 1–31

Verma A, Saini JK, Singh HB, Wiley J (2021) Phytomicrobiome interactions and sustainable agriculture. Wiley

Vieira A, Moura M, Silva L (2021) Soil metagenomics in grasslands and forests–a review and bibliometric analysis. Appl Soil Ecol 167:104047

Wang H, Qian X, Zhang L, Xu S, Li H, Xia X, Dai L, Xu L, Yu J, Liu X (2018) A method of high throughput monitoring crop physiology using chlorophyll fluorescence and multispectral imaging. Front Plant Sci 9:407

Wang Y, Zhang Y-L, Zhang R, Li C-X, Lai X-J, Ma X-R (2025) Effect of ecological restoration on physicochemical properties, bacterial and fungal communities of red sand soil in Longquan Mountain Forest Park. Bact Fungal Communities Red Sand Soil Longquan Mt For Park

Wasilewski T, Kamysz W, Gębicki J (2024) AI-assisted detection of biomarkers by sensors and biosensors for early diagnosis and monitoring. Biosensors 14

Wich SA, Koh LP (2018) Conservation drones: mapping and monitoring biodiversity. Oxford University Press

Wilson AD (2013) Diverse applications of electronic-nose technologies in agriculture and forestry. Sensors 13:2295–2348

Wu Z, Pullman ME (2020) Sustainable food logistics systems. In: The Oxford handbook of supply chain management

Yang C, Westbrook JK, Suh CP, Martin DE, Hoffmann WC, Lan Y, Fritz BK, Goolsby JA (2014) An airborne multispectral imaging system based on two consumer-grade cameras for agricultural remote sensing. Remote Sens 6(6):5257–5278

Zakari RY, Shafik W, Kalinaki K, Iheaturu CJ (2024) 15 Internet of Things (IoFT) of forestry technologies and applications in forest management. Adv IoT Technol Appl Ind 4 Digit Econ: 275

Concluding Chapter: Climate Change and Mango Production—Synthesis, Synergy, and Pathways Forward

Introduction

Mango (*Mangifera indica* L.), revered as the "King of Fruits," is a linchpin of tropical agriculture, nourishing millions and sustaining economies across the globe. Yet, this vital crop faces unprecedented threats from climate change, including shifting temperature regimes, erratic rainfall, and escalating biotic pressures. This book, *Climate Change and Mango Production: Potential Adaptation and Mitigation Options*, synthesizes 32 chapters of cutting-edge research, offering a mosaic of strategies to safeguard mango cultivation. This concluding chapter weaves these threads into a cohesive narrative, highlighting synergies across disciplines and charting a roadmap for resilient mango systems in an uncertain climate future.

Synthesis of Key Themes

The chapters of this book collectively underscore the profound and multifaceted impacts of climate change on mango production systems, while offering actionable strategies for adaptation and mitigation. Below is a synthesis of the major themes that emerge from the contributions:

Climate Change Impacts on Mango Physiology and Production

Chapters 2–7 elucidate how rising temperatures and erratic precipitation disrupt mango physiology, altering flowering patterns, nutrient uptake, and fruit quality. Key physiological processes such as photosynthesis, nutrient uptake, flowering, and

S. A. H. Naqvi et al. (eds.), *Climate Change and Mango Production*,
https://doi.org/10.1007/978-3-031-87035-4

fruit set are compromised under heat stress, leading to reduced yields, poor fruit quality (e.g., sunburn, cracking), and delayed phenological cycles. For instance, prolonged heat stress impairs photosynthesis and heat waves impair pollination by desiccating floral structures, while Chap. 5 links irregular rainfall to uneven ripening and postharvest losses (Chap. 7), while elevated CO_2 levels influence secondary metabolite production (Chap. 2). These physiological shifts are compounded by soil health degradation (Chap. 12) and pathogen proliferation (Chap. 17), underscoring the need for holistic adaptation strategies.

Innovative Adaptation and Mitigation Strategies

- **Microbiome Engineering**: Chapter 1 highlights rhizosphere microbiota as a frontline defense, enhancing nutrient uptake and stress tolerance. Synergies with Chap. 18's polyembryonic rootstocks demonstrate how host-microbe interactions can be harnessed for disease resistance.
- **Water and Soil Management**: Chapters 10 and 12 advocate for smart irrigation and organic amendments, which align with Chap. 13's postharvest innovations to reduce losses in water-scarce regions.
- **Nanotechnology and Disease Control**: Nanoparticles (Chaps. 21, 31) offer targeted pathogen suppression, complementing biocontrol agents (Chap. 28) and resistant cultivars (Chap. 26) for integrated pest management.

Soil health (Chap. 12) and microbiome engineering (Chap. 1) emerge as central pillars for climate-resilient mango cultivation. The rhizosphere's microbial communities PGPR, AMF, and nitrogen-fixing bacteria enhance nutrient availability, water retention, and disease suppression. Chapter 18 further demonstrates how polyembryonic rootstocks with robust root systems improve resistance to soilborne pathogens like *Fusarium* and *Ceratocystis*. However, soil degradation under climate change (e.g., salinity, nutrient leaching) threatens these benefits, necessitating organic amendments and precision nutrient management (Chaps. 8 and 20).

Disease and Pest Dynamics in a Warming World and Technological Leapfrogging

Climate change exacerbates fungal, bacterial, and insect pest pressures. Chapters 17, 22, and 23 detail how warmer, humid conditions favor pathogens like *Colletotrichum gloeosporioides* (anthracnose) and *Ceratocystis fimbriata* (MSDS), while Chap. 24 identifies expanding ranges of fruit flies and borers. Notably, Chapter 19 demonstrates that exogenous chemical applications can mitigate floral diseases in heat-stressed cultivars like Chaunsa Summer Bahisht, but sustainable alternatives (e.g., nanoparticles in Chap. 21) are critical to reducing chemical dependency.

From microbiome engineering (Chap. 1) to AI-driven disease surveillance (Chap. 32), technological advancements offer transformative solutions. Hydroponic grafting (Chap. 9), climate-smart irrigation (Chap. 10), and biodegradable fruit bagging (Chap. 15) improve resource efficiency, while nanotechnology (Chaps. 21, 31) and CRISPR-based breeding (implied in Chap. 28) target pathogen resistance. Postharvest innovations like solar drying and IoT-enabled cold chains (Chap. 13) reduce losses and enhance market access.

Socio-economic Vulnerabilities and Equity

The socio-economic impacts of climate change on smallholder farmers (Chap. 11) are stark. Erratic yields, rising input costs, and export restrictions due to phytosanitary issues (Chap. 26) disproportionately affect communities in Pakistan, India, and Brazil. Gender disparities in labor and decision-making further compound vulnerabilities, underscoring the need for inclusive policies (Chap. 14) and farmer-centric technologies (Chap. 10).

Policy and Institutional Frameworks

Effective governance is critical for scaling resilience strategies. Chapter 14 critiques fragmented policies in major mango-producing nations and advocates for integrated frameworks that align subsidies, extension services, and climate data systems. Case studies from Brazil and Mexico (Chap. 26) highlight the role of public–private partnerships in phytosanitary compliance and value chain optimization.

Interconnections Across Chapters

The chapters are deeply interconnected, revealing a cohesive narrative about systemic resilience:

- **Microbiome Engineering ↔ Soil Health ↔ Disease Resistance**
  The rhizosphere's microbial consortia (Chap. 1) enhance nutrient cycling and suppress soilborne pathogens (Chaps. 12 and 18), reducing reliance on chemical fungicides (Chap. 23). This synergy is critical for organic systems and aligns with nanoparticle applications (Chap. 21) for integrated disease management.
- **Grafting ↔ Climate Adaptation ↔ Socio-Economic Resilience**
  Heat-tolerant rootstocks (Chap. 9) improve orchard survival in drought-prone regions (Chap. 4), directly supporting farmer livelihoods (Chap. 11). However, scaling grafted saplings require clean nursery protocols (Chap. 16) and policy incentives (Chap. 14).

- **Precision Agriculture ↔ Pest Management ↔ Global Trade**
  AI-driven pest forecasting (Chap. 32) enables targeted interventions, reducing pesticide use and meeting export phytosanitary standards (Chap. 26). This links to postharvest innovations (Chap. 13) that maintain fruit quality during transport.
- **Secondary Metabolites ↔ Stress Tolerance ↔ Nutritional Quality**
  Climate-induced shifts in secondary metabolites (Chap. 2) may enhance mango resilience but alter nutritional profiles (Chap. 6). Breeding programs (Chap. 28) must balance stress tolerance with consumer preferences for taste and health benefits.

Current and Future Perspectives: Gaps and Emerging Challenges

Despite progress, critical gaps remain:

Research Gaps

- **Long-Term Microbiome Dynamics**: While Chap. 1 highlights bioinoculants, their persistence under extreme weather is poorly understood.
- **Gene-Environment Interactions**: Molecular mechanisms linking heat stress to malformation disease (Chap. 28) require genomic and transcriptomic exploration.
- **Socio-Technical Barriers**: Farmer adoption of technologies like IoT (Chap. 10) is hindered by digital literacy and infrastructure gaps (Chap. 11).

Emerging Challenges

- **Pathogen Evolution**: Climate-driven genetic plasticity in *Ceratocystis* (Chap. 22) and *Fusarium* (Chap. 28) may outpace current resistance strategies.
- **Water-Energy-Food Nexus**: Competing demands for irrigation (Chap. 10) and renewable energy (Chap. 13) in water-scarce regions like Turkey (Chap. 4) necessitate integrated planning.
- **Carbon Footprint**: Mango value chains (Chap. 26) must address emissions from cold storage and chemical inputs to meet net-zero targets.

Future Directions

- **Climate-Resilient Breeding**: Leveraging polyembryonic germplasm (Chap. 18) and CRISPR to develop cultivars with dual tolerance to heat and salinity.
- **Circular Economy Models**: Recycling agricultural waste into biochar (Chap. 12) or biodegradable packaging (Chap. 15).
- **One Health Approaches**: Linking mango microbiome health to human nutrition and ecosystem stability.

Practical Implications

For Farmers and Practitioners

- Adopt microbial consortia and organic mulches to enhance soil moisture and suppress diseases.
- Implement fruit bagging and drip irrigation to reduce heat stress and water use.
- Use AI-based apps (Chap. 32) for real-time pest alerts and irrigation scheduling.

For Policymakers

- Prioritize climate-smart subsidies for bioinoculants, solar dryers, and resilient rootstocks.
- Strengthen phytosanitary infrastructure to meet EU and US import standards (Chap. 26).
- Fund farmer cooperatives to scale innovations like nanotechnology (Chap. 21) and collective bargaining.

For Researchers

- Develop multi-omics databases to link mango genotypes with climate resilience traits.
- Validate nanoparticle safety and field efficacy across diverse agroecologies.
- Model pest-pathogen-climate interactions using machine learning (Chap. 29).

Final Thoughts

This book bridges the gap between climate science and mango agronomy, offering a roadmap for resilience in the face of existential threats. Its contributions are threefold:

1. **Holistic Understanding**: By integrating soil science, pathology, socio-economics, and technology, it rejects siloed approaches in favor of systemic solutions.
2. **Actionable Innovation**: From microbiome engineering to policy frameworks, it provides scalable tools for immediate and long-term adaptation.
3. **Global Relevance**: Case studies from Turkey, Pakistan, and Brazil contextualize strategies for both emerging and established mango economies.

This book transcends a mere compilation of chapters; it is a manifesto for climate-resilient mango systems. By interlacing microbiome science, nanotechnology, policy, and tradition, we chart a viable path toward sustaining the "King of Fruits." The solutions exist, their implementation hinges on our collective resolve. As climate change reshapes our world, let this synthesis ignite a global movement where mango orchards thrive not despite the climate but because of our ingenuity and unity.

As climate change accelerates, the mango industry must embrace adaptive co-management; a fusion of traditional knowledge and cutting-edge science. The path forward demands collaboration across disciplines, borders, and sectors to ensure that the "King of Fruits" remains a symbol of abundance, not fragility, in the Anthropocene.

The future of mango cultivation is not a solitary struggle, it is a shared journey toward resilience, equity, and abundance.

Editors

Syed Atif Hasan Naqvi, St. Catherines, ON, Canada
Shakeel Ahmad, Multan, Punjab, Pakistan
Mukhtar Ahmed, Rawalpindi, Punjab, Pakistan

The manufacturer's authorised representative in the EU is Springer Nature Customer Service Centre GmbH, Europaplatz 3, 69115 Heidelberg, Germany. If you have any concerns regarding our products, please contact ProductSafety@springernature.com

Printed and bound by CPI Group (UK) Ltd, Croydon, CR0 4YY
01/10/2026
02225922-0019